JAY WITHGOTT • MATTHEW LAPOSATA

ESSENTIAL ENVIRONMENT

THE SCIENCE BEHIND THE STORIES

Fourth Edition

PEARSON

Boston   Columbus   Indianapolis   New York   San Francisco   Upper Saddle River   Amsterdam
Cape Town   Dubai   London   Madrid   Milan   Munich   Paris   Montréal   Toronto   Delhi
Mexico City   São Paulo   Sydney   Hong Kong   Seoul   Singapore   Taipei   Tokyo

VP/Editor-in-Chief: Beth Wilbur
Executive Director of Development: Deborah Gale
Executive Editor: Chalon Bridges
Project Editor/Developmental Editor: Nora Lally-Graves
Editorial Assistant: Rachel Brickner
Editorial Media Manager: Lee Ann Doctor
Marketing Manager: Lauren Rodgers
Managing Editor, Environmental Science: Gina M. Cheselka
Project Manager, Science: Wendy Perez
Art Production Manager: Connie Long

Operations Specialist: Maura Zaldivar
Photo Researcher: Maureen Spuhler
Composition/Full Service: Element LLC
Production Editor: Heidi Allgair
Illustrations: Imagineering
Interior Design: Yin Ling Wong
Cover Design: Yvo Riezebos Design
Cover Photograph: Photolibrary/Patrick Neri; *The buildings of downtown Seattle seen from across Elliot Bay and Gas Works Park during sunset, WA.*

Library of Congress Cataloging-in-Publication Data
Withgott, Jay.
 Essential environment : the science behind the stories / Jay Withgott, Matthew Laposata. —4th ed.
   p. cm.
 Includes bibliographical references and index.
 ISBN-13: 978-0-321-75290-1
 ISBN-10: 0-321-75290-2
 1. Environmental sciences. I. Laposata, Matthew. II. Title.
 GE105.B74 2011
 363.7—dc22   2011011755

ABOUT OUR SUSTAINABILITY INITIATIVES

This book is carefully crafted to minimize environmental impact. The materials used to manufacture this book originated from sources committed to responsible forestry practices. The paper is FSC® certified. The binding, cover, and paper come from facilities that minimize waste, energy consumption, and the use of harmful chemicals.

Pearson closes the loop by recycling every out-of-date text returned to our warehouse. We pulp the books, and the pulp is used to produce items such as paper coffee cups and shopping bags. In addition, Pearson aims to become the first climate neutral educational publishing company.

The future holds great promise for reducing our impact on the Earth's environment, and Pearson is proud to be leading the way. We strive to publish the best books with the most up-to-date and accurate content, and to do so in ways that minimize our impact on the Earth.

FSC
www.fsc.org
MIX
Paper from responsible sources
FSC® C084269

PEARSON

ISBN-10: 0-321-75290-2
ISBN-13: 978-0-321-75290-1

ABOUT THE AUTHORS

With this edition we welcome Dr. Matthew Laposata as an author. Professor of environmental science at Kennesaw State University in Georgia, Matt teaches and coordinates his university's environmental science courses while actively engaging in outside projects to promote environmental science education. Matt's ideas, energy, and commitment to effective teaching have already enlivened our book and strengthened its content and approach. Please welcome him to our author team!

JAY WITHGOTT has authored *Essential Environment* and its parent volume, *Environment: The Science behind the Stories*, since their inception. In dedicating himself to these books, he works to keep abreast of a diverse and rapidly changing field and continually seeks to develop new and better ways to help today's students learn environmental science.

As a researcher, Jay has published papers in ecology, evolution, animal behavior, and conservation biology in journals ranging from *Evolution* to *Proceedings of the National Academy of Sciences*. As an instructor, he has taught university lab courses in ecology and other disciplines. As a science writer, Jay has authored articles for journals and magazines including *Science, New Scientist, BioScience, Smithsonian,* and *Natural History*. By combining his scientific training with prior experience as a newspaper reporter and editor, he strives to make science accessible and engaging for general audiences. Jay holds degrees from Yale University, the University of Arkansas, and the University of Arizona.

Jay lives with his wife, biologist Susan Masta, in Portland, Oregon.

MATTHEW LAPOSATA is a professor of environmental science at Kennesaw State University (KSU). He holds a bachelor's degree in biology education from Indiana University of Pennsylvania, a master's degree in biology from Bowling Green State University, and a doctorate in ecology from The Pennsylvania State University.

Matt is the coordinator of KSU's two-semester general education science sequence titled *Science, Society, and the Environment*, which enrolls roughly 7,000 students a year. He focuses exclusively on introductory environmental science courses and has enjoyed teaching and interacting with thousands of nonscience majors during his career. He is an active scholar in environmental science education and has received grants from state, federal, and private sources to develop and evaluate innovative curricular materials. His scholarly work has received numerous awards, including the Georgia Board of Regents' highest award for the Scholarship of Teaching and Learning.

He resides in suburban Atlanta with his wife, Lisa, and children, Lauren, Cameron, and Saffron.

CONTENTS

iv

PREFACE

Dear Student,

You are coming of age at a unique and momentous time in history. Within your lifetime, our global society must chart a promising course for a sustainable future. The stakes could not be higher.

Today we live long lives enriched with astonishing technologies, in societies more free, just, and equal than ever before. We enjoy wealth on a scale our ancestors could hardly have dreamed of. Yet we have purchased these wonderful things at a price. By exploiting Earth's resources and ecological services, we are depleting our planet's bank account and running up its credit card.

Your future is being shaped by the phenomena you will learn about in your environmental science course. Environmental science gives us a big-picture understanding of the world and our place within it. Environmental science also offers hope and solutions, revealing ways to address problems. Environmental science is not just some subject you study in college. It provides you basic literacy in the foremost issues of the 21st century, and it relates to everything around you for your entire life.

We have written this book because today's students will shape tomorrow's world. At this unique moment in history, the decisions and actions of your generation are key to achieving a sustainable future for our civilization. The many challenges that face us can seem overwhelming, but you should feel encouraged and motivated. Remember that each dilemma is also an opportunity. For every problem that human carelessness has created, human ingenuity can devise an answer. Now is the time for innovation, creativity, and the fresh perspectives that a new generation can offer. Your own ideas and energy *will* make a difference. You are the solution!

JAY WITHGOTT AND MATTHEW LAPOSATA

Dear Instructor,

You perform one of our society's most important jobs: educating today's students—the citizens and leaders of tomorrow—on the fundamentals of the world around them, the nature of science, and the most vital issues of our time. We have written this book to assist you in this endeavor because we feel that the vital role of environmental science in today's world makes it imperative to engage, educate, and inspire a broad audience of students.

In *Essential Environment: The Science behind the Stories,* we strive to implement the very best in modern teaching approaches and to show how science can inform human efforts to design a sustainable society. We aim to maintain a balanced approach and to encourage critical thinking as we flesh out the social debate over environmental issues. As we assess the challenges facing our civilization and our planet, we focus on providing hope and solutions.

As environmental science has grown, so has the length of textbooks that cover it. With this volume, we aim to meet the needs of introductory courses that favor a more succinct and affordable book. We have distilled the most essential content from our full-length book, *Environment: The Science behind the Stories,* now in its fourth edition. We have streamlined our material and carefully crafted our writing to make *Essential Environment* every bit as readable, informative, and engaging as its parent volume.

NEW TO THIS EDITION

In this fourth edition of *Essential Environment,* we have incorporated the most current information from a fast-moving field. Moreover, a number of major changes new to this edition enhance our presentation while strengthening our commitment to teach science in an engaging and accessible way.

➤ **Chapter sequence.** We split Chapter 9 from the third edition into two chapters: coverage of forests and protected areas (Chapter 9) and coverage of cities and urban sustainability (Chapter 18). This allows for more focused attention to these topics. We also modified the sequence of chapters in our Foundations section to better match the preferences of most instructors, placing economics and policy after the natural sciences but retaining these as vital foundations of environmental science.

➤ **EnvisionIt.** These new full-page photo essays are designed to draw in visual learners and to help all students envision vital concepts. Some *EnvisionIt* pages help students visualize the scale of a phenomenon. Others bring vibrant life to the human side of an environmental issue or show how issues affect people in cultures throughout the world. Because today's students are so visually driven, we expect this new feature will help to better engage them in the book and in your course.

➤ **FAQ.** By highlighting students' frequently asked questions, this new feature pinpoints widely held misconceptions and points students toward what the evidence from environmental science tells us.

➤ **New Central Case Studies.** Eight of our 17 *Central Case Studies* are new to this edition, giving you a wealth of fresh stories and new ways to frame issues in environmental science. In the new case studies, students learn about Chesapeake Bay and its restoration, Iowa's no-till agriculture, the science and the debate over bisphenol A, water issues on the Mississippi River, air pollution control efforts in Los Angeles and Tehran, and Germany's use of feed-in tariffs to promote solar energy. The *Deepwater Horizon* oil spill now frames our discussion of fossil fuel energy.

➤ **New Science behind the Story features.** Fully 13 of our 18 *Science behind the Story* features are new to this edition, giving you a more current and exciting selection of scientific studies to highlight. Students will follow researchers as they radio-track trash, uncover the impacts of bottled water, apply forensics in conservation biology, measure the effects of enriched carbon dioxide on forests, restore mine sites, and study everything from ozone depletion to urban ecology to organic agriculture.

➤ **FSC-certified paper.** One of our new *Central Case Studies* describes how our book uses sustainable paper from Michigan's Upper Peninsula, certified by the Forest Stewardship Council. Students learn about the process used to make the very textbook they are holding in their hands while they discover how certification programs and corporate responsibility can help drive sustainability efforts.

➤ **Enhanced photos, art, and graphs.** We have strengthened our visual presentation of material throughout the text. This edition includes 160 new photos and 85 new graphs and illustrations, while 80 additional figures have been revised to reflect current data or for better clarity or pedagogy.

➤ **An exciting new online platform.** With this edition we are thrilled to gain a new online learning and assessment platform called *Mastering Environmental Science*. Powerful yet easy to use, you can employ *Mastering* to assess student learning outside the classroom. After consulting with instructors, we apportioned our features and activities between the print book and the *Mastering* platform in a way that makes best use of the advantages of each medium. As a result, you will find certain popular features from previous editions of the print book now on *Mastering*. These include *Interpreting Graphs and Data* exercises, the interactive *GraphIt!* program, *Causes and Consequences* exercises, *Viewpoints* essays, and more.

➤ **Video Field Trips.** Brand-new on the *Mastering Environmental Science* site is a series of *Video Field Trips*. These brief videos are a wonderful resource, especially for courses unable to take students into the field. The videos include a short visit to a landfill, wastewater treatment plant, organic farm, eWaste site, prescribed fire burn, solar power sites and more.

EXISTING FEATURES

We have also retained the major features that made the first three editions of our book unique and that are proving so successful in classrooms across North America:

➤ **Central Case Studies integrated throughout the text.** We integrate each chapter's *Central Case Study* into the main text, weaving information and elaboration throughout the chapter. In this way, compelling stories about real people and real places help to teach general concepts by giving students a tangible framework with which to incorporate novel ideas.

➤ **The Science behind the Story.** Because we strive to engage students in the scientific process of testing and discovery, we feature *The Science behind the Story* in each chapter. By guiding students through key research efforts, this feature shows not merely *what* scientists discovered, but *how* they discovered it.

➤ **Weighing the Issues.** Our *Weighing the Issues* questions aim to help develop the critical-thinking skills students need to navigate multifaceted issues at the juncture of science, policy, and ethics. These questions serve as stopping points for students to reflect on what they have read, wrestle with complex dilemmas, and engage in spirited classroom discussion.

➤ **End-of-chapter features.** The features that conclude each chapter target particular student needs. *Testing Your Comprehension* provides concise study questions on main topics. *Seeking Solutions* encourages broader creative thinking aimed at finding solutions, and "Think It Through" questions place students in a scenario and empower them to make decisions to resolve problems. *Calculating Ecological Footprints* enables students to quantify the impacts of their own choices and measure how individual impacts scale up to the societal level.

➤ **An emphasis on solutions.** For many students, today's deluge of environmental dilemmas can cause them to feel that there is no hope or that they cannot personally make a difference. We have aimed to counter this impression by drawing out innovative solutions being developed around the world. While being careful not to paint too rosy a picture of the challenges that lie ahead, we endeavor to instill hope and encourage action.

Essential Environment: The Science behind the Stories has grown from our experiences in teaching, research, and writing. We have been guided in our efforts by input from hundreds of instructors across North America who have served as reviewers and advisors. The participation of so many learned, thoughtful, and committed experts and educators has improved this volume in countless ways.

We sincerely hope that our efforts are worthy of the immense importance of our subject matter. We invite you to let us know how well we have achieved our goals and where you feel we have fallen short. Please write to us in care of our editor Chalon Bridges (chalon.bridges@pearson.com). We value your feedback and are eager to know how we can serve you better.

JAY WITHGOTT AND MATTHEW LAPOSATA

INSTRUCTOR SUPPLEMENTS

Instructor Resource DVD with TestGen Computerized Testbank (0321753216)

This powerful media package is organized chapter by chapter and includes all the teaching resources in one convenient location. You'll find videos, PowerPoint presentations, Active Lecture questions to facilitate class discussions (for use with or without clickers), and an image library that includes all art and tables from the text. Test questions available in both Word and TestGen files offer hundreds of assessment options including unique graph and figure interpretation questions.

Instructor Guide (0321753224)

This comprehensive resource contains chapter objectives, lecture outlines, key terms, teaching tips, additional resources, and suggested answers for the *Weighing the Issues* and *Science behind the Stories* features that appear in the text.

Mastering Environmental Science (www.masteringenvironmentalscience.com)

Mastering Environmental Science from Pearson has been designed and refined with a single purpose in mind: to help educators create that moment of understanding with their students. The Mastering online homework and tutoring system delivers self-paced tutorials that provide individualized coaching, focus on your course objectives, and are responsive to each student's progress. The Mastering system helps instructors maximize class time with customizable, easy-to-assign, and automatically graded assessments that motivate students to learn outside of class and arrive prepared for lecture. By complementing your teaching with our engaging technology and content, you can be confident your students will arrive at that moment—the moment of true understanding.

Blackboard Premium for Essential Environment (0321753232)

ACKNOWLEDGMENTS

A textbook is the product of *many* more minds and hearts than one might guess from the names on the cover. The two of us are exceedingly fortunate to be supported and guided by a tremendous publishing team and by a small army of experts in environmental science who have generously shared their time and expertise.

First and foremost, we would like to thank our executive editor, Chalon Bridges. Chalon's commitment to quality educational publishing is simply inspirational. By her own example she motivates our team to relish the challenge of taking a successful and well-received book and making it still better. Project editor Nora Lally-Graves ably guided us through this edition's revision, improving the book's content, handling countless logistics, and helping our new author team to function seamlessly. Working closely with Chalon and Nora on this edition has been a joy for us, and we appreciate their caring dedication, patience, and good humor. We also thank editorial director Michael Young and executive director of development Deborah Gale, as well as our editor-in-chief Beth Wilbur for her strong and steady support of this book through its four editions.

Photo researcher Maureen Spuhler helped acquire quality photos, Sally Peyrefitte provided meticulous copy-editing of our text, and editorial assistant Rachel Brickner was always there with timely and effective help. Once the manuscript was ready, Wendy Perez and Gina Cheselka saw it through to production, while Connie Long handled our art program. We want to offer a big thank-you to Heidi Allgair, Cindy Miller, and the rest of the staff at Element LLC for a tremendous job putting the book together. We also thank Tim Flem, Mary Ann Murray, Susan Teahan, Elaine Soares, Travis Amos, and Maureen Eide for their contributions to this and earlier editions.

Our thanks also go to Danielle DuCharme for revising our *Instructor Guide,* Thomas Pliske for revising the *Test Bank*, and Heidi Marcum for revising our PowerPoint slides. For their work on the *Mastering Environmental Science* website and our media supplements, we thank Lee Ann Doctor, Sean O'Connor, and Katie Foley. A special thanks to Eric Flagg for his tremendous *Video Field Trips.*

We give a big thanks to marketing manager Lauren Garritson for her dedicated work marketing the book. And of course, the many field representatives who help communicate our vision and deliver our product to instructors are absolutely vital, and we deeply appreciate their tireless work and commitment.

Lastly, Jay gives loving thanks to his wife, Susan, who has endured each edition's preparation with patience and sacrifice, providing support and sustenance throughout. Matt thanks the family, friends, and colleagues who enrich his existence. He is grateful for his children giving him three reasons to care deeply about the future and, most importantly, he thanks his wife, Lisa, for gracing his life with a woman more beautiful, talented, and insightful than he deserves.

We dedicate this book to today's students, who will shape tomorrow's world.

JAY WITHGOTT AND MATTHEW LAPOSATA

REVIEWERS

We wish to express special thanks to the dedicated reviewers who shared their time and expertise to help make this edition the best it could be. Nearly 600 instructors and outside experts have reviewed material for the previous three editions of this book and the first four editions of this book's parent volume, where they are acknowledged in full. Below we acknowledge those who contributed particularly to this fourth edition of *Essential Environment*, in most cases with multiple chapter reviews. If the thoughtfulness and thoroughness of these reviewers are any indication, we feel confident that the teaching of environmental science is in excellent hands!

Matthew Abbott, *Des Moines Area Community College* and *Iowa Lakes Community College*

Terrence Bensel, *Allegheny College*
Anne Bower, *Philadelphia University*
Tait Chirenje, *Stockton College*
Erica Kipp, *Pace University*
Ned J. Knight, *Linfield College*
James Kubicki, *The Pennsylvania State University*
Kurt M. Leuschner, *College of the Desert*
Jeffrey Mahr, *Georgia Perimeter College*
Troy Mutchler, *Kennesaw State University*
Virginia D. Rivers, *Truckee Meadows Community College*
Kimberly Schulte, *Georgia Perimeter College*
Erin Seiling, *Lenoir-Rhyne University*
Julie Stoughton, *University of Nevada, Reno*
Todd T. Tracy, *Northwestern College*
Lorne Wolfe, *Georgia Southern University*

Pre-Revision Reviewers

Abbed Babaei, *Cleveland State University*
Charles R. Bomar, *University of Wisconsin—Stout*
C. Lee Burras, *Iowa State University*
Richard Clements, *Chattanooga State Technical Community College*
Carol A. Hoban, *Kennesaw State University*
Leslie Kanat, *Johnson State College*
Susan W. Karr, *Carson–Newman College*
Erica Kosal, *North Carolina Wesleyan College*

Heidi Marcum, *Baylor University*
Chris Migliaccio, *Miami Dade College*
Roger del Moral, *University of Washington*
Eric Sanden, *University of Wisconsin—River Falls*
Kimberly Schulte, *Georgia Perimeter College*
Roy Sofield, *Chattanooga State Technical Community College*
Patricia A. Terry, *University of Wisconsin—Green Bay*
Danielle M. Wirth, *Des Moines Area Community College*

1 Science and Sustainability: An Introduction to Environmental Science

Upon completing this chapter, you will be able to:

➤ Define the term *environment* and describe the field of environmental science

➤ Explain the importance of natural resources and ecosystem services to our lives

➤ Discuss the consequences of population growth and resource consumption

➤ Describe the steps of the scientific method

➤ Understand the nature and importance of science, and characterize aspects of the process of science

➤ Compare and contrast various approaches in environmental ethics

➤ Diagnose and illustrate major pressures on the global environment

➤ Articulate the concepts of sustainability and sustainable development

Our Island, Earth

OUR ISLAND, EARTH

Viewed from space, our home planet resembles a small blue marble suspended in a vast inky-black void. Earth may seem enormous to us as we go about our lives on its surface, but the astronaut's view reveals that Earth and its systems are finite and limited. From this perspective, it becomes clear that as our population and our consumption of resources increase, so does our capacity to alter our planet and damage the very systems that keep us alive.

Our environment surrounds us

A photograph of Earth offers a revealing perspective, but it cannot convey the complexity of our environment. Our **environment** consists of all the living and nonliving things around us. It includes the continents, oceans, clouds, and ice caps you can see in the photo of Earth from space, as well as the animals, plants, forests, and farms that comprise the landscapes surrounding us. In a more inclusive sense, it encompasses the structures, urban centers, and living spaces that people have created. In its broadest sense, our environment also includes the complex webs of social relationships and institutions that shape our daily lives.

People commonly use the term *environment* in the first, most narrow sense—to mean a nonhuman or "natural" world apart from human society. This usage is unfortunate, because it masks the important fact that people exist within the environment and are part of nature. As one of many species on Earth, we share dependence on a healthy planet. The limitations of language make it all too easy to speak of "people and nature," or "humans and the environment," as though they were separate and did not interact. However, the fundamental insight of environmental science is that we are part of the "natural" world and that our interactions with the rest of it matter a great deal.

Environmental science explores our interactions with the world

Understanding our relationship with the world around us is vital because we depend utterly on our environment for air, water, food, shelter, and everything else essential for living. Moreover, we modify our environment. Many of our actions have enriched our lives, bringing us better health, longer life spans, and greater material wealth, mobility, and leisure time—but they have also often degraded the natural systems that sustain us. Impacts such as air and water pollution, soil erosion, and species extinction compromise our well-being, pose risks to human life, and jeopardize our ability to build a society that will survive and thrive in the long term.

Environmental science is the study of how the natural world works, how our environment affects us, and how we affect our environment. We need to understand how we interact with our environment so that we can devise solutions to our most pressing challenges. It can be daunting to reflect on the sheer magnitude of environmental dilemmas that confront us today, but these problems also bring countless opportunities for creative solutions.

Environmental scientists study the issues most centrally important to our world and its future. Right now,

global conditions are changing more quickly than ever. Right now, we are gaining scientific knowledge more rapidly than ever. And right now, the window of opportunity for acting to solve problems is still open. With such bountiful challenges and opportunities, this moment in history is an exciting time to be alive—and to be studying environmental science.

We rely on natural resources

An island by definition is finite and bounded, and its inhabitants must cope with limitations in the materials they need. On our island, Earth, human beings, like all living things, ultimately face environmental constraints. Specifically, there are limits to many of our **natural resources,** the various substances and energy sources we take from our environment and that we rely on to survive. Natural resources that are replenished over short periods are known as **renewable natural resources.** Some renewable resources, such as sunlight, wind, and wave energy, are perpetually renewed and essentially inexhaustible. Others, such as timber, water, and soil, renew themselves over months, years, or decades. In contrast, **nonrenewable natural resources,** such as minerals and crude oil, are in finite supply and are formed much more slowly than we use them. Once we deplete a nonrenewable resource, it is no longer available.

We can view the renewability of natural resources as a continuum (**FIGURE 1.1**). Renewable resources such as timber, water, and soil can be depleted if we use them faster than they are replenished. For example, overpumping groundwater can deplete underground aquifers and turn lush landscapes into deserts. Populations of animals and plants we harvest from the wild may vanish if we overharvest them.

We rely on ecosystem services

If we think of natural resources as "goods" produced by nature, then it is also true that Earth's natural systems provide "services" on which we depend. Our planet's ecological systems purify air and water, cycle nutrients, regulate climate, pollinate plants, and receive and recycle our waste. Such essential services are commonly called **ecosystem services.** Ecosystem services arise from the normal functioning of natural systems and are not meant for our benefit, yet we could not survive without them. Later we will examine the countless and profound ways that ecosystem services support our lives and civilization (pp. 36, 90, 94–95).

Just as we can deplete natural resources if we take too many of them, we can degrade ecosystem services by depleting resources, destroying habitat, or generating pollution. In recent years, our depletion of nature's goods and our disruption of nature's services have intensified, driven by rising affluence and a human population that grows larger every day.

Population growth amplifies our impact

For nearly all of human history, fewer than a million people populated Earth at any one time. Today our population has grown beyond 7 billion people—several thousand times more! **FIGURE 1.2** shows just how recently and suddenly this monumental change has come about.

Renewable natural resources
- Sunlight
- Wind energy
- Wave energy
- Geothermal energy

- Fresh water
- Forest products
- Agricultural crops
- Soils

Nonrenewable natural resources
- Crude oil
- Natural gas
- Coal
- Copper, aluminum, and other metals

FIGURE 1.1 ▲ Natural resources lie along a continuum from perpetually renewable **(left)** to nonrenewable **(right)**. Perpetually renewable, or inexhaustible, resources, such as sunlight and wind energy, will always be there for us. Renewable resources such as timber, soils, and fresh water may be replenished on intermediate time scales, if we are careful not to deplete them. Nonrenewable resources, such as oil and coal, exist in limited amounts that could one day be gone.

Two phenomena triggered remarkable increases in population size. The first was our transition from a hunter-gatherer lifestyle to an agricultural way of life. This change began around 10,000 years ago and is known as the **agricultural revolution.** As people began to grow crops, raise domestic animals, and live sedentary lives on farms and in villages, they found it easier to meet their nutritional needs. As a result, they began to live longer and to produce more children.

The second notable phenomenon, known as the **industrial revolution,** began in the mid-1700s. It entailed a shift from rural life, animal-powered agriculture, and handcrafted goods to an urban society provisioned by the mass production of factory-made goods and powered by **fossil fuels** (nonrenewable energy sources including oil, coal, and natural gas (pp. 328–335)). Industrialization brought technological advances and improvements in sanitation and medicine, and it enhanced agricultural production through the use of fossil-fuel-powered equipment and synthetic pesticides and fertilizers (p 136).

The factors driving population growth have brought us better lives in many ways. But as our world fills up with people, population growth has begun to threaten our well-being. We must ask how well the planet can accommodate 7 billion of us—or the 9 billion forecast by 2050. Already our sheer numbers, unparalleled in history, are putting unprecedented stress on natural systems and the availability of resources.

Resource consumption exerts social and environmental pressures

Population growth is unquestionably at the root of many environmental concerns, but the growth in resource consumption also plays a role. As the industrial revolution enhanced the material affluence of many of the world's people, it considerably increased our consumption of natural resources and manufactured goods.

The "tragedy of the commons" When publicly accessible resources are open to unregulated exploitation, they inevitably become overused and, as a result, are damaged or depleted. So argued the late Garrett Hardin of the University of California at Santa Barbara in his 1968 essay in the journal *Science,* titled "The Tragedy of the Commons."

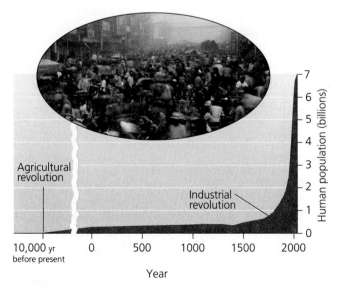

FIGURE 1.2 ▲ For almost all of human history, our population was low and relatively stable. It increased after the agricultural revolution and then skyrocketed as a result of the industrial revolution. Our growing population has given rise to congested urban areas, such as this city **(inset)** in Java, Indonesia. Data compiled from U.S. Census Bureau, U.N. Population Division, and other sources.

Basing his argument on a scenario described in a 19th-century pamphlet, Hardin explained that in a public pasture, or "common," open to unregulated grazing, each person who grazes animals will be motivated by self-interest to increase the number of his or her animals in the pasture. Because no single person owns the pasture, no one has incentive to expend effort taking care of it, and everyone takes what he or she can until the resource is depleted. This is known as the **tragedy of the commons**. Ultimately, overgrazing will cause the pasture's food production to collapse.

Some argue that private ownership best addresses this problem. Others point to cases in which people sharing a common resource have voluntarily organized and cooperated to enforce its responsible use. Still others maintain that the dilemma justifies government regulation of the use of resources held in common by the public, from grazing land and forests to clean air and water.

 WEIGHING THE ISSUES

The Tragedy of the Commons Imagine you make your living by fishing. You are free to boat anywhere and set out as many lines and traps as you like, and your catches have been good. However, the fishing grounds are getting crowded, and you find yourself competing with more people for fewer fish. Catches decline year by year, leaving you and the other fishers with catches too meager to support your families. Some call for dividing the waters and selling access to individuals plot by plot. Others implore the government to regulate how many fish can be caught. Still others want to team up, set quotas themselves, and prevent newcomers from entering the market. What do you think is the best way to combat this tragedy of the commons and restore the fishery, and why?

Our ecological footprint As global affluence has risen, human society has consumed more and more of the planet's resources. We can quantify resource consumption using the concept of the ecological footprint, developed in the 1990s by environmental scientists Mathis Wackernagel and William Rees. An **ecological footprint** expresses environmental impact in terms of the cumulative area of biologically productive land and water required to provide the resources a person or population consumes and to dispose of or recycle the waste the person or population produces (**FIGURE 1.3**). It measures the total area of Earth's biologically productive surface that a given person or population "uses" once all direct and indirect impacts are totaled up.

For humanity as a whole, Wackernagel and his colleagues calculate that our species is now using 50% more of the planet's resources than are available on a sustainable basis. That is, we are depleting renewable resources by using them 50% faster than they are being replenished—like drawing the principal out of a bank account rather than living off the interest. This excess use has been termed **overshoot**, because we have overshot, or surpassed, Earth's capacity to sustainably support us (**FIGURE 1.4**). Moreover, people from wealthy nations such

FIGURE 1.3 ▲ An *ecological footprint* represents the total area of biologically productive land and water needed to produce the resources and dispose of the waste for a given person or population. The footprint of an average citizen of an affluent nation is much larger than the physical area in which the person lives day to day. Adapted from an illustration by Philip Testemale in Wackernagel, M., and W. Rees, 1996. *Our ecological footprint: Reducing human impact on the Earth*. Gabriola Island, British Columbia: New Society Publishers.

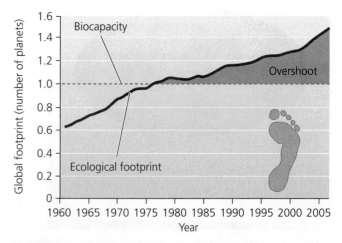

FIGURE 1.4 ▲ The global ecological footprint of the human population is over 2.5 times larger than it was a half-century ago and now exceeds what Earth can bear in the long run, scientists have calculated. Data indicate that we have already overshot Earth's biocapacity—its capacity to support us—by 50%. That is, we are using renewable natural resources 50% faster than they are being replenished. Data from WWF International, 2010. *Living planet report 2010*. Published in 2010 by WWF-World Wide Fund for Nature.

as the United States have much larger ecological footprints than do people from poorer nations. If all the world's people consumed resources at the rate of U.S. citizens, we would need the equivalent of four-and-a-half planet Earths.

Environmental science can help us avoid past mistakes

It remains to be seen what consequences resource consumption and population growth will have for today's global society, but we have historical evidence that civilizations can crumble when pressures from population and consumption overwhelm resource availability. Easter Island is a classic case (see **THE SCIENCE BEHIND THE STORY**, pp. 6–7).

Many great civilizations have fallen after degrading their environments, and each has left devastated landscapes in its wake: the Greek and Roman empires; the Angkor civilization of Southeast Asia; and the Maya, Anasazi, and other civilizations of the New World. In Iraq and other regions of the Middle East, areas that are barren desert today were lush enough to support the origin of agriculture when ancient civilizations thrived there. In his 2005 book *Collapse*, scientist and author Jared Diamond analyzed existing research and formulated general reasons why civilizations succeed and persist, or fail and collapse. Success and persistence, he argued, depend largely on how societies interact with their environments and on how they respond to problems.

In today's globalized society, the stakes are higher than ever because our environmental impacts are global. If we cannot forge sustainable solutions to our problems, then the resulting societal collapse will be global. Fortunately, environmental science holds keys to building a better world. By studying environmental science, you will learn to evaluate the changes happening around us and to think critically and creatively about actions to take in response.

THE NATURE OF ENVIRONMENTAL SCIENCE

Environmental scientists aim to comprehend how Earth's natural systems function, how these systems influence people, and how we are influencing these systems. Many environmental scientists are motivated by a desire to develop solutions to environmental problems. These solutions (such as new technologies, policy decisions, or resource management strategies) are *applications* of environmental science. The study of such applications and their consequences is, in turn, also part of environmental science.

Environmental science is interdisciplinary

Studying our interactions with our environment is a complex endeavor that requires expertise from many disciplines, including ecology, Earth science, chemistry, biology, geography, economics, political science, demography, ethics, and others. Environmental science is thus an **interdisciplinary** field—one that borrows techniques from multiple

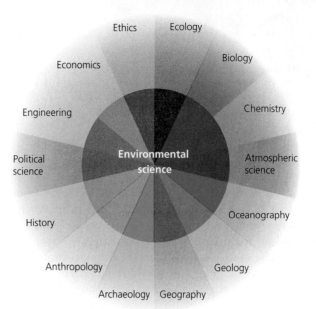

FIGURE 1.5 ▲ Environmental science is an interdisciplinary pursuit, involving input from many different established fields of study across the natural sciences and social sciences.

disciplines and brings their research results together in a broad synthesis (**FIGURE 1.5**).

Traditional established disciplines are valuable because their scholars delve deeply into topics, uncovering new knowledge and developing expertise in particular areas. In contrast, interdisciplinary fields are valuable because their practitioners consolidate and synthesize the specialized knowledge from many different disciplines and make sense of it in a broad context to better serve the multifaceted interests of society.

Environmental science is especially broad because it encompasses not only the **natural sciences** (disciplines that examine the natural world), but also the **social sciences** (disciplines that address human interactions and institutions). Most environmental science programs focus predominantly on the natural sciences, whereas programs that incorporate the social sciences extensively often use the term **environmental studies**. Whichever approach one takes, these fields reflect many diverse perspectives and sources of knowledge.

Just as an interdisciplinary approach to studying issues can help us better understand them, an integrated approach to addressing environmental problems can produce effective solutions for society. As one example, we used to add lead to gasoline to make cars run more smoothly, even though researchers knew that lead emissions from tailpipes caused health problems, including brain damage and premature death. In 1970 air pollution was severe, and motor vehicles accounted for 78% of U.S. lead emissions. Over the following years, environmental scientists, engineers, medical researchers, and policymakers all merged their knowledge and skills into a process that eventually brought about a ban on leaded gasoline. By 1996 all gasoline sold in the United States was unleaded, and the nation's largest source of atmospheric lead emissions had been completely eliminated.

Easter Island's immense statues

The Lesson of Easter Island

Easter Island is one of the most remote spots on the globe, located in the Pacific Ocean 3,750 km (2,325 mi) from South America and 2,250 km (1,395 mi) from the nearest inhabited island. When the first European explorers reached the island (today called Rapa Nui) in 1722, they found a barren landscape populated by fewer than 2,000 people, who lived in caves and eked out a marginal existence from a few meager crops. However, explorers also noted that the desolate island featured hundreds of gigantic statues of carved stone, evidence that a sophisticated civilization had once inhabited the island.

Historians and anthropologists wondered how people without wheels or ropes, on an island without trees, could have moved statues 10 m (33 ft) high weighing 90 metric tons (99 tons) as far as 10 km (6.2 mi) from the quarries where they were chiseled to the coastal sites where they were erected. The explanation, scientists discovered, was that the island did not always lack trees.

Indeed, scientific research tells us that the island had once been lushly forested and had supported a prosperous society of 6,000 to 30,000 people. Tragically, this once-flourishing civilization overused its resources and cut down all its trees, destroying itself in a downward spiral of starvation and conflict. Today Easter Island stands as a parable and a warning for what can happen when a population consumes too much of the limited resources that support it.

To explore the mystery of Easter Island's past, scientists have used various methods. Some, such as British scientist John Flenley, have excavated sediments from the bottom of the island's lakes, drilling cores deep into the mud and examining ancient grains of pollen preserved there. Pollen grains vary from one plant species to another, so scientists can reconstruct, layer by layer, the history of vegetation in a region through time. By analyzing pollen grains under scanning electron microscopes, Flenley and other researchers found that when Polynesian people arrived (likely between A.D. 300 and 900), the island was covered with a species of palm tree related to the Chilean wine palm, a tall and thick-trunked tree.

Moreover, archaeologists located ancient palm nut casings in caves and crevices, and a geologist found carbon-lined channels in the soil that matched root channels typical of the Chilean wine palm. Scientists deciphering the island people's script on stone tablets discerned characters etched in the form of palm trees.

By studying pollen and the remains of wood from charcoal, scientists such as French archaeologist Catherine Orliac found that at least 21 other species of plants, many of them trees, had also been common but are now completely gone. The island had clearly supported a diverse forest. However, starting around A.D. 750, tree populations declined and ferns and grasses became more common, according to pollen analysis from one lake site. By A.D. 950, the trees were largely gone, and around A.D. 1400 overall pollen levels plummeted, indicating a dearth of vegetation.

The same sequence of events occurred two centuries later at the other two lake sites, which were higher and more remote from village areas. Researchers first hypothesized that the forest loss was due to climate change, but evidence instead supported the hypothesis that the people had gradually denuded their own island.

Environmental science is not the same as environmentalism

Although many environmental scientists are interested in solving problems, it would be incorrect to confuse environmental science with environmentalism, or environmental activism. They are *not* the same.

Environmental science is the pursuit of knowledge about the workings of the environment and our interactions with it. In contrast, **environmentalism** is a social movement dedicated to protecting the natural world—and, by extension, people—from undesirable changes brought about by human actions. Although environmental scientists search for solutions to environmental problems, they strive to keep their research rigorously objective and free from ideology, personal values, and preconceptions. Remaining open to whatever conclusions the data demand is a hallmark of the effective scientist.

THE NATURE OF SCIENCE

Modern scientists describe **science** as a systematic process for learning about the world and testing our understanding of it. The term *science* is also used to refer to the accumulated body of knowledge that arises from this dynamic process of questioning, observation, testing, and discovery.

Knowledge gained from science can be applied to address societal needs—for instance, to develop technology

The haunting statues of Easter Island (Rapa Nui) were erected by a sophisticated civilization that collapsed after depleting its resource base and devastating its island environment.

Remains from charcoal fires aged by radiocarbon dating show that besides crops and birds, early islanders feasted on the bounty of the sea, including porpoises, fish, sharks, turtles, octopus, and shellfish. But analysis of islanders' diets in the later years indicated that the people consumed little seafood. With the trees gone, the islanders could no longer build the great double canoes their proud Polynesian ancestors had used for centuries to fish and travel among islands.

As resources declined, archaeologists found, the islanders began keeping their main domesticated food animal, chickens, in stone fortresses with entrances designed to prevent theft. The once prosperous and peaceful civilization fell into clan warfare, as revealed by unearthed weapons, skeletons, and skulls with head wounds.

Is the story of Easter Island as unique and isolated as the island itself, or does it hold lessons for our world today? Like the Easter Islanders, we are all stranded together on an island with limited resources. Earth may be vastly larger and richer in resources than Easter Island, but Earth's human population is also much greater.

The Easter Islanders must have seen that they were depleting their resources, but it seems that they could not stop. Whether we can learn from the history of Easter Island and act more wisely to conserve the resources of our island, Earth, is entirely up to us.

The trees provided fuelwood, building material for houses and canoes, fruit to eat, fiber for clothing, and, presumably, logs with which to move the stone statues. By hiring groups of men to recreate the feat, anthropologists experimentally tested hypotheses about how the islanders moved their monoliths down from the quarries. The methods that have worked involve using numerous tree trunks as rollers or sleds, along with great quantities of rope. The only likely source of rope on the island would have been the fibrous inner bark of the hauhau tree, a species that today is near extinction.

With the trees gone, rain would have eroded soil away—a phenomenon confirmed by data from the lake bottoms, where large quantities of sediment accumulated. Erosion of the islanders' agricultural land would have lowered yields of bananas, sugar cane, and sweet potatoes, leading to starvation and population decline.

Archaeological evidence supports such a scenario of environmental degradation and civilization decline. Analysis of 6,500 bones by archaeologist David Steadman has shown that at least 31 species of birds nested on Easter Island and served as a food source for the islanders. Today, only one native bird species is left.

or to inform policy and management decisions (**FIGURE 1.6**). Many scientists are motivated by the potential to develop useful applications, whereas others are motivated simply by a desire to understand how the world works.

Scientists test ideas by critically examining evidence

Science is all about asking and answering questions. Scientists examine how the world works by making observations, taking measurements, and testing whether their ideas are supported by evidence. The effective scientist thinks critically and does not simply accept conventional wisdom from others. The scientist becomes excited by novel ideas but is skeptical and judges ideas by the strength of evidence that supports them. In these ways, scientists are good role models for the rest of us, because every one of us can benefit from learning to think critically in our everyday lives.

A great deal of scientific work is **observational science** or **descriptive science,** research in which scientists gather basic information about organisms, materials, systems, or processes that are not well known or that cannot be manipulated in experiments. In this approach, researchers explore new frontiers of knowledge by observing and measuring phenomena to gain a better understanding of them. Such research is common in traditional fields such as astronomy, paleontology, and taxonomy, as well as in newer, fast-growing fields such as molecular biology and genomics.

(a) Methanol-powered fuel-cell car

(b) Prescribed burning

FIGURE 1.6 ▲ Scientific knowledge can be applied in engineering and technology and in policy and management decisions. Energy-efficient automobiles **(a)** are technological advances made possible by materials and energy research. Prescribed burning **(b)**, shown here in the Ouachita National Forest, Arkansas, is a management practice to restore healthy forests that is informed by scientific research into forest ecology.

Once enough general information is known about a subject, scientists can begin posing more specific questions that seek deeper explanations about how and why things are the way they are. At this point they may pursue **hypothesis-driven science,** research that proceeds in a targeted and structured manner, using experiments to test hypotheses within a framework traditionally known as the scientific method.

The scientific method is a traditional approach to research

The **scientific method** is a technique for testing ideas with observations. There is nothing mysterious about the scientific method; it is merely a formalized version of the way any of us might naturally use logic to resolve a question. Because science is an active, creative process, an innovative scientist may sometimes find good reason to depart from the traditional scientific method. Moreover, scientists in different fields may approach their work differently because they deal with dissimilar types

of information. However, scientists of all persuasions broadly agree on fundamental elements of the process of scientific inquiry. The scientific method (**FIGURE 1.7**) typically consists of the steps outlined below.

Make observations Advances in science typically begin with the observation of some phenomenon that the scientist wishes to explain. Observations set the scientific method in motion and also function throughout the process.

Ask questions Curiosity is a fundamental human characteristic. Just observe the explorations of babies or young children in a new environment—they want to touch, taste, watch, and listen to anything that catches their attention, and as soon as they can speak, they begin asking questions. Scientists, in this respect, are kids at heart. Why are certain plants or animals less common today than they once were? Why are storms becoming more severe and flooding more frequent? What is causing excessive growth of algae in local ponds? When pesticides poison fish or frogs, are people also affected? All of these are questions environmental scientists ask.

Develop a hypothesis Scientists address their questions by devising explanations they can test. A **hypothesis** is a statement that attempts to explain a phenomenon or answer a

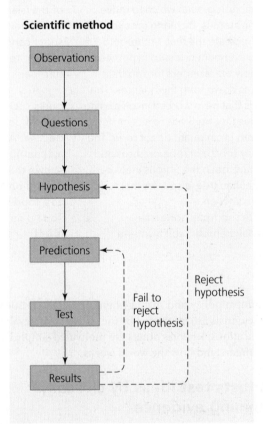

FIGURE 1.7 ▲ The scientific method is the traditional experimental approach that scientists use to learn how the world works. This diagram is a simplified generalization that, although useful for instructive purposes, cannot convey the true dynamic and creative nature of science. Moreover, researchers from different disciplines may pursue their work in ways that vary legitimately from this model.

scientific question. For example, a scientist investigating why algae are growing excessively in local ponds might observe chemical fertilizers being applied on farm fields nearby. The scientist might then state a hypothesis as follows: "Agricultural fertilizers running into ponds cause the amount of algae in the ponds to increase."

Make predictions   The scientist next uses the hypothesis to generate **predictions,** specific statements that can be directly and unequivocally tested. In our algae example, a researcher might predict: "If agricultural fertilizers are added to a pond, the quantity of algae in the pond will increase."

Test the predictions   Scientists test predictions one at a time by gathering evidence that could potentially refute the prediction and thus disprove the hypothesis. An **experiment** is an activity designed to test the validity of a prediction or a hypothesis. It involves manipulating **variables,** or conditions that can change.

For example, a scientist could test the prediction linking algal growth to fertilizer by selecting two identical ponds and adding fertilizer to one while leaving the other in its natural state. In this example, fertilizer input is an **independent vari-**

FIGURE 1.8 ▼ Dr. Jennifer Smith of the Scripps Institution of Oceanography in San Diego uses a quadrat with a digital camera to photograph sites along a transect of a coral reef at a remote atoll in the South Pacific. Data from analysis of the photos will help her test hypotheses about how human impacts affect the condition and community structure of coral reefs.

able, a variable the scientist manipulates, whereas the quantity of algae that results is the **dependent variable,** one that depends on the fertilizer input. If the two ponds are identical except for a single independent variable (fertilizer input), then any differences that arise between the ponds can be attributed to that variable. Such an experiment is known as a **controlled experiment** because the scientist controls for the effects of all variables except the one whose effect he or she is testing. In our example, the pond left unfertilized serves as a **control,** an unmanipulated point of comparison for the manipulated **treatment** pond.

Whenever possible, it is best to *replicate* one's experiment; that is, to stage multiple tests of the same comparison of control and treatment. Our scientist could perform a replicated experiment on, say, 10 pairs of ponds, adding fertilizer to one of each pair.

Analyze and interpret results   Scientists record **data,** or information, from their studies (**FIGURE 1.8**). They particularly value *quantitative* data (information expressed using numbers) because numbers provide precision and are easy to compare. The scientist running the fertilization experiment, for instance, might quantify the area of water surface covered by algae in each pond or might measure the dry weight of algae in a certain volume of water taken from each.

However, even with the precision that numbers provide, a scientist's results may not be clear-cut. Data from treatments and controls may vary only slightly, or replicates may yield different results. The researcher must therefore analyze the data using statistical tests. With these mathematical methods, scientists can determine objectively and precisely the strength and reliability of patterns they find.

If experiments disprove a hypothesis, the scientist will reject the hypothesis and may formulate a new one to replace it. If experiments fail to disprove a hypothesis, this lends support to the hypothesis but does not *prove* it is correct. The scientist may choose to generate new predictions to test the hypothesis in different ways and further assess its likelihood of being true. Thus, the scientific method loops back on itself, often giving rise to repeated rounds of hypothesis revision and new experimentation (see Figure 1.7).

If repeated tests fail to reject a particular hypothesis, evidence in favor of it accumulates, and the researcher may eventually conclude that the hypothesis is well supported. Ideally, the scientist would want to test all possible explanations. For instance, our researcher might formulate an additional hypothesis, proposing that algae increase in fertilized ponds because chemical fertilizers diminish the numbers of fish or invertebrate animals that eat algae. It is possible, of course, that both hypotheses could be correct and that each may explain some portion of the initial observation that local ponds were experiencing algal blooms.

We test hypotheses in different ways

An experiment in which the researcher actively chooses and manipulates the independent variable is known as a *manipulative experiment*. A manipulative experiment provides the strongest type of evidence a scientist can obtain because it can reveal causal relationships, showing that changes in an independent variable cause changes in a dependent variable. In practice, however, we cannot run manipulative experiments for all questions, especially for processes that operate at large spatial scales or on long time scales. For example, in studying the effects of global climate change (Chapter 14), we cannot add carbon dioxide to 10 treatment planets and 10 control planets and then compare the results! Thus, in environmental science, it is common for scientists to run *natural experiments* that compare how dependent variables are expressed in naturally different contexts, and to search for *correlation,* or statistical association among variables.

For instance, let's suppose our scientist studying algae surveys 50 ponds, 25 of which happen to be fed by fertilizer runoff from nearby farms and 25 of which are not. Let's say he or she finds seven times more algal growth in the fertilized ponds. The scientist would conclude that algal growth is correlated with fertilizer input; that is, that one tends to increase along with the other.

This type of evidence is weaker than the causal demonstration that manipulative experiments can provide, but sometimes a natural experiment is the only feasible approach for a subject of immense scale, such as an ecosystem or a planet. Because many questions in environmental science are complex and exist on large scales, they must be addressed with correlative data. As such, environmental scientists cannot always provide clear-cut, black-and-white answers to questions from policymakers and the public. Nonetheless, good correlative studies can make for strong science, and they preserve the real-world complexity that manipulative experiments often sacrifice. Whenever possible, scientists try to integrate both natural and manipulative experiments to gain the advantages of each.

The scientific process continues beyond the scientific method

Scientific work takes place within the context of a community of peers. To have an impact, a researcher's work must be published and made accessible to this community. Thus, the scientific method is embedded within a larger process involving the scientific community as a whole (**FIGURE 1.9**).

Peer review When a researcher's work is done and the results analyzed, he or she writes up the findings and submits them to a journal (a scholarly publication in which scientists share their work). The journal's editor asks several other scientists who specialize in the subject area to examine the manuscript, provide comments and criticism (generally anonymously), and judge whether the work merits publication in the journal. This procedure, known as **peer review,** is an essential part of the scientific process.

Peer review is a valuable guard against faulty science contaminating the literature on which all scientists rely. However, because scientists are human, personal biases and politics can sometimes creep into the review process. Fortunately, just as individual scientists strive to remain objective in conducting their research, the scientific community does its best to ensure fair review of all work. Winston Churchill once called democracy the worst form of government, except for all the others that had been tried. The same might be said about peer review: It is an imperfect system, yet it is the best we have.

Conference presentations Scientists frequently present their work at professional conferences, where they interact

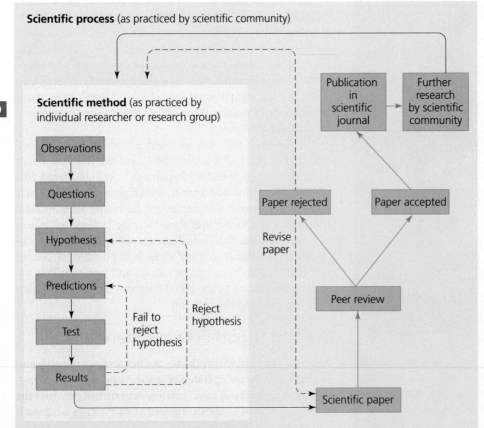

Scientific process (as practiced by scientific community)

Scientific method (as practiced by individual researcher or research group)

Observations → Questions → Hypothesis → Predictions → Test → Results

Fail to reject hypothesis

Reject hypothesis

Scientific paper

Peer review

Paper rejected — Revise paper

Paper accepted

Publication in scientific journal → Further research by scientific community

FIGURE 1.9 ◄ The scientific method **(inner yellow box)** followed by individual researchers or research teams exists within the context of the overall process of science at the level of the scientific community **(outer green box)**. This process includes peer review and publication of research, acquisition of funding, and the elaboration of theory through the cumulative work of many researchers.

with colleagues and receive informal comments on their research. Such feedback can help improve a researcher's work before it is submitted for publication.

Grants and funding   To fund their research, most scientists need to spend enormous amounts of time requesting grant money from private foundations or from government agencies such as the National Science Foundation. Grant applications undergo peer review just as scientific papers do, and competition for funding is generally intense.

Scientists' reliance on funding sources can occasionally lead to conflicts of interest. A researcher who obtains data showing his or her funding source in an unfavorable light may feel reluctant to publish the results for fear of losing funding—or worse yet, may be tempted to doctor the results. This situation can arise, for instance, when an industry funds research to test its products for safety or environmental impact. Most scientists resist these pressures, but when you are critically assessing a scientific study, it is always a good idea to note where the researchers obtained their funding.

Repeatability   The careful scientist may test a hypothesis repeatedly in various ways, and after the research results are published, other scientists may seek to reproduce the results in their own experiments. Scientists are inherently cautious about accepting a novel hypothesis, so the more a result can be reproduced by different research teams, the more confidence scientists will have that it provides the correct explanation for an observed phenomenon.

Theories   If a hypothesis survives repeated testing by numerous research teams and continues to predict experimental outcomes and observations accurately, it may be incorporated into a theory. A **theory** is a widely accepted, well-tested explanation of one or more cause-and-effect relationships that has been extensively validated by a great amount of research. Whereas a hypothesis is a simple explanatory statement that may be disproven by a single experiment, a theory consolidates many related hypotheses that have been supported by a large body of data.

Note that scientific use of the word *theory* differs from popular usage of the word. In everyday language, when we say something is "just a theory" we are suggesting it is a speculative idea without much substance. Scientists, however, mean just the opposite when they use the term. To them, a theory is a conceptual framework that explains a phenomenon and has undergone extensive and rigorous testing, such that confidence in it is extremely strong.

For example, Darwin's theory of evolution by natural selection (pp. 46–48) has been supported and elaborated by many thousands of studies over 150 years of intensive research. Research has shown repeatedly and in great detail how plants and animals change over generations, or evolve, expressing characteristics that best promote survival and reproduction. Because of its strong support and explanatory power, evolutionary theory is the central unifying principle of modern biology. Other prominent scientific theories include atomic theory, cell theory, big bang theory, plate tectonics, and general relativity.

Science goes through "paradigm shifts"

As the scientific community accumulates data in a given area of research, interpretations may change. Thomas Kuhn's 1962 book *The Structure of Scientific Revolutions* argued that science goes through periodic revolutions: dramatic upheavals in thought, in which one **paradigm,** or dominant view, is abandoned for another. For example, before the 16th century, scientists believed that Earth was at the center of the universe. Their data on the movements of planets fit that concept quite well, yet the idea eventually was disproved by Nicolaus Copernicus, who showed that placing the sun at the center of the solar system explained the planetary data even better.

Another paradigm shift occurred in the 1960s, when geologists accepted plate tectonics (pp. 228–229). By this time, evidence for the movement of continents and the action of tectonic plates had accumulated and become overwhelmingly convincing. Such paradigm shifts demonstrate the strength and vitality of science, showing it to be a process that refines and improves itself through time.

Understanding how science works is vital to assessing how scientific interpretations improve through time as information accrues. This process is especially relevant in environmental science, a young field that is changing rapidly as we learn vast amounts of new information. However, to understand and address environmental problems, and to assess our actions and their consequences, we need more than science. We also need ethics. Science does not take place in a vacuum—it is influenced by the worldviews and cultural backgrounds of the scientists who practice it. Cultural influences also guide how engineers, managers, policymakers, and citizens apply scientific knowledge. Thus, our examination of ethics below (and of economics and policy in Chapter 5) will help us learn how values shape human behavior and how information from the sciences is interpreted and put to use in our society.

ENVIRONMENTAL ETHICS

Ethics is a branch of philosophy that involves the study of good and bad, of right and wrong. The term *ethics* can also refer to the set of moral principles or values held by a person or a society. Ethicists help clarify how people judge right from wrong by elucidating the criteria, standards, or rules that people use in making these judgments. Such criteria are grounded in values—for instance, promoting human welfare, maximizing individual freedom, or minimizing pain and suffering.

People of different cultures or with different worldviews may differ in their values and thus in the specific actions they consider to be right or wrong. This is why some ethicists are **relativists,** who believe that ethics do and should vary with social context. However, different human societies show a remarkable extent of agreement on what moral standards are appropriate. Thus, many ethicists are **universalists,** who maintain that there exist objective notions of right and wrong that hold across cultures and contexts. For both relativists and universalists, ethics is not just descriptive, but *prescriptive;* it tells us how we *ought to* behave.

Ethical standards are the criteria that help differentiate right from wrong. One classic ethical standard is the *categorical imperative* proposed by German philosopher Immanuel Kant, which advises us to treat others as we would prefer to be treated ourselves. In Christianity this is called the "golden rule," and most of the world's religions teach this same lesson. Another standard is the principle of *utility,* elaborated by British philosophers Jeremy Bentham and John Stuart Mill. The utilitarian principle holds that something is right when it produces the greatest practical benefits for the most people. We all employ such ethical standards as tools for making countless decisions in our everyday lives.

Environmental ethics pertains to people and the environment

The application of ethical standards to relationships between people and nonhuman entities is known as **environmental ethics.** Our interactions with our environment frequently give rise to ethical questions that can be difficult to resolve. Consider some examples:

1. Is the present generation obligated to conserve resources for future generations? If so, how much should we sacrifice?
2. Can we justify exposing some communities to a disproportionate share of pollution? If not, what actions are warranted to prevent this?
3. Are humans justified in driving species to extinction? If destroying a forest would drive extinct an insect species few people have heard of but would create jobs for 10,000 people, would that action be ethically admissible? What if it were an owl species? What if only 100 jobs would be created?

Answers to such questions depend partly on what ethical standard(s) a person adopts. They also depend on the breadth and inclusiveness of the person's domain of ethical concern. A person who feels responsibility for the welfare of insects would answer the third question very differently from a person whose domain of ethical concern ends with people. Three loosely conceived categories summarize differences among personal domains of ethical concern. These three ethical perspectives, or *worldviews,* are anthropocentrism, biocentrism, and ecocentrism (**FIGURE 1.10**).

Anthropocentrism describes a human-centered view of our relationship with the environment. An anthropocentrist denies or ignores the notion that nonhuman entities can have rights. An anthropocentrist measures the costs and benefits of actions solely according to their impact on people. For example, if development of a mining project would provide significant economic benefits while doing little harm to aesthetics or human health, an anthropocentrist would conclude it was a worthwhile venture, even if it would destroy many plants and animals. Conversely, if protecting the area from development would provide greater economic, spiritual, or other benefits to people, an anthropocentrist would favor its protection. In the anthropocentric perspective, anything not providing benefit to people is considered to be of negligible value.

In contrast, **biocentrism** ascribes value to certain living things or to the biotic realm in general. In this perspective, human life and nonhuman life both have ethical standing. In the case of a mining proposal, a biocentrist might oppose mine development if it would destroy many plants and animals, even if it would generate economic benefits and pose no threat to human health.

Ecocentrism judges actions in terms of their effects on whole ecological systems, which consist of living and nonliving elements and the relationships among them. An ecocentrist values the well-being of entire species, communities, or ecosystems (we

FIGURE 1.10 ▲ We can categorize people's ethical perspectives as anthropocentric, biocentric, or ecocentric. An anthropocentrist grants ethical standing only to human beings and judges actions solely in terms of their effects on people. A biocentrist values and considers all living things, human and otherwise. An ecocentrist extends ethical consideration to living and nonliving components of the environment and takes a holistic view of the connections among these components, valuing the larger functional systems of which they are a part.

will study these in Chapters 2–4) over the welfare of a given individual. Implicit in this view is that preserving systems generally protects their components, whereas just protecting certain components may not safeguard the entire system. Ecocentrism is a more holistic perspective than biocentrism or anthropocentrism. It encompasses a wider variety of entities and seeks to preserve the connections that tie them together into functional systems.

Conservation and preservation arose with the 20th century

With the onset of the industrial revolution, more people began adopting biocentric and ecocentric worldviews. In the 19th and 20th centuries, worldviews of people in the United States evolved quickly as the nation pushed west, urbanized, and exploited the continent's resources, boosting affluence and dramatically altering the landscape in the process.

A key voice for restraint during this period of rapid growth and change was **John Muir** (1838–1914), a Scottish immigrant to the United States who made California's Yosemite Valley his wilderness home. Although Muir chose to live in solitude in his beloved Sierra Nevada for long stretches of time, he also became politically active and won fame as a tireless advocate for the preservation of wilderness (**FIGURE 1.11**).

Muir was motivated by the rapid deforestation he witnessed throughout North America and by his belief that the natural world should be treated with the same respect we give to cathedrals. He promoted the **preservation** ethic, which holds that we should protect our environment in a pristine, unaltered state. Muir argued that nature deserved protection for its own inherent value (an ecocentric argument), but he also maintained that nature promoted human happiness (an anthropocentric

FIGURE 1.11 ▼ A pioneering advocate of the preservation ethic, John Muir is also remembered for his efforts to protect the Sierra Nevada from development and for his role in founding the Sierra Club, a leading environmental organization. Here Muir **(right)** is shown with President Theodore Roosevelt in Yosemite National Park. After his 1903 wilderness camping trip with Muir, the president instructed his interior secretary to increase protected areas in the Sierra Nevada.

FIGURE 1.12 ▲ Gifford Pinchot, the first chief of what would become the U.S. Forest Service, was a leading proponent of the conservation ethic. The conservation ethic holds that people should use natural resources but should strive to ensure the greatest good for the greatest number of people for the longest time.

argument based on the principle of utility). "Everybody needs beauty as well as bread," he wrote, "Places to play in and pray in, where nature may heal and give strength to body and soul alike."

Some of the factors that motivated Muir also inspired the forester **Gifford Pinchot** (1865–1946; **FIGURE 1.12**), who founded what would become the U.S. Forest Service and served as its chief in Theodore Roosevelt's administration. Like Muir, Pinchot opposed the deforestation and unregulated development of North American lands. However, Pinchot took a more anthropocentric view of how and why we should value nature. He espoused the **conservation** ethic, which holds that people should put natural resources to use but that we have a responsibility to manage them wisely. The conservation ethic employs a utilitarian standard, stating that we should allocate resources in a way that provides the greatest good to the greatest number of people for the longest time. Whereas preservation aims to preserve nature for its own sake and for our aesthetic and spiritual benefit, conservation promotes the prudent, efficient, and sustainable extraction and use of natural resources for the benefit of present and future generations.

Pinchot and Muir came to represent different branches of the American environmental movement, and their contrasting ethical approaches often pitted them against one another on policy issues of the day. Nonetheless, they both represented reactions against a prevailing "development ethic," which holds that people are and should be masters of nature and which promotes economic development without regard to its negative consequences. Both Pinchot and Muir left legacies that reverberate today in the various ethical approaches to environmentalism.

WEIGHING THE ISSUES

Preservation and Conservation With which ethic do you identify more—preservation or conservation? Think of a forest or wetland or other important natural resource in your region. Give an example of a situation in which you might adopt a preservation ethic and an example of one in which you might adopt a conservation ethic. Are there conditions under which you'd follow neither, but instead adopt a "development ethic"?

Aldo Leopold's land ethic inspires many people

As a young forester and wildlife manager, **Aldo Leopold** (1887–1949; **FIGURE 1.13**) began his career in the conservationist camp, having graduated from Yale Forestry School, which Pinchot had helped found just as Roosevelt and Pinchot were advancing conservation on the national stage. As a forest manager in Arizona and New Mexico, Leopold embraced the government policy of shooting predators, such as wolves, to increase populations of deer and other game animals.

At the same time, Leopold followed the advance of ecological science. He eventually ceased to view certain species as "good" or "bad" and instead came to see that healthy ecological systems depend on protecting all their interacting parts. Drawing an analogy to mechanical maintenance, he wrote, "to keep every cog and wheel is the first precaution of intelligent tinkering."

It was more than science that pulled Leopold from an anthropocentric perspective toward a more holistic one. One day he shot a wolf, and when he reached the animal, Leopold was transfixed by "a fierce green fire dying in her eyes." The experience remained with him for the rest of his life and helped lead him to an ecocentric ethical outlook. Years later, as a University of Wisconsin professor, Leopold argued that people should view themselves and "the land" as members of the same community and that we are obligated to treat the land in an ethical manner. In his 1949 essay "The Land Ethic," he wrote:

> All ethics so far evolved rest upon a single premise: that the individual is a member of a community of interdependent parts. . . . The land ethic simply enlarges the boundaries of the community to include soils, waters, plants, and animals, or collectively: the land. . . . A land ethic changes the role of *Homo sapiens* from conqueror of the land-community to plain member and citizen of it. . . . It implies respect for his fellow-members, and also respect for the community as such.

Leopold intended that the land ethic would help guide decision making. "A thing is right," he wrote, "when it tends to preserve the integrity, stability, and beauty of the biotic community. It is wrong when it tends otherwise." Leopold died before seeing "The Land Ethic" and his best-known book, *A Sand County Almanac,* in print, but today many view him as the most eloquent and important philosopher of environmental ethics.

Environmental justice seeks fair treatment for all people

Our society's domain of ethical concern has been expanding from rich to poor and from majority races and ethnic groups to minority ones. This ethical expansion involves applying a standard of fairness and equality, and it has given rise to the environmental justice movement. **Environmental justice** involves the fair and equitable treatment of all people with respect to environmental policy and practice, regardless of their income, race, or ethnicity.

The environmental justice movement has been fueled by the perception that poor people tend to be exposed to a greater share of pollution, hazards, and environmental degradation than are richer people. Environmental justice advocates also note that racial and ethnic minorities tend to suffer more exposure to hazards than whites. Indeed, studies across North America repeatedly document that poor and nonwhite communities each bear heavier burdens of air pollution, lead poisoning, pesticide exposure, toxic waste exposure, and workplace hazards. This is thought to occur because lower-income and minority communities often have less access to information on environmental health risks, less political power with which to protect their interests, and less money to spend on avoiding or alleviating risks. Environmental justice proponents also sometimes blame institutionalized racism and inadequate government policies.

A protest in the early 1980s by African Americans in Warren County, North Carolina, against a toxic waste dump in their community is widely seen as the beginning of the movement (**FIGURE 1.14**). The state had chosen to establish the dump in the county with the highest percentage of African Americans.

Likewise, white residents of the Appalachian region have long been the focus of environmental justice concerns. Mountaintop coal mining practices (p. 240) in this economically neglected region provide some jobs to local residents, but also pollute water, bury streams, degrade forests,

FIGURE 1.13 ▲ Aldo Leopold, a wildlife manager and pioneering environmental philosopher, articulated a new relationship between people and the environment. In his essay "The Land Ethic," he called on people to include the environment in their ethical framework.

and cause flooding. Low-income residents of affected Appalachian communities have historically had little political power to voice complaints over the impacts of these mining practices.

Today, although our economies have grown, the gaps between rich and poor have widened. And despite much progress toward racial equality in Western societies, significant inequities remain. Although environmental laws have proliferated, minorities and the poor still suffer substandard environmental conditions (**FIGURE 1.15**). Yet today more people are fighting environmental hazards in their communities and winning.

One success story is in California's San Joaquin Valley. The poor, mostly Latino, farm workers in this region who help harvest much of the U.S. food supply also suffer some of the nation's worst air pollution. Industrial agriculture produces pesticide emissions, dairy feedlot emissions, and windblown dust from eroding farmland, yet this pollution was not being regulated. Valley residents enlisted the help of several organizations, including the Center on Race, Poverty, and the Environment, a San Francisco–based environmental justice law firm, and succeeded in convincing California regulators to enforce Clean Air Act provisions and California legislators to pass new laws regulating agricultural emissions.

WEIGHING THE ISSUES

Environmental Justice Consider the place where you grew up. Where were the factories, waste dumps, and polluting facilities located, and who lived closest to them? Who lives nearest them in the town or city that hosts your campus? Do you think the concerns of environmental justice advocates are justified? If so, what could be done to ensure that poor communities are no more polluted than wealthy ones?

FIGURE 1.14 ▼ Communities of poor people and people of color have suffered more than their share of environmental hazards, a situation that has given rise to the environmental justice movement. The movement gained prominence with this protest against a toxic waste dump in Warren County, North Carolina.

FIGURE 1.15 ▲ Hurricane Katrina revealed our ongoing need for environmental justice because the people affected most by the storm and its aftermath were poor and nonwhite. These girls are playing in the Lower Ninth Ward of New Orleans, where many homes were destroyed and water remained unsafe to drink long afterwards. Their mother had moved back here after Katrina destroyed her home, but poverty forced her to accept donated gutted housing once the Federal Emergency Management Agency cut off payments.

Just as wealthy people often impose their pollution on poorer people, wealthy nations do the same to poorer nations. For instance, the millions of tons of hazardous waste that we in developed nations produce in our factories, power plants, and incinerators must go somewhere (p. 394). Proper disposal is expensive, so companies often find it cheaper to pay cash-strapped nations to take the waste—and cheaper still to dump it illegally. In developing nations with lax environmental and health regulations, workers and residents are often uninformed of or unprotected against the dangers from this waste. An international treaty, the Basel Convention, prohibits the international export of waste, but trade and illegal dumping continue. Although 169 nations have ratified the treaty, the United States, the world's largest exporter, has not. Environmental justice at all levels is a key component in pursuing the environmental, economic, and social goals of the modern drive for sustainability.

SUSTAINABILITY AND THE FUTURE OF OUR WORLD

Recall the ethical question posed earlier (p. 12): Is the present generation obligated to conserve resources for future generations? This question cuts to the core of **sustainability,** a guiding principle of modern environmental science and a concept you will encounter throughout this book.

Sustainability means living within our planet's means, such that Earth and its resources can sustain us—and all life—for the foreseeable future. Sustainability means leaving our children and grandchildren a world as rich and full as the world we live in now. It means conserving Earth's resources so that our descendants may enjoy them as we have. It means developing solutions that work in the long term. Sustainability requires maintaining fully functioning ecological systems, because we cannot sustain human civilization without sustaining the natural systems that nourish it.

We can think of our planet's resources as a bank account. If we deplete resources, we draw down the bank account. However, we can choose instead to use the interest and leave the principal intact so that we can continue using the interest far into the future. Currently we are drawing down Earth's **natural capital,** its accumulated wealth of resources. Recall (p. 4) that one research group estimates that we are withdrawing our planet's natural capital 50% faster than it is being replenished. To live off *nature's interest*—its replenishable resources—is sustainable. To draw down resources faster than they are replaced is to eat into *nature's capital,* and we cannot get away with this for long.

Population and consumption drive environmental impact

Humanity is placing an ever-greater burden on Earth's systems. We add over 200,000 people to the planet *each day,* and the ongoing growth of human population amplifies nearly all of our environmental impacts (Chapter 6). Our consumption of resources has risen even faster than our population. The modern rise in affluence has been a positive development for humanity, and our conversion of the planet's natural capital has made life more pleasant for us so far. However, like rising population, rising per capita consumption magnifies the demands we make on our environment.

Moreover, the world's citizens have not benefited equally from the overall rise in affluence. Today the 20 wealthiest nations boast over 55 times the per capita income of the 20 poorest nations—nearly three times the gap that existed just four decades ago. The ecological footprint of the average citizen of a developed nation such as the United States is considerably larger than that of the average resident of a developing country (**FIGURE 1.16**).

WEIGHING THE ISSUES

Ecological Footprints What do you think accounts for the variation in sizes of per capita ecological footprints among societies? Do you think that nations with larger footprints have an ethical obligation to reduce their environmental impact, so as to leave more resources available for nations with smaller footprints? Why or why not?

Our growing population and consumption are intensifying the many environmental impacts we examine in this book, including erosion and other impacts from agriculture (Chapter 7), deforestation (Chapter 9), toxic substances (Chapter 10), mineral extraction and mining impacts (Chapter 11), fresh water depletion (Chapter 12), fisheries declines (Chapter 12), air and water pollution (Chapters 12 and 13), waste generation (Chapter 17), and, of course, global climate change (Chapter 14). These impacts degrade our health and quality of life, and they alter the ecosystems and landscapes in which we all live (**FIGURE 1.17**). They are also driving the loss of Earth's biodiversity (Chapter 8)—perhaps our greatest problem, because extinction is irreversible; once a species becomes extinct, it is lost forever.

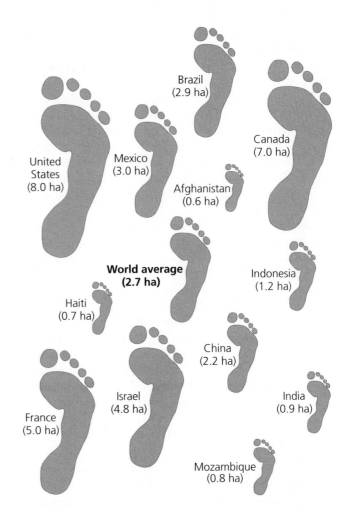

FIGURE 1.16 ▲ The citizens of some nations have much larger ecological footprints than the citizens of others. Shown are ecological footprints for average citizens of several developed and developing nations, along with the world's average per capita footprint of 2.7 hectares. One hectare (ha) = 2.47 acres. Data are for 2007, from Global Footprint Network, 2010.

The most comprehensive assessment of the condition of the world's ecological systems and their capacity to continue supporting us was completed in 2005, when over 2,000 leading environmental scientists from nearly 100 nations completed the **Millennium Ecosystem Assessment.** The Millennium Ecosystem Assessment makes clear that our degradation of environmental systems is having negative impacts on all of us, but that with care and diligence we can still turn many of these trends around.

Sustainable solutions abound

Humanity's challenge is to develop solutions that enhance our quality of life while protecting and restoring the environment that supports us. How we tackle this challenge will largely determine the nature of our lives in the 21st century and beyond. Fortunately, many workable solutions are at hand. For instance:

▶ Renewable energy sources (Chapter 16) are being developed to replace fossil fuels. In extracting fossil fuels, we have been splurging on a one-time, short-lived

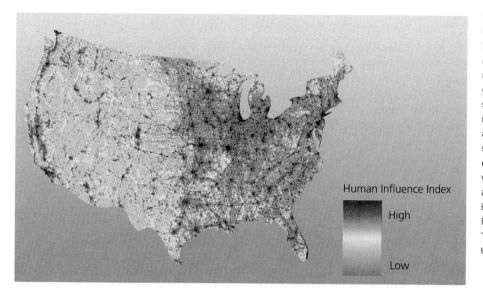

FIGURE 1.17 ◀ Human activity has heavily influenced much of the United States, especially in cities, on farms, and across the eastern portion of the nation. This map summarizes influence on terrestrial ecosystems by human settlement, roads and transportation networks, nighttime light pollution, and agriculture and other land use. It demonstrates that we live in a highly modified environment and suggests we would be wise to carefully nurture natural systems and manage remaining resources. Used by permission of the Center for International Earth Science Information Network (CIESIN), The Earth Institute, Columbia University. © 2008.

Human Influence Index

High

Low

bonanza, and researchers estimate that we have depleted nearly half the world's oil supplies (Chapter 15). Today scientists, engineers, and policymakers are working to develop alternative energy options and to increase energy efficiency.

▶ In response to agricultural impacts, scientists and others have developed and promoted soil conservation, high-efficiency irrigation, and organic agriculture (Chapter 7).

▶ Legislation and technological advances have reduced the pollution emitted by industry and automobiles in wealthier countries (Chapters 5 and 13).

▶ Conservation biologists are helping to protect habitat, slow extinction, and safeguard endangered species (Chapter 8).

▶ Recycling is helping to conserve resources and relieve waste disposal problems (Chapter 17).

▶ Governments, businesses, and individuals are taking steps to reduce emissions of the greenhouse gases that drive climate change (Chapter 14).

Sustainable development involves environmental protection, economic well-being, and social justice

Today's search for sustainable solutions centers on **sustainable development,** the use of resources in a manner that satisfies our current needs but does not compromise the future availability of resources. Sustainable development aims to enhance people's quality of life while preserving environmental quality. The modern drive for sustainable development arose from the recognition that society's poorer people often suffer the most from environmental degradation. This realization led advocates of environmental protection, advocates of economic development, and advocates of social justice to begin working together toward common goals. Increasingly, sustainable development efforts by governments, businesses, industries, organizations, and individuals everywhere—from students on campus (pp. 414–415) to international representatives at the United Nations (**FIGURE 1.18**)—are generating sustainable solutions that satisfy a

triple bottom line by meeting environmental, economic, and social goals simultaneously.

Sustainability and the triple bottom line require that we limit our environmental impact while promoting economic well-being and social equity. These aims oblige us to make an ethical commitment to our fellow citizens and to future generations. They also require that we apply knowledge from the sciences to help devise ways to limit our impact and maintain the environmental systems on which we depend.

The question "How can we develop in a sustainable way?" may be the single most important question we face. Environmental science holds the key to addressing this question. Because so much remains to be studied and done, and because it is so central to our modern world, environmental science will remain an exciting frontier for you to explore as a student today and as an informed citizen throughout your life.

FIGURE 1.18 ▼ Former South African President Thabo Mbeki hugs a boy who performed in the welcoming ceremony of the United Nations–sponsored World Summit on Sustainable Development. Held in Johannesburg, South Africa, in 2002, the summit hosted 10,000 delegates from 200 nations who set sustainable development goals. Sustainability requires that each generation leave enough resources for future generations to live as well or better.

➤ CONCLUSION

Finding effective ways of living peacefully, healthfully, and sustainably on our diverse and complex planet requires a solid ethical grounding and a thorough scientific understanding of natural and social systems. Environmental science helps us understand our intricate relationship with our environment and informs our attempts to solve and prevent environmental problems. Many of today's trends may worry us, but others give us reason to hope—and identifying a problem is the first step toward devising a solution. Solving environmental problems can move us toward health, longevity, peace, and prosperity. Science in general, and environmental science in particular, can aid us in our efforts to develop balanced, workable, sustainable solutions and to create a better world for ourselves and our children.

TESTING YOUR COMPREHENSION

1. How and why did the agricultural revolution affect human population size? How and why did the industrial revolution affect human population size? Explain some social consequences and some environmental impacts that have resulted.

2. What is the *tragedy of the commons*? Explain how the concept might apply to an unregulated industry that is a source of water pollution.

3. What is *environmental science*? Name several disciplines involved in environmental science.

4. Contrast the two meanings of *science*. Name three applications of science.

5. Describe the scientific method. What is its typical sequence of steps? What needs to occur before a researcher's results are published? Why is this process important?

6. What does the study of ethics encompass? Differentiate two classic ethical standards, the categorical imperative and the principle of utility. What is environmental ethics?

7. Compare and contrast anthropocentrism, biocentrism, and ecocentrism.

8. Differentiate the preservation ethic from the conservation ethic. Explain the contributions of John Muir and Gifford Pinchot in the history of environmental ethics.

9. Describe Aldo Leopold's land ethic. How did Leopold define the "community" to which ethical standards should be applied?

10. What is *sustainable development,* and why is it important? What is meant by the *triple bottom line*?

SEEKING SOLUTIONS

1. Resources such as soils, timber, fresh water, food crops, and biodiversity are renewable if we use them in moderation but can become nonrenewable if we overexploit them (see Figure 1.1, p. 3). For each of these five resources, describe one way in which we sometimes overexploit them, and one thing we could do to conserve them. In supplying your answers, feel free to look ahead and peruse coverage of these issues throughout this book.

2. Why do you think the Easter Islanders did not or could not stop themselves from stripping their island of its trees? What similarities do you perceive between Easter Island and the modern history of our society? What differences do you see between their predicament and ours?

3. What environmental problem do *you* feel most acutely yourself? Do you think there are people in the world who do not view your issue as a problem? Who might they be, and why might they take a different view?

4. Describe your ethical perspective, or worldview, as it pertains to your relationship with the environment. How do you think your culture has influenced your worldview? How do you think your personal experience has influenced it? Do you feel that you fit into any particular category discussed in this chapter? Why or why not?

5. **THINK IT THROUGH** You have become head of a major funding agency that grants money to researchers pursuing work in environmental science. You must give your staff several priorities to determine what types of scientific research to fund. What environmental problems would you most like to see addressed with research? Describe the research you think would need to be completed so that workable solutions to these problems could be developed. What else besides scientific information would be needed to develop sustainable solutions?

CALCULATING ECOLOGICAL FOOTPRINTS

Mathis Wackernagel and his colleagues at the Global Footprint Network have continued to refine the method of calculating ecological footprints—the amount of biologically productive land and water required to produce the energy and natural resources we consume and to absorb the wastes we generate. According to their most recent data, there are 1.8 hectares (4.4 acres) available for each person in the world, yet we use on average 2.7 ha (6.7 acres) per person, creating a global ecological deficit, or overshoot (p. 4), of 50%.

Compare the ecological footprints of each nation listed in the table. Calculate their proportional relationships to the world population's average ecological footprint and to the area available globally to meet our ecological demands.

Nation	Ecological footprint (hectares per person)	Proportion relative to world average footprint	Proportion relative to world area available
Bangladesh	0.6	0.2 (0.6 + 2.7)	0.3 (0.6 + 1.8)
Tanzania	1.2		
Colombia	1.9		
Thailand	2.4		
Mexico	3.0		
Sweden	5.9		
United States	8.0		
World average	2.7	1.0 (2.7 + 2.7)	1.5 (2.7 + 1.8)
Your personal footprint (see question 4)			

Data from *Living planet report 2010*. WWF International, Zoological Society of London, and Global Footprint Network.

1. Why do you think the ecological footprint for people in Bangladesh is so small?
2. Why is it so large for people in the United States?
3. Based on the data in the table, how do you think average per capita income affects ecological footprints?
4. Go to an online footprint calculator such as the one at http://www.myfootprint.org or http://www.footprintnetwork.org/en/index.php/GFN/page/personal_ footprint, and take the test to determine your own personal ecological footprint. Enter the value you obtain in the table, and calculate the other values as you did for each nation. How does your footprint compare to those of the average person in the United States? How does it compare to that of people from other nations? Name three actions you could take to reduce your footprint.

Go to **www.masteringenvironmentalscience.com** for homework assignments, practice quizzes, Pearson eText, and more.

2 Environmental Systems: Matter, Energy, and Ecosystems

Upon completing this chapter, you will be able to:

➤ Describe the nature of environmental systems

➤ Explain and apply the fundamentals of environmental chemistry

➤ Describe the molecular building blocks of organisms

➤ Differentiate among the types of energy and explain the basics of energy flow

➤ Distinguish photosynthesis from respiration and summarize their importance to living things

➤ Define *ecosystem* and evaluate how living and nonliving entities interact in ecosystem-level ecology

➤ Outline the fundamentals of landscape ecology and ecological modeling

➤ Assess ecosystem services and how they benefit our lives

➤ Describe how water, carbon, phosphorus, and nitrogen cycle through the environment

An oysterman unloads his catch on the shores of the Chesapeake Bay

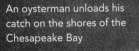

The Vanishing Oysters of the Chesapeake Bay

"I'm 60. Danny's 58. We're the young ones."
—Grant Corbin, Oysterman in Deal Island, Maryland

"The Bay continues to be in serious trouble. And it's really no question why this is occurring. We simply haven't managed the Chesapeake Bay as a system the way science tells us we must."
—Will Baker, President, Chesapeake Bay Foundation

A visit to Deal Island, Maryland, on the Chesapeake Bay reveals a situation that unfortunately is all too common in modern America. The island, which was once bustling with productive industries and growing populations, is suffering. Economic opportunities in the community are few, and its populace is increasingly "graying" as more and more young people leave to find work elsewhere. In 1930, Deal Island had a population of 1,237 residents. In 2000 it held a mere 578—and only one household in five included children.

Unlike other parts of the country with similar stories of economic decline, the demise of Deal Island and other bayside towns was not caused by the closing of a local factory, steel mill, or corporate headquarters. It was caused by people decimating the Chesapeake Bay oyster fishery.

The Chesapeake Bay was once a thriving system of interacting plants, animals, and microbes. Healthy populations of blue crabs, scallops, and fish such as giant sturgeon, striped bass, and shad inhabited the bay. Nutrients carried to the bay by thousands of streams in the bay's roughly 168,000-km^2 (64,000-mi^2) **watershed** (the land area that funnels water to the bay through rivers) nourished fields of underwater grasses that provided food and refuge to juvenile fish, shellfish, and crabs. Hundreds of millions of oysters kept the bay's water clear by filtering nutrients and *phytoplankton* (microscopic photosynthetic algae, protists, and cyanobacteria that drift near the surface) from the water column.

Although oysters had been eaten locally for some time, the intensive harvest of bay oysters for export began in the 1830s, and by the 1880s the bay boasted the world's largest oyster fishery.

People flocked to the Chesapeake to work on oystering ships or in canneries, shucking houses (where oysters are separated from their shells), dockyards, and shipyards. Bayside towns prospered along with the oyster industry and developed a unique maritime culture that defined the region.

But by 2010 the bay's oysters had been reduced to a mere 1% of their historical abundance, and the oyster industry was all but ruined. Perpetual overharvesting, habitat destruction, virulent oyster diseases, and water pollution had nearly eradicated this economically and ecologically important species from bay waters. The monetary losses associated with this fishery collapse have been staggering, costing the economies of Maryland and Virginia an estimated $4 billion in lost economic activity from 1980 to 2010 alone.

One of the biggest impacts in recent decades on oysters is the pollution of the bay with high levels of the nutrients nitrogen and phosphorus from agricultural fertilizers, animal manure, stormwater runoff, and atmospheric compounds produced by fossil fuel combustion. Elevated levels of these nutrients cause the number of phytoplankton in the bay to explode as they are no longer held in check by extensive oyster filtration of bay waters. When phytoplankton die, settle to the bay bottom, and are decomposed by bacteria, this depletes oxygen in the water (a condition called **hypoxia**) and creates "dead zones" in the bay. Grasses, oysters, and other immobile organisms perish in dead zones when deprived of oxygen. Crabs, fish, and other mobile organisms are forced to flee to habitats where oxygen levels are higher, but they face smaller food supplies and increased predation pressure. Collectively, hypoxia affects numerous components of the Chesapeake Bay system and contributes to the bay's listing on the Environmental Protection Agency's list of highly polluted waters.

Recent events have, at long last, given reason for hope for the recovery of the Chesapeake Bay system. The EPA agreed in 2010, for the first time in the region, to hold bay states to strict pollutant "budgets" whose aim is to reduce inputs of nitrogen and phosphorus into Chesapeake Bay by one-third of current levels by 2025. Further, oyster restoration efforts are finally showing promise (see **THE SCIENCE BEHIND THE STORY,** pp. 34–35), and aquaculture shows potential for reinvigorating the oyster industry in the Chesapeake. If these efforts prove successful and we can begin to restore the bay to health, Deal Island and other communities may once again enjoy prosperity on the scenic shores of the Chesapeake. ■

EARTH'S ENVIRONMENTAL SYSTEMS

Understanding the rise and fall of the oyster industry in the Chesapeake Bay, like many other human impacts on the environment, involves comprehending environmental systems and how they function. Our planet's environment consists of complex networks of interlinked systems. These systems include processes that shape the land, air, water, and climate; ecological webs of relationships among species; and the interaction of living organisms with the nonliving entities around them. Earth's systems also include the cycles of key chemical elements and compounds that support life and regulate climate. We depend on these systems for our very survival.

Taking a "systems approach" is helpful in environmental science because so many issues are multifaceted and interconnected. This type of approach poses challenges, however, because systems often show behavior that is difficult to understand and predict. Even so, environmental scientists are rising to the challenge of studying systems holistically, helping us to develop comprehensive solutions to complicated problems such as those faced in the Chesapeake Bay.

Systems involve feedback loops

A **system** is a network of relationships among components that interact with and influence one another through the exchange of energy, matter, or information. Systems receive inputs of energy, matter, or information; process these inputs; and produce outputs. For example, the Chesapeake Bay receives inputs of fresh water, sediments, nutrients, and pollutants from the rivers that empty into it. Oystermen, fishermen, and crabbers harvest some of the Bay system's output: matter and energy in the form of seafood. This output subsequently becomes input to the human economic system and to the digestive systems of the many people who consume the seafood.

Sometimes a system's output can serve as input to that same system, a circular process known as a **feedback loop**. Feedback loops are of two types, negative and positive. In a **negative feedback loop** (**FIGURE 2.1A**) output that results from a system moving in one direction acts as input that moves the system in the other direction. Input and output essentially neutralize one another's effects, stabilizing the system. A thermostat, for instance, stabilizes a room's temperature by turning the furnace on when the room gets cold and shutting it off when the room gets hot. Similarly, negative feedback regulates our body temperature. If we get too hot, our sweat glands pump out moisture that evaporates to cool us down, or we may move from sun to shade. If we get too cold, we shiver, creating heat, or we move into the sun or put on more clothing. Most systems in nature involve negative feedback loops. Negative feedback loops enhance stability, and in the long run, only those systems that are stable will persist.

Positive feedback loops have the opposite effect. Rather than stabilizing a system, they drive it further toward an extreme. One positive feedback cycle of great concern to environmental scientists today involves the melting of glaciers and sea ice in the Arctic due to global warming (pp. 308–309). Ice and snow, being white, reflect sunlight and keep surfaces cool. But if the climate warms enough to melt the ice and snow, darker surfaces of land and water are exposed, and these darker surfaces absorb more sunlight. This absorption of light warms the surface, causing further melting, which in turn exposes more dark surface area, leading to further warming (**FIGURE 2.1B**). Runaway cycles of positive feedback are rare in nature, but they are common in natural systems altered by human impact, and they can destabilize those systems.

Environmental systems interact

Natural systems can be categorized in many different ways. For instance, scientists divide Earth's major components into structural spheres. The **lithosphere** (p. 227) contains the rock and sediment beneath our feet, in the planet's uppermost layers. The **atmosphere** (p. 279) is composed of the air surrounding our planet.

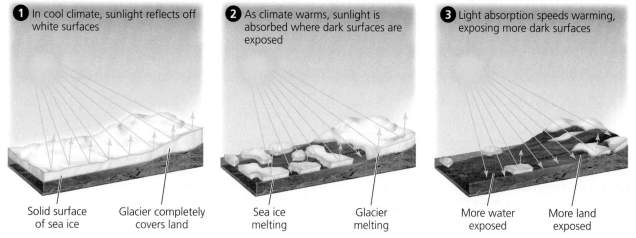

(a) Negative feedback

FIGURE 2.1 ◀ Negative feedback loops **(a)** exert a stabilizing influence on systems and are common in nature. The human body's response to heat and cold involves a negative feedback loop. Positive feedback loops **(b)** have a destabilizing effect on systems and push them toward extremes. As Arctic glaciers and sea ice melt because of global warming, darker surfaces are exposed, which absorb more sunlight, causing further warming and further melting.

❶ In cool climate, sunlight reflects off white surfaces

❷ As climate warms, sunlight is absorbed where dark surfaces are exposed

❸ Light absorption speeds warming, exposing more dark surfaces

Solid surface of sea ice   Glacier completely covers land   Sea ice melting   Glacier melting   More water exposed   More land exposed

(b) Positive feedback

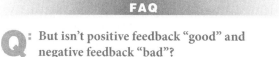

FAQ

Q: But isn't positive feedback "good" and negative feedback "bad"?

A: Understanding negative and positive feedback in systems can be difficult, because it goes against the the way we use those terms in everyday language. In daily life, positive feedback (such as a complimentary comment on a paper written for a course in school) is something that makes us feel good, whereas negative feedback (such as criticism on a paper) may make us feel bad. In essence, we have been trained to view positive feedback as a stabilizing force ("Keep up the good work, and you'll get a good grade") and negative feedback as a destabilizing force ("You need to change your approach if you're going to succeed").

In environmental systems, it's the opposite! Negative feedback works against change in systems, and in doing so it enhances stability, typically keeping conditions within ranges beneficial to organisms. Positive feedback exerts destabilizing effects that push conditions in systems to extremes, threatening organisms adapted to the system's normal conditions. Thus, negative feedback in environmental systems keeps conditions stable for living things whereas positive feedback can harm them.

The **hydrosphere** (p. 250) encompasses all water—salt or fresh, liquid, ice, or vapor—in surface bodies, underground, and in the atmosphere. The **biosphere** (p. 52) consists of all the planet's living organisms and the abiotic (nonliving) portions of the environment with which they interact. Categorizing environmental systems in this manner can help make Earth's dazzling complexity comprehensible, but it's important to remember that most natural systems overlap or interact.

The Chesapeake Bay and the rivers that empty into it are an example of interacting systems. On a map, these rivers are a branched and braided network of water channels surrounded by farms, cities, and forests (**FIGURE 2.2**). But where are this system's boundaries? For a scientist interested in runoff and the flow of water, sediment, or pollutants, it may make the most sense to view the Chesapeake Bay's watershed as a system. However, for a scientist interested in the Bay's dead zones, it may be best to view the watershed together with the Chesapeake Bay as the system of interest, because their interaction is central to the problem. In environmental science, identifying the boundaries of systems depends on the questions being addressed.

The dead zones in the Chesapeake Bay are due to the extremely high levels of nitrogen and phosphorus delivered to its waters from the six states in its watershed and the 15 states in its *airshed* (the geographic area that produces air pollutants that are likely to end up in a waterway). In

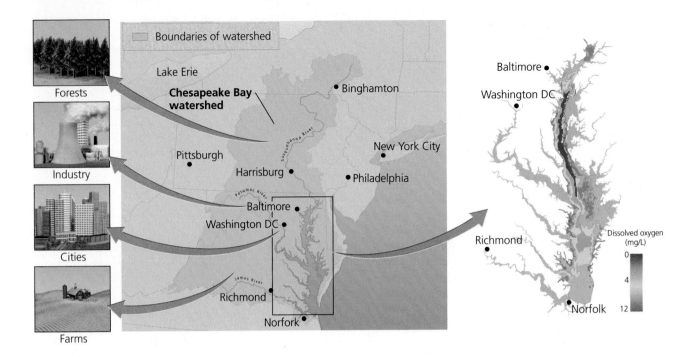

FIGURE 2.2 ▲ The Chesapeake Bay watershed encompasses 168,000 km² (64,000 mi²) of land area in six states and the District of Columbia. Tens of thousands of streams carry water, sediment, and pollutants from a variety of sources downriver to the Chesapeake, where nutrient pollution has given rise to large areas of hypoxic waters. The zoomed-in map **(at right)** shows dissolved oxygen concentrations in the Chesapeake Bay in summer 2007. Oysters, crabs, and fish typically require a minimum of 3 mg/L of oxygen and are therefore excluded from large portions of the bay where oxygen levels are too low. Figure at right adapted from Chesapeake Bay Record Dead Zone Map, Chesapeake Bay Foundation.

2007, the Bay received an estimated 127 million kg (281 million lb) of nitrogen and 8.3 million kg (18.2 million lb) of phosphorus, with roughly one-third of nitrogen inputs from atmospheric sources. Agriculture was a major source of these nutrients, contributing 38% of the nitrogen **(FIGURE 2.3A)** and 45% of the phosphorus **(FIGURE 2.3B)** entering the bay.

Elevated nitrogen and phosphorus inputs cause phytoplankton—microscopic algae and other organisms drifting near the surface—to flourish. High densities lead

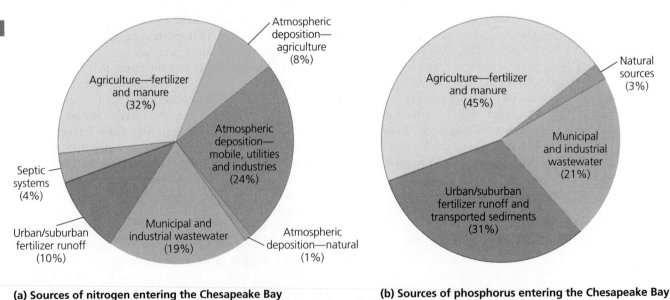

(a) Sources of nitrogen entering the Chesapeake Bay

(b) Sources of phosphorus entering the Chesapeake Bay

FIGURE 2.3 ▲ The Chesapeake Bay receives inputs of nitrogen **(a)** and phosphorus **(b)** from many sources in its watershed. Data from Chesapeake Bay Program Watershed Model Phase 4.3 (Chesapeake Bay Program Office, 2009).

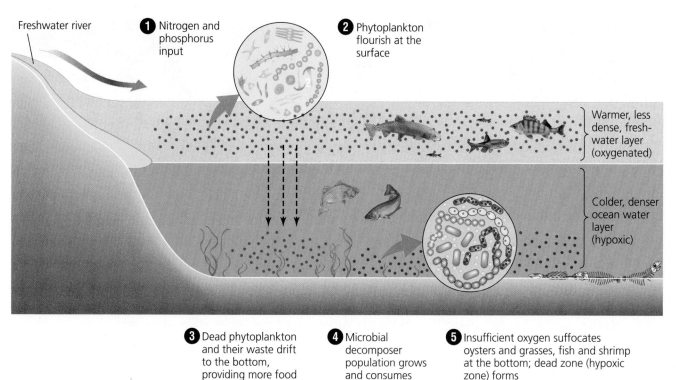

Freshwater river

1 Nitrogen and phosphorus input

2 Phytoplankton flourish at the surface

Warmer, less dense, fresh-water layer (oxygenated)

Colder, denser ocean water layer (hypoxic)

3 Dead phytoplankton and their waste drift to the bottom, providing more food for bacteria to decompose

4 Microbial decomposer population grows and consumes more oxygen

5 Insufficient oxygen suffocates oysters and grasses, fish and shrimp at the bottom; dead zone (hypoxic zone) forms

FIGURE 2.4 ▲ Excess nitrogen and phosphorus causes eutrophication in aquatic systems such as the Chesa-peake Bay. Coupled with stratification (layering) of water, eutrophication can severely deplete dissolved oxygen. Nutrients from river water **1** boost growth of phytoplankton **2**, which die and are decomposed at the bottom by bacteria **3**. Stability of the surface layer prevents deeper water from absorbing oxygen to replace oxygen con-sumed by decomposers **4**, and the oxygen depletion suffocates or drives away bottom-dwelling marine life **5**. This process gives rise to hypoxic zones like those in the bay.

to elevated mortality in phytoplankton populations, and dead phytoplankton settle to the bottom. The remains of dead phytoplankton are joined on the bottom by the waste products of *zooplankton,* tiny creatures that feed on phy-toplankton. This abundance of organic material causes an explosion in populations of bacterial decomposers, which deplete the oxygen in bottom waters while consuming this material. Deprived of oxygen, organisms will flee if they can or will suffocate if they cannot. Oxygen replenishes slowly at the bottom because fresh water entering the bay from rivers remains naturally stratified in a layer at the surface and mixes slowly with the denser, saltier bay water. This limits the amount of oxygenated surface water that reaches the bottom-dwelling life that needs it. This process of nutri-ent overenrichment, blooms of algae, increased production of organic matter, and subsequent ecosystem degradation is known as **eutrophication** (**FIGURE 2.4**). Eutrophication tends to be driven by increases in nitrogen in marine en-vironments and by increases in phosphorus in freshwater environments, though both contribute to eutrophication in all waters.

Increased nutrient pollution from farms, cities, and industries has led to the development of over 400 document-ed hypoxic dead zones globally as of 2008 (**FIGURE 2.5**), in-cluding that of the Chesapeake Bay as well as a large dead zone that forms each year in the Gulf of Mexico off the

Louisiana coast near the mouth of the Mississippi River (p. 267). Fisheries in these regions have seen reduced catches of seafood, and decreased economic activity, because of human-induced dead zones. The increase in the number of dead zones—there were 162 documented in the 1980s and 49 in the 1960s—re-flects how the activities of people are changing the chemistry of waters around the world. Let's look more closely at how chem-istry is involved in important issues in environmental science.

CHEMISTRY AND THE ENVIRONMENT

Chemistry plays a central role in the environmental challeng-es facing the Chesapeake Bay. Understanding how too much nitrogen or phosphorus in one part of a system can lead to too little oxygen in another requires a good working knowledge of chemistry.

Indeed, examine any environmental issue, and you will likely discover chemistry playing a key role. Chemistry is crucial to understanding how environmental chemicals af-fect the health of humans and wildlife, how air pollutants cause acid precipitation, and how synthetic chemicals thin the ozone layer. To appreciate the importance of chemistry in environmental science, we must begin with a grasp of the fundamentals.

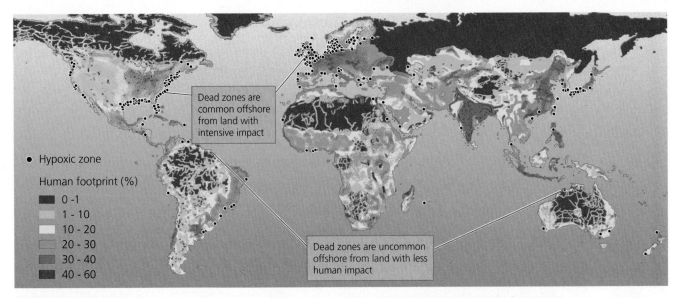

FIGURE 2.5 ▲ Over 400 marine dead zones have been recorded across the world. These dead zones (shown by dots in the map) occur mostly offshore from areas of land with the greatest human ecological footprints (here, expressed on a scale of 0 to 100, with higher numbers indicating bigger human footprints). Data from Diaz, R., and R. Rosenberg, 2008. Spreading dead zones and consequences for marine ecosystems. *Science* 321: 926–929. Reprinted with permission from AAAS.

Atoms and elements are chemical building blocks

All material in the universe that has mass and occupies space is termed **matter**. Matter exists in the universe as a solid, liquid, or gas. Matter may be transformed from one type of substance into others, but it cannot be created or destroyed. This principle is referred to as the **law of conservation of matter**. In environmental science, this principle helps us understand that the amount of matter stays constant as it is recycled in nutrient cycles and ecosystems (pp. 31, 36–41). It also makes it clear that we cannot simply wish away "undesirable" matter, such as nuclear waste or toxic pollutants.

The **nitrogen, phosphorus,** and **oxygen** that play key roles in the Chesapeake Bay's predicament are each elements. An **element** is a fundamental type of matter, a chemical substance with a given set of properties, that cannot be broken down into substances with other properties in chemical reactions. Chemists currently recognize 92 elements occurring in nature, as well as more than 20 others that have been artificially created. Elements needed in large quantities

by organisms, such as **carbon**, nitrogen, and calcium, are called **nutrients**. Each element is assigned an abbreviation, or chemical symbol. The *periodic table of the elements* (see **APPENDIX: Periodic Table of Elements**) summarizes information on the elements in a comprehensive way.

Elements are composed of **atoms**, the smallest components that maintain the chemical properties of the element. An atom's **protons** (positively charged particles) and **neutrons** (particles lacking electrical charge) are in its nucleus. The atoms of each element have a defined number of protons, called the *atomic number*. (Carbon, for instance, has six protons; thus, its atomic number is 6.) An atom's nucleus is surrounded by **electrons** (negatively charged particles), which balance the positive charge of the protons (**FIGURE 2.6**).

Although all atoms of a given element contain the same number of protons, they do not necessarily contain the same number of neutrons. Atoms with differing numbers of neutrons are referred to as **isotopes** (**FIGURE 2.7A**). Isotopes are denoted by their elemental symbol preceded by the *mass number*, or combined number of protons and neutrons in the atom. For example, ^2H (deuterium) is an

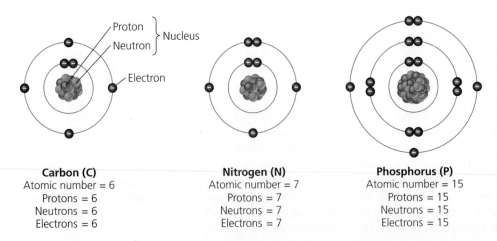

Carbon (C)
Atomic number = 6
Protons = 6
Neutrons = 6
Electrons = 6

Nitrogen (N)
Atomic number = 7
Protons = 7
Neutrons = 7
Electrons = 7

Phosphorus (P)
Atomic number = 15
Protons = 15
Neutrons = 15
Electrons = 15

FIGURE 2.6 ◄ In an atom, protons and neutrons remain in the nucleus, and electrons move around the nucleus. Each chemical element has its own particular number of protons. Carbon possesses 6 protons, nitrogen 7, and phosphorus 15. These schematic diagrams are meant to clearly show and compare numbers of electrons for these three elements. In reality, however, electrons do *not* orbit the nucleus in rings as shown; they move through space in more complex ways.

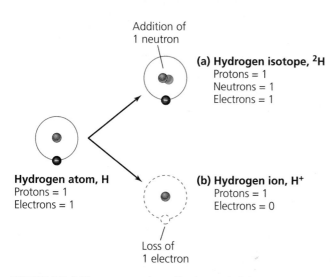

Addition of
1 neutron

(a) Hydrogen isotope, ^{2}H
Protons = 1
Neutrons = 1
Electrons = 1

Hydrogen atom, H
Protons = 1
Electrons = 1

(b) Hydrogen ion, H$^+$
Protons = 1
Electrons = 0

Loss of
1 electron

FIGURE 2.7 ▲ The mass number of hydrogen is 1, because a typical atom of this element contains 1 proton and 0 neutrons. Deuterium (hydrogen-2, or ^{2}H), a different isotope of hydrogen **(a)**, contains a neutron as well as a proton and thus has greater mass than a typical hydrogen atom; its mass number is 2. Shown in **(b)** is the hydrogen ion, H$^+$. By losing its electron, it gains a positive charge.

isotope of **hydrogen** with one neutron (and one proton) in the nucleus rather than the more common ^{1}H ("normal" hydrogen), which contains zero neutrons and one proton. Because they differ slightly in mass, isotopes of an element differ slightly in their behavior.

Although elements cannot be broken down by chemical reactions, some isotopes are *radioactive* and "decay," changing their chemical identity as they shed subatomic particles and emit high-energy radiation. *Radioisotopes* decay into lighter and lighter radioisotopes, until they become *stable isotopes*, isotopes that are not radioactive. Each radioisotope decays at a rate determined by that isotope's *half-life*, the amount of time it takes for one-half the atoms in a given sample to give off radiation and decay. Different radioisotopes have very different half-lives, ranging from fractions of a second to billions of years. The radioisotope uranium-235 (^{235}U) is our society's source of energy for commercial nuclear power (p. 346). It decays into a series of daughter isotopes, eventually forming lead-207 (^{207}Pb), and has a half-life of about 700 million years.

Atoms may also gain or lose electrons to become **ions**, electrically charged atoms or combinations of atoms (**FIGURE 2.7B**). Ions are denoted by their elemental symbol followed by their ionic charge. For instance, a common ion used by mussels and clams to form shells is Ca^{2+}, a calcium atom that has lost two electrons and so has a charge of positive 2.

Atoms bond to form molecules and compounds

Because of attractions between their electrons, atoms can bond together and form **molecules**, combinations of two or more atoms. Common molecules containing only a single element include those of oxygen gas (O$_2$) and nitrogen gas (N$_2$), both of which are abundant in air. As shown in these examples, scientists use a **chemical formula** (such as O$_2$ and N$_2$) as a shorthand way to indicate the type and number of

atoms in the molecule. A molecule composed of atoms of two or more different elements is called a **compound**. Water is a compound composed of two hydrogen atoms bonded to one oxygen atom (H$_2$O). Another compound is **carbon dioxide**, consisting of one carbon atom bonded to two oxygen atoms (CO$_2$).

Ions of differing charge bind with one another to form compounds with **ionic bonds**. A crystal of table salt, sodium chloride (NaCl), is held together by ionic bonds between the positively charged sodium ions (Na$^+$) and the negatively charged chloride ions (Cl$^-$). Atoms that lack an electrical charge combine by "sharing" electrons in **covalent bonds**. For example, two atoms of hydrogen share their electrons when they bind together to form hydrogen gas (H$_2$).

Elements, molecules, and compounds can also come together in **solutions** without chemically bonding. Air in the atmosphere is a solution formed of constituents such as nitrogen, oxygen, water, carbon dioxide, **methane** (CH$_4$), and **ozone** (O$_3$). Human blood, ocean water, plant sap, and metal alloys (p. 236) such as brass are all solutions.

Hydrogen ions determine acidity

In any aqueous solution, a small number of water molecules split apart, each forming a hydrogen ion (H$^+$) and a hydroxide ion (OH$^-$). The product of hydrogen and hydroxide ion concentrations is always 10^{-14}; as one increases, the other decreases. Pure water contains equal numbers of these ions, each at a concentration of 10^{-7}, and we say that this water is **neutral**. Solutions in which the H$^+$ concentration is greater than the OH$^-$ concentration are **acidic**, whereas solutions in which the OH$^-$ concentration is greater than the H$^+$ concentration are **basic**.

The pH scale (**FIGURE 2.8**) was devised to quantify the acidity or basicity of solutions. It runs from 0 to 14. Pure wa-

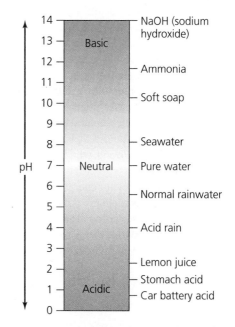

FIGURE 2.8 ▲ The pH scale measures how acidic or basic (alkaline) a solution is. The pH of pure water is 7, the midpoint of the scale. Acidic solutions have higher hydrogen ion concentrations and lower pH, whereas basic solutions have lower hydrogen ion concentrations and higher pH.

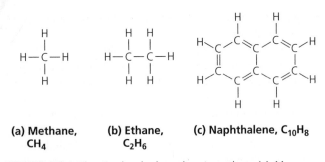

(a) Methane,
CH₄

(b) Ethane,
C₂H₆

(c) Naphthalene, C₁₀H₈

FIGURE 2.9 ▲ The simplest hydrocarbon is methane (a). Many hydrocarbons consist of linear chains of carbon atoms with hydrogen atoms attached; the shortest of these is ethane (b). The air pollutant naphthalene (c) is a ringed hydrocarbon.

ter has a pH of 7 (neutral), solutions with pH less than 7 are acidic, and solutions with pH greater than 7 are basic. Each step on the scale represents a tenfold difference in hydrogen ion concentration. Thus, a substance with pH of 6 contains 10 times as many hydrogen ions as a substance with pH of 7, and 100 times as many hydrogen ions as a substance with pH of 8. Figure 2.8 shows pH for a number of common substances. Most biological solutions have a pH between 6 and 8, and substances that are very acidic (battery acid) or very basic (sodium hydroxide) are harmful to living things. The acidification of soils and water from acid rain (pp. 291–294) in the northeastern and midwestern United States is just one example of how pH changes caused by human activities can affect organisms in the environment.

Matter is composed of organic and inorganic compounds

Beyond their need for water, living things also depend on organic compounds. *Organic compounds* consist of carbon atoms joined by bonds, and they may include other elements, such as hydrogen, nitrogen, oxygen, sulfur, and phosphorus. Carbon's unusual ability to build elaborate molecules by linking carbon molecules to one another in chains, rings, and other structures has resulted in millions of different organic compounds. Because of the diversity of organic compounds and their importance in living organisms, chemists differentiate organic compounds from *inorganic compounds*, which lack carbon–carbon bonds.

One class of organic compound that is important in environmental science is **hydrocarbons**, which consist solely of atoms of carbon and hydrogen. Hydrocarbons make up the fossil fuels we combust for so many of our energy needs (Chapter 15). The simplest hydrocarbon is methane (CH₄), the key component of natural gas; it has one carbon atom covalently bonded to four hydrogen atoms (**FIGURE 2.9**). Crude oil is a complex mixture of hundreds of types of hydrocarbons.

Macromolecules are building blocks of life

Just as the carbon atoms in hydrocarbons may be strung together in chains, other organic compounds sometimes combine to form long chains of repeated molecules. Some of these chains, called *polymers*, play key roles as building blocks of life. Three types of polymers are essential to life: proteins,

nucleic acids, and carbohydrates. Along with lipids (which are not polymers), these four types of essential molecules are referred to as **macromolecules** because of their large size.

Proteins are made up of long chains of organic molecules called *amino acids*. Organisms combine up to 20 types of amino acids into long chains to build proteins. Proteins comprise the majority of each organism's matter and serve many functions in living things. Some help produce tissues and provide structural support. For example, animals use proteins to generate skin, hair, muscles, and tendons. Some proteins help store energy, and others transport substances. Some function in the immune system, defending the organism against foreign attackers. Others act as hormones, molecules that serve as chemical messengers within an organism. Proteins can also serve as enzymes, which are molecules that catalyze, or promote, certain chemical reactions.

Nucleic acids direct the production of proteins. The two nucleic acids—*deoxyribonucleic acid (DNA)* and *ribonucleic acid (RNA)*—carry hereditary information for organisms and are responsible for passing traits from parents to offspring. Nucleic acids are composed of series of nucleotides, each of which contains a sugar molecule, a phosphate group, and a nitrogenous base. DNA includes four types of nucleotides and can be pictured as a ladder twisted into a spiral, giving the molecule a shape called a double helix (**FIGURE 2.10**). Regions of DNA coding for particular proteins that perform particular functions are called **genes**.

Carbohydrates include simple sugars and large molecules comprised of chemically bonded simple sugars. Glucose (C₆H₁₂O₆) fuels living cells and serves as a building block for complex carbohydrates, such as starch. Plants use starch to store energy, and animals eat plants to acquire starch. Plants and animals also use complex carbohydrates to build structure. Insects and crustaceans form hard shells from the

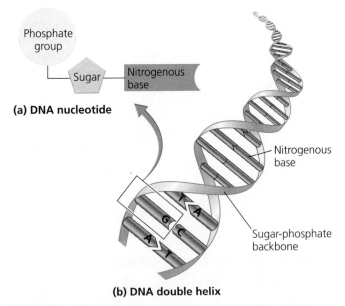

Phosphate group

Sugar

Nitrogenous base

(a) DNA nucleotide

Nitrogenous base

Sugar-phosphate backbone

(b) DNA double helix

FIGURE 2.10 ▲ Nucleic acids encode genetic information in the sequence of nucleotides (a), small molecules that pair together like rungs of a ladder. DNA includes four types of nucleotides, each with a different nitrogenous base: adenine (A), guanine (G), cytosine (C), and thymine (T). Adenine (A) pairs with thymine (T), and cytosine (C) pairs with guanine (G). In RNA, thymine is replaced by uracil (U). DNA twists into the shape of a double helix (b).

carbohydrate chitin. Cellulose, the most abundant organic compound on Earth, is a complex carbohydrate found in the cell walls of leaves, bark, stems, and roots.

Lipids are a chemically diverse group of compounds, classified together because they do not dissolve in water. Lipids include fats and oils (for energy storage), phospholipids (for membranes), waxes (for structure), and steroids (for hormone production).

Organisms use cells to compartmentalize macromolecules

All living things are composed of **cells**, the most basic unit of organismal organization. Organisms range in complexity from single-celled bacteria to plants and animals that contain millions of cells. Cells vary greatly in size, shape, and function. Biologists classify organisms into two groups based on the structure of their cells. The cells of *eukaryotes* (plants, animals, fungi, and protists) contain a membrane-enclosed nucleus and various membrane-enclosed organelles that perform specific functions. *Prokaryotes* (bacteria and archaea) are generally single-celled, and their cells lack membrane-enclosed organelles and a nucleus.

ENERGY FUNDAMENTALS

To create and maintain organized complexity, whether of a cell, an organism, or an ecological system, requires energy. Energy is needed to organize matter into complex forms, to build and maintain cellular structure, to power interactions among species, and to drive the geologic forces that shape our planet. Energy is somehow involved in nearly every biological, chemical, and physical phenomenon.

But what, exactly, is energy? **Energy** is an intangible phenomenon that can change the position, physical composition, or temperature of matter. Scientists differentiate two types of energy: **potential energy**, energy of position; and **kinetic energy**, energy of motion. Consider river water held behind a dam. Prevented from moving downstream, the water accumulates potential energy. When the dam gates are opened, this potential energy is converted to kinetic energy as the water rushes downstream.

Such energy transfers take place at the atomic level every time a chemical bond is broken or formed. **Chemical energy** is potential energy held in the bonds between atoms. Converting a molecule that has high-energy bonds (such as the carbon–carbon bonds of hydrocarbons in crude oil) into molecules with lower-energy bonds (such as the bonds in water or carbon dioxide) releases energy by changing potential energy into kinetic energy, and produces motion, action, or heat. Just as our automobile engines split the hydrocarbons of gasoline to release chemical energy and generate movement, our bodies split glucose molecules from our food for the same purpose (**FIGURE 2.11**).

Energy is always conserved, but it changes in quality

Although energy can change from one form to another, it cannot be created or destroyed. Just as matter is conserved, the total energy in the universe remains constant and thus is said to be conserved. Scientists have dubbed this principle the **first law of thermodynamics**. The potential energy of the water behind a dam will equal the kinetic energy of its eventual movement down the riverbed. Similarly, burning converts the chemical potential energy in a log of firewood to an equal amount of energy produced as heat and light.

Although the overall amount of energy is conserved in any process of energy transfer, the **second law of thermodynamics** states that the nature of energy will change from a more-ordered state to a less-ordered state, if no force counteracts this tendency. That is, systems tend to move toward increasing disorder, or *entropy*. For instance, a log of firewood—the highly organized and structurally complex product of many years of tree growth—transforms in a campfire to carbon ash, smoke, and gases such as carbon dioxide and water vapor, as well as the light and heat of the flame. With the help of oxygen, the complex biological polymers making up the wood are converted into a disorganized assortment of

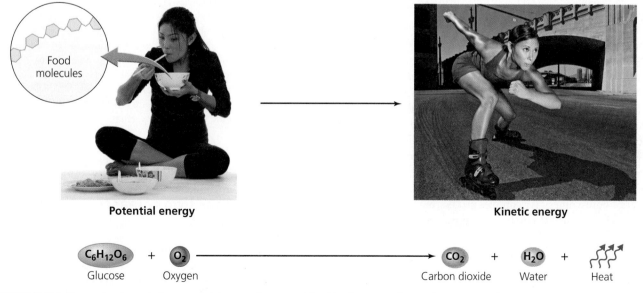

Potential energy   **Kinetic energy**

$$C_6H_{12}O_6 + O_2 \longrightarrow CO_2 + H_2O + \text{Heat}$$

Glucose   Oxygen   Carbon dioxide   Water   Heat

FIGURE 2.11▲ Energy is released when potential energy is converted to kinetic energy. Potential energy stored in sugars (such as glucose) in the food we eat, combined with oxygen, becomes kinetic energy when we exercise, releasing carbon dioxide, water, and heat as by-products.

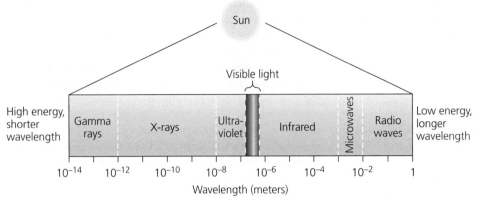

FIGURE 2.12 ◄ The sun emits radiation from many portions of the electromagnetic spectrum. Visible light makes up only a small proportion of this energy. Some radiation that reaches our planet is reflected back; some is absorbed by air, land, and water; and a small amount powers photosynthesis (see Figure 14.1, p. 301).

rudimentary molecules and heat and light energy. When energy transforms from a more-ordered to less-ordered state, it cannot accomplish tasks as efficiently. For example, the energy available in ash (a less-ordered state of wood) is far lower than that available in a log of firewood (the more-ordered state of wood). Living organisms are able to resist entropy through regular inputs of energy from food sources and photosynthesis. Once death occurs and those energy inputs cease, an organism undergoes decomposition and loses its highly organized structure.

Light energy from the sun powers most living systems

The energy that powers Earth's ecological systems comes primarily from the sun. The sun releases radiation from large portions of the electromagnetic spectrum, although our atmosphere filters much of this out and we see only some of this radiation as visible light (**FIGURE 2.12**).

Some organisms use the sun's radiation directly to produce their own food. Such organisms, called **autotrophs** or **producers**, include green plants, algae, and bacteria called cyanobacteria. Autotrophs turn light energy from the sun into chemical energy through a process called **photosynthesis** (**FIGURE 2.13**). In photosynthesis, sunlight powers a series of chemical reactions that convert carbon dioxide and water into sugars, transforming energy from the sun into high-quality energy (sugars) the organism can use. It is an example of moving toward a state of lower entropy, and as such it requires a substantial input of outside energy.

Photosynthesis occurs within cell organelles called *chloroplasts*, where the light-absorbing pigment *chlorophyll* (which is what makes plants green) uses solar energy to initiate a series of chemical reactions called *light reactions*. During these reactions, water molecules split and react to form hydrogen ions (H^+) and molecular oxygen (O_2), thus creating the oxygen that we breathe. The light reactions also produce small, high-energy molecules that are used to fuel reactions in the *Calvin cycle*, where carbon atoms from carbon dioxide are linked together to manufacture sugars. Photosynthesis is a complex process, but the overall reaction can be summarized in the following equation:

$$6CO_2 + 6H_2O + \text{the sun's energy} \longrightarrow C_2H_{12}O_6 + 6O_2$$
$$\text{(sugar)}$$

Thus in photosynthesis, green plants draw up water from the ground through their roots, absorb carbon dioxide from the air through their leaves, and harness the energy in sunlight to create sugars (such as $C_6H_{12}O_6$) for their growth and maintenance. As

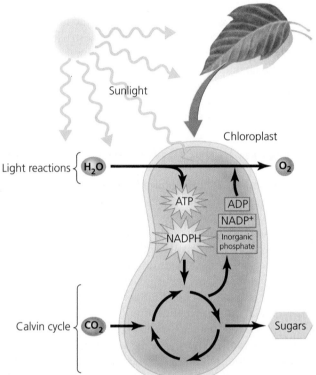

FIGURE 2.13 ▲ In photosynthesis, autotrophs such as plants, algae, and cyanobacteria use sunlight to convert carbon dioxide and water into sugars and oxygen. This schematic diagram summarizes the complex sets of chemical reactions that take place within chloroplasts. In the light reactions, water is converted to oxygen in the presence of sunlight, creating high-energy molecules (ATP and NADPH). These molecules help drive reactions in the Calvin cycle, in which carbon dioxide is used to produce sugars. Molecules of ADP, NADP+, and inorganic phosphate produced in the Calvin cycle in turn help power the light reactions, creating an endless loop.

a by-product, they release the oxygen that we, and all other animals, breathe.

Cellular respiration releases chemical energy

Organisms make use of the chemical energy created by photosynthesis in a process known as **cellular respiration**. To release the chemical energy of glucose, cells use oxygen to break the high-energy chemical bonds in glucose ($C_6H_{12}O_6$) and reform its starting materials, water (H_2O) and carbon dioxide (CO_2). The energy released during this process is used to form new chemical bonds or to perform other tasks within cells. The net equation for cellular respiration is the opposite of that for photosynthesis:

$$C_6H_{12}O_6 + 6O_2 \longrightarrow 6CO_2 + 6H_2O + \text{energy}$$
$$\text{(sugar)}$$

However, the energy gained per glucose molecule in respiration is only two-thirds of the amount of energy required to synthesize a glucose molecule in photosynthesis—a prime example of the second law of thermodynamics in action.

Respiration occurs in autotrophs and also in **heterotrophs**, or **consumers**, organisms that gain their energy by feeding on the biomass of other organisms. Heterotrophs include most animals, as well as the fungi and microbes that decompose organic matter. In most ecological systems, plants, algae, or cyanobacteria form the base of a food chain through which energy passes to heterotrophs (pp. 68–70).

ECOSYSTEMS

Let's now apply our knowledge of chemistry and energy to see how energy, matter, and nutrients move through the living and nonliving environment. An **ecosystem** consists of all organisms and nonliving entities that occur and interact in a particular area. Animals, plants, water, soil, nutrients—all these and more help comprise ecosystems.

Energy flows and matter cycles through ecosystems

The ecosystem concept originated with scientists who recognized that biological entities are tightly intertwined with the chemical and physical aspects of their environment. For instance, in the Chesapeake Bay **estuary** (a water body where rivers flow into the ocean, mixing fresh water with salt water), aquatic organisms are intimately affected by the flow of water, sediment, and nutrients from the rivers that feed the bay and from the land that feeds those rivers. In turn, the photosynthesis, respiration, and decomposition of these organisms influence the chemical composition of the Chesapeake's waters.

Ecologists soon began analyzing ecosystems as an engineer might analyze the operation of a machine. In this view, ecosystems are systems that receive inputs of energy, process and transform that energy while cycling matter internally, and produce a variety of outputs (such as heat, water flow, and animal waste products) that can feed into other ecosystems.

Energy flows in one direction through ecosystems; it arrives mostly as radiation from the sun, powers the system, and exits in the form of heat (**FIGURE 2.14A**). Matter, in contrast, is generally recycled within ecosystems (**FIGURE 2.14B**). Energy and matter are passed among organisms (producers,

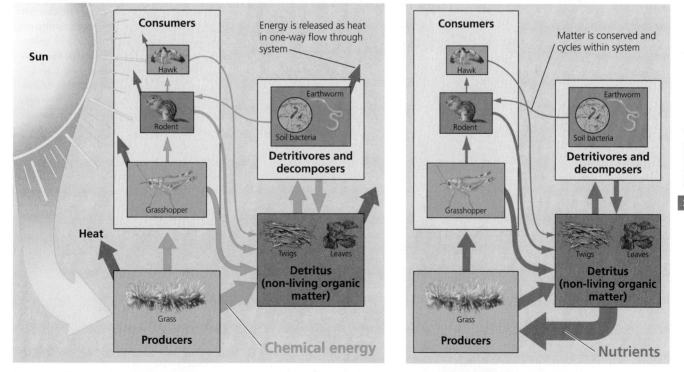

(a) Energy flowing through an ecosystem

(b) Matter cycling within an ecosystem

FIGURE 2.14 ▲ Energy enters, flows through, and exits an ecosystem. In **(a)**, light energy from the sun (**yellow arrow**) drives photosynthesis in producers, which begins the transfer of chemical energy (**green arrows**) among trophic levels (p. 68) and detritus. Energy exits the system through respiration in the form of heat (**red arrows**). In contrast, matter cycles within an ecosystem. In **(b)**, **blue arrows** show the movement of nutrients among trophic levels and detritus. In both diagrams, box sizes represent relative magnitudes of energy or matter content, and arrow widths represent relative magnitudes of energy or matter transfer. Such magnitudes may vary tremendously from one ecosystem to another. For simplicity, various abiotic components (such as water, air, and inorganic soil content) of ecosystems are omitted from these schematic diagrams. (We will revisit the flow of energy and matter among organisms in greater detail in Chapter 4 (pp. 68–71).)

consumers, and decomposers) through food web relationships (pp. 70–71). Matter is recycled because when organisms die and decay, their nutrients remain in the system. In contrast, most energy that organisms take in is eventually lost to the environment through respiration.

Energy is converted to biomass

Energy flow in most ecosystems begins with the sun's radiation. As autotrophs convert solar energy to the energy of chemical bonds in sugars during photosynthesis, they perform *primary production*. Specifically, the total amount of chemical energy produced by autotrophs is termed *gross primary production*. Autotrophs use most of this production to power their own metabolism by cellular respiration, however. The energy that remains after respiration and is used to generate biomass (such as leaves, stems, and roots; p. 69) ecologists call **net primary production**. Net primary production is equal to gross primary production minus cellular respiration.

Ecosystems vary in the rate at which autotrophs convert energy to biomass. This rate is termed *productivity*, and ecosystems whose producers convert solar energy to biomass rapidly are said to have high **net primary productivity**. Freshwater wetlands, tropical forests, and coral reefs tend to have the highest net primary productivities, whereas deserts, tundra, and open ocean tend to have the lowest (**FIGURE 2.15A**). These differences in net primary productivity among ecosystem types result in distinct geographic patterns across the globe (**FIGURE 2.15B**). In terrestrial ecosystems, net primary productivity tends to increase with temperature and precipitation. In aquatic ecosystems, net primary productivity tends to rise with light and the availability of nutrients. The limiting nature of nutrients in waters such as the Chesapeake is why additions of nitrogen and phosphorus from people's activities have such significant effects on the system.

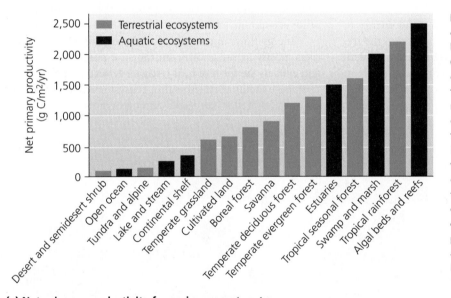

(a) Net primary productivity for major ecosystem types

FIGURE 2.15 ◄ Freshwater wetlands, tropical forests, coral reefs, and algal beds show high net primary productivities on average **(a)**, whereas deserts, tundra, and the open ocean show low values. A world map created from satellite data **(b)** shows that on land, net primary productivity varies geographically with temperature and precipitation. In the world's oceans, net primary productivity is highest around the margins of continents, where nutrients (of both natural and human origin) run off from land. Data in (a) from Whittaker, R.H., 1975. *Communities and ecosystems*, 2nd ed. New York: MacMillan. Map in (b) from satellite data presented by Field, C.B., et al., 1998. Primary production of the biosphere: Integrating terrestrial and oceanic components. *Science* 281: 237–240. Reprinted with permission from AAAS.

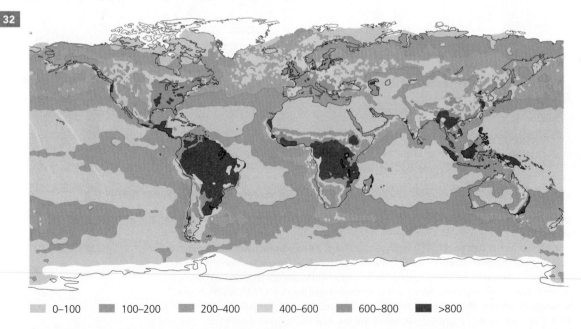

| 0–100 | 100–200 | 200–400 | 400–600 | 600–800 | >800 |

(b) Global map of net primary productivity

Ecosystems interact across landscapes

Ecosystems vary widely in size. An ecosystem can be as small as a puddle of water containing algae and tadpoles or as large as forest, lake, or bay that supports a diversity of habitats and species. In general, the term refers to a system of moderate geographic extent that is somewhat self-contained. For example, the tidal marshes in the Chesapeake where river water empties into the bay are an ecosystem, as are the sections of the bay dominated by oyster reefs. However, ecosystems that border one another may interact extensively. For instance, rivers, tidal marshes, and open waters in estuaries all interact, as do forests and grasslands where they converge. Areas where ecosystems meet may consist of transitional zones called *ecotones*, in which elements of each ecosystem mix.

WEIGHING THE ISSUES

Ecosystems Where You Live Think about the area where you live. How would you describe this area's ecosystems? How do these systems interact with one another? If one ecosystem were greatly disturbed (say, if a wetland or forest were replaced by a shopping mall), what might be the impacts on nearby natural systems?

Because components of different ecosystems may intermix, ecologists often find it useful to view these systems on larger geographic scales that encompass multiple ecosystems. In such a broad-scale approach, called **landscape ecology**, scientists study how landscape structure affects the abundance, distribution, and interaction of organisms. Taking a view across the landscape is important in studying birds that migrate long distances, mammals that move seasonally between mountains and valleys, and fish such as salmon that swim upriver from the ocean to reproduce. A landscape-level approach is also useful for planning sustainable cities and regional development (p. 404). These studies have been greatly aided by satellite imaging and *geographic information systems (GIS)*—computer software that takes multiple types of data (for instance, on geology, hydrology, vegetation, animal species, and human development) and layers them together on a common set of geographic coordinates.

For a landscape ecologist, a landscape is made up of *patches* (of ecosystems, communities, or habitat) arrayed spatially in a *mosaic* (**FIGURE 2.16**). Landscape ecology is of great interest to **conservation biologists** (p. 174), scientists who study the loss, protection, and restoration of biodiversity (see **The Science behind the Story**, pp. 34–35). Populations of organisms have specific habitat requirements and so occupy

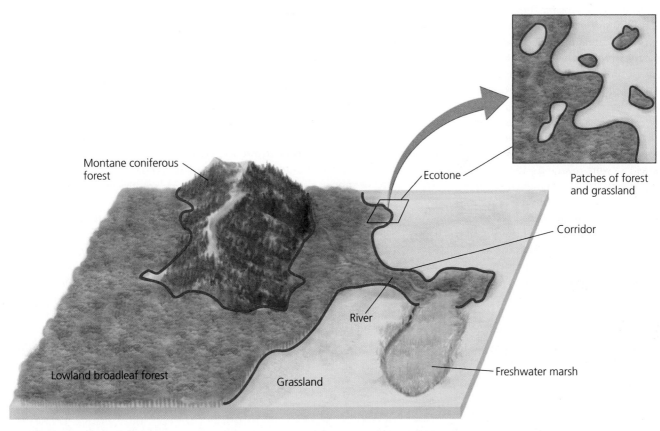

FIGURE 2.16 ▲ Landscape ecology deals with spatial patterns above the ecosystem level. This generalized diagram of a landscape shows a mosaic of patches of five ecosystem types (three terrestrial types, a marsh, and a river). Thick red lines indicate ecotones. A stretch of lowland broadleaf forest running along the river serves as a corridor connecting the large region of forest on the left to the smaller patch of forest alongside the marsh (allowing forest animals to move between patches). The inset shows a magnified view of the forest-grassland ecotone and how it consists of patches on a smaller scale.

David Schulte, U.S. Army Corps
of Engineers

34

THE SCIENCE BEHIND THE STORY

"Turning the Tide" for Native Oysters in the Chesapeake Bay

In 2001, the Eastern oyster (*Crassostrea virginica*) was in dire trouble in the Chesapeake Bay. Populations of oysters in the Chesapeake (whose name is derived from an Algonquin word *Chesepiook*, meaning "great shellfish bay") were a mere 1% of their historical abundance. Restoration efforts had largely failed. The bay oyster industry, once the largest in the world, had collapsed as a result of 200 years of overharvesting, poor water quality, virulent oyster diseases

spread throughout the bay by transplanted oysters, and the destruction of the reefs that are preferred oyster habitat. Nonetheless, when scientists or resource managers proposed to help rebuild oyster populations by significantly restricting oyster harvests or establishing oyster reef "sanctuaries," these initiatives were typically defeated by the politically powerful oyster industry.

With the collapse of the native oyster fishery and with political obstacles blocking restoration projects for native oysters, support grew among the oyster industry, state resource managers, and some scientists for the introduction of Suminoe oysters (*Crassostrea ariakensis*) from Asia. This species seemed well suited for conditions in the bay and showed resistance to the parasitic diseases that were ravaging native oysters. Proponents argued that Suminoe oyster introduction would reestablish thriving oyster populations in the bay, revitalizing the decimated oyster fishery and improving water quality (because as oysters feed, they filter phytoplankton and sediments from the water column).

Filter-feeding by oysters is an important ecological service in the bay because it reduces phytoplankton densities, clarifies waters, and supports the growth of underwater grasses that

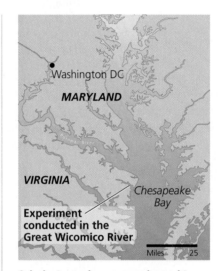

Schulte's study was conducted in the Great Wicomico River in Virginia in the lower Chesapeake Bay.

provide food and refuge for waterfowl and young crabs. Because introductions of invasive species can have profound ecological impacts (p. 75), the Army Corps of Engineers was directed to coordinate an environmental impact statement (EIS, p. 100) on oyster restoration approaches in the Chesapeake.

It was in this politically charged, high-stakes environment that Dave Schulte, a scientist with the Corps and doctoral student at the Virginia Institute of Marine Sciences, set out to determine whether there was a viable approach to

restoring native oyster populations. The work he and his team began would help turn the tide in favor of native oysters in the bay's restoration efforts.

One of the biggest impacts on native oysters was the destruction of oyster reefs by a century of intensive oyster harvesting. Oysters settle and grow best on the shells of other oysters, and over long periods this process forms reefs (underwater outcrops of living oysters and oyster shells) that solidify and become as hard as stone. Throughout the bay, massive reefs that at one time had jutted out of the water at low tide had been reduced to rubble on the bottom from a century of repeated scouring by metal dredgers used by oyster harvesting ships. The key, Schulte realized, was to construct artificial reefs like those that once existed, to get oysters off the bottom—away from smothering sediments and hypoxic waters—and up into the plankton-rich upper waters.

Armed with the resources available to the Corps, he opted to take a landscape ecology approach (p. 33) and restore patches of reef habitat on nine complexes of reefs covering a total of 35.3 hectares (87 acres) in an oyster sanctuary near the mouth of the Great Wicomico River (**see map**) in the lower Chesapeake Bay. This was a very

suitable patches across the landscape. If habitat patches are highly fragmented and isolated (pp. 169, 201–202), the populations in those patches may perish. Accordingly, establishing corridors of habitat (p. 203; also see Figure 2.16) to link patches is one approach that conservation biologists pursue as they attempt to maintain biodiversity.

Modeling aids our understanding of ecosystems

Another way in which ecologists seek to make sense of the natural systems they study is by working with models. In science, a **model** is a simplified representation of a complicated

A water cannon blows oysters shells off a barge and onto the river bottom to create an artificial oyster reef for the experiment.

different approach from the smaller-scale restoration efforts of the past.

Artificial reefs of two heights were constructed in 2004 (**see photograph**), and oysters were allowed to colonize the reefs, safe from harvesting. Oyster populations on the constructed reefs were sampled in 2007, and the results were stunning. The reef complex supported an estimated 185 million oysters, a number nearly as large as the wild population of 200 million oysters estimated to live on the remaining degraded habitat in all of Maryland's waters.

Higher reefs supported an average of over 1,000 oysters per square meter—four times more than the lower reefs and 170 times more than unrestored bottom (**see graph**). Like natural reefs, the constructed reefs began to solidify, providing a firm foundation for the settlement of new oysters. In 2009, Schulte's research made a splash when his team published its findings in the journal *Science*, bringing international attention to their study.

After reviewing eight alternative approaches to oyster restoration that involved one or more oyster species, the Corps advocated an approach that avoided the introduction of non-native oysters. Instead it proposed a combination of native oyster restoration, a temporary moratorium on oyster harvests (accompanied by a compensation program for the oyster industry), and enhanced support for oyster aquaculture in the bay region.

Schulte's restoration project cost roughly $3 million and will require substantial investments if it is to be repeated elsewhere in the bay. This is particularly true in upper portions of the bay, where oyster reproduction levels are lower (requiring restored reefs to be "seeded" with oysters), water conditions are poorer, and oysters are less resistant to disease. Many scientists contend that expanded reef restoration efforts are worth the cost because they enhance oyster populations and provide a vital service to the bay through water filtering. Some scientists also see value in promoting oyster farming, in which restoration efforts would be

supported by businesses instead of taxpayers.

Regardless of how they will be funded, protected sites for oyster restoration efforts are being established. In 2010, Maryland proposed creating 3,640 hectares (9,000 acres) of new oyster sanctuaries—25% of existing oyster reefs in state waters—where restoration projects like Schulte's could be replicated. This movement toward increased protection for oyster populations, coupled with findings of growing resistance to disease in bay oysters, has given new hope that native oysters may once again thrive in the bay that bears their name.

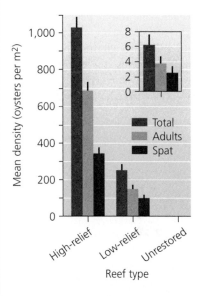

Reef height had a profound effect on the density of adult oysters and spat (newly settled oysters). Schulte's work suggested that native oyster populations could rebound in portions of the bay if they were provided elevated reefs and protected from harvest. Data from Schulte, D.M., R.P. Burke, and R.N Lipicus, 2009. Unprecedented restoration of a native oyster metapopulation. *Science* 325: 1124–1128. Reprinted with permission from AAAS.

natural process, designed to help us understand processes and make predictions.

Ecological modeling is the practice of constructing and testing models that aim to explain and predict how ecological systems function. These models are grounded in actual data and based on hypotheses about how system components interact. Ecological modeling is extremely useful in large, intricate systems that are difficult to isolate and study. For example, ecological models are useful for understanding the flow of nutrients into the Chesapeake Bay and in predicting the responses of oysters and underwater grasses to changing water conditions.

Ecosystems provide vital services

All life on our planet, including human life, depends on healthy, functioning ecosystems. When Earth's ecosystems function normally and undisturbed, they provide goods and services that we could not survive without. Examples of such **ecosystem services** include regulating atmospheric gases, precipitation, and temperature; providing people with food and natural resources; pollinating plants and controlling crop pests; preventing soil erosion; filtering wastes and pollutants; and providing recreational opportunities and aesthetic, artistic, spiritual, and educational value (**FIGURE 2.17**).

When human activities impair ecosystem functioning, we must devote resources in an attempt to provide these services ourselves, requiring extra investments of time and money. For example, when we eliminate insect predators from agricultural fields by using intensive farming practices, we impair the ecological service of pest control that nature provides. Farmers must then apply synthetic pesticides to reduce pest populations, costing them resources and increasing exposures to these chemicals among people and wildlife.

One of the most important ecosystem services is the cycling of nutrients. Through the processes that take place within and among ecosystems, the chemical elements and compounds that we need—water, carbon, nitrogen, phosphorus, and many more—cycle through our environment in intricate ways.

BIOGEOCHEMICAL CYCLES

Just as nitrogen and phosphorus from fertilizer on Pennsylvania farm fields end up in Chesapeake Bay oysters on our dinner plates, all nutrients move through the environment in complex and fascinating ways. Whereas energy enters an ecosystem from the sun, flows from one organism to another, and is dissipated to the atmosphere as heat, the physical matter of an ecosystem is circulated over and over again.

Nutrients circulate in biogeochemical cycles

Nutrients move through ecosystems in **nutrient cycles** (or **biogeochemical cycles**) that circulate elements or molecules through the lithosphere, atmosphere, hydrosphere, and biosphere. A carbon atom in your fingernail today might have been part of the muscle of a cow a year earlier, may have resided in a blade of grass a month before that, and may have been part of a dinosaur's tooth 100 million years ago. After we die, the nutrients in our bodies will spread widely through the environment, eventually being incorporated by an untold number of organisms far into the future.

Nutrients move from one *pool*, or *reservoir*, to another, remaining for varying amounts of time (*residence time*) in

FIGURE 2.17 ▲ Ecological processes naturally provide countless *ecosystem services*. Our society, indeed our very survival, depends on these services.

each. The dinosaur, the grass, the cow, and your body are each reservoirs for carbon atoms. When a reservoir releases more materials than it accepts, it is called a *source*, and when a reservoir accepts more materials than it releases, it is called a *sink*. **FIGURE 2.18** illustrates these concepts in a simple manner. The rate at which materials move between reservoirs is termed a *flux*, and the flux between any given pair of pools

can change over time. As we will see in the following sections, human activities affect the cycling of nutrients by altering fluxes, residence times, and the relative amounts of nutrients in reservoirs.

The water cycle influences all other cycles

Water is so integral to life that we frequently take it for granted. The essential medium for many biochemical reactions, water plays key roles in nearly every environmental system, including each nutrient cycle. Water transports nutrients, sediments, and pollutants from the continents to the oceans via rivers, streams, and surface runoff. Nutrients can then be carried thousands of miles on ocean currents. Water also brings atmospheric pollutants from the air back down to the surface when they dissolve in falling rain or snow. These activities make the water cycle, or **hydrologic cycle** (**FIGURE 2.19**), an integral part of nutrient cycling on Earth.

The oceans are the largest reservoir in the hydrologic cycle, holding more than 97% of all water on Earth. The fresh water we depend on for our survival accounts for the remaining water, and two-thirds of this small amount is tied up in

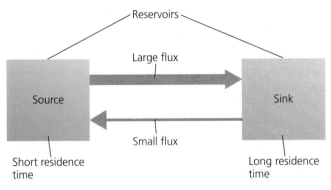

FIGURE 2.18 ▲ The main components of a biogeochemical cycle are reservoirs (places where materials are stored) and fluxes (rates at which materials move among reservoirs). A source releases more materials than it accepts, and a sink accepts more materials than it releases.

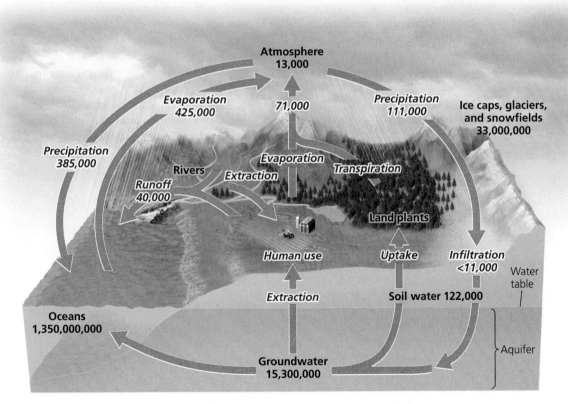

FIGURE 2.19 ▲ The water cycle, or hydrologic cycle, summarizes the many routes that water molecules take as they move through the environment. Gray arrows represent fluxes among reservoirs, or pools, for water. The water cycle is a system unto itself but also plays key roles in other biogeochemical cycles. Oceans hold 97% of our planet's water, whereas most fresh water resides in groundwater and icecaps. Water vapor in the atmosphere condenses and falls to the surface as precipitation; then it evaporates from land and transpires from plants to return to the atmosphere. Water flows downhill into rivers, eventually reaching the oceans. In the figure, pool names are printed in black type, and numbers in black type represent pool sizes expressed in units of cubic kilometers (km^3). Processes, printed in italic red type, give rise to fluxes, printed in italic red type and expressed in km^3 per year. Data from Schlesinger, W.H., 1997. *Biogeochemistry: An analysis of global change*, 2nd ed. London: Academic Press.

glaciers, snowfields, and ice caps (p. 250). Thus, considerably less than 1% of the planet's water is in a form that we can readily use—groundwater, surface fresh water, and rain from atmospheric water vapor.

Water moves from oceans, lakes, ponds, rivers, and moist soil into the atmosphere by **evaporation**, the conversion of a liquid to gaseous form. Water also enters the atmosphere by **transpiration**, the release of water vapor by plants through their leaves. Transpiration and evaporation act as natural processes of distillation, because water escaping into the air as a gas leaves behind its dissolved substances. Water returns from the atmosphere to Earth's surface as **precipitation** when water vapor condenses and falls as rain or snow. Precipitation may be taken up by plants and used by animals, but much of it flows as **runoff** into streams, rivers, lakes, ponds, and oceans.

Some water soaks down through soil and rock through a process called **infiltration** to recharge underground reservoirs known as **aquifers**. Aquifers are spongelike regions of rock and soil that hold **groundwater**, water found underground beneath layers of soil. The uppermost level of groundwater held in an aquifer is referred to as the **water table**. Some aquifers hold groundwater for short periods of time, whereas others contain quite ancient water.

Human activity has affected nearly every flux, reservoir, and residence time in the water cycle. By damming rivers, we have slowed the movement of water from the land to the sea, and we increase evaporation by holding river waters in reservoirs. We have removed natural vegetation by clear-cutting and developing land, which increases surface runoff, decreases infiltration and transpiration, and promotes soil erosion. Our withdrawals of surface water and groundwater for agriculture, industry, and domestic uses have depleted rivers, lakes, and streams and have lowered water tables. And by emitting into the atmosphere pollutants that dissolve in water droplets, we have changed the chemical nature of precipitation, in effect sabotaging the natural distillation process that evaporation and transpiration provide. (We will revisit the water cycle, water resources, and human impacts in more detail in Chapter 12).

The carbon cycle circulates a vital organic nutrient

The **carbon cycle** describes the routes that carbon atoms take through the environment (**FIGURE 2.20**). Because carbon forms the backbone of essential biological molecules, its cycling is of great importance. Producers, including terrestrial and aquatic plants, algae, and cyanobacteria, pull carbon dioxide out of the atmosphere and out of surface water to use in photosynthesis. Autotrophs and the heterotrophs that consume them use carbohydrates produced in photosynthesis to fuel their respiration and for structural growth, releasing carbon back into the atmosphere and waters as CO_2. Because plants are a sizable reservoir of carbon and CO_2 is a greenhouse gas (p. 300), researchers are attempting to measure the amount of CO_2 that various plants sequester to evaluate the potential of vegetation in combating global climate change.

Earth's oceans function extensively in the carbon cycle. Oceans absorb carbon-containing compounds from the atmosphere, terrestrial runoff, undersea volcanoes, and the detritus of marine organisms. Ocean waters are the second largest reservoir of carbon on Earth. The largest reservoir of carbon, sedimentary rock (p. 230), is formed in oceans and freshwater wetlands. When organisms in these habitats die, their remains can settle in sediments, and as layers of sediment accumulate, the older layers are buried more deeply and experience high pressure for long periods. Over time, these conditions can convert the soft tissues of dead organisms into fossil fuels—coal, oil, and natural gas—deep underground (p. 329). The shells and skeletons of aquatic organisms, which are rich in calcium carbonate ($CaCO_3$), are converted by these high pressures underground to sedimentary rock, such as limestone. Although any given carbon atom spends a relatively short time in the atmosphere, carbon trapped in sedimentary rock may reside there for hundreds of millions of years. Carbon trapped in sedimentary rocks and fossil fuel deposits may eventually be released into the oceans or atmosphere by geologic processes such as uplift, erosion, and volcanic eruptions. It also reenters the atmosphere when we extract and burn fossil fuels.

The largest human impact on the carbon cycle is through our use of fossil fuels as an energy source. It is estimated that since the middle of the 18th century, people have added over 250 billion metric tons (276 billion tons) of carbon dioxide to the atmosphere through the combustion of coal, oil, and natural gas—greatly increasing the flux of carbon from the lithosphere to the atmosphere and shortening the residence time of carbon in fossil fuel deposits.

Moreover, when people burn forests and fields to clear land for agriculture, the carbon in wood and leaves is released to the atmosphere. Because these cleared sites have less vegetation, photosynthesis removes less carbon dioxide from the atmosphere than before. As a result of vegetation clearing and fossil fuel combustion, scientists estimate that today's atmospheric carbon dioxide reservoir is the largest that Earth has experienced in the past 800,000 years—likely in the past 20 million years (p. 302)—and is a driving force behind global climate change (Chapter 14).

The nitrogen cycle involves specialized bacteria

Nitrogen makes up 78% of our atmosphere by mass and is the sixth most abundant element on Earth. It is an essential ingredient in the proteins, DNA, and RNA that build our bodies and, like phosphorus, is an essential nutrient for plant growth. Thus the **nitrogen cycle** (**FIGURE 2.21**) is of vital importance to us and to all other organisms. Despite its abundance in the air, nitrogen gas (N_2) is chemically inert and cannot cycle out of the atmosphere and into living organisms without assistance from lightning, highly specialized bacteria, or human intervention. However, once nitrogen undergoes the right kind of chemical change, it becomes biologically active and available to the organisms that need it, and it can act as a potent fertilizer.

To become biologically available, inert nitrogen gas (N_2) must be "fixed," or combined with hydrogen in nature to form ammonia (NH_3), whose water-soluble ions of ammonium (NH_4^+) can be taken up by plants. **Nitrogen fixation** can be accomplished in two ways: by the intense energy of lightning

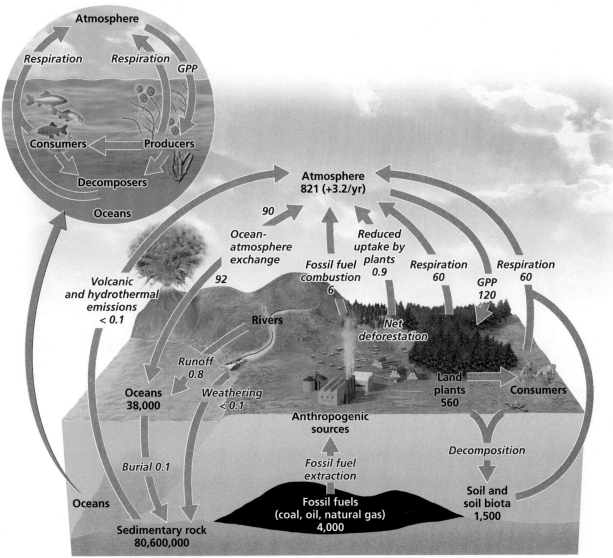

FIGURE 2.20 ▲ The carbon (C) cycle summarizes the many routes that carbon atoms take as they move through the environment. Gray arrows represent fluxes among reservoirs, or pools, for carbon. In the carbon cycle, plants use carbon dioxide from the atmosphere for photosynthesis (gross primary production, or "GPP" in the figure). Carbon dioxide is returned to the atmosphere through respiration by plants, their consumers, and decomposers. The oceans sequester carbon in their water and in deep sediments. The vast majority of the planet's carbon is stored in sedimentary rock. In the figure, pool names are printed in black type, and numbers in black type represent pool sizes expressed in petagrams (units of 10^{15} g) of C. Processes, printed in italic red type, give rise to fluxes, printed in italic red type and expressed in petagrams of C per year. Data from Schlesinger, W.H., 1997. *Biogeochemistry: An analysis of global change*, 2nd ed. London: Academic Press.

strikes, or by particular types of **nitrogen-fixing bacteria** that inhabit the top layer of soil. These bacteria live in a mutualistic relationship (p. 68) with many types of plants, including soybeans and other legumes, providing them nutrients by converting nitrogen to a usable form. Other types of specialized bacteria then perform a process known as **nitrification**, converting ammonium ions first into nitrite ions (NO_2^-), then into nitrate ions (NO_3^-). Plants can take up these ions, which also become available after atmospheric deposition on soils or in water or after application of nitrate-based fertilizer.

Animals obtain the nitrogen they need by consuming plants or other animals. Decomposers obtain nitrogen from dead and decaying plant and animal matter and from the urine and feces of animals. Once decomposers process the nitrogen-rich compounds, they release ammonium ions, making these available to nitrifying bacteria to convert again to nitrates and nitrites. The next step in the nitrogen cycle occurs when **denitrifying bacteria** convert nitrates in soil or water to gaseous nitrogen. Denitrification thereby completes the cycle by releasing nitrogen back into the atmosphere as a gas.

We have greatly influenced the nitrogen cycle

Historically, nitrogen fixation was a *bottleneck*, a step that limited the flux of nitrogen out of the atmosphere into water-soluble forms. Once people discovered how to fix nitrogen on massive scales, a process called **industrial fixation**, we

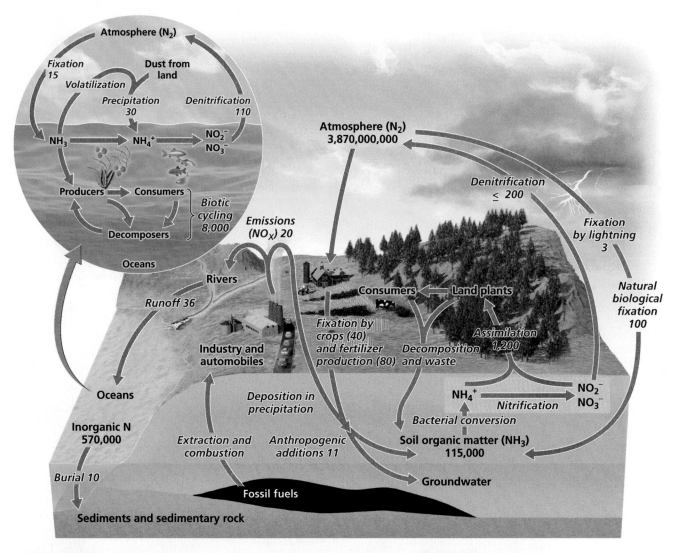

FIGURE 2.21 ▲ The nitrogen (N) cycle summarizes the many routes that nitrogen atoms take as they move through the environment. Gray arrows represent fluxes among reservoirs, or pools, for nitrogen. In the nitrogen cycle, specialized bacteria play key roles in "fixing" atmospheric nitrogen and converting it to chemical forms that plants can use. Other types of bacteria convert nitrogen compounds back to the atmospheric gas N_2. In the oceans, inorganic nitrogen is buried in sediments, whereas nitrogen compounds are cycled through food webs as they are on land. In the figure, pool names are printed in black type, and numbers in black type represent pool sizes expressed in teragrams (units of 10^{12} g) of N. Processes, printed in italic red type, give rise to fluxes, printed in italic red type and expressed in teragrams of N per year. Data from Schlesinger, W.H., 1997. *Biogeochemistry: An analysis of global change*, 2nd ed. London: Academic Press.

accelerated its flux into other reservoirs. Today, our species is fixing at least as much nitrogen artificially as is being fixed naturally, and we are overwhelming nature's denitrification abilities. A 2008 study reported that people apply a staggering 150 million metric tons (165 million tons) of fixed nitrogen to Earth's land area each year.

Human alteration of the nitrogen cycle has had profound impacts beyond the eutrophication of aquatic ecosystems. Oddly enough, the overapplication of nitrogen-based fertilizers can strip the soil of other vital nutrients, such as calcium and potassium, reducing soil fertility. Additionally, burning fossil fuels, forests, or fields generates nitrogenous compounds in the atmosphere that act as greenhouse gases (Chapter 14), form acid precipitation (p. 291), and create photochemical smog (p. 288).

The phosphorus cycle circulates a limited nutrient

Sedimentary rocks are the largest reservoir in the phosphorus cycle (**FIGURE 2.22**). Environmental concentrations of phosphorus available to organisms tend to be very low, because weathering (p. 137), which releases phosphate ions (PO_4^{3-}) into water, is the only process that makes phosphorus available for uptake. Phosphorus is a key component of cell membranes, DNA, RNA, and a number of vital biochemical compounds, so its rarity in the environment makes it a limiting factor for plant growth. Aquatic producers take up phosphates from surrounding waters, whereas terrestrial producers take up phosphorus from soil water through their roots. Phosphorus is incorporated into the tissues of producers, and consumers obtain their required phosphorus by

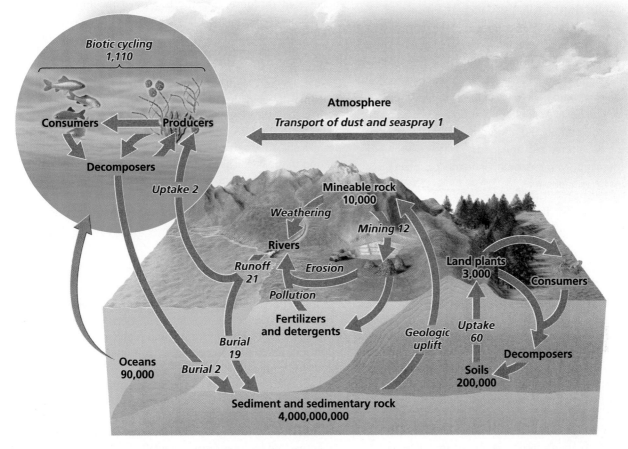

FIGURE 2.22 ▲ The phosphorus (P) cycle summarizes the many routes that phosphorus atoms take as they move through the environment. Gray arrows represent fluxes among reservoirs, or pools, for phosphorus. Most phosphorus resides underground in rock and sediment, but the phosphorus cycle moves this element through the soil, the oceans, and freshwater and terrestrial ecosystems. Rocks containing phosphorus are uplifted geologically and slowly weathered away. Small amounts of phosphorus cycle through food webs, where this nutrient is often a limiting factor for plant growth. In the figure, pool names are printed in black type, and numbers in black type represent pool sizes expressed in teragrams (units of 10^{12} g) of P. Processes, printed in italic red type, give rise to fluxes, printed in italic red type and expressed in teragrams of P per year. Data from Schlesinger, W.H., 1997. *Biogeochemistry: An analysis of global change*, 2nd ed. London: Academic Press.

eating the tissues of other organisms. Phosphorus from dead organisms or waste products in the oceans can precipitate into solid form and settle to the bottom in sediments, which are eventually compressed into sedimentary rock.

Human activities increase phosphorus concentrations in surface waters through substantial runoff of the phosphorus-rich fertilizers we apply to lawns and farmlands. A 2008 study determined that an average hectare of land in the Chesapeake Bay region received a net input of 4.52 kg (10 lb) of phosphorus per year, promoting phosphorus accumulation in soils, runoff into waterways, and phytoplankton blooms and hypoxia in the bay. People also add phosphorus to waterways through releases of treated wastewater rich in phosphates from domestic use of phosphate detergents.

Tackling nutrient enrichment requires diverse approaches

With our reliance on synthetic fertilizers for food production and fossil fuels for energy, nutrient enrichment of ecosystems will be a challenge for many years to come.

But there are a number of approaches available to control nutrient pollution in the Chesapeake Bay watershed, Mississippi River watershed, and other waterways affected by eutrophication:

▶ Reducing fertilizer use on farms and lawns

▶ Changing the timing of fertilizer application to minimize rainy-season runoff

▶ More effectively managing manure applications to farmland to reduce nutrient runoff

▶ Planting and maintaining vegetation "buffers" around streams that trap nutrient and sediment runoff

▶ Using artificial wetlands to filter stormwater and farm runoff

▶ Restoring nutrient-absorbing wetlands along waterways

▶ Improving technologies in sewage treatment plants to enhance nitrogen and phosphorus capture

▶ Restoring frequently flooded lands to reduce runoff

▶ Reducing fossil fuel combustion to minimize atmospheric inputs of nitrogen to waterways

Nutrient Pollution and Its Financial Impacts A sizable amount of the nitrogen and phosphorus that enters the Chesapeake Bay originates from farms and other sources far from the bay, yet it is people living near the bay, such as oystermen, who bear many of the negative impacts. Who do you believe should be responsible for addressing this problem? Should environmental policies on this issue be developed and enforced by state governments, the federal government, both, or neither? Explain the reasons for your answer.

After 25 years of failed pollution control agreements and nearly $6 billion spent on cleanup efforts, the Chesa-peake Bay Foundation (CBF), a nonprofit organization dedicated to conserving the bay, sued the Environmental Protection Agency in January 2009 for failing to use its available powers under the Clean Water Act to clean up the bay. The CBF's lawsuit focused media attention on the plight of the bay, its ongoing water quality issues, and its depleted fisheries—and spurred action. In May 2009, President Obama directed the EPA and other federal agencies to establish a comprehensive plan for the Chesapeake, and in May 2010 the EPA and the CBF announced a settlement in which the EPA agreed to provide aggressive pollution regulation in the bay. Thanks to the efforts of concerned citizens and advocacy organizations like the CBF, the 17 million people living in the Chesapeake Bay watershed have reason to hope that the Chesapeake Bay tomorrow may be healthier than it is today.

➤ CONCLUSION

Earth hosts many interacting systems, and the way we perceive them depends on the questions we ask. Life interacts with its abiotic environment in ecosystems, systems through which energy flows and materials are recycled. Understanding the biogeochemical cycles that describe the movement of nutrients within and among ecosystems is crucial, because human activities are causing significant changes in the ways these cycles function.

Understanding energy, energy flow, and chemistry enhances our comprehension of how organisms interact with one another, how they relate to their nonliving environment, and how environmental systems function. Energy and chemistry are tied to nearly every significant process in environmental science. Moreover, applications of chemistry can provide solutions to environmental problems involving agricultural practices, water resources, air quality, energy policy, and environmental health.

Thinking in terms of systems is important in understanding how Earth works, so that we may learn not to disrupt its processes and how to mitigate any disruptions we cause. By studying the environment from a systems perspective and by integrating scientific findings with the policy process, people who care about the Chesapeake Bay are working today to solve their pressing problems.

TESTING YOUR COMPREHENSION

1. Which type of feedback loop is most common in nature, and which more commonly results from human action? For either type of feedback loop, can you think of an example that was not mentioned in the text?

2. Describe how hypoxic conditions can develop in aquatic ecosystems such as the Chesapeake Bay.

3. Differentiate an ion from an isotope.

4. Describe the two major forms of energy, and give examples of each. Compare and contrast the first law of thermodynamics and the second law of thermodynamics.

5. What substances are produced by photosynthesis? By cellular respiration?

6. Describe the typical movement of energy through an ecosystem. Describe the typical movement of matter through an ecosystem.

7. List five ecosystem services provided by functioning ecosystems, and rank them according to your perceived value of each.

8. What role do each of the following play in the carbon cycle?
   ➤ Cars
   ➤ Photosynthesis
   ➤ The oceans
   ➤ Earth's crust

9. Distinguish the function performed by nitrogen-fixing bacteria from that performed by denitrifying bacteria.

10. How has human activity altered the hydrologic cycle? The carbon cycle? The phosphorus cycle? The nitrogen cycle? To what environmental problems have these changes given rise?

SEEKING SOLUTIONS

1. Can you think of an example of an environmental problem not mentioned in this chapter that a good knowledge of chemistry could help us solve? Explain your answer.

2. Consider the ecosystem(s) that surround(s) your campus. How is each affected by human activities?

3. For a conservation biologist interested in sustaining populations of each organism below, why would it be

helpful to take a landscape ecology perspective? Explain your answer in each case.

- ▶ A forest-breeding warbler that suffers poor nesting success in small fragmented forest patches
- ▶ A bighorn sheep that must move seasonally between mountains and lowlands
- ▶ A toad that lives in upland areas but travels cross-country to breed in localized pools each spring

4. A simple change in the flux between just two reservoirs in a single nutrient cycle can potentially have major consequences for ecosystems and, indeed, for the globe. Explain how this can be, using one example from the carbon cycle and one example from the nitrogen cycle.

5. **THINK IT THROUGH** You are an oysterman in the Chesapeake Bay, and your income is decreasing because hypoxic zones are making it harder to harvest oysters. One day your senator comes to town, and you have a one-minute audience with her. What steps would you urge her to take in Washington, D.C., to try to help alleviate the bay's water quality problems and bring back the oyster fishery?

Now suppose that you are a Pennsylvania farmer who has learned that the federal government is insisting that you use 30% less fertilizer on your crops each year to reduce nutrient inputs to the Chesapeake. You know that in good growing years you could do without that fertilizer, and you'd be glad not to have to pay for it. But in bad growing years, you need the fertilizer to ensure a harvest so that you can continue making a living. And you must apply the fertilizer each spring before you know whether it will be a good or bad year. What would you tell your senator when she comes to town?

CALCULATING ECOLOGICAL FOOTPRINTS

In ecological systems, a rough rule of thumb is that when energy is transferred from plants to plant-eaters or from prey to predator, the efficiency is only about 10% (pp. 68–69). Much of this inefficiency is a consequence of the second law of thermodynamics. Another way to think of this is that eating 1 calorie of material from an animal is the ecological equivalent of eating 10 calories of plant material.

Humans are considered omnivores because we can eat both plants and animals. The choices we make about what to eat have significant ecological impacts. With this in mind, calculate the ecological energy requirements for four different diets, each of which provides a total of 2,000 dietary calories per day.

Diet	Source of calories	Number of calories consumed	Ecologically equivalent calories	Total ecologically equivalent calories
100% plant	Plant			
0% animal	Animal			
90% plant	Plant	1,800	1,800	3,800
10% animal	Animal	200	2,000	
50% plant	Plant			
50% animal	Animal			
0% plant	Plant			
100% animal	Animal			

1. How many ecologically equivalent calories would it take to support you for a year, for each of the four diets listed?

2. How does the ecological impact from a diet consisting strictly of animal products (meat, eggs, milk, and other dairy products) compare with that of a strictly vegetarian diet? How many additional ecologically equivalent calories do you consume each day by including as little as 10% of your calories from animal sources?

3. What percentages of the calories in your own diet do you think come from plant versus animal sources? Estimate the ecological impact of your diet, relative to a strictly vegetarian one.

4. Describe some challenges of providing food for the growing human population, especially as people in many poorer nations develop a taste for an American-style diet rich in animal protein and fat.

3 Evolution, Biodiversity, and Population Ecology

Upon completing this chapter, you will be able to:

➤ Explain the process of natural selection and cite evidence for this process

➤ Describe the ways in which evolution influences biodiversity

➤ Discuss reasons for species extinction and mass extinction events

➤ List the levels of ecological organization

➤ Outline the characteristics of populations that help predict population growth

➤ Assess logistic growth, carrying capacity, limiting factors, and other fundamental concepts in population ecology

Golden toads at Monteverde

Striking Gold in a Costa Rican Cloud Forest

"What a terrible feeling to realize that within my own lifetime, a species of such unusual beauty, one that I had discovered, should disappear from our planet."
—Dr. Jay M. Savage, on the Golden Toad

"To keep every cog and wheel is the first precaution of intelligent tinkering."
—Aldo Leopold

During a 1963 visit to Central America, biologist Jay M. Savage heard rumors of a previously undocumented toad living in Costa Rica's mountainous Monteverde region. The elusive amphibian, according to local residents, was best known for its brilliant golden color. Savage was told the toad was hard to find because it appeared only during the early part of the region's rainy season.

Monteverde means "green mountain" in Spanish, and the name couldn't be more appropriate. The village of Monteverde sits beneath the verdant slopes of the Cordillera de Tilarán, mountains that receive over 400 cm (157 in.) of annual rainfall. Some of the lush forests above Monteverde, which begin at an altitude of around 1,500 m (4,920 ft, just under a mile high), are known as *cloud forests* because much of their moisture arrives with low-moving clouds that blow inland from the Caribbean Sea.

Monteverde's cloud forest was not fully explored at the time of Savage's first visit, and researchers who had been there described the area as pristine, with a rich bounty of ferns, liverworts, mosses, clinging vines, orchids, and other organisms that thrive in cool, misty environments. Savage knew that such conditions create ideal habitat for many toads and other amphibians.

In May of 1964, Savage organized an expedition into the muddy mountains above Monteverde to try to document the existence of the undescribed toad species in its natural habitat. Late on the afternoon of May 14, he and his colleagues found what they were looking for. Approaching the mountain's crest,

they spotted bright orange patches on the forest's black floor. In one area only 5 m (16 ft) in diameter, they counted 200 male golden toads searching for females at a breeding congregation. Savage gave the creature the scientific name *Bufo periglenes* (literally, "the brilliant toad").

The discovery received international attention, making a celebrity of the tiny toad and making a travel destination of its mountain home, which was soon protected within the Monteverde Cloud Forest Preserve.

At the time, no one knew that the Monteverde ecosystem was about to be transformed. No one foresaw that the oceans and atmosphere would begin warming because of global climate change (Chapter 14) and cause Monteverde's moisture-bearing clouds to rise, drying the forest. No one anticipated the spread of a lethal fungal disease called chytridiomycosis, caused by the pathogen *Batrachochytrium dendrobatidis*, that would infect frogs and toads even in the most pristine environments. And no one could guess that within 25 years the glamorous, newly discovered golden toad would vanish from the Earth. ∎

EVOLUTION AS THE WELLSPRING OF EARTH'S BIODIVERSITY

The golden toad was new to science, and countless species still await discovery. Even so, scientists understand quite well how the world became populated with the remarkable diversity of organisms we see today. Research shows that our planet has progressed from a stark world inhabited solely by microbes to a lush cornucopia of 1.8 million (and likely millions more) species (**FIGURE 3.1**).

A **species** is a particular type of organism or, more precisely, a population or group of populations whose members share certain traits and can freely breed with one another and produce fertile offspring. A **population** is a group of individuals of a particular species that live in a particular area. Over eons of time, our planet's species and populations have been molded by the process of biological evolution, giving us the vibrant abundance of life that enriches Earth today.

Evolution in the broad sense means change over time, and biological evolution consists of genetic change in populations of organisms across generations. Changes in genes (p. 28) often lead to modifications in the appearance, functioning, or behavior of organisms from generation to generation through time. Biological evolution results from random genetic changes, and it may proceed randomly or may be directed by natural selection. **Natural selection** is the process by which inherited characteristics that enhance survival and reproduction are passed on more frequently to future generations than those that do not, thus altering the genetic makeup of populations through time.

Evolution by natural selection is one of the best-supported and most illuminating concepts in all of science. From a scientific standpoint, evolutionary theory is indispensable, because it is the foundation of modern biology. Understanding evolution is also vital for a full appreciation of environmental science. Perceiving how organisms adapt to their environments and change over time is crucial for comprehending ecology and for understanding the history of life. Evolutionary processes influence many aspects of environmental science, including agriculture, pesticide resistance, medicine, and environmental health.

Natural selection shapes organisms and diversity

In 1858, **Charles Darwin** and **Alfred Russel Wallace** each independently proposed the concept of natural selection as a mechanism for evolution and as a way to explain the great variety of living things. Both Darwin and Wallace were exceptionally keen naturalists from England who had studied plants and animals in such exotic locales as the Galápagos Islands (Darwin) and the Malay Archipelago (Wallace). In the century and a half since then, many thousands of scientists have refined our understanding of natural selection and evolution.

Natural selection is a simple concept that offers an astonishingly powerful explanation for patterns evident in nature. The idea of natural selection follows logically from a few straightforward premises that are readily apparent to anyone who observes the life around us:

▸ Organisms face a constant struggle to survive and reproduce.

▸ Organisms tend to produce more offspring than can survive.

▸ Individuals of a species vary in their characteristics.

Variation is due to differences in genes, the environments within which genes are expressed, and the interactions between genes and environment. As a result of this variation,

(a) Resplendent quetzal

(b) Heliconia flower

(c) Harlequin frog

(d) Scutellerid bug

FIGURE 3.1 ▲ Much of our planet's biological diversity resides in tropical rainforests. Monteverde's cloud-forest community includes organisms such as this **(a)** resplendent quetzal (*Pharomachrus mocinno*), **(b)** heliconia (*Heliconia wagneriana*), **(c)** harlequin frog (*Atelopus varius*), and **(d)** scutellerid bug (*Pachycoris torridus*).

some individuals of a species will be better suited to their environment than others and will be better able to reproduce.

Many characteristics are passed from parent to offspring through genes, and a parent that produces many offspring will pass on more genes to the next generation than a parent that produces few or no offspring. In the next generation, therefore, the genes of better-adapted individuals will be more prevalent than those of individuals that are less well adapted. From one generation to another through time, characteristics, or traits, that lead to better and better reproductive success in a given environment will evolve in the population. A trait that promotes success is called an **adaptation.**

Selection acts on genetic variation

For an organism to pass a trait along to future generations, genes in the organism's DNA (p. 28) must code for the trait. In an organism's lifetime, its DNA will be copied millions of times by millions of cells. In all this copying and recopying, sometimes a mistake is made. Accidental changes in DNA, called **mutations,** give rise to genetic variation among individuals. If a mutation occurs in a sperm or egg cell, it may be passed on to the next generation. Most mutations have little effect, but some can be deadly, whereas others can be beneficial. Those that are not lethal provide the genetic variation on which natural selection acts.

Genetic variation is also generated as organisms mix their genetic material through sexual reproduction. When organisms reproduce sexually, a portion of each parent's genes contributes to the genes of the offspring. This process produces novel combinations of genes, generating variation among individuals.

During the evolutionary process, natural selection does not simply weed out unfit individuals. It also helps to elaborate and diversify traits that may lead to the formation of new species and whole new types of organisms (**FIGURE 3.2**).

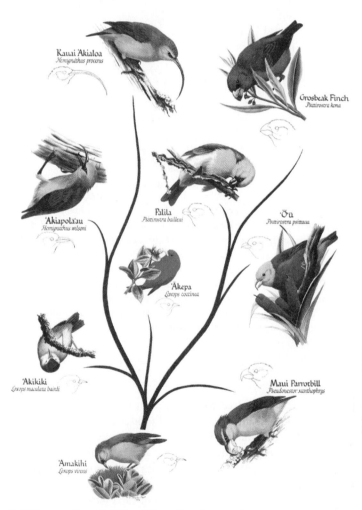

(a) Divergent evolution of Hawaiian honeycreepers

(b) Convergent evolution of a cactus in Arizona (top) and a spurge (euphorb) in the Canary Islands (bottom)

FIGURE 3.2 ▲ Natural selection can cause closely related species to diverge in appearance **(a)** as they adapt to selective pressures in different environments. In the group of birds known as Hawaiian honeycreepers, closely related species have adapted to different food resources and habitats, as indicated by the diversity in their plumage colors and the shapes of their bills. Natural selection can also cause distantly related species to converge in appearance **(b)** if the selective pressures of their environments are similar. A classic case of such *convergent evolution* involves many cacti of the Americas and euphorbs of Africa. Plants in each family independently adapted to arid environments through the evolution of tough succulent stems to hold water, thorns to keep thirsty animals away, and leafless photosynthetic stems to reduce surface area and water loss.

Environmental conditions determine what pressures natural selection will exert, and these conditions change over time. Genes and environments constantly interact as species engage in a perpetual process of adapting to the changing conditions around them. However, this process requires a great deal of time, and a species cannot adapt at once every time conditions change. For example, the current warming of Earth's climate (Chapter 14) is occurring too rapidly for most species to adapt.

FAQ

Q: Isn't evolution based on just one man's beliefs?

A: Because Charles Darwin contributed so much to our early understanding of evolution, many people assume the concept itself hinges on his ideas. But scientists and lay-people had been observing nature and puzzling over fossils for a long time, and the notion of evolution was being discussed long before Darwin. Once he and Alfred Russel Wallace independently proposed the concept of natural selection, scientists finally gained a precise and feasible mechanism to explain how and why organisms change across generations. Later, geneticists discovered Gregor Mendel's research and worked out how traits are inherited—and modern evolutionary biology was born. Twentieth-century scientists Fisher, Wright, Dobzhansky, Simpson, Mayr, and others ran experiments and developed sophisticated mathematical models, documenting phenomena with extensive evidence and making evolutionary biology one of science's strongest fields. Since then, evolutionary research by thousands of scientists has driven our understanding of biology and has facilitated spectacular advances in agriculture and medicine.

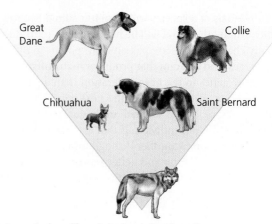

(a) Ancestral wolf and derived dog breeds

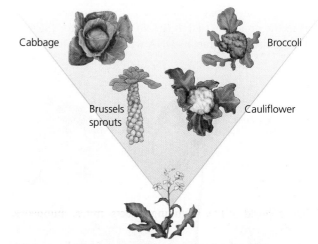

(b) Ancestral *Brassica oleracea* and derived crops

FIGURE 3.3 ▲ Selection imposed by people (selective breeding, or artificial selection) has resulted in the numerous breeds of dogs **(a)**. By starting with the gray wolf (*Canis lupus*) as the ancestral wild species, and by breeding like with like and selecting for the traits we prefer, we have evolved breeds as different as Great Danes and Chihuahuas. By this same process we have created the immense variety of crop plants we depend on for sustenance **(b)**. Cabbage, brussels sprouts, broccoli, and cauliflower were all generated from a single ancestral species, *Brassica oleracea*.

Evidence of selection is all around us

The results of natural selection are all around us, visible in every adaptation of every organism. In addition, numerous lab experiments (mostly with fast-reproducing organisms such as bacteria and fruit flies) have demonstrated rapid evolution of traits via natural selection.

The evidence for selection that may be most familiar to us is that which Darwin himself cited prominently in his work 150 years ago: our breeding of domesticated animals. In our dogs, cats, and livestock, we have conducted selection under our own direction. We have chosen animals with traits we like and bred them together, while not breeding animals with variants we do not like. Through such *selective breeding,* we have been able to augment particular traits we prefer.

Consider the great diversity of dog breeds (**FIGURE 3.3A**). People generated every type of dog alive today by starting with a single ancestral species and selecting for desired traits as in-

dividuals were bred together. From Great Dane to Chihuahua, all dogs can interbreed and produce viable offspring, yet breeders maintain striking differences among them by allowing only like individuals to breed with like. This process of selection conducted under human direction is termed **artificial selection.**

Artificial selection has given us the many crop plants we depend on for food, all of which people domesticated from wild ancestors and carefully bred over years, centuries, or millennia (**FIGURE 3.3B**). Through selective breeding, we have created corn with larger, sweeter kernels; wheat and rice with larger and more numerous grains; and apples, pears, and oranges with better taste. We have diversified single types into many—for instance, breeding variants of the plant *Brassica oleracea* to create broccoli, cauliflower, cabbage, and brussels sprouts. Our entire agricultural system is based on artificial selection. Thus, we depend on a working understanding of evolution for the very food we eat every day.

Evolution generates biological diversity

Evolution has generated our planet's remarkable diversity of life. But what do we mean by "diversity"? **Biological diversity,** or **biodiversity** for short, refers to the variety of life across all levels of biological organization, including the diversity of species, their genes, their populations, and their communities (we will introduce communities shortly: p. 52 and Chapter 4).

Scientists have described about 1.8 million species, but many more remain undiscovered or unnamed. Estimates for the total number of species in the world range up to 100 million, with many of them thought to occur in tropical forests. In this light, the discovery of a new toad species in Costa Rica seems less surprising. Although Costa Rica covers a tiny fraction (0.01%) of Earth's surface area, it is home to 5–6% of all species known to science. And of the 500,000 species scientists estimate exist in the country, only 87,000 (17%) have been inventoried and described.

Tropical rainforests such as Costa Rica's, however, are by no means the only places rich in biodiversity. Step outside anywhere, even in a major city, and you will find numerous species within easy reach. Plants poke up from cracks in asphalt in every city in the world, and even Antarctic ice harbors microbes. A handful of backyard soil may contain an entire miniature world of life, including insects, mites, millipedes, nematode worms, plant seeds, fungi, and millions of bacteria. (We will examine Earth's biodiversity in detail in Chapter 8.)

Speciation produces new types of organisms

How did Earth come to have so many species? Whether there are 1.8 million or 100 million, such large numbers require scientific explanation. The process by which new species are generated is termed **speciation.** Speciation can occur in a number of ways, but most biologists consider the main mode of species formation to be *allopatric speciation,* species formation due to the physical separation of populations over some geographic distance. To understand allopatric speciation, begin by picturing a population of organisms. Individuals within the population possess many similarities that unify them as a species because they are able to reproduce with one another and share genetic information. However, if the population is broken up into two or more isolated populations, individuals from one population cannot reproduce with individuals from the others.

When a mutation arises in the DNA of an organism in one of these isolated populations, it cannot spread to the other populations. Over time, each population will independently accumulate its own set of mutations. Eventually, the populations may diverge, or grow different enough, that their members can no longer mate with one another. Once this has happened, there is no turning back; the populations cannot interbreed, and they have embarked on their own independent evolutionary trajectories as separate species (**FIGURE 3.4**). The populations will continue diverging in their characteristics as

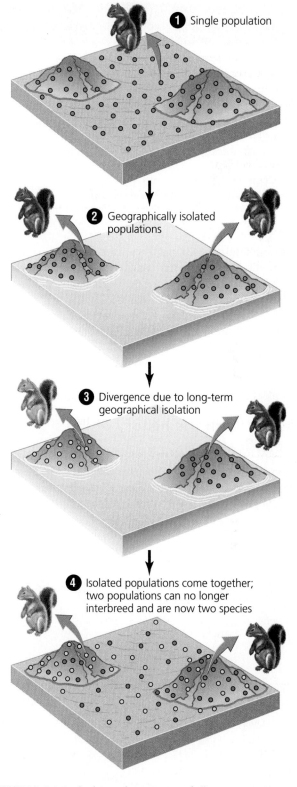

1 Single population

2 Geographically isolated populations

3 Divergence due to long-term geographical isolation

4 Isolated populations come together; two populations can no longer interbreed and are now two species

FIGURE 3.4 ▲ In the long, slow process of allopatric speciation, some geographical barrier splits a population. In this diagram, two mountaintops 1 are turned into islands by rising sea level 2, isolating populations of squirrels. Each isolated population accumulates its own independent set of genetic changes over time, until individuals become genetically distinct and unable to breed with individuals from the other population 3. The two populations now represent separate species and will remain so even if the geographical barrier is removed and the new species intermix 4.

chance mutations accumulate that cause the populations to become different in random ways. If environmental conditions happen to be different for the populations, then natural selection may accelerate the divergence.

The geographic isolation of populations that can lead to allopatric speciation may occur in various ways (**TABLE 3.1**). In each case, populations must remain isolated for a long time, generally thousands of generations. Through the speciation process, over millions of years, single species can generate multiple species, each of which can in turn generate more.

We can learn the history of life's diversification

Innumerable speciation events have generated complex patterns of diversity at levels above the species level. Evolutionary biologists study such patterns, examining how groups of organisms arose and how they evolved the characteristics they show. These scientists represent life's history by using branching, treelike diagrams called **phylogenetic trees** (**FIGURE 3.5**). Similar to family genealogies, phylogenetic trees depict relationships among species, major groups of species, populations, or genes. In such trees of species, each branching point represents a speciation event. Scientists construct these trees by analyzing patterns of similarity among present-day organisms and inferring which groups share similarities because they are related.

Scientists also decipher life's history by studying fossils. As organisms die, some are buried by sediment. Under certain conditions, the hard parts of their bodies—such as bones, shells, and teeth—may be preserved as sediments are compressed into rock (p. 230). Minerals replace the organic material, leaving behind a **fossil,** an imprint in stone of the dead organism (**FIGURE 3.6**). In countless locations across the world, geologic processes over millions of years have buried sedimentary rock layers and later brought them to the surface, revealing assemblages of fossilized plants and animals from different time periods. By aging the rock layers that contain fossils, scientists can learn when particular organisms lived. The cumulative body of fossils worldwide is known as the **fossil record.**

Life's history, as revealed by phylogenetic trees and by the fossil record, is complex, but a few big-picture trends are apparent. In its 3.5 billion years on Earth, life has evolved complex structures from simple ones, and large sizes from small ones. However, simplicity and small size have also evolved when favored by natural selection, and it is easy to argue that Earth still belongs to the bacteria and other microbes, some of them little changed over eons.

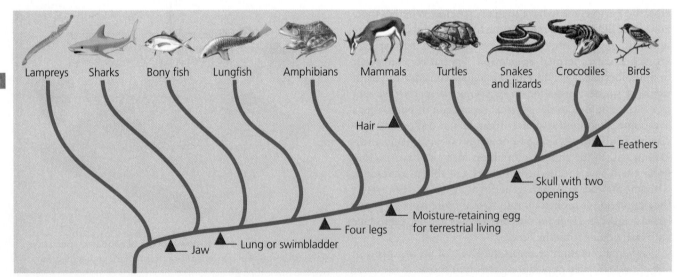

FIGURE 3.5 ▲ Phylogenetic trees show the history of life's divergence. Similar to family genealogies, these diagrams illustrate relationships among groups of organisms, as inferred from comparisons among present-day creatures. The tree shown here is a greatly simplified representation of relationships among groups of vertebrates—just one small portion of the huge and complex "tree of life." Each branch results from a speciation event, and as you follow the tree upward from its trunk to the tips of its branches you proceed forward in time, tracing life's history. By mapping traits onto phylogenetic trees, biologists can study how traits have evolved over time. In this diagram, major traits are mapped using arrows to indicate when they originated. For instance, all vertebrates "above" the point at which jaws are indicated have jaws, whereas lampreys diverged before jaws originated and thus lack them.

FIGURE 3.6 ▲ The fossil record helps reveal the history of life on Earth. The numerous fossils of trilobites indicate that these animals, now extinct, were abundant in the oceans from roughly 540 million to 250 million years ago.

Even fans of microbes, however, must marvel at the exquisite adaptations of animals, plants, and fungi: The heart that beats so reliably for an animal's entire lifetime that we take it for granted. The complex organ system to which the heart belongs. The stunning plumage of a peacock in full display. The ability of each and every plant on the planet to lift water and nutrients from the soil, gather light from the sun, and turn it into food. The staggering diversity of beetles and other insects. The human brain and its ability to reason. All these and more have resulted as the process of evolution has generated new species and whole new branches on the tree of life.

Although speciation generates Earth's biodiversity, it is only one part of the equation, because the vast majority of species that once lived are now gone—from creatures of the deep past, such as the trilobite in Figure 3.6, to recent ones, such as Monteverde's golden toad. The disappearance of a species from Earth is called **extinction.** From studying the fossil record, paleontologists (scientists who study life's history) calculate that the average time a species spends on Earth is 1–10 million years. The number of species in existence at any one time is equal to the number added through speciation minus the number removed by extinction.

Some species are more vulnerable to extinction than others

In general, extinction occurs when environmental conditions change so rapidly or severely that a species cannot adapt because natural selection simply does not have enough time to

work. All manner of events can cause extinction—climate change, the rise and fall of sea level, the arrival of new species, severe weather events, and more. In general, small populations and species narrowly specialized on some particular resource or way of life are most vulnerable to extinction from environmental change.

The golden toad was a prime example of a vulnerable species. It was **endemic** to the Monteverde cloud forest, meaning that it occurred nowhere else on the planet. Endemic species face relatively high risks of extinction because all their members belong to a single, sometimes small, population. At the time of its discovery, the golden toad was known from only a 4-km^2 (1,000-acre) area of Monteverde. It also required very specific conditions to breed successfully. During the spring at Monteverde, water collects in shallow pools within the network of tree roots that span the cloud forest's floor. The golden toad gathered to breed in these root-bound reservoirs, and it was here that Jay Savage and his companions collected their specimens. Monteverde provided ideal habitat for the golden toad, but the extent of that habitat was minuscule—so any environmental stresses that deprived the toad of the resources it needed to survive might doom the entire world population of the species.

Earth has seen several episodes of mass extinction

Most extinction occurs gradually, one species at a time. The rate at which this type of extinction occurs is referred to as the *background extinction rate.* However, Earth has seen five events of staggering proportions that killed off massive numbers of species at once. These episodes, called **mass extinction events,** have occurred at widely spaced intervals in Earth history and have wiped out 50–95% of our planet's species each time (p. 166).

The best-known mass extinction occurred 65 million years ago and brought an end to the dinosaurs (although birds are modern representatives of dinosaurs). Evidence suggests that the impact of a gigantic asteroid caused this event, called the Cretaceous-Tertiary, or K-T, event. Still more severe was the mass extinction at the end of the Permian period 250 million years ago (see **APPENDIX: Geologic Time Scale** for Earth's geologic periods). Paleontologists estimate that 75–95% of all species may have perished during this event, described by one researcher as "the mother of all mass extinctions." Scientists do not yet know what caused the end-Permian extinction event, but hypotheses include massive volcanism, an asteroid impact, methane releases and global warming, or some combination of factors.

The sixth mass extinction is upon us

Many biologists have concluded that Earth is currently entering its sixth mass extinction event—and that we are the cause. Changes to Earth's natural systems set in motion by human population growth, development, and resource depletion have driven many species extinct and are threatening countless more. The alteration and destruction of natural habitats, the hunting and harvesting of species, and the introduction of species from one place to another where they can harm native species—these processes and more are causing us to lose the biological diversity that makes Earth such a unique planet (pp. 168–171).

This loss affects people directly, because other organisms provide us with life's necessities—food, fiber, medicine, and ecosystem services (pp. 2, 36, 90, 94–95) without which we cannot survive. Species extinction may well be the biggest environmental problem we face, because the loss of a species is irreversible.

Amphibians such as the golden toad are disappearing faster than just about any other type of organism. According to the most recent scientific assessments, 40% of frog, toad, and salamander species are in decline, 30% are in danger of extinction, and nearly 170 species have vanished just within the last few years or decades. Some species are disappearing in remote and pristine areas, suggesting that infection by chytrid fungus—in particular, a pathogen named *Batrachochytrium dendrobatidis*—could be responsible. Many researchers think a variety of causes are affecting amphibians at once in a "perfect storm" of impacts.

LEVELS OF ECOLOGICAL ORGANIZATION

The extinction of species, their generation through speciation, and other evolutionary mechanisms and patterns play key roles in ecology. **Ecology** is the study of interactions among organisms and between organisms and their environments. It is often said that ecology provides the stage upon which the play of evolution unfolds. The two are intertwined in many ways.

We study ecology at several levels

Life occurs in a hierarchy of levels, from atoms, molecules, and cells (pp. 26–29) up through the **biosphere,** which is the cumulative total of living things on Earth and the areas they inhabit. *Ecologists* are scientists who study relationships on the higher levels of this hierarchy (**FIGURE 3.7**), namely on the levels of the organism, population, community, ecosystem, and biosphere.

At the level of the organism, the science of ecology describes relationships between the organism and its physical environment. Organismal ecology helps us understand, for example, what aspects of the golden toad's environment were important to it, and why. In contrast, **population ecology** examines the dynamics of population change and the factors that affect the distribution and abundance of members of a population. It helps us understand why populations of some species (such as the golden toad) decline while populations of others (such as ourselves) increase. **Community ecology** (Chapter 4) focuses on patterns of species diversity and on interactions among species. In the case of Monteverde, it allows us to study how the golden toad interacted with other species of its cloud-forest community. Ecology at the ecosystem level (Chapter 2) reveals patterns, such as the flow of energy and nutrients, by studying living and nonliving components of systems in conjunction. As new technologies allow scientists to learn more about the complex dynamics of natural systems

Levels of Ecological Organization		
	Biosphere	The sum total of living things on Earth and the areas they inhabit
	Ecosystem	A functional system consisting of a community, its nonliving environment, and the interactions between them
	Community	A set of populations of different species living together in a particular area
	Population	A group of individuals of a species that live in a particular area
	Organism	An individual living thing

FIGURE 3.7 ▲ Life exists in a hierarchy of levels. Ecology includes the study of the organismal, population, community, and ecosystem levels and, increasingly, the level of the biosphere.

on a global scale, many ecologists are expanding their horizons to the biosphere as a whole.

Habitat, niche, and specialization are key concepts in ecology

At the level of the organism, each individual relates to its environment in ways that tend to maximize its survival and reproduction. One key relationship involves the specific environment in which an organism lives, its **habitat.** A species' habitat consists of the living and nonliving elements around it, including rock, soil, leaf litter, humidity, plant life, and more. The golden toad lived in a habitat of cloud forest—specifically, on the moist forest floor, using seasonal pools for breeding and burrows for shelter.

Each organism thrives in certain habitats and not in others, leading to nonrandom patterns of *habitat use*. Mobile organisms actively select habitats in which to live from among the range of options they encounter, a process called *habitat selection*. In the case of plants and rooted animals (such as sea anemones), whose young disperse and settle passively, patterns of habitat use result from success in some habitats and failure in others.

Habitats are scale dependent. A tiny soil mite may use less than a square meter of soil in its lifetime. A vulture, elephant, or whale, in contrast, may traverse miles upon miles of air, land, or water each day. Species also may have different habitat needs in different seasons. Many migratory birds use distinct breeding, wintering, and migratory habitats.

The criteria by which organisms favor some habitats over others can vary greatly. The soil mite may assess habitats in terms of the chemistry, moisture, and texture of the soil and the percentage and type of organic matter. The vulture may ignore not only soil but also topography and vegetation, focusing solely on the abundance of dead animals that it scavenges for food. Every species assesses habitats differently because every species has different needs.

Habitat use is important in environmental science because the availability and quality of habitat are crucial to an organism's well-being. Indeed, because habitats provide everything an organism needs, including nutrition, shelter, breeding sites, and mates, the organism's very survival depends on the availability of suitable habitats. Often this need results in conflict with people who want to alter or develop a habitat for their own purposes.

Another way in which an organism relates to its environment is through its niche. A species' **niche** reflects its use of resources and its functional role in a community. This includes its habitat use, its consumption of certain foods, its role in the flow of energy and matter, and its interactions with other organisms. The niche is a multidimensional concept, a kind of summary of everything an organism does. The pioneering ecologist Eugene Odum once wrote that "habitat is the organism's address, and the niche is its profession."

Organisms vary in the breadth of their niche. Species with narrow breadth, and thus very specific requirements, are said to be **specialists.** Those with broad tolerances, able to use a wide array of resources, are **generalists.** Specialists succeed by being extremely good at the things they do, but they are vulnerable when conditions change and threaten the habitat or resource on which they have specialized. Generalists succeed by being able to live in many different places and withstand variable conditions, but they may not thrive in any one situation as much as a specialist does. An organism's habitat use, niche, and degree of specialization each reflect adaptations of the species and are products of natural selection.

POPULATION ECOLOGY

Individuals of a species inhabiting a particular area make up a population. Species may consist of multiple populations that are geographically isolated from one another. This is the case with a species characteristic of Monteverde—the resplendent quetzal (*Pharomachrus mocinno*), considered one of the world's most spectacular birds (see Figure 3.1a). Although it ranges from southernmost Mexico to Panama, the resplendent quetzal is adapted to high-elevation tropical forest and is absent from surrounding low-elevation areas. Moreover, human development has destroyed much of its montane cloud-forest habitat. Thus, the species today exists in isolated populations scattered across Central America.

In contrast, the human species is a consummate generalist, and we have spread into nearly every corner of the planet. As a result, it is difficult to define a distinct human population on anything less than the global scale. Some would maintain that in the ecological sense of the word, all 7 billion of us comprise one population.

Several characteristics help us predict population dynamics

All populations—from humans to quetzals to golden toads—show characteristics that help population ecologists predict the future dynamics of the population.

Population size   Expressed as the number of individual organisms present at a given time, **population size** may increase, decrease, undergo cyclical change, or remain the same over time. Extinctions are generally preceded by population declines. As late as 1987, scientists documented a golden toad population at Monteverde in excess of 1,500 individuals, but in 1988 researchers sighted only 10 toads, and in 1989 they found just a single individual. By 1990, the species had disappeared.

The passenger pigeon (*Ectopistes migratorius*), also now extinct, illustrates the extremes of population size (**FIGURE 3.8**). Not long ago it was the most abundant bird in North America; flocks of passenger pigeons literally darkened the skies. In the early 1800s, ornithologist Alexander Wilson watched a flock of 2 billion birds 390 km (240 mi) long that took 5 hours to fly over and sounded like a tornado. Passenger pigeons nested in gigantic colonies in the forests of the upper Midwest and southern Canada. Once people began cutting the forests, however, the birds made easy targets for market hunters, who gunned down thousands at a time and shipped them to market by the wagonload. By the end of the 19th century, the passenger pigeon population had declined to such a low number that they could not form the colonies they apparently needed in order to breed. In 1914, the last passenger pigeon on Earth died in the Cincinnati Zoo, bringing the continent's most numerous bird species to extinction within just a few decades.

Population density   The flocks and breeding colonies of passenger pigeons showed high population density, another attribute that ecologists assess to understand populations. **Population density** describes the number of individuals in a population per unit area. For instance, the 1,500 golden toads counted in 1987 within 4 km^2 indicated a density of 375 toads per km^2. In general, large organisms have low population densities because each individual requires many resources—and thus a great deal of area—to survive.

High population density makes it easier for organisms to group together and find mates, but it can also lead to competition and conflict if space, food, or mates are in limited supply. Overcrowded organisms may become vulnerable to the predators that feed on them, and close contact among individuals can increase the transmission of infectious disease. For these reasons, organisms sometimes leave an area when densities become too high. In contrast, at low population densities, organisms benefit from more space and resources but may find it harder to locate mates and companions.

Overcrowding at high population densities in small remnants of habitat is thought to have doomed Monteverde's harlequin frog (*Atelopus varius*; see Figure 3.1c), an amphibian that disappeared at the same time as the golden toad. The harlequin frog is a habitat specialist, favoring

(a) Passenger pigeon

(b) 19th-century lithograph of pigeon hunting in Iowa

FIGURE 3.8 ▲ The passenger pigeon **(a)** was once North America's most numerous bird, and its flocks literally darkened the skies when millions of birds passed overhead **(b)**. However, human cutting of forests and hunting drove the species to extinction within a few decades.

"splash zones," areas alongside rivers and streams that receive spray from waterfalls and rapids. As Monteverde's climate grew warmer and drier in the 1980s and 1990s, water flow decreased, and many streams dried up (see **THE SCIENCE BEHIND THE STORY,** pp. 56–57). Splash zones grew smaller and fewer, and harlequin frogs were forced to cluster together in what remained of the habitat. Researchers J. Alan Pounds and Martha Crump recorded frog population densities up to 4.4 times higher than normal, with more than 2 frogs per meter (3.3 ft) of stream. Such crowding made the frogs vulnerable to predator attack, assault from parasitic flies, and transmission of disease pathogens such as chytrid fungus.

Population distribution It was not simply the harlequin frog's density, but also its distribution in space that led to its demise at Monteverde. **Population distribution,** or **population dispersion,** describes the spatial arrangement of organisms in an area. Ecologists define three distribution types: random, uniform, and clumped (**FIGURE 3.9**). In a *random distribution,* individuals are located haphazardly in space in no particular pattern. This type of distribution can occur when the resources an organism needs are widespread and other organisms do not strongly influence where members of a population settle.

A *uniform distribution* is one in which individuals are evenly spaced. This can occur when individuals hold territories or otherwise compete for space. In a desert where water is scarce, each plant needs a certain amount of space for its roots to gather moisture. Plants may even poison one another's roots as a means of competing for space. As a result, each surviving plant may be equidistant from others.

In a *clumped distribution,* the pattern most common in nature, organisms arrange themselves according to the availability of the resources they need. Many desert plants grow in patches around springs or along washes that flow with water

after rainstorms. During their mating season, golden toads were found clumped at seasonal breeding pools. Humans, too, exhibit clumped distribution, as people frequently aggregate together in urban centers. Clumped distributions often indicate that species are seeking certain habitats or resources that are themselves clumped.

Sex ratio A population's **sex ratio** is its proportion of males to females, and this can influence whether the population will increase or decrease in size over time. In monogamous species (in which each sex takes a single mate), a 1:1 sex ratio maximizes population growth, whereas an unbalanced ratio leaves many individuals of one sex without mates. Most species are not monogamous, however, so sex ratios may vary from one species to another.

Age structure Populations generally consist of individuals of different ages. **Age structure,** or **age distribution,** describes the relative numbers of organisms of each age within a population (pp. 119–120). By combining this information with data on the reproductive potential of individuals in different age classes, a population ecologist can predict how the population may grow or shrink.

For many plants and animals that continue growing in size as they age, older individuals reproduce more. A tree that is large because it is old can produce more seeds, and a fish that is large because it is old may produce more eggs. In some animals, such as birds, the experience they gain with age often makes older individuals better breeders.

Human beings are unusual because we often survive past our reproductive years. A human population made up largely of older (post-reproductive) individuals will tend to decline over time, whereas a population with many young people (of reproductive or pre-reproductive age) is likely to increase. (We will use diagrams to explore these ideas further in Chapter 6 as we study human populations.)

(a) Random

(b) Uniform

(c) Clumped

FIGURE 3.9 ▲ Individuals in a population can spatially distribute themselves in three fundamental ways. In a random distribution **(a)**, organisms are dispersed at random through the environment. In a uniform distribution **(b)**, individuals are spaced evenly, at equal distances from one another. Territoriality can result in such a pattern. In a clumped distribution **(c)**, individuals occur in patches, concentrated more heavily in some areas than in others. Habitat selection, clustered resource patterns, or flocking to avoid predators can result in such a pattern.

Populations may grow, shrink, or remain stable

Now that we have outlined some key attributes of populations, we are ready to take a quantitative view of population change by examining some simple mathematical concepts used by population ecologists and by **demographers** (scientists who study human populations). Population growth, or decline, is determined by four factors:

- ▶ Births within the population (*natality*)
- ▶ Deaths within the population (*mortality*)
- ▶ *Immigration* (arrival of individuals from outside the population)
- ▶ *Emigration* (departure of individuals from the population)

Births and immigration add individuals to a population, whereas deaths and emigration remove individuals. To measure a population's **growth rate,** the rate of change in a population's size per unit time, we calculate the birth rate plus the immigration rate, minus the death rate plus the emigration rate, each expressed as the number per 1,000 individuals per year:

$$(\text{birth rate} + \text{immigration rate}) - (\text{death rate} + \text{emigration rate}) = \text{growth rate}$$

The resulting number tells us the net change in a population's size per 1,000 individuals per year. For example, a population with a birth rate of 18 per 1,000/yr, a death rate of 10 per 1,000/yr, an immigration rate of 5 per 1,000/yr, and an emigration rate of 7 per 1,000/yr would have a growth rate of 6 per 1,000/yr:

$$(18/1{,}000 + 5/1{,}000) - (10/1{,}000 + 7/1{,}000) = 6/1{,}000$$

Thus, a population of 1,000 in one year will reach 1,006 in the next. If the population is 1,000,000, it will reach 1,006,000 the next year. Such population increases are often expressed as percentages, which we can calculate using the following formula:

$$\text{growth rate} \times 100\%$$

Thus, a growth rate of 6/1,000 would be expressed as:

$$6/1{,}000 \times 100\% = 0.6\%$$

By measuring population growth in terms of percentages, scientists can compare increases and decreases in populations of far different sizes. They can also project changes that will occur in the population over longer periods, much as you might calculate the amount of interest your savings account will earn over time.

Unregulated populations increase by exponential growth

When a population increases by a fixed percentage each year, it is said to undergo **exponential growth.** Imagine you put money in a savings account at a fixed interest rate and leave it untouched for years. As the principal accrues interest and grows larger, you earn still more interest, and the sum grows by escalating amounts each year. The reason is that a fixed percentage of a small number makes for a small increase, but that same percentage of a large number produces a large increase. Thus, as savings accounts (or populations) become larger, each incremental increase likewise gets larger. Such acceleration is a characteristic of exponential growth.

THE SCIENCE BEHIND THE STORY

Climate Change, Disease, and the Amphibians of Monteverde

Dr. J. Alan Pounds (left) with Dr. Luis Coloma, looking for harlequin frogs

Soon after the golden toad's disappearance, scientists began to investigate the potential role of global climate change (Chapter 14) in driving cloud-forest species toward extinction. They noted that the period from July 1986 to June 1987 was the driest on record at Monteverde, with unusually high temperatures and record-low stream flows. These conditions had caused the golden toad's breeding pools to dry up in the spring of 1987, likely killing nearly all of the eggs and tadpoles in the pools.

By reviewing reams of weather data, scientists found that the number of dry days and dry periods each winter in the Monteverde region had increased between 1973 and 1998. Because amphibians breathe and absorb moisture through their skin, they are susceptible to dry conditions. Based on these facts, herpetologists J. Alan Pounds and Martha Crump in 1994 hypothesized that hot, dry conditions were to blame for high adult mortality and breeding problems among golden toads and other amphibians.

Throughout this period, scientists worldwide were realizing that the atmosphere and oceans were warming because of human release of carbon dioxide and other greenhouse gases (Chapter 14). With this in mind, Pounds and others reviewed the scientific literature on ocean and atmospheric science to analyze the effects on Monteverde's local climate due to warming patterns in the ocean regions around Costa Rica.

Warmer oceans, the researchers found, caused clouds to pass over at higher elevations, where they were no longer in contact with the trees. Once the cloud forest's moisture supply was pushed upward, the forest began to dry out (**see first figure**).

In a 1999 paper in the journal *Nature*, Pounds and two colleagues reported that climate modification was causing local changes at the species,

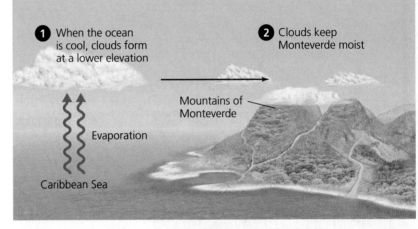

(a) Cool ocean conditions

① When the ocean is cool, clouds form at a lower elevation

② Clouds keep Monteverde moist

Mountains of Monteverde

Evaporation

Caribbean Sea

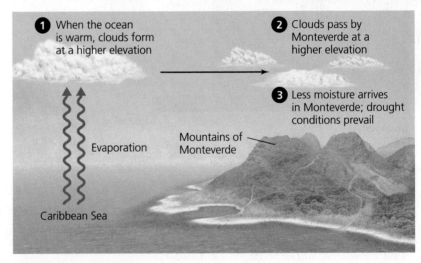

(b) Warm ocean conditions

① When the ocean is warm, clouds form at a higher elevation

② Clouds pass by Monteverde at a higher elevation

③ Less moisture arrives in Monteverde; drought conditions prevail

Mountains of Monteverde

Evaporation

Caribbean Sea

Monteverde's cloud forest gets its moisture from clouds that sweep inland from the oceans. When ocean temperatures are cool (a), the clouds keep Monteverde moist. Warmer ocean conditions (b) resulting from climate change cause clouds to form at higher elevations and pass over the mountains, drying the forest.

population, and community levels. They argued that higher clouds and decreasing moisture could explain the disappearance of the golden toad and harlequin frog, and also the concurrent population crashes and subsequent disappearance of 20 other species of frogs and toads from the Monteverde region.

Moreover, whole communities were being altered. As the montane forests dried out, drought-tolerant species of birds and reptiles shifted upslope, and moisture-dependent species were stranded at the mountaintops by a rising tide of aridity. If a species has nowhere to go, then extinction may result.

Pounds and his colleagues expanded the story further in *Nature* in 2006. Although clouds had risen higher in the sky, the extra moisture evaporating from warming oceans was increasing cloud cover overall, blocking sunlight during the day and trapping heat at night. As a result, at Monteverde and other tropical locations, daytime and nighttime temperatures were becoming more similar.

Such conditions are optimal for chytrid fungi, pathogens that can lethally infect amphibians. In recent years the chytrid fungus *Batrachochytrium dendrobatidis* is thought to have contributed to the likely extinction of 67 of the world's 113 species of harlequin frogs. At Monteverde and elsewhere, Pounds's team argued, climate change is promoting disease epidemics that are driving extinct many of the world's amphibians.

Other researchers agreed that chytrid fungus is a major threat but disputed a connection to climate change. In 2008, one team led by biologist Karen Lips of Southern Illinois University, Carbondale, reanalyzed the Pounds team's data and also mapped amphibian declines and inferred how the non-native and invasive chytrid fungus had spread rapidly in waves across Central and South America in recent years (**see second figure**). The analysis by Lips and her team suggests

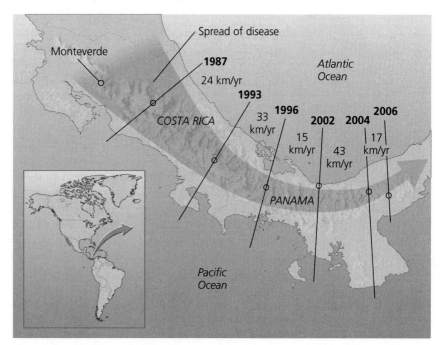

Dr. Karen Lips's research team used known dates of decline (years in figure) in harlequin frog populations at sites in Costa Rica and Panama to infer the spread of a wave of infection (arrow) by chytrid fungus across the region. By their analysis, chytrid reached Monteverde before 1987 and was well into Panama by 2000. Adapted from Lips, K.R., et al., 2008. Riding the wave: Reconciling the roles of disease and climate change in amphibian declines. *PLoS Biology* 6: 441–454.

that the dramatic amphibian die-offs at Monteverde and elsewhere are directly due to the arrival of this devastating pathogen and do not necessarily involve climate change.

Later that year, a team led by Jason Rohr of the University of South Florida set out to test the Pounds hypothesis (that climate change was promoting chytrid infection) versus the Lips hypothesis (that chytrid was spreading regardless of climate change). Rohr's group tested for statistical correlations (p. 10) among climate variables, chytrid spread, and the timing of the 67 suspected extinctions of harlequin frog species. This team found a clear spatial pattern in the sequence of extinctions of harlequin frogs that was consistent with the spread of disease, but the researchers could not attribute this beyond doubt to chytrid.

Rohr's team found that harlequin frog extinctions were correlated with rising tropical air temperatures, but the researchers questioned whether this was causal, given that increased extinctions were also correlated with just about every other variable that rose between 1970 and 1990, from human population growth to regional beer and banana production. The researchers concluded that because climate change has been shown to influence many biological systems, it is likely a factor in amphibian declines, but that we do not yet have solid evidence for this causal link.

Clearly more research into the effects of both disease and climate change on frogs is needed—and fast. Biologists today are racing to find the answers, hoping to save many of the world's amphibian species from extinction.

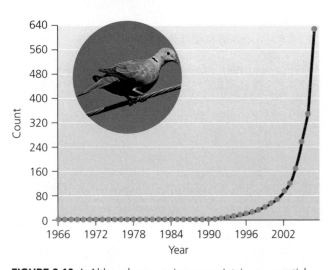

FIGURE 3.10 ▲ Although no species can maintain exponential growth indefinitely, some may grow exponentially for a time when colonizing an unoccupied environment or exploiting an unused resource. The Eurasian collared dove (*Streptopelia decaocto*) is currently spreading across the North American continent, propelled by exponential growth. Data from Sauer, J.R., et al., 2008. *The North American Breeding Bird Survey, Results and Analysis 1966–2007.* v. 5.15.2008. USGS Patuxent Wildlife Research Center, Laurel, MD.

We can visualize changes in population size by using population growth curves. The J-shaped curve in **FIGURE 3.10** shows exponential increase. Populations of organisms increase exponentially unless they meet constraints. Each organism reproduces by a certain amount, and as populations get larger, there are more individuals reproducing by that amount. If there are no external limits, ecologists theoretically expect exponential growth.

Normally, exponential growth occurs in nature only when a population is small, competition is minimal, and environmental conditions are ideal for the organism in question. Most often, these conditions occur when the organism is introduced to a new environment that contains abundant resources to exploit. Mold growing on a piece of bread or fruit and bacteria colonizing a recently dead animal are cases in point. Plants colonizing regions during primary succession (p. 74) after glaciers recede or volcanoes erupt may also grow exponentially. A current example of exponential growth in the United States is the Eurasian collared dove (see Figure 3.10). Unlike its extinct relative the passenger pigeon, this species arrived here from Europe and has spread across North America in a matter of years. The Eurasian collared dove thrives in human-disturbed areas and apparently has not (yet) encountered anything to limit its population growth.

Limiting factors restrain population growth

Exponential growth rarely lasts long. If even a single species in Earth's history had increased exponentially for very many generations, it would have blanketed the planet's surface! Instead, every population eventually is constrained by **limiting factors**—physical, chemical, and biological attributes of the environment that restrain population growth. The interaction of these factors determines the **carrying capacity,** the

maximum population size of a species that a given environment can sustain.

Ecologists use the curve in **FIGURE 3.11** to show how an initial exponential increase is slowed and eventually brought to a standstill by limiting factors. Called the **logistic growth curve,** it rises sharply at first but then begins to level off as the effects of limiting factors become stronger. Eventually the force of these factors—collectively termed *environmental resistance*—stabilizes the population size at its carrying capacity. One day soon, the Eurasian collared dove population in North America will slow in its growth and then reach carrying capacity. Populations of other European birds that spread across North America in recent decades, such as the house sparrow and the European starling, have apparently peaked and are today stabilizing.

For animals in terrestrial environments, limiting factors include the availability of food, water, mates, shelter, and suitable breeding sites; temperature extremes; prevalence of disease; and abundance of predators. Plants are often limited by amounts of sunlight and moisture and the type of soil chemistry, in addition to disease and to attack from plant-eating animals. In aquatic systems, limiting factors include salinity, sunlight, temperature, dissolved oxygen, fertilizers, and pollutants.

A population's density can enhance or diminish the impact of certain limiting factors. Recall that high population density can help organisms find mates but can also increase competition and the risk of predation and disease. Such factors are said to be **density-dependent** factors, because their influence rises and falls with population density. The logistic growth curve in Figure 3.11 represents the effects of density dependence. The larger the population size, the stronger the effects of environmental resistance.

Density-independent factors are limiting factors whose influence is not affected by population density. Temperature extremes and catastrophic events such as floods, fires, and landslides are examples of density-independent factors because

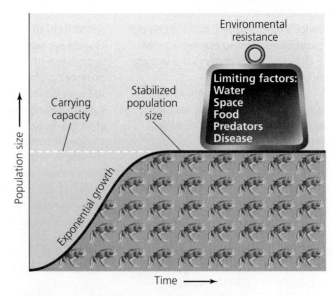

FIGURE 3.11 ▲ The logistic growth curve shows how population size may increase rapidly at first, then grow more slowly, and finally stabilize at a carrying capacity. Carrying capacity is determined both by the *biotic potential* of the organism and by various external limiting factors, collectively termed *environmental resistance.*

they can eliminate large numbers of individuals without regard to their density.

The logistic curve is a simplified model, and real populations can behave differently. Some may cycle above and below the carrying capacity. Others may overshoot the carrying capacity and then crash, destined either for extinction or recovery (**FIGURE 3.12**).

Carrying capacities can change

Because environments are complex and ever-changing, carrying capacity can vary. If a fire destroys a forest, for example, the carrying capacities for most forest animals will decline, whereas those for certain species that benefit from fire will increase. Our own species has proved capable of intentionally altering our environment so as to reduce environmental resistance and raise our carrying capacity. When our ancestors began to build shelters and use fire for heating and cooking, they reduced the environmental resistance of areas with cold climates and were able to expand into new territory. As limiting factors are overcome (through development of new technologies or through natural environmental change), the carrying capacity for a species may increase. People have managed so far to increase the planet's carrying capacity for our species, but we have done so by appropriating immense proportions of the planet's resources. In the process, we have reduced the carrying capacities for countless other organisms that rely on those same resources.

WEIGHING THE ISSUES

Carrying Capacity and Human Population Growth As we have seen (pp. 2–3), the global human population has surpassed 7 billion, and we have far exceeded our planet's historic carrying capacity for people. What factors increased Earth's carrying capacity for us? Are there limiting factors for the human population? What might they be? Do you think we can keep raising our carrying capacity in the future? Might Earth's carrying capacity for us decrease?

Reproductive strategies vary among species

Limiting factors from an organism's environment provide only half the story of population regulation. The other half comes from the attributes of the organism itself. For example, organisms differ in their *biotic potential*, or ability to produce offspring. A fish with a short gestation period that lays thousands of eggs at a time has high biotic potential, whereas a whale with a long gestation period that gives birth to a single calf at a time has low biotic potential.

Giraffes, elephants, humans, and other large animals with low biotic potential produce relatively few offspring during their lifetimes. Species that take this approach to reproduction require a long time to gestate and raise each offspring, but the considerable energy and resources they devote to caring for and protecting them helps give these few offspring a high

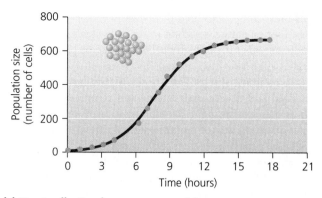

(a) Yeast cells, *Saccharomyces cerevisiae*

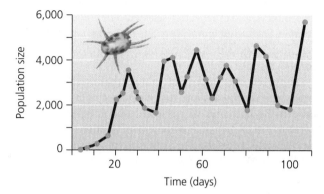

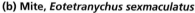
(b) Mite, *Eotetranychus sexmaculatus*

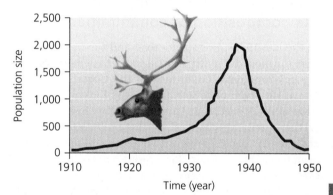
(c) St. Paul reindeer, *Rangifer tarandus*

FIGURE 3.12 ▲ Population growth in nature may depart from the logistic growth curve in various ways. Yeast cells from an early lab experiment show logistic growth **(a)** that closely matches the theoretical model. Some organisms, such as the mite shown in **(b)**, show cycles in which population fluctuates above and below the carrying capacity. Populations that rise too fast and deplete resources may crash just as suddenly **(c)**, like the population of reindeer introduced to the Bering Sea island of St. Paul. Data from: (a) Pearl, R., 1927. The growth of populations. *Quarterly Review of Biology* 2: 532–548; (b) Huffaker, C.B., 1958. Experimental studies on predation: Dispersion factors and predator-prey oscillations. *Hilgardia* 27: 343–383, Figure 7. © 1958 by the Regents of the University of California; (c) Adapted from Scheffer, Victor B., 1951. The rise and fall of a reindeer herd. *Scientific Monthly* 73: 356–362, Fig. 1. Reprinted with permission from AAAS.

likelihood of survival. Such species are said to be **K-selected** (so named because their populations tend to stabilize over time near carrying capacity, commonly symbolized as *K*). Because their populations stay close to carrying capacity, these organisms must compete to hold their own in a crowded world. In these species, natural selection favors individuals that invest in producing offspring of high quality that can be good competitors.

In contrast, species that are **r-selected** have high biotic potential and devote their energy and resources to producing as many offspring as possible in a relatively short time. Their offspring do not require parental care after birth, so r-strategists simply leave their survival to chance. The abbreviation *r* denotes the rate at which a population increases in the absence of limiting factors. Populations of r-selected species fluctuate greatly, such that they are often well below carrying capacity. This is why natural selection in these species favors traits that lead to rapid population growth. Many fish, plants, frogs, insects, and others are r-selected. The golden toad is one example. Each adult female laid 200–400 eggs, and its tadpoles spent just 5 weeks in the breeding pools, unprotected, metamorphosing into adults.

However, it is important to note that *these are two extremes on a continuum* and that most species fall somewhere between the extremes of r-selected and K-selected species. Moreover, many organisms show combinations of traits that do not clearly correspond to a place on the continuum. A redwood tree (*Sequoia sempervirens*), for instance, is large and long-lived, yet it produces many small seeds and offers no parental care.

Changes in populations influence the composition of communities

In the late 1980s, the golden toad and the harlequin frog were the most diligently studied species affected by changing environmental conditions in the Costa Rican cloud forest. However, once scientists began looking at populations of other species at Monteverde, they began to notice further troubling changes. By the early 1990s, not only had golden toads, harlequin frogs, and other organisms been pushed from their cloud-forest habitat into apparent extinction, but also many species from lower, drier habitats had begun to appear at Monteverde. By the year 2000, 15 dry-forest species had moved into the cloud forest and begun to breed.

Meanwhile, population sizes of several cloud-forest bird species declined. After 1987, 20 of 50 frog species vanished from one part of Monteverde, and ecologists later reported more disappearances, including those of two lizards native to the cloud forest. Scientists hypothesized that changes in climate were causing population fluctuations and disease outbreaks and were unleashing changes in the composition of the community (see **The Science behind the Story,** pp. 56–57).

FIGURE 3.13 ▲ Costa Rica has protected many of its diverse natural areas, stimulating the nation's economy through ecotourism in the process. Here, visitors experience a walkway through the forest canopy in one of the nation's parks.

Conservation can help address biodiversity loss

Changes in populations and communities have been taking place naturally as long as life has existed, but today human development, resource extraction, and population pressure are speeding the rate of change and altering the types of change. Science is crucial to helping us understand our impacts, but the phenomena that threaten biodiversity have complex social, economic, and political roots, and we must understand these aspects if we are to develop solutions.

Fortunately, millions of people around the world are taking action to safeguard biodiversity and the ecological and evolutionary processes that make Earth such a unique place (pp. 174–181). Costa Ricans have been confronting the challenges to their nation's biodiversity, and their actions show what even a small country of modest means can do. Despite efforts like the Monteverde Cloud Forest Reserve, established in 1972, by the 1990s Costa Rica was losing its forests faster than any nation in the world. The nation's leaders resolved to step up land protection efforts and expand their national park system. Today Costa Rica protects more than 12% of its area in national parks, and devotes a further 13% to other types of wildlife and conservation reserves. As a result, people from around the world flock to Costa Rica to visit these natural areas (**FIGURE 3.13**) in a phenomenon called **ecotourism**. Ecotourism draws over 1 million visitors to Costa Rica each year, employs thousands of Costa Ricans, and pumps a billion dollars per year into the country's economy. By working to protect habitat, safeguard endangered species, and recover their populations, Costa Rica and its citizens are now reaping ecological and economic benefits from their conservation efforts.

➤ CONCLUSION

The golden toad and other organisms of the Monteverde cloud forest have helped illuminate the fundamentals of evolution and population ecology that are integral to environmental science. The evolutionary processes of natural selection, speciation, and extinction help determine Earth's biodiversity. Understanding

how ecological processes function at the population level is crucial to protecting biodiversity threatened by the mass extinction event that many biologists maintain is already underway. Population ecology also informs the study of human populations, another key endeavor in environmental science (Chapter 6).

TESTING YOUR COMPREHENSION

1. Explain the premises and logic that support the concept of natural selection.

2. Describe two examples of evidence for selection.

3. How does allopatric speciation occur?

4. Name two organisms that have gone extinct, and give a probable reason for each extinction.

5. What is the difference between a species and a population? Between a population and a community?

6. Contrast the concepts of habitat and niche.

7. List and describe each of the five major population characteristics discussed in this chapter. Explain how each shapes population dynamics.

8. Can any species undergo exponential growth forever? Explain your answer.

9. Describe how limiting factors relate to carrying capacity.

10. Explain the difference between K-selected species and r-selected species. Can you think of examples of each that were not mentioned in the chapter?

SEEKING SOLUTIONS

1. In what ways has artificial selection changed people's quality of life? Give examples. Can you imagine a way in which artificial selection could be used to improve our quality of life further? Can you imagine a way it could be used to reduce our environmental impact?

2. What types of species are most vulnerable to extinction, and what kinds of factors threaten them? What species in your region are threatened with extinction? What reasons lie behind their endangerment?

3. Do you think the human species can continue raising its global carrying capacity? How so, or why not? Do you think we *should* try to keep raising our carrying capacity? Why or why not?

4. Describe the evidence suggesting that changes in temperature and precipitation led to the extinction of the golden toad and to population crashes for other amphibians at Monteverde. Why do scientists also think that disease played a role? What do you think could be done to help make future such declines in other species less likely?

5. **THINK IT THROUGH** You are a population ecologist studying animals in a national park, and policymakers are asking for advice on how to apportion the government's limited conservation funds. How would you rate the following three species, from most vulnerable (and thus most in need of conservation attention) to least vulnerable? Give reasons for your choices.

   ▸ A bird with an even (1:1) sex ratio that is a habitat generalist
   ▸ A salamander endemic to the park that lives in high-elevation forest
   ▸ A fish that specializes on a few types of invertebrate prey and has a large population size

CALCULATING ECOLOGICAL FOOTPRINTS

Americans love their coffee. In 2010, coffee consumption in the United States reached nearly 3.3 billion pounds (out of about 17.5 billion pounds produced globally). Next to petroleum, coffee is the most valuable (legal) commodity on the world market, and the United States is its leading importer. Given this information, estimate the coffee consumption rates in the table.

Most coffee is produced on large tropical plantations, where coffee is the only tree species and is grown in full sun where natural forests have been cut. However, approximately 2% of coffee is produced in smaller groves where coffee trees are intermingled with other species under a partial rainforest canopy. These *shade-grown* coffee plantations maintain greater habitat diversity for tropical rainforest wildlife.

	Population	Pounds of coffee per day	Pounds of coffee per year
You (or the average American)	1	0.029	10.6
Your class			
Your hometown			
Your state			
United States			

Data from International Coffee Organization, and O'Brien, T.G., and M.F. Kinnaird, 2003. Caffeine and conservation. *Science* 300: 587.

1. What percentage of global coffee production is consumed in the United States? If U.S. coffee drinkers consumed only shade-grown coffee, how much would shade-grown production need to increase to meet demand?

2. How much extra would you be willing to pay per pound for shade-grown coffee as opposed to standard coffee, if you knew that your money would help to prevent habitat loss or extinction for animals such as resplendent quetzals, endangered Costa Rican squirrel monkeys, and the many songbirds that migrate between Latin America and North America?

3. If everyone in the United States were willing to pay as much extra per pound for shade-grown coffee as you are, how much additional money would that provide for conservation of biodiversity in the tropics each year?

MasteringENVIRONMENTALSCIENCE™

Go to **www.masteringenvironmentalscience.com** for homework assignments, practice quizzes, Pearson eText, and more.

4 Species Interactions and Community Ecology

Upon completing this chapter, you will be able to:

➤ Compare and contrast the major types of species interactions

➤ Characterize feeding relationships and energy flow, using them to construct trophic levels and food webs

➤ Distinguish characteristics of a keystone species

➤ Characterize succession and notions of community change

➤ Perceive and predict the potential impacts of invasive species in communities

➤ Explain the goals and methods of restoration ecology

➤ Describe and illustrate the terrestrial biomes of the world

Aggregation of zebra mussels

Black and White, and Spread All Over: Zebra Mussels Invade the Great Lakes

"We are seeing changes in the Great Lakes that are more rapid and more destructive than any time in [their] history."
—Andy Buchsbaum, National Wildlife Federation

"When you tear away the bottom of the food chain, everything that is above it is going to be disrupted."
—Tom Nalepa, National Oceanic and Atmospheric Administration

Things had been looking up for the Great Lakes. The pollution-fouled waters of Lake Erie and the other Great Lakes shared by Canada and the United States had become gradually cleaner in the years following the U.S. Clean Water Act of 1970 and a binational agreement in 1972. As government regulation brought industrial discharges under control, people once again began to use the lakes for recreation, and populations of fish rebounded.

Then the zebra mussel arrived. Black-and-white-striped shellfish the size of a dime, zebra mussels attach to hard surfaces and feed on algae by filtering water through their gills. This mollusk, given the scientific name *Dreissena polymorpha*, is native to the Caspian Sea, Black Sea, and Azov Sea in western Asia and eastern Europe. In 1988, it was discovered in North American waters at Lake St. Clair, which connects Lake Erie with Lake Huron. People had brought it to this continent by accident when ships arriving from Europe discharged ballast water containing the mussels or their larvae.

Within just two years of their discovery, zebra mussels had multiplied and reached all five of the Great Lakes. The next year, these invaders entered New York's Hudson River to the east, and the Illinois River at Chicago to the west. From the Illinois River and its canals, they soon reached the Mississippi River, giving them access to a vast watershed covering 40% of the United States. In just three more years, they spread to 19 U.S. states and two Canadian prov-

inces. By 2010, zebra mussels had colonized waters in 30 U.S. states.

How could a mussel spread so quickly? The zebra mussel's larval stage is well adapted for long-distance dispersal. Its tiny larvae drift freely for several weeks, traveling as far as the currents take them. Adults that attach themselves to boats and ships may be transported from one place to another, even to isolated lakes and ponds well away from major rivers. Moreover, in North America the mussels encountered none of the particular species of predators, competitors, and parasites that had evolved with them in the Old World and limited their population growth there.

Why all the fuss? Zebra mussels clog water intake pipes at factories, power plants, municipal water supplies, and wastewater treatment facilities (**FIGURE 4.1**). At one Michigan power plant, workers counted 700,000 mussels per square meter of pipe surface. Great densities of these organisms can damage boat engines, degrade docks, foul fishing gear, and sink buoys that ships use for navigation. Through such

FIGURE 4.1 ▲ In addition to their many ecological impacts, zebra mussels can clog water intake pipes of power plants and industrial facilities.

impacts, zebra mussels cost Great Lakes economies an estimated $5 billion in the first decade of the invasion, and they continue to impose costs of hundreds of millions of dollars each year.

Zebra mussels also have severe impacts on ecological systems. They eat primarily phytoplankton (p. 21), the microscopic photosynthetic algae, protists, and cyanobacteria that drift in open water. Because each mussel filters a liter or more of water every day, zebra mussels consume so much phytoplankton that they can deplete populations. Phytoplankton is the foundation of the Great Lakes food web, so its depletion is bad news for zooplankton (p. 25), the tiny aquatic animals that eat phytoplankton—and for the fish that eat both. Researchers are finding that water bodies with zebra mussels have fewer zooplankton and open-water fish than water bodies without them. Zebra mussels also suffocate native mollusks by attaching to their shells.

However, zebra mussels also benefit some bottom-feeding invertebrates and fish—at least in the short term. By filtering algae and organic matter from open water and depositing nutrients in feces that sink, they shift the community's nutrient balance to the bottom and benefit the species that feed there. Once they clear the water, sunlight penetrates more deeply, spurring the growth of large-leafed underwater plants and algae. In the long term, however, eutrophication (pp. 25, 267) may ensue, bringing harm to the system.

In the past several years, scientists have noted a surprising new twist: One invader is being displaced by another. The quagga mussel (*Dreissena bugensis*), a close relative of the zebra mussel from Ukraine, is spreading through the Great Lakes and beyond. This species, named after an extinct zebra-like animal, appears to be replacing the zebra mussel in many locations. Scientists are only beginning to understand what consequences this shift may have for ecological communities. ∎

SPECIES INTERACTIONS

By interacting with many species in a variety of ways, zebra mussels and quagga mussels have set in motion an array of changes in the ecological communities they have invaded. Interactions among species are the threads in the fabric of communities. Ecologists organize species interactions into several fundamental categories (**TABLE 4.1**).

Competition can occur when resources are limited

When multiple organisms seek the same limited resource, their relationship is said to be one of **competition**. Competing organisms do not usually fight with one another directly and physically. Competition is generally more subtle and indirect, taking place as organisms vie with one another to procure resources. Such resources include food, water, space, shelter, mates, sunlight, and more. Competitive interactions can take place between members of the same species (*intraspecific competition*) or between members of different species (*interspecific competition*).

If one species is a very effective competitor, it may exclude other species from resource use entirely. This has occurred in parts of the Great Lakes as zebra mussels have displaced native mussels, and it is happening now as quagga mussels begin to displace zebra mussels. Alternatively, competing species may be able to coexist, adapting over evolutionary time as natural selection (pp. 46–48) favors individuals that use slightly different resources or that use shared resources in different ways. For example, if two bird species eat the same type of seeds, natural selection might favor one species coming to specialize on larger seeds and the other coming to specialize on smaller seeds. Or one bird species might become more active in the morning and the other more active in the evening, thus minimizing interference. This process is called **resource partitioning** because the species partition, or divide, the resource they use in common by specializing in different ways (**FIGURE 4.2**).

In competitive interactions, each participant exerts a negative effect on other participants by taking resources the others could have used. This is reflected in the two minus signs shown for competition in Table 4.1. In other types of interactions, some participants benefit while others are harmed; that is, one species exploits the other (note the +/− interactions in Table 4.1). Such *exploitative* interactions include predation, parasitism, and herbivory (**FIGURE 4.3**).

TABLE 4.1 Effects of Species Interactions on Their Participants

Type of interaction	Effect on Species 1	Effect on Species 2
Competition	−	−
Predation, parasitism, herbivory	+	−
Mutualism	+	+
"+" denotes a positive effect; "−" denotes a negative effect.		

White-breasted nuthatch climbs down trunk looking for insects

Yellow-bellied sapsucker drills rows of holes and consumes sap and insects stuck in sap

Pileated woodpecker digs deeply into wood to find large insects

Brown creeper climbs up trunk looking for tiny insects

FIGURE 4.2 ▲ When species compete, they tend to partition resources, each specializing on a slightly different resource or way of attaining a shared resource. Various types of birds—including the woodpeckers, creeper, and nuthatch shown here—feed on insects from tree trunks, but they use different portions of the trunk, seeking different foods in different ways.

Predators kill and consume prey

Every living thing needs to procure food, and for most animals, that means eating other living organisms. **Predation** is the process by which individuals of one species, a *predator*, hunt, capture, kill, and consume individuals of another species, the *prey* (see Figure 4.3a). Interactions among predators and prey structure the food webs we will examine shortly, and they help shape community composition by influencing the relative numbers of predators and prey.

Zebra mussels consume the smaller types of zooplankton, and this predation has caused zooplankton populations to decline by up to 70% in Lake Erie and the Hudson River. (Also contributing to this decline is the fact that zebra mussels, by eating phytoplankton, compete with zooplankton for food.) Most predators are also prey, however, and zebra mussels have become a food source for a variety of North American fish, ducks, muskrats, and crayfish.

Predation can sometimes drive population dynamics. An increase in the population size of prey creates more food for predators, which may survive and reproduce more effectively as a result. As the predator population rises, intensified predation drives down the population of prey. Diminished numbers of prey in turn cause some predators to starve, so that the predator population declines. This allows the prey population to begin rising again, starting the cycle anew (**FIGURE 4.4**).

Predation also has evolutionary ramifications. Individual predators that are more adept at capturing prey will likely live longer, healthier lives and be better able to provide for their offspring. Natural selection (pp. 46–48) will thereby lead to the evolution of adaptations that make predators better hunters. Prey face an even stronger selective pressure—the risk of immediate death. As a result, predation pressure has driven the evolution of an elaborate array of defenses against being eaten (**FIGURE 4.5**).

Parasites exploit living hosts

Organisms can exploit other organisms without killing them. **Parasitism** is a relationship in which one organism, the *parasite*, depends on another, the *host*, for nourishment or some other benefit while simultaneously doing the host harm (see Figure 4.3b). Unlike predation, parasitism usually does not result in an organism's immediate death.

Some types of parasites are free-living and come into contact with their hosts infrequently. For example, the cuckoos of Eurasia and the cowbirds of the Americas lay their eggs in other birds' nests and let the host birds raise the parasite's young. Many other parasites live inside their hosts. These parasites include disease pathogens such as the protists that cause malaria and dysentery, as well as animals such as tapeworms

FIGURE 4.3 ▼ In predation (a), a predator kills and eats prey, just as this fire-bellied snake is devouring a frog. In parasitism (b), a parasite benefits by doing harm to its host, just as these sea lampreys gain nourishment by sucking blood from fish in the Great Lakes. In herbivory (c), animals feed on plants, just as this larva (caterpillar) of the death's head hawk moth is feeding on leaves in Europe.

(a) Predation

(b) Parasitism

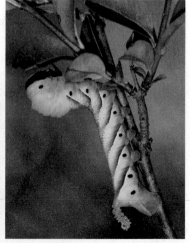

(c) Herbivory

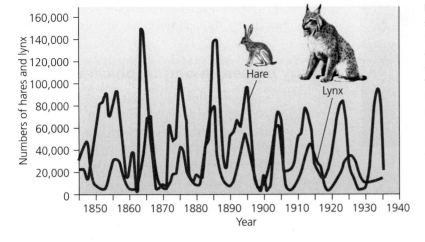

FIGURE 4.4 ◄ Predator-prey systems occasionally show paired cycles, in which increases and decreases in the population of one species help drive increases and decreases in the population of the other. A classic case is that of hares and the lynx that prey on them in Canada. These data come from fur-trapping records of the Hudson Bay Company and represent numbers of each animal trapped. Data from MacLulich, D.A., 1937. Fluctuation in the numbers of varying hare (*Lepus americanus*). *Univ. Toronto Stud. Biol. Ser.* 43, Toronto: University of Toronto Press.

that inhabit their hosts' digestive tracts. Other parasites live on the exterior of their hosts. For example, the sea lamprey (*Petromyzon marinus*) is a tube-shaped vertebrate that grasps the bodies of fish with a suction-cup mouth and a rasping tongue, sucking their blood for days or weeks (see Figure 4.3b). Sea lampreys invaded the Great Lakes from the Atlantic Ocean after people dug canals to connect the lakes for shipping, and the lampreys soon devastated economically important fisheries of chubs, lake herring, whitefish, and lake trout.

Many insects parasitize other insects, often killing them in the process, and are called *parasitoids*. Various species of parasitoid wasps lay eggs on caterpillars. When the eggs hatch, the wasp larvae burrow into the caterpillar's tissues and slowly consume them. The wasp larvae then metamorphose into adults and fly from the body of the dying caterpillar.

Just as predators and prey evolve in response to one another, so do parasites and hosts, in a process termed *coevolution*. Hosts and parasites can become locked in a duel of escalating adaptations, known as an *evolutionary arms race*. Like rival nations racing to stay ahead of one another in military technology, host and parasite may repeatedly evolve new responses to the other's latest advance. In the long run, though, it may not be in a parasite's best interest to do its host too much harm. A parasite might leave more offspring in the next generation—and thus be favored by natural selection—if it allows its host to live longer.

Herbivores exploit plants

One of the most common exploitative interactions is **herbivory**, which occurs when animals feed on the tissues of plants (see Figure 4.3c). Insects that feed on plants are the most widespread type of herbivore, and just about every plant in the world is attacked by some type of insect. In most cases, herbivory does not kill a plant outright but may affect its growth and reproduction.

Like animal prey, plants have evolved an impressive arsenal of defenses against the animals that feed on them. Many plants produce chemicals that are toxic or distasteful to herbivores. Others arm themselves with thorns, spines, or irritating hairs. In response, herbivores may evolve ways to overcome these defenses, and the plant and the animal may embark on an evolutionary arms race.

(a) Cryptic coloration **(b) Warning coloration** **(c) Mimicry**

FIGURE 4.5 ▲ Natural selection to avoid predation has resulted in many fabulous adaptations. Some prey hide from predators by *crypsis*, or camouflage, such as this gecko **(a)** on tree bark. Other prey species are brightly colored to warn predators that they are toxic or distasteful, such as this yellowjacket **(b).** Still others fool predators with mimicry. This caterpillar **(c),** when disturbed, swells and curves its tail end and shows false eyespots, making it look like a snake's head.

FIGURE 4.6 ▲ In mutualism, individuals of different species benefit one another. Hummingbirds visit flowers to gather nectar, and in the process transfer pollen between flowers, helping the plant to reproduce.

Mutualists benefit one another

Unlike exploitative interactions, **mutualism** is a relationship in which two or more species benefit from interacting with one another. Generally, each partner provides some resource or service that the other needs. Many mutualistic relationships—like many parasitic relationships—occur between organisms that live in close physical contact. Physically close association between interacting species (whether in mutualistic or parasitic interactions) is called **symbiosis.**

Thousands of terrestrial plant species depend on mutualisms with fungi; plant roots and some fungi together form symbiotic associations called mycorrhizae. In these relationships, the plant provides energy and protection to the fungus, and the fungus helps the plant absorb nutrients from the soil. In the ocean, coral polyps, the tiny animals that build coral reefs (pp. 258–259), share beneficial arrangements with algae known as zooxanthellae. The coral provide housing and nutrients for the algae in exchange for a steady supply of food that the algae produce through photosynthesis (p. 30). You, too, are part of a symbiotic mutualism. Your digestive tract is filled with microbes that help you digest food—microbes you are providing a place to live. Without these mutualistic microbes, none of us would likely survive for long.

Not all mutualists live in close proximity. **Pollination** (**FIGURE 4.6**), an interaction vital to agriculture and our food supply (pp. 148–149), involves free-living organisms that may encounter each other only once. Insects, birds, bats, and other creatures transfer pollen (containing male sex cells) from flower to flower, fertilizing ovaries (containing female sex cells) that grow into fruits with seeds. Most pollinating animals visit flowers for their nectar, a reward the plant uses to entice them. The pollinators receive food, and the plants are pollinated and reproduce. Various types of bees pollinate 73% of our crops, one expert has estimated—from soybeans to potatoes to tomatoes to beans to cabbage to oranges.

ECOLOGICAL COMMUNITIES

A **community** is an assemblage of organisms living in the same area at the same time. Members of a community interact with one another in the ways discussed above, and these species interactions help determine the structure, function, and species composition of communities. *Community ecologists* are interested in which species coexist, how they interact, how communities change through time, and why these patterns exist.

Energy passes among trophic levels

Some of the most important interactions among community members involve who eats whom. As organisms feed on one another, matter and energy move through the community from one **trophic level**, or rank in the feeding hierarchy, to another (**FIGURE 4.7**; and compare Figure 2.14, p. 31).

Producers *Producers,* or *autotrophs* ("self-feeders," p. 30), comprise the first trophic level. Terrestrial green plants, cyanobacteria, and algae capture solar energy and use photosynthesis to produce sugars (p. 30).

Consumers Organisms that consume producers are known as *primary consumers* and comprise the second trophic level. Herbivorous grazing animals, such as deer and grasshoppers, are primary consumers. The third trophic level consists of *secondary consumers*, which prey on primary consumers. Wolves that prey on deer are considered secondary consumers, as are rodents and birds that prey on grasshoppers. Predators that feed at still higher trophic levels are known as *tertiary consumers.* Examples of tertiary consumers include hawks and owls that eat rodents that have eaten grasshoppers.

Detritivores and decomposers Detritivores and decomposers consume nonliving organic matter. *Detritivores,* such as millipedes and soil insects, scavenge the waste products or dead bodies of other community members. *Decomposers,* such as fungi and bacteria, break down leaf litter and other nonliving matter into simpler constituents that can be taken up and used by plants. These organisms enhance the topmost soil layers (pp. 137–138) and play essential roles as the community's recyclers, making nutrients from organic matter available for reuse by living members of the community.

In Great Lakes communities, phytoplankton are the main producers, floating freely and photosynthesizing with sunlight that penetrates the upper layer of the water. Zooplankton are primary consumers, feeding on the phytoplankton. Phytoplankton-eating fish are primary consumers, and zooplankton-eating fish are secondary consumers. At higher trophic levels are tertiary consumers such as larger fish and birds that feed on plankton-eating fish. (The left side of Figure 4.7 shows these relationships in a very generalized form.) Zebra mussels and quagga mussels, by eating both phytoplankton and zooplankton, function on multiple trophic levels. When an organism dies and sinks to the bottom, detritivores scavenge its tissues, and decomposers recycle its nutrients.

Energy, numbers, and biomass decrease at higher trophic levels

At each trophic level, organisms use energy in cellular respiration (pp. 30–31), and most energy ends up being given off as heat. Only a small portion of the energy is transferred to the next trophic level through predation, herbivory, or parasitism. A rough rule of thumb is that each trophic level

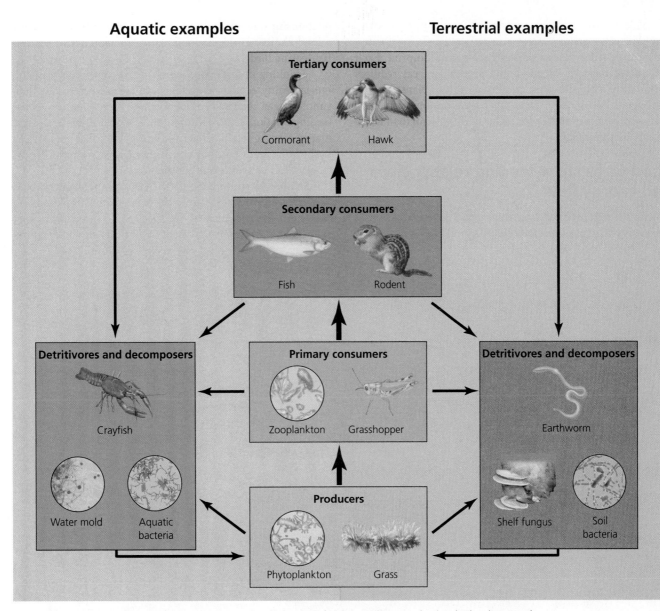

FIGURE 4.7 ▲ Ecologists organize species hierarchically by their feeding rank, or trophic level. The diagram shows aquatic **(left)** and terrestrial **(right)** examples at each level. Arrows indicate the direction of energy flow. Producers produce food by photosynthesis, primary consumers (herbivores) feed on producers, secondary consumers eat primary consumers, and tertiary consumers eat secondary consumers. Communities can have more or fewer trophic levels than in this example. Detritivores and decomposers feed on nonliving organic matter and the remains of dead organisms from all trophic levels, and they "close the loop" by returning nutrients to the soil or the water column for use by producers.

contains just 10% of the energy of the trophic level below it (although the actual proportion can vary greatly).

This pattern can be visualized as a pyramid (**FIGURE 4.8**). The pattern also tends to hold for the numbers of organisms at each trophic level; generally, fewer organisms exist at higher trophic levels than at lower ones. A grasshopper eats many plants in its lifetime, a rodent eats many grasshoppers, and a hawk eats many rodents. Thus, for every hawk in a community there must be many rodents, still more grasshoppers, and an immense number of plants. And because the difference in numbers of organisms among trophic levels tends to be large, the same pyramid-like relationship also often holds true for **biomass**, the collective mass of living matter in a given place and time.

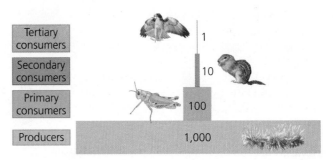

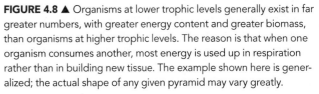

FIGURE 4.8 ▲ Organisms at lower trophic levels generally exist in far greater numbers, with greater energy content and greater biomass, than organisms at higher trophic levels. The reason is that when one organism consumes another, most energy is used up in respiration rather than in building new tissue. The example shown here is generalized; the actual shape of any given pyramid may vary greatly.

CHAPTER 4   Species Interactions and Community Ecology

69

This pyramid pattern illustrates why eating at lower trophic levels—being a vegetarian rather than a meat-eater, for instance—decreases a person's ecological footprint. Each amount of meat or other animal product we eat requires the input of a considerably greater amount of plant material (see Figure 7.22, p. 152). Thus, when we eat animal products, we use up far more energy for each calorie we gain than when we eat plant products.

Food webs show feeding relationships and energy flow

As energy is transferred from lower trophic levels to higher ones, it is said to pass up a *food chain*, a linear series of feeding relationships. Plant, grasshopper, rodent, and hawk make up a food chain—as do phytoplankton, zooplankton, fish, and fish-eating birds.

Thinking in terms of food chains is conceptually useful, but in reality ecological systems are far more complex than simple linear chains. A more accurate representation of the feeding relationships in a community is a **food web,** a visual map of feeding relationships and energy flow that uses arrows to show the many paths by which energy passes among organisms as they consume one another.

FIGURE 4.9 shows a food web from a temperate deciduous forest of eastern North America. It is greatly simplified and leaves out the vast majority of species and interactions that occur. Note, however, that even within this simple diagram,

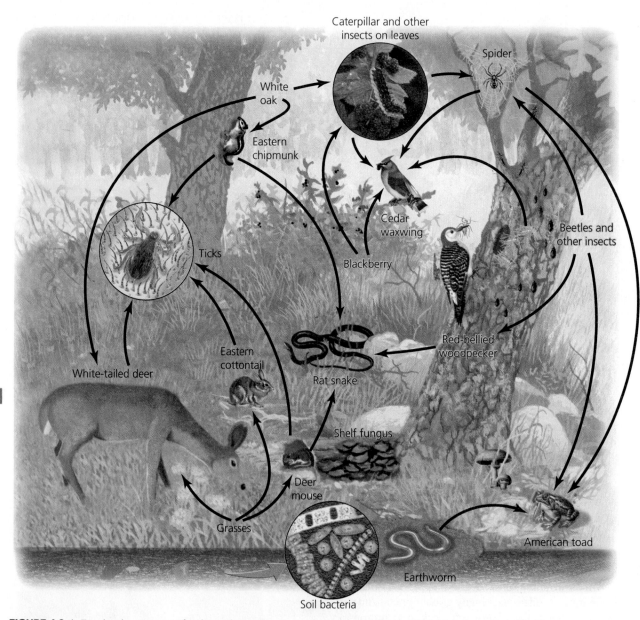

FIGURE 4.9 ▲ Food webs represent feeding relationships in a community. This food web pertains to eastern North America's temperate deciduous forest and includes organisms on several trophic levels. In a food web diagram, arrows lead from one organism to another to indicate the direction of energy flow as a result of predation, parasitism, or herbivory. For example, an arrow leads from the grass to the cottontail rabbit to indicate that cottontails consume grasses. The arrow from the cottontail to the tick indicates that parasitic ticks derive nourishment from cottontails. Like most food web diagrams, this one is a simplification, because the actual community contains many more species and interactions than can be shown.

we can pick out a number of different food chains involving different sets of species.

A Great Lakes food web would involve the phytoplankton that photosynthesize near the water's surface, the zooplankton that eat them, fish that eat phytoplankton and zooplankton, larger fish that eat the smaller fish, and lampreys that parasitize the fish. It would include a number of native mussels and clams and, since 1988, the zebra mussels and quagga mussels that are displacing them. This food web would include diving ducks that used to feed on native bivalves and now prey on the mussels. It would also show that crayfish and other bottom-dwelling invertebrates feed from the refuse of the mussels. Finally, the food web would include underwater plants and macroscopic algae, whose growth is enhanced as the non-native mussels filter out phytoplankton, allowing sunlight to penetrate more deeply into the water column. (Jump ahead to see Figure 4.13a (p. 76) for an illustration of some of these effects.)

Overall, zebra and quagga mussels alter the Great Lakes food web essentially by shifting productivity from open-water regions to *benthic* (bottom) and *littoral* (nearshore) regions. In so doing, the mussels help benthic and littoral fish and make life harder for open-water fish (see **THE SCIENCE BEHIND THE STORY,** pp. 72–73).

Some organisms play bigger roles in communities than others

"Some animals are more equal than others," George Orwell wrote in his classic novel *Animal Farm*. Although Orwell was making wry sociopolitical commentary, his remark hints at a truth in ecology. In communities, ecologists have found, some species exert greater influence than do others. A species that has strong or wide-reaching impact far out of proportion to its abundance is often called a **keystone species.** A keystone is the wedge-shaped stone at the top of an arch that holds the structure together. Remove the keystone, and the arch will collapse (**FIGURE 4.10A**). In an ecological community, removing a keystone species will have ripple effects that substantially alter the food web.

Often, large-bodied secondary or tertiary consumers at the tops of food chains are considered keystone species. Top predators control populations of herbivores, which otherwise would multiply and could greatly modify the plant community (**FIGURE 4.10B**). Thus, predators at high trophic levels can indirectly promote populations of organisms at low trophic levels by keeping species at intermediate trophic levels in check, a phenomenon ecologists refer to as a **trophic cascade.** In the United States, for example, government bounties long promoted the hunting of wolves and mountain lions, which were largely

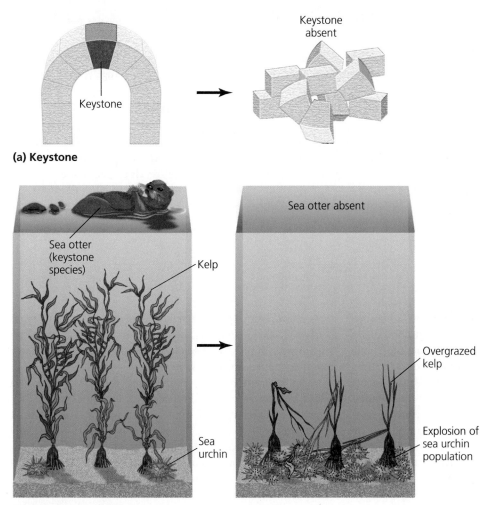

(a) Keystone

(b) A keystone species

FIGURE 4.10 ◀ A keystone is the wedge-shaped stone at the top of an arch that holds its structure together **(a)**. A keystone species, such as the sea otter, is one that exerts great influence on a community's composition and structure **(b)**. Sea otters consume sea urchins, which eat kelp in marine nearshore environments of the Pacific. When otters are present, they keep urchin numbers down, allowing lush underwater forests of kelp to grow and provide habitat for many other species. When otters are absent, urchin populations increase and devour the kelp, destroying habitat and depressing species diversity.

Determining Zebra Mussels' Impacts on Fish Communities

Dr. David Strayer samples aquatic invertebrates

When zebra mussels appeared in the Great Lakes, people feared for sport fisheries and predicted that fish population declines could cost billions of dollars. The mussels would deplete the phytoplankton and zooplankton that fish depended on, people reasoned.

However, food webs are complicated systems, and disentangling them to infer the impacts of any one species is fraught with difficulty. Thus, even after 15 years, there was no solid evidence of widespread harm to fish populations.

So, aquatic biologist David Strayer of the Institute of Ecosystem Studies in Millbrook, New York, joined Kathryn Hattala and Andrew Kahnle of New York State's Department of Environmental Conservation (DEC). They mined datasets on fish populations in the Hudson River, which zebra mussels had invaded in 1991.

Strayer and other researchers had been studying aspects of the community for years. Their data showed that after zebra mussels invaded the Hudson:

- Biomass of phytoplankton had fallen by 80%.
- Biomass of small zooplankton had fallen by 76%.
- Biomass of large zooplankton had fallen by 52%.

Zebra mussels increased filter-feeding in the community 30-fold, depleting phytoplankton and small zooplankton and leaving larger zooplankton with less phytoplankton to eat. Overall, the zooplankton and invertebrate animals of the open water (which are eaten by open-water fish) declined by 70%.

However, Strayer had also found that *benthic*, or bottom-dwelling, invertebrates in shallow water (especially in the nearshore, or *littoral*, zone) had increased notably, because the mussels' shells provide habitat structure and their feces provide nutrients.

These contrasting trends in the benthic shallows and the open deep water led Strayer's team to hypothesize that zebra mussels would harm open-water fish that ate plankton but would help littoral-feeding fish. They predicted that following the zebra mussel invasion, larvae and juveniles of six common open-water fish species

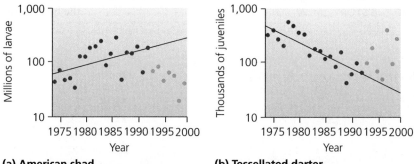

(a) American shad   **(b) Tessellated darter**

Larvae of American shad **(a)**, an open-water fish, had been increasing in abundance before zebra mussels invaded (red points and trend line). After zebra mussels invaded, shad larvae decreased (orange points). Juveniles of the tessellated darter **(b)**, a littoral fish, had been decreasing in abundance before zebra mussels invaded (red points and trend line). After zebra mussels invaded, they increased (orange points). *Source:* Strayer, D., et al., 2004. Effects of an invasive bivalve (*Dreissena polymorpha*) on fish in the Hudson River estuary. *Canadian J. Fisheries and Aquatic Sciences* 61: 924–941. © 2004. Reprinted by permission of NRC Research Press.

exterminated by the middle of the 20th century. In the absence of these predators, deer populations have grown unnaturally dense and have overgrazed forest-floor vegetation and eliminated tree seedlings, causing major changes in forest structure.

The removal of top predators across the United States was an uncontrolled large-scale experiment with unintended consequences, but ecologists have verified the keystone species concept in controlled experiments. Classic research by marine biologist Robert Paine established that the predatory sea star (starfish) *Pisaster ochraceus* influences the community composition of intertidal organisms (pp. 257–258) on the Pacific coast of North America. When *Pisaster* is present in this community, species diversity is high, with various types of barnacles, mussels, and algae. When *Pisaster* is removed, the mussels it preys on become numerous and displace other species, suppressing species diversity.

Animals at high trophic levels, such as wolves, sea stars, and sea otters (see Figure 4.10), can be keystone species, but other types of organisms also may exert strong community-wide effects. "Ecosystem engineers" physically modify the environment shared by community members. Beavers build dams across streams, flooding acres of dry land and turning them to swamp. Prairie dogs dig burrows that aerate the soil and serve as homes for other animals.

would decline in number, decline in growth rate, and shift downriver toward saltier water, where mussels are absent. Conversely, they predicted that larvae and juveniles of 10 littoral fish species would increase in number, increase in growth rate, and shift upriver to regions of greatest zebra mussel density.

To test their predictions, the researchers analyzed data from fish surveys carried out by DEC scientists and consultants over 26 years, spanning periods before and after the zebra mussel's arrival. Strayer's team compared data on abundance, growth, and distribution of young fish before and after 1991.

The results supported their predictions. Larvae and juveniles of open-water fish, such as American shad, blueback herring, and alewife, tended to decline in abundance in the years after zebra mussels were introduced **(see first figure, part (a))**. Those of littoral fish, such as tessellated darter, bluegill, and largemouth bass, tended to increase **(see first figure, part (b))**.

Growth rates showed the same trend: Open-water fish grew more slowly after zebra mussels invaded, whereas littoral fish grew more quickly. In terms of distribution in the 248-km (154-mi) stretch of river studied, open-water fish shifted downstream toward areas with fewer zebra mussels, whereas littoral fish shifted upstream toward areas with more zebra mussels **(see second figure)**. Overall, the data supported the hypothesis that the fish community would respond to changes in food resources caused by zebra mussels. The results were published in 2004 in the *Canadian Journal of Fisheries and Aquatic Sciences*.

As Strayer and his colleagues continue their research, they learn more and more. In 2008 they published a broader analysis of the Hudson's food web, showing that although littoral species benefited from zebra mussels, the fact that the mussels clarified the water made littoral species more susceptible to variation in clarity due to sediment input at times of high water flow.

Strayer and others also recently showed that populations of native mussels and clams in the Hudson had crashed after the zebra mussel invaded (likely as a result of competition for food), but that starting in 2000, these native bivalves, instead of going extinct, suddenly stabilized and persisted at about 4–22% of their pre-invasion population sizes. The researchers could not determine the reason for this turnaround, but they suggested that perhaps the native species would continue to play a role in the community.

Research such as this helps illuminate the impacts that particular species interactions have on communities as a whole. In this case, the research may also help fisheries biologists to better manage commercially and recreationally important fish populations in the Hudson River and other areas invaded by zebra mussels.

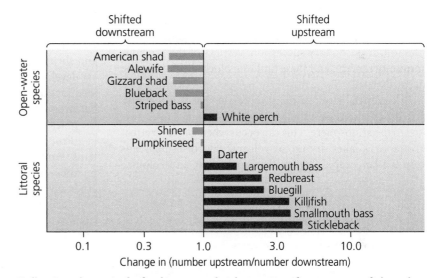

Following the arrival of zebra mussels, the young of open-water fish such as American shad, blueback herring, and alewife tended to shift downstream toward areas with fewer zebra mussels. Young of littoral fish, such as killifish, bluegill, and largemouth bass, tended to shift upstream toward areas with more zebra mussels. Source: Strayer, D., et al., 2004. Effects of an invasive bivalve (*Dreissena polymorpha*) on fish in the Hudson River estuary. *Canadian J. Fisheries and Aquatic Sciences* 61: 924–941. © 2004. Reprinted by permission of NRC Research Press.

Ants disperse seeds, redistribute nutrients, and selectively protect or destroy different insects and plants near their colonies. Less conspicuous organisms toward the bottoms of food chains may have even greater impact. Remove the fungi that decompose dead matter, or the insects that control plant growth, or the phytoplankton that are the base of the marine food chain, and a community may change very rapidly indeed. However, because there are usually more species at lower trophic levels, it is less likely that any single one of them alone has wide influence. Often if one species is removed, other species that remain may be able to perform many of its functions.

Communities respond to disturbance in different ways

The removal of a keystone species is just one of many types of disturbance that can modify the composition, structure, or function of an ecological community. Over time, a community may experience natural disturbances ranging from gradual phenomena such as climate change to sudden events such as fires, floods, landslides, or hurricanes. Today, human impacts are major sources of disturbance for ecological communities worldwide—from habitat alteration to pollution to the introduction of non-native species such as the zebra mussel.

Communities are dynamic systems and may respond to disturbance in several ways. A community that resists change and remains stable despite disturbance is said to show **resistance** to the disturbance. Alternatively, a community may show **resilience,** meaning that it changes in response to disturbance but later returns to its original state. Or, a community may be modified by disturbance permanently and may never return to its original state.

Succession follows severe disturbance

If a disturbance is severe enough to eliminate all or most of the species in a community, the affected site may then undergo a somewhat predictable series of changes that ecologists have traditionally called **succession**. In the conventional view of this process, there are two types of succession. **Primary succession** follows a disturbance so severe that no vegetation or soil life remains from the community that occupied the site. In primary succession, a biotic community is built essentially from scratch. In contrast, **secondary succession** begins when a disturbance dramatically alters an existing community but does not destroy all life or all organic matter in the soil. In secondary succession, vestiges of the previous community remain, and these building blocks help shape the process.

At terrestrial sites, primary succession takes place after a bare expanse of rock, sand, or sediment becomes newly exposed to the atmosphere. This can occur when glaciers retreat (**FIGURE 4.11**), lakes dry up, or volcanic lava or ash spreads across the landscape. Species that arrive first and colonize the new substrate are referred to as **pioneer species**. Pioneer species are well adapted for colonization, having traits such as spores or seeds that can travel long distances.

The pioneers best suited to colonizing bare rock are the mutualistic aggregates of fungi and algae known as *lichens*. In lichens, the algal component provides food and energy via photosynthesis while the fungal component takes a firm hold on rock and captures moisture. As lichens grow, they secrete acids that break down the rock surface. This begins the formation of soil, and soon small plants, insects, and worms find the rocky outcrops more hospitable. As new organisms arrive, they provide more nutrients and habitat. As time passes, larger plants establish themselves, the amount of vegetation increases, and species diversity rises.

Secondary succession begins when a fire, a hurricane, logging, or farming removes much of the biotic community. Consider a farmed field in eastern North America that has been abandoned (**FIGURE 4.12**). After farming ends, the site will be colonized by pioneer species of grasses, herbs, and forbs that disperse well or were already in the vicinity. Soon, shrubs and fast-growing trees such as aspens begin to grow from the field. As time passes, pine trees rise above these trees and shrubs, forming a pine-dominated forest. This pine forest develops an understory of hardwood trees, because pine seedlings do not grow well under mature pines, whereas some hardwood seedlings do. Eventually the hardwoods outgrow the pines, creating a hardwood forest.

Processes of succession occur in many ecological systems. For instance, ponds may undergo succession as algae, microbes, plants, and zooplankton grow, reproduce, and die, gradually filling the water body with organic matter. The pond acquires additional organic matter and sediments from streams and runoff, and it can eventually fill in, becoming a bog (p. 254) or even a terrestrial system.

In the traditional view of succession described here, the transitions between stages lead to a *climax community,* which remains in place with little modification until some disturbance restarts succession. Early ecologists felt that each region had its own characteristic climax community, determined by climate.

Communities may undergo shifts

Today, ecologists recognize that community change is far more variable and less predictable than early models of succession suggested. Conditions at one stage may promote progression to another stage, or organisms may, through competition, inhibit a community's progression to another stage. The trajectory of change can vary greatly according to chance factors, such as which particular species happen to gain an early foothold. And communities are not determined solely by climate, but vary with soil conditions and other factors from one time or place to another.

Moreover, once a community is disturbed and changes are set in motion, there is no guarantee that the community will ever return to its original state. Sometimes communities may undergo a **phase shift,** or **regime shift,** in which the overall character of the community fundamentally changes. This can occur if some crucial climate threshold is passed, a keystone species is lost, or an exotic species invades. For instance, in

FIGURE 4.11 ▼ As this Norwegian glacier retreats, small plants (foreground) begin the process of primary succession.

Time ➞

FIGURE 4.12 ▲ Secondary succession occurs after a disturbance, such as a fire, landslide, or farming, removes most vegetation from an area. Here is shown a typical series of changes in a plant community of eastern North America following the abandonment of a farmed field.

recent years many coral reef communities have undergone a phase shift and become dominated by algae, often after people had overharvested populations of fish or turtles that eat algae. Once algae overgrows a reef, the community may never shift back to its original coral-dominated state. Phase shifts make it clear that we cannot count on being able to reverse damage caused by human disturbance. Instead, some of the changes we set in motion may become permanent.

FAQ

Q: If we disturb a community, won't it return to its original state later if we just leave the area alone?

A: Probably not, if the disturbance has been substantial. For example, if soil has become compacted, or water sources have dried up, then the plant species that grew at the site originally may no longer be able to grow. Different plant species may take their place—and among them, a different suite of animal species may find habitat. Because species interact and because they rely on particular habitat conditions, a change in one aspect of a community can lead to a cascade of other changes. Sometimes a whole new community may arise. For instance, in some grasslands, livestock grazing and fire suppression have led shrubs and trees to invade, changing the grasslands to shrublands. And we have seen (p. 71) how removing sea otters can lead to the destruction of kelp forest communitites. In the past, people didn't realize how permanent such changes could be, because we tended to view natural systems as static, predictable, and liable to return to equilibrium. But today ecologists recognize that these systems are highly dynamic and can sometimes undergo rapid, extreme, and long-lasting change.

Invasive species pose new threats to community stability

Traditional concepts of communities involve species native to an area. But what if a new species arrives from elsewhere? And what if this non-native (also called *alien* or *exotic*) species turns *invasive,* spreading widely and becoming dominant in a community? Such **invasive species** can alter a community substantially and are one of the central ecological forces in today's world.

Most exotic species do not turn invasive, but those that do are generally species that people have introduced, intentionally or by accident, from elsewhere in the world. Global trade helped spread zebra and quagga mussels, which were inadvertently transported in the ballast water of cargo ships. To maintain stability at sea, ships take water into their hulls as they begin their voyage and then discharge that water at their destination. Decades of unregulated exchange of ballast water has ferried countless species across the oceans.

Introduced species may become invasive when limiting factors (p. 58) that regulate their population growth are absent. Plants and animals brought to a new area may leave behind the predators, parasites, and competitors that had exploited them in their native land. If few or no organisms in the new environment prey on the introduced species, parasitize it, or compete with it, then the introduced species may thrive and spread. As the species proliferates, it may exert diverse influences on other community members (**FIGURE 4.13**).

For example, various grasses introduced in the American West for ranching have overrun entire regions, pushing out native vegetation. Fish introduced into streams for sport compete with and exclude native fish. Hundreds of island-dwelling animals and plants worldwide have been driven extinct by the goats, pigs, and rats introduced by human colonists. (We will examine more examples in Chapter 8 [p. 170].)

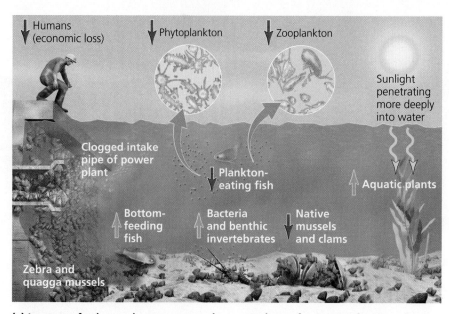

(a) Impacts of zebra and quagga mussels on members of a Great Lakes nearshore community

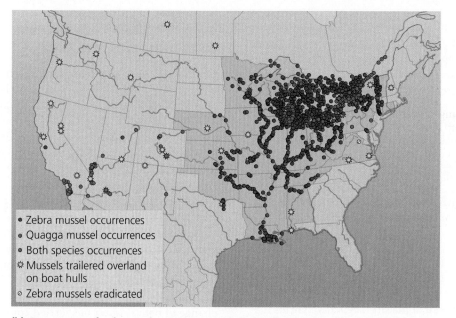

- Zebra mussel occurrences
- Quagga mussel occurrences
- Both species occurrences
- Mussels trailered overland on boat hulls
- Zebra mussels eradicated

(b) Occurrence of zebra and quagga mussels in North America, as of June 2011

FIGURE 4.13 ▲ The zebra mussel and the quagga mussel are biological invaders that are modifying ecological communities. By filtering phytoplankton and small zooplankton from open water, they exert impacts on other species **(a)**, both negative (red downward arrows) and positive (green upward arrows). The map in **(b)** shows known occurrences of the zebra mussel (red dots) and quagga mussel (green dots) in North America as of June 2011. In just two decades, they spread from a small area of the Great Lakes to waterways across the continent. People have transported them inadvertently on boat hulls (yellow stars) as far west as California. Populations in two locations have been eradicated. Source (b): U.S. Geological Survey.

Ecologists generally view the impacts of invasive species—and introduced species in general—as overwhelmingly negative. Yet many people enjoy the beauty of introduced ornamental plants in their gardens. Some organisms are introduced intentionally to control pest populations through biocontrol (pp. 147–148). And some introduced species that have turned invasive provide benefits to our economy, such as the European honeybee, which pollinates many of our crops (p. 149). Whatever view one takes, the impacts of invasive species on native species and ecological communities are significant, and they grow year by year with the increasing mobility of people and the globalization of our society.

Are Invasive Species All Bad? Some ethicists question the notion that all invasive species should automatically be considered bad. If we introduce a non-native species to a community and it greatly modifies the community, do you think that is a bad thing? What if it drives another species extinct? What if the invasive species arrived on its own, rather than through human intervention? What if it provides economic services to our society, as the European honeybee does? What ethical standard(s) (pp. 11–12) would you apply to determine whether we should battle or accept a given invasive species?

Scientific research and media attention to zebra and quagga mussels helped put invasive species on the map as a major environmental and economic problem. In 1990, the U.S. Congress passed legislation that led to the National Invasive Species Act of 1996. Among other things, this law mandated that ships dump their freshwater ballast at sea and exchange it with salt water before entering the Great Lakes.

Since then, funding has become widely available for the control and eradication of invasive species. Managers have tried to control the spread of zebra mussels by removing them manually, applying toxic chemicals, drying them out, depriving them of oxygen, introducing predators and diseases, and stressing them with heat, sound, electricity, carbon dioxide, and ultraviolet light. However, most of these are localized and short-term fixes unable to make a dent in the huge populations at large in the environment. With one invasive species after another, managers are finding that control and eradication measures are so difficult and expensive that preventive strategies (such as ballast water regulations) represent a much better investment.

To prevent invasions, it helps to be able to predict where a species might spread. By analyzing the biology of the organism, scientists can try to model the environmental conditions in which it will thrive. In 2007, researchers applied knowledge of how zebra and quagga mussels use calcium in water to create their shells to predict where the mussels might do best. The researchers mapped low-risk and high-risk regions across North America—and for the most part these conformed to the areas of actual spread so far.

Altered communities can be restored

Invasive species are adding to the tremendous transformations that people have already forced on natural landscapes and communities through habitat alteration, deforestation, pollution, climate change, the hunting of keystone species, and other activities. With so much of Earth's landscape altered by human impact, many communities and ecosystems are severely degraded. Because ecological systems support our civilization and all of life, when degraded systems cease to function, it threatens our health and well-being.

This realization has given rise to the science of **restoration ecology**. Restoration ecologists research the historical conditions of ecological communities as they existed before our industrialized civilization altered them. They then try to devise ways to restore some of these areas to an earlier condition. In some cases, the intent is primarily to restore the functionality of the system—to reestablish a wetland's ability to filter pollutants and recharge groundwater, for example, or a forest's ability to cleanse the air, build soil, and provide habitat for wildlife. In other cases, the aim is to return a community all the way to its natural "pre-settlement" condition. Either way, the science of restoration ecology informs the practice of **ecological restoration,** the actual on-the-ground efforts to carry out these visions and restore communities.

For instance, nearly all the tallgrass prairie in the United States was converted to agriculture in the 19th century. Now, people are restoring patches of prairie by planting native plants, weeding out invaders and competitors, and introducing controlled fire to mimic the fires that historically maintained this community (**FIGURE 4.14**).

The world's largest restoration project is the ongoing effort to restore parts of the Florida Everglades. The Everglades, a 7,500-km^2 (4,700-mi^2) system of marshes and seasonally flooded grasslands, has been drying out for decades because

FIGURE 4.14 ▶ USDA Forest Service ecologists from the Midewin National Tallgrass Prairie inspect native grasses in a prairie restoration area on the site of the former Joliet Arsenal in Illinois.

Restoring "Natural" Communities

Practitioners of ecological restoration in North America often aim to restore communities to their natural state. But what is meant by "natural"? Does it mean the state of the community before industrialization? Before Europeans came to the New World? Before any people laid eyes on the community?

Let's say Native Americans altered a forest community 8,000 years ago by burning the underbrush regularly to improve hunting, and continued doing so until Europeans arrived 400 years ago and cut down the forest for farming. Today the area's inhabitants want to restore the land to its "natural" forested state. Should restorationists try to recreate the forest of the Native Americans, or the forest that existed before Native Americans arrived? What values do you think underlie the desire for restoration?

the water that feeds it has been managed for flood control and overdrawn for irrigation and development. Economically important fisheries have suffered greatly as a result, and populations of wading birds have dropped by 90–95%. The 30-year, $7.8-billion restoration project intends to restore water by undoing damming and diversions of 1,600 km (1,000 mi) of canals, 1,150 km (720 mi) of levees, and 200 water control structures. Because the Everglades provide drinking water for millions of Florida citizens, as well as considerable tourism revenue, restoring its ecosystem services (pp. 2, 36, 90, 94–95) should prove economically beneficial as well as ecologically valuable. (We will explore ecological restoration further in Chapter 8 [pp. 180–181].)

As our population grows and development spreads, ecological restoration is becoming an increasingly vital conservation strategy. However, restoration is difficult, time-consuming, and expensive, and it is not always successful. It is therefore best, whenever possible, to protect natural systems from degradation in the first place, so that restoration does not become necessary.

EARTH'S BIOMES

Across the world, each location is home to different sets of species, leading to endless variety in community composition. However, communities in far-flung places often share strong similarities in their structure and function. This allows us to classify communities into broad types. A **biome** is a major regional complex of similar communities—a large-scale ecological unit recognized primarily by its dominant plant type and vegetation structure. The world contains a number of biomes, each covering large geographic areas (**FIGURE 4.15**).

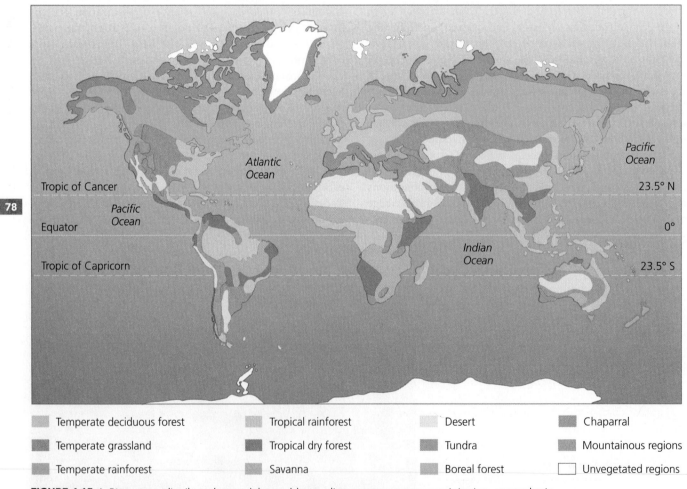

Temperate deciduous forest	Tropical rainforest	Desert	Chaparral
Temperate grassland	Tropical dry forest	Tundra	Mountainous regions
Temperate rainforest	Savanna	Boreal forest	Unvegetated regions

FIGURE 4.15 ▲ Biomes are distributed around the world according to temperature, precipitation, atmospheric and oceanic circulation patterns, and other factors.

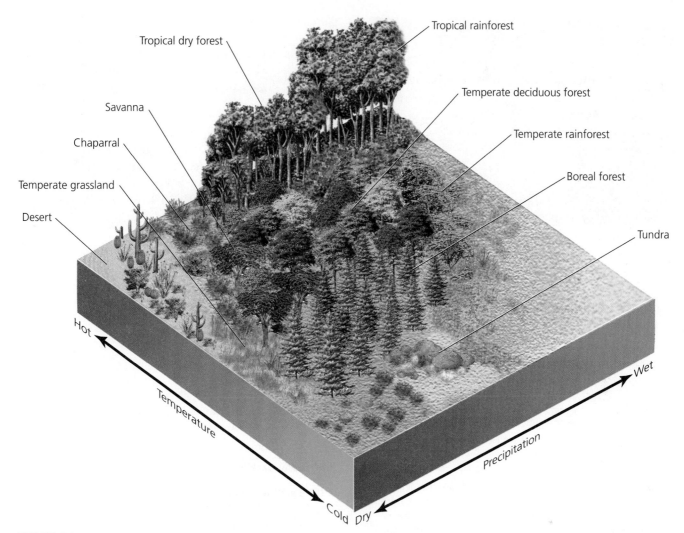

Tropical dry forest

Tropical rainforest

Savanna

Temperate deciduous forest

Chaparral

Temperate rainforest

Temperate grassland

Boreal forest

Desert

Tundra

Hot

Wet

Temperature

Precipitation

Cold Dry

FIGURE 4.16 ▲ As precipitation increases, vegetation generally becomes taller and more luxuriant. As temperature increases, types of plant communities change. Together, temperature and precipitation are the main factors determining which biome occurs in a given area.

Climate influences the locations of biomes

Which biome covers each portion of the planet depends on a variety of abiotic factors, including temperature, precipitation, atmospheric and oceanic circulation patterns, and soil characteristics. Among these factors, temperature and precipitation exert the greatest influence (**FIGURE 4.16**). Because biome type is largely a function of climate, and because temperature and precipitation are among the best indicators of climate, scientists often use *climate diagrams*, or *climatographs*, to depict such information.

Global climate patterns cause biomes to occur in large patches in different parts of the world. For instance, temperate deciduous forest occurs in eastern North America, Europe, and eastern China. Note in Figure 4.16 how patches representing the same biome tend to occur at similar latitudes. This is due to Earth's north–south gradient in temperature and to atmospheric circulation patterns (p. 281).

Aquatic and coastal systems also show biome-like patterns

In our discussion of biomes, we will focus solely on terrestrial systems because the biome concept has traditionally been developed for and applied to terrestrial systems. However, areas equivalent to biomes also exist in the oceans, along coasts, and in freshwater systems. One might consider the shallows along the world's coastlines to represent one aquatic system, the continental shelves another, and the open ocean, the deep sea, coral reefs, and kelp forests as still others. Many coastal systems—such as salt marshes, rocky intertidal communities, mangrove forests, and estuaries—share both terrestrial and aquatic components. And freshwater systems such as those of the Great Lakes are widely distributed throughout the world.

Unlike terrestrial biomes, aquatic systems are shaped not by air temperature and precipitation, but by water temperature, salinity, dissolved nutrients, wave action, currents, depth, light levels, and type of substrate (e.g., sandy, muddy, or rocky bottom). Marine communities are also more clearly

(a) Temperate deciduous forest

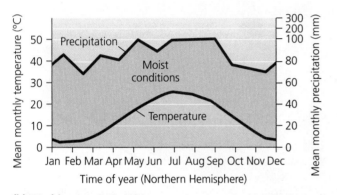

(b) Washington, D.C., USA

FIGURE 4.17 ▲ Temperate deciduous forests (**a**) experience fairly stable seasonal precipitation but varied seasonal temperatures. Scientists use climate diagrams (**b**) to illustrate an area's average monthly precipitation and temperature. In these diagrams, the x axis marks months of the year, and paired y axes denote temperature and precipitation. The curves indicate precipitation (blue) and temperature (red) from month to month. When the precipitation curve lies above the temperature curve, as is the case year-round in the temperate deciduous forest around Washington, D.C., the region experiences relatively "moist" conditions, indicated with green coloration. Climatograph here and in Figures 4.18 through 4.26 adapted from Breckle, S.W., and H. Walter, trans. by G. and D. Lawlor, 2002. *Walter's vegetation of the Earth: The ecological systems of the geo-biosphere*, 4th ed. Originally published by Eugen Ulmer KG, 1999, used by permission.

delineated by their animal life than by their plant life. (We will examine freshwater, marine, and coastal systems in the greater detail they deserve in Chapter 12.)

We can divide the world into roughly ten terrestrial biomes

Temperate deciduous forest The **temperate deciduous forest** (**FIGURE 4.17**) that dominates the landscape around the southern Great Lakes is characterized by broad-leafed trees that are deciduous, meaning that they lose their leaves each fall and remain dormant during winter, when hard freezes would endanger leaves. These midlatitude forests occur in Europe and eastern China as well as in eastern

(a) Temperate grassland

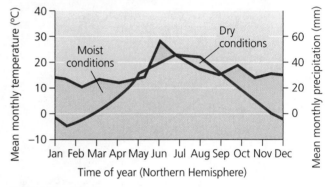

(b) Odessa, Ukraine

FIGURE 4.18 ▲ Temperate grasslands experience seasonal temperature variation and too little precipitation for many trees to grow. This climatograph indicates "moist" climate conditions (**green**), as well as "dry" climate conditions (in **yellow**, when the temperature curve is above the precipitation curve). Climatograph adapted from Breckle, S.W., 2002.

North America—all areas where precipitation is spread relatively evenly throughout the year.

Soils of the temperate deciduous forest are relatively fertile, but this biome consists of far fewer tree species than in tropical rainforests. Oaks, beeches, and maples are a few of the most common trees in these forests. Some typical animals of the temperate deciduous forest of eastern North America are shown in Figure 4.9 (p. 70).

Temperate grassland As we travel westward from the Great Lakes, temperature differences between winter and summer become more extreme, rainfall diminishes, and we find **temperate grasslands** (**FIGURE 4.18**). This is because the limited precipitation in the Great Plains region can support grasses more easily than trees. Also known as *steppe* or *prairie*, temperate grasslands were once widespread throughout parts of North and South America and much of central Asia. Today people have converted most of the world's grasslands for agriculture. Characteristic vertebrates of the North American grasslands include American bison, prairie dogs, pronghorn antelope, and ground-nesting birds such as meadowlarks and prairie chickens.

(a) Temperate rainforest

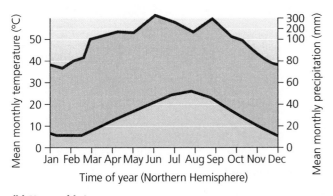

(b) Nagasaki, Japan

FIGURE 4.19 ▲ Temperate rainforests receive a great deal of precipitation and feature moist, mossy interiors. Climatograph adapted from Breckle, S.W., 2002.

(a) Tropical rainforest

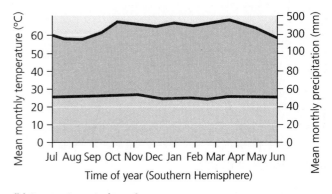

(b) Bogor, Java, Indonesia

FIGURE 4.20 ▲ Tropical rainforests, famed for their biodiversity, grow under constant, warm temperatures and a great deal of rain. Climatograph adapted from Breckle, S.W., 2002.

Temperate rainforest Further west in North America, the topography becomes varied, and biome types intermix. The coastal Pacific Northwest region, with its heavy rainfall, features **temperate rainforest** (**FIGURE 4.19**). Coniferous trees such as cedars, spruces, hemlocks, and Douglas fir grow very tall in the temperate rainforest, and the forest interior is shaded and damp. Moisture-loving animals such as the bright yellow banana slug are common.

The soils of temperate rainforests are fertile but are susceptible to landslides and erosion if forests are cleared. Logging has eliminated most old-growth trees in these forests, driving species such as the endangered spotted owl (p. 176) toward extinction. Local people often support further timber extraction, but they also suffer the consequences of overharvesting.

Tropical rainforest In tropical regions we see the same pattern found in temperate regions: Areas of high rainfall grow rainforests, areas of intermediate rainfall support dry or deciduous forests, and areas of lower rainfall are dominated by grasses. However, tropical biomes differ from their temperate counterparts in other ways because

they are closer to the equator and therefore warmer on average year-round. For one thing, they hold far greater biodiversity.

Tropical rainforest (**FIGURE 4.20**) is found in Central America, South America, Southeast Asia, west Africa, and other tropical regions and is characterized by year-round rain and uniformly warm temperatures. Tropical rainforests have dark, damp interiors, lush vegetation, and highly diverse communities, with more species of insects, birds, amphibians, and other animals than any other biome.

These forests are not dominated by single species of trees, as are forests closer to the poles, but instead consist of very high numbers of tree species intermixed, each at a low density. Any given tree may be draped with vines, enveloped by strangler figs, and loaded with epiphytes (orchids and other plants that grow in trees), such that trees occasionally collapse under the weight of all the life they support.

Despite this profusion of life, tropical rainforests have poor, acidic soils that are low in organic matter. Nearly all nutrients in this biome are contained in the plants, not in the soil. An unfortunate consequence is that once tropical

(a) Tropical dry forest

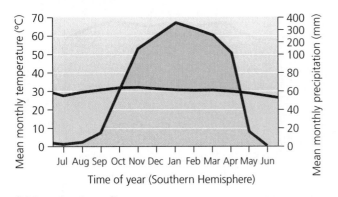

(b) Darwin, Australia

FIGURE 4.21 ▲ Tropical dry forests experience significant seasonal variation in precipitation and relatively stable warm temperatures. Climatograph adapted from Breckle, S.W., 2002.

(a) Savanna

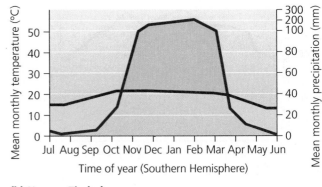

(b) Harare, Zimbabwe

FIGURE 4.22 ▲ Savannas are grasslands with clusters of trees. They experience slight seasonal variation in temperature but significant variation in rainfall. Climatograph adapted from Breckle, S.W., 2002.

rainforests are cleared, the nutrient-poor soil can support agriculture for only a short time (pp. 139–140). As a result, farmed areas are abandoned quickly, and the soil and vegetation recover slowly.

Tropical dry forest Tropical areas that are warm year-round but where rainfall is lower overall and highly seasonal give rise to **tropical dry forest**, or *tropical deciduous forest* (**FIGURE 4.21**), a biome widespread in India, Africa, South America, and northern Australia. Wet and dry seasons each span about half a year in tropical dry forest. Organisms that inhabit tropical dry forest have adapted to seasonal fluctuations in precipitation and temperature. For instance, plants are deciduous and often leaf out and grow profusely with the rains, then drop their leaves during the driest times of year.

Rains during the wet season can be extremely heavy and, coupled with erosion-prone soils, can lead to severe soil loss where people have cleared forest. Across the globe, we have converted a great deal of tropical dry forest to agriculture. Clearing for farming or ranching is made easier by the fact

that vegetation heights are much lower and canopies less dense than in tropical rainforest.

Savanna Drier tropical regions give rise to **savanna** (**FIGURE 4.22**), tropical grassland interspersed with clusters of acacias or other trees. The savanna biome is found today across stretches of Africa, South America, Australia, India, and other dry tropical regions. Precipitation in savannas usually arrives during distinct rainy seasons and concentrates grazing animals near widely spaced water holes. Common herbivores on the African savanna include zebras, gazelles, and giraffes, and the predators of these grazers include lions, hyenas, and other highly mobile carnivores.

Desert Where rainfall is sparse, **desert** (**FIGURE 4.23**) forms. The driest biome on Earth, most deserts receive well under 25 cm (9.8 in.) of precipitation per year, much of it during isolated storms months or years apart. Some deserts, such as Africa's Sahara and Namib deserts, are mostly bare sand dunes, whereas others, such as the Sonoran Desert of Arizona and northwest Mexico, receive more rain and are more

(a) Desert

(b) Cairo, Egypt

FIGURE 4.23 ▲ Deserts are dry year-round, but they are not always hot. The temperature curve is consistently above the precipitation curve in this climatograph of Cairo, Egypt, indicating that the region experiences "dry" conditions all year. Climatograph adapted from Breckle, S.W., 2002.

(a) Tundra

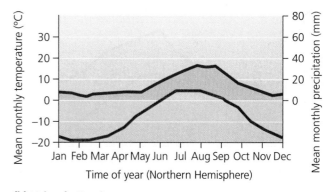

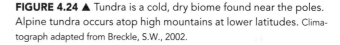

(b) Vaigach, Russia

FIGURE 4.24 ▲ Tundra is a cold, dry biome found near the poles. Alpine tundra occurs atop high mountains at lower latitudes. Climatograph adapted from Breckle, S.W., 2002.

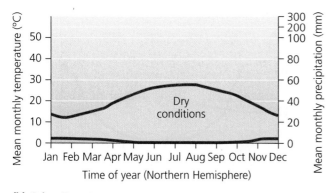

heavily vegetated. Desert soils can be quite saline and are sometimes known as lithosols, or stone soils, for their high mineral and low organic-matter content. Because deserts have low humidity and relatively little vegetation to insulate them from temperature extremes, sunlight readily heats them in the daytime, but daytime heat is quickly lost at night. As a result, temperatures vary widely from day to night and across seasons of the year.

Desert animals and plants show many adaptations to deal with a harsh climate. Most reptiles and mammals, such as rattlesnakes and kangaroo mice, are active in the cool of night, and many Australian desert birds are nomadic, wandering long distances to find areas of recent rainfall and plant growth. Many desert plants have thick, leathery leaves to reduce water loss or green trunks that enable the plant to photosynthesize without leaves, which lose water. The spines of cacti and other desert plants protect them from being eaten by herbivores desperate for the precious water they hold. Such traits have evolved by convergent evolution in deserts across the world (see Figure 3.2b, p. 47).

Tundra Nearly as dry as desert, **tundra** (**FIGURE 4.24**) occurs at very high latitudes along the northern edges of Russia, Canada, and Scandinavia. Extremely cold winters with little daylight and summers with lengthy days characterize this landscape of lichens and low, scrubby vegetation without trees. The great seasonal variation in temperature and day length results from this biome's high-latitude location, angled toward the sun in summer and away from the sun in winter.

Because of the cold climate, underground soil remains permanently frozen and is called *permafrost*. During winter, the surface soil freezes as well; then, when the weather warms, it melts and produces pools of water that make ideal habitat for mosquitoes and other insects. The swarms of insects benefit bird species that migrate long distances to breed during the brief but productive summer. Caribou also migrate to the tundra to breed and then leave for the winter. Only a few animals, such as polar bears and musk oxen, can survive year-round in this extreme climate.

Tundra also occurs as *alpine tundra* at the tops of tall mountains in temperate and tropical regions. Here,

(a) Boreal forest

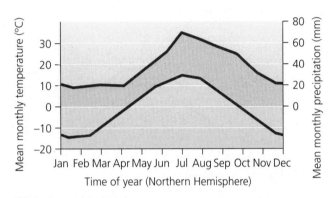

(b) Archangelsk, Russia

FIGURE 4.25 ▲ Boreal forest is characterized by long, cold winters, relatively cool summers, and moderate precipitation. Climatograph adapted from Breckle, S.W., 2002.

(a) Chaparral

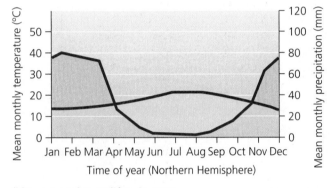

(b) Los Angeles, California, USA

FIGURE 4.26 ▲ Chaparral is a highly seasonal biome dominated by shrubs, influenced by marine weather, and dependent on fire. Climatograph adapted from Breckle, S.W., 2002.

high elevation creates conditions similar to those of high latitude.

Boreal forest The **boreal forest**, often called *taiga* (**FIGURE 4.25**), stretches in a broad band across much of Canada, Alaska, Russia, and Scandinavia. It consists of a few species of evergreen trees, such as black spruce, that dominate large stretches of forest interspersed with occasional bogs and lakes. These forests develop in cooler, drier regions than do temperate forests, and they experience long, cold winters and short, cool summers. Soils are typically nutrient-poor and somewhat acidic. As a result of the strong seasonal variation in day length, temperature, and precipitation, many organisms compress a year's worth of feeding and breeding into a few warm, wet months. Year-round residents of boreal forest include mammals such as moose, wolves, bears, lynx, and rodents. This biome also hosts many insect-eating birds that migrate from the tropics to breed during the brief, intensely productive summer season.

Chaparral In contrast to the boreal forest's broad, continuous distribution, **chaparral** (**FIGURE 4.26**) is limited to fairly small patches widely flung around the globe. Chaparral consists mostly of evergreen shrubs and is densely thicketed. This biome is also highly seasonal, with mild, wet winters and warm, dry summers. This type of climate is induced by oceanic influences and is often termed "Mediterranean." In addition to ringing the Mediterranean Sea, chaparral occurs along the coasts of California, Chile, and southern Australia. Chaparral communities experience frequent fire, and their plant species are adapted to resist fire or even to depend on it for germination of their seeds.

➤ CONCLUSION

The natural world is so complex that we can visualize it in many ways and at various scales. Dividing the world's communities into major types, or biomes, is informative at the broadest geographic scales. Understanding how communities function at more local scales requires understanding how species interact with one another. Species interactions such as competition, predation, parasitism, and mutualism give rise to effects that are both weak and strong, direct and indirect. Feeding relationships can be represented with the concepts of trophic levels and food webs, and particularly influential species are sometimes called keystone species. People alter ecological communities, partly by introducing non-native species that sometimes turn invasive. But increasingly, through ecological restoration, we are also attempting to undo the changes we have caused.

TESTING YOUR COMPREHENSION

1. How does competition promote resource partitioning?

2. Contrast the several types of exploitative species interactions. How do predation, parasitism, and herbivory differ?

3. Give examples of symbiotic and nonsymbiotic mutualisms. Describe at least one way in which mutualisms affect your daily life.

4. Using the concepts of trophic levels and energy flow, explain why the ecological footprint of a vegetarian is smaller than that of a meat eater.

5. Differentiate a food chain from a food web. Which best represents the reality of communities, and why?

6. What is meant by the term *keystone species*, and what types of organisms are most often considered keystone species?

7. Explain primary succession. How does it differ from secondary succession?

8. What is restoration ecology? Why is it an important scientific pursuit in today's world?

9. What factors most strongly influence the type of biome that forms in a particular place on land? What factors determine the type of aquatic system that may form in a given location?

10. Draw a typical climate diagram for a tropical rainforest. Label all parts of the diagram, and describe all of the types of information an ecologist could glean from such a diagram. Now draw a climate diagram for a desert. How does it differ from the rainforest climatograph, and what does this tell you about how the two biomes differ?

SEEKING SOLUTIONS

1. Suppose you spot two species of birds feeding side by side, eating seeds from the same plant. You begin to wonder whether competition is at work. Describe how you might design scientific research to address this question. What observations would you try to make at the outset? Would you try to manipulate the system to test your hypothesis that the two birds are competing? If so, how?

2. Spend some time outside on your campus, in your yard, or in the nearest park or natural area. Find at least 10 species of organisms (plants, animals, or others), and observe each one long enough to watch it feed or to make an educated guess about how it derives its nutrition. Now, using Figure 4.9 as a model, draw a simple food web involving all the organisms you observed.

3. Can you think of one organism not mentioned in this chapter as a keystone species that you believe may be a keystone species? For what reasons do you suspect this? How could you experimentally test whether an organism is a keystone species?

4. Why do scientists consider invasive species to be a problem? What makes a species "invasive," and what ecological effects can invasive species have?

5. **THINK IT THROUGH** A federal agency has put you in charge of devising responses to the zebra mussel invasion. Based on what you know from this chapter, how would you seek to control this species' spread and reduce its impacts? What strategies would you consider pursuing immediately, and for which strategies would you commission further scientific research? For each of your ideas, name one benefit or advantage, and identify one obstacle it might face in being implemented. What additional steps might you suggest to deal with the unfolding quagga mussel invasion?

CALCULATING ECOLOGICAL FOOTPRINTS

In 2005, environmental scientists David Pimentel, Rodolfo Zuniga, and Doug Morrison of Cornell University reviewed estimates for the economic and ecological costs inflicted by introduced and invasive species in the United States. They found that approximately 50,000 species have been introduced in the United States and that these account for over $120 billion in economic costs each year. These costs include direct losses and damage as well as costs required for control of the invasive species. (The researchers did not try to quantify monetary estimates for losses of biodiversity, ecosystem services, and aesthetics, which they say would drive total costs several times higher.) Calculate values missing from the table to determine the number of introduced species of each type of organism and the annual cost that each inflicts on our economy.

Group of organism	Percentage of total introduced	Number of species introduced	Percentage of total annual costs	Annual economic costs
Plants	50.0	25,000	27.2	
Microbes	40.0		20.2	
Arthropods	9.0		15.7	
Fish	0.28		4.2	
Birds	0.19		1.5	$1.9 billion
Mollusks	0.18		1.7	
Reptiles and amphibians	0.11		0.009	
Mammals	0.04	20	29.4	$37.5 billion
TOTAL	100	50,000	100	$127.4 billion

Data from Pimentel, D., R. Zuniga, and D. Morrison, 2005. Update on the environmental and economic costs associated with alien-invasive species in the United States. *Ecological Economics* 52: 273–288.

1. Of the 50,000 species introduced into the United States, half are plants. Describe two ways in which non-native plants might be brought to a new environment. How might we help prevent non-native plants from establishing in new areas and posing threats to native communities?

2. Organisms that damage crop plants are the most costly of introduced species. Weeds, pathogenic microbes, and arthropods that attack crops together account for half of the costs documented by Pimentel and his colleagues. What steps can we—farmers, governments, and all of us as a society—take to minimize the impacts of invasive species on crops?

3. How might your own behavior influence the influx and ecological impacts of non-native species such as those listed above? Name three things you could personally do to help reduce the impacts of invasive species.

Go to **www.masteringenvironmentalscience.com** for homework assignments, practice quizzes, Pearson eText, and more.

5 Environmental Economics and Environmental Policy

Upon completing this chapter, you will be able to:

➤ Describe principles of economic theory and summarize their implications for the environment

➤ Compare the concepts of economic growth, economic health, and sustainability

➤ Explain the approaches of environmental economics and ecological economics

➤ Describe the aims of environmental policy and assess its societal context

➤ Discuss the history of U.S. environmental policy and identify major U.S. environmental laws

➤ Characterize the institutions involved with international environmental policy and describe how nations handle transboundary issues

➤ Outline the environmental policy process and evaluate its effectiveness

➤ Discuss the role of science in the policy process

➤ Contrast the different approaches to environmental policy

Contaminated beach near Tijuana River mouth

San Diego and Tijuana: Pollution Problems and Policy Solutions

"Never doubt that a small group of thoughtful, committed citizens can change the world—indeed it is the only thing that ever has."
—Anthropologist Margaret Mead

"It is the continuing policy of the Federal Government . . . to create and maintain conditions under which man and nature can exist in productive harmony and fulfill the social, economic, and other requirements of present and future generations of Americans."
—National Environmental Policy Act

The beaches south of San Diego boast some of the world's best waves for surfing. These days, however, most surfers avoid the temptation. For it is here that the heavily polluted Tijuana River flows across the international border from Mexico and empties into the Pacific Ocean, disgorging millions of gallons of untreated wastewater.

"When it rains, I call it the sewage tsunami," says surfer and environmentalist Serge Dedina. "For 40 square miles, from Imperial Beach to Coronado, there is a brown plume as far as the eye can see."

Such incidents occur when heavy rains overwhelm the ability of sewage treatment plants to process wastewater. San Diego's coastal waters receive stormwater runoff when rains wash pollutants into local rivers. Across the border in the Mexican city of Tijuana, the aging, leaky sewer system becomes clogged with debris, causing raw sewage to overflow into the streets and, eventually, into the Tijuana River.

Winding northwestward through the arid landscape of northern Baja California, Mexico, the Tijuana River crosses the U.S. border south of San Diego (FIGURE 5.1A). A river's **watershed** consists of all the land from which water drains into the river, and the Tijuana River's watershed covers 4,500 km² (1,750 mi²) and is home to 2 million people of two nations. The Tijuana River watershed is a transboundary watershed (so named because it crosses a political boundary—in this case, an international border), with approximately 70% of its area in Mexico. On the Mexican side of the border, the river and its tributaries are lined with farms,

apartments, shanties, and factories, as well as leaky sewage treatment plants and toxic dump sites. Rains wash pollutants from all these sources into the Tijuana River and eventually onto U.S. and Mexican beaches (FIGURE 5.1B).

Although pollution has flowed in the Tijuana River for decades, the problem grew worse in recent years as the region's population boomed, outstripping the capacity of sewage treatment facilities. As impacts intensified, people on both sides of the border pressed policymakers for action. As a result, Mexico and the United States worked together to construct a wastewater treatment plant to handle excess waste from Tijuana. The South Bay International Wastewater Treatment Plant (IWTP) began operating just north of the border in 1997 and treats up to 95 million L (25 million gal) of wastewater each day. Unfortunately, the facility reached its capacity within three years because Tijuana's population grew so quickly, and excess wastewater began flowing downriver.

Since then, beach closures and pollution-related health advisories have been commonplace. Garbage carried by the river litters the beaches. "Every day I find broken glass, balloons, or can pop-tops. I've

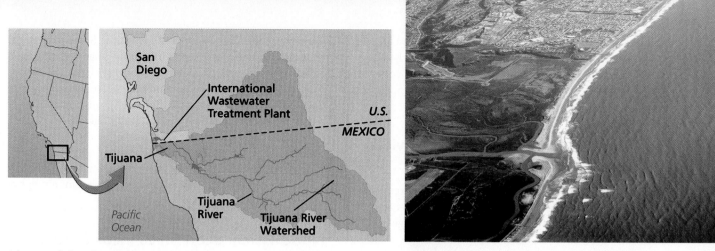

(a) Map of the Tijuana River watershed

(b) Wastewater enters the ocean near Tijuana

FIGURE 5.1 ▲ The Tijuana River winds northwestward from Mexico into California just south of San Diego, draining 4,500 km² (1,750 mi²) of land in its watershed (colored green in map) **(a)**. Pollution entering the river affects people on both sides of the border and sometimes creates a visible brown plume of wastewater entering the Pacific Ocean **(b)**. The photo shows an aerial view from the north, with Tijuana in the background.

even found hypodermic needles. It's really sad," one resident of Imperial Beach told her local newspaper.

Mexican residents of the Tijuana River watershed suffer more pollution because most live in poverty relative to their U.S. neighbors. Close to one-third of Tijuana's homes are not connected to a sewer system, and in poor neighborhoods such as Loma Taurina, river pollution directly affects people's day-to-day lives by contaminating water for drinking and washing and by promoting risks of disease, especially when flooding occurs. The rise of U.S.-owned factories, or *maquiladoras*, on the Mexican side of the border also has contributed to pollution, through the direct disposal of industrial waste and by attracting thousands of new workers to the already crowded region. In many ways, economic inequities spanning the border region have aggravated its problems with water pollution.

As we explore environmental economics and environmental policy in this chapter, we will periodically return to the Tijuana River watershed and see how people are making progress by using science and policy to help address the region's pollution challenges. ∎

ECONOMICS: APPROACHES AND ENVIRONMENTAL IMPLICATIONS

Many environmental problems share the mix of impacts we see in the Tijuana River watershed—harming human health, altering ecological systems, inflicting economic damage, and contributing to inequities among people. In the Tijuana River watershed, pollution affects the region's economies—while economic inequities, in turn, worsen pollution. Sewage-tainted water carries pathogens (organisms that cause illness), posing health risks and leading to higher medical costs. Untreated

wastewater lowers concentrations of dissolved oxygen, killing economically valuable fish and shellfish. Pollution and beach closures reduce recreation and tourism both in Mexico and in southern California, whose beaches each year host 175 million visitors who spend over $1.5 billion. As a result, finding ways to reduce pollution will help the region economically.

Economics is the study of how people decide to use resources to provide goods and services in the face of demand for them. By this definition, environmental problems are also economic problems that vary with population and per capita resource consumption. Indeed, the word *economics* and the word *ecology* come from the same Greek root, *oikos*, meaning "household." Economists traditionally have studied the household of human society, while ecologists study the broader household of all life.

Several types of economies exist

An **economy** is a social system that converts resources into **goods**, material commodities manufactured for and bought by individuals and businesses; and **services**, work done for others as a form of business. The oldest type of economy is the **subsistence economy**. People in subsistence economies meet their daily needs by subsisting on what they can gather from nature or produce on their own, rather than working for wages and then purchasing life's necessities.

A second type of economy is the **capitalist market economy**. In this system, interactions among buyers and sellers determine which goods and services are produced, how much is produced, and how these are produced and distributed. Capitalist economies contrast with state socialist economies, or **centrally planned economies**, in which government determines how to allocate resources. In reality, today's capitalist and socialist economies have borrowed much from one another and are in fact hybrid systems (often termed **mixed economies**).

In modern mixed economies, governments typically intervene in the market for several reasons: (1) to eliminate unfair advantages held by single buyers or sellers; (2) to provide social services, such as national defense, medical care, and education; (3) to provide "safety nets" for the elderly, victims of natural disasters, and so on; (4) to manage the commons (p. 3–4); and (5) to reduce pollution and other threats to health and quality of life.

Economies rely on goods and services from the environment

Economies receive inputs (such as natural resources) from the environment, process them in complex ways that enable human society to function, and then discharge outputs (such as waste) into the environment. Although these interactions between human economies and the nonhuman environment are readily apparent, traditional economic schools of thought have long overlooked the importance of these connections. Indeed, most mainstream economists still adhere to a worldview that largely ignores the environment (**FIGURE 5.2A**)—and this worldview continues to drive most policy decisions. However, modern economists belonging to the fast-growing fields of environmental economics and ecological economics (pp. 93–94) explicitly recognize that human economies are subsets of the environment and depend crucially upon it for natural resources and ecosystem services (**FIGURE 5.2B**).

Economic activity uses natural resources (pp. 2–3), the substances and forces we need to survive: the sun's energy, the fresh water we drink, the trees that provide us lumber, the rocks that provide us metals, and the fossil fuels that power our machines. We can think of natural resources as "goods" produced by nature.

Environmental systems also naturally function in a manner that supports economies. Earth's ecological systems purify air and water, form soil, cycle nutrients, regulate climate, pollinate plants, and recycle the waste generated by our economic activity. Such essential processes, called **ecosystem services** (pp. 2, 36), support the life that makes our economic activity possible.

While our environment enables economic activity by providing ecosystem goods and services, economic activity can affect the environment in return. When we deplete natural resources and generate pollution, we often degrade the capacity of ecological systems to function. In fact, the Millennium Ecosystem Assessment (p. 16) concluded in 2005 that 15 of 24 ecosystem services its scientists surveyed globally were being degraded or used unsustainably. The degradation of ecosystem services can in turn disrupt economies, as we see along the Tijuana River, where pollution depresses people's economic opportunities. Ecological degradation is harming poor people more than wealthy people, according to the Millennium Ecosystem Assessment. As a result, restoring ecosystem services stands as a prime avenue for alleviating poverty.

Adam Smith proposed an "invisible hand"

When economics began to develop as a discipline in the 18th century, many philosophers argued that individuals acting in their own self-interest would harm society (as in the tragedy of the commons, p. 3–4). However, Scottish philosopher **Adam Smith** (1723–1790) believed that self-interested behavior could benefit society, as long as the behavior was constrained by the rule of law and private property rights and operated within fairly competitive markets. Known today as a founder of **classical economics**, Smith felt that when people are free to pursue their own economic self-interest in a competitive marketplace, the marketplace will behave as if guided by "an invisible hand" that leads their actions to benefit society as a whole. Smith's philosophy remains a pillar of free-market thought today.

Neoclassical economics incorporates psychology and cost-benefit analysis

Economists subsequently adopted more quantitative approaches as they aimed to explain human behavior. **Neoclassical economics** examines the psychological factors underlying consumer choices, explaining market prices in terms of consumer preferences for units of particular commodities. In neoclassical economic theory, buyers desire the lowest possible price, whereas sellers desire the highest possible price. As a result of this conflict, a compromise price is reached, and the "right" quantities of commodities are bought and sold. This balance is often phrased in terms of *supply*, the amount of a product offered for sale at a given price, and *demand*, the amount of a product people will buy at a given price if free to do so (**FIGURE 5.3**).

To evaluate an action or decision, neoclassical economists often use **cost-benefit analysis**. In this approach, economists total up estimated costs for a proposed action and compare these to the sum of benefits estimated to result from the action. If benefits exceed costs, the action should be pursued; if costs exceed benefits, it should not. Given a choice of alternative actions, the one with the greatest excess of benefits over costs should be chosen.

This reasoning seems eminently logical, but problems often arise because not all costs and benefits can be easily identified, defined, or quantified. For example, it may be easy to tally up the costs of installing equipment to reduce pollution, yet difficult to assess the effects of pollution on people's health or lifestyles. Moreover, monetary values can often be assigned more easily to economic benefits (such as jobs created by a factory) than to environmental costs (such as long-term health impacts of the factory's pollution on a community), so economic benefits tend to be overrepresented in cost-benefit analyses. As a result, environmental advocates often feel these analyses are biased in favor of economic development and against environmental protection.

Neoclassical economics has profound implications for the environment

Today's capitalist market systems operate largely in accord with the principles of neoclassical economics. These systems have generated unprecedented material wealth for our societies. Alas, four fundamental assumptions of neoclassical economics often contribute to environmental degradation.

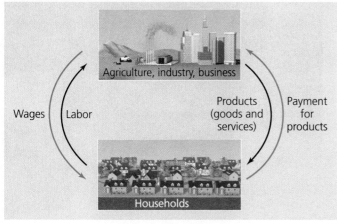

(a) Conventional view of economic activity

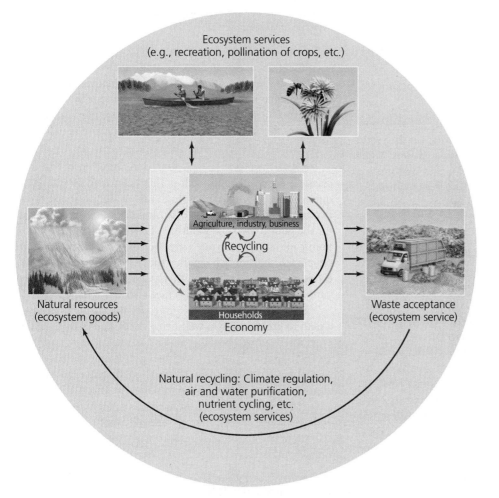

(b) Economic activity as viewed by environmental and ecological economists

Are resources infinite or substitutable? One assumption is that natural resources and human resources (such as workers) are either infinite or largely substitutable and interchangeable. This implies that once we have used up a resource, we should be able to find a replacement for it. Certainly it is true that many resources can be replaced. However, some cannot. Nonrenewable resources (such as fossil fuels) can truly be depleted, and many renewable resources (such as forest products) can also be used up if we exploit them faster than they are replenished.

Should we discount the future? Second, neoclassical economics grants an event in the future less value than one in the present. In economic terminology, future effects are

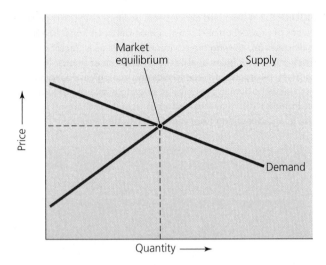

FIGURE 5.3 ▲ In a supply-and-demand graph, the demand curve indicates the quantity of a given good (or service) that consumers desire at each price, and the supply curve indicates the quantity produced at each price. The market automatically moves toward an equilibrium point at which supply equals demand.

FIGURE 5.4 ▲ River pollution creates external costs. This woman washing clothes in the Tijuana River suffers pollution from factories upstream, and her use of detergents causes pollution for people living downstream.

"discounted." Short-term costs and benefits are granted more importance than long-term costs and benefits, causing us to ignore the long-term consequences of policy decisions. Many environmental problems unfold gradually, and discounting causes us to downplay the impacts on future generations of the pollution we create and the resources we deplete today.

Are all costs and benefits internal? A third assumption of neoclassical economics is that all costs and benefits associated with an exchange of goods or services are borne by individuals engaging directly in the transaction. In other words, it is assumed that the costs and benefits are "internal" to the transaction, experienced by the buyer and seller alone. However, many transactions affect other members of society. For example, pollution from a *maquiladora* along the Tijuana River can harm people living downstream. In such a case, members of society not involved in producing the pollution end up paying its costs. When market prices do not take the social, environmental, or economic costs of pollution into account, then taxpayers bear the burden of paying them. Costs of a transaction that affect people other than the buyer or seller are known as **external costs** (**FIGURE 5.4**). External costs commonly include the following:

▸ Human health problems

▸ Property damage

▸ Declines in desirable features of the environment, such as fewer fish in a stream

▸ Aesthetic damage, such as from air pollution or clear-cutting

▸ Stress and anxiety experienced by people downstream or downwind from a pollution source

▸ Declining real estate values resulting from these problems

By ignoring external costs, economies create a false impression of the consequences of particular choices and unjustly subject people to the impacts of transactions in which they did not participate. External costs comprise one reason governments develop environmental legislation and regulations.

Is all growth good? A fourth assumption of the neoclassical economic approach is that economic growth is required to keep employment high and maintain social order. Economic growth, it is argued, should create opportunities for the poor to become wealthier. By making the overall economic pie larger, everyone's slice becomes larger, even if some people still have much smaller slices than others. In today's economies, economic growth has become the quantitative yardstick by which progress is measured.

To the extent that economic growth is a means to an end—a path to greater human well-being—it is a good thing. However, when growth becomes an end in itself it may no longer be the best route toward well-being. Sociologists have even coined a word for the way material goods often fail to bring contentment to people affluent enough to afford them: *affluenza*. Moreover, critics of the growth paradigm fear that the endless pursuit of economic growth will eventually destroy our economic system, because resources to support growth are ultimately limited.

How sustainable is economic growth?

Our global economy is seven times the size it was just half a century ago. All measures of economic activity—trade, rates of production, amount and value of goods manufactured—are higher than ever before. This has brought many people much greater material wealth (although not equitably, and gaps between rich and poor are wide and growing).

The modern-day United States exemplifies the view that "more and bigger" are always better. Spurred on by advertising and the increased availability of goods due to technological advances and expanded global trade, Americans have embarked on a frenzy of consumption unparalleled in history.

The dramatic rise in per-person consumption has numerous consequences (**FIGURE 5.5**).

Economic growth stems from two sources: (1) an increase in inputs to the economy (e.g., greater inputs of labor and natural resources) and (2) improvements in the efficiency of production due to better technologies and approaches (i.e., ideas and equipment that enable us to produce more goods with fewer inputs). This second approach—whereby we produce more with less—is often termed *economic development*.

As our population and consumption rise, it is becoming clearer that we cannot sustain growth forever using the first approach. Nonrenewable resources on Earth are finite in quantity, and there are limits on the rates that we can harvest many renewable resources. As for the second approach, we have used technological innovation to push back the limits on growth time and again. More-efficient technologies for extracting minerals, fossil fuels, and groundwater allow us to exploit these resources more fully with less waste. Automated farm machinery, fertilizers, and chemical pesticides enable us to grow more food per unit area of land (p. 136). Better machinery in our factories speeds our manufacturing. We continue to make computer chips more powerful while also making them smaller and using fewer raw materials. In all these ways, we are producing more goods and services while using fewer resources.

Can we conclude, then, that improvements in technology will allow us to overcome all our environmental limitations and continue economic growth indefinitely? Surely we can innovate and achieve further efficiency and economic growth without depleting our resource base—but ultimately, nonrenewable resources are finite, and renewable resources can be exploited only at limited rates. If our population and consumption continue to grow and we do not shift to full reuse and recycling, we will inevitably deplete resources and put ever-greater demands on our capacity to innovate.

Ecological economists argue that civilizations do not, in the long run, overcome their environmental limitations. Ecological economists apply principles of ecology and systems science (Chapter 2) to the analysis of economic systems, and they view natural systems as good models for making human economies sustainable. In nature, every population faces limiting factors and a carrying capacity (p. 58), and systems generally operate in self-renewing cycles, not in a linear manner. Many ecological economists advocate economies that do not grow and do not shrink, but rather are stable. Such **steady-state economies** are intended to mirror natural systems. Ecological economists maintain that quality of life should continue to rise in a steady-state economy, as a result of technological advances and behavioral changes (such as greater

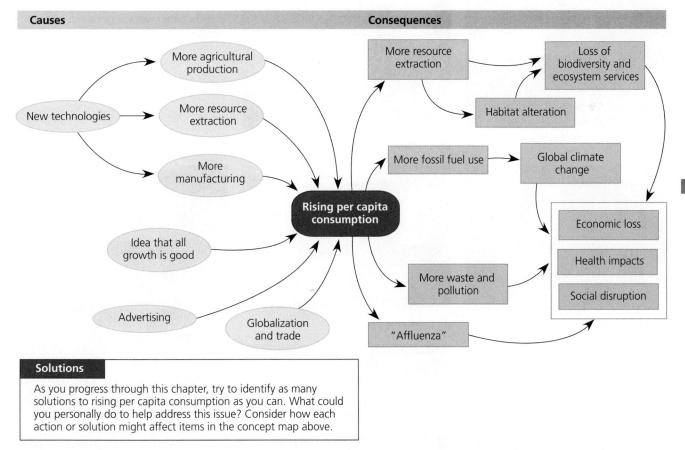

FIGURE 5.5 ▲ The rise in per-person consumption of goods and services stems from multiple causes (**ovals on left**) and results in a diversity of environmental, social, and economic consequences (**boxes on right**). Arrows in this concept map lead from causes to consequences. Note that items grouped within an outlined box do not necessarily share any special relationship; the outlined box is intended merely to streamline the figure.

use of recycling) that enhance sustainability. They argue that wealth and human well-being can continue to increase after economic growth has leveled off.

Environmental economists tend to agree with ecological economists that economies are unsustainable if population growth is not reduced and resources are not used more efficiently, but they argue that, with effort, we can accomplish these changes and attain sustainability within our current economic systems. By modifying the principles of neoclassical economics to address environmental challenges, environmental economists maintain that we can keep our economies growing and that technology can continue to improve efficiency. Thus, whereas ecological economists call for revolution, environmental economists call for reform. One approach environmental economists take is to assign monetary values to ecosystem goods and services, so as to better integrate them into traditional cost-benefit analyses.

We can assign monetary value to ecosystem goods and services

Ecosystem services are said to have **nonmarket values**, values not usually included in the market price of a good or service (**TABLE 5.1** and **FIGURE 5.6**). For example, the aesthetic and recreational pleasure we obtain from natural landscapes is something of real value. Yet because we do not pay money for this, its value is hard to quantify and appears in no traditional measures of economic worth. Or consider Earth's water cycle (p. 37), by which rain fills our reservoirs with drinking water, rivers give us hydropower and flush away our waste, and water evaporates, purifying itself of contaminants and readying itself to fall again as rain. This natural cycle is vital to our very existence, yet because we do not quantify its value, markets impose no financial penalties when we disturb it.

To resolve this dilemma, environmental and ecological economists have sought ways to assign values to ecosystem services. In one technique, economists use surveys to determine how much people are willing to pay to protect or restore a resource. In another approach, they measure the money, time, or effort people expend to travel to parks for recreation. Economists also compare housing prices for similar homes in different environmental settings to infer the dollar value of landscapes, views, and peace and quiet. They may also measure the cost required to restore natural systems that have been damaged, to replace those systems' functions with technology, or to reduce harm from pollution.

FIGURE 5.6 ◄ Accounting for nonmarket values such as those shown here may help us to make better environmental and economic decisions. See Table 5.1 for details.

(a) Use value

(b) Existence value

(c) Option value

(d) Aesthetic value

(e) Scientific value

(f) Educational value

(g) Cultural value

TABLE 5.1 Values That Modern Market Economies Generally Do Not Address

Nonmarket value	Is the worth we ascribe to things . . .
Use value	that we use directly
Existence value	simply because they exist, even though we may never experience them directly (e.g., an endangered species in a far-off place)
Option value	that we do not use now but might use later
Aesthetic value	for their beauty or emotional appeal
Scientific value	that may be the subject of scientific research
Educational value	that teach us about ourselves or the world
Cultural value	that sustain or help define our culture

In 1997, a research team headed by environmental economist Robert Costanza set out to calculate the total economic value of all the services that oceans, forests, wetlands, and other ecosystems provide across the world. Costanza's team combed the scientific literature and evaluated over 100 studies that used various methods to estimate dollar values for 17 major ecosystem services such as water purification, climate regulation, plant pollination, and pollution cleanup (**FIGURE 5.7**). To improve the accuracy of estimates, the researchers reevaluated the data using multiple techniques. They then multiplied average estimates for each ecosystem by the global area occupied by each. Their analysis, reported in the journal *Nature*, calculated that Earth's biosphere in total provides at least $33 trillion ($46 trillion in 2011 dollars) worth of ecosystem services each year—more than the GDP of all nations combined!

In a follow-up study in 2002, Costanza joined Andrew Balmford and 17 other colleagues to compare the benefits and costs of preserving natural systems versus converting wild lands for agriculture, logging, or fish farming. After reviewing many studies, they reported in the journal *Science* that a global network of nature reserves covering 15% of Earth's land surface and 30% of the ocean would be worth between $4.4 and $5.2 trillion. This amount is 100 times greater than the value of those areas were they to be converted for direct exploitative human use—demonstrating, in their words, that "conservation in reserves represents a strikingly good bargain."

Businesses are responding to sustainability concerns

As economists rethink old assumptions and as more consumers and investors express preferences for sustainable products and services, many industries, businesses, and corporations are finding that they can make money by "greening" their operations.

Some companies have cultivated an eco-conscious image from the start, such as Ben & Jerry's (ice cream), Patagonia (outdoor apparel), Seventh Generation (household products), and Credo (formerly Working Assets; phone service). In recent years, however, corporate sustainability has gone mainstream, and some of the world's largest corporations have joined in, including Ford Motor Company, Toyota, McDonald's, Starbucks, IKEA, Dow, Dupont, BASF, Intel, and Wal-Mart (**FIGURE 5.8**). Hewlett-Packard runs programs to reuse and recycle used toner cartridges, electronics, and plastics. Nike, Inc., collects millions of used sneakers each year and recycles their materials to create surfaces for basketball courts, tennis courts, and running tracks. The Gap, Inc., built a sustainably designed headquarters building, cut energy use in its stores and distribution centers, and promotes alternative transportation for its employees. Today many corporations are finding ways to increase energy

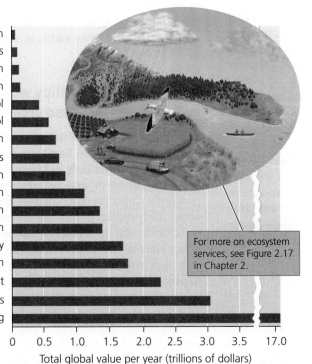

For more on ecosystem services, see Figure 2.17 in Chapter 2.

FIGURE 5.7 ◀ Environmental economists in 1997 estimated the value of the world's ecosystem services at more than $33 trillion ($46 trillion in 2011 dollars). Shown are subtotals for each major type of ecosystem service. The $33 trillion figure is an underestimate because it does not include values from ecosystems for which adequate data were unavailable. Data from Costanza, R., et al., 1997. The value of the world's ecosystem services and natural capital. *Nature* 387: 253–260.

FIGURE 5.8 ▲ A Wal-Mart cashier bags compact fluorescent bulbs in reusable canvas bags for a customer. Wal-Mart provides the most celebrated recent example of corporate greening efforts. The world's largest retailer has launched a program to sell organic and sustainable products, reduce packaging and use recycled materials, enhance fuel efficiency in its truck fleet, reduce energy use in its stores, power itself with renewable energy, cut carbon dioxide emissions, and preserve one acre of natural land for every acre developed. It is also developing a "sustainability index" to rate its products and inform consumers. Although many remain skeptical of Wal-Mart's commitment to sustainability, the corporation's vast reach and its ability to persuade suppliers to alter their practices in order to retain its business mean that any change Wal-Mart enacts could have far-reaching impacts.

efficiency, reduce toxic substances, increase the use of recycled materials, and minimize greenhouse gas emissions. In doing so, they often find that they reduce costs and increase profit.

Of course, corporations exist to make money for their shareholders, so they cannot be expected to pursue goals that do not turn a profit. Moreover, many corporate greening efforts are more rhetoric than reality, and corporate **greenwashing** may mislead consumers into thinking a company is acting more sustainably than it is. For instance, the bottled water industry advertises its products with words such as "pure" and "natural." Images of forests and alpine springs lead us to believe that bottled water is cleaner and healthier for us to drink—when in reality the water is often less safe, the plastic bottles are a major source of waste, oil is used to manufacture and transport the bottles, and the industry depletes aquifers in local communities (pp. 268–269).

In the end, corporate actions hinge on consumer behavior. It is up to all of us in our roles as consumers to encourage trends in sustainability by rewarding those businesses that truly promote sustainable solutions.

Markets can fail

When they do not reflect the full costs and benefits of actions, markets are said to fail. **Market failure** occurs when markets do not take into account the environment's positive contributions to economies (such as ecosystem services) or when they do not reflect the negative impacts of economic activity on the environment or on people (external costs). Traditionally, market failure has been countered by government intervention. Governments can restrain corporate behavior through laws and regulations. They can tax environmentally harmful activities. Or, they can design economic incentives that use market mechanisms to promote fairness, resource conservation, and economic sustainability. We will now examine these approaches in our discussion of environmental policy.

ENVIRONMENTAL POLICY: AN OVERVIEW

Economic analysis and scientific research can help us determine when resources are being depleted, ecosystem services are being degraded, or people's quality of life is declining. Once a society reaches broad agreement that such a problem exists, it may persuade its leaders to try to resolve the problem through the making of policy. **Policy** consists of a formal set of general plans and principles intended to address problems and guide decision making in specific instances. **Public policy** is policy made by governments, including those at the local, state, federal, and international levels. Public policy consists of laws, regulations, orders, incentives, and practices intended to advance societal well-being. **Environmental policy** is policy that pertains to human interactions with the environment. It generally aims to regulate resource use or reduce pollution in order to promote human welfare and/or protect natural systems.

Forging effective environmental policy requires input from science, ethics, and economics. Science provides information and analysis needed to identify and understand problems and devise solutions. Ethics and economics offer criteria to assess problems and to help clarify how society might like to address them. Government interacts with citizens, organizations, and the private sector in various ways to formulate policy (**FIGURE 5.9**).

Environmental policy addresses issues of fairness and resource use

Market capitalism is driven by incentives for short-term economic gain rather than long-term social and environmental stability. It provides little motivation for businesses or individuals to behave in ways that minimize environmental impact or that equalize costs and benefits among parties. As we noted, such *market failure* has traditionally been viewed as justification for government intervention. Environmental policy aims to protect environmental quality and the natural resources people use, and also to promote equity or fairness in people's use of resources.

The tragedy of the commons Policy to protect resources held and used in common by the public is intended to safeguard them from depletion or degradation. As environmental scientist Garrett Hardin explained in his 1968

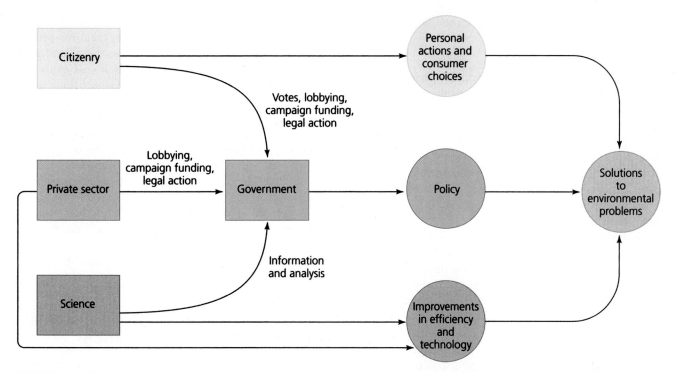

FIGURE 5.9 ▲ Policy plays a central role in how we as a society address environmental problems. Voters, the private sector, and lobbying groups representing various interests influence government representatives. Scientific research also informs government decisions. Governmental representatives and agencies formulate policy that aims to address problems. Public policy—along with improvements in technology and efficiency and personal actions and consumer purchasing choices—can produce lasting solutions to environmental problems.

essay "The Tragedy of the Commons" (pp. 3–4), a resource held in common that is accessible to all will eventually become overused and degraded. Therefore, he argued, it is in our best interest to develop guidelines for the use of such resources. In Hardin's example of a common pasture, guidelines might limit the number of animals each individual can graze or might require pasture users to pay to restore and manage the shared resource. These two concepts—restriction of use and active management—are central to environmental policy today.

Free riders   A second reason to develop policy for publicly held resources is the **free rider** predicament. Let's say a community on a river suffers from water pollution that emanates from 10 different factories. The problem could in theory be solved if every factory voluntarily agreed to reduce its own pollution. However, once they all begin reducing their pollution, it becomes tempting for any one of them to stop doing so. Such a factory, by avoiding the sacrifices others are making, would in essence get a "free ride." If enough factories take a free ride, the whole effort will collapse. Because of the free rider problem, private voluntary efforts are generally less effective than efforts mandated by public policy, which help ensure that all parties sacrifice equitably.

External costs   Environmental policy also aims to promote fairness by eliminating external costs, ensuring that some parties do not use resources in ways that harm

others. For example, a factory may reap greater profits by discharging waste freely into a river and avoiding paying for waste disposal or recycling. Its actions, however, impose external costs (water pollution, decreased fish populations, aesthetic degradation, health risks) on downstream users of the river. U.S.-owned *maquiladoras* in the Tijuana River watershed dump waste that affects Mexican families downstream (see Figure 5.4). Likewise, wastewater from the growing number of people living in the watershed further pollutes the river, imposing external costs on families farther downstream and beachgoers in Mexico and California.

THE ISSUES

Do We Really Need Environmental Policy? Many free-market advocates contend that environmental laws and regulations are an undesirable government intrusion into private affairs. Adam Smith (p. 90) argued that individuals benefit society by pursuing their own self-interest. Do you agree? Can you describe a situation in which an individual acting in self-interest either (a) benefits society by addressing an environmental problem, or (b) harms society by causing an environmental problem? How might policy help in either situation? What are some advantages and disadvantages of environmental laws and regulations?

U.S. ENVIRONMENTAL LAW AND POLICY

The United States provides a good focus for understanding environmental policy in constitutional democracies worldwide, for several reasons. First, the United States has pioneered innovative environmental policy. Second, U.S. policies have served as models—of both success and failure—for other nations and international government bodies. Third, the United States exerts a great deal of influence on the affairs of other nations. Finally, understanding U.S. environmental policy at the federal level helps us understand it at local, state, and international levels.

The three branches of the U.S. federal government—legislative, executive, and judicial—are each involved in aspects of environmental policy. Once **legislation**, or statutory law, is passed by Congress and signed into law by the president, its implementation and enforcement is assigned to an administrative agency within the executive branch. Administrative agencies are the source of a great deal of policy, in the form of **regulations**, specific rules intended to help achieve the objectives of the more broadly written statutory law. Besides issuing regulations, administrative agencies monitor compliance with laws and regulations and enforce them when they are violated. The judicial branch interprets law as needed in response to suits in the courts.

The structure of the federal government is mirrored at the state level with governors, legislatures, judiciaries, and agencies. State laws cannot violate principles of the U.S. Constitution, and if state and federal laws conflict, federal laws take precedence. Many states with dense urban populations, such as California, New York, New Jersey, and Massachusetts, have strong environmental laws and well-funded environmental agencies. Citizens of such states put more emphasis on safeguarding environmental quality because they have witnessed extensive environmental degradation in the past.

To safeguard public health in locations such as San Diego's beaches, California's state legislators have required state environmental health officials to set standards and test waters for bacterial contamination. Officials issue an advisory, or warning, when bacterial concentrations in nearshore waters exceed health limits established by California law. As we proceed through our discussion of federal policy, keep in mind that important environmental policy is also created at the state and local levels.

Early U.S. environmental policy promoted development

The laws that comprise U.S. environmental policy were created largely in three periods. Laws enacted during the first period, from the 1780s to the late 1800s, accompanied the westward expansion of the nation and were intended mainly to promote settlement and the use of the continent's abundant natural resources (**FIGURE 5.10**).

Among these early laws were the *General Land Ordinances of 1785* and *1787*, which gave the federal government the right to manage unsettled lands and created a grid system for surveying them and readying them for private ownership. Between 1785 and the 1870s, the federal government promoted settlement on lands it had expropriated from Native Americans, and it doled out these lands to its citizens. Western settlement provided these citizens with means to achieve prosperity while relieving crowding in Eastern cities. It expanded the geographical reach of the United States at a time when the young nation was still jostling with European powers for control of the continent. It also wholly displaced the millions of Native Americans who had long inhabited these lands. U.S. environmental policy of this era reflected the public perception that Western lands were practically infinite and inexhaustible in natural resources. Laws encouraged settlers, entrepreneurs, and land speculators to move west, and this hastened the closing of the frontier.

The second wave of U.S. environmental policy encouraged conservation

In the late 1800s, as the continent became more populated and its resources were increasingly exploited, public perception and government policy toward natural resources began to shift. Laws of this period aimed to alleviate some of the environmental impacts associated with westward expansion.

In 1872, Congress designated Yellowstone as the world's first national park. In 1891, Congress passed a law authorizing the president to create "forest reserves" in order to prevent overharvesting and protect forested watersheds. In 1903, President Theodore Roosevelt created the first national wildlife refuge. These acts enabled the creation of a national park system, national forest system, and national wildlife refuge system that still stand as global models (pp. 193, 200). These developments reflected a new understanding that the continent's resources were exhaustible and required legal protection.

Land management policies continued through the 20th century, addressing soil conservation in the Dust Bowl years (p. 141) and wilderness preservation through the Wilderness Act of 1964 (p. 200), which sought to preserve pristine lands "untrammeled by man, where man himself is a visitor who does not remain."

The third wave responded to pollution

Further social changes in the 20th century gave rise to the third major period of U.S. environmental policy. In a more densely populated nation driven by technology, heavy industry, and intensive resource consumption, Americans found themselves better off economically but living with dirtier air, dirtier water, and more waste and toxic chemicals. During the 1960s and 1970s, several events triggered increased awareness of environmental problems and brought about a shift in public policy.

A landmark event was the 1962 publication of *Silent Spring*, a book by American scientist and writer Rachel Carson (**FIGURE 5.11**). *Silent Spring* awakened the public to the nega-

(a) Settlers in Custer County, Nebraska, circa 1860

FIGURE 5.10 ◄ The Homestead Act of 1862 allowed settlers **(a)** to claim, for a $16 fee, 65 ha (160 acres) of public land by living there for five years and cultivating the land or building a home. The General Mining Act of 1872 legalized and promoted mining **(b)** by private individuals on public land for just $5 per acre, subject to local customs, with no government oversight. Although the Timber Culture Act of 1873 promoted tree planting on settled lands, elsewhere the timber industry was allowed to clear-cut the nation's ancient trees **(c)** with little government policy to limit logging or encourage replanting or conservation.

(c) Loggers felling an old-growth tree, Washington

tive ecological and health effects of pesticides and industrial chemicals (p. 210). The book's title refers to Carson's warning that pesticides might kill so many birds that few would be left to sing in springtime.

Ohio's Cuyahoga River (**FIGURE 5.12**) also drew attention to pollution hazards. The Cuyahoga was so polluted with oil and industrial waste that the river actually caught fire near Cleveland a number of times in the 1950s and 1960s. This spectacle, coupled with an oil spill offshore from Santa Barbara, California, in 1969, moved the public to urge Congress and the president to do more to protect the environment. The first Earth Day in 1970 helped to galvanize public support for action to address pollution problems.

Today, largely because of environmental policies enacted since the 1960s, public health is better protected and the nation's air and water are considerably cleaner. All of us alive today owe a great deal to the dedicated people who designed policy to tackle pollution problems during this period.

(b) Nineteenth-century mining operation, Lynx Creek, Alaska

FIGURE 5.11 ▲ Scientist and writer Rachel Carson illuminated the problem of pollution from DDT and other pesticides in her 1962 book, *Silent Spring*.

NEPA gives citizens input into policy decisions

Besides Earth Day, two federal actions marked 1970 as the dawn of the modern era of environmental policy in the United States. On January 1, 1970, President Richard Nixon signed the **National Environmental Policy Act (NEPA)** into law. NEPA created an agency called the Council on Environmental Quality and required that an **environmental impact statement (EIS)** be prepared for any major federal action that might significantly affect environmental quality. An EIS is a

FIGURE 5.12 ▼ In a spectacular display of the need for better control over water pollution, Ohio's Cuyahoga River caught fire multiple times in the 1950s and 1960s. The Cuyahoga was so polluted with oil and industrial waste that the river would burn for days at a time.

report of results from studies that assess the potential environmental impacts that would likely result from development projects undertaken or funded by the federal government.

NEPA's effects have been far-reaching. The EIS process forces government agencies and businesses that contract with them to evaluate environmental impacts before proceeding with a new dam, highway, or construction project. Although the EIS process generally does not halt such projects, it can serve as an incentive to minimize environmental damage. NEPA also grants ordinary citizens input in the policy process by requiring that EISs be made publicly available and that policymakers solicit and consider public comment on them.

Creation of the EPA marked a shift in environmental policy

Six months after signing NEPA into law, Nixon issued an executive order calling for a new integrated approach to environmental policy. "The Government's environmentally related activities have grown up piecemeal over the years," the order stated. "The time has come to organize them rationally and systematically." Nixon's order moved elements of agencies regulating water quality, air pollution, solid waste, and other issues into the newly created **Environmental Protection Agency (EPA)**. The order charged the EPA with conducting and evaluating research, monitoring environmental quality, setting and enforcing standards for pollution levels, assisting the states in meeting standards and goals, and educating the public.

FAQ

Q: Isn't the EPA an advocate for the environment?

A: Like all administrative agencies, the EPA is part of the executive branch and operates in line with the policies of the presidential administration in power at the time. As such, the EPA under a conservative president may function very differently from the EPA under a liberal one. Indeed, sometimes the agency may impede environmental regulations! The EPA employs many career scientists who carry out careful scientific research and make scientifically informed policy recommendations. They advise administrators appointed by the president, however, and policy decisions are ultimately made by these politically appointed administrators.

Other prominent laws followed

Ongoing public demand for a cleaner environment during this period resulted in a number of key laws that remain fundamental to U.S. environmental policy (**FIGURE 5.13**). For river pollution problems like those of the Tijuana River, a crucial law has been the Clean Water Act of 1977. Prior to passage of federal laws such as the Clean Water Act, pollution problems were left largely to local and state governments or

Key Environmental Protection Laws, 1963–1980

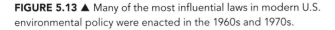

1963	Clean Air Act
1964	Wilderness Act
1965–1966	Federal Water Pollution Control Act, Solid Waste Disposal Act
1968	Wild and Scenic Rivers Act
1970	National Environmental Policy Act
1971–1972	Marine Mammal Protection Act, Federal Pesticide Act
1973	Endangered Species Act
1974	Safe Drinking Water Act
1976	Toxic Substances Control Act
1977–1978	Clean Water Act, Soil and Water Conservation Act
1980	CERCLA ("Superfund")

FIGURE 5.13 ▲ Many of the most influential laws in modern U.S. environmental policy were enacted in the 1960s and 1970s.

were addressed through lawsuits. The flaming waters of the Cuyahoga, however, indicated to many people that tough legislation was needed. Thanks to restrictions on pollutants by the Federal Water Pollution Control Acts of 1965 and 1972, and then the Clean Water Act, U.S. waterways finally began to recover. These laws regulated the discharge of wastes, especially from industry, into rivers and streams. The Clean Water Act also aimed to protect wildlife and establish a system for granting permits for the discharge of pollutants. Today thousands of federal, state, and local laws and regulations help protect health and environmental quality in the United States and abroad.

The social context for policy evolves over time

Historians suggest that major advances in environmental policy occurred in the 1960s and 1970s because (1) evidence of environmental problems became widely and readily apparent, (2) people could visualize policies to deal with the problems, and (3) the political climate was ripe, with a supportive public and leaders who were willing to act. In addition, photographs from the space program allowed humanity to see, for the first time ever, images of the Earth from space (see photos on pp. 1 and 418). It is hard for us today to comprehend the power of those images at the time, but they revolutionized many people's worldviews by making us aware of the finite nature of our planet.

By the 1990s, the political climate in the United States had changed. Although public support for the goals of environmental protection remained high, many citizens and policy experts began to feel that the legislative and regulatory means used to achieve these goals often imposed economic burdens on businesses or individuals. Increasingly, attempts were made to roll back or weaken environmental laws, culminating in an array of efforts by the George W. Bush administration and the Republican-controlled Congresses in power from 1994 through 2006.

As advocates of environmental protection watched their hard-won gains eroding, many began to propose new perspectives and strategies. In a provocative 2004 essay titled "The Death of Environmentalism," political consultants Michael Shellenberger and Ted Nordhaus argued that environmental advocates needed to stop simply offering technical policy fixes and instead needed to appeal to people's core values and articulate a positive, inspiring vision for the future. These suggestions opened a spirited discussion, and in 2008 Barack Obama embraced a similar approach in his presidential campaign.

As the United States' international leadership in environmental policy has waned in recent decades, other nations have enhanced their attention to environmental issues. The 1992 Earth Summit at Rio de Janeiro, Brazil, and the 2002 World Summit on Sustainable Development in Johannesburg, South Africa, were the largest diplomatic conferences ever held, unifying leaders from 200 nations around the idea of sustainable development (pp. 17, 414). Internationally, we are embarking on a new wave of environmental policy, one focused on sustainability and sustainable development. This approach aims to safeguard natural systems while raising living standards for the world's people. Moreover, the pressing issue of global climate change (Chapter 14) has come to dominate much of the world's discussion over environmental policy (**FIGURE 5.14**). As we continue to feel the social, economic, and ecological effects of environmental degradation,

FIGURE 5.14 ▼ College students and activists at the 2009 Power Shift event in Washington, D.C., urge U.S. leaders to enact policies to help bring the atmosphere's carbon dioxide concentration back down to 350 parts per million. This was one of several major events in recent years that expressed grassroots support for addressing global climate change through the political process.

environmental policy and the search for sustainable solutions will become central parts of governance and everyday life for all of us in the years ahead.

INTERNATIONAL ENVIRONMENTAL POLICY

Environmental systems pay no heed to political boundaries. For instance, most of the world's major rivers cross international borders. As a result, environmental problems like those along the Tijuana River are frequently international in scope. Because U.S. law has no authority in Mexico or any other nation outside the United States, international law is vital to solving transboundary problems.

International law includes customary law and conventional law

International law known as **customary law** arises from long-standing practices, or customs, held in common by most cultures. International law known as **conventional law** arises from conventions, or treaties, into which nations enter. One example is the Montreal Protocol, a 1987 accord among more than 160 nations to reduce the emission of airborne chemicals that thin the ozone layer (pp. 290–291). Another example is the Kyoto Protocol to reduce greenhouse gas emissions that contribute to global climate change (pp. 318–319). **TABLE 5.2** shows a selection of major environmental treaties.

The South Bay International Wastewater Treatment Plant that treats wastewater from the Tijuana River watershed was built as a result of a 1990 treaty between Mexico and the United States. Many social, economic, and environmental issues along the U.S.–Mexican border are influenced by the 1994 **North American Free Trade Agreement (NAFTA)** (see **THE SCIENCE BEHIND THE STORY,** pp. 106–107).

Several organizations shape international environmental policy

A number of international organizations act to influence the policy and behavior of nations by providing funding, applying political or economic pressure, and/or directing media attention.

The United Nations   Founded in 1945 and including representatives from all nations of the world, the **United Nations (U.N.)** seeks "to maintain international peace and security; to develop friendly relations among nations; to cooperate in solving international economic, social, cultural and humanitarian problems and in promoting respect for human rights and fundamental freedoms; and to be a centre for harmonizing the actions of nations in attaining these ends."

Headquartered in New York City, the U.N. plays an active role in international environmental policy by sponsoring conferences, coordinating treaties, and publishing research. An agency within it, the *United Nations Environment Programme (UNEP)*, promotes sustainability with research and outreach activities that provide information to policymakers and scientists throughout the world.

The World Bank   Established in 1944 and based in Washington, D.C., the **World Bank** is one of the largest sources of funding for economic development. This institution shapes environmental policy through its funding of dams, irrigation infrastructure, and other major projects. In fiscal year 2010, the World Bank provided $59 billion in loans and support for projects designed to benefit the poorest people in the poorest countries.

Despite its admirable mission, the World Bank has frequently been criticized for funding projects that cause more environmental problems than they solve, such as dams that

TABLE 5.2 Some Major International Environmental Treaties

Convention or Protocol	Year it came into force	Nations that have ratified it	U.S. status
Convention on International Trade in Endangered Species of Wild Fauna and Flora (CITES) (p. 177)	1975	175	Ratified
Ramsar Convention on Wetlands of International Importance	1975	159	Ratified
Protocol on Substances that Deplete the Ozone Layer (Montreal Protocol), of the Vienna Convention for the Protection of the Ozone Layer (pp. 290–291)	1989	196	Ratified
Basel Convention on the Control of Transboundary Movements of Hazardous Wastes and Their Disposal (pp. 15, 394)	1992	172	Signed but has not ratified
Convention on Biological Diversity (p. 177)	1993	168	Signed but has not ratified
Stockholm Convention on Persistent Organic Pollutants (p. 223)	2004	152	Signed but has not ratified
Kyoto Protocol, of the U.N. Framework Convention on Climate Change (pp. 318–319)	2005	184	Signed but has not ratified

flood valuable forests and farmlands in order to provide electricity. Providing for the needs of growing populations in poor nations while minimizing damage to the environmental systems on which people depend can be a tough balancing act. Environmental scientists today agree that sustainable development must be the guiding principle for such efforts.

The European Union   The **European Union (EU)** seeks to promote Europe's unity and its economic and social progress (including environmental protection) and to "assert Europe's role in the world." The EU can sign binding treaties on behalf of its 27 member nations and can enact regulations that have the same authority as national laws. The EU's European Environment Agency addresses waste management, noise pollution, water pollution, air pollution, habitat degradation, and natural hazards. The EU also seeks to remove trade barriers among member nations. It has classified some nations' environmental regulations as barriers to trade, arguing that the stricter environmental laws of some northern European nations limit the import and sale of environmentally harmful products from other member nations.

The World Trade Organization   Based in Geneva, Switzerland, the **World Trade Organization (WTO)** represents multinational corporations and promotes free trade by reducing obstacles to international commerce and enforcing fairness among nations in trading practices. Whereas the United Nations and the European Union have limited influence over nations' internal affairs, the WTO has authority to impose financial penalties on nations that do not comply with its directives. These penalties can affect environmental policy.

Like the EU, the WTO has interpreted some national environmental laws as unfair barriers to trade. For instance, in 1995 the U.S. EPA issued regulations requiring cleaner-burning gasoline in U.S. cities, following Congress's amendments of the Clean Air Act. Brazil and Venezuela filed a complaint with the WTO, saying the new rules discriminated against the petroleum they exported to the United States, which did not burn as cleanly. The WTO agreed, ruling that even though the South American gasoline posed a threat to human health in the United States, the EPA rules were an illegal trade barrier. The ruling forced the United States to weaken its regulations. Not surprisingly, critics have frequently charged that the WTO aggravates environmental problems.

International treaties to promote commerce allow industries and corporations to weaken environmental protection laws if they view them as barriers to trade. Chapter 11 of NAFTA allows an investor in one country to sue another country if its laws hinder the investor's ability to make profits. Canada's cattle industry demanded $300 million from U.S. taxpayers for banning Canadian beef after mad cow disease was found in it. A Canadian producer of a chemical in MTBE, a gasoline additive that California banned to protect public health, sued California for $1 billion in lost profits. A U.S. company forced Mexico to pay it $16 million after local residents refused to let it reopen a toxic waste dump in their neighborhood. Not all such courtroom challenges succeed, but NAFTA's Chapter 11 cases have discouraged states and nations from passing new environmental protection laws.

Nongovernmental organizations   A number of **nongovernmental organizations (NGOs)** have become international in scope and exert influence over international environmental policy. Some, such as the Nature Conservancy, focus on accomplishing conservation objectives on the ground (in its case, purchasing and managing land and habitat for rare species) without becoming politically involved. Other groups, such as Conservation International, the World Wide Fund for Nature, Greenpeace, and Population Connection, attempt to shape policy through research, education, lobbying, or protest.

Trade Barriers and Environmental Protection   If Nation A has stricter laws for environmental protection than Nation B, and if these laws restrict the ability of Nation B to export its goods to Nation A, then by the policy of the WTO and the EU, Nation A's environmental protection laws can be overruled in the name of free trade. Do you think this is right? What if Nation A is a wealthy industrialized country and Nation B is a poor developing country that needs every economic boost it can get?

SCIENCE AND THE ENVIRONMENTAL POLICY PROCESS

In constitutional democracies such as the United States, every person has a political voice and can make a difference. However, money wields influence, and some people and organizations are far more influential than others. We will explore some of these dynamics as we examine the main steps of the policymaking process and the role that science plays in policy.

Policy results from a stepwise process

Environmental policy involves a multiple-step process that requires initiative, dedication, and the support of many people (**FIGURE 5.15**).

❶ **Identify a problem**   The first step in the policy process is to identify an environmental problem. This requires curiosity, observation, record keeping, and an awareness of our relationship with the environment—so scientific inquiry and data collection play key roles. For example, assessing the contamination of San Diego- and Tijuana-area beaches required detecting the contamination, recognizing the ecological and health impacts of untreated wastewater, and understanding water flow dynamics among the beaches, the Pacific Ocean, and the Tijuana River watershed.

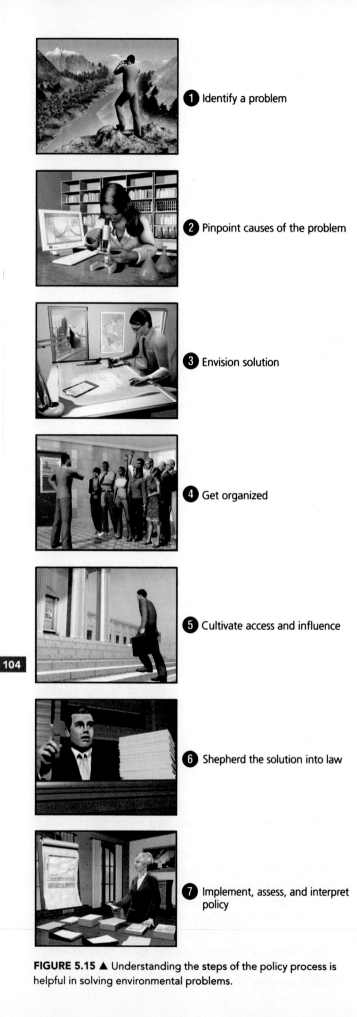

1 Identify a problem

2 Pinpoint causes of the problem

3 Envision solution

4 Get organized

5 Cultivate access and influence

6 Shepherd the solution into law

7 Implement, assess, and interpret policy

FIGURE 5.15 ▲ Understanding the steps of the policy process is helpful in solving environmental problems.

2 Pinpoint causes of the problem The next step in the policy process is to discover specific causes of the problem, and this often requires scientific research. A person seeking causes for the Tijuana River's pollution might notice that pollution intensified once U.S.-based companies began opening *maquiladoras* on the Mexican side of the border. Advocates of the *maquiladora* system argue that these factories provide much-needed jobs while keeping companies' costs down by paying Mexican workers low wages. Critics argue that the factories are waste-generating, water-guzzling polluters whose transboundary nature makes them difficult to regulate.

3 Envision a solution The better one can pinpoint causes of a problem, the more effectively one can envision solutions to it. Science plays a vital role here too, although solutions often rely on social or political action. In San Diego, citizen activists wanted Tijuana to enforce its own pollution laws more effectively—something that, once visualized, began to happen when San Diego city employees started training and working with their Mexican counterparts to keep hazardous waste out of the sewage treatment system.

4 Get organized When it comes to influencing policy, organizations are generally more effective than individuals. Yet small coalitions and even individual citizens who are motivated, informed, and organized can solve environmental problems. San Diego-area resident Lori Saldaña provides an example. Concerned about the Tijuana River's pollution, Saldaña reviewed plans for the international wastewater treatment plant that the U.S. government proposed to build. She concluded that it would merely shift pollution from the river to the ocean, where sewage would be released 5.6 km (3.5 mi) offshore. Working with her local Sierra Club chapter, Saldaña protested the plant's design and participated in the lawsuit that forced the EPA to conduct further studies and eventually implement design changes. After a decade of activism and work on a commission on border issues, Saldaña ran for the California State Assembly in 2004 and won, becoming the representative from California's 76th district.

5 Cultivate access and influence The next step in the policy process entails gaining access to policymakers who have the clout to enact change. People gain access and influence through lobbying and campaign contributions. Anyone can spend time or money trying to change an elected official's mind, but this is much more difficult for an ordinary citizen than for the professional lobbyists employed by businesses and organizations seeking a voice in politics. Supporting a candidate's reelection efforts with money is another way to make one's voice heard, and any individual can donate money to political campaigns.

Both of these methods were employed by a private-sector consortium seeking to contract with the U.S. government to build a wastewater treatment plant to supplement the existing international plant. Backers of this effort, the Bajagua Project, lobbied government officials and gave campaign contributions to California congressional representatives, whom they hoped would support their cause. Following years of such efforts, in 2006 the International Boundary and Water Commission agreed to support the Bajagua Project—although the commission changed its mind two years later.

❻ Shepherd the solution into law The next step is to prepare a bill, or draft law, that embodies the desired solutions. Anyone can draft a bill, but members of the House and Senate must introduce the bill and shepherd it from subcommittee through full committee and on to passage by the full Congress (**FIGURE 5.16**). If it passes through all of these steps and gains the president's signature, the bill becomes law, but it can die in countless fashions along the way.

❼ Implement, assess, and interpret policy Following a law's enactment, administrative agencies (such as the EPA) implement regulations. Policymakers evaluate the policy's successes and failures and may revise the policy as necessary. Moreover, courts interpret law in response to lawsuits, and much environmental policy has lived or died by judicial interpretation. As societal and environmental conditions change, the policy process may circle back to its first step as fresh problems are identified, and the process may begin anew.

Science plays a role in policy but can be misused

A nation's strength depends on its commitment to science, and this is why governments devote a portion of our taxes to fund scientific research. The more information a policymaker can glean from scientific research, the better policy he or she will be able to craft.

Unfortunately, sometimes policymakers allow ideology alone to determine policy on scientific matters. In 2004, the nonpartisan Union of Concerned Scientists released a statement that faulted the George W. Bush administration for ignoring scientific advice; manipulating scientific information for political ends; censoring, suppressing, and editing reports from government scientists; placing people who were unqualified or had clear conflicts of interest in positions of power; and misleading the public by misrepresenting scientific knowledge. More than 12,000 American scientists signed on to this statement. Many government scientists working on politically sensitive issues such as climate change or endangered species protection said they had found their work suppressed or discredited and their jobs threatened. Many chose self-censorship.

For these reasons, most American scientists greeted the election of Barack Obama with relief, as Obama spoke of "restoring scientific integrity to government" and ensuring "that scientific data [are] never distorted or concealed to serve a political agenda." Of course, either political party can politicize science, and as of 2011, the Union of Concerned Scientists was faulting the Obama administration for failing to fully live up to its promises.

Whenever taxpayer-funded science is suppressed or distorted for political ends—by the right or the left—we all lose. Abuses of power generally come to light only when brave government scientists risk their careers to alert the public and when journalists work hard to uncover and publicize these issues. We cannot simply take for granted that science will play a role in policy. Scientifically literate citizens of a democracy need to stay vigilant and help make sure our government representatives are making proper use of the tremendous scientific assets we have at our disposal.

APPROACHES TO ENVIRONMENTAL POLICY

When most people think of environmental policy, what comes to mind are major laws, such as the Clean Water Act, or government regulations, such as those specifying what a factory can and cannot dump into a river. However, environmental policy is far more diverse.

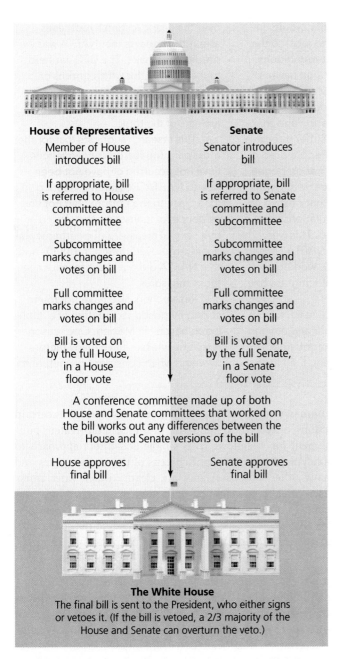

House of Representatives

Member of House introduces bill

If appropriate, bill is referred to House committee and subcommittee

Subcommittee marks changes and votes on bill

Full committee marks changes and votes on bill

Bill is voted on by the full House, in a House floor vote

Senate

Senator introduces bill

If appropriate, bill is referred to Senate committee and subcommittee

Subcommittee marks changes and votes on bill

Full committee marks changes and votes on bill

Bill is voted on by the full Senate, in a Senate floor vote

A conference committee made up of both House and Senate committees that worked on the bill works out any differences between the House and Senate versions of the bill

House approves final bill

Senate approves final bill

The White House
The final bill is sent to the President, who either signs or vetoes it. (If the bill is vetoed, a 2/3 majority of the House and Senate can overturn the veto.)

FIGURE 5.16 ▲ Before a bill becomes U.S. law, it must clear hurdles in both legislative bodies. If the bill passes the House and Senate, a conference committee works out differences between House and Senate versions before the bill is sent to the president. The president may then sign or veto the bill.

Workers assemble television circuit boards for Panasonic in a *maquiladora* in Tijuana, Mexico.

Assessing the Environmental Impacts of NAFTA

A number of international treaties address environmental issues (see Table 5.2). But treaties not aimed specifically at environmental concerns may still have major environmental consequences. The North American Free Trade Agreement (NAFTA) is one such treaty.

Mexico, the United States, and Canada signed NAFTA to promote free trade among them. NAFTA, which came into force in 1994, eliminated trade barriers such as tariffs on imports and exports. Nations erect tariffs in order to raise prices on foreign goods so that foreign industries won't drive domestic industries out of business—but if nations can agree to mutually eliminate tariffs, it makes goods cheaper for everyone.

Many people worried that NAFTA threatened to undermine protections for workers and the environment. For instance, if a nation's regulations to protect environmental quality or worker safety are viewed as a barrier to trade or investment, under NAFTA those regulations can potentially be overturned.

Moreover, many people felt that industries, motivated to decrease

costs and increase profits, would move their factories (and jobs) to the nation with the weakest regulations. This could create "pollution havens." Many people thought Mexico would be overrun by *maquiladoras* seeking to profit from lax regulation and that Mexico would thereby suffer intensive pollution (**see figure**). People predicted that once such a migration began, this could lead to a "race to the bottom" whereby all three nations would begin gutting their regulations in an attempt to lure business. This was part of the reason that NAFTA caused so many blue-collar U.S. workers to fear that their jobs would migrate to Mexico.

In response to these fears, two side agreements for labor and environmental concerns were negotiated. The environmental agreement, the North

American Agreement on Environmental Cooperation, set up a tri-national Commission on Environmental Cooperation (CEC). The CEC has monitored NAFTA's effects on the environment over the years, testing hypotheses about the impacts that NAFTA was predicted to have. The CEC has held four symposia, for which dozens of researchers have published over 50 research papers analyzing different aspects of the topic.

This research suggests that for the most part, the feared consequences have not occurred or have not been due to NAFTA. Researchers testing for a "race to the bottom" did not find evidence for it. Instead, they found that several measures of environmental quality improved during the 1990s and that NAFTA did not seem to influence these measures.

Researchers also have not found strong evidence for creation of a "pollution haven" in Mexico. One reason is that businesses have many factors to weigh when considering whether to

We follow three types of policy approaches

Today's environmental policy can follow a variety of strategies within three major approaches (**FIGURE 5.17**).

Lawsuits in the courts Prior to the legislative push of recent decades, most environmental policy questions in the United States were addressed with lawsuits in the courts. Individuals suffering external costs from pollution would sue polluters, one case at a time. The courts sometimes punished polluters by ordering them to stop their operations or pay damages to the affected individuals. However, as industrialization proceeded and population grew denser, pollution became harder to avoid, and judges were reluctant to hinder industry. People began to view legislation and regulation as more effective means of protecting public health and safety.

Command-and-control policy Most environmental laws of recent decades, and most regulations enforced by agencies today, use a **command-and-control** approach. In the

command-and-control approach, an agency prohibits certain actions, or sets rules, standards, or limits, and threatens punishment for violations. This simple and direct approach to policymaking has brought citizens of the United States and many other nations cleaner air, cleaner water, safer workplaces, healthier neighborhoods, and many other improvements in quality of life. The relatively safe, healthy, comfortable lives most of us enjoy today owe much to the command-and-control environmental policy of the past few decades.

Economic policy tools Despite the successes of command-and-control policy, many people have grown disenchanted with the top-down, sometimes heavy-handed, nature of an approach that dictates particular solutions to problems. As a result, political scientists, economists, and policymakers today are exploring alternative approaches that aim to channel the innovation and economic efficiency of market capitalism in directions that benefit the public. Such economic policy tools use financial incentives to promote desired outcomes, discourage undesired outcomes, and encourage

Maquiladoras in the border areas of Mexico flourished in the years after NAFTA, employing many people but creating substantial pollution. Perhaps the worst pollution occurred at the site shown here. At this abandoned lead recycling plant, *Metales y Derivados*, the U.S. owner left 6,600 tons of hazardous waste amid a working-class Tijuana neighborhood.

to protect threatened species and habitats.

Moreover, consumer demand in the United States and Canada for sustainably produced products helped encourage improvement in Mexico. Many Mexican coffee farmers converted to sustainable plantations, and researchers found that *maquiladoras* selling products north of the border made more environmental improvements than those selling only in Mexico.

However, despite all this, environmental impacts in Mexico grew worse after NAFTA. For instance, air and water pollution from *maquiladoras* in border areas such as Tijuana increased greatly. Yet, researchers determined that this was due not to a pollution haven or race to the bottom, but to accelerated economic growth in Mexico. Consumption and pollution from economic growth simply outpaced the country's ability to enhance regulation and environmental protection.

Overall, CEC-sponsored research in the years since NAFTA has shown that trade liberalization can lead to environmental improvements, but only if policymakers pay close attention to trends and are ready to make adjustments. Opportunities for creating win-win policies that benefit both trade and environmental quality exist, researchers say, and are ready to be seized.

relocate, and environmental regulations are merely one of these. Ironically, the one clear case of a pollution haven occurred not in Mexico, but in Canada! Exports of hazardous waste (mostly from steel and chemical factories) from the United States to Canada more than quadrupled soon after NAFTA went into effect. Disposal was cheaper in Canada, with fewer regulations and liability concerns. Canada responded by tightening its regulations.

CEC-sponsored researchers also tested hypotheses that NAFTA might enhance sustainability by facilitating the spread of environmentally superior technology, products, and approaches among nations. Researchers found plenty of examples. Canada began using the EPA's Energy Star program (p. 344). Mexico banned the pesticide DDT. Chemical manufacturers set up a transnational program for reducing hazards. And the three nations began conferring on how

CHAPTER 5 Environmental Economics and Environmental Policy

107

private entities competing in a marketplace to innovate and produce new or better solutions at lower cost.

Each of these three major approaches has strengths and weaknesses, and each is best suited to different conditions. The approaches may also be used together. For instance, government regulation is often needed to frame market-based efforts, and citizens can use the courts to ensure that regulations are enforced. Let's now explore several types of economic policy tools.

Green taxes discourage unsustainable activities

In taxation, money passes from private parties to the government, which reapportions it in services to benefit the public. Taxing undesirable activities helps to "internalize" external costs by making them part of the cost of doing business. Taxes on environmentally harmful activities and products are called **green taxes**. When a business pays a green tax, it is essentially reimbursing the public for environmental damage it causes.

Under green taxation, a firm owning a factory that pollutes a waterway would pay taxes on the amount of pollution it discharges—the more pollution the higher the tax payment. This gives firms a financial incentive to reduce pollution while allowing them the freedom to decide how to do so. One polluter might choose to invest in technologies to reduce its pollution if doing so is less costly than paying the taxes. Another polluter might choose to pay the taxes instead—funds the government might then apply toward mitigating pollution in some other way.

Green taxes have yet to gain widespread support in the United States, although similar "sin taxes" on cigarettes and alcohol are tools of U.S. social policy. Taxes on pollution have been widely instituted in Europe, where many nations have adopted the **polluter-pays principle**, which specifies that the party that pollutes should be held responsible for covering the costs of its impacts. Today there is debate about whether we should implement *carbon taxes*—taxes on gasoline, coal-based electricity, and fossil-fuel-intensive products—in order to fight climate change (p. 321).

PROBLEM: Pollution from factory harms people's health

FIGURE 5.17 ▲ For any given environmental problem, such as pollution from a factory, we may consider three major types of policy approaches: ❶ seeking compensation through lawsuits, ❷ limiting pollution through command-and-control legislation and regulation, and ❸ reducing pollution through market-based or other economic strategies.

SOLUTIONS: Three policy approaches

❶ Sue factory in court seeking damages and/or injunction.

EPA

❷ Government regulation restricts emissions allowed.

❸ Market-based approaches: factories that pollute less outcompete polluting factory through permit-trading, avoiding green taxes, collecting subsidies, or selling ecolabeled products. Polluting factory must find ways to cut emissions to survive in marketplace.

Green taxation provides incentive for industry to lower emissions not merely to a level specified in a regulation, but to still lower levels. However, green taxes do have drawbacks. One is that businesses will most likely pass on their tax expenses to consumers.

Subsidies promote certain activities

Another type of economic policy tool is the **subsidy**, a government giveaway of money or resources that is intended to encourage a particular industry or activity. Subsidies take many forms, and one is the *tax break*. Relieving the tax burden on an industry, firm, or individual assists it by reducing its expenses.

Subsidies can be used to promote environmentally sustainable activities, but all too often they are used to prop up unsustainable ones. In the United States, subsidies for timber extraction (pp. 193–194), grazing (p. 144), and mineral extraction (pp. 241–242) on public lands all benefit private parties while often degrading publicly held resources. Fossil fuel industries have benefited as well. From 2002 to 2008, the U.S. government gave $72 billion of its citizens' money—$240 per person—to fossil fuel corporations, while spending only $29 billion on renewable energy efforts. About $54 billion of the fossil fuel subsidies were in the form of tax breaks.

In total, the world's governments spend roughly $1.45 *trillion* each year on subsidies judged to be harmful to the environment and to the economy, according to environmental scientists Norman Myers and Jennifer Kent—an amount larger than the economies of all but five nations. The average U.S. taxpayer pays $2,000 per year in environmentally harmful subsidies, plus $2,000 more through increased prices for goods and through degradation of ecosystem services, Myers and Kent estimate.

Permit trading can save money and produce results

In the innovative market-based approach known as **permit trading**, the government creates a market in permits for an environmentally harmful activity, and companies, utilities, or industries are allowed to buy, sell, or trade rights to conduct the activity. For instance, to decrease emissions of air pollutants, a government might grant emissions permits and set up an **emissions trading** system. In a **cap-and-trade** emissions trading system, the government first determines the overall amount of pollution it will accept (i.e., it caps the total amount of pollution allowed) and then issues permits to polluters that allow them each to emit a certain fraction of that amount. Polluters may exchange these permits with other polluters, and each year the government may reduce the amount of overall emissions allowed (Figure 14.23, p. 321).

Suppose, for example, you are a plant owner with permits to release 10 units of pollution, but you find that you can become more efficient and release only 5 units instead. You then have a surplus of permits, which might be very valuable to some other plant owner who is having trouble reducing pollution or who wants to expand production. In such a case, you can sell your extra permits. Doing so generates income for you and meets the needs of the other plant, while the total amount of pollution does not increase. By providing companies an economic incentive to find ways to reduce emissions, permit trading can reduce expenses for both industry and the public relative to a conventional regulatory system.

A cap-and-trade system for sulfur dioxide has been in place in the United States since 1995, established by amendments to the Clean Air Act (p. 283) that mandated lower emissions of this air pollutant, which contributes to acid deposition (pp. 291–294). Since then, sulfur dioxide emissions from sources in the program have declined by 64% (**FIGURE 5.18**), sulfate deposition has been reduced, and air quality and visibility have improved. The cuts were attained at much less cost than was predicted, with no apparent effect on electricity supply or economic growth. Savings from the permit trading system are estimated at billions of dollars per year, and the EPA calculates that the program's benefits outweigh its costs by about 40 to 1.

Currently, European nations are operating a market in carbon emissions in an effort to address climate change (p. 321). Some U.S. industries take part in carbon trading through the Chicago Climate Exchange, and emissions trading programs are being established by coalitions of U.S. states (p. 321).

 THE ISSUES

A License to Pollute? Some environmental advocates oppose emissions trading because they view it as giving polluters "a license to pollute." How do you feel about emissions trading as a means of reducing air pollution? Would you favor command-and-control regulation instead? What advantages and disadvantages do you see in each approach?

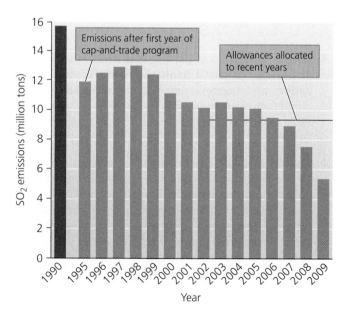

FIGURE 5.18 ▲ Emissions of sulfur dioxide from sources participating in the emissions trading program mandated by the 1990 Clean Air Act amendments have fallen 64% since 1990. As of 2009, emissions throughout the United States had dropped well below the amount allocated in permits (black line). Data from U.S. Environmental Protection Agency.

FIGURE 5.19 ▲ Ecolabeling gives consumers information on products with low environmental impact and enables consumers to encourage more sustainable business practices through their purchasing decisions. Organic juices are just one example of ecolabeled products that have become widely available in the marketplace.

Ecolabeling empowers consumers

Another strategy that uses the marketplace to counteract market failure allows consumers to play the key role. When manufacturers designate on their labels how their products were grown, harvested, or manufactured, this approach—called **ecolabeling**—tells consumers which brands use environmentally benign processes (**FIGURE 5.19**). By preferentially buying ecolabeled products, consumers provide businesses a powerful incentive to switch to more sustainable processes. One early example of this was labeling cans of tuna as "dolphin-safe," indicating that the methods used to catch the tuna avoid the accidental capture of dolphins. Other common examples include labeling recycled paper (p. 387), organic foods (pp. 154–157), and lumber harvested through sustainable forestry (pp. 197–199).

Market incentives also operate at the local level

You may have already taken part in transactions involving financial incentives as policy tools. Many municipalities charge residents for waste disposal according to the amount of waste they generate. Other cities place taxes or disposal fees on items whose safe disposal is costly, such as tires and motor oil. Still others give rebates to residents who buy water-efficient appliances, because the rebates cost the city less than upgrading its wastewater treatment system. Likewise, power companies sometimes offer discounts to customers who buy high-efficiency lightbulbs and appliances, because doing so is cheaper for the utilities than expanding the generating capacity of their plants.

The creative use of economic policy tools is growing at all levels, while command-and-control regulation and legal action in the courts continue to play vital roles in environmental policymaking. As a result, we have a variety of effective policy strategies we can consider as we seek sustainable solutions to our society's challenges.

➤ CONCLUSION

Environmental policymaking is a problem-solving pursuit that makes use of science, ethics, and economics and that requires an astute understanding of the political process. Conventional command-and-control approaches of legislation and regulation remain the most common policy approaches, but innovative economic policy tools are increasingly being developed. These tools, as well as efforts of environmental and ecological economists to quantify the value of ecosystem services, are helping to bring economic approaches to bear on environmental protection and resource conservation. Equating economic well-being with economic growth, as economists and policymakers traditionally have, suggests that economic well-being entails a trade-off with environmental quality. However, as modern scientific, economic, and political efforts proceed, they are helping us to see that safeguarding environmental quality helps to promote our quality of life.

TESTING YOUR COMPREHENSION

1. Name two key ways in which human economies are linked to the natural environment.

2. Describe four ways in which critics maintain that neoclassical economic approaches can worsen environmental problems.

3. Compare and contrast the views of neoclassical economists, environmental economists, and ecological economists, particularly regarding the issue of economic growth.

4. What are ecosystem services? Give several examples. Describe how some economists have tried to assign monetary worth to nonmarket values and ecosystem services.

5. Describe and critique three common justifications for environmental policy. Explain the concept of external costs, and state why it is relevant to environmental policy.

6. Summarize the differences among the first, second, and third waves of environmental policy in U.S. history. Describe two current priorities in international environmental policy.

7. What did the National Environmental Policy Act accomplish? Briefly describe the origin and duties of the U.S. Environmental Protection Agency.

8. List the steps of the environmental policy process, from identification of a problem through the implementation, assessment, and interpretation of policy.

9. Compare and contrast the three major approaches to environmental policy: lawsuits, command-and-control policy, and economic policy tools. Describe an advantage and disadvantage of each.

10. Differentiate among a green tax, a subsidy, a tax break, and an emissions permit.

SEEKING SOLUTIONS

1. Do you think that a steady-state economy is a practical alternative to our current approach that prioritizes economic growth? Why or why not?

2. Do you think we should attempt to quantify and assign market values to ecosystem services and other entities that have only nonmarket values? Why or why not?

3. Reflect on the causes for the transitions in U.S. history from one type of environmental policy to another. Now peer into the future, and think about how life might be different in 25, 50, or 100 years. What would you speculate about the environmental policy of the future? What issues might it address? Do you predict we will have more or less environmental policy?

4. Compare the roles of the United Nations, the European Union, the World Bank, the World Trade Organization, and nongovernmental organizations. If you could gain the support of just one of these institutions for a policy you favored, which would you choose? Why?

5. **THINK IT THROUGH** You have just been elected to the U.S. House of Representatives. People in your district suffer an unusual amount of water pollution from untreated municipal wastewater, chemical discharges from factories, and oil spillage from commercial and recreational boats. What policy approaches would you choose to pursue to search for solutions to these problems? Give reasons for your choices.

CALCULATING ECOLOGICAL FOOTPRINTS

The League of Conservation Voters is a nongovernmental organization that tracks the voting records of U.S. policymakers on environmental issues and provides information to citizens. It is an advocacy organization, and you may or may not agree with its positions, but its website provides an easy way to track the votes of your Congressional representatives on bills dealing with environmental policy. Go to the LCV's website at http://www.lcv.org. Under "Scorecard," click on the link for the most recent year. Find the representatives for your state, and click on their names to explore their voting records. The Y's and N's show how they voted on key bills. To read descriptions of the bills, click on the links for those bills.

Voting Records of Your Congressional Representatives

	Name of Congressperson	How he/she voted on your bill of interest
A member of the U.S. House of Representatives from your state		
A U.S. senator from your state		
A member of the Congressional leadership		

1. After exploring the voting records of your state's representatives in the House, find one vote that you disagree with. In the first row of the table, write the representative's name and describe his or her vote on your bill of interest. What do you disagree with, and why?

2. Next, explore the voting records of your state's two U.S. senators, and find one vote that you disagree with. In the second row of the table, write the senator's name and describe his or her vote on your bill of interest. What do you disagree with, and why?

3. For one of these issues, outline a letter that you would write to your Congressional representative explaining why you feel he or she should support your position on the issue.

4. The League of Conservation Voters advocates particular positions. Did you find cases in which you would take a position that differs from theirs? If so, what did you disagree with, and why?

5. Select one of the bills you see described, and explain how you think passage of the bill might influence the ecological footprint of people in your state.

Go to **www.masteringenvironmentalscience.com** for homework assignments, practice quizzes, Pearson eText, and more.

Human Population

Upon completing this chapter, you will be able to:

➤ Perceive the scope of human population growth

➤ Evaluate how human population, affluence, and technology affect the environment

➤ Explain and apply the fundamentals of demography

➤ Outline and assess the concept of demographic transition

➤ Describe how wealth and poverty, the status of women, and family-planning affect population growth

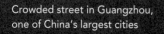

Crowded street in Guangzhou,
one of China's largest cities

China's One-Child Policy

"Population growth is analogous to a plague of locusts. What we have on this earth today is a plague of people."
—TED TURNER, Media Magnate and Supporter of the United Nations Population Fund

"As you improve health in a society, population growth goes down. . . . Before I learned about it, I thought it was paradoxical."
—BILL GATES, CHAIR, Microsoft Corporation

The People's Republic of China is the world's most populous nation, home to one-fifth of the 7 billion people living on Earth at the start of 2012.

When Mao Zedong founded the country's current regime 63 years earlier, roughly 540 million people lived in a mostly rural, war-torn, impoverished nation. Mao believed population growth was desirable, and under his leadership China grew and changed. By 1970, improvements in food production, food distribution, and public health allowed China's population to swell to 790 million people. At that time, the average Chinese woman gave birth to 5.8 children in her lifetime.

However, the country's burgeoning population and its industrial and agricultural development were eroding the nation's soils, depleting its water, leveling its forests, and polluting its air. Chinese leaders realized that the nation might not be able to feed its people if their numbers grew much larger. They saw that continued population growth could exhaust resources and threaten the stability and progress of Chinese society. The government decided in 1970 to institute a population control program that prohibited most Chinese couples from having more than one child.

The program began with education and outreach efforts encouraging people to marry later and have fewer children. Along with these efforts, the Chinese government increased the accessibility of contraceptives and abortion. By 1975, China's annual population growth rate had dropped from 2.8% to 1.8%.

To further decrease birth rates, in 1979 the government took the more drastic step of instituting a system of rewards and punishments to enforce a one-child limit. One-child families received better access to schools, medical care, housing, and government jobs. Mothers with only one child were given longer maternity leaves. Families with more than one child, meanwhile, were subjected to monetary fines, employment discrimination, and social scorn and ridicule. In some cases, the fines exceeded half of a couple's annual income.

Population growth rates dropped still further, but public resistance to the policy was simmering. Beginning in 1984, the one-child policy was loosened, strengthened, and then loosened again, as government leaders sought to maximize population control while minimizing public opposition. Today the one-child program is less strict than in past years and applies mostly to families in urban areas. Many farmers and ethnic minorities in rural areas are exempted, because success on the farm often depends on having multiple children.

In enforcing its policies, China has been conducting one of the largest and most controversial social experiments in history. In purely quantitative terms, the experiment has been a success: The nation's growth rate is now down to 0.5%, making it easier for the country to deal with its many social, economic, and environmental challenges.

However, the one-child policy has also produced unintended consequences. Traditionally, Chinese culture has valued sons because they carry on the family name, assist with farm labor in rural areas, and care for aging parents. Daughters, in contrast, will most likely marry and leave their parents, as the culture dictates. As a result, they cannot provide the same benefits to their parents as will sons. Thus, faced with being limited to just one child, many Chinese couples prefer a son to a daughter. Tragically, this has led to selective abortion, killing of female infants, an unbalanced sex ratio, and a black-market trade in teenaged girls for young men who cannot find wives.

Further problems are expected in the near future, including an aging population and a shrinking workforce. Moreover, China's policies have elicited intense criticism worldwide from people who oppose government intrusion into personal reproductive choices.

As other nations become more and more crowded, might their governments also feel forced to turn to drastic policies that restrict individual freedoms? In this chapter, we examine human population dynamics worldwide, consider their causes, and assess their consequences for the environment and our society. ∎

FAQ

Q: How big is a billion?

A: Human beings have trouble conceptualizing huge numbers. As a result, we often fail to recognize the true magnitude of a number such as 7 billion. Although we know that a billion is bigger than a million, we tend to view both numbers as impossibly large and therefore similar in size. For example, guess (without calculating) how long it would take a banker to count out $1 million if she did it at a rate of a dollar a second for 8 hours a day, 7 days a week. Now guess how long it would take to count $1 billion at the same rate. The difference between your estimate and the answer may surprise you. Counting $1 million would take a mere 35 days, whereas counting $1 billion would take 95 years! Living 1 million seconds takes only 12 days, while living for 1 billion seconds requires more than 31 years. You couldn't live for 7 billion seconds if you tried, because that would take 221 years. Examples like these can help us appreciate the "b" in "billion."

OUR WORLD AT SEVEN BILLION

While China works to slow its population growth, numbers of people continue to rise in most nations. Most population growth is occurring in poverty-stricken developing nations that are ill-equipped to handle it. India (**FIGURE 6.1**) is on course to surpass China as the world's most populous nation in coming decades. Although the *rate* of human population growth is slowing, our absolute numbers are still increasing. Thus, our growth rate remains positive, and in 2011, our global population will surpass 7 billion.

FIGURE 6.1 ▲ Population growth is driving our global numbers to 7 billion and beyond. India is on course to overtake China and become the world's most populous nation. In this photo, Indian women wait in line for immunizations for their babies.

The human population is growing rapidly

Our global population grows by over 80 million people each year, a number equivalent to the combined populations of California, Texas, and New York. We add 2.6 people to the planet every second and could fill two large college football stadiums (capacity 110,000 each) with the number of humans added to the population every day. It took until after 1800, virtually all of human history, for our population to reach 1 billion. Yet we've added 6 billion people to the global population since then, with the last three installments of 1 billion people taking only 12 years each (**FIGURE 6.2**) No previous generation has ever lived amid so many other people.

What accounts for our unprecedented growth? Exponential growth—the increase in a quantity by a fixed percentage per unit time—accelerates increase in population size over time, just as compound interest accrues in a savings account (p. 55). The reason, you will recall, is that a fixed percentage of a small number makes for a small increase, but that same percentage of a large number produces a large increase. Thus, even if the growth *rate* remains steady, population *size* will increase by greater increments with each successive generation.

For much of the 20th century, the growth rate of the human population rose from year to year. This rate peaked at 2.1% during the 1960s and has declined to 1.2% since then. Although 1.2% may sound small, exponential growth endows small numbers with large consequences. A hypothetical population starting with one man and one woman that grows at 1.2% gives rise to a population of 2,939 after only 40 generations and 112,695 after 60 generations. Also, although the

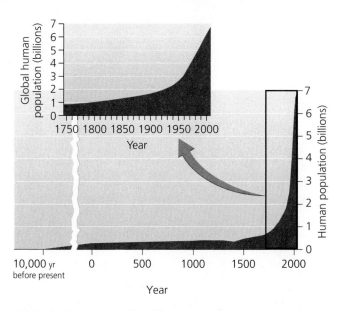

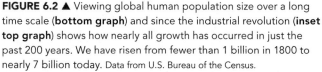

FIGURE 6.2 ▲ Viewing global human population size over a long time scale (**bottom graph**) and since the industrial revolution (**inset top graph**) shows how nearly all growth has occurred in just the past 200 years. We have risen from fewer than 1 billion in 1800 to nearly 7 billion today. Data from U.S. Bureau of the Census.

global growth rate is 1.2%, rates vary widely from region to region and are highest in developing nations (**FIGURE 6.3**).

At a 1.2% annual growth rate, a population doubles in size in only 58 years. We can roughly estimate doubling times with a handy rule of thumb. Just take the number 70 (which is 100 times 0.7, the natural logarithm of 2) and divide it by the annual percentage growth rate: 70/1.2 = 58.3. China's current growth rate of 0.5% means it would take roughly 140 years (70/0.5 = 140) for its population to double. Had China not instituted its one-child policy and had its growth rate remained unchecked at 2.8%, it would have taken only 25 years (70/2.8 = 25) to double in size.

To help you visualize just how fast our world population has grown—and will continue to grow in your lifetime—take a look at the **ENVISIONIT** feature on p. 116. In other chapters you will similarly find an EnvisionIt page that uses visual images to bring to life important concepts in environmental science.

Is there a limit to human population growth?

Our spectacular growth in numbers has resulted largely from technological innovations, improved sanitation, better medical care, increased agricultural output, and other factors that have brought down death rates and infant mortality rates. Birth rates have not declined as much, so births have outpaced deaths for many years now, leading to population growth. But can the human population continue to grow indefinitely?

Environmental factors set limits on the growth of populations (p. 58), but environmental scientists who have tried to pin a number to the human carrying capacity (p. 59) have come up with wildly differing estimates. The most rigorous estimates range from 1–2 billion people living prosperously in a healthy environment to 33 billion people living in extreme poverty in a degraded world of intensive cultivation without natural areas.

The difficulty in estimating the carrying capacity for humans is that we are a particularly successful species, and we have repeatedly overcome predicted limits on growth by developing new technologies and ways of securing resources. For example, British economist Thomas Malthus (1766–1834) argued in his influential work *An Essay on the Principle of Population* (1798) that if society did not reduce its birth rate, then rising death rates would reduce the population through war, disease, and starvation. Although his contention was reasonable at the time, agricultural improvements in the 19th century increased food supplies, and his prediction did not come to pass. Similarly, biologist Paul Ehrlich predicted in his 1968 book *The Population Bomb* that human population growth would soon outpace food production

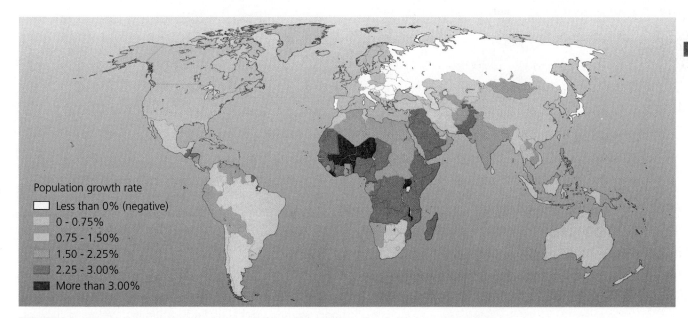

FIGURE 6.3 ▲ Population growth rates vary greatly from place to place. Population is growing fastest in poorer nations of the tropics and subtropics but is now beginning to decrease in some northern industrialized nations. Shown are natural rates of population change (p. 122) as of 2010. Data from Population Reference Bureau, 2010. *2010 World population data sheet.* By permission.

1900

Earth has become more crowded as our global population has grown.

1950

From 1900 to 1950 numbers grew by 50%.

2000

Then our numbers exploded: In 2000 our population was 2.4 times greater than in 1950, an increase of 140%.

2050

For every two people alive in 2000, there will be three in 2050. Will enough natural resources remain to provide our children the standard of living we enjoy today?

You Can Make a Difference

➤ Support women's rights and education in developing nations.

➤ Even as population grows, reducing your resource consumption will lessen your ecological footprint.

and unleash massive famine and conflict in the latter 20th century. However, thanks to the way the "Green Revolution" (p. 136) increased food production in developing regions in the decades after his book, Ehrlich's dire forecasts did not fully materialize.

Does this mean human innovation will always find a way to support our population? Under the view that many economists hold, resource depletion due to population increase is not a problem if new resources can be found or created to replace depleted ones (p. 91). In contrast, environmental scientists recognize that few resources are actually created by people and that not all resources can be replaced once they are depleted. For example, once a species is extinct, we cannot replicate its exact function in an ecosystem. As the human population continues to climb, we may yet continue to find ways to raise our carrying capacity. But given our knowledge of population ecology and logistic growth (pp. 58–59), we have no reason to presume that human numbers can increase forever.

But is human population growth really a problem? Even if resource substitution could hypothetically enable our population to grow indefinitely, could we maintain the *quality* of life we desire for ourselves and our descendants? Unless the availability and quality of all resources keeps pace forever with population growth, the average person in the future will have less space in which to live, less food to eat, and less material wealth than the average person does today. Thus population growth is indeed a problem if it depletes resources, stresses social systems, and degrades the natural environment, such that our quality of life declines (**FIGURE 6.4**).

Population is one of several factors that affect the environment

One widely used formula gives us a handy way to think about population and other factors that affect environmental quality. Nicknamed the **IPAT model**, it is a variation of a formula proposed in 1974 by Paul Ehrlich and John Holdren. The IPAT model represents how our total impact (I) on the environment results from the interaction among population (P), affluence (A), and technology (T):

$$I = P \times A \times T$$

Increased population intensifies impact on the environment as more individuals take up space, use natural resources, and gen-

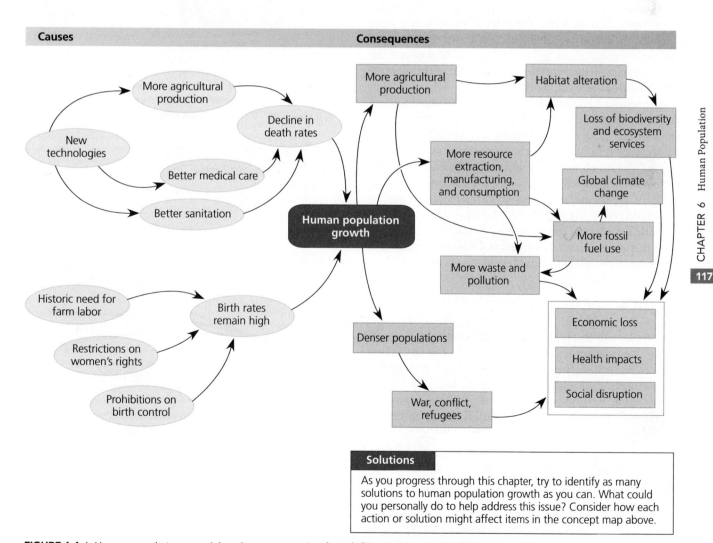

FIGURE 6.4 ▲ Human population growth has diverse causes (**ovals on left**) and consequences (**boxes on right**), as we shall see throughout this chapter. The consequences of rapid population growth are generally negative for society and the environment. Arrows in this concept map lead from causes to consequences. Note that items grouped within outlined boxes do not necessarily share any special relationship; the outlined boxes are merely to streamline the figure.

erate waste. Increased affluence magnifies environmental impact through the greater per capita resource consumption that generally has accompanied enhanced wealth. Technology that enhances our abilities to exploit minerals, fossil fuels, old-growth forests, or ocean fisheries generally increases impact, but technology to reduce smokestack emissions, harness renewable energy, or improve manufacturing efficiency can decrease impact.

We might also add a fourth factor, sensitivity (S), to the equation to denote how sensitive a given environment is to human pressures. For instance, the arid lands of western China are more sensitive to human disturbance than the moist regions of southeastern China. Plants grow more slowly in the arid west, making the land more vulnerable to deforestation and soil degradation. Thus, adding an additional person to western China has more environmental impact than adding one to southeastern China. We could refine the IPAT equation further by adding terms for the effects of social institutions such as education, laws and their enforcement, stable and cohesive societies, and ethical standards that promote environmental well-being. Such factors all affect how population, affluence, and technology translate into environmental impact.

Modern-day China shows how all elements of the IPAT formula can combine to cause tremendous environmental impact in little time. Although China boasts the world's fastest growing economy, the country is battling unprecedented environmental challenges brought about by rapid economic development. Intensive agriculture has expanded westward out of the country's moist rice-growing areas, causing farmland to erode and blow away, much like the Dust Bowl tragedy that befell the U.S. heartland in the 1930s (p. 141). China has overpumped aquifers and has drawn so much water for irrigation from the Yellow River that the once-mighty waterway now dries up in many stretches (p. 262). The nation now faces urban air pollution and massive traffic jams from rapidly rising numbers of automobiles. In August 2010, for example, a 100-km (60-mi) traffic jam formed on the outskirts of Beijing and persisted for more than 10 days! As the world's developing countries try to attain the material prosperity that industrialized nations enjoy, China is a window into what much of the rest of the world could soon become.

DEMOGRAPHY

It is a fallacy to think of people as being somehow outside nature. We exist within our environment as one species out of many. As such, all the principles of population ecology that apply to toads, frogs, and passenger pigeons (Chapter 3) apply to humans as well. The application of principles from population ecology to the study of statistical change in human populations is the focus of **demography**.

Demography is the study of human population

Demographers study population size, density, distribution, age structure, sex ratio, and rates of birth, death, immigration, and emigration of people, just as population ecologists study these characteristics in other organisms. Each of these characteristics is useful for predicting population dynamics and environmental impacts.

Population size   Our global human population of more than 7 billion is spread among over 200 nations with populations ranging up to China's 1.34 billion, India's 1.21 billion, and the 312 million of the United States (**FIGURE 6.5**). The United Nations Population Division estimates that by the year 2050, the global population will surpass 9 billion (**FIGURE 6.6**). However, population size alone—the absolute number of individuals—doesn't tell the whole story. Rather, a population's environmental impact depends on its density, distribution, and composition (as well as on affluence, technology, and other factors outlined earlier).

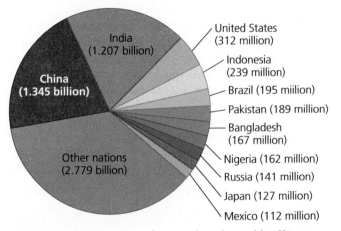

FIGURE 6.5 ▲ Almost one in five people in the world is Chinese, and more than one of every six live in India. Three of every five people live in one of the 11 nations that have populations above 100 million. Data are projected for mid-2011, based on Population Reference Bureau, 2010. *2010 world population data sheet.*

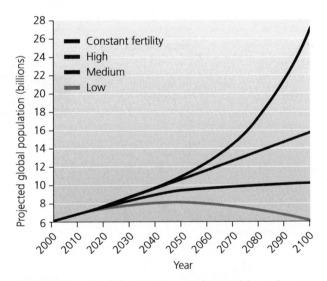

FIGURE 6.6 ▲ The United Nations predicts world population growth. In the latest projection (made in 2010), population is estimated to reach 11.0 billion in the year 2050 if fertility rates remain constant at 2005–2010 levels (top line in graph). However, U.N. demographers expect fertility rates to continue falling, so they arrived at a best guess (medium scenario) of 9.3 billion for 2050. In the high scenario, if women on average have 0.5 child more than in the medium scenario, population will reach 10.6 billion in 2050. In the low scenario, if women have 0.5 child fewer than in the medium scenario, the world will contain 8.1 billion people in 2050. Adapted by permission from Population Division of the Department of Economic and Social Affairs of the United Nations Secretariat, 2011. *World population prospects: The 2010 revision,* http://esa.un.org/unpp, Fig 1. © United Nations, 2011.

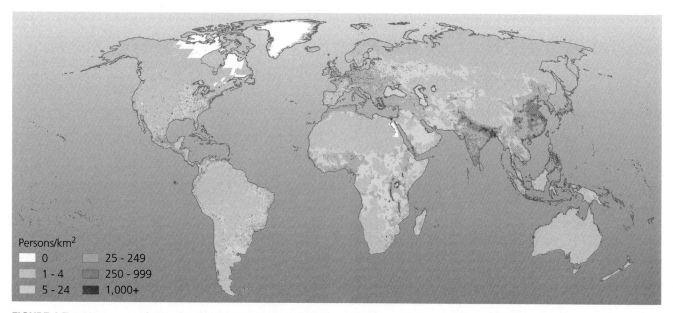

FIGURE 6.7 ▲ Human population density varies tremendously from one region to another. Arctic and desert regions have the lowest population densities, whereas areas of India, Bangladesh, and eastern China have the densest populations. The World: Population Density, 2000. Center for International Earth Science Information Network (CIESIN), Columbia University; and Centro Internacional de Agricultura Tropical (CIAT). 2005. Gridded Population of the World Version 3 (GPWv3). Palisdes, NY: Socioeconomic Data and Applications Center (SEDAC), Columbia University. By permission.

Population density and distribution   People are distributed unevenly across our planet. In ecological terms, our distribution is clumped (pp. 54–55) at all spatial scales. At the global scale (**FIGURE 6.7**), population density is highest in regions with temperate, subtropical, and tropical climates and lowest in regions with extreme-climate biomes, such as desert, rainforest, and tundra. Human population is dense along seacoasts and rivers, and less dense away from water. At more local scales, we cluster together in cities and towns.

This uneven distribution means that certain areas bear more environmental impact than others. Just as the Yellow River experiences pressure from Chinese farmers, the world's other major rivers all receive more than their share of human impact. At the same time, some areas with low population density are sensitive (a high S value in our revised IPAT model) and thus vulnerable to impact. Deserts and arid grass-lands, for instance, are easily degraded by agriculture and ranching that commandeers too much water.

Age structure   Age structure describes the relative numbers of individuals of each age class within a population (p. 54). These data are valuable for predicting future dynamics of human populations. A population made up mostly of individuals past reproductive age will tend to decline over time. In contrast, a population with many individuals of reproductive age or pre-reproductive age is likely to increase. A population with an even age distribution will likely remain stable as births keep pace with deaths.

Age structure diagrams, often called population pyramids, are visual tools scientists use to illustrate age structure (**FIGURE 6.8**). The width of each horizontal bar represents the number of people in each age class. A pyramid with a wide

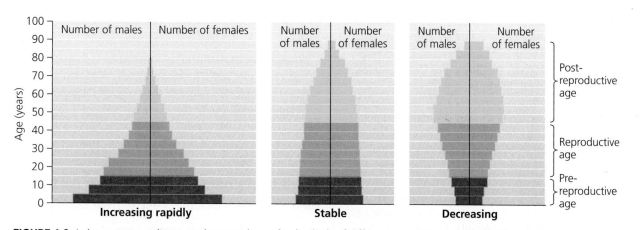

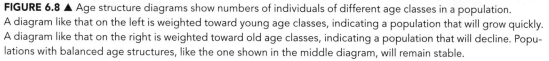

FIGURE 6.8 ▲ Age structure diagrams show numbers of individuals of different age classes in a population. A diagram like that on the left is weighted toward young age classes, indicating a population that will grow quickly. A diagram like that on the right is weighted toward old age classes, indicating a population that will decline. Populations with balanced age structures, like the one shown in the middle diagram, will remain stable.

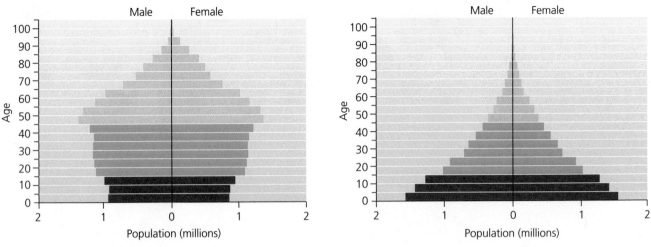

(a) Age pyramid of Canada in 2010

(b) Age pyramid of Madagascar in 2010

FIGURE 6.9 ▲Canada **(a)** shows a fairly balanced age structure, with relatively even numbers of individuals in various age classes. Madagascar **(b)** shows an age distribution heavily weighted toward young people. Madagascar's population growth rate (2.9%) is over seven times greater than Canada's (0.4%). Data from Population Division of the Department of Economic and Social Affairs of the United Nations Secretariat, 2011. *World population prospects: The 2010 revision,* http://esa.un.org/unpp. © United Nations, 2011. By permission.

base denotes a large proportion of people who have not yet reached reproductive age—and this indicates a population soon capable of rapid growth.

As an example, compare age structures for Canada and Madagascar (**FIGURE 6.9**). Madagascar's large concentration of individuals in young age groups portends a great deal of reproduction. Not surprisingly, Madagascar has a much greater population growth rate than Canada.

Today, populations are aging in many nations, including the United States. The global median age today is 28, but it will be 38 in the year 2050. China's population policies have radically changed its age structure. In 1970 the median age in China was 20; by 2050 it will be 45. In 1970, China had more children under age 5 than people over 60, but by 2050 there will be 12 times more people over 60 than under age 5! This dramatic shift in age structure (**FIGURE 6.10**) will challenge China's economy, health care system, families, and military forces because fewer working-age people will be available to support social programs to assist the rising number of older people. However, a changing age structure may also lead to increased benefits to society from volunteer efforts by productive retirees. In many ways both good and bad, "graying" populations will affect societies in China, the United States, and elsewhere throughout our lifetimes.

Sex ratios The ratio of males to females also can affect population dynamics. The naturally occurring sex ratio at birth in human populations features a slight preponderance of males; for every 100 female infants born, about 106 male infants are born. This phenomenon is an evolutionary adaptation to the fact that males are slightly more prone to death during any given year of life. It tends to ensure that the ratio of men to women will be approximately equal when people reach reproductive age. Thus, a slightly uneven sex ratio at birth may be beneficial. However, a greatly distorted ratio can lead to problems.

In recent years, demographers have witnessed an unsettling trend in China: The ratio of newborn boys to girls has become strongly skewed. In the 2000 census, 120 boys were reported born for every 100 girls. Some provinces reported sex ratios as high as 138 boys for every 100 girls. The leading hypothesis for these unusual sex ratios is that some parents, having learned the gender of their fetuses by ultrasound, are selectively aborting female fetuses.

China's skewed sex ratio may further lower population growth rates. However, it has proved tragic for the "missing girls." It is also having the undesirable social consequence of leaving many Chinese men single. This, in turn, has resulted in a grim new phenomenon. In parts of rural China, teenaged girls are being kidnapped and sold to families in other parts of the country as brides for single men.

WEIGHING THE ISSUES

China's Reproductive Policy Consider the benefits as well as the problems associated with a reproductive policy such as China's. Do you think a government should be able to enforce strict penalties for citizens who fail to abide by such a policy? If you disagree with China's policy, what alternatives can you suggest for dealing with the resource demands of a rapidly growing population?

Population change results from birth, death, immigration, and emigration

Rates of birth, death, immigration, and emigration determine whether a population grows, shrinks, or remains stable. The formula for measuring population growth (p. 55)

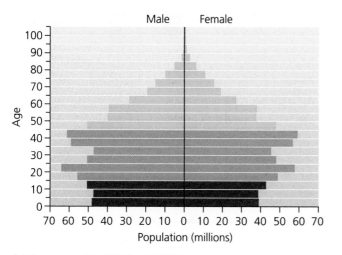

Male   Female

(a) Age pyramid of China in 2010

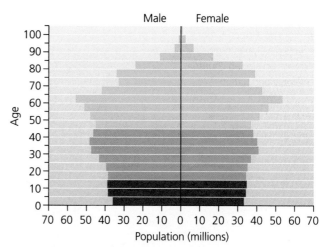

Male   Female

(b) Projected age pyramid of China in 2050

(c) Young female factory workers in Hong Kong

(d) Elderly Chinese

FIGURE 6.10 ▲ As China's population ages, older people will outnumber the young. Population pyramids show the predicted graying of the Chinese population from 2010 **(a)** to 2050 **(b)**. Today's children may, as working-age adults **(c)**, face pressures to support greater numbers of elderly citizens **(d)** than has any previous generation. Data in (a) and (b) from Population Division of the Department of Economic and Social Affairs of the United Nations Secretariat, 2011. *World population prospects: The 2010 revision,* http://esa.un.org/unpp. © United Nations, 2011. By permission.

pertains to people: Birth and immigration add individuals to a population, whereas death and emigration remove individuals. Technological advances during the last 200 years led to a dramatic decline in human death rates, widening the gap between birth rates and death rates and resulting in the global human population expansion.

In recent decades, reductions in birth rates around the world have led to an overall decline in the global growth rate (**FIGURE 6.11**). Note, however, that although the rate of growth is slowing, the absolute size of the population continues to increase. Even though our growth rate is getting smaller year by year, we continue to add over 80 million people to the planet each year.

Total fertility rate influences population growth

One key statistic demographers calculate to examine a population's potential for growth is the **total fertility rate (TFR)**, the average number of children born per woman during her lifetime. **Replacement fertility** is the TFR that keeps the

size of a population stable. For humans, replacement fertility roughly equals a TFR of 2.1. (Two children replace the mother and father, and the extra 0.1 accounts for the risk of a child dying before reaching reproductive age.) If the TFR drops below 2.1, population size (in the absence of immigration) will shrink.

Various factors have driven TFR downward in many nations in recent years. Historically, people tended to conceive many children, which helped ensure that at least some would survive. Today's improved medical care has reduced infant mortality rates, making it less necessary to bear multiple children. Urbanization has also driven TFR down; whereas rural families need children to contribute to farm labor, in urban areas children are usually excluded from the labor market, are required to go to school, and impose economic costs on their families. Moreover, if a government provides some form of social security, as most do these days, parents need fewer children to support them in their old age. Finally, with greater educational opportunities and changing roles in society, women tend to shift into the labor force, putting less emphasis on child rearing.

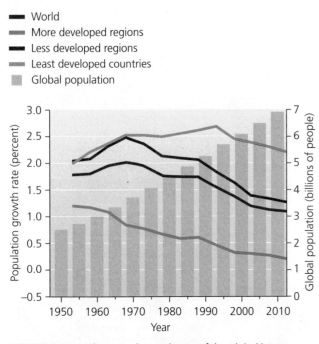

FIGURE 6.11 ▲ The annual growth rate of the global human population (**dark blue line**) peaked in the late 1960s and has declined since then. Growth rates of developed regions (**green line**) have fallen since 1950, while those of developing regions (**red line**) have fallen since the global peak in the late 1960s. For the world's least developed countries (**orange line**), growth rates began to fall in the 1990s. Although growth *rates* are declining, global population size (**gray bars**) is still growing about the same amount each year, because smaller percentage increases of ever-larger numbers produce roughly equivalent additional amounts. Data from Population Division of the Department of Economic and Social Affairs of the United Nations Secretariat, 2011. *World population prospects: The 2010 revision*, http://esa.un.org/unpp. © United Nations, 2011. By permission.

TABLE 6.1 Total Fertility Rates for Major Regions	
Region	Total fertility rate (TFR)
Africa	4.7
Australia and the South Pacific	2.5
Latin America and the Caribbean	2.3
Asia	2.2
North America	2.0
Europe	1.6

Data from Population Reference Bureau, 2010. 2010 World population data sheet.

All these factors have come together in Europe, where TFR has dropped from 2.6 to 1.6 in the past half-century. Nearly every European nation now has a fertility rate below the replacement level, and populations are declining in 17 of 45 European nations. In 2010, Europe's overall annual **natural rate of population change** (change due to birth and death rates alone, excluding migration) was between 0.0% and 0.1%. Worldwide by 2010, 72 countries had fallen below the replacement fertility of 2.1. These countries make up roughly 45% of the world's population and include China (with a TFR of 1.5). **TABLE 6.1** shows TFRs of major continental regions.

Many nations have experienced the demographic transition

Many nations with lowered birth rates and TFRs are experiencing a common set of interrelated changes. In countries with good sanitation, effective health care, and reliable food supplies, more people than ever before are living long lives. As a result, over the past 50 years the life expectancy (the time a person can expect to live) for the average person worldwide has increased from 46 to 69 years, with much of this increase attributed to reductions in infant mortality rates. Societies going through these changes are generally those that have undergone urbanization and industrialization and have generated personal wealth for their citizens.

To make sense of these trends, demographers developed a concept called the **demographic transition**. This is a model of economic and cultural change first proposed in the 1940s and 1950s by demographer Frank Notestein to explain the declining death rates and birth rates that have occurred in Western nations as they industrialized. Notestein believed nations move from a stable pre-industrial state of high birth and death rates to a stable post-industrial state of low birth and death rates (**FIGURE 6.12**). Industrialization, he proposed, causes these rates to fall by first decreasing mortality and then lessening the need for large families. Parents thereafter choose to invest in quality of life rather than quantity of children. Because death rates fall before birth rates fall, a period of net population growth results. Thus, under the demographic transition model, population growth is seen as a temporary phenomenon that occurs as societies move from one stage of development to another.

The pre-industrial stage The first stage of the demographic transition model is the **pre-industrial stage**, characterized by conditions that have defined most of human history. In pre-industrial societies, death rates and birth rates are both high. Death rates are high because disease is widespread, medical care rudimentary, and food supplies unreliable and difficult to obtain. Birth rates are high because people must compensate for infant mortality by having several children. In this stage, children are valuable as workers who can help meet a family's basic needs. Populations within the pre-industrial stage are not likely to experience much growth, which is why the human population was relatively stable until the industrial revolution.

Industrialization and falling death rates Industrialization initiates the second stage of the demographic transition, known as the **transitional stage**. This transition from the pre-industrial stage to the industrial stage is generally characterized by declining death rates due to increased food production and improved medical care. Birth rates in the transitional stage remain high, however, because people have not yet grown used to the new economic and social conditions. As a result, population growth surges.

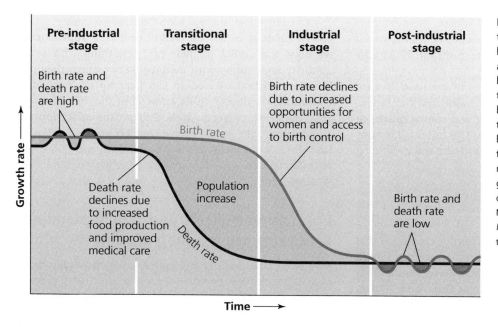

FIGURE 6.12 ◄ The demographic transition models a process that has taken some populations from a pre-industrial stage of high birth rates and high death rates to a post-industrial stage of low birth rates and low death rates. In this diagram, the wide green area between the two curves illustrates the gap between birth and death rates that causes rapid population growth during the middle portion of this process. Adapted from Kent, M., and K. Crews, 1990. *World population: Fundamentals of growth.* By permission of the Population Reference Bureau.

The industrial stage and falling birth rates The third stage in the demographic transition is the **industrial stage**. Industrialization increases opportunities for employment outside the home, particularly for women. Children become less valuable, in economic terms, because they do not help meet family food needs as they did in the pre-industrial stage. If couples are aware of this, and if they have access to birth control, they may choose to have fewer children. Birth rates fall, closing the gap with death rates and reducing population growth.

The post-industrial stage In the final stage, the **postindustrial stage**, both birth and death rates have fallen to low and stable levels. Population sizes stabilize or decline slightly. The society enjoys the fruits of industrialization without the threat of runaway population growth.

Is the demographic transition a universal process?

The demographic transition has occurred in many European countries, the United States, Canada, Japan, and several other developed nations over the past 200 to 300 years. It is a model that may or may not apply to all developing nations as they industrialize now and in the future. On the one hand, note in Figure 6.12 how growth rates fell first for industrialized nations, then for less developed nations, and finally for least developed nations, suggesting that it may merely be a matter of time before all nations experience the transition. On the other hand, some developing nations may already be suffering too much from the impacts of large populations to replicate the developed world's transition. Cultures that place greater value on childbirth or grant women fewer freedoms may never complete the transition.

Moreover, natural scientists estimate that we would need the natural resources of four-and-a-half planet Earths if everyone lived like a modern American. Thus, whether developing nations (which include the vast majority of the planet's people) pass through the demographic transition is one of the most important questions for the future of our civilization and Earth's environment.

POPULATION AND SOCIETY

Demographic transition theory links the quantitative study of how populations change with the societal factors that influence (and are influenced by) population dynamics. Let's examine a few of these societal factors more closely.

Family planning is a key approach for controlling population growth

Perhaps the greatest single factor enabling a society to slow its population growth is the ability of women and couples to engage in **family planning**, the effort to plan the number and spacing of one's children. Family-planning programs and clinics offer information and counseling to potential mothers and fathers on reproductive issues. An important component of family planning is **birth control**, the effort to control the number of children one bears, particularly by reducing the frequency of pregnancy. This relies on **contraception**, the deliberate attempt to prevent pregnancy despite sexual intercourse. Common methods of modern contraception in use today include condoms, spermicide, hormonal treatments (birth control pill/hormone injection), intrauterine devices (IUDs), and permanent sterilization through tubal ligation or vasectomy. Many family-planning organizations aid clients by offering free or discounted contraceptives.

Worldwide in 2010, 55% of women (aged 15–49) reported using contraceptives, with rates of use varying widely among nations. China, at 86%, had the highest rate of contraceptive use of any nation. Six European nations showed rates of contraceptive use of 70% or more, as did Australia,

Brazil, Canada, Costa Rica, Cuba, Micronesia, New Zealand, Nicaragua, Paraguay, Puerto Rico, Thailand, the United States, and Uruguay. At the other end of the spectrum, 17 African nations had rates below 10%.

Low usage rates for contraceptives in some societies are due to religious doctrine or cultural influences that hinder family planning, denying counseling and contraceptives to people who might otherwise use them. This can result in family sizes that are larger than the parents desire and to elevated rates of population growth.

Empowering women reduces fertility rates

Today, many social scientists and policymakers recognize that for population growth to slow and stabilize, women should be granted equal power to men in societies worldwide. This would have many benefits: Studies show that where women are freer to decide whether and when to have children, fertility rates fall, and the resulting children are better cared for, healthier, and better educated.

One benefit of equal rights for women is the ability to make reproductive decisions. In many societies, men restrict women's decision-making abilities, including decisions as to how many children they will bear. Fertility rates have dropped most noticeably in nations where women have gained improved access to contraceptives and to family planning (see **THE SCIENCE BEHIND THE STORY**, pp. 128–129).

Equality for women also involves expanding educational opportunities for them, because in many nations girls are discouraged from pursuing an education or kept out of school altogether. Over two-thirds of the world's people who cannot read are women. And data clearly show that as women receive educational opportunities, fertility rates decline (**FIGURE 6.13**). Education helps more women pursue careers, delay childbirth, and have a greater say in reproductive decisions.

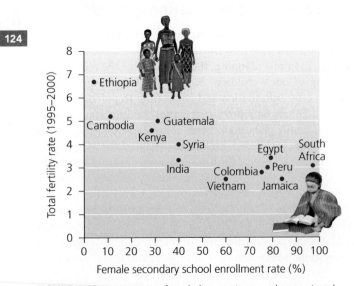

FIGURE 6.13 ▲ Increasing female literacy is strongly associated with reduced birth rates in many nations. Data from McDonald, M., and D. Nierenberg, 2003. Linking population, women, and biodiversity. *State of the world 2003*. Washington, D.C.: Worldwatch Institute.

In a physiological sense, access to family planning and improved rights gives women control over their **reproductive window**, the period of their life, beginning with sexual maturity and ending with menopause, in which they may become pregnant. A woman can bear up to 25 children within this window (**FIGURE 6.14A**), but she may choose to delay the birth of her first child to pursue education and employment. She may also use contraception to delay her first child, space births within the window, and "close" her reproductive window after achieving her desired family size (**FIGURE 6.14B**).

Population policies and family-planning programs are working around the globe

Data show that funding and policies that encourage family planning can lower population growth rates in all types of nations, even those that are least industrialized. No nation has pursued a sustained population control program as intrusive as China's, but other rapidly growing nations have implemented less restrictive programs.

India was the first nation to implement population control policies. However, when some policymakers introduced forced sterilization in the 1970s, the resulting outcry brought down the government. Since then, India's efforts have been more modest and far less coercive, focusing on family planning and reproductive health care. However, today a number of Indian states also run programs of incentives and disincentives promoting a "two-child norm," and current debate centers on whether this is a just and effective approach.

The government of Thailand has reduced birth rates and slowed population growth. In the 1960s, Thailand's growth rate was 2.3%, but today it stands at 0.6%. This decline was achieved without a one-child policy, resulting instead from an education-based approach to family planning and the increased availability of contraceptives. Brazil, Mexico, Iran, Cuba, and many other developing countries have instituted active programs to reduce their population growth that entail setting targets and providing incentives, education, contraception, and reproductive health care. These programs are working: Studies show that nations with such programs have lower fertility rates than similar nations without them (**FIGURE 6.15**).

In 1994, the United Nations hosted a milestone conference on population and development in Cairo, Egypt, at which 179 nations endorsed a platform calling on all governments to offer their citizens universal access to reproductive health care within 20 years. The conference marked a turn away from older notions of top-down population policy geared toward pushing contraception and lowering population to preset targets. Instead, it urged governments to offer better education and health care and to address social needs that affect population from the bottom up (such as alleviating poverty, disease, and sexism). However, worldwide funding for family planning fell by more than one-third in the decade following the Cairo conference, slowing progress on these initiatives.

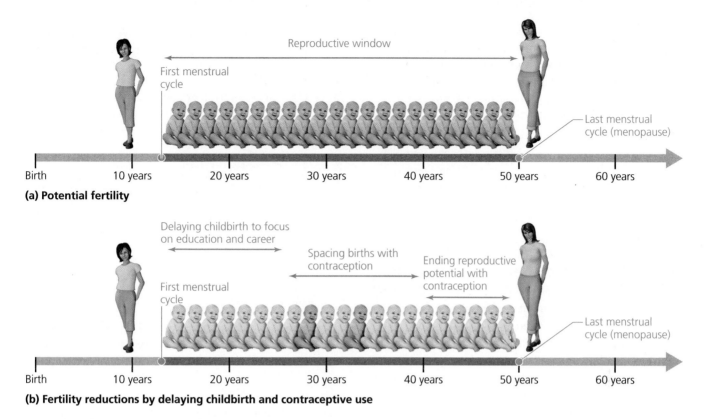

First menstrual
cycle

Reproductive window

Last menstrual
cycle (menopause)

Birth | 10 years | 20 years | 30 years | 40 years | 50 years | 60 years

(a) Potential fertility

Delaying childbirth to focus
on education and career

Spacing births with
contraception

Ending reproductive
potential with
contraception

First menstrual
cycle

Last menstrual
cycle (menopause)

Birth | 10 years | 20 years | 30 years | 40 years | 50 years | 60 years

(b) Fertility reductions by delaying childbirth and contraceptive use

FIGURE 6.14 ▲ **(a)** Women can potentially have very high fertility within their "reproductive window" but **(b)** can reduce the number of births by delaying the birth of their first child to pursue education and career and by using contraception to space pregnancies or to end their reproductive window at the time of their choosing.

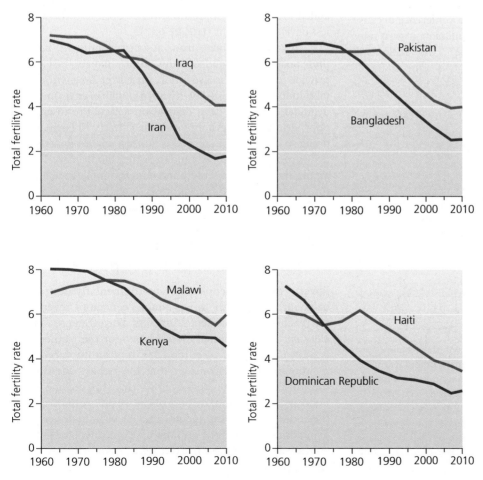

FIGURE 6.15 ◄ Data from four pairs of neighboring countries demonstrate the effectiveness of family planning in reducing fertility rates. In each case, the nation that invested in family planning and (in some cases) made other reproductive rights, education, and health care more available to women (**blue lines**) reduced its total fertility rate (TFR) more than its neighbor (**red lines**). Data from Population Reference Bureau and Harrison, P., and F. Pearce, 2000. *AAAS Atlas of population and environment*, edited by the American Association for the Advancement of Science, © 2000 by the American Association for the Advancement of Science. Used by permission of the publisher, University of California Press.

Should the United States Abstain from International Family Planning? Over the years, the United States has joined 180 other nations in providing millions of dollars to the United Nations Population Fund (UNFPA), which advises governments on family planning, sustainable development, poverty reduction, reproductive health, and AIDS prevention in many nations, including China. Starting in 2001, the Bush administration withheld funds, saying that U.S. law prohibits funding any organization that "supports or participates in the management of a program of coercive abortion or involuntary sterilization" and claiming that the Chinese government has been implicated in both. Many nations criticized the U.S. decision, and the European Union offered UNFPA additional funding to offset the loss of U.S. contributions. Once President Obama came to office, he reinstated funding, asking Congress to allocate $60 million in 2009. What do you think U.S. policy should be? Should the United States fund family-planning efforts in other nations? What conditions, if any, should it place on the use of such funds?

Reducing poverty lowers fertility

Over half the world's people live below the internationally defined poverty line of U.S. $2 per day. The alleviation of poverty was a prime target of the Cairo conference because poorer societies tend to show higher population growth rates than do wealthier societies (**FIGURE 6.16**). This relationship

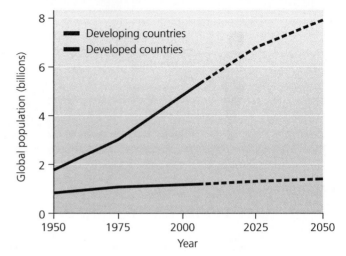

FIGURE 6.17 ▲ Over 99% of the next 1 billion people added to Earth's population will be born into the less developed, poorer parts of the world. Dashed portions of the lines indicate projected future trends. Data from Population Division of the Department of Economic and Social Affairs of the United Nations Secretariat, 2011. *World population prospects: The 2010 revision*, http://esa.un.org/unpp. © United Nations, 2011. By permission.

operates in both directions: Poverty exacerbates population growth, and rapid population growth worsens poverty.

These trends also influence the distribution of people on the planet. In 1960, 70% of the world's population lived in developing nations. As of 2010, 82% of all people live in these countries. Moreover, fully 99% of the next billion people to be added to the global population will be born into these poor, less developed regions (**FIGURE 6.17**).

This is unfortunate from a social standpoint, because these people will be added to the nations that are least able to provide for them. It is also unfortunate from an environmental standpoint, because poverty often results in environmental degradation. People who depend on agriculture in areas of poor farmland, for instance, may need to try to farm even if doing so degrades the soil and is not sustainable. Poverty also drives people to cut forests and to deplete biodiversity. For example, impoverished settlers and miners hunt large mammals for "bush meat" in Africa's forests, including the great apes that are now heading toward extinction.

Expanding wealth increases the environmental impact per person

Poverty can lead people into environmentally destructive behavior, but wealth can produce even more severe and far-reaching environmental impacts. The affluence of a society such as the United States, Japan, or the Netherlands is built on levels of resource consumption unprecedented in human history. Consider that the richest one-fifth of the world's people possess over 80 times the income of the poorest one-fifth and use 86% of the world's resources (**FIGURE 6.18**). The environmental impact of human activities depends not only on the number of people involved but also on the way those people live (recall the *A* for *affluence* in the IPAT equation).

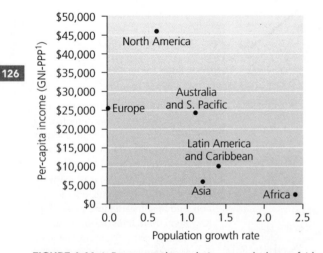

FIGURE 6.16 ▲ Poverty and population growth show a fairly strong correlation, despite the influence of many other factors. Regions with the most rapid population growth tend to show lower per capita incomes. Per capita income is here measured in GNI PPP, or "gross national income in purchasing power parity," a measure that standardizes income among nations by converting it to "international" dollars, which indicate the amount of goods and services one could buy in the United States with a given amount of money. Data from Population Reference Bureau, 2010. *2010 World population data sheet.*

Average per U.S. resident:
- $46,970 annual income
- 31,000 m² of land
- 19.0 metric tons of CO_2 emitted per year

(a) A family living in the United States

Average per resident of India:
- $2,960 annual income
- 2,800 m² of land
- 1.1 metric tons of CO_2 emitted per year

(b) A family living in India

FIGURE 6.18 ◀ A typical U.S. family **(a)** may own a large house with a wealth of material possessions. A typical family in a developing nation such as India **(b)** may live in a small, sparsely furnished dwelling with few material possessions and little money or time for luxuries. Compared with the average resident of India, the average U.S. resident enjoys 16 times more income, uses 11 times more land, and emits 17 times more carbon dioxide emissions. Data from Population Reference Bureau, 2009. *2009 World Population Data Sheet.*

We previously explored the concept of the *ecological footprint*, the cumulative amount of Earth's surface area required to provide the raw materials a person or population consumes and to dispose of or recycle the waste produced (pp. 4, 16). Individuals from affluent societies leave considerably larger per capita ecological footprints (see Figure 1.16, p. 16). In this sense, the addition of one American to the world has as much environmental impact as the addition of 3.5 Chinese, 9 Indians, or 13 Afghans. This fact reminds us that the "population problem" does not lie solely with the developing world!

Indeed, just as population is rising, so is consumption. We have seen that humanity's global ecological footprint surpassed Earth's capacity to support us in 1970s (p. 4) and our species is now living 50% beyond its means (**FIGURE 6.19**). The rising consumption that is accompanying the rapid industrialization of China, India, and other populous nations makes it all the more urgent for us to find this path to global sustainability.

If humanity's overarching goal is to generate a high standard of living and quality of life for all people, then developing nations must find ways to slow their population growth. However, those of us living in the developed world must also be willing to reduce our consumption. Earth does not hold enough resources to sustain all 7 billion of us at the current North American standard of living, nor can we venture out and bring home extra planets. We must make the best of the one place that supports us all. ◀

THE SCIENCE BEHIND THE STORY

Fertility Decline in Bangladesh

R esearch in developing nations indicates that poverty and overpopulation can create a vicious cycle, in which poverty promotes high fertility and high fertility obstructs economic development. Are there policy steps that such countries can take to bring down fertility rates? Scientific analysis of family-planning programs in the South Asian nation of Bangladesh suggests that there are.

Dr. James Phillips of Columbia University

Bangladesh is one of the poorest and most densely populated countries on the planet. Its 166 million people live in an area about the size of Wisconsin, and 45% of them live below the poverty line. With few natural resources and 1,150 people per km² (nearly 3,000 per mi²—2.5 times the population density of New Jersey), limiting population growth is critically important. Back in 1976, Bangladeshi president Ziaur Rahman declared, "If we cannot do something about population, nothing else that we accomplish will matter much."

Since then, Bangladesh has made striking progress in slowing its population growth. Despite stagnant economic development, low literacy rates, poor health care, and limited rights for women, the nation's total fertility rate (TFR) has dropped markedly. In the 1970s, the average woman in Bangladesh gave birth to nearly 7 children over the course of her life. Today, the TFR is 2.4 (**see first figure**).

Researchers hypothesized that family-planning programs were responsible for Bangladesh's rapid reduction in TFR. Because conducting an experiment to test such a hypothesis is difficult, some researchers took advantage of a natural experiment (pp. 9–10). By comparing Bangladesh to countries that are socioeconomically similar but have had less success in lowering TFR, such as Pakistan (see Figure 6.15, p. 125), researchers concluded that Bangladesh succeeded because of aggressive, well-funded outreach efforts that were sensitive to the values of its traditional society.

However, no two nations are identical, so it is difficult to draw firm conclusions from such broad-scale comparisons. This is why the Matlab Family Planning and Health Services Project, in the isolated rural area of Matlab, Bangladesh,

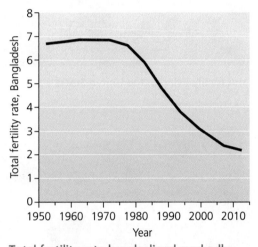

Total fertility rate has declined markedly in Bangladesh in the past 40 years, in part because of family-planning programs. Data from Population Division of the Department of Economic and Social Affairs of the United Nations Secretariat, 2011. *World population prospects: The 2010 revision*, http://esa.un.org/unpp. © United Nations, 2011. By permission.

has become one of the best-known experiments in family planning in developing countries.

- ▬▬ Ecological footprint
- ▪▰▪ Projected ecological footprint
- ---- Biocapacity

FIGURE 6.19 ◀ The global ecological footprint of the human population is estimated to be 50% greater than what Earth can bear in the long run. If population and consumption continue to rise (orange dashed line), we will increase our ecological deficit, or degree of overshoot, until systems give out and populations crash. If, instead, we pursue a path to sustainability (red dashed line), we can eventually repay our ecological debt and sustain our civilization. Adapted from WWF International, 2010. *Living Planet Report 2010.* Published by WWF-World Wide Fund for Nature. © 2010 WWF (panda.org). Zoological Society of London, and Global Footprint Network.

The Matlab Project was an intensive outreach program run collaboratively by the Bangladeshi government and international aid organizations. Each household in the project area received biweekly visits from local women offering counseling, education, and free contraceptives. Compared to a similar government-run program in a nearby area, the Matlab Project featured more training, more services, and more frequent visits. In both areas, a highly organized data collection system gave researchers detailed information about births, deaths, and health-related behaviors such as contraceptive use. The result was an experiment comparing the Matlab Project with the government-run project.

When Matlab Project director James Phillips and his colleagues reviewed a decade's worth of data in 1988, they found that fertility rates had declined in both areas. The decline appeared to be due almost entirely to a rise in contraceptive use, because other factors—such as the average age of marriage—remained the same. Phillips and his colleagues also found the declines to be significantly greater in the Matlab area than in the government-run area. These findings suggested that high-intensity outreach efforts could affect TFR even in the absence of significant improvements in women's status, education, or economic development.

Why was the outreach program successful? One hypothesis was that visits from health care workers helped

In the Matlab Project, Bangladeshi households received visits from local women offering counseling, education, and free contraceptives.

convince local women that small families are desirable. However, in 1999, Mary Arends-Kuenning, a University of Michigan graduate student, and her colleagues reported that there was no relationship between the number of visits made by outreach workers and women's perception of the ideal family size, either in Matlab or nearby areas. Ideal family size declined equally in all areas. Instead of creating new demand for birth control, the Matlab Project appears to have helped women convert an already-existing desire for fewer children into behaviors, such as contraceptive use, that reduce fertility.

Bangladesh's ability to rein in fertility rates despite unfavorable social and economic conditions bodes well for impoverished nations facing explosive population growth. However, further reductions may require political and socioeconomic changes that are difficult to implement in traditional, resource-strapped countries such as Bangladesh. Nonetheless, scientific research from Matlab has illuminated the impacts of family-planning programs on fertility, helping to inform population control efforts in Bangladesh and elsewhere.

► CONCLUSION

Today's human population is larger than at any time in the past. Our growing population and our growing consumption affect the environment and our ability to meet the needs of all the world's people. The great majority of children born today are likely to live their lives in conditions far less healthy and prosperous than most of us in the industrialized world are accustomed to.

However, there are at least two major reasons to be encouraged. First, although global population is still rising, the rate of growth has decreased nearly everywhere, and some countries are even seeing population declines. Most developed nations have passed through the demographic transition, showing that it is possible to lower death rates while stabilizing population and creating more prosperous

societies. Second, progress has been made in expanding rights for women worldwide. Although there is still a long way to go, women are obtaining better education, more economic independence, and more ability to control their reproductive decisions. Aside from the clear ethical progress these developments entail, they are helping to slow population growth.

Human population cannot continue to rise forever. The question is how it will stop rising: Will it be through the gentle and benign process of the demographic transition, through restrictive governmental intervention such as China's one-child policy, or through the miserable Malthusian checks of disease and social conflict caused by overcrowding and competition for scarce resources? Moreover, sustainability demands a further challenge—that we stabilize our population size in time to avoid destroying the natural systems that support our economies and societies. We are indeed a special species. We are the only one to achieve the dominance to alter much of Earth's landscape and even its climate system. We are also the only species with the intelligence needed to halt the increase in our own numbers before we destroy the very systems on which we depend.

TESTING YOUR COMPREHENSION

1. What is the approximate current human global population? How many people are being added to the population each day?

2. Why has the human population continued to grow despite environmental limitations? Give one example of an innovation that increased the carrying capacity for humans.

3. Contrast the views of environmental scientists with those of economists and policymakers regarding whether population growth is a problem. Name several reasons why population growth is commonly viewed as a problem.

4. Explain the IPAT model. How can technology either increase or decrease environmental impact? Provide at least two examples.

5. Describe how demographers use size, density, distribution, age structure, and sex ratio of a population to estimate how it may change. How does each of these factors help determine the impact of human populations on the environment?

6. What is the total fertility rate (TFR)? Why is the replacement fertility for humans approximately 2.1? How is Europe's TFR affecting its natural rate of population change?

7. Why have fertility rates fallen in many countries?

8. How does the demographic transition model explain the increase in population growth rates in recent centuries? How does it explain the recent decrease in population growth rates in many countries?

9. Why are the empowerment of women and the pursuit of gender equality viewed as important to controlling population growth? Describe the aim of family-planning programs.

10. Why do poorer societies have higher population growth rates than wealthier societies? How does poverty affect the environment? How does affluence affect the environment?

SEEKING SOLUTIONS

1. China's reduction in birth rates is leading to significant change in the nation's age structure. Review Figure 6.10, which shows that the population is growing older, leading to the top-heavy population pyramid for the year 2050. What effects might this ultimately have on Chinese society? What steps could be taken in response?

2. Apply the IPAT model to the example of China provided in the chapter. How do population, affluence, technology, and ecological sensitivity affect China's environment? Now consider your own country or your own state. How do population, affluence, technology, and ecological sensitivity affect your environment? How can we minimize the environmental impacts of growth in the human population?

3. Do you think that all of today's developing nations will complete the demographic transition and come to enjoy a permanent state of low birth and death rates? Why or why not? What steps might we as a global society take to help ensure that they do? Now think about developed nations such as the United States and Canada. Do you think these nations will continue to lower and stabilize their birth and death rates in a state of prosperity? What factors might affect whether they do so?

4. **THINK IT THROUGH** India's prime minister puts you in charge of that nation's population policy. India has a population growth rate of 1.6% per year, a TFR of 2.7, a 49% rate of contraceptive use, and a population that is 71% rural. What policy steps would you recommend, and why?

5. **THINK IT THROUGH** Now suppose that you have been tapped to design population policy for Germany. Germany is losing population at an annual rate of 0.2%, has a TFR of 1.3, a 66% rate of contraceptive use, and a population that is 73% urban. What policy steps would you recommend, and why?

CALCULATING ECOLOGICAL FOOTPRINTS

A nation's population size and the affluence of its citizens each influence its resource consumption and environmental impact. As of 2010, the world's population passed 6.9 billion, and average per capita income was $10,030 per year, and the latest estimate for the world's average ecological footprint was 2.7 hectares (ha) per person. The sampling of data in the table will allow you to explore patterns in how population, affluence, and environmental impact are related.

Nation	Population (millions of people)	Affluence (per capita income, in GNI PPP)[1]	Personal impact (per capita footprint, in ha/person)	Total impact (national footprint, in millions of ha)
Belgium	10.8	$34,760	8.0	86
Brazil	193.3	$10,070	2.9	
China	1,338.1	$6,020	2.2	
Ethiopia	85.0	$870	1.1	
India	1,188.8	$2,960	0.9	
Japan	127.4	$35,220	4.7	
Mexico	110.6	$14,270	3.0	
Russia	141.9	$15,630	4.4	
United States	309.6	$46,970	8.0	2,477

[1]GNI PPP is "gross national income in purchasing power parity," a measure that standardizes income among nations by converting it to "international" dollars, which indicate the amount of goods and services one could buy in the United States with a given amount of money.

Data Sources: Population and affluence data are from Population Reference Bureau, 2010. World population data sheet 2010. Footprint data are for 2007, from WWF International, Zoological Society of London, and Global Footprint Network. Living Planet Report 2010.

1. Calculate the total impact (national ecological footprint) for each country.

2. Draw a graph illustrating per capita impact (on the *y* axis) versus affluence (on the *x* axis). What do the results show? Explain why the data look the way they do.

3. Draw a graph illustrating total impact (on the *y* axis) in relation to population (on the *x* axis). What do the results suggest to you?

4. Draw a graph illustrating total impact (on the *y* axis) in relation to affluence (on the *x* axis). What do the results suggest to you?

5. You have just used three of the four variables in the IPAT equation. Now give one example of how the T (technology) variable could potentially increase the total impact of the United States, and one example of how it could potentially decrease the U.S. impact.

7 Soil, Agriculture, and the Future of Food

Upon completing this chapter, you will be able to:

➤ Explain the challenges of feeding a growing human population

➤ Compare and contrast traditional, industrial, and sustainable agricultural approaches

➤ Identify the goals, methods, and consequences of the Green Revolution

➤ Explain the importance of soils to agriculture

➤ Analyze the types and causes of soil erosion and land degradation

➤ Explain the principles of soil conservation and provide solutions to soil erosion and soil degradation

➤ Compare and contrast approaches to irrigation, fertilization, and pest management in industrial and sustainable agriculture

➤ Describe the science of genetically modified food and evaluate the debate over its use

➤ Assess the benefits and drawbacks of feedlots and aquaculture

➤ Describe organic agriculture and summarize its recent growth

Iowa farmers Nate Ronsiek (left) and Paul "Butch" Schroeder (right)

Iowa's Farmers Practice No-Till Agriculture

"The nation that destroys its soil destroys itself."
—U.S. President Franklin D. Roosevelt

"There are two spiritual dangers in not owning a farm. One is the danger of supposing that breakfast comes from the grocery, and the other that heat comes from the furnace."
—Aldo Leopold, Conservationist and philosopher

Iowa farmers Paul "Butch" Schroeder and his brother David know that their livelihoods depend on keeping their soil healthy and productive. The Schroeder brothers farm 1,200 ha (3,000 acres) in western Iowa, where rich prairie soils have historically made for bountiful grain harvests.

Yet the Schroeders also know that repeated cycles of plowing and planting since farmers first settled the region have diminished the soil's productivity. They know that much of the topsoil—the valuable surface layer richest in organic matter and nutrients—has been lost to erosion: washed away by water and blown away by wind. Turning the earth by tilling (plowing, disking, harrowing, or chiseling) aerates the soil and works weeds and old crop residue into the soil to nourish it, but tilling also leaves the surface bare, allowing wind and water to erode away precious topsoil.

As a result, the Schroeder brothers abandoned the conventional practice of tilling the soil after harvests and instead turned to **no-till** farming. Rather than plowing after each harvest, they began leaving crop residues atop their fields, keeping the soil covered with plant material at all times. To plant the next crop, they cut a thin, shallow groove into the soil surface, dropped in seeds, and covered them. By planting seeds of the new crop through the residue of the old, less soil erodes away, organic material accumulates, and the soil soaks up more water—all of which encourages better plant growth. On portions of their land where some tilling is required, they practice **conservation tillage**, an approach involving limited

NORTH AMERICA

Iowa

Atlantic Ocean

Pacific Ocean

SOUTH AMERICA

tilling—only as much as is needed.

The Schroeders have found that no-till farming saves time and money as well. By forgoing tilling, they reduce the number of passes they need to make on the tractor, which saves fuel, time, effort, and wear and tear on equipment.

Butch and David Schroeder also practice other conservation measures on their land. They take soil samples to determine how much fertilizer different areas need, so as not to overapply it. They employ planting methods designed to reduce erosion. They forgo farming on lands that are vulnerable or have poor soil, retiring them as part of the U.S. government's Conservation Reserve Program, which pays farmers to take highly erodible land out of cultivation. All these measures save farmers money while protecting environmental quality and nurturing soil as an investment for sustainable yields in the future.

Future investments are important when a family farm is passed down to the next generation. In Sioux County in northwest Iowa, 26-year-old Nate Ronsiek has taken charge of the farm his late father Vince left to him. Nate and his wife Rachel are putting into practice the strategies for conservation that Nate's father taught him, as well as those he learned as a student at Kansas State University.

"Nate has been farming a short time, but he's doing a lot of things right on his farm that are saving him money and improving the environment," says Greg Marek, a conservationist with the local Natural Resource Conservation Service office.

To raise grain for their 65 stock cows, Ronsiek practices no-till farming. He wanted to test the approach for himself, so he worked with agricultural extension agents from Iowa State University to conduct experiments on his own land and compare results from no-till fields and conventional fields. After three years, his no-till fields produced as much corn as the conventional fields while requiring less time and money. Ronsiek says he can clearly see how water infiltrates better in the no-till fields and how those fields suffer less erosion.

"I'm excited," Ronsiek says. "In future growing seasons yields are expected to increase—profit potential, too. What I don't spend on fuel and time for conventional tillage field trips I can invest elsewhere on the farm."

By enhancing soil conditions and reducing erosion, no-till techniques are benefiting Iowa's people and environment as well, cutting down on pollution in its air, waterways, and ecosystems. Similar effects are being felt elsewhere in the world where no-till methods are being applied.

No-till farming is not a panacea everywhere. But in suitable regions, proponents say it can help make agriculture sustainable. We will need sustainable agriculture if we are to feed the world's human population while protecting the natural environment, including the soils that vitally support our production of food. ■

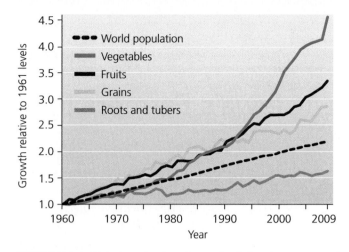

FIGURE 7.1 ▲ Global production of most foods has risen more quickly than world population over the past half-century. This means that we have produced more food per person each year. Trend lines show cumulative increases relative to 1961 levels (for example, a value of 2.0 means twice the 1961 amount). Population is measured in number of people, and food is measured by weight. Data from U.N. Food and Agriculture Organization (FAO).

every 5 seconds, somewhere in the world, a child starves to death. Many people are undernourished because they are too poor to purchase the food they need, but political obstacles, conflict, and inefficiencies in distribution contribute significantly to hunger as well.

Globally, the number of people suffering from undernourishment fell steadily from 1970 through the late 1990s (**FIGURE 7.2**). The percentage of undernourished people in developing countries, where hunger issues are most pronounced, fell even more steeply. These encouraging trends were reversed because of rising food prices in the 2000s and the economic slump of 2008–2009, and both the number and percentage of hungry people increased. But in 2010 the

THE RACE TO FEED THE WORLD

Thanks to the efforts of farmers such as the Schroeder brothers and Nate Ronsiek, our ability to produce food has grown even faster than global population over the past half-century (**FIGURE 7.1**). Improving people's quality of life by producing more food per person is a monumental achievement of which humanity can be proud.

However, not all of the 7 billion people alive today have enough to eat, and we expect to add an additional 2 billion people to the human population by 2050. Providing **food security**, the guarantee of an adequate, safe, nutritious, and reliable food supply available to all people at all times, will therefore be one of our greatest challenges in coming decades.

We face undernutrition, overnutrition, and malnutrition

Despite our rising food production, 925 million people worldwide suffer from **undernutrition**, receiving fewer calories than the minimum dietary energy requirement. As a result,

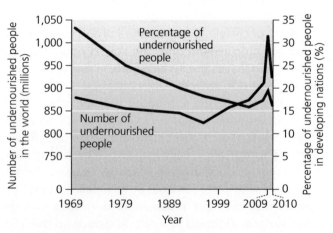

FIGURE 7.2 ▲ The number of people suffering undernourishment in the world declined from 1970 through the late 1990s, while the percentage of people in developing nations who were undernourished fell still more. However, both of these indicators of hunger increased recently in response to economic conditions. Data from Food and Agriculture Organization of the United Nations, Rome, 2010. *The state of food insecurity in the world, 2010.* Rome.

number of undernourished people declined from the previous year, in large part because of more favorable economic conditions in developing countries. Moreover, hunger levels, as a percentage of population, are substantially lower than the rates that existed 40 years ago.

Although 1 billion people lack access to nutritious foods, many others consume too many calories each day and suffer from **overnutrition**. Overnutrition leads to unhealthy weight gain, which leads to cardiovascular disease, diabetes, and other health problems. In the United States, more than three in five adults are technically overweight, and more than one in four are obese. But overnutrition is a global problem. The World Health Organization estimates that in 2008, 1.5 billion adults worldwide were overweight and that at least 500 million were obese. It predicts that by 2015 those numbers will rise to 2.3 billion and 700 million.

Just as the *quantity* of food a person eats is important for health, so is the *quality* of food. **Malnutrition**, a shortage of nutrients the body needs, occurs when a person fails to obtain a complete complement of vitamins and minerals. Malnutrition can easily lead to disease (**FIGURE 7.3**). For example, people who eat a diet that is high in starch but deficient in protein or essential amino acids (p. 28) can develop *kwashiorkor*.

FIGURE 7.3 ▲ Millions of children, including this child with his mother in Somalia, suffer from forms of malnutrition, such as kwashiorkor and marasmus.

Children who have recently stopped receiving protein from breast-feeding are most at risk for developing kwashiorkor, which causes bloating of the abdomen, deterioration and discoloration of hair, mental disability, immune suppression, developmental delays, and reduced growth. Protein deficiency together with a lack of calories can lead to *marasmus*, which causes wasting or shriveling among millions of children in the developing world. Dietary conditions such as anemia (a lack of dietary iron which causes fatigue and developmental disabilities), iodine deficiency (which causes swelling of the thyroid gland and brain damage), and vitamin A deficiency (which leads to blindness) are also alarmingly prevalent.

Some biofuels reduce food supplies

In recent years, some well-intentioned efforts to promote renewable energy have had unintended consequences that have affected people's access to nutritious foods. **Biofuels** (pp. 358-361) are fuels derived from organic materials and used in internal combustion engines as replacements for petroleum. In the United States, ethanol produced from corn (pp. 359-361) is the primary biofuel.

When government subsidies for ethanol were expanded in 2007, U.S. ethanol production nearly doubled as new ethanol facilities opened and farmers began selling their corn for ethanol instead of for food. In Iowa, for example, over half of the corn grown in the state is used for ethanol production.

The use of corn for fuel production caused a scarcity of corn worldwide, and prices for basic foods skyrocketed. The situation worsened when prices for other staple grains also rose because farmers shifted fields formerly devoted to food crops into biofuel production (Chapter 16). For low-income people, the steep rise in food prices was frightening. Thousands staged protests, and riots erupted in many nations. This food crisis contributed to the increase in global hunger shown in Figure 7.2.

THE CHANGING FACE OF AGRICULTURE

If we are to increase food security for all people, we will need to examine the ways we produce our food and how we can make them more sustainable. As the human population has grown, so have the amounts of land and resources we devote to agriculture. We can define **agriculture** as the practice of raising crops and livestock for human use and consumption.

We obtain most of our food and fiber from **cropland**, land used to raise plants for human use, and from **rangeland**, land used for grazing livestock. Agriculture currently covers 38% of Earth's land surface. Of this land, 26% is rangeland and 12% is cropland.

Industrial agriculture is a recent human invention

During most of the human species' 160,000-year existence, we were hunter-gatherers, depending on wild plants and animals for our food and fiber. Then about 10,000 years ago, as glaciers retreated and the climate warmed, people in some cultures began to raise plants from seed and to domesticate animals.

For thousands of years, the work of cultivating, harvesting, storing, and distributing crops was performed by human and animal muscle power, along with hand tools and simple machines—an approach known as **traditional agriculture**. Traditional farmers typically plant **polycultures** ("many types") that mix different crops in small plots of farmland, such as the Native American farming systems that mixed maize, beans, squash, and peppers. Traditional agriculture is still practiced today but has rapidly been overtaken by newer methods of farming that increase crop yields.

After thousands of years of practicing traditional agriculture, the industrial revolution (p. 3) introduced large-scale mechanization and fossil fuel combustion to agriculture just as it did to industry. Farmers replaced horses and oxen with machinery that provided faster and more powerful means of cultivating, harvesting, transporting, and processing crops. Such **industrial agriculture** also boosted yields by intensifying irrigation and introducing synthetic fertilizers, while the advent of chemical pesticides reduced herbivory by crop pests and competition from weeds. Today, industrial agriculture is practiced on over 25% of the world's cropland and dominates areas such as Iowa.

However, the use of machinery created a need for highly organized approaches to farming, leading to vast areas being planted with single crops in orderly, straight rows. Such **monocultures** ("one type") make farming more efficient, but they provide fewer habitats in farm fields, which reduces biodiversity. Moreover, when all plants in a field are genetically similar, as in monocultures, all are equally susceptible to viral diseases, fungal pathogens, or insect pests that can spread quickly from plant to plant.

Industrial agriculture's widespread use of genetically similar crop varieties has led to efforts to conserve the wild relatives of crop plants and crop varieties indigenous to various regions, because they may contain genes we will one day need to introduce into our commercial crops. Seeds of these species are stored in some 1,400 **seed banks**, institutions that preserve seed types, around the world, including a "doomsday seed vault" established in 2008 on the island of Spitsbergen in Arctic Norway. Many agricultural scientists feel that we also need to protect the genetic integrity of wild relatives of crop plants by preventing genetically modified versions of these plants from exchanging genes with them. These scientists want to avoid such genetic "contamination" of wild populations so that we preserve the natural gene combinations of plants that are well adapted to their environments.

Our use of monocultures also contributes to a narrowing of the human diet. Globally, 90% of the food we consume now comes from just 15 crop species and eight livestock species—a drastic reduction in diversity from earlier times. Only 30% of the maize varieties that grew in Mexico in the 1930s exist today. In the United States, many fruit and vegetable crops have decreased in diversity by 90% in less than a century.

The Green Revolution boosted production—and exported industrial agriculture

The desire for greater quantity and quality of food for our growing population led in the mid- and late-20th century to the **Green Revolution**. This agricultural "revolution" introduced new technology, crop varieties, and farming practices to the developing world and drastically increased food production in these nations.

The transfer of the technology to the developing world that marked the Green Revolution began in the 1940s, when the late U.S. agricultural scientist **Norman Borlaug** introduced Mexico's farmers to a specially bred type of wheat (**FIGURE 7.4**). This strain of wheat produced large seed heads, was resistant to diseases, was short in stature to resist wind, and produced high yields. Within two decades of planting this new crop, Mexico tripled its wheat production and began exporting wheat. The stunning success of this program inspired similar projects around the world. Borlaug—who won the Nobel Peace Prize for his work—took his wheat to India and Pakistan and helped transform agriculture there.

Soon many developing countries were doubling, tripling, or quadrupling their yields using selectively bred strains of wheat, rice, corn, and other crops from industrialized nations. These crops dramatically increased yields and helped millions avoid starvation. When Borlaug died in 2009 at age 95, he was widely celebrated as having "saved more lives than anyone in history."

The effects of industrial agriculture have been mixed

Industrial agriculture has allowed food production to keep pace with our growing population, but it has many adverse environmental and social impacts. On the positive side, high-input industrial agriculture succeeded dramatically in producing higher crop yields from each hectare of land and reducing pressure to develop natural areas for new farmland. Between 1961 and 2008, for example, food production rose 150% and population rose 100%, while area converted for agriculture increased only 10%.

On the negative side, the intensive application of water, fossil fuels, inorganic fertilizers, and synthetic pesticides worsened pollution, topsoil losses, and soil quality. Industri-

FIGURE 7.4 ▼ Norman Borlaug holds examples of the wheat variety he bred that helped launch the Green Revolution. The high-yielding, disease-resistant wheat helped boost agricultural productivity in many developing countries.

ial agriculture also requires far more energy than traditional agriculture. From 1900 to 2000, as industrial agriculture expanded around the world, humans increased energy inputs into agriculture by 80 times while expanding the world's cultivated area by just 33%. Industrial agriculture also displaces low-income farmers who cannot afford the advanced technologies it requires. This can force poor rural people to migrate to urban areas, as is occurring today in many regions of the developing world (pp. 400-401).

WEIGHING THE ISSUES

The Green Revolution and Population

In the 1960s, India's population was skyrocketing, and its traditional agriculture was not producing enough food to support the growth. By adopting Green Revolution agriculture, India sidestepped mass starvation. However, Norman Borlaug called his Green Revolution methods "a temporary success in man's war against hunger and deprivation," something to give us breathing room in which to deal with what he called the "Population Monster." Indeed, in the years since intensifying its agriculture, India has added several hundred million more people and continues to suffer widespread poverty and hunger.

Do you think the Green Revolution has solved problems, deferred problems, or created new ones? Have the benefits of the Green Revolution outweighed its costs?

Sustainable agriculture reduces environmental impacts

We have achieved our impressive growth in food production by devoting more fossil fuel energy to agriculture; intensifying our use of irrigation, fertilizers, and pesticides; planting and harvesting more frequently; cultivating more land; and developing (through crossbreeding and genetic engineering) more productive crop and livestock varieties. However, many of these practices have degraded soils, polluted waters, and affected biodiversity.

We cannot simply keep expanding agriculture into new areas, because land suitable and available for farming is running out. Instead, we must find ways to improve the efficiency of food production in areas already under cultivation. Industrial agriculture in some form seems necessary to feed our planet's 7 billion people, but many experts feel we will be better off in the long run by raising animals and crops in ways that are less polluting, are less resource-intensive, and cause less impact on natural systems. To this end, many farmers and agricultural scientists are creating agricultural systems that better mimic the way a natural ecosystem functions.

Sustainable agriculture describes agriculture that maintains the healthy soil, clean water, and genetic diversity essential to long-term crop and livestock production. It is agriculture that can be practiced in the same way far into the future. Treating agricultural systems as ecosystems is a key aspect of sustainable agriculture, and this general lesson applies regardless of location, scale, or the crop involved.

One key component of making agriculture sustainable is reducing the fossil-fuel-intensive inputs we devote to ag-

riculture and decreasing the pollution these inputs cause. Fossil fuels power farm machinery and the pumps that draw irrigation water. They are also used as raw materials in the synthesis of some fertilizers and synthetic pesticides. *Low-input agriculture* describes agriculture that uses lesser amounts of pesticides, fertilizers, growth hormones, antibiotics, water, and fossil fuel energy than are used in industrial agriculture. Such approaches also seek to reduce food production costs by allowing nature to provide valuable ecosystem services, such as fertilization and pest control, that farmers using industrial agricultural methods must pay for themselves.

In the sections that follow, we will examine how modern food production affects the basic elements of agriculture—maintaining soil fertility, watering and fertilizing crops, and controlling agricultural pests—and how more sustainable methods can reduce these impacts while maintaining high crop yields. We will begin with soils, the foundation of agriculture.

SOILS

Soil is not merely lifeless dirt; it is a complex system consisting of disintegrated rock, organic matter, water, gases, nutrients, and microorganisms. Healthy soil is vital for agriculture, for forests (Chapter 9), and for the functioning of Earth's natural systems. Productive soil is a renewable resource. Once depleted, soil may renew itself over time, but this renewal generally occurs very slowly. If we abuse soil through careless or uninformed practices, we can greatly reduce its ability to support life for long time periods.

We generally overlook the startling complexity of soil. By volume, soil consists very roughly of 50% mineral matter and up to 5% organic matter. The rest consists of pore space taken up by air or water. The organic matter in soil includes living and dead microorganisms as well as decaying material derived from plants and animals. The soil ecosystem supports a diverse collection of bacteria, fungi, protists, worms, insects, and burrowing animals (**FIGURE 7.5**). The composition of a region's soil can have as much influence on its ecosystems as do climate, latitude, and elevation.

Soil forms slowly

Soil formation begins when the lithosphere's parent material is exposed to the effects of the atmosphere, hydrosphere, and biosphere (pp. 22–23). **Parent material** is the base geologic material in a particular location. It can be hardened lava or volcanic ash; rock or sediment deposited by glaciers; wind-blown dunes; sediments deposited in riverbeds, floodplains, lakes, and the ocean; or **bedrock**, the continuous mass of solid rock that makes up Earth's crust. Parent material is broken down by **weathering**, the physical, chemical, and biological processes that convert large rock particles into smaller particles.

Once weathering has produced fine particles, biological activity contributes to soil formation through the deposition, decomposition, and accumulation of organic matter. As plants, animals, and microbes die or deposit waste, this material is incorporated amid the weathered rock particles, mixing with minerals. The deciduous trees of temperate forests, for example, drop their leaves each fall, making leaf litter available to the detritivores and decomposers (p. 68) that break it

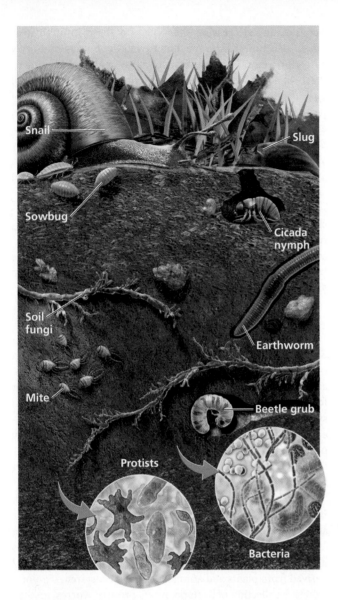

FIGURE 7.5 ▲ Soil is a complex mixture of organic and inorganic components and is full of living organisms whose actions help to keep it fertile. In fact, entire ecosystems exist in soil. Most soil organisms, from bacteria to fungi to insects to earthworms, decompose organic matter. Many, such as earthworms, also help to aerate the soil.

down and incorporate its nutrients into the soil. In decomposition, complex organic molecules are broken down into simpler ones, including those that plants can take up through their roots. Partial decomposition of organic matter creates *humus*, a dark, spongy, crumbly mass of material made up of complex organic compounds. Soils with high humus content hold moisture well and are productive for plant life.

A soil profile consists of layers known as horizons

As wind, water, and organisms move and sort the fine particles that weathering creates, distinct layers eventually develop. Each layer of soil is termed a **horizon**, and the cross-section as a whole, from surface to bedrock, is known as a **soil profile**.

The simplest way to categorize soil horizons is to recognize A, B, and C horizons corresponding respectively to topsoil,

subsoil, and parent material. However, soil scientists often recognize at least three additional horizons, including an O horizon (litter layer) that consists primarily of organic matter from organisms (**FIGURE 7.6**). Soils from different locations vary, and few soil profiles contain all six of these horizons, but any given soil contains at least some of them.

Generally, the degree of weathering and the concentration of organic matter decrease as one moves downward in a soil profile. Minerals are generally transported downward as a result of **leaching**, the process whereby solid particles suspended or dissolved in liquid are transported to another location. In some soils, minerals may be leached so rapidly that plants are deprived of nutrients.

A crucial horizon for agriculture and ecosystems is the A horizon, or **topsoil**. Topsoil consists mostly of inorganic mineral components such as weathered substrate, with organic matter and humus from above mixed in. Topsoil is the portion of the soil that is most nutritive for plants, and it takes its loose texture, dark coloration, and strong water-holding capacity from its humus content. The O and A horizons are home to most of the countless organisms that give life to soil. Topsoil is vital for agriculture, but agriculture practiced unsustainably

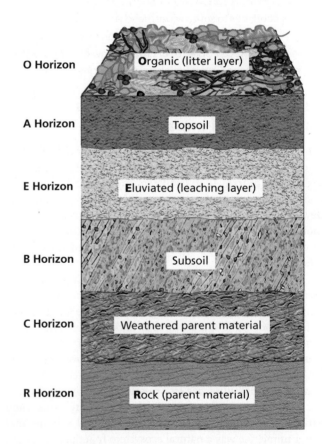

FIGURE 7.6 ▲ Mature soil consists of layers, or horizons, that have different compositions and characteristics. Uppermost is the **O horizon**, or litter layer (O = organic), consisting mostly of organic matter deposited by organisms. Below it lies the **A horizon**, or topsoil, consisting of some organic material mixed with mineral components. Minerals and organic matter tend to leach out of the **E horizon** (E = eluviation, or leaching) into the **B horizon**, or subsoil, where they accumulate. The **C horizon** consists largely of weathered parent material and overlies an **R horizon** (R = rock) of pure parent material.

over time will deplete organic matter, reducing the soil's fertility and ability to hold water. When a farmer practices no-till farming, he or she essentially creates an O horizon of crop residue to cover the topsoil and then plants seeds of the new crop through this O horizon into the protected topsoil layer.

Regional differences in soil traits affect agriculture

Soil characteristics vary from place to place, and are deeply affected by variables like climate. In the Amazon, the enormous amount of rain that falls leaches minerals and nutrients out of the topsoil and E horizon. Anything not captured by plants is taken quickly down to the water table, below the reach of plants' roots. Warm temperatures speed the decomposition of leaf litter and the uptake of nutrients by plants, so only small amounts of humus remain in the thin topsoil layer. As a result,

when tropical forest is cleared for farming, cultivation quickly depletes the soil's fertility.

The traditional form of agriculture in tropical forested areas is *swidden* agriculture, in which the farmer cultivates a plot for one to a few years and then moves on to clear another plot, leaving the first to grow back to forest (**FIGURE 7.7A**). At low population densities this can be sustainable, but with today's dense human populations, soils may not be allowed enough time to regenerate. As a result, intensive agriculture has degraded the soils of many tropical areas.

On the Iowa prairie, however, there is less rainfall, and therefore less leaching, which keeps nutrients within reach of plants' roots. Plants return nutrients to the topsoil when they die, maintaining its fertility. The thick, rich topsoil of temperate grasslands can be farmed repeatedly with minimal loss of fertility if techniques such as no-till farming and conservation tillage are used (**FIGURE 7.7B**).

(a) Tropical swidden agriculture on nutrient-poor soil

(b) Industrial agriculture on Iowa's rich topsoil

FIGURE 7.7 ◄ In tropical forested areas such as Surinam **(a)**, the traditional form of farming is *swidden* agriculture. In this practice, forest is cut, the plot is farmed for one to a few years, and the farmer then moves on to clear another plot. Frequent movement is necessary because tropical rainforest soils **(see inset)** are nutrient-poor and easily depleted. On the Iowa prairie **(b)**, less rainfall means that nutrients are not leached from the topsoil. Organic matter accumulates, forming rich soil that can be sustainably farmed. Note the thick, dark layer of topsoil in the inset for the Iowa farm.

Q: What is "slash-and-burn" agriculture?

A: Soils in tropical forests are not well suited for cultivating crops because they contain relatively low levels of plant nutrients. But although the soil is nutrient-poor, there are large amounts of nutrients tied up in the forest's lush vegetation. When farmers develop plots of tropical rainforest for agriculture, they enrich the soil by cutting down and burning the plants on the site. The nutrient-rich ash is then tilled into the soil, providing sufficient fertility to grow crops. The practice is therefore called *slash-and-burn* agriculture.

The nutrients from the ash, however, are usually depleted in only one to a few years. Lacking the resources to purchase synthetic fertilizer, poor farmers simply move deeper into the forest, repeat the process on a new plot of forested land, and cause further impacts on these productive and biologically diverse ecosystems.

MAINTAINING HEALTHY SOILS

Throughout the world, especially in drier regions, it has gotten more difficult to raise crops and graze livestock. Soils have deteriorated in quality and declined in productivity—a process termed **soil degradation**. Each year, our planet gains 80 million people yet loses 5–7 million ha (12–17 million acres, about the size of West Virginia) of productive cropland to degradation. The common causes include soil erosion, nutrient depletion, water scarcity, salinization (p. 145), waterlogging (p. 145), chemical pollution, changes in soil structure and pH, and loss of organic matter from the soil. Over the past 50 years, scientists estimate that soil degradation has reduced potential rates of global grain production on cropland by 13%.

Erosion can degrade ecosystems and agriculture

Erosion is the removal of material from one place and its transport toward another by the action of wind or water (**FIGURE 7.8**). *Deposition* occurs when eroded material arrives at a new location and is deposited. Erosion and deposition are natural processes that in the long run can help create soil. Flowing water can deposit freshly eroded sediment rich in nutrients across river valleys and deltas, producing fertile and productive soils. However, erosion often is a problem locally for ecosystems and agriculture because it generally occurs much more quickly than soil is formed. Erosion also tends to remove topsoil, the most valuable soil layer for living things. In general, steeper slopes, greater precipitation intensities, and sparser vegetative cover all lead to greater water erosion.

People have made land more vulnerable to erosion through three widespread practices:

▶ Overcultivating fields through poor planning or excessive tilling

▶ Overgrazing rangeland with more livestock than the land can support

▶ Clearing forests on steep slopes or with large clear-cuts (pp. 194-195)

Soil erosion is a global problem

In today's world, humans are the primary cause of erosion, and we have accelerated it to unnaturally high rates. In a 2004 study, geologist Bruce Wilkinson concluded that human activities move over 10 times more soil than all other natural processes on the surface of the planet combined. A 2007 study by soil scientist David Montgomery found an even greater degree of human impact. Montgomery's study also pointed toward a solution, revealing that land farmed under conservation approaches erodes at slower rates than land under conventional farming.

More than 19 billion ha (47 billion acres) of the world's croplands suffer from erosion and other forms of soil degradation resulting from human activities. In the past decade, China lost an area of arable farmland the size of Indiana. Farmlands in the United States lose roughly 5 tons of soil for every ton of grain harvested. In Africa, soil degradation over the next 40 years could reduce crop yields by half. Couple these declines in soil quality and crop yields with the rapid population growth occurring in many of these areas, and we begin to see why some observers describe the future of agriculture as a crisis situation.

Desertification reduces productivity of arid lands

Much of the world's population lives and farms in *drylands*, arid and semi-arid environments that cover about 40% of Earth's land surface. Precipitation in these regions is too meager to meet the demand for water from growing human populations, so drylands are prone to desertification. **Desertification** describes a form of land degradation in which more than 10% of productivity is lost as a result of erosion, soil compaction, forest removal, overgrazing, drought, salinization, climate change, water depletion, and

FIGURE 7.8 ▼ Water erosion, like that occurring here on farmland in Iowa, can readily remove soil from areas where soil is exposed.

other factors. Most such degradation results from wind and water erosion.

By some estimates, desertification endangers the food supply or well-being of more than 1 billion people in over 100 countries, and costs tens of billions of dollars in income each year through reduced productivity. A 2007 United Nations report estimated that desertification could worsen as climate change alters rainfall patterns, displacing as many as 50 million people in 10 years. China alone loses $6.5 billion annually from desertification. In its western reaches, desert areas are expanding and joining one another because of overgrazing from over 400 million goats, sheep, and cattle.

As a result of desertification, in recent years gigantic dust storms from denuded land in China have blown across the Pacific Ocean to North America, and dust storms from Africa's Sahara Desert have blown across the Atlantic Ocean to the Caribbean Sea (see Figure 13.6, p. 283). Such massive dust storms occurred in the United States during the Dust Bowl days of the early 20th century, when desertification shook American agriculture and society to their roots.

The Dust Bowl prompted the United States to fight erosion

Prior to large-scale cultivation of North America's Great Plains, native prairie grasses of this temperate grassland region held soils in place. In the late 19th and early 20th centuries, many homesteading settlers arrived in Oklahoma, Texas, Kansas, New Mexico, and Colorado with hopes of making a living there as farmers. Farmers grew abundant wheat, and ranchers grazed many thousands of cattle, sometimes expanding onto unsuitable land and contributing to erosion by removing native grasses and altering soil structure.

In the early 1930s, a drought exacerbated the ongoing human impacts, and the region's strong winds began to erode millions of tons of topsoil. Dust storms traveled up to 2,000 km (1,250 mi), blackening rain and snow as far away as New York and Washington, D.C. Some areas lost 10 cm (4 in.) of topsoil in a few years. The most-affected region in the southern Great Plains became known as the **Dust Bowl**, a term now also used for the historical event itself (**FIGURE 7.9**). The "black blizzards" of the Dust Bowl forced thousands of farmers off their land.

In response, the U.S. government, along with state and local governments, increased support for research into soil conservation measures. The U.S. Congress passed the Soil Conservation Act of 1935, establishing the Soil Conservation Service (SCS). This new agency worked closely with farmers to develop conservation plans for individual farms.

The SCS (currently named the *Natural Resources Conservation Service*) served as a model for other nations that established their own soil conservation agencies. In South America, no-till agriculture has exploded in popularity and now covers a large proportion of farmland in Argentina, Brazil, and Paraguay. The shift to no-till farming across this vast region came about largely through local grassroots organization by farmers, with the help of agronomists and government extension agents who provided them with information and resources.

(a) Kansas dust storm, 1930s

(b) Dust Bowl region

FIGURE 7.9 ▲ Drought combined with poor agricultural practices brought devastation and despair to millions of U.S. farmers in the 1930s in the Dust Bowl region of the southern Great Plains. The photo **(a)** shows towering clouds of dust approaching houses near Dodge City, Kansas, in a 1930s dust storm. The map **(b)** shows the Dust Bowl region, with darker colors indicating the areas most affected.

Sustainable agriculture begins with soil management

A number of farming techniques can reduce the impacts of conventional cultivation on soils and combat soil degradation (**FIGURE 7.10**).

Crop rotation In **crop rotation**, farmers alternate the type of crop grown in a given field from one season or year to the next (**FIGURE 7.10A**). Rotating crops can return nutrients to the soil, break cycles of disease associated with continuous cropping, and minimize the erosion that can come from letting fields lie fallow. Many U.S. farmers rotate their fields between wheat or corn and soybeans from one year to the next. Soybeans are legumes, plants that have specialized bacteria on their roots that fix nitrogen (pp. 38–39), revitalizing soil that the previous crop had partially depleted of nutrients. Crop rotation also reduces insect pests; if an insect is adapted to feed and lay eggs on one crop, planting a different type of crop will leave its offspring with nothing to eat.

In a practice similar to crop rotation, many no-till farmers such as those in Iowa plant cover crops, such as nitrogen-

(a) Crop rotation

(b) Contour farming

(c) Terracing

(d) Intercropping

(e) Shelterbelts

(f) No-till farming

FIGURE 7.10 ▲ The world's farmers have adopted various strategies to conserve soil. Rotating crops such as soybeans and corn **(a)** helps restore soil nutrients and reduce impacts of crop pests. Contour farming **(b)** reduces erosion on hillsides. Terracing **(c)** minimizes erosion in steep mountainous areas. Intercropping **(d)** can reduce soil loss while maintaining soil fertility. Shelterbelts **(e)** protect against wind erosion. In **(f)**, corn grows up from amid the remnants of a "cover crop" used in no-till agriculture.

replenishing clover, to prevent erosion during times of year when the main crops are not growing.

Contour farming   Water running down a hillside with little vegetative cover can easily carry soil away, so farmers have developed several methods for cultivating slopes. **Contour farming** (**FIGURE 7.10B**) consists of plowing furrows

sideways across a hillside, perpendicular to its slope and following the natural contours of the land. In contour farming, the side of each furrow acts as a small dam that slows runoff and captures eroding soil. Contour farming is most effective on gradually sloping land with crops that grow well in rows. Iowa farmers Butch and David Schroeder practice contour farming on all their fields. They also plant buffer strips of vegetation along the borders of their fields and along nearby streams, which further protects against erosion and water pollution.

Terracing On extremely steep terrain, terracing (**FIGURE 7.10C**) is the most effective method for preventing erosion. Terraces are level platforms, sometimes with raised edges, that are cut into steep hillsides to contain water from irrigation and precipitation. **Terracing** transforms slopes into series of steps like a staircase, enabling farmers to cultivate hilly land without losing huge amounts of soil to water erosion.

Intercropping Farmers may also minimize erosion by **intercropping**, planting different types of crops in alternating bands or other spatially mixed arrangements (**FIGURE 7.10D**). Intercropping helps slow erosion by providing more ground cover than does a single crop. Like crop rotation, intercropping reduces vulnerability to insects and disease and, when a nitrogen-fixing legume is used, replenishes the soil. In Brazil, some no-till farmers intercrop cover crops with food crops such as maize, soybeans, wheat, onions, cassava, grapes, tomatoes, tobacco, and orchard fruit.

Shelterbelts A widespread technique to reduce erosion from wind is to establish **shelterbelts**, or *windbreaks* (**FIGURE 7.10E**). These are rows of trees or other tall plants that are planted along the edges of fields to slow the wind. On the Great Plains, fast-growing species such as poplars are often used. Shelterbelts can be combined with intercropping by planting mixed crops in rows surrounded by or interspersed with rows of trees that provide fruit, wood, or protection from wind.

Conservation tillage *Conservation tillage* describes an array of approaches that reduce the amount of tilling relative to conventional farming. No-till farming is the ultimate form of conservation tillage. To plant using the no-till method (**FIGURE 7.10F**), a tractor pulls a "no-till drill" that cuts furrows through the O horizon of dead weeds and crop residue and the upper levels of the A horizon. The device drops seeds into the furrow and closes the furrow over the seeds, minimizing disturbance to the soil.

By increasing organic matter and soil biota while reducing erosion, no-till farming and conservation tillage can improve soil quality and combat global climate change by storing carbon in soils. In the United States today, nearly one-quarter of farmland is under no-till cultivation, and over 40% is farmed using conservation tillage (**FIGURE 7.11**). According to U.S. government figures, erosion rates in the United States declined from 9.1 tons/ha (3.7 tons/acre) in 1982 to 5.9 tons/ha (2.4 tons/acre) in 2003, thanks to conservation tillage.

No-till and conservation tillage methods have become most widespread in subtropical and temperate South America. In Brazil, Argentina, and Paraguay, over half of all cropland is now under no-till cultivation. In this part of the world, heavy rainfall promotes erosion, causing tilled soils to lose organic matter and nutrients, and hot weather can overheat tilled soil. Thus, the no-till approach is especially helpful here, and results have exceeded those in the United States: Crop yields increased (in some cases they nearly doubled), erosion was reduced, soil quality was enhanced, and pollution declined, all while costs to farmers dropped by roughly 50%.

Critics of no-till farming in the United States have noted that this approach often requires substantial use of chemical herbicides (because weeds are not physically removed from fields) and synthetic fertilizer (because other plants take up a significant portion of the soil's nutrients). In many industrialized countries, this has indeed been the case. Proponents, however, point out that in developing regions of South America, farmers have departed from the industrialized model by relying more heavily on *green manures* (dead plants as fertilizer) and by rotating fields with cover crops, including nitrogen-fixing legumes. The manures and legumes nourish the soil, and cover crops reduce weeds by taking up space the weeds might occupy. Although this approach is often not practical for large-scale intensive agriculture, farmers are educating themselves on the available approaches and choosing those that are best for their farms.

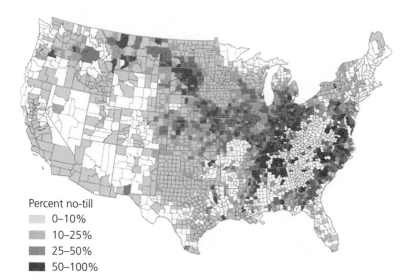

Percent no-till
- 0–10%
- 10–25%
- 25–50%
- 50–100%

FIGURE 7.11 ◀ Across the United States, red and orange colors on this county-by-county map show that no-till methods are practiced most in the Midwest and along the central and southern Atlantic seaboard. Data from Conservation Technology Information Center, National Crop Residue Management Survey, 2004.

Grazing practices can contribute to soil degradation

We have focused in this chapter largely on the cultivation of crops as a source of impacts on soils and ecosystems, but raising livestock also has impacts. Humans keep a total of 3.4 billion cattle, sheep, and goats that graze primarily on grasses on the open range. As long as livestock populations do not exceed a range's carrying capacity (p. 58) and do not consume grass faster than it can regrow, grazing may be sustainable. However, when too many livestock eat too much of the plant cover, it impedes plant regrowth and prevents the replacement of biomass, and the result is **overgrazing**.

When livestock remove too much plant cover, soil is exposed and made vulnerable to erosion. In a positive feedback cycle (pp. 22–23), soil erosion makes it difficult for vegetation to regrow, a problem that perpetuates the lack of cover and gives rise to more erosion (**FIGURE 7.12**). Moreover, non-native weedy plants that are unpalatable to livestock may outcompete native vegetation in the new, modified environment.

Too many livestock trampling the ground can also compact soils and alter their structure. Soil compaction makes it more difficult for water to infiltrate, for soils to be aerated, for plants' roots to expand, and for roots to conduct cellular respiration (pp. 30–31). All of these effects further decrease the growth and survival of native plants.

As a cause of soil degradation worldwide, overgrazing is equal to cropland agriculture, and it is a greater cause of desertification. Fully 70% of the world's rangeland is classified as degraded, and lost productivity on these lands is estimated at $23.3 billion per year. Rangelands in the western United States have often been overgrazed by a "tragedy of the commons" process (pp. 3–4), but today increasing numbers of ranchers are working cooperatively with government agencies, environmental scientists, and even environmental advocates to find ways to ranch more sustainably and safeguard the health of the land.

Agricultural subsidies affect soil degradation

In theory, the marketplace should discourage people from farming and grazing using intensive methods that degrade the land they own if such practices are not profitable. But land degradation often unfolds gradually, whereas farmers and ranchers generally cannot afford to go without profits in the short term, even if they know conservation is in their long-term interests.

Many nations spend billions of dollars in government subsidies to support agriculture. Roughly one-fifth of the income of the average U.S. farmer comes from subsidies. Proponents of such subsidies stress that the uncertainties of weather make profits and losses from farming unpredictable from year to year. To persist, these proponents say, an agricultural system needs some way to compensate farmers for bad years. This may be the case, but subsidies can encourage people to cultivate land that would otherwise not be farmed; to produce more food than is needed, driving down prices for other producers; and to practice methods that further degrade the land. Thus, opponents of subsidies argue that subsidizing environmentally destructive agricultural practices is unsustainable. They suggest that a better model is for farmers to buy insurance to protect themselves against short-term production failures.

A number of U.S. and international programs promote soil conservation

While government subsidies may promote soil degradation, the U.S. Congress has also enacted a number of provisions promoting soil conservation through the farm bills it passes every five to six years. Many of these provisions require farmers to adopt soil conservation plans and practices before they can receive government subsidies.

The **Conservation Reserve Program**, established in the 1985 farm bill, pays farmers to stop cultivating highly erodible cropland and instead place it in conservation reserves

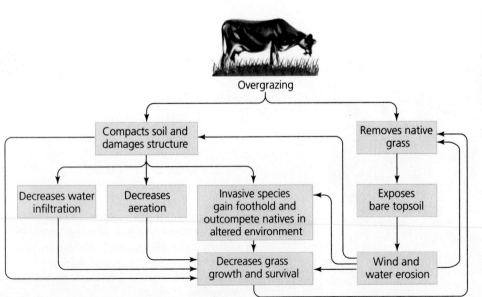

FIGURE 7.12 ◀ When grazing by livestock exceeds the carrying capacity of rangelands and their soil, it can set in motion a series of consequences and positive feedback loops that degrade soils and grassland ecosystems.

planted with grasses and trees. Lands under the Conservation Reserve Program now cover an area nearly the size of Iowa, and the United States Department of Agriculture (USDA) estimates that each dollar invested in this program saves nearly 1 ton of topsoil. Besides reducing erosion, the Conservation Reserve Program generates income for farmers, improves water quality, and provides habitat for wildlife. Nationwide, the government pays farmers about $1.8 billion per year for the conservation of lands totaling between 12–16 million ha (30–40 million acres).

Congress reauthorized and expanded the Conservation Reserve Program in the farm bills of 1996, 2002, and 2008. Butch and David Schroeder are among the many farmers in Iowa and across the nation who have utilized this program, taking land with damaged and erodible soil out of production. The 2008 bill limited total conservation reserve lands to 32 million acres but also funded 14 other programs for the conservation of grasslands, wetlands, wildlife habitat, and other resources. Internationally, the United Nations promotes soil conservation and sustainable agriculture through a variety of programs led by the Food and Agriculture Organization (FAO).

WATERING AND FERTILIZING CROPS

Soil degradation is not the only impact of agriculture on the environment. We affect natural systems, sometimes far away from farm fields, when we supply plants with supplemental water and nutrients to boost crop yields using unsustainable approaches.

Irrigation boosts productivity but can damage soil

Plants require water for optimum growth, and people have long supplemented the water that crops receive from rainfall. The artificial provision of water to support agriculture is known as **irrigation**. By irrigating crops, people maintain high yields in times of drought and turn previously dry and unproductive regions into fertile farmland.

Worldwide, irrigated acreage has increased along with the adoption of industrial farming methods, and 70% of the fresh water withdrawn by humans is applied to crops. In some cases, withdrawing water for irrigation has depleted aquifers and dried up rivers and lakes (pp. 262–263).

In some locations, excessive irrigation has degraded soils. **Waterlogging** occurs when over-irrigation causes the water table to rise to the point that water drowns plant roots, depriving them of access to gases and essentially suffocating them. A more frequent problem is **salinization**, the buildup of salts in surface soil layers. In dryland areas where precipitation is minimal and evaporation rates are high, the evaporation of water from the soil's A horizon may pull water with dissolved salts up from lower horizons. When the water evaporates at the surface, those salts remain on the soil, often turning the soil surface white. Salinization now inhibits production on one-fifth of all irrigated cropland globally, costing more than $11 billion each year.

Sustainable approaches to irrigation maximize efficiency

One of the most effective ways to reduce water use in agriculture is to better match crops and climate. Many arid regions have been converted into productive farmland through extensive irrigation, often with the support of government subsidies that make irrigation water artificially inexpensive. Some farmers in these areas cultivate crops that require large amounts of water, such as rice and cotton. This leads to extensive water losses from evaporation in the arid climate. Choosing other crops that require far less water, such as beans or wheat, could enable these areas to remain agriculturally productive and yet greatly reduce water use.

Another approach is to embrace new technologies that improve water use efficiency in irrigation. Currently, irrigation efficiency worldwide is low, as plants end up using only 43% of the water that we apply—the rest evaporates or soaks into the soil away from plant roots (**FIGURE 7.13A**). New drip irrigation systems that deliver water directly to plant roots can increase efficiencies to over 90% (**FIGURE 7.13B**). Such systems

(a) Conventional irrigation

(b) Drip irrigation

FIGURE 7.13 ▲ Currently, plants take up less than half the water we apply in irrigation. Conventional methods **(a)** lose a great deal of water to evaporation. In more-efficient irrigation approaches, water is precisely targeted to plants. In drip irrigation systems, such as this one watering grape vines in California **(b)**, hoses are arranged so that water drips directly onto the plants.

were once quite expensive to install and were largely used by farmers in wealthier nations, but are becoming cheap enough that more farmers in developing countries will be able to afford them. Drip systems are not utilized in the vast fields of monocultures like those profiled in the Central Case Study, but they are useful on smaller plots of farmland and with perennial plants, such as fruit trees.

Fertilizers boost crop yields but can be overapplied

Plants require nitrogen, phosphorus, and potassium to grow, as well as smaller amounts of more than a dozen other nutrients. Leaching and uptake by plants removes these nutrients from soil, and if soils come to contain too few nutrients, crop yields decline. Therefore, we go to great lengths to enhance nutrient-limited soils by adding **fertilizer**, substances that contain essential nutrients.

There are two main types of fertilizers. **Inorganic fertilizers** are mined or synthetically manufactured mineral supplements. **Organic fertilizers** consist of the remains or wastes of organisms and include animal manure, crop residues, fresh vegetation (*green manure*), and *compost*, a mixture produced when decomposers break down organic matter, including food and crop waste, in a controlled environment.

Historically, humans relied on organic fertilizers to replenish soil nutrients. But during the latter half of the 20th century, farmers in industrialized and Green Revolution regions widely embraced the use of inorganic fertilizers (**FIGURE 7.14**). This use has greatly boosted our global food production, but the overapplication of inorganic fertilizers is causing increasingly severe pollution problems. Inorganic fertilizers are generally more susceptible to leaching than organic fertilizers and more readily contaminate groundwater supplies. Nutrients from these fertilizers can also have impacts far beyond the boundaries of the fields. We saw how nitrogen and phosphorus runoff from farms in the Midwest

and other sources spurs phytoplankton blooms in the Gulf of Mexico near the mouth of the Mississippi River and creates an oxygen-depleted "dead zone" that kills animal and plant life (Chapter 2). Such eutrophication (pp. 23–25, 267–268) occurs at countless river mouths, lakes, and ponds throughout the world. Air pollution can also arise from components of some nitrogen fertilizers that evaporate into the air, contributing to the formation of photochemical smog (pp. 288–289) and acid deposition (pp. 291–294).

Sustainable fertilizer use involves monitoring and targeting nutrients

Sustainable approaches to fertilizing crops target the delivery of nutrients to plant roots and avoid the overapplication of fertilizer. This is accomplished in many ways. Farmers using drip irrigation systems can add fertilizer to irrigation water, thereby releasing it only above plant roots. No-till or conservation tillage systems often inject fertilizer along with seeds, concentrating it near the developing plant. Farmers can also avoid overapplication by regularly monitoring soil nutrient content and applying fertilizer only when soil nutrient levels are too low, as was done by the Schroeder brothers in Iowa. Strips of vegetation planted at the edges of fields and along streams can capture nutrient runoff along with eroded soils.

Sustainable agriculture also embraces the use of organic fertilizers, as they can provide some benefits that inorganic fertilizers cannot. Organic fertilizers provide not only nutrients, but also organic matter that improves soil structure, nutrient retention, and water-retaining capacity. The use of organic fertilizers is not without cost, though. For instance, when manure is applied in amounts needed to supply sufficient nitrogen for a crop, it may introduce excess phosphorus that can run off into waterways. Accordingly, sustainable approaches do not rely solely on organic fertilizers, but rather integrate them with the targeted delivery of nutrients using inorganic fertilizer.

CONTROLLING PESTS

Throughout the history of agriculture, the insects, fungi, viruses, rats, and weeds that eat or compete with our crop plants have taken advantage of the ways we cluster food plants into agricultural fields. Pests and weeds pose an especially great threat to monocultures, where a pest adapted to specialize on the crop can move easily from plant to plant.

What people term a *pest* is any organism that damages crops that are valuable to us. What we term a *weed* is any plant that competes with our crops. There is nothing inherently malevolent in the behavior of a pest or a weed. These organisms are simply trying to survive and reproduce, but they affect farm productivity when doing so.

We have developed thousands of chemical pesticides

To suppress pests and weeds, people have developed thousands of chemicals to kill insects (*insecticides*), plants (*herbicides*), and fungi (*fungicides*). Such poisons are collectively termed **pesticides**. The highly modified ecosystems of industrial farming limit the ability of natural mechanisms

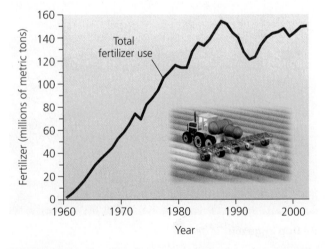

FIGURE 7.14 ▲ Use of synthetic inorganic fertilizers has risen sharply over the past half-century and now stands at over 147 million metric tons annually. (The temporary drop during the early 1990s was due to economic decline in countries of the former Soviet Union following that nation's dissolution.) Data from Food and Agriculture Organization of the United Nations (FAO).

to control pest populations. Hence, as industrial agriculture grew in use, farmers turned to chemical means to control agricultural pests.

Three-quarters of all pesticides are applied on agricultural land. Since 1960, pesticide use has risen fourfold worldwide. Usage in industrialized nations has leveled off in the past two decades, but it continues to rise in the developing world. Exposure to synthetic pesticides can have health consequences for people and other organisms under some circumstances (Chapter 10), so their use in food production can have far-reaching effects.

Pests evolve resistance to pesticides

Despite the toxicity of these chemicals, their effectiveness tends to decline with time as pests evolve resistance to them. Recall from our discussion of natural selection (pp. 46–49) that organisms within populations vary in their traits. Because most insects, weeds, and microbes can occur in huge numbers, it is likely that a small fraction of individuals may by chance already have genes that enable them to metabolize and detoxify a pesticide (p. 213). These individuals will survive exposure to the pesticide, while individuals without these genes will not. If an insect that is genetically resistant to an insecticide survives and mates with other resistant individuals,

the genes for pesticide resistance will be passed to their offspring. As resistant individuals become more prevalent in the pest population, insecticide applications will cease to be effective and the population will increase in number (**FIGURE 7.15**).

In many cases, industrial chemists are caught up in an evolutionary arms race (p. 67) with the pests they battle, racing to increase or retarget the toxicity of their chemicals while the armies of pests evolve ever-stronger resistance to their efforts. Because we seem to be stuck in this cyclical process, it has been nicknamed the "pesticide treadmill." As of 2011, among arthropods (insects and their relatives) alone, there were more than 9,900 known cases of resistance by 586 species to over 330 insecticides. Hundreds more weed species and plant diseases have evolved resistance to herbicides and other pesticides. Many species, including insects such as the green peach aphid, Colorado potato beetle, and diamondback moth, have evolved resistance to multiple chemicals.

Biological control pits one organism against another

Because of pesticide resistance, toxicity to nontarget organisms, and human health risks from some synthetic chemicals, agricultural scientists increasingly battle pests and weeds with organisms that eat or infect them. This more

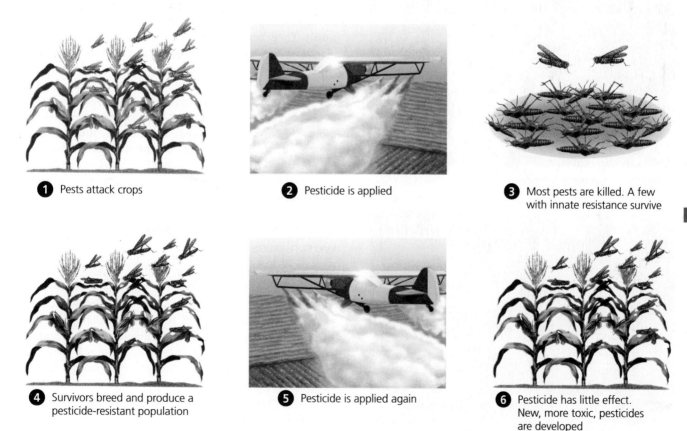

1 Pests attack crops

2 Pesticide is applied

3 Most pests are killed. A few with innate resistance survive

4 Survivors breed and produce a pesticide-resistant population

5 Pesticide is applied again

6 Pesticide has little effect. New, more toxic, pesticides are developed

FIGURE 7.15 ▲ Through the process of natural selection (pp. 46–49), crop pests may evolve resistance to the poisons we apply to kill them. When a pesticide is applied to an outbreak of insect pests, it may kill all individuals except those few with an innate immunity, or resistance, to the poison (resistant individuals are colored red in the diagram). Those surviving individuals may establish a population with genes for resistance to the poison. Future applications of the pesticide may then be ineffective, forcing us to develop a more potent poison or an alternative means of pest control.

pest control strategy, called **biological control** or **biocontrol**, operates on the principle that "the enemy of one's enemy is one's friend." For example, parasitoid wasps (p. 67) are natural enemies of many caterpillars. These wasps lay eggs on a caterpillar, and the larvae that hatch from the eggs feed on the caterpillar, eventually killing it. Parasitoid wasps are frequently used as biocontrol agents and have often succeeded in controlling pests and reducing chemical pesticide use.

One classic case of successful biological control is the introduction of the cactus moth, *Cactoblastis cactorum*, from Argentina to Australia in the 1920s to control invasive prickly pear cactus that was overrunning rangeland. Within just a few years, the moth managed to free millions of hectares of rangeland from the cactus (**FIGURE 7.16**).

However, biocontrol approaches entail risks. Biocontrol organisms are sometimes more difficult to manage than chemical controls, because they cannot be "turned off" once they are initiated. Further, biocontrol organisms have in some cases become invasive and harmed nontarget organisms. Following the cactus moth's success in Australia, for exam-

ple, it was introduced in other countries to control non-native prickly pear. Moths introduced to Caribbean islands spread to Florida on their own and are now eating their way through rare native cacti in the southeastern United States. If these moths reach Mexico and the southwestern United States, they could decimate many native and economically important species of prickly pear there. Because of concerns about unintended impacts, researchers study biocontrol proposals carefully before putting them into action, and government regulators must approve these efforts.

Integrated pest management combines varied approaches to pest control

As it became clear that both chemical and biocontrol approaches pose risks, agricultural scientists and farmers began developing more sophisticated strategies, trying to combine the best attributes of each approach. **Integrated pest management (IPM)** incorporates numerous techniques, including biocontrol, use of chemicals when needed, close monitoring of populations, habitat alteration, crop rotation, transgenic crops, alternative tillage methods, and mechanical pest removal.

IPM has become popular in many parts of the world that are embracing sustainable agriculture techniques. Indonesia stands as an exemplary case. This nation had subsidized pesticide use heavily for years, but its scientists came to understand that pesticides were actually making pest problems worse. They were killing the natural enemies of the brown planthopper, which began to devastate rice fields as its populations exploded. Concluding that pesticide subsidies were costing money, causing pollution, and apparently decreasing yields, the Indonesian government in 1986 banned the import of 57 pesticides, slashed pesticide subsidies, and promoted IPM. Within just four years, pesticide production fell by half, imports fell by two-thirds, and subsidies were phased out (saving $179 million annually). Rice yields rose 13% with IPM, and since then the approach has spread to dozens of other nations.

(a) Before cactus moth introduction

(b) After cactus moth introduction

FIGURE 7.16 ▲ In a classic case of biocontrol, larvae of the cactus moth, *Cactoblastis cactorum*, were used to clear non-native prickly pear cactus from millions of hectares of rangeland in Queensland, Australia. These photos from the 1920s show an Australian ranch before **(a)** and after **(b)** introduction of the moth.

Pollinators are beneficial "bugs" worth preserving

Managing insect pests is such a major issue in agriculture that it is easy to fall into a habit of thinking of all insects as somehow bad or threatening. But in fact, most insects are harmless to agriculture, and some are absolutely essential. The insects that pollinate crops are among the most vital, yet least understood and appreciated, factors in our food production (**FIGURE 7.17**). Pollinators are the unsung heroes of agriculture.

Pollination (p. 68) is the process by which male sex cells of a plant (pollen) fertilize female sex cells of a plant; it is the botanical version of sexual intercourse. Pollinators are animals that move pollen from one flower to another. Flowers are, in fact, evolutionary adaptations that function to attract pollinators. The sugary nectar and protein-rich pollen in flowers serve as rewards to lure these sexual intermediaries, and the sweet smells and bright colors of flowers are signals to advertise these rewards.

FIGURE 7.17 ▲ North American farmers hire beekeepers to bring hives (in white boxes in photo) of European honeybees (inset, on apple blossom) to their crops when it is time for flowers to be pollinated. Recently, honeybees have suffered devastating epidemics, making it increasingly important to conserve native species of pollinators.

Our staple grain crops are derived from grasses and are wind-pollinated, but 800 types of cultivated plants rely on bees and other insects for pollination. Overall, native species of bees in the United States alone are estimated to provide $3 billion of pollination services each year to crop agriculture.

Preserving the biodiversity of native pollinators is especially important today because the domesticated workhorse of pollination, the honeybee (*Apis mellifera*), is in decline. In recent years, two accidentally introduced parasitic mites have swept through honeybee populations, decimating hives and pushing many beekeepers toward financial ruin. On top of this, starting in 2006, entire hives inexplicably began dying off. In each of the last several years, up to one-third of all honeybees in the United States have vanished from what is being called *colony collapse disorder*. Scientists are racing to discover the cause of this mysterious syndrome. Leading hypotheses are insecticide exposure, an unknown new parasite, or a combination of stresses that weaken bees' immune systems and destroy social communication within the hive. We must be very careful when trying to control the "bad" bugs that threaten the plants we value so that we do not kill the "good" insects as well.

GENETICALLY MODIFIED FOOD

The Green Revolution enabled us to feed a greater number and proportion of the world's people, but relentless population growth is demanding still more innovation. A new set of potential solutions began to arise in the 1980s and 1990s as advances in genetics enabled scientists to directly alter the genes of organisms, including crop plants and livestock. The genetic modification of organisms that provide us food holds promise to enhance nutrition and the efficiency of agriculture while lessening impacts on the planet's environmental systems. However, genetic modification may also pose risks that are not yet well understood. This possibility has given rise to anxiety and protest by consumer advocates, small farmers, environmental activists, and critics of big business.

Foods can be genetically modified

The genetic modification of crops and livestock is one type of genetic engineering. **Genetic engineering** is any process whereby scientists directly manipulate an organism's genetic material in the laboratory by adding, deleting, or changing segments of its DNA (p. 28). **Genetically modified (GM) organisms** are organisms that have been genetically engineered using **recombinant DNA**, which is DNA that has been patched together from the DNA of multiple organisms (**FIGURE 7.18**).

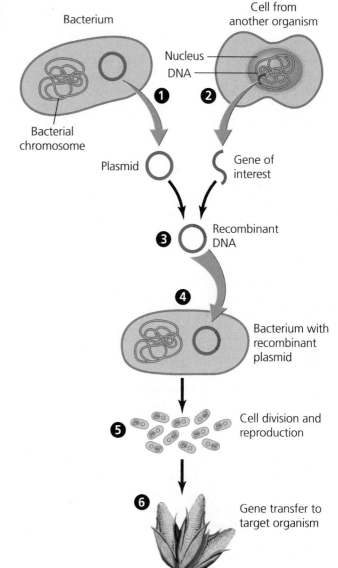

FIGURE 7.18 ▲ To create recombinant DNA, scientists first isolate *plasmids* ❶ small circular DNA molecules, from a bacterial culture. DNA containing a gene of interest ❷ is then removed from another organism. Scientists insert the gene of interest into the plasmid to form recombinant DNA ❸. This recombinant DNA enters new bacteria ❹, which then reproduce ❺, generating many copies of the desired gene. The desired gene is then transferred to many individuals of the target plant or animal ❻. The individuals that successfully incorporate the desired gene are then cultured and grown. If all goes as planned, the new gene will be expressed in the genetically modified organism as a desirable trait, such as rapid growth or high nutritional content in a food crop.

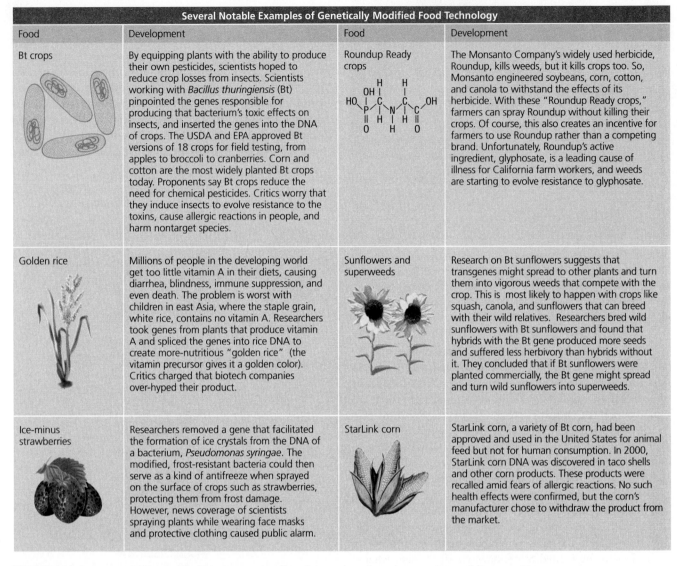

Several Notable Examples of Genetically Modified Food Technology			
Food	Development	Food	Development
Bt crops	By equipping plants with the ability to produce their own pesticides, scientists hoped to reduce crop losses from insects. Scientists working with *Bacillus thuringiensis* (Bt) pinpointed the genes responsible for producing that bacterium's toxic effects on insects, and inserted the genes into the DNA of crops. The USDA and EPA approved Bt versions of 18 crops for field testing, from apples to broccoli to cranberries. Corn and cotton are the most widely planted Bt crops today. Proponents say Bt crops reduce the need for chemical pesticides. Critics worry that they induce insects to evolve resistance to the toxins, cause allergic reactions in people, and harm nontarget species.	Roundup Ready crops	The Monsanto Company's widely used herbicide, Roundup, kills weeds, but it kills crops too. So, Monsanto engineered soybeans, corn, cotton, and canola to withstand the effects of its herbicide. With these "Roundup Ready crops," farmers can spray Roundup without killing their crops. Of course, this also creates an incentive for farmers to use Roundup rather than a competing brand. Unfortunately, Roundup's active ingredient, glyphosate, is a leading cause of illness for California farm workers, and weeds are starting to evolve resistance to glyphosate.
Golden rice	Millions of people in the developing world get too little vitamin A in their diets, causing diarrhea, blindness, immune suppression, and even death. The problem is worst with children in east Asia, where the staple grain, white rice, contains no vitamin A. Researchers took genes from plants that produce vitamin A and spliced the genes into rice DNA to create more-nutritious "golden rice" (the vitamin precursor gives it a golden color). Critics charged that biotech companies over-hyped their product.	Sunflowers and superweeds	Research on Bt sunflowers suggests that transgenes might spread to other plants and turn them into vigorous weeds that compete with the crop. This is most likely to happen with crops like squash, canola, and sunflowers that can breed with their wild relatives. Researchers bred wild sunflowers with Bt sunflowers and found that hybrids with the Bt gene produced more seeds and suffered less herbivory than hybrids without it. They concluded that if Bt sunflowers were planted commercially, the Bt gene might spread and turn wild sunflowers into superweeds.
Ice-minus strawberries	Researchers removed a gene that facilitated the formation of ice crystals from the DNA of a bacterium, *Pseudomonas syringae*. The modified, frost-resistant bacteria could then serve as a kind of antifreeze when sprayed on the surface of crops such as strawberries, protecting them from frost damage. However, news coverage of scientists spraying plants while wearing face masks and protective clothing caused public alarm.	StarLink corn	StarLink corn, a variety of Bt corn, had been approved and used in the United States for animal feed but not for human consumption. In 2000, StarLink corn DNA was discovered in taco shells and other corn products. These products were recalled amid fears of allergic reactions. No such health effects were confirmed, but the corn's manufacturer chose to withdraw the product from the market.

FIGURE 7.19 ▲ As genetically modified foods were developed, a number of products ran into public opposition or trouble in the marketplace. A selection of such cases serves to illustrate some of the issues that proponents and opponents of genetically modified foods have been debating.

The goal is to place genes that produce certain proteins and code for certain desirable traits (such as rapid growth, disease resistance, or high nutritional content) into organisms lacking those traits. An organism that contains DNA from another species is called a **transgenic** organism, and the genes that have moved between them are called **transgenes**.

The creation of transgenic organisms is one type of **biotechnology**, the material application of biological science to create products derived from organisms. Biotechnology has helped us develop medicines, clean up pollution, understand the causes of cancer and other diseases, dissolve blood clots after heart attacks, and make better beer and cheese. **FIGURE 7.19** shows several notable developments in GM foods. The stories behind them illustrate both the promises and pitfalls of food biotechnology.

The genetic alteration of plants and animals by people is nothing new; through artificial selection (p. 48) we have influenced the genetic makeup of our livestock and crop plants for thousands of years. However, as critics are quick to point out, the techniques geneticists use to create GM organisms differ from traditional selective breeding in several ways. For one, selective breeding mixes genes from individuals of the same or similar species, whereas scientists creating recombinant DNA routinely mix genes of organisms as different as viruses and crops, or spiders and goats. For another, selective breeding deals with whole organisms living in the field, whereas genetic engineering works with genetic material in the lab. Third, traditional breeding selects from combinations of genes that come together on their own, whereas genetic engineering creates the novel combinations directly. Thus, traditional breeding changes organisms through the process of selection (pp. 46–49), whereas genetic engineering is more akin to the process of mutation (p. 47).

Biotechnology is transforming the products around us

In just three decades, GM foods have gone from science fiction to mainstream agriculture. Most GM crops today are engineered to resist herbicides, so that farmers can apply herbicides to kill weeds without having to worry about killing their crops. Other crops are engineered to resist insect attack. Some are modified for both types of resistance. Resistance to herbicides and pests enables large-scale commercial farmers to grow crops more efficiently. As a result, sales of GM seeds to these farmers in the United States and other countries have risen quickly.

In 2010, GM varieties comprised 86% of the U.S. corn harvest and 93% of the soybean and cotton crops. This is incredible growth, considering that GM crop varieties have been commercially planted only since 1996. Worldwide, three of every four soybean plants are now transgenic, as is one of every four corn plants, one of every five canola plants, and half of all cotton plants. Given the diversity of foods that contain corn products and the overwhelming majority of the U.S. corn crop that is GM varieties, it is extremely likely that you consume GM foods on a daily basis.

Soybeans account for most of the world's GM crops (**FIGURE 7.20A**). Of the 29 nations growing GM crops in 2010, six (the United States, Brazil, Argentina, India, Canada, and China) accounted for 92% of production, with the United States alone growing nearly half the global total

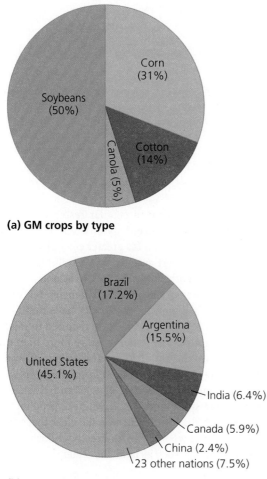

(a) GM crops by type

(b) GM crops by nation

FIGURE 7.20 ▲ Of the world's genetically modified crops **(a)**, soybeans constitute the majority so far. Of the world's nations **(b)**, the United States devotes the most land area to GM crops. Data are for 2010, from the International Service for the Acquisition of Agri-Biotech Applications.

(**FIGURE 7.20B**). The market value of GM crops in 2009 was estimated at $10.5 billion.

What are the impacts of GM crops?

As GM crops were adopted and as biotech business expanded, many citizens, scientists, and policymakers became concerned. Some feared the new foods might be dangerous for people to eat. Others worried that pests would evolve resistance to the supercrops and become "superpests" or that transgenes would be transferred from crops to other plants and turn them into "superweeds." Some worried that transgenes might become integrated into closely related wild relatives of crops.

Because GM technology is rapidly changing, and because its large-scale introduction into our environment is recent, there remains much that we don't yet know. Millions of Americans eat GM foods every day without outwardly obvious signs of harm, and direct evidence for negative ecological effects is limited so far. However, it is still too early to dismiss concerns about environmental impacts without further scientific research.

Data indicate that the effect of GM crops on pesticide use has been mixed. The adoption of insect-resistant GM crops appears to reduce the use of chemical insecticides, but farmers planting herbicide-tolerant crops tend to use more herbicide because their crops can withstand it. A 2010 study by the Institute of Science in Society calculated that as GM crops expanded in the United States between 1996 and 2008, insecticide use declined by 64 million pounds, but herbicide use increased by 383 million pounds.

Genetically modified crops have the potential to advance agriculture sustainably by promoting no-till farming with herbicide-resistant crops, engineering crops with high drought tolerance for use in arid regions, and developing high-yield crops that can feed our growing population without converting additional natural areas to agriculture.

However, GM crops are expensive to create, and crops with traits that might benefit poor, small-scale farmers of developing countries (such as increased nutrition, drought tolerance, and salinity tolerance) have not been widely commercialized, perhaps because corporations have less economic incentive to do so. Often, a company's GM crops are tolerant to herbicides that the same company manufactures and profits from (e.g., Monsanto's Roundup Ready crops). Whereas the Green Revolution was a largely public venture, the "gene revolution" promised by GM crops is largely driven by the financial interests of corporations selling proprietary products. This has driven widespread concern that the global food supply is being dominated by a few large agrobiotech corporations that develop GM technologies.

The future of GM foods seems likely to hinge on social, economic, legal, and political factors as well as scientific ones—and these factors vary in different nations. European consumers have expressed widespread unease about GM technologies, and European governments mandate that all foods containing GM organisms be labeled as such. In contrast, U.S. consumers have largely accepted the GM crops approved by U.S. agencies, generally without even realizing that the majority of their food contains GM products.

The world's large, rapidly industrializing nations (such as Brazil, India, and China) are now aggressively pursuing GM crops, even as ethical, economic, and political debates over the costs and benefits of these foods continue. Although many experts feel we should adopt the **precautionary principle** (the idea that one should not undertake new action until the ramifications of that action are well understood) and proceed with caution on GM foods, the demands placed on food supplies by a growing human population may accelerate their adoption around the world.

RAISING ANIMALS FOR FOOD: LIVESTOCK, POULTRY, AND AQUACULTURE

Food from cropland agriculture makes up a large portion of the human diet, but most of us also eat animal products. Just as farming methods have changed over time, so have the ways we raise animals for food.

As wealth and global commerce have increased, so has humanity's consumption of meat, milk, eggs, and other animal products (**FIGURE 7.21**). The world population of domesticated animals raised for food rose from 7.2 billion animals to 25.2 billion animals between 1961 and 2009. Most of these animals are chickens. Global meat production has increased fivefold since 1950, and per capita meat consumption has doubled. The United Nations Food and Agriculture Organization (FAO) estimates that as more developing nations go through the demographic transition (pp. 122–123) and become wealthier, total meat consumption will nearly double by the year 2050.

Our food choices are also energy choices

What we choose to eat has ramifications for how we use energy and the land that supports agriculture. Recall our discussions of trophic levels and pyramids of energy (pp. 68–71). Every time energy moves from one trophic level to the next,

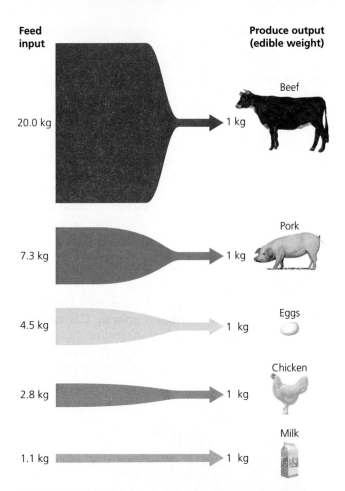

FIGURE 7.22 ▲ Different animal food products require different amounts of input of animal feed. Chickens must be fed 2.8 kg of feed for each 1 kg of resulting chicken meat, for instance, whereas 20 kg of feed must be provided to cattle to produce 1 kg of beef. Data from Smil, V., 2001. *Feeding the world: A challenge for the twenty-first century.* Cambridge, MA: MIT Press.

as much as 90% is used up in cellular respiration. For this reason, eating meat is far less energy-efficient than relying on a vegetarian diet and leaves a far greater ecological footprint.

Some animals convert grain feed into milk, eggs, or meat more efficiently than others (**FIGURE 7.22**). Scientists have calculated relative energy-conversion efficiencies for different types of animals. Such energy efficiencies have ramifications for land use because land and water are required to raise food for the animals, and some animals require more than others. **FIGURE 7.23** shows the area of land and weight of water required to produce 1 kg (2.2 lb) of food protein for milk, eggs, chicken, pork, and beef. Producing eggs and chicken meat requires the least space and water, whereas producing beef requires the most. Such differences make clear that when we choose what to eat, we are also indirectly choosing how to make use of resources such as land and water.

Feedlot agriculture has benefits and costs

In traditional agriculture, livestock are kept by farming families near their homes or are grazed on open grasslands by nomadic herders or sedentary ranchers. These traditions

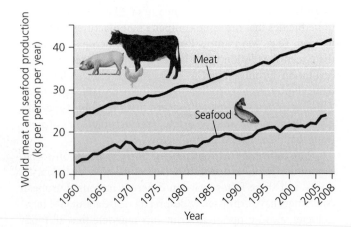

FIGURE 7.21 ▲ Per capita consumption of meat from farmed animals has risen steadily worldwide, as has per capita consumption of seafood (marine and freshwater, harvested and farmed). Data from U.N. Food and Agriculture Organization (FAO).

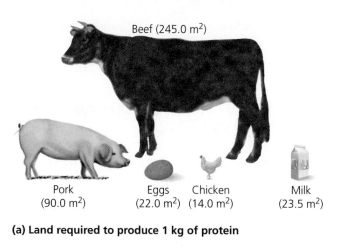

Beef (245.0 m²)

Pork   Eggs   Chicken   Milk
(90.0 m²)   (22.0 m²) (14.0 m²)   (23.5 m²)

(a) Land required to produce 1 kg of protein

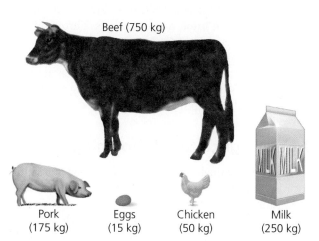

Beef (750 kg)

Pork   Eggs   Chicken   Milk
(175 kg)   (15 kg)   (50 kg)   (250 kg)

(b) Water required to produce 1 kg of protein

FIGURE 7.23 ▲ Producing different types of animal products requires different amounts of land and water. Raising cattle for beef requires by far the most land and water of all animal products. Data from Smil, V., 2001. *Feeding the world: A challenge for the twenty-first century.* Cambridge, MA: MIT Press.

survive, but the advent of industrial agriculture has brought a new method. **Feedlots**, also known as *factory farms* or *concentrated animal feeding operations (CAFOs)*, are essentially huge warehouses or pens designed to deliver energy-rich food to animals living at extremely high densities (**FIGURE 7.24**).

Today nearly half the world's pork and most of its poultry come from feedlots.

Feedlot operations allow for economic efficiency and greater food production, which makes meat affordable to more people. Feedlots also offer one overarching benefit for environmental quality: Taking cattle and other livestock off rangeland and concentrating them in feedlots reduces the grazing impacts they would otherwise exert across large portions of the landscape.

Intensified animal production through the industrial feedlot model has some negative effects, though. Forty-five percent of our global grain production goes to feed livestock and poultry. This elevates the price of staple grains and endangers food security for the very poor. Livestock produce prodigious amounts of manure and urine, and their waste can pollute surface water and groundwater. The crowded conditions under which animals are often kept necessitate heavy use of antibiotics to control disease. The overuse of antibiotics can cause microbes to become resistant to the antibiotics (just as pests become resistant to pesticides; p. 147), making these drugs less effective. Livestock are also a major source of greenhouse gases that lead to climate change (pp. 300-301). A comprehensive FAO report in 2006 concluded that livestock agriculture contributes 9% of our carbon dioxide emissions, 37% of our methane emissions, and 65% of our nitrous oxide emissions—altogether, 18% of the emissions driving climate change, a larger share than automobile transportation!

Proper management of feedlots can minimize their impacts, however, and both the EPA and the states regulate U.S. feedlots. Manure from feedlots, for example, can be applied to cropland as a fertilizer, reducing the need for inorganic fertilizer.

We raise seafood with aquaculture

Besides growing plants as crops and raising animals on rangelands and in feedlots, we rely on aquatic organisms for food. Wild fish populations are plummeting throughout the world's oceans as increased demand and new technologies have led us to overharvest most marine fisheries (pp. 273–275). This means that raising fish and shellfish on "fish

FIGURE 7.24 ▼ Most meat eaten in the United States comes from animals raised in feedlots, or factory farms. These locations house thousands of chickens **(a)** or cattle **(b)** at high densities. The animals are dosed liberally with antibiotics to control disease.

(a) Chicken factory farm in Arkansas

(b) Cattle feedlot in California

farms" may be the only way to meet our growing demand for these foods.

We call the cultivation of aquatic organisms for food in controlled environments **aquaculture**, and people pursue aquaculture with over 220 freshwater and marine species (**FIGURE 7.25**). Many aquatic species are grown in open water in large, floating net-pens. Others are raised in ponds or holding tanks. Aquaculture is the fastest-growing type of food production; in the past 20 years, global output has increased fivefold. Most widespread in Asia, aquaculture today produces $80 billion worth of food and provides three-quarters of the freshwater fish and over half of the shellfish that we eat.

Aquaculture helps reduce fishing pressure on overharvested and declining wild stocks. Furthermore, aquaculture consumes fewer fossil fuels and provides a safer work environment than does commercial fishing. Fish farming can also be remarkably energy-efficient, producing as much as 10 times more fish per unit area than is harvested from waters of the continental shelf and up to 1,000 times more than is harvested from the open ocean.

Along with its benefits, aquaculture has disadvantages. Aquaculture can produce prodigious amounts of waste, both from the farmed organisms and from the feed that goes uneaten and decomposes in the water. Like feedlot animals, commercially farmed fish often are fed grain, and this affects food supplies for people. In other cases, farmed fish are fed fish meal made from wild ocean fish such as herring and anchovies, whose harvest may place additional stress on wild fish populations.

If farmed aquatic organisms escape into ecosystems where they are not native (as several carp species have done in U.S. waters), they may spread disease to native stocks or may outcompete native organisms for food or habitat. Salmon genetically modified for rapid growth, for exam-

ple, could introduce genes into wild salmon populations or spread disease if they were to escape from aquaculture facilities.

THE GROWTH OF SUSTAINABLE AGRICULTURE

Industrial agriculture has allowed food production to keep pace with our growing population, but it involves many adverse environmental and social impacts. These range from the degradation of soils to reliance on fossil fuels to problems arising from pesticide use, genetic modification, and intensive feedlot and aquaculture operations. Although intensive commercial agriculture may help alleviate certain environmental pressures, it often worsens others. Throughout this chapter, we have seen sustainable approaches to agriculture that maintain high crop yields, minimize resource inputs into food production, and lessen the environmental impacts of farming. Let's now take a closer look at the growth of sustainable agriculture around the world.

Organic agriculture is booming

One type of sustainable agriculture is **organic agriculture**, which is agriculture that uses no synthetic fertilizers, insecticides, fungicides, or herbicides. In 1990, the U.S. Congress passed the Organic Food Production Act to establish national standards for organic products and facilitate their sale. Under this law, in 2000 the USDA issued criteria by which crops and livestock could be officially certified as organic, and these standards went into effect in 2002 as part of the National Organic Program. California, Washington, and Texas established stricter state guidelines for labeling foods organic, and today many U.S. states and over 70 nations have laws spelling out organic standards.

For farmers, organic farming can bring a number of benefits: lower input costs, enhanced income from higher-value produce, and reduced chemical pollution and soil degradation (see **THE SCIENCE BEHIND THE STORY**, pp. 156–157). Transitioning to organic agriculture does involve some risk, however, as farmers must use organic approaches for three years before their products can be certified as organic and sold at prices commanded by organic foods. The main obstacle for consumers to organic foods is price. Organic products tend to be 10–30% more expensive than conventional ones, and some (such as milk) can cost twice as much. However, enough consumers are willing to pay more for organic products that grocers and other businesses are making them more widely available.

Today, three of four Americans buy organic food at least occasionally, and more than four of five retail groceries offer it. U.S. consumers spent $26.7 billion on organic food in 2010, amounting to 4% of all food sales (**FIGURE 7.26A**). Worldwide, sales of organic food tripled between 2000 and 2009, when sales surpassed $54 billion.

Production is increasing along with demand (**FIGURE 7.26B**). Although organic agriculture takes up less than 1%

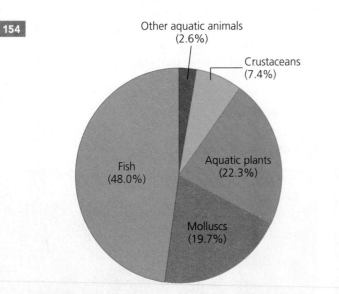

FIGURE 7.25 ▲ Aquaculture involves many types of fish, but also a wide diversity of other marine and freshwater organisms. Data from U.N. Food and Agriculture Organization (FAO).

The average grocery store item travels 1,400 miles from its origin to your shopping cart.

This long-distance transport consumes oil and contributes to pollution and climate change.

You Can Make a Difference

➤ Ask your campus's food services director to buy locally grown produce and locally made products.

➤ Shop at farmer's markets, or help to start one on campus.

➤ Partner with farmers in community-supported agriculture programs.

Locally grown produce and locally made products cut down on the carbon footprint of our food.

Heirloom Tomatoes $3 50 /lb

155

Does Organic Farming Work?

Dr. Paul Mäder, Research Institute of Organic Agriculture, Switzerland

Organic farming puts fewer synthetic chemicals into the soil, air, and water, but can it produce large enough crop yields to feed the world's people?

To search for the answer we can turn to Switzerland, a leader in organic farming. One in every nine hectares of agricultural land here is managed organically—the fourth-highest rate in the world—and Swiss scientists have taken the lead in examining the benefits and costs of organic farming. The longest-running and most-cited studies of organic agriculture come from experimental farms

at Therwil, Switzerland, established back in 1977. Long-term studies are rare and valuable in agriculture and in ecology, because they can reveal slow processes or subtle effects that get swamped out by variation in shorter-term studies.

At the Swiss research site, wheat, potatoes, and other crops are grown in plots cultivated in different treatments:

▶ Conventional farming using large amounts of chemical pesticides, herbicides, and inorganic fertilizer

▶ Conventional farming that also uses organic fertilizer (cattle manure)

▶ Organic farming using only manure, mechanical weeding, and plant extracts to control pests

▶ Organic farming that also adds natural boosts, such as herbal extracts in compost

Researchers record crop yields at harvest each year. They analyze the soil regularly, measuring nutrient content, pH, structure, and other variables. They also measure the biological diversity and activity of microbes and invertebrates in the soil. Such indicators of soil quality help researchers assess the potential for long-term farm productivity.

In 2002, Paul Mäder and colleagues from two Swiss research institutes reported in the journal *Science* results from 21 years of data. Over this time, the organic fields yielded 80% of what the conventional fields produced. Organic crops of winter wheat yielded 90% of the conventional yield. Organic potato crops averaged 58–66% of conventional yields because of nutrient deficiency and disease.

Although the organic plots produced 20% less, they did so while receiving 35–50% less fertilizer than the

conventional fields and 97% fewer pesticides. Thus, Mäder's team concluded, the organic plots were highly efficient and represent "a realistic alternative to conventional farming systems."

How can organic fields produce decent yields without synthetic agricultural chemicals? The answer lies in the soil. Mäder's team found that soil in the organic plots had better structure, better supplies of some nutrients, and much more microbial activity and invertebrate biodiversity (**first figure**). Differences were visible just by looking at the fields (**second figure**).

The Swiss data reflect those found by scientists elsewhere. For instance, U.S. researchers comparing organic and conventional farms in North Dakota and Nebraska found that organic farming produced soils with more microbial life, earthworm activity, water-holding capacity,

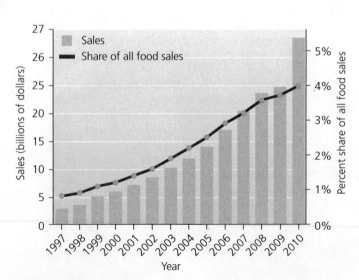

(a) Sales of organic food

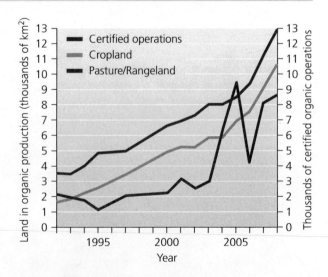

(b) Extent of organic agriculture

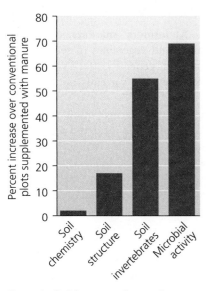

Organic fields outperformed conventional fields supplemented with manure in their soil quality, including averaged values for soil chemistry (six variables), soil structure (three variables), soil invertebrates (five variables), and microbial activity (six variables). Organic fields outperformed conventional fields without manure (not shown) still more. Data from Mäder, P., et al., 2002. Soil fertility and biodiversity in organic farming. *Science* 296: 1694–1697.

topsoil depth, and naturally occurring nutrients. Given such differences in soil quality, researchers expect that organic fields should perform better and better relative to conventional fields as time goes by—in other words, that they are more sustainable.

Studies are continuing at the Swiss plots to determine whether this is true. As one example, in 2009, Jens Leifeld and two colleagues at a Zurich research institute analyzed soil carbon content after 27 years. They found that soil carbon had decreased in all the treatments, but that it had declined most in conventional plots using inorganic fertilizer or no fertilizer.

(a) Organic plot

(b) Conventional plot

Organic plots **(a)** showed more weeds but more earthworm castings and healthier soil structure than conventional plots **(b)** in the Swiss experiment.

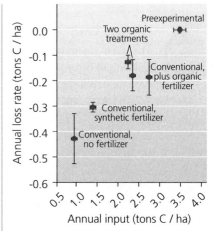

Organic plots lost less soil carbon over 27 years of farming than did conventional plots with no fertilizer or only synthetic fertilizer. Conventional plots with organic fertilizer supplements retained just as much soil carbon as the organic plots. Data from Leifeld, J., et al., 2009. Consequences of conventional versus organic farming on soil carbon: Results from a 27-year field experiment. *Agronomy Journal* 101: 1204–1218.

Conventional plots that supplemented synthetic fertilizer with cow manure, however, did just as well as the organic plots in retaining soil carbon (**third figure**).

As more long-term data are published and new studies begin, we are learning more and more about the benefits of organic agriculture for soil quality and about how we might improve conventional methods to maximize crop yields while protecting the long-term sustainability of agriculture.

FIGURE 7.26 ◀ Sales of organic food in the United States **(a)** have increased rapidly in the past decade, both in total dollar amounts (bars) and as a percentage of the overall food market (line). Since the mid-1990s in the United States **(b)**, acreage devoted to organic crops and livestock has each quadrupled, and the number of certified operations has more than tripled. (a) Adapted by permission from Willer, Helga, and Lukas Kilcher (Eds.), (2009). *The world of organic agriculture, statistics and emerging trends* 2009. FiBL-IFOAM Report. IFOAM, Bonn; FiBL, Frick; ITC, Geneva. Data used with permission from Organic Trade Association OTA: Manufacturer Survey 2007. Additional, updated data provided by Organic Trade Association. (b) Data from USDA Economic Research Service.

of agricultural land worldwide (35 million ha, or 86.5 million acres, in 2008), this area is rapidly expanding. Two-thirds of this area is in developed nations, and farmers in all 50 U.S. states and 1.4 million farmers in more than 150 nations practice organic farming commercially to some extent. Contrary to common misconception, organic farming can be performed on both big and small farms, as long as the criteria for organic certification are satisfied.

Government policies have aided organic farming. In the United States, the 2008 Farm Bill set aside $112 million over five years for organic agriculture, and the government helps to defray certification expenses. The European Union adopted a policy in 1993 to support farmers financially during conversion to organic agriculture. Once conversion is complete, studies suggest that reduced inputs and higher market prices can make organic farming at least as profitable for the farmer as conventional methods.

Locally supported agriculture is growing

Apart from organic methods, another component of the move toward sustainable agriculture is an attempt to reduce fossil fuel use from the long-distance transport of food (see **ENVISIONIT**, p. 155). The average food product sold in a U.S. supermarket travels at least 2,300 km (1,400 mi) between the farm and the grocery. Because of the travel time, supermarket produce is often chemically treated to preserve freshness and color.

In response, increasing numbers of farmers and consumers in developed nations are supporting local small-scale agriculture. Farmers' markets are springing up throughout North America as people rediscover the joys of fresh, locally grown produce. At **farmers' markets**, consumers buy meats and fresh fruits and vegetables in season from local producers. These markets generally offer a wide choice of organic items and unique local varieties not found in supermarkets.

Some consumers are even partnering with local farmers in a phenomenon called **community-supported agriculture (CSA)**. In a CSA program, consumers pay farmers in advance for a share of their yield, usually a weekly delivery of produce. Consumers get fresh seasonal produce, and farmers get a guaranteed income stream up front to invest in their crops—a welcome alternative to taking out loans and being at the mercy of the weather.

Sustainable agriculture provides a roadmap for the future

The best approach for making an agricultural system sustainable is to mimic the way a natural ecosystem functions. Ecosystems are sustainable because they operate in cycles and are internally stabilized with negative feedback loops. In this way they provide a useful model for agriculture.

The approaches embraced by the Schroeder brothers and Nate Ronsiek exemplify how agricultural systems can be more efficiently integrated with surrounding natural ecosystems and reduce the environmental impacts from food production. Efforts like these, together with other approaches farmers around the world are taking to help make agriculture sustainable, will be crucial for all of us as we progress through the coming century.

➤ CONCLUSION

Our species has a 10,000-year history with agriculture, and over this time methods of food production have changed substantially. Many practices of intensive industrial agriculture exert substantial negative environmental impacts. At the same time, the increased production had boosted food supplies and helped to relieve certain pressures on land or resources.

What is certain is that if our planet is to support 9 billion people by mid-century without further degrading the soil, water, pollinators, and other resources and ecosystem services that support our food production, we must find ways to shift to sustainable agriculture. Approaches such as biological pest control, organic agriculture, pollinator conservation, preservation of native crop diversity, sustainable aquaculture, and likely some degree of careful and responsible genetic modification of food may all be parts of the game plan we will need to set in motion to achieve a sustainable future.

TESTING YOUR COMPREHENSION

1. Describe patterns in global food security from 1970 to 2010. Name two nutritional deficiencies commonly seen in the modern world.

2. Compare and contrast the methods used in traditional and industrial agriculture. How does sustainable agriculture differ from industrial agriculture?

3. How are soil horizons created? List and describe the major horizons in a typical soil profile. What is the general pattern of distribution of organic matter in a typical profile?

4. Name three human activities that can promote soil erosion. Describe several farming techniques (such as terracing and no-till farming) that can help reduce the risk of erosion.

5. Explain how over-irrigation can damage soils and reduce crop yields.

6. How do fertilizers boost crop growth? How can large amounts of fertilizer added to soil also end up in water supplies and the atmosphere? How do sustainable agriculture approaches reduce fertilizer runoff?

7. Explain how pesticide resistance occurs.

8. How is a transgenic organism created? How is genetic engineering different from traditional agricultural breeding? How is it similar?

9. Compare and contrast the land and water required to produce 1 kg of beef, pork, chicken, eggs, and milk.

10. What are some economic benefits of aquaculture? What are some negative environmental impacts?

SEEKING SOLUTIONS

1. How do you think a farmer can best help to conserve soil? How do you think a scientist can best help to conserve soil? How do you think a national government can best help to conserve soil?

2. Assess several ways in which high-input industrial agriculture can be beneficial for the environment and several ways in which it can be detrimental. Now suggest several ways in which we might modify industrial agriculture to mitigate its environmental impacts.

3. What factors make for an effective biological control strategy of pest management? What risks are involved in biocontrol? If you had to decide whether to use biocontrol against a particular pest, what questions would you want to have answered before you decide?

4. **THINK IT THROUGH** You are the head of an international granting agency that assists farmers with soil conservation and sustainable agriculture. You have $10 million to disburse. Your agency's staff has decided that the funding should go to (1) farmers in an arid area of Africa prone to salinization, (2) farmers in a fast-growing area of Indonesia where swidden agriculture is practiced, (3) farmers in Argentina practicing no-till agriculture, and (4) farmers in a dryland area of Mongolia undergoing desertification. What types of projects would you recommend funding in each of these areas, how would you apportion your funding among them, and why?

5. **THINK IT THROUGH** You are a USDA official and must decide whether to allow the planting of a new genetically modified strain of cabbage that produces its own pesticide and has twice the vitamin content of regular cabbage. What questions would you ask of scientists before deciding whether to approve the new crop? What scientific data would you want to see? Would you also consult nonscientists or consider ethical, economic, and social factors?

CALCULATING ECOLOGICAL FOOTPRINTS

As food production became more industrialized during the 20th century, more and more energy has been expended to store food and ship it to market. In the United States today, food travels an average of 1,400 miles from the field to your table. The price you pay for the food covers the cost of this long-distance transportation, which in 2004 was approximately $1 per ton per mile. Assuming that the average person eats 2 pounds of food per day, calculate the food transportation costs for each category in the table below.

Consumer	Daily cost	Annual cost
You	$1.40	$511
Your class		
Your town		
Your state		
United States		

Pirog, R., and A. Benjamin, 2003. Checking the food odometer: Comparing food miles for local versus conventional produce sales to Iowa institutions. Leopold Center for Sustainable Agriculture, Iowa State University, Ames.

1. What specific challenges to environmental sustainability are imposed by a food production and distribution system that relies on long-range transportation to bring food to market?

2. A 2003 study noted that locally produced food traveled only 50 miles or so to market, thus saving 96% of the transportation costs. Locally grown foods may be fresher and cause less environmental impact as they are brought to market, but what are the disadvantages to you as a consumer in relying on local food production? Do you think the advantages outweigh those disadvantages?

3. What has happened to gasoline prices recently? How would future increases in the price of gas affect your answers to the preceding questions?

Go to **www.masteringenvironmentalscience.com** for homework assignments, practice quizzes, Pearson eText, and more.

8 Biodiversity and Conservation Biology

Upon completing this chapter, you will be able to:

➤ Characterize the scope of biodiversity on Earth

➤ Contrast the background extinction rate with periods of mass extinction

➤ Evaluate the primary causes of biodiversity loss

➤ Specify the benefits of biodiversity

➤ Assess the science and practice of conservation biology

➤ Analyze efforts to conserve threatened and endangered species

➤ Compare and contrast conservation efforts above the species level

A Siberian tiger in the Sikhote-Alin Mountains

Saving the Siberian Tiger

"Future generations would be truly saddened that this century had so little foresight, so little compassion, such lack of generosity of spirit for the future that it would eliminate one of the most dramatic and beautiful animals this world has ever seen."
—GEORGE SCHALLER, Wildlife Biologist, on the tiger

"If you kill a tiger, you can buy a motorbike."
—ANONYMOUS POACHER, on selling tiger parts

Historically, tigers roamed widely across Asia from Turkey to northeast Russia to Indonesia. However, people have driven the majestic striped cats from nearly all of their range. Today, tigers are exceedingly rare and are sliding toward extinction. Just over 3,000 tigers survive, down from 100,000 a century ago.

Tigers of the subspecies known as the Siberian tiger are the largest cats in the world. These regal animals today find their last refuge in the forests of the remote Sikhote-Alin Mountains of the Russian Far East. For thousands of years the Siberian tiger coexisted with the region's native people and held a prominent place in their lore. These people viewed it as a guardian of the mountains and forests, and they rarely killed a tiger unless it had preyed on a person.

The Russians who moved into the region in the early 20th century had no such cultural traditions. They hunted tigers relentlessly for sport and hides, and the tiger population dipped to perhaps just 20–30 animals. In response, the Russian government banned the hunting of tigers, and the population began to recover. However, poachers started killing tigers illegally to sell their body parts to China and other Asian countries, where they are used in traditional medicine and as alleged aphrodisiacs. Meanwhile, logging, road building, and agriculture degraded and fragmented tiger habitat, providing easy access for still more poachers.

International conservation groups got involved just in time, working with Russian biologists to save the dwindling tiger population. One such group was the Hornocker Wildlife Institute, now part of the Wildlife Conservation Society (WCS). In 1992 the group helped launch the Siberian Tiger Project, devoted to studying and conserving the tiger and its habitat. The team put together a plan to protect the tiger, began educating people on the animal's value, and worked closely with people who live near the big cats.

Today, WCS biologists track tigers with radio-collars, monitor their movements and health, determine causes of death when they die, and study aspects of the tiger's ecosystem. They also work with the region's people and help fund local wildlife officials to deter and capture poachers.

Thanks to such efforts, the Siberian tiger population stabilized, even while the world's other tiger populations were declining. The last range-wide survey, in 2005, found between 428 and 502 Siberian tigers in the wild, while 1,500 more survived in zoos and captive breeding programs. However, government funding and law enforcement to deter poaching were reduced, and data since 2005 suggest that tiger numbers are falling yet again.

Many dedicated scientists, conservationists, and policymakers continue trying to save these endangered animals. In November 2010, leaders of the 13 nations where tigers still survive met at a historic summit in St. Petersburg, Russia, marking the first time that multiple heads of state had ever convened to focus on saving a single species of wild animal. Russian prime minister Vladimir Putin, Chinese premier Wen Jiabao, World Bank president Robert Zoellick, and actor Leonardo DiCaprio were among the luminaries participating in the conference. At this International Tiger Forum, the leaders signed a declaration that set in motion a strategic multinational plan called the Global Tiger Recovery Program.

This program aims to double the tiger population by 2022 (the next "Year of the Tiger" by the Chinese zodiac) by protecting habitat, cracking down on poaching, and addressing illegal trade in pelts and body parts. National governments, conservation organizations, and the World Bank promised millions of dollars, although more is needed—an estimated $350 million over the first 5 years of the program. Representatives of the 13 nations planned to work out details of financing during 2011.

Some proponents of tiger conservation criticized the program, worrying that funding would not be adequate, that specific measures to reduce demand for tiger body parts were not spelled out, and that proposed actions were not focused enough. Nonetheless, by demonstrating support for tiger conservation at the highest political level, the summit gave tiger conservation efforts a clear boost. The struggle to save the tiger from imminent extinction is one of numerous efforts around the world today to stem the loss of our planet's priceless biological diversity. ∎

OUR PLANET OF LIFE

Our rising human population and resource consumption are putting ever-greater pressure on the flora and fauna of our planet, from tigers to tiger beetles. We are diminishing Earth's diversity of life, the very quality that makes our planet so special.

Biodiversity encompasses multiple levels

Biological diversity, or **biodiversity** (p. 49), describes the variety of life across all levels of biological organization, including the diversity of species, their genes, their populations, and their communities. Biodiversity is a concept as multifaceted as life itself, and biologists employ different working definitions according to their own aims and philosophies. Yet scientists agree that the concept applies across the major levels in the organization of life (**FIGURE 8.1**). The level that is easiest to visualize and most commonly used is species diversity.

Ecosystem diversity

Species diversity

Genetic diversity

FIGURE 8.1 ▲ The concept of biodiversity encompasses several levels in the hierarchy of life. Species diversity **(middle frame of the figure)** refers to the number or variety of species. Genetic diversity **(bottom frame)** refers to variation in DNA composition among individuals within a species. Ecosystem diversity **(top frame)** and related concepts refer to variety at levels above the species level, such as ecosystems, communities, habitats, or landscapes.

Species diversity A **species** (p. 46) is a distinct type of organism, a set of individuals that uniquely share certain characteristics and can breed with one another and produce fertile offspring. Biologists use differing criteria to distinguish one species from another. Some emphasize characteristics shared because of common ancestry, whereas others emphasize the ability to interbreed. In practice, however, scientists generally agree on species identities.

We can express **species diversity** in terms of the number or variety of species in a particular region. One component of species diversity is *species richness*, the number of species. Another is *evenness* or *relative abundance*, the extent to which species differ in numbers of individuals.

Speciation (pp. 49–50) generates new species, whereas extinction (p. 51) diminishes species richness. Immigration, emigration, and local extinction may change species richness locally, but only speciation and extinction change it globally.

Taxonomists classify species by their similarity into a hierarchy of categories meant to reflect evolutionary relationships. Related species are grouped together into *genera* (singular: *genus*); related genera are grouped into families; and so on (**FIGURE 8.2**). Every species is given a two-part Latin or Latinized scientific name denoting its genus and species. The tiger, *Panthera tigris*, is similar to the world's other species of large cats, such as the jaguar (*Panthera onca*), the leopard (*Panthera pardus*), and the African lion (*Panthera leo*). These four species are closely related in evolutionary terms, and this is indicated by the genus name they share, *Panthera*. They are more distantly related to cats in other genera such as the cheetah (*Acinonyx jubatus*) and the bobcat (*Felis rufus*), although all cats are classified together in the family Felidae.

Biodiversity exists below the species level in the form of *subspecies*, populations of a species that occur in separate geographic areas and differ in some characteristics. Subspecies arise by the same processes that drive speciation (pp. 49–50) but result when divergence stops short of forming separate species. Scientists denote subspecies with a third part of the scientific name. The Siberian tiger, *Panthera tigris altaica*, is one of five (or perhaps only four) subspecies of tiger still surviving (**FIGURE 8.3**). Tiger subspecies differ in color, coat thickness, stripe patterns, and size. For example, *Panthera tigris altaica* is taller at the shoulder than the Bengal tiger (*Panthera tigris tigris*) of India and Nepal, and it has a thicker coat and larger paws.

Genetic diversity Scientists designate subspecies when they recognize substantial, genetically based differences among individuals from different populations of a species. However, all species consist of individuals that vary genetically from one another to some degree, and this genetic diversity is an important component of biodiversity. **Genetic diversity** encompasses the differences in DNA composition (p. 28) among individuals.

Genetic diversity provides the raw material for adaptation to local conditions. A diversity of genes for coat thickness in tigers allowed natural selection (pp. 46–48) to favor genes for thin fur in Bengal tigers living in warm regions, and genes for thick fur in Siberian tigers living in cold regions. In the long term, populations with more genetic diversity may be more likely to persist, because their variation better enables them to cope with environmental change.

Populations with little genetic diversity are vulnerable to environmental change because they may happen to lack ge-

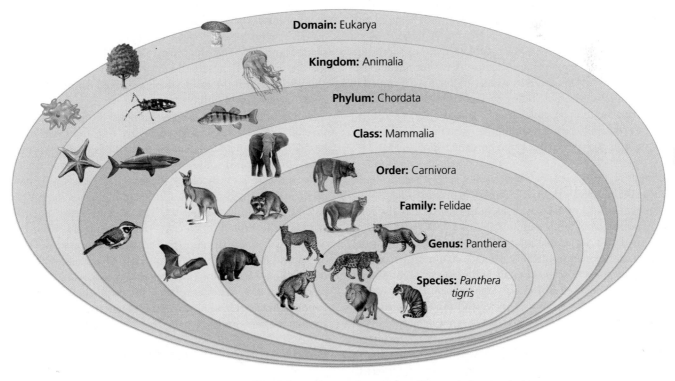

FIGURE 8.2 ▲ Taxonomists classify organisms using a hierarchical system meant to reflect evolutionary relationships. Species that are similar in their appearance, behavior, and genetics (because they share recent common ancestry) are placed in the same genus. Organisms of similar genera are placed within the same family. Families are placed within orders, orders within classes, classes within phyla, phyla within kingdoms, and kingdoms within domains. For instance, tigers belong to the class Mammalia, along with elephants, kangaroos, and bats. However, the differences among these species, which have evolved and diverged over millions of years, are great enough that they are placed in different orders, families, and genera.

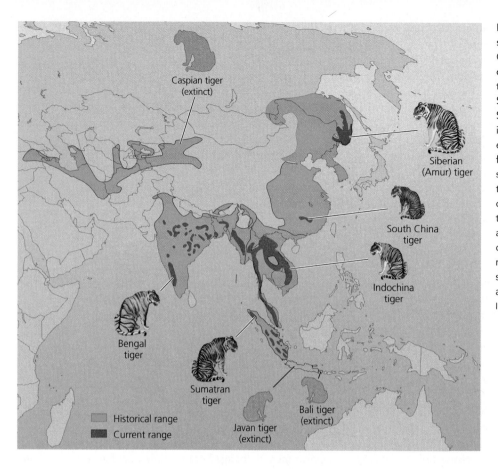

FIGURE 8.3 ◄ Three of the eight subspecies of tiger—the Bali, Javan, and Caspian tigers—were driven extinct during the 20th century. Today only the Siberian, Bengal, Indochina, and Sumatran tigers persist, while the South China tiger has not been seen in 25 years and may be extinct. Deforestation, hunting, and other pressures from people have caused tigers of all subspecies to disappear from 93% of the geographic range they historically occupied. Researchers estimate that the majority of surviving individuals are crowded into less than half of 1% of the species' original range. This map contrasts the ranges of the eight subspecies in the years 1800 **(orange)** and 2000 **(red)**. Data from the Tiger Information Center.

netic variants that would help them adapt to novel conditions. Populations with low genetic diversity may also be more vulnerable to disease and may suffer *inbreeding depression*, which occurs when genetically similar parents mate and produce weak or defective offspring. Scientists have sounded warnings over low genetic diversity in species that have dropped to low population sizes, including cheetahs, bison, and elephant seals, but the full consequences of reduced diversity in these species remain to be seen. Diminished genetic diversity in our crop plants is a prime concern to humanity (p. 136).

Ecosystem diversity Biodiversity encompasses levels above the species level, as well. *Ecosystem diversity* refers to the number and variety of ecosystems, but biologists may also refer to the diversity of biotic communities or habitats within some specified area. If the area is large, scientists may also consider the geographic arrangement of habitats, communities, or ecosystems at the landscape level, including the sizes and shapes of patches and the connections among them. Under any of these concepts, a seashore of rocky and sandy beaches, forested cliffs, offshore coral reefs, and ocean waters would hold far more biodiversity than the same acreage of a monocultural cornfield. A mountain slope whose vegetation changes with elevation from desert to hardwood forest to conifer forest to alpine meadow would hold more biodiversity than an equal-sized area consisting of only desert, forest, or meadow.

Many species await discovery

Scientists often express biodiversity in terms of species richness because that component is most easily measured and is a

good gauge for overall biodiversity. Yet we still are profoundly ignorant of the number of species that exist in the world. So far, scientists have described about 1.8 million species of plants, animals, and microorganisms. However, estimates for the number that actually exist range from 3 million to 100 million, with the most widely accepted estimates in the neighborhood of 14 million.

Our knowledge of species numbers is incomplete for several reasons. First, many species are tiny and easily overlooked. These include bacteria, nematodes (roundworms), fungi, protists, and soil-dwelling arthropods. Second, many organisms are so difficult to identify that ones thought to be identical sometimes turn out, once biologists look more closely, to be multiple species. Third, some areas of Earth remain little explored. We have barely sampled the ocean depths, hydrothermal vents (p. 260), or the canopies and soils of tropical forests. As one example, a 2005 expedition to the remote Foja Mountains of New Guinea discovered over 40 new species of vertebrates, plants, and butterflies in less than a month, while research in marine waters nearby turned up another 50 new species.

Biodiversity is unevenly distributed

Some taxonomic groups hold more species than others. In this respect, insects show a staggering predominance over all other forms of life (**FIGURE 8.4**). Within insects, about 40% are beetles, and beetles alone outnumber all non-insect animals and all plants. No wonder the 20th-century British biologist J.B.S. Haldane famously quipped that God must have had "an inordinate fondness for beetles."

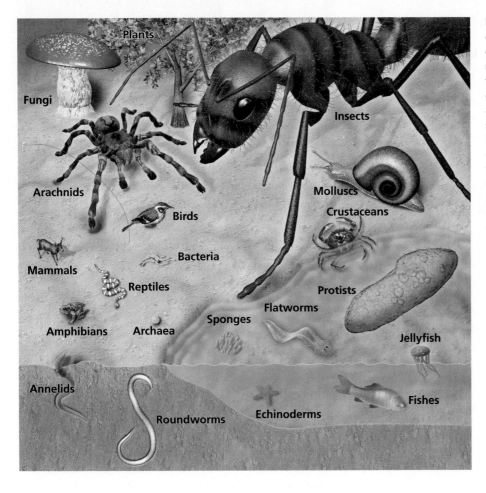

FIGURE 8.4 ◄ This illustration shows organisms scaled in size to the number of species known from each major taxonomic group. This gives a visual sense of the disparity in species richness among groups. However, because most species are not yet discovered or described, some groups (such as bacteria, archaea, insects, nematodes, protists, fungi, and others) may contain far more species than we now know of. Data from Groombridge, B., and M.D. Jenkins, 2002. *Global biodiversity: Earth's living resources in the 21st century.* UNEP-World Conservation Monitoring Centre. Cambridge, U.K.: Hoechst Foundation.

Living things are distributed unevenly across our planet, as well. For instance, species richness generally increases as one nears the equator. This pattern of variation with latitude is called the *latitudinal gradient*, and hypotheses abound to explain it. A leading idea is that greater amounts of solar energy, heat, and humidity at tropical latitudes lead to more plant growth, making areas nearer the equator more productive and able to support more animals. The relatively stable climates of equatorial regions, in turn, discourage single species from dominating ecosystems and, instead, allow numerous species to coexist. Whereas variable environmental conditions favor generalists (species that can tolerate a wide range of circumstances), stable conditions favor specialists (species that do particular things especially well). Another proposed explanation for the latitudinal gradient is that glaciation events repeatedly forced organisms toward tropical latitudes, leaving the polar and temperate regions relatively species-poor.

The latitudinal gradient influences the species diversity of Earth's biomes (pp. 78–84). Tropical dry forests and rainforests support far more species than tundra and boreal forests, for instance. At smaller scales, diversity varies with habitat type. Structurally diverse habitats tend to allow for more ecological niches (p. 53) and support greater species richness and evenness. For instance, forests generally support more diversity than grasslands.

For any given area, species diversity tends to increase with diversity of habitats, because each habitat supports a somewhat different set of organisms. Thus, ecotones (transi-tion zones where habitats intermix; p. 33) often support high biodiversity. Because human disturbance can sometimes increase habitat diversity, species diversity may rise in disturbed areas. However, this is true only at local scales. At larger scales, human disturbance decreases diversity because specialists disappear when habitats are homogenized and because species that rely on large expanses of habitat disappear when habitats are fragmented.

EXTINCTION AND BIODIVERSITY LOSS

Biodiversity at all levels is being lost to human impact, most irretrievably in the extinction of species. **Extinction** (p. 51) occurs when the last member of a species dies and the species ceases to exist. The disappearance of a particular population from a given area, but not the entire species globally, is referred to as **extirpation**. The tiger has been extirpated from most of its historic range (see Figure 8.3), but it is not yet extinct. Extirpation is an erosive process that can, over time, lead to extinction.

Human impact is responsible for most cases of extirpation and extinction today, but these processes also occur naturally, albeit at a much slower rate. If species did not naturally go extinct, we would be up to our ears in dinosaurs, trilobites, and millions of other creatures that vanished from Earth long before we appeared. Paleontologists estimate that roughly 99% of all species that ever lived are now extinct, and that the remaining 1% comprises the wealth of species on our planet today.

Most extinctions preceding the appearance of human beings occurred one by one for independent reasons, at a pace referred to as the *background extinction rate* (p. 51). By studying traces of organisms preserved in the fossil record (p. 50), scientists infer that for mammals and marine animals, each year, on average, 1 species out of every 1–10 million vanished.

Earth has experienced five mass extinction episodes

Extinction rates rose far above this background rate at several points in Earth's history. In the past 440 million years, our planet experienced five major episodes of **mass extinction** (p. 51; **FIGURE 8.5**). Each event eliminated more than one-fifth of life's families and at least half its species. The most severe episode occurred at the end of the Permian period (see **APPENDIX: Geologic Time Scale** for Earth's geologic periods). At this time, 248 million years ago, close to 90% of all species went extinct. The best-known episode occurred 65 million years ago at the end of the Cretaceous period, when an apparent asteroid impact brought an end to the dinosaurs and many other groups.

If current trends continue, the modern era, known as the Quaternary period, may see the extinction of more than half

FIGURE 8.6 ▲ The ivory-billed woodpecker was one of North America's most majestic birds and lived in old-growth forests throughout the southeastern United States. Forest clearing and timber harvesting eliminated the mature trees it needed for food, shelter, and nesting, and this symbol of the South appeared to go extinct. In recent years, fleeting, controversial observations in Arkansas, Louisiana, and Florida have raised hopes that the species persists, but proof has been elusive.

of all species. Although similar in scale to previous mass extinctions, today's ongoing mass extinction is different in two primary respects. First, we are causing it. Second, we will suffer as a result.

We are setting the sixth mass extinction in motion

Over just the past few centuries, we have recorded hundreds of instances of species extinction caused by people. Among North American birds in the past two centuries alone, we have driven into extinction the Carolina parakeet, great auk, Labrador duck, passenger pigeon (pp. 53–54), probably the Bachman's warbler and Eskimo curlew, and possibly the ivory-billed woodpecker (**FIGURE 8.6**). Several more species, including the whooping crane, Kirtland's warbler, and California condor (p. 177), teeter on the brink of extinction.

However, species extinctions caused by people precede written history. Archaeological evidence shows that in case after case, a wave of extinctions followed close on the heels of human arrival on islands and continents (**FIGURE 8.7**). After Polynesians reached Hawaii, half its birds went extinct. Birds, mammals, and reptiles vanished following human arrival on many other oceanic islands, including large land masses such as New Zealand and Madagascar. Dozens of species of large vertebrates died off in Australia after people arrived roughly 50,000 years ago, and North America lost 33 genera of large mammals once people arrived more than 10,000 years ago.

Today, species loss is accelerating as our population growth and resource consumption put increasing strain on

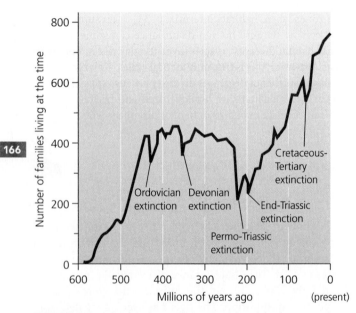

FIGURE 8.5 ▲ The fossil record shows evidence for five episodes of mass extinction during the past half-billion years of Earth history. At the end of the Ordovician, Devonian, Permian, Triassic, and Cretaceous periods, 50–95% of the world's species appear to have gone extinct. (This graph shows families, not species, which is why the drops appear less severe.) Each time, biodiversity later rebounded to equal or higher levels, but each rebound required millions of years. Data from Raup, D.M., and J.J. Sepkoski, 1982. Mass extinctions in the marine fossil record. *Science* 215:1501–1503. Reprinted with permission from AAAS.

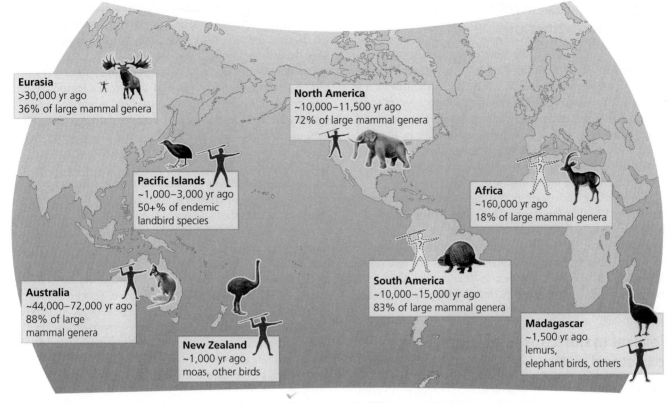

FIGURE 8.7 ▲ This map shows for each region the time of human arrival and the extent of the recent extinction wave. Illustrated are representative extinct megafauna from each region. The human hunter icons are sized according to the degree of evidence that human hunting was a cause of extinctions; larger icons indicate more certainty that humans (as opposed to climate change or other factors) were the cause. Data for South America and Africa are so far too sparse to be conclusive, and future archaeological and paleontological research could well alter these interpretations. Adapted from Barnosky, A.D., et al., 2004. Assessing the causes of late Pleistocene extinctions on the continents. *Science* 306: 70–75; and Wilson, E.O., 1992. *The diversity of life*. Cambridge, MA: Belknap Press.

habitats and wildlife. In 2005, scientists with the Millennium Ecosystem Assessment (p. 16) calculated that the current global extinction rate is 100 to 1,000 times greater than the background rate. They projected that the rate would increase tenfold or more in future decades.

To monitor endangered species, the International Union for Conservation of Nature (IUCN) maintains the **Red List**, an updated list of species facing high risks of extinction. The 2010 Red List reported that 21% (1,131) of mammal species, 13% (1,240) of bird species, and 30% (1,898) of amphibian species are threatened with extinction. Among other major groups (for which assessments are not complete), 17% to 73% of species are judged to be at high risk of extinction. In the United States alone over the past 500 years, 237 animals and 30 plants are known to have gone extinct. For all these figures, the actual numbers are without doubt greater than the known numbers.

Among the 1,131 mammals facing possible extinction is the tiger, which despite—or perhaps because of—its tremendous size and reputation as a fierce predator, is one of the most endangered large animals on the planet. In 1950, eight tiger subspecies existed (see Figure 8.3). Today, three are extinct. The Bali tiger went extinct in the 1940s, the Caspian tiger in the 1970s, and the Javan tiger in the 1980s. The South China tiger has not been seen in 25 years and little of its habitat remains, so scientists fear it too will soon be extinct, if it is not already.

Q: If a mass extinction is happening, why don't I see species going extinct all around me?

A: There are two reasons that most of us don't personally sense the scale of biodiversity loss. First, if you live in a town or city, the plants and animals you see from day to day are generalist species that thrive in disturbed areas. In contrast, the species in trouble are those that rely on less-disturbed habitats, and you may need to go further afield to find them.

Second, a human lifetime is very short! The loss of populations and species over the course of our lifetime may seem a slow process to us, but on Earth's timescale it is sudden—almost instantaneous. Because each of us is born into a world that has already lost many species, we don't recognize what's already been lost. Likewise, our grandchildren won't appreciate what we've lost in our lifetimes. Each human generation experiences just a portion of the overall phenomenon, so we have difficulty sensing the big picture. Nonetheless, researchers who study biology and naturalists who spend their time outdoors *are* seeing a great deal of biodiversity loss—and that's precisely why they feel so passionate about it.

Biodiversity loss involves more than extinction

Extinction is only part of the story of biodiversity loss. The larger part involves declining population sizes. As a species' numbers decline, its geographic range often shrinks as it is extirpated from parts of its range. Thus, many species today are less numerous and occupy less area than they once did. Tigers numbered well over 100,000 worldwide in the 19th century but number only 3,000 to 3,500 today. Such declines mean that genetic diversity and ecosystem diversity, as well as species diversity, are being lost.

To measure and quantify this degradation, scientists at the World Wildlife Fund and the United Nations Environment Programme (UNEP) developed a metric called the *Living Planet Index*. This index summarizes trends in the populations of 2,544 vertebrate species that are sufficiently monitored to provide reliable data. Between 1970 and 2007, the Living Planet Index fell by roughly 30% (**FIGURE 8.8**), driven primarily by biodiversity losses in tropical regions.

Several major causes of biodiversity loss stand out

Scientists have identified four primary causes of population decline and species extinction: habitat loss, pollution, overharvesting, and invasive species. Global climate change (Chapter 14) now is becoming the fifth. Each of these causes is intensified by human population growth and by our increasing per capita consumption of resources.

Habitat loss Habitat loss is the single greatest cause of biodiversity loss today. It is the primary cause of population declines in 83% of threatened mammals and 85% of threatened birds, according to UNEP data. For example, the prairies native to North America's Great Plains are today almost entirely converted to agriculture. Less than 1% of prairie habitat remains. As a result, grassland bird populations have declined by an estimated 82–99%.

Habitat destruction has occurred widely in nearly every biome (**FIGURE 8.9**). Over half of the world's temperate forests, grasslands, and shrublands had been converted by 1950 (mostly for agriculture). Across Asia, scientists estimate that 40% of the tiger's remaining habitat has disappeared just in the last decade. Today habitat is being lost most rapidly in tropical rainforests, tropical dry forests, and savannas.

Because organisms are adapted to the habitats in which they live, any major change in their habitat is likely to render it less suitable for them. Many human activities alter, degrade, or destroy habitat. Farming replaces diverse natural communities with simplified ones composed of only a few plant species. Grazing modifies the structure and species composition of grasslands. Either type of agriculture can lead to desertification. Clearing forests removes the food, shelter, and other resources that forest-dwelling organisms need to survive. Hydroelectric dams turn rivers into reservoirs upstream and affect water conditions and floodplain communities downstream. Urban sprawl supplants natural ecosystems, driving many species from their homes.

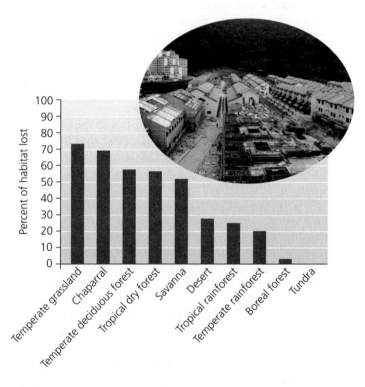

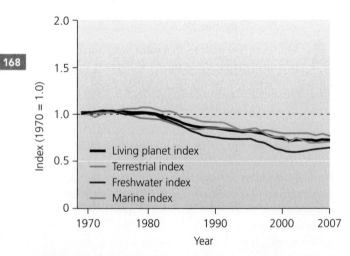

FIGURE 8.8 ▲ The Living Planet Index serves as an indicator of the state of global biodiversity. Index values summarize trends for 7,953 populations of 2,544 vertebrate species. Between 1970 and 2007, the Living Planet Index fell by roughly 30%. The index for terrestrial species fell by 25%; for freshwater species, 35%; and for marine species, 24%. Most losses are in tropical regions, where the index has declined by 60%. In contrast, temperate areas are recovering, showing an improvement of 29%. Data from World Wide Fund for Nature, 2010. *The Living Planet Report, 2010.* Gland, Switzerland.

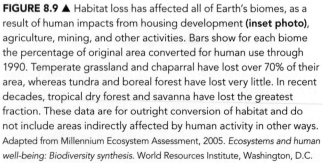

FIGURE 8.9 ▲ Habitat loss has affected all of Earth's biomes, as a result of human impacts from housing development **(inset photo)**, agriculture, mining, and other activities. Bars show for each biome the percentage of original area converted for human use through 1990. Temperate grassland and chaparral have lost over 70% of their area, whereas tundra and boreal forest have lost very little. In recent decades, tropical dry forest and savanna have lost the greatest fraction. These data are for outright conversion of habitat and do not include areas indirectly affected by human activity in other ways. Adapted from Millennium Ecosystem Assessment, 2005. *Ecosystems and human well-being: Biodiversity synthesis.* World Resources Institute, Washington, D.C.

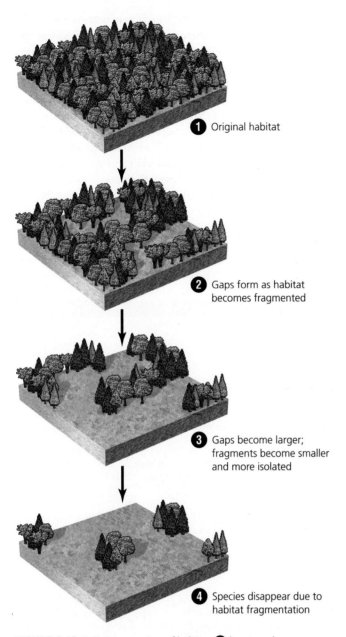

FIGURE 8.10 ▲ Fragmentation of habitat ❶ begins when gaps are created ❷ within a natural habitat. As development proceeds, these gaps expand ❸, join together, and eventually come to dominate the landscape ❹, stranding islands of habitat in their midst. As habitat becomes fragmented, fewer populations can persist, and numbers of species in the fragments decline.

❶ Original habitat

❷ Gaps form as habitat becomes fragmented

❸ Gaps become larger; fragments become smaller and more isolated

❹ Species disappear due to habitat fragmentation

Of course, human habitat alteration benefits some species. Animals such as house sparrows, pigeons, gray squirrels, rats, and cockroaches thrive in cities and suburbs. However, the species that benefit from our presence are relatively few; for every species that wins, more lose. Furthermore, the species that do well in our midst tend to be weedy generalists that are in little danger of disappearing any time soon.

Habitat loss occurs most commonly through gradual, piecemeal degradation such as **habitat fragmentation** (**FIGURE 8.10**). When farming, logging, road building, or development intrude into a forest, they break up a continuous expanse of forest habitat into an array of fragments, or patches. As habitat fragmentation proceeds across a landscape, animals and plants adapted to the forest habitat disappear from one fragment after

another. In response to habitat fragmentation, conservation biologists design landscape-level strategies to prioritize areas to be preserved (pp. 201–203).

Pollution Pollution harms organisms in many ways. Air pollution (Chapter 13) degrades forest ecosystems. Water pollution (Chapter 12) impairs fish and amphibians. Agricultural runoff (including fertilizers, pesticides, and sediments; Chapters 2 and 7) harms many terrestrial and aquatic species. Heavy metals, polychlorinated biphenyls (PCBs), endocrine-disrupting compounds, and other toxic chemicals poison people and wildlife (Chapter 10). Plastic garbage in the ocean can strangle, drown, or choke marine creatures (pp. 270–271). The effects of oil and chemical spills on wildlife (pp. 267, 338–340) are dramatic and well known. However, although pollution is a substantial threat, it tends to be less significant than public perception holds it to be, and it is far less influential than habitat loss.

Overharvesting For most species, hunting or harvesting by people will not in itself pose a threat of extinction, but for species like the Siberian tiger, it can. Large in size, few in number, long-lived, and raising few young in its lifetime—a classic K-selected species (p. 59)—the Siberian tiger is just the type of animal to be vulnerable to hunting. The advent of Russian hunting nearly drove the animal extinct, whereas decreased hunting after World War II allowed the population to increase. By the 1980s, the Siberian tiger population was likely up to 250 individuals. The political freedom that came with the Soviet Union's breakup in 1989, however, brought with it a freedom to harvest Siberia's natural resources, including the tiger, without regulations or rules, and poachers illegally killed at least 180 Siberian tigers between 1991 and 1996. This coincided with an economic expansion in many Asian countries, where tiger penises are believed to boost human sexual performance and where tiger bones, claws, whiskers, and other body parts are used to try to treat a variety of health problems (**FIGURE 8.11**). Although no proof of

FIGURE 8.11 ▼ Body parts from tigers are sold as traditional medicines and aphrodisiacs in some Asian cultures. Poachers are illegally killing tigers to satisfy the surging market demand for these items. Here a street vendor in northern China displays tiger body parts for sale.

their effectiveness has been demonstrated, sale of body parts from one tiger fetches at least $15,000 on the black market—a powerful economic temptation for poachers in poor regions.

Hunting has reduced the populations of many K-selected animals. The Atlantic gray whale was driven extinct, and several other whales remain threatened or endangered. Gorillas and other primates that are killed for their meat may face extinction soon. Thousands of sharks are killed each year simply so their fins can be used in soup. Today the oceans contain only 10% of the large animals they once did (p. 274).

To combat overharvesting, governments have passed laws, signed treaties, and strengthened anti-poaching efforts. Scientists have begun using genetic analyses to expose illegal hunting and wildlife trade. For instance, DNA testing can reveal the geographic origins of elephant ivory and determine whether whale meat sold in markets came from animals caught illegally (see **THE SCIENCE BEHIND THE STORY**, pp. 178–179).

Invasive species Our introduction of non-native species to new environments, where some may become invasive (pp. 75–77), also displaces native species (**FIGURE 8.12**). Some introductions are accidental. Examples include aquatic organisms transported in the ballast water of ships (such as zebra mussels; Chapter 4), animals that escape from the pet trade, and weeds whose seeds cling to our socks as we travel from place to place. Other introductions are intentional. People have long brought food crops and animals with them as they colonized new places, and today we continue international trade in exotic pets and ornamental plants.

Most organisms introduced to new areas perish, but the few types that survive may do very well, especially if they are freed from the predators and parasites that attacked them back home or from the competitors that had limited their access to resources. Once released from the limiting factors (p. 58) of predation, parasitism, and competition, an introduced species

Invasive Species		
Species	**Invasive in...**	**Effects**
Gypsy moth (*Lymantria dispar*) 	Northeastern United States (Native to Eurasia)	In the 1860s, a scientist introduced the gypsy moth to Massachusetts in the belief that it might help produce a commercial-quality silk. The moth failed to start a silk industry, and instead spread through the northeastern United States, where its outbreaks defoliate trees over large regions every few years.
European starling (*Sturnus vulgaris*) 	North America (Native to Europe)	The bird was first introduced to New York City in the late 19th century by Shakespeare devotees intent on bringing every bird mentioned in Shakespeare's plays to America. It only took 75 years for starlings to spread to all corners of North America, becoming one of the continent's most abundant birds. Starlings are thought to outcompete native birds for nest holes.
Cheatgrass (*Bromus tectorum*) 	Western United States (Native to Eurasia)	In just 30 years after its introduction to Washington state in the 1890s, cheatgrass has spread across much of the western United States. It crowds out other plants, uses up the soil's nitrogen, and burns readily. Fire kills many of the native plants, but not cheatgrass, which grows back even stronger amid the lack of competition.
Brown tree snake (*Boiga irregularis*) 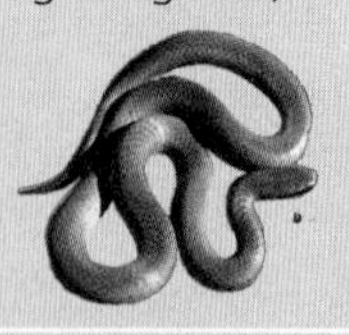	Guam (Native to Southeast Asia)	Nearly all native forest bird species on the South Pacific island of Guam have disappeared. The culprit is the brown tree snake, brought to the island inadvertently as stowaways in cargo bays of military planes in World War II. Guam's birds had not evolved with tree snakes, and so had no defenses against the snake's nighttime predation. The snakes have spread to other islands where they are repeating their ecological devastation. The arrival of this snake is the greatest fear of conservation biologists in Hawaii.
Kudzu (*Pueraria montana*) 	Southeastern United States (Native to Japan)	Kudzu is a vine that can grow 30 m (100 ft) in a single season. The U.S. Soil Conservation Service introduced kudzu in the 1930s to help control erosion. Adaptable and extraordinarily fast-growing, kudzu has taken over thousands of hectares of forests, fields, and roadsides.
Asian long-horned beetles (*Anoplophora glabripennis*) 	United States (Native to Asia)	Having arrived in imported lumber in the 1990s, these beetles burrow into trees and interfere with the trees' ability to absorb and process water and nutrients. They may wipe out the majority of hardwood trees in an area. Several U.S. cities, including Chicago and Seattle, have cleared thousands of trees after detecting these invaders.
Rosy wolfsnail (*Euglandina rosea*) 	Hawaii (Native to Southeastern United States and Latin America)	In the 1950s, well-meaning scientists introduced the rosy wolfsnail to Hawaii to prey upon and reduce the population of another invasive species, the giant African land snail. Within a few decades, however, the carnivorous rosy wolfsnail had instead driven more than half of Hawaii's native species of banded tree snails to extinction.

FIGURE 8.12 ▲ Invasive species are species that thrive in areas where they are introduced, often harming native species. This chart shows a few of the many thousands of invasive species.

FIGURE 8.13 ▲ As Arctic warming melts the sea ice from which they hunt seals, polar bears must swim farther for food. A lawsuit brought by environmental groups forced the U.S. Fish and Wildlife Service in 2008 to list the polar bear as threatened under the Endangered Species Act as a result of climate change.

may proliferate and displace native species. Invasive species cause billions of dollars in economic damage each year.

Climate change The preceding four types of human impacts affect biodiversity in discrete places and times. In contrast, our manipulation of Earth's climate (Chapter 14) is having global impacts. As our emissions of greenhouse gases from fossil fuel combustion cause temperatures to warm worldwide, we modify weather patterns and increase the frequency of extreme weather events.

Extreme weather events such as droughts increase stress on populations, and warming temperatures force organisms to shift their geographic ranges toward the poles and higher in altitude. Some species will not be able to adapt. Mountaintop organisms (such as the cloud-forest fauna at Monteverde in Chapter 3) cannot move further upslope to escape warming temperatures, so they may perish. Trees may not be able to move poleward fast enough. As ranges shift, animals and plants may find themselves among new communities of prey, predators, and parasites to which they are not adapted. In the Arctic, where warming has been greatest, the polar bear (**FIGURE 8.13**) has been listed as a threatened species under the U.S. Endangered Species Act (p. 176) because thawing ice hinders its ability to hunt seals (p. 314). All in all, scientists predict that a 1.5–2.5 °C (2.7–4.5 °F) global temperature rise could put 20–30% of the world's plants and animals at increased risk of extinction.

All five of these causes of biodiversity loss are intensified by human population growth and rising per capita resource consumption. As researchers gain a solid scientific understanding of the causes of biodiversity loss, we are also coming to appreciate its consequences (**FIGURE 8.14**). Moreover, these causes may interact, resulting in impacts greater than the sum of their parts. The current collapse of amphibian populations throughout the world provides an example. Today entire populations of frogs, toads, and salamanders are vanishing without a trace. Nearly 2,500 of the 6,300 known species are in decline, and roughly 170 species stud-

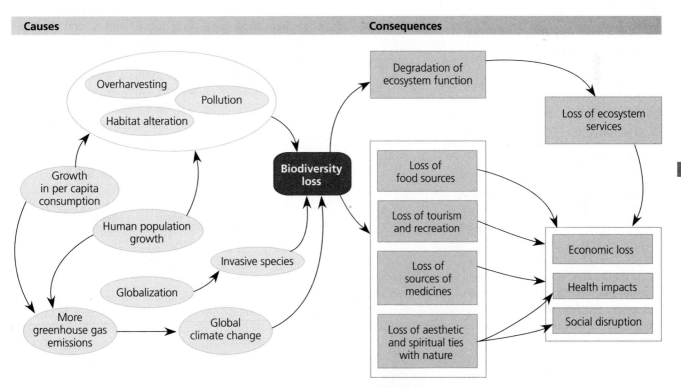

FIGURE 8.14 ▲ The loss of biodiversity stems from a variety of causes (**ovals on left**) and results in a number of consequences (**boxes on right**) for ecological systems and human well-being. Arrows in this concept map lead from causes to consequences. Note that items grouped within outlined boxes do not necessarily share any special relationship; the outlined boxes are intended merely to streamline the figure.

Solutions

As you progress through this chapter, try to identify as many solutions to biodiversity loss as you can. What could you personally do to help address this issue? Consider how each action or solution might affect items in the concept map above.

ied just years or decades ago are thought to be gone. As these creatures disappear before our eyes, scientists are racing to discover the reasons, and they have found evidence implicating habitat destruction, chemical pollution, disease, invasive species, and climate change. In some cases, researchers surmise that factors in combination are multiplying one another's effects.

BENEFITS OF BIODIVERSITY

Biodiversity loss matters from an ethical perspective, because many people feel that organisms have an intrinsic right to exist. However, losing biodiversity is also a problem for human society because of the many tangible, pragmatic ways that biodiversity benefits people and supports our society.

Biodiversity provides ecosystem services

Contrary to popular opinion, some things in life can indeed be free—as long as we protect the ecological systems that provide them. Intact forests provide clean air and water, and they buffer hydrologic systems against flooding and drought. Native crop varieties provide insurance against disease. Wildlife can attract tourism and boost economies. Intact ecosystems provide these and other valuable processes, known as *ecosystem services* (pp. 2, 36, 90, 95), for all of us, free of charge. According to UNEP, biodiversity:

- Provides food, fuel, fiber, and shelter.
- Purifies air and water.
- Detoxifies and decomposes wastes.
- Stabilizes Earth's climate.
- Moderates floods, droughts, and temperatures.
- Cycles nutrients and renews soil fertility.
- Pollinates plants, including many crops.
- Controls pests and diseases.
- Maintains genetic resources for crop varieties, livestock breeds, and medicines.
- Provides cultural and aesthetic benefits.
- Gives us the means to adapt to change.

In these ways, organisms and ecosystems support vital processes that people cannot replicate or would need to pay for if nature did not provide them. The annual economic value of just 17 of these ecosystem services has been estimated at more than $46 trillion per year (p. 95).

Biodiversity helps maintain ecosystem function

Ecological research shows that biodiversity tends to enhance the *stability* of communities and ecosystems. Research also finds that biodiversity tends to increase the *resilience* (p. 74) of ecological systems—their ability to

weather disturbance, bounce back from stress, or adapt to change. Thus, when we lose biodiversity this can diminish a natural system's ability to function and to provide services to our society.

Will the loss of a few species really make much difference in an ecosystem's ability to function? Consider a metaphor first offered by Paul and Anne Ehrlich (pp. 115, 117): The loss of one rivet from an airplane's wing—or two, or three—may not cause the plane to crash. But as more rivets are removed the structure will be compromised, and eventually the loss of just one more rivet will cause it to fail.

Research suggests that removing a top predator such as a tiger can indeed have a strong impact, because top predators are often keystone species (pp. 71–72). A single tiger may prey on many herbivores, each of which may consume many plants—so the removal of a species like the tiger can have consequences that multiply as they cascade down the food chain.

"Ecosystem engineers" (pp. 72–73) such as ants and earthworms can be every bit as influential as keystone species, so the loss of an ecosystem engineer from a system can likewise set major changes in motion. Ecosystems are complex, and it can be difficult to predict which species are most important. Thus, many people prefer to apply the precautionary principle (pp. 152, 222) in the spirit of Aldo Leopold (p. 14), who advised, "To keep every cog and wheel is the first precaution of intelligent tinkering."

Biodiversity enhances food security

Biodiversity provides the food we eat. Throughout history, people have used 7,000 plant species and several thousand animal species for food. Today nutritional experts worry that industrial agriculture has narrowed our diet. Globally, we now get 90% of our food from just 15 crop species and 8 livestock species, and this lack of diversity leaves us vulnerable to failures of particular crops. In a world where 1 billion people go hungry and more are malnourished, we can improve food security (the guarantee of an adequate, safe, nutritious, and reliable food supply; p. 134) by finding sustainable ways to harvest or farm novel or underutilized wild species and rare crop varieties (**FIGURE 8.15**). For example, the babassu palm of the Amazon produces more vegetable oil than any other plant. The serendipity berry generates a sweetener 3,000 times sweeter than table sugar. Several salt-tolerant grasses and trees are so hardy that farmers can irrigate them with salt water to produce animal feed, a vegetable oil substitute, and other products.

Moreover, crop relatives and wild ancestors of crops hold reservoirs of genetic diversity (p. 136) that can help save our monocultural crops from catastrophe when we transfer helpful genes by crossbreeding or genetic engineering. We have already received tens of billions of dollars' worth of disease resistance from the wild relatives of potatoes, wheat, barley, and other crops.

Organisms provide drugs and medicines

People have made medicines from plants for centuries, and many of today's pharmaceuticals are derived from

Food Security and Biodiversity: Potential new food sources		
Species	Native to...	Potential uses and benefits
Amaranths (three species of *Amaranthus*)	Tropical and Andean America	Grain and leafy vegetable; livestock feed; rapid growth, drought resistant
Buriti palm (*Mauritia flexuosa*)	Amazon lowlands	"Tree of life" to Amerindians; vitamin-rich fruit; pith as source for bread; palm heart from shoots
Maca (*Lepidium meyenii*)	Andes Mountains	Cold-resistant root vegetable resembling radish, with distinctive flavor; near extinction
Babirusa (*Babyrousa babyrussa*)	Indonesia: Moluccas and Sulawesi	A deep-forest pig; thrives on vegetation high in cellulose and hence less dependent on grain
Capybara (*Hydrochoeris hydrochoeris*)	South America	World's largest rodent; meat esteemed; easily ranched in open habitats near water
Vicuna (*Lama vicugna*)	Central Andes	Threatened species related to llama; source of meat, fur, and hides; can be profitably ranched
Chachalacas (*Ortalis*, many species)	South and Central America	Tropical birds; adaptable to human habitations; fast-growing

FIGURE 8.15 ▲ By protecting biodiversity, we enhance food security. The wild species shown here are a fraction of the many plants and animals that could supplement our food supply. Adapted from Wilson, E.O., 1992. *The diversity of life*. Cambridge, MA: Belknap Press.

Medicines and Biodiversity: Natural sources of pharmaceuticals		
Plant	Drug	Medical application
Pineapple (*Ananas comosus*)	Bromelain	Controls tissue inflammation
Autumn crocus (*Colchicum autumnale*)	Colchicine	Anticancer agent
Yellow cinchona (*Cinchona ledgeriana*)	Quinine	Antimalarial
Common thyme (*Thymus vulgaris*)	Thymol	Cures fungal infection
Pacific yew (*Taxus brevifolia*)	Taxol	Anticancer (especially ovarian cancer)
Velvet bean (*Mucuna deeringiana*)	L-Dopa	Parkinson's disease suppressant
Common foxglove (*Digitalis purpurea*)	Digitoxin	Cardiac stimulant

FIGURE 8.16 ▲ By protecting biodiversity, we enhance our ability to treat illness. Shown are just a few of the plants found to provide chemical compounds of medical benefit. Adapted from Wilson, E.O., 1992. *The diversity of life*. Cambridge, MA: Belknap Press.

The world's biodiversity holds an even greater treasure chest of medicines still to be discovered. Yet with every species that goes extinct, we lose one more opportunity to find cures for cancer, AIDS, or other maladies (**FIGURE 8.17**).

FIGURE 8.17 ▲ We lost opportunities for medical advances when two species of gastric brooding frogs went extinct shortly after their discovery in Australia's rainforests. Females of these bizarre frogs raised their young inside their stomachs, where the young apparently exuded substances to neutralize their mother's stomach acids. Any such substance could be of immense value for treating human stomach ulcers, which affect 25 million U.S. citizens. When both frog species went extinct in the 1980s, they took their secrets with them forever.

chemical compounds from wild plants (**FIGURE 8.16**). The rosy periwinkle produces compounds that treat Hodgkin's disease and a deadly form of leukemia. Had this plant from Madagascar become extinct, these two fatal diseases would have claimed far more victims. In Australia, a rare species of cork, *Duboisia leichhardtii*, provides hyoscine, a compound that physicians use to treat cancer, stomach disorders, and motion sickness. The Pacific yew of North America's Pacific Northwest produces a compound that forms the basis for the anti-cancer drug taxol. Each year, pharmaceutical products owing their origin to wild species generate up to $150 billion in sales and save thousands of human lives.

Biodiversity boosts economies through tourism and recreation

Many people like to travel to observe wildlife and explore natural areas, and in so doing they create economic opportunities for residents living near protected natural areas. Visitors spend money at local businesses, hire local people as guides, and support parks that employ local residents. Such **ecotourism** (p. 60) can bring jobs and income to areas that otherwise might be poverty-stricken.

Ecotourism has become a vital source of income for Costa Rica, with its rainforests; Australia, with its Great Barrier Reef; Belize, with its reefs, caves, and rainforests; and Kenya and Tanzania, with their savanna wildlife. The United States, too, benefits from ecotourism; its national parks draw millions of visitors from around the world. Although too much development for ecotourism can damage the natural assets that draw people, ecotourism can serve as a powerful financial incentive for nations, states, and local communities to preserve natural areas and reduce impacts on the landscape and on native species.

People value and seek out connections with nature

Not all of biodiversity's benefits to people can be expressed in the hard numbers of economics or the practicalities of food and medicine. Some scientists and philosophers argue that people find a deeper value in biodiversity. Harvard University biologist and author Edward O. Wilson has popularized the notion of **biophilia**, asserting that human beings have an instinctive love for nature and feel an emotional bond with other living things (**FIGURE 8.18**). Wilson and others cite as evidence of biophilia our affinity for parks and wildlife, our love for pets, the high value of real estate with a view of natural landscapes, and our interest in hiking, bird-watching, fishing, hunting, backpacking, and similar outdoor pursuits.

In a 2005 book, writer Richard Louv adds that as today's children are increasingly deprived of outdoor experiences and

direct contact with wild organisms, they suffer what he calls "nature-deficit disorder." Although it is not a medical condition, Louv argues that this alienation from biodiversity and nature damages childhood development and may lie behind many of the emotional and physical problems young people in developed nations face today.

Do we have ethical obligations toward other species?

Aside from all of biodiversity's pragmatic benefits, many people feel that living organisms have an inherent right to exist. In this view, biodiversity conservation is justified on ethical grounds alone. As Maurice Hornocker wrote when he and his associates first established the Siberian Tiger Project, "Saving the most magnificent of all the cat species and one of the most endangered should be a global responsibility. . . . If they aren't worthy of saving, then what are we all about? What *is* worth saving?"

Today many people are engaged in efforts to save vanishing species. The search for solutions to today's biodiversity crisis is dynamic and exciting, and scientists are developing innovative approaches to maintaining the diversity of life on Earth.

CONSERVATION BIOLOGY: THE SEARCH FOR SOLUTIONS

Today, more and more scientists and citizens perceive a need to stop the loss of biodiversity. In his 1994 autobiography, *Naturalist*, E.O. Wilson wrote:

> When the [20th] century began, people still thought of the planet as infinite in its bounty. [Yet] in one lifetime, exploding human populations have reduced wildernesses to threatened nature reserves. Ecosystems and species are vanishing at the fastest rate in 65 million years. Troubled by what we have wrought, we have begun to turn in our role from local conqueror to global steward.

Conservation biology arose in response to biodiversity loss

The urge to act as responsible stewards of natural systems, and to use science as a tool in that endeavor, helped spark the rise of conservation biology. **Conservation biology** is a scientific discipline devoted to understanding the factors, forces, and processes that influence the loss, protection, and restoration of biological diversity.

Conservation biologists aim to develop solutions to such problems as habitat degradation and species loss (see **ENVISIONIT**, p. 175). Conservation biology is thus an applied and goal-oriented science, with implicit values and ethical standards. Conservation biologists integrate an understanding of evolution and extinction with ecology and the dynamic nature of environmental systems. They use field data, lab data, theory, and experiments to study our impacts on other organisms. They also design, test, and implement ways to alleviate human impact.

FIGURE 8.18 ▼ An Indonesian girl peers into a flower of *Rafflesia arnoldii*, the largest flower in the world. The concept of biophilia holds that human beings have an instinctive love and fascination for nature and a deep-seated desire to affiliate with other living things.

Sampling insects in Madagascar

Conservation biologists are racing to save species from decline and extinction.

Tracking rhinos in Africa

They work in the field, in the lab, at zoos, and with local people in areas that need protecting ...

... striving to recover populations of plants and animals threatened by habitat loss and other causes.

Blood sample from Seychelles Magpie Robin

Sea turtle with transmitter, Hong Kong

Sampling plants in Florida

Radio tracking birds in Spain

You Can Make a Difference

➤ Preserve or restore wildlife habitat in your yard, in your community, or on your campus.

➤ Volunteer time or money to organizations working to save species and habitat.

➤ Buy shade-grown coffee, organic produce, and other wildlife-friendly products.

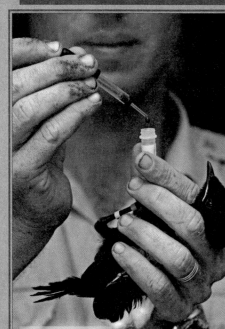

At the genetic level, *conservation geneticists* ask how small a population can become and how much genetic variation it can lose before running into problems such as inbreeding depression (p. 164). By determining a population's *minimum viable population size*, conservation geneticists can help wildlife managers decide how vital it may be to increase the population. Studies of genes, populations, and species inform conservation efforts with habitats, communities, ecosystems, and landscapes. Because small and isolated subpopulations are most vulnerable to extirpation, conservation biologists pay special attention to them. By examining how organisms disperse from one habitat patch to another, and how their genes flow among subpopulations, conservation biologists try to learn how likely a population is to persist or succumb in the face of habitat change or other threats.

Endangered species are a focus of conservation efforts

The primary legislation for protecting biodiversity in the United States is the **Endangered Species Act**. Passed in 1973, the Endangered Species Act (ESA) forbids the government and private citizens from taking actions that destroy endangered species or their habitats. The ESA also forbids trade in products made from endangered species. The aim is to prevent extinctions, stabilize declining populations, and enable populations to recover. As of 2011, there were 1,061 species in the United States listed as "endangered" and 313 more listed as "threatened," the status considered one notch less severe than endangered. For about half of these species, government agencies are running recovery plans to protect them and stabilize or increase their populations.

The ESA has had a number of notable successes. Following the 1973 ban on the pesticide DDT and years of effort by wildlife managers, the bald eagle, peregrine falcon, brown pelican, and other birds have recovered and are no longer listed as endangered (**FIGURE 8.19**). Intensive management programs with other species, such as the red-cockaded woodpecker, have held populations steady in the face of continued pressure on habitat. In fact, roughly 40% of declining populations have been stabilized.

This success comes despite the fact that the U.S. Fish and Wildlife Service and the National Marine Fisheries Service, the agencies responsible for upholding the ESA, are perennially underfunded for the job. Reauthorization of the ESA faced stiff opposition from the Republican Congresses in power from 1994 to 2006. Efforts to weaken the ESA by stripping it of its ability to safeguard habitat were narrowly averted in 2006 after 5,700 scientists sent Congress a letter of protest.

Polls repeatedly show that most Americans support protecting endangered species. Yet some opponents feel that the ESA places more value on the life of an endangered organism than it does on the livelihood of a person. This was a common perception in the Pacific Northwest in the 1990s, when protection for the northern spotted owl slowed logging in old-growth rainforest and loggers began to fear for their jobs. In addition, many landowners worry that fed-

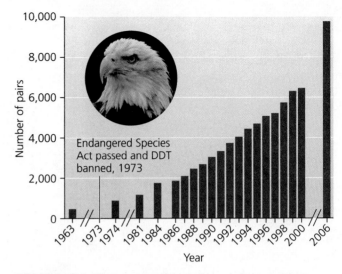

FIGURE 8.19 ▲ The recovery of the bald eagle is a success story of the U.S. Endangered Species Act. The national symbol of the United States was close to extinction in the Lower 48 states in the 1960s. Following its protection under the ESA and a ban on the pesticide DDT in 1973, the eagle population began a long rebound. With its Lower-48 population around 10,000 pairs in 2007, the bald eagle was declared recovered and was removed from the Endangered Species List. Data from U.S. Fish and Wildlife Service, based on annual volunteer eagle surveys. The paucity of data after 2000 is because surveys began to be discontinued once it became clear that the eagle was recovering.

eral officials will restrict the use of private land on which threatened or endangered species are found. This has led to a practice described as "shoot, shovel, and shut up" among landowners who want to conceal the presence of such species on their land.

In fact, however, the ESA has stopped few development projects—and a number of provisions of the ESA and its amendments promote cooperation with landowners. *Habitat conservation plans* and *safe harbor agreements* are arrangements that allow landowners to harm species in some ways if they improve habitat for them in others.

Today a number of nations have laws protecting species, although they are not always effective. When Canada enacted its *Species at Risk Act* in 2002, the Canadian government was careful to stress cooperation with landowners and provincial governments, rather than presenting the law as a mandate from the national government. Environmental advocates and many scientists protested that the law was too weak and failed to protect habitat adequately.

In Russia, the government issued Decree 795 in 1995, creating a Siberian tiger conservation program and declaring the tiger a natural and national treasure. In 2007 it established a national park in the Sikhote-Alin Mountains to help protect the tiger, and the next year Prime Minister Vladimir Putin personally visited scientists in the field as they tagged a tiger. However, state funding for tiger conservation remained so meager that the Wildlife Conservation Society felt it necessary to help pay for Russians to enforce their own anti-poaching laws. Whether government support increases in the wake of Russia's hosting of the 2010 International Tiger Forum in St. Petersburg remains to be seen.

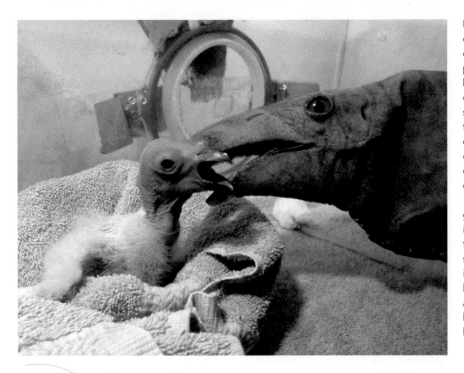

FIGURE 8.20 ◄ To save the California condor from extinction, biologists are raising chicks in captivity. Feeding them with hand puppets designed to look and feel like the heads of adult condors, the biologists shield each chick from all contact with humans, so that when the bird is grown it does not feel an attachment to people. Condors had declined because people killed them, they collided with electrical wires, and they succumbed to lead poisoning after scavenging carcasses of animals killed with lead shot. By 1982, only 22 condors remained, and biologists decided to take all the birds into captivity, in hopes of boosting their numbers and then releasing them. The ongoing program is succeeding. As of 2011, there were 189 birds in captivity and 181 birds living in the wild, having been released at sites in California, Arizona, and Baja California. Several pairs have begun nesting, and so far 24 chicks have been raised in the wild.

Conservation efforts include international treaties

The United Nations has facilitated several international treaties to protect biodiversity. The 1973 **Convention on International Trade in Endangered Species of Wild Fauna and Flora (CITES)** protects endangered species by banning the international transport of their body parts. When nations enforce it, CITES can protect tigers and other rare species whose body parts are traded internationally.

In 1992, leaders of many nations agreed to the **Convention on Biological Diversity**. This treaty embodies three goals: to conserve biodiversity, to use biodiversity in a sustainable manner, and to ensure the fair distribution of biodiversity's benefits. Among its accomplishments, the treaty has prompted nations to augment protected reserves, has enhanced global markets for shade-grown coffee and other crops grown without removing forests, has ensured that African nations share in the economic benefits of ecotourism from wildlife preserves, and has replaced pesticide-intensive farming practices with sustainable ones in some rice-producing Asian nations. Yet the treaty's overall goal—"to achieve, by 2010, a significant reduction of the current rate of biodiversity loss at the global, regional, and national level"—was not met. Fully 193 nations have become parties to the Convention on Biological Diversity. The only ones choosing *not* to do so are tiny Andorra, the Vatican, and the United States.

Captive breeding, reintroduction, and cloning are being pursued

In the effort to save threatened and endangered species, zoos and botanical gardens have become centers for **captive breeding**, in which individuals are bred and raised in controlled conditions with the intent of reintroducing them into the wild. One example is the program to save the California condor, North America's largest bird (**FIGURE 8.20**). The successful program to reintroduce gray wolves to Yellowstone National Park has proven popular with the American public but has met stiff resistance from ranchers, who fear the wolves will attack their livestock. In Arizona and New Mexico, a wolf reintroduction program is making headway, but a number of wolves have been shot.

China is considering a reintroduction program for the Siberian tiger. The Chinese government says it is preparing 600 captive Siberian tigers for release into the far northeastern portion of the country. Critics note that forests there are so fragmented that efforts should focus on improving habitat first.

One new idea for saving species from extinction is to create individuals by cloning them. In this technique, DNA (p. 28) from an endangered species is inserted into a cultured egg without a nucleus, and the egg is implanted into a female of a closely related species that acts as a surrogate mother. Several mammals have been cloned in this way, with mixed results. Some scientists even talk of recreating extinct species from DNA recovered from preserved body parts. Indeed, in 2009 a subspecies of Pyrenean ibex (a type of mountain goat) was cloned from cells taken from the last surviving individual, which had died in 2000. The cloned baby ibex died shortly after birth. Even if cloning can succeed from a technical standpoint, however, such efforts are not an adequate response to biodiversity loss. Without ample habitat and protection in the wild, having cloned animals in a zoo does little good.

Forensics is being used to protect species

Forensic science, or **forensics**, involves the scientific analysis of evidence to make an identification or answer a question relating to a crime or an accident. Conservation biologists are now employing forensics to protect species at risk from illegal harvesting. By analyzing DNA from organisms or their tissues sold at

Dr. C. Scott Baker of Oregon State University running genetic analyses in a Japanese hotel room

Using Forensics to Uncover Illegal Whaling

As any television buff knows, forensic science is a crucial tool in solving mysteries and fighting crime. In recent years conservation biologists have been using forensics to unearth secrets and catch bad guys in the multibillion-dollar illegal global wildlife trade. One such detective story comes from the Pacific Ocean and Japan.

The meat from whales has long been a delicacy in Japan and some other nations. Whaling ships decimated populations of most species of whales in the 20th century through overhunting, and the International Whaling Commission (IWC) outlawed commercial whaling worldwide beginning in 1986. Yet whale meat continues to be sold at market to wealthy consumers today (**see photo**). This meat comes legally from several sources:

1. From scientific hunts. Japan and several other nations negotiated with the IWC to continue to hunt limited numbers of whales for research purposes, and this meat may be sold afterwards.

2. From whales killed accidentally when caught in fishing nets meant for other animals (*bycatch*, p. 274).

3. Possibly from stockpiles frozen before the IWC's moratorium.

However, conservation biologists long suspected that much of the whale meat on the market was actually caught illegally for the purpose of selling for food and that fleets from Japan and other nations were killing more whales than international law allowed. Once DNA sequencing technology was developed, scientists could use this tool to find out.

The detectives in this story are conservation geneticists C. Scott Baker, Stephen Palumbi, Frank Cipriano, and their colleagues. For close to two decades they have been traveling to Asia on what have

Whale meat is sold at this market in South Korea.

amounted to top-secret grocery shopping trips.

It began in 1993 when Baker and Palumbi bought samples of whale meat—all labeled simply as *kujira*, the generic Japanese term for whale meat—from a number of markets in Japan and sequenced DNA from these samples. Law forbids the export of whale meat, so the researchers had to run their analyses in their hotel rooms with portable genetic kits. Once they were back home in the United States, they compared their data with sequences from known whale species.

By analyzing which samples matched which, they concluded that they had sampled meat from nine minke whales, four fin whales, one humpback whale, and two dolphins. Moreover, because subspecies of whales from different oceans differ genetically, the researchers were able to analyze the genetic variation in their samples and learn that one fin whale came from the Atlantic whereas the other three were from the Pacific, and that eight of the nine minke whales came from the Southern Hemisphere.

Because several of these species and/or subspecies were off-limits to hunting, the data suggested that some of the meat had been hunted, processed, or traded illegally. Baker and Palumbi concluded in a 1994 paper in the journal *Science* that "the existence of legal whaling serves as a cover for the sale of illegal whale products." They urged that the international

market, researchers can often determine the species or subspecies and, sometimes, its geographic origin. This can help detect illegal activity, enhancing the enforcement of laws protecting wildlife (see **THE SCIENCE BEHIND THE STORY**, above).

One example involves African elephants killed for ivory from their tusks. Trade in ivory has been banned globally in an effort to stop the slaughter of elephants. After airport customs agents seized 6.5 tons of tusks in Singapore in 2002, researchers led by Samuel Wasser of the University of Washington analyzed DNA from the tusks to determine the geographic origin of the elephants that were killed. The researchers sought to find out whether the tusks belonged to savanna elephants killed in

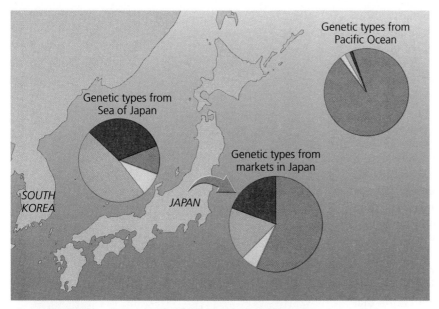

The distribution of genetic types from minke whale meat in Japanese markets (center pie chart) shows evidence both of whales from the Sea of Japan (left pie chart) and of whales from the open Pacific Ocean (right pie chart). Note how proportions of the types from the market are intermediate between those of each geographic area of ocean. Adapted with permission from Lukoschek, V., et al., 2009. High proportion of protected minke whales sold on Japanese markets due to illegal, unreported, or unregulated exploitation. *Animal Conservation* 12:385–395.

and found evidence of 12 species or subspecies of whales, along with orcas, porpoises, and dolphins—and even sheep and horses! Seven of the whale species were internationally protected, and together these constituted 10% of the whale meat for sale in Japanese markets.

Genetic analyses of minke whale samples from Japan's markets also indicated that a surprisingly large percentage came from animals from the Sea of Japan, where Korea and Japan harvested them as fishing bycatch. One-third of the meat on the market was coming from the Sea of Japan, meaning that four times as many whales were being killed there as Japan was reporting (**see top figure**). The research team calculated that Japan and Korea together were taking so many minke whales from the Sea of Japan that they would eventually wipe out the population (**see bottom figure**).

In 2007, Baker led a team that combined genetic forensics with ecological methods to estimate numbers of individual whales whose meat was passing through Korean markets. They inferred that meat from 827 minke whales had passed through South Korea's market in five years. The nation had reported catching only 458 minke whales as fishing bycatch, leading the researchers to conclude that the remainder had been taken illegally.

The governments of Japan and South Korea have tried to refute these findings. Yet the technology and approaches that turn scientists into forensic detectives are now influencing the debate and the negotiation over whaling policy at the international level.

community monitor catches more closely.

Two years later, Baker, Palumbi, and Cipriano presented results from markets in South Korea and Japan. Again their genetic sleuthing revealed a diversity of whale species, and they stated that their data were "difficult to reconcile" with records of legal catches (scientific whaling by Japan and fishing bycatch by South Korea) reported by these nations to the IWC. Among the whales they detected were two specimens of what seemed to be a subspecies or species of whale new to science.

In 2000, the team analyzed 655 samples labeled as whale meat from Japanese and South Korean markets

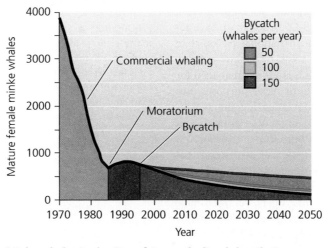

Minke whales in the Sea of Japan declined sharply (orange color in graph) until the 1986 moratorium on their capture. Population models forecast that bycatch of 150, 100, or even 50 minke whales per year would be enough to prevent the population's recovery. Data suggest that actual bycatch from the Sea of Japan has been close to 150 per year. Adapted from Baker, C.S., et al., 2000. Predicted decline of protected whales based on molecular genetic monitoring of Japanese and Korean markets. *Proc. Roy. Soc. Lond. B* 267:1191–1199.

Zambia (the origin of the shipment), or forest elephants from other locations. The DNA matched known samples from Zambian elephants, revealing that many more elephants were being killed there than Zambia's government had realized. In response, the government replaced its wildlife director and began imposing harsher sentences on poachers and ivory smugglers.

Some species act as "umbrellas," protecting habitat and communities

Scientists know that protecting species does little good if the larger ecological systems they rely on are not also sustained. Yet no law or treaty exists to protect communities or ecosystems. Therefore, conservation biologists often use particular species as tools to conserve habitats, communities, and ecosystems. Such species are called *umbrella species* because they serve as a kind of umbrella to protect many other species. Umbrella species often are large animals that roam great distances, such as the Siberian tiger. Because such animals require large areas, meeting their habitat needs helps meet those of thousands of less charismatic species that might never elicit as much public interest.

Environmental advocacy organizations have found that using large, charismatic vertebrates as spearheads for biodiversity conservation is an effective strategy. This approach of promoting particular *flagship species* is evident in the longtime symbol of the World Wide Fund for Nature (World Wildlife Fund in North America), the panda. The panda is a large endangered mammal requiring sizeable stands of undisturbed bamboo forest. Its lovable appearance has made it a favorite with the public—and an effective vehicle for soliciting support for conservation efforts that protect far more than just the panda.

Protected areas conserve biodiversity at the ecosystem level

Although most legislation, funding, and resources for biodiversity conservation go toward single-species approaches, our practice of preserving areas of undeveloped land in parks and protected areas helps to conserve habitats, communities, ecosystems, and landscapes (pp. 199–203). So far we have set aside 12% of the world's land area in national parks, state parks, provincial parks, wilderness areas, biosphere reserves, and other protected areas. Many such lands are managed for recreation, water quality protection, or other purposes, not for biodiversity, and many suffer from illegal logging, poaching, or resource extraction. Yet these areas offer animals and plants a degree of protection from human persecution, and some are large enough to preserve whole natural systems that otherwise would be fragmented, degraded, or destroyed.

Protected areas alone may not be enough. India has established reserves to protect its remaining tigers, yet tigers have disappeared from at least two of these reserves. Today a major challenge is to provide linkages among protected areas across the landscape, so that isolated populations of wide-ranging species like tigers can intermix—and so that organisms can move in response to climate change as it alters habitats within protected areas.

Biodiversity hotspots pinpoint regions of high diversity

One international approach oriented around geographic regions, rather than single species, is that of **biodiversity hotspots**. A hotspot is an area that supports an especially great number of species that are **endemic** (p. 51) to the region: that is, found nowhere else in the world (**FIGURE 8.21**). To qualify as a hotspot, a location must harbor at least 1,500 endemic plant species (0.5% of the world's total). In addition, a hotspot must have already lost 70% of its habitat as a result of human impact, and be in danger of losing more.

FIGURE 8.21 ▲ The ring-tailed lemur is a primate that is endemic to the island of Madagascar. Over 2,000 individuals survive in zoos worldwide thanks to captive breeding, but the lemur's natural habitat is fast disappearing. Madagascar has lost over 90% of its forests because of human population growth, poverty, and resource extraction. One recent president encouraged conservation and ecotourism, but when his government fell, illegal logging resumed, destroying large areas of protected forest. Many lemurs were killed and sold for their meat, and timber was exported to meet demand from wealthy nations.

The nonprofit group Conservation International maps 34 biodiversity hotspots (**FIGURE 8.22**). The ecosystems of these areas together once covered 15.7% of the planet's land surface but today, because of habitat loss, cover only 2.3%. This small amount of land is the exclusive home for half the world's plant species and 42% of all terrestrial vertebrate species. The hotspot concept gives incentive to focus on these areas, where the greatest number of unique species can be protected per unit effort.

We can restore degraded ecosystems

Protecting natural areas before they become degraded is the best way to safeguard biodiversity and ecological systems. However, we can often restore degraded natural systems to some semblance of their former condition through **ecological restoration** (p. 77). Ecological restoration projects aim not just to bring back populations of animals and plants, but also to reestablish the processes that make ecosystems function. By restoring complex natural systems such as Illinois prairies or the Florida Everglades (pp. 77–78), restoration

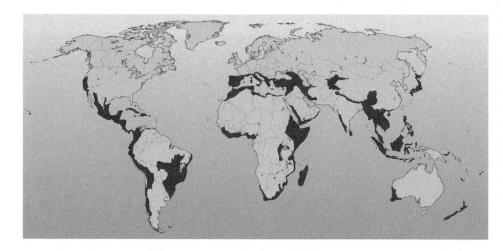

FIGURE 8.22 ◄ Some areas of the world possess exceptionally high numbers of species found nowhere else. Many conservation biologists support prioritizing habitat preservation in these areas, dubbed *biodiversity hotspots*. Shown in red are the 34 biodiversity hotspots mapped by Conservation International. Only about 15% of the area in red is actually habitat for these species; most is developed. Data from Conservation International.

ecologists aim to recreate systems that filter pollutants, cleanse water and air, build soil, and recharge groundwater, providing habitat for wildlife and services for people.

Today's highest-profile restoration project is an effort to restore the vast marshes of southern Iraq. In the 1990s, Iraqi ruler Saddam Hussein drained the marshes, aiming to devastate the region's people, whom he viewed as disloyal. Today ecologists from many nations have joined the people of the marshes in a multimillion-dollar restoration effort. The project was able to restore natural water flow to 75% of the region, so that vegetation grew back and wildlife and people began returning. The effort was on the verge of success when drought descended and Turkey and Syria began diverting water upriver for their own purposes. As of 2011, ecologists and human rights supporters alike were searching for ways to complete this ambitious restoration project.

WEIGHING THE ISSUES

Single-Species Conservation? What would you say are some advantages of focusing on conserving single species, versus trying to conserve habitats, communities, ecosystems, or landscapes? What are some disadvantages? Which do you think is the better approach, or should we pursue both?

Community-based conservation is growing

Helping people, wildlife, and ecosystems at the same time, as the Iraqi marsh restoration project intends, is the focus of many current efforts in conservation biology. In past decades, conservationists from developed nations, in their zeal to preserve ecosystems in other nations, often neglected the needs of people in the areas they wanted to protect. Developing nations came to view this as a kind of neocolonialism. Today this has largely changed, and many conservation biologists actively engage local people in efforts to protect land and wildlife (**FIGURE 8.23**).

This cooperative approach, called **community-based conservation**, is being pursued to help protect tigers in many places. In India, multiple projects offer education, health care, and development assistance to communities living amid the

shrinking habitat of the Bengal tiger. In Cambodia, people who used to hunt Indochinese tigers are being retrained and paid salaries as forest guards to protect the animals from poachers, or as wildlife technicians to help with science and monitoring. In Russia, the Wildlife Conservation Society is working with local hunters to reduce poaching of Siberian tigers and to increase populations of deer and other animals that are prey for both tigers and hunters. The WCS is also establishing a market in the West for sustainably harvested products from the region that are certified "tiger-friendly." Proceeds from sales of the products aim to supplement the incomes of up to 1,000 local people by 12–25%.

Making conservation beneficial for local people requires hard work, investment, and trust on all sides. While setting aside land may deprive local people of short-term access to exploitable resources, it also helps ensure that these resources will not be used up or sold to foreign corporations, but can instead be sustainably managed. Moreover, parks and reserves draw ecotourism that supports local economies. Community-based conservation has not always been successful, but in a world of rising human population, we will require locally based management for biodiversity that sustainably meets people's needs.

FIGURE 8.23 ▲ In community-based conservation, conservation biologists partner with local people, empowering them to conserve wildlife and habitat in their own region. Here, Costa Rican schoolgirls plant trees in a park in their nation's capital, San Jose.

➤ CONCLUSION

Data from scientists worldwide confirm what any naturalist who has watched the habitat change in his or her hometown already knows: From amphibians to tigers, biological diversity is being lost rapidly and visibly within our lifetimes. This erosion of biodiversity threatens to result in a mass extinction event equivalent to those of the geologic past. Habitat alteration, invasive species, pollution, overharvesting of biotic resources, and climate change are the primary causes of biodiversity loss. This loss matters, because human society cannot function without biodiversity's many pragmatic benefits. Conservation biologists are rising to the challenge of conducting science aimed at saving endangered species, protecting their habitats, recovering populations, and preserving and restoring natural ecosystems. The innovative strategies these scientists are pursuing hold promise to slow the loss of biodiversity on Earth.

TESTING YOUR COMPREHENSION

1. What is biodiversity? Describe three levels of biodiversity.

2. What are the five primary causes of biodiversity loss? Give one specific example of each.

3. List three invasive species, and describe their impacts.

4. Define the term *ecosystem services*. Give three examples of ecosystem services that people would have a hard time replacing if their natural sources were eliminated.

5. What is the relationship between biodiversity and food security? Between biodiversity and pharmaceuticals? Give three examples of potential benefits of biodiversity conservation for food security and medicine.

6. Describe three reasons why people suggest biodiversity conservation is important.

7. Name two successful accomplishments of the U.S. Endangered Species Act. Now name two reasons some people have criticized it.

8. Describe how captive breeding can help with endangered species recovery, and give an example. Now explain why cloning will never be, in itself, an effective response to species loss.

9. What is the difference between an umbrella species and a keystone species? Could one species be both an umbrella species and a keystone species?

10. What is a biodiversity hotspot? Describe community-based conservation.

SEEKING SOLUTIONS

1. Many arguments have been advanced for the importance of preserving biodiversity. Which argument do you think is most compelling, and why? Which argument do you think is least compelling, and why?

2. Some people declare that we shouldn't worry about endangered species because extinction has always occurred. How would you respond to this view?

3. Advocates of biodiversity preservation from developed nations have long pushed to set aside land in biodiversity-rich regions of developing nations. Leaders of developing nations have responded by accusing these advocates of neo-colonialism. "Your nations attained prosperity and power by overexploiting their environments decades or centuries ago," these leaders ask, "so why should we now sacrifice our development by setting aside our land and resources?" What would you say to these leaders? What would you say to the environmental advocates? Do you see ways that both preservation and development goals might be reached?

4. **THINK IT THROUGH** You are an influential legislator in a nation that has no endangered species act, and you want to introduce legislation to protect your nation's vanishing biodiversity. Consider the U.S. Endangered Species Act and Canada's Species At Risk Act, as well as international efforts such as CITES and the Convention on Biological Diversity. What strategies would you write into your legislation? How would your law be similar to and different from each of these efforts?

5. **THINK IT THROUGH** As a citizen and resident of your community, and a parent of two young children, you attend a town meeting called to discuss the proposed development of a shopping mall and condominium complex. The development would eliminate a 100-acre stand of forest, the last sizeable forest stand in your town. The developers say the forest loss will not matter because plenty of 1-acre stands still exist scattered throughout the area. Consider the development's possible impacts on the community's biodiversity, children, and quality of life. What will you choose to tell your fellow citizens and the town's decision-makers at this meeting, and why?

CALCULATING ECOLOGICAL FOOTPRINTS

Of the five major causes of biodiversity loss discussed in this chapter, habitat alteration arguably has the greatest impact. In their 1996 book introducing the ecological footprint concept, authors Mathis Wackernagel and William Rees present a consumption/land use matrix for an average North American. Each cell in the matrix lists the number of hectares of land of that type required to provide for the different categories of a person's consumption (food, housing, transportation, consumer goods, and services). Of the 4.27 hectares required to support this average person, 0.59 hectares are forest, with most (0.40 hectares) being used to meet the housing demand. Using this information, calculate the missing values in the table.

	Hectares of forest used for housing	Total forest hectares used
You	0.40	0.59
Your class		
Your state		
United States		

Data from Wackernagel, M., and W. Rees, 1996. Our ecological footprint: Reducing human impact on the earth. *British Columbia, Canada: New Society Publishers.*

1. Approximately two-thirds of the forests' productivity is consumed for housing. To what use(s) would you speculate that most of the other third is put?

2. If the harvesting of forest products exceeds the sustainable harvest rate, what will be the likely consequence for the forest? For communities surrounding the forest?

3. What impacts would you expect on biodiversity in each of the following cases, and why?
   a. The cutting of small plots of forest within a large forest
   b. The clear-cutting (p. 194) of an entire forest
   c. The clear-cutting of an entire forest followed by planting of a monocultural plantation of young trees

Go to **www.masteringenvironmentalscience.com** for homework assignments, practice quizzes, Pearson eText, and more.

9 Forests, Forest Management, and Protected Areas

Upon completing this chapter, you will be able to:

➤ Summarize the ecological and economic contributions of forests

➤ Outline the history and current scale of deforestation

➤ Assess aspects of forest management and describe methods of harvesting timber

➤ Identify federal land management agencies and the lands they manage

➤ Recognize types of parks and protected areas, and evaluate issues involved in their design

A worker examines paper being produced at the mill in Escanaba, Michigan.

Certified Sustainable Paper in Your Textbook

"FSC is the high bar in forest certification, because its standards protect forests of significant conservation value and consistently deliver meaningful improvements in forest management on the ground."
—KERRY CESAREO, DEPUTY DIRECTOR, WWF-US Forests Program

"As a company and as individuals, we must be able to look ourselves in the mirror and know that we are truly doing what is right and good for the environment."
—RICK WILLETT, PRESIDENT AND CEO, NewPage Corporation

As you turn the pages of this textbook, you are handling paper made from trees that were grown, managed, harvested, and processed using certified sustainable practices.

If you were to trace the paper in this book back to its origin, you would find yourself standing in a diverse mixed forest of aspen, birch, beech, maple, spruce, and pine in the Upper Peninsula of Michigan. This sparsely populated region near the Canadian border, flanked by Wisconsin, Lake Michigan, and Lake Superior, remains heavily forested, despite supplying timber to our society for nearly 200 years.

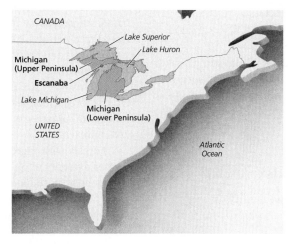

The trees cut to make this book's paper were selected for harvest based on a sustainable management plan designed to avoid depleting the forest of its mature trees or degrading the ecological functions the forest performs. The logs were then transported to a nearby pulp and paper mill at Escanaba, a community of 13,000 people on the shore of Lake Michigan. The mill is Escanaba's largest employer, providing jobs to about 1,100 residents.

At the Escanaba mill, the wood is chipped and then fed into a digester where the chips are cooked with chemicals to break down the wood's molecular bonds. Millworkers then bleach, wash, and screen the wood's cellulose fibers, and mix them with water and dye. This mixture is poured onto a moving mat, where heavy rollers press it into thin sheets. The sheets of newly formed paper are then dried on heated rollers and made ready to receive a variety of coatings. The paper is wound into immense reels, later to be cut into sheets. In this way the Escanaba mill produces about 770,000 tons of paper each year. It recycles chemicals and water used in the process, and it combusts discarded waste to help power the mill.

At every stage in this process, independent third-party inspectors from the Forest Stewardship Council (FSC) examine the practices being used to ensure that they meet the FSC's strict criteria for sustainable forest management and paper production. The Forest Stewardship Council is an organization that officially certifies forests, companies, and products that meet sustainability standards. FSC-certified timber harvesting operations on Michigan's Upper Peninsula are required to protect rare species and sensitive habitats, safeguard water sources, control erosion, minimize pesticide use, and maintain the diversity of the forest and its ability to regenerate after harvesting. FSC certification is the best way for consumers of forest products to know that they are supporting sustainable practices that protect forests.

The Escanaba mill is one of 11 mills run by the New-Page Corporation, based in Ohio. NewPage's mills (in Kentucky, Maine, Maryland, Michigan, Minnesota,

Wisconsin, and Nova Scotia) together produce 4.4 million tons of coated paper each year for books, magazines, newspapers, food packaging, and more. Only some of this paper is FSC-certified, but NewPage seeks out suppliers of wood from private land who are certified at least to the less rigorous standards of the Sustainable Forestry Initiative®. NewPage states that it does not use wood from old-growth forests, rainforests, or forests of exceptional conservation value.

FSC certification for the paper in your textbook was made possible because Pearson Education, the publisher of this book, is striving to follow sustainable practices. Pearson supported FSC paper for this book at the request of your authors and editors because our entire team feels that an environmental science textbook should walk its talk. So, as you flip through this book, you can feel satisfied that you are doing a small part to help safeguard the world's forests by supporting sustainable forestry practices. ■

FOREST ECOSYSTEMS AND FOREST RESOURCES

Forests currently cover 31% of Earth's land surface (**FIGURE 9.1**). They provide habitat for countless organisms; help maintain soil, air, and water quality; and play key roles in our planet's biogeochemical cycles (pp. 36–41). Forests have also long provided humanity with wood for fuel, construction, paper, and more.

Many kinds of forests exist

A **forest** is any ecosystem with a high density of trees. Most of the world's forests occur as *boreal forest* (p. 84), a biome that stretches across Canada, Scandinavia, and Russia; or as *tropical rainforest* (p. 81), a biome in South and Central America, equatorial Africa, Indonesia, and southeast Asia. *Temperate deciduous forests* (p. 80), *temperate rainforests* (p. 81), and *tropical dry forests* (p. 82) also cover large regions.

Within each forest biome, the nature of the plant community varies from region to region because of differences in soil and climate. As a result, ecologists and forest managers classify forests into *forest types*, categories defined by their predominant tree species (**FIGURE 9.2**). The eastern United States contains 10 forest types ranging from spruce-fir to oak-hickory to longleaf-slash pine. The western United States contains 13 forest types ranging from Douglas fir and hemlock–sitka spruce forests of the moist Pacific Northwest to ponderosa pine and pinyon-juniper woodlands of the dry interior. Michigan's Upper Peninsula hosts a diversity of forest types. Near Escanaba, spruce-fir forest intermixes with forests of aspen, birch, maple, and beech, and areas of white, red, and jack pine.

Forests are ecologically complex

Because of their structural complexity and their capacity to provide many niches for organisms, forests comprise some of the richest ecosystems for biodiversity (**FIGURE 9.3**). Insects, birds, mammals, and other animals subsist on the leaves, fruits, and seeds that trees produce, or find shelter in the cavities of tree trunks. Tree canopies are full of life, and understory shrubs and groundcover plants provide food and shelter for yet more organisms. The leaves, stems, and roots of plants are colonized by an extensive array of fungi and microbes. On the forest floor, leaf litter nourishes the soil, and a multitude of soil organisms helps to decompose plant material and cycle nutrients.

Forests with a greater diversity of plants tend to host a greater diversity of organisms overall. In general, old-growth forests host more biodiversity than younger forests, because older forests contain more structural diversity and thus more microhabitats and resources for more species.

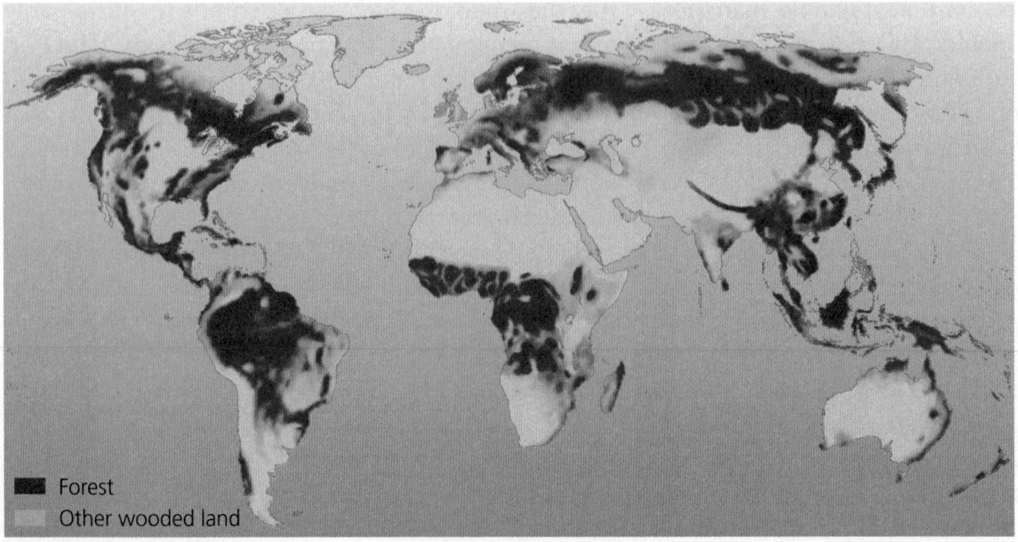

FIGURE 9.1 ◄ Forests cover 31% of Earth's land surface. Most widespread are boreal forests in the north and tropical forests in South America, Africa, and Southeast Asia. Areas classified as "wooded land" support trees at sparser densities. Data from Food and Agricultural Organization of the United Nations, 2010. *Global forest resources assessment 2010.*

Forest
Other wooded land

(a) Maple-beech-birch forest, Michigan's Upper Peninsula

(b) Oak-hickory forest, West Virginia

(c) Ponderosa pine forest, northern Arizona

(d) Redwood forest, coastal northern California

FIGURE 9.2 ▲ Among the 23 forest types found in the continental United States are **(a)** maple-beech-birch forest, **(b)** oak-hickory forest, **(c)** ponderosa pine forest, and **(d)** redwood forest.

Forests provide ecosystem services

Besides hosting biodiversity, forests supply vital ecosystem services (pp. 2, 36, 90, 95; **FIGURE 9.4**). As plants grow, their roots stabilize the soil and help to prevent erosion. When rain falls, leaves and leaf litter slow runoff by intercepting water. This prevents flooding, helps water soak into the ground to nourish roots and recharge aquifers, reduces soil erosion, and helps keep streams and rivers clean. Forest plants also filter pollutants and purify water as they take it up from the soil and release it to the atmosphere in transpiration (p. 38). Plants produce the oxygen that we breathe, regulate moisture and precipitation patterns, and moderate climate. Trees' roots draw minerals up from deep soil layers and deliver them to surface soil layers where other plants can use them. Plants also return organic material to the topsoil in the form of litter.

By performing all these ecological functions, forests are indispensable for our survival. Forests also enhance our quality of life by providing people with cultural, aesthetic, health, and recreation values (pp. 94–95). People seek out forests for adventure and for spiritual solace alike—to admire beautiful trees, to hike, to observe wildlife, to enjoy clean air, and for many other reasons.

Of all the services that forests provide, their storage of carbon has elicited great interest as nations debate how to control global climate change (Chapter 14). Because trees absorb carbon dioxide from the air during photosynthesis (p. 30) and then store carbon in their tissues, forests serve as a major reservoir for carbon. In fact, scientists estimate that the plants and soil of the world's forests store more carbon than the entire atmosphere contains. When plant matter is burned or when it decomposes, carbon dioxide is released—and thereafter less vegetation remains to soak it up. Therefore, when we cut forests, we release CO_2 to the atmosphere and we worsen global climate change. The more forests we preserve or restore, the better we can address climate change.

Forests provide valuable resources

Carbon storage and other ecosystem services make forests priceless to our society, but forests also provide many economically valuable resources. Among these are plants for

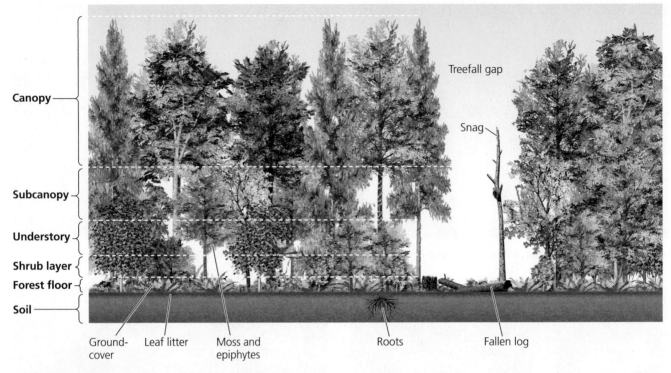

Canopy

Subcanopy

Understory

Shrub layer

Forest floor

Soil

Treefall gap

Snag

Ground-cover Leaf litter Moss and epiphytes Roots Fallen log

FIGURE 9.3 ▲ In this cross-section of a mature forest, the crowns of the largest trees form the canopy, and trees beneath them form the shaded subcanopy and understory. Shrubs and groundcover grow just above the forest floor, which may be covered in leaf litter teeming with invertebrate animals. Vines, mosses, lichens, and epiphytes cover portions of trees and the forest floor. Snags (dead trees) provide food and nesting sites for woodpeckers and other animals, and fallen logs nourish the soil. Treefall gaps caused by fallen trees let light through the canopy and create openings in the forest, allowing early successional plants to grow.

O_2

CO_2

Store carbon

Support biodiversity

Produce oxygen

Provide fuel wood, lumber, paper, medicines, dyes, foods, fibers

H_2O

Purify water, filter pollution

Return organic matter to soil

Slow runoff, prevent flooding

Provide health, beauty, recreation

Transport minerals to soil surface

Stabilize soil, prevent erosion

FIGURE 9.4 ▲ Forests provide us a diversity of priceless ecosystem services, as well as resources that we can harvest.

medicines, dyes, and fibers; animals, plants, and mushrooms for food; and, of course, wood from trees. For millennia, wood has fueled our fires, keeping us warm and well fed. We have used lumber to construct the houses that keep us sheltered. Using timber, we built the ships that carried people and cultures between continents. And wood from forest trees gave us paper, the medium of the first information revolution.

In recent decades, industrial harvesting has allowed us to extract more timber than ever before. The extraction of forest resources has been instrumental in helping us achieve the standard of living we now enjoy. Most commercial logging today takes place in Canada, Russia, and other nations with large expanses of boreal forest, and in tropical nations with large areas of rainforest, such as Brazil and Indonesia. In the United States, most logging takes place in pine plantations of the South and conifer forests of the West.

FOREST LOSS

When trees are removed more quickly than they can regrow, it leads to deforestation. **Deforestation**, the clearing and loss of forests, has altered landscapes and ecosystems across much of our planet. In the time it takes you to read this sentence, 2 hectares (5 acres) of tropical forest will have been cleared. As we alter and eliminate forests, we lose biodiversity, worsen climate change, and disrupt the ecosystem services that support our societies. The alteration, fragmentation, and outright loss of forested land represent one of our society's primary challenges (see **ENVISIONIT**, p. 189).

Each year Earth loses 18 million acres of forest—an area the size of South Carolina.

Satellite photo of Rondonia, Brazil, 1975

Satellite photo of same area, 2001

As roads are built into the Amazon rainforest, small settlers and large corporations cut the forest and convert the land for cattle ranching and soy production. Development follows wherever roads are built.

Forest loss drives biodiversity loss, climate change, erosion, and other problems.

Cattle on burned and cleared land

A soy farm leaves a tiny forest fragment.

Often people can farm the land for only a few years before the soil gives out and they have to cut more forest.

Settlers in the Amazon

You Can Make a Difference

➤ Buy certified sustainable wood products.

➤ Eat less meat (raising livestock requires more land than raising crops).

➤ Contribute to forest protection efforts in your region and around the world.

189

Agriculture and demand for wood puts pressure on forests

We all use wood and other forest products, and we all rely on food and fiber grown on cropland and rangeland—much of which occupies lands where forests once stood. To make way for agriculture and to extract wood products, people have been clearing forests for millennia.

Although forest clearing has fed our civilization's growth, forest loss has negative consequences, especially as human population grows. Deforestation causes biodiversity loss, soil degradation, and desertification (p. 140), and it adds carbon dioxide to the atmosphere, contributing to global climate change.

In 2010, the U.N. Food and Agriculture Organization (FAO) released its latest *Global Forest Resources Assessment*, a report based on remote sensing data from satellites, analysis from forest experts, questionnaire responses, and statistical modeling. The assessment concluded that we are eliminating 13 million ha (32 million acres) of forest across the world each year. Subtracting annual regrowth from this amount makes for an annual net loss of 5.2 million ha (12.8 million acres)—an area about half the size of Kentucky or twice the size of Massachusetts. The good news is that this rate (for the decade 2000–2010) is lower than it was in the 1990s, when 8.3 million ha (20.5 million acres) were lost worldwide each year.

Forests are being felled most quickly in the tropical rainforests of Latin America and Africa. Developing nations in these regions are striving to expand areas of settlement for their burgeoning populations and to boost their economies by extracting natural resources and selling them abroad. Moreover, many people in these societies harvest fuelwood for their daily cooking and heating needs (pp. 357–358). In contrast, some areas of Europe and North America are slowly gaining forest cover as they recover from past deforestation.

We deforested much of North America

As the United States and Canada expanded westward across North America over the past 400 years, forests were cleared for timber and farmland. The vast deciduous forests of the East were cleared by the mid-19th century, making way for countless small farms. Timber from these forests built the cities of the Atlantic seaboard and the upper Midwest. As a farming economy shifted to an industrial one, wood was used to stoke the furnaces of industry. Logging operations moved south, where vast pine woodlands and bottomland hardwood forests were cleared and replaced with pine plantations. Once mature trees were removed from these areas, timber companies moved west, cutting the continent's biggest trees in the Rocky Mountains, the Sierra Nevada, the Cascade Mountains, and the Pacific Coast ranges. Exploiting forest resources helped American society to develop, but we were not harvesting forests sustainably. Instead, we were depleting our store of renewable resources for the future.

By the 20th century, very little **primary forest**—natural forest uncut by people—remained in the lower 48 U.S. states, and today even less is left (**FIGURE 9.5**). Nearly all the large oaks and maples found in eastern North America today, and even most redwoods of the California coast, are merely *second-growth* trees: trees that sprouted after old-growth timber was cut. Such second-growth trees characterize **secondary forest**.

The fortunes of loggers have risen and fallen with the availability of big trees. As each region was denuded of timber, the industry declined, and timber companies moved on while local loggers lost their jobs. If the remaining ancient trees of North America—most in British Columbia and Alaska—are cut, many loggers will be out of jobs once again. Their employers will move on to nations of the developing world, as many already have.

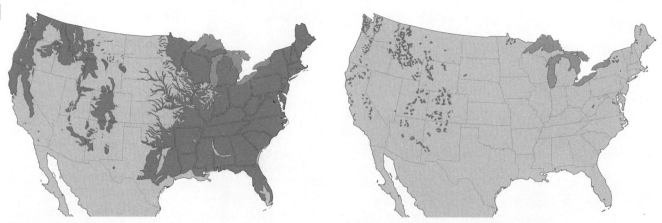

(a) 1620: Areas of primary (uncut) forest

(b) Today: Areas of primary (uncut) forest

FIGURE 9.5 ▲ When Europeans first were colonizing North America **(a)**, much of the continent was covered in primary forest, shown in green. Today, nearly all this primary forest is gone **(b)**, having been cut to make way for agriculture and to provide timber. Much of the landscape has become reforested with secondary forest, which generally contains smaller trees of different species compositions. *Sources:* (a) adapted from Greeley, W.B., 1925. The relation of geography to timber supply, *Economic Geography* 1:1–11; and (b) map by George Draffan, www.endgame.org.

Forests are being lost rapidly in developing nations

Uncut primary forests still remain in many developing countries, and these nations are in the position the United States and Canada enjoyed a century or two ago: having a resource-rich frontier that they can develop. Today's advanced technology allows these countries to exploit their resources and push back their frontiers even faster than occurred in North America. As a result, deforestation is rapid in places such as Brazil, Indonesia, and West Africa. Tropical forests in these areas are home to far more biodiversity than the temperate forests of North America.

Developing nations are often desperate enough for economic development and foreign capital that they impose few or no restrictions on logging. Often they allow their timber to be extracted by foreign multinational corporations, which pay fees to the developing nation's government for a **concession**, or right to extract the resource. Once a concession is granted, the corporation has little incentive to manage forest resources sustainably. Most economic benefits are short term, reaped not by local residents but by the foreign corporation. Local people may receive temporary employment, but once the timber is gone they no longer have the forest and the ecosystem services it had provided. Moreover, much of the wood extracted in developing nations is exported to Europe and North America. In this way, our consumption of high-end furniture and other wood products in developed nations can fuel forest destruction in poorer nations.

Throughout Southeast Asia and Indonesia today, vast areas of tropical rainforest are being cleared to establish plantations of oil palms (**FIGURE 9.6**). Oil palm fruit produces palm oil, which is used in snack foods, soaps, and cosmetics, and as a biofuel. In Indonesia, oil palm plantations have displaced over 6 million ha (15 million acres) of rainforest. Clearing for plantations encourages further development and eases access for people to enter the forest and conduct logging illegally. The palm oil boom represents a conundrum for environmental advocates. Many people eager to fight climate change had urged the development of biofuels (pp. 358–361) to replace fossil fuels. Yet grown at the large scale that our society is demanding, monocultural plantations of biofuel crops such as oil palms are causing severe environmental impacts by displacing natural forests.

FIGURE 9.6 ▲ Oil palm plantations like this one in Borneo are replacing primary forest across large areas of Southeast Asia and Indonesia. Forest clearing for plantations promotes further development, illegal logging, and forest degradation. Since 1950, the immense island of Borneo **(maps at bottom)** has lost most of its forest cover. Data from Radday, M., WWF-Germany, 2007. Designed by Hugo Ahlenius, UNEP/GRID-Arendal. Extent of deforestation in Borneo 1950–2001, and projection towards 2020. http://maps.grida.no/go/graphic/extent-of-deforestation-in-borneo-1950–2005-and-projection-towards-2020.

WEIGHING THE ISSUES

Logging Here or There Suppose you are an activist protesting the logging of old-growth trees near your hometown. Now let's say you know that if the protest is successful, the company will move to a developing country and cut its primary forest instead. Would you still protest the logging in your hometown? Would you pursue any other approaches?

Solutions are emerging

New solutions are being proposed to address deforestation in developing nations. Some conservation organizations are buying concessions and using them to preserve forest rather than cut it down. Others are pursuing community-based conservation (p. 181).

In Indonesia, NewPage Corporation, the supplier of this textbook's paper, is funding a three-year project called POTICO (Palm Oil, Timber, and Carbon Offsets) to reduce deforestation and illegal logging. In this project, the nonprofit World Resources Institute (WRI) is working with palm oil companies that own concessions to clear primary rainforest, and the WRI will steer them instead to land that is already logged and degraded. WRI will then protect the forests that were slated for conversion or else allow FSC-certified sustainable forestry in them. Because primary forest stores more carbon than oil palm plantations, these land swaps can reduce Indonesia's greenhouse gas emissions and qualify for credit via carbon offsets (p. 322).

Carbon offsets are central to emerging international plans to curb deforestation and climate change together. Forest loss accounts for 12–25% of the world's greenhouse gas emissions—as much as all the world's vehicles emit. Thus, at recent international climate conferences (p. 319), negotiators have outlined a program called *Reducing Emissions from Deforestation and Forest Degradation (REDD)*, whereby wealthy industrialized nations would pay poorer developing nations to conserve forest. Under this plan, poor nations would gain income while rich nations would receive carbon credits to offset their emissions.

FOREST MANAGEMENT

Our demand for forest resources is rising, so we need to take care in managing the forests from which we take resources. *Foresters* are professionals who manage forests through the practice of **forestry**. Foresters must balance our society's demand for forest products against the central importance of forests as ecosystems. Today, sustainable forest management practices are spreading as informed consumers demand sustainably produced products. Just as your textbook uses FSC-certified paper from sustainably managed forests, more and more forest products are now made using certified sustainable practices.

Debates over how to manage forest resources reflect broader questions about how to manage natural resources in general. Resources such as fossil fuels and many minerals are nonrenewable, whereas resources such as the sun's energy are perpetually renewable. Between these extremes lie resources, such as timber, that are renewable if they are not exploited too rapidly (pp. 2–3). Besides timber, these resources include soils, fresh water, rangeland, wildlife, and fisheries.

Resource management describes our use of strategies to manage and regulate the harvest of potentially renewable resources. Sustainable resource management involves harvesting these resources in ways that do not deplete them. Resource managers are guided by research in the natural sciences, as well as by social, political, and economic factors.

Resource managers follow several strategies

A key question in managing resources is whether to focus strictly on the resource of interest or to look more broadly at the environmental system of which it is a part. Taking a broader view often helps avoid degrading the system and thereby helps sustain the resource in the long term.

Maximum sustainable yield Traditionally, a guiding principle in resource management has been **maximum sustainable yield**. Its aim is to achieve the maximum amount of resource extraction without depleting the resource from one harvest to the next. Recall the logistic growth curve (see Figure 3.11, p. 58), which shows how a population grows most quickly at an intermediate size—specifically, at one-half of carrying capacity (p. 58). A fisheries manager aiming for maximum sustainable yield will prefer to keep fish populations at intermediate levels so that they rebound quickly after

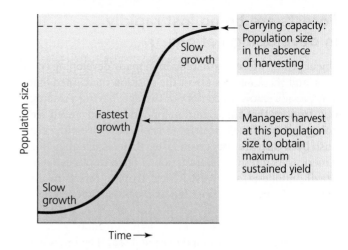

FIGURE 9.7 ▲ Using the concept of maximum sustainable yield, resource managers maximize the amount of resource harvested while keeping the harvest sustainable in perpetuity. For a wildlife population or fisheries stock that grows according to a logistic growth curve, managers aim to keep the population at half the carrying capacity, because populations grow fastest at intermediate sizes.

each harvest. Doing so should result in the greatest amount of fish harvested over time, while sustaining the population indefinitely (**FIGURE 9.7**).

This management approach, however, keeps the fish population at only half its carrying capacity—well below the size it would attain in the absence of fishing. Reducing a population in this way will likely affect other species and alter the food web dynamics of the community. From an ecological point of view, management for maximum sustainable yield may set in motion significant changes.

In forestry, maximum sustainable yield argues for cutting trees shortly after they go through their fastest stage of growth. Because trees often grow most quickly at intermediate ages, trees are generally cut long before they have grown as large as they would in the absence of harvesting. This practice maximizes timber production over time, but it alters forest ecology and eliminates habitat for species that rely on mature trees.

Ecosystem-based management Because of these dilemmas, more and more managers today espouse **ecosystem-based management**, which aims to minimize impact on the ecological processes that provide the resource. Many certified sustainable forestry plans protect certain forested areas, restore ecologically important habitats, and consider patterns at the landscape level (p. 33), allowing timber harvesting while preserving the ecological processes and functional integrity of the forest ecosystem. It can be challenging, however, to determine how best to implement this type of management. Ecosystems are complex, and our understanding of how they operate is limited. Thus, ecosystem-based management has come to mean different things to different people.

Adaptive management Some management actions will succeed, and some will fail. **Adaptive management** involves systematically testing different approaches and aiming to

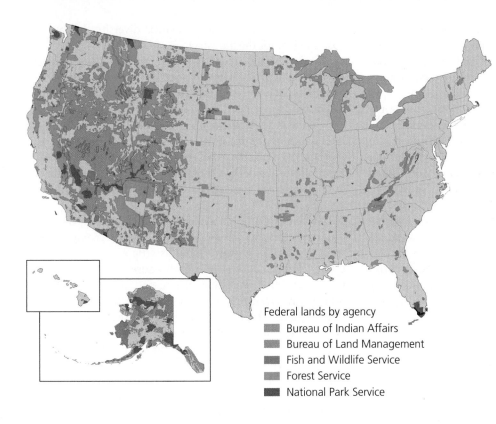

FIGURE 9.8 ◀ Federal agencies own and manage well over 250 million ha (600 million acres) of land in the United States. These include national forests, national parks, national wildlife refuges, Native American reservations, and Bureau of Land Management lands. Data from United States Geological Survey.

Federal lands by agency
- Bureau of Indian Affairs
- Bureau of Land Management
- Fish and Wildlife Service
- Forest Service
- National Park Service

improve methods through time. For managers, it entails monitoring the results of one's practices and adjusting them as needed, based on what is learned. Adaptive management is intended as a fusion of science and management, because hypotheses about how best to manage resources are explicitly tested.

We extract timber from public and private lands

We began managing forest resources in the United States a century ago in response to rampant deforestation and widespread fear of a "timber famine." This led the federal government to form a system of forest reserves: public lands set aside to grow trees, produce timber, protect water quality, and serve as insurance against scarcities of lumber. Today the U.S. **national forest** system consists of 77 million ha (191 million acres) managed by the U.S. Forest Service and covering over 8% of the nation's land area (**FIGURE 9.8**). The U.S. Forest Service was established in 1905 under Gifford Pinchot, whose conservation ethic (p. 13) meant managing the forests for "the greatest good of the greatest number in the long run." Pinchot believed that the nation should extract and use resources from its public lands, but that wise and careful management of timber resources was imperative.

Today, most timber harvesting in the United States takes place on private land owned by the timber industry or by small landholders (**FIGURE 9.9**). Timber companies also extract timber from the national forests and from publicly

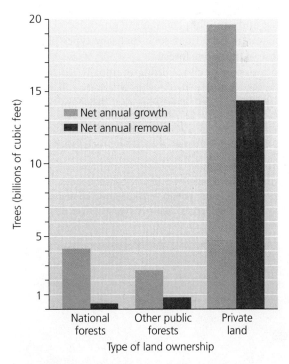

FIGURE 9.9 ▲ As the United States recovers from deforestation, trees (measured in cubic feet of wood biomass) are growing faster than they are being removed, particularly on national forests and other public lands. However, forests that regrow after logging often differ substantially from the forests that were removed. "Private land" in this figure combines land owned by the timber industry and by small landholders. Data from USDA Forest Service, 2008. *Forest resources of the United States, 2007.* U.S. Department of Agriculture, Washington, D.C.

held state forests. U.S. Forest Service employees manage timber sales and build roads to provide access for logging companies, which sell the timber they harvest for profit. In this way, taxpayers subsidize private timber harvesting on public land (p. 108).

Timber extraction from U.S. national forests rose in the 1950s as the nation underwent a postwar economic boom, paper consumption rose, and the population expanded into newly built suburban homes. Since the 1980s, harvests from national forests have declined as economic trends shifted and forest management philosophy evolved. At present, in an average year, about 2.1% of U.S. forest acreage is cut for timber. Tree regrowth now outpaces tree removal on national forests by 11 to 1 (see Figure 9.9). However, even when regrowth outpaces removal, the character of forests may change. Once primary forest is replaced by younger secondary forest or a single-species plantation, the resulting community may be very different from the original forest, and generally it is ecologically less valuable.

We harvest timber by several methods

Most timber has been harvested by **clear-cutting**, in which all trees in an area are cut at once (**FIGURE 9.10**). Clear-cutting is cost-efficient, and to some extent it can mimic natural disturbance events such as fires, tornadoes, or windstorms. However, the ecological impacts of clear-cutting are considerable. An entire ecological community is removed, soil erodes away, and sunlight penetrates to ground level, changing microclimatic conditions such that new types of plants replace those of the native forest.

Public aversion to clear-cutting eventually led foresters and the timber industry to develop alternative harvesting methods (**FIGURE 9.11**). For example, selection systems preserve much of a forest's structural diversity (but are less cost-efficient for the industry).

Today, North America's timber industry focuses on production from plantations of fast-growing tree species planted in single-species monocultures (p. 136). Because all trees in a given stand are planted at the same time, the stands are *even-aged*, with all trees the same age (**FIGURE 9.12**). Stands are cut after a certain number of years (called the *rotation time*), and the land is replanted with seedlings. Plantation forestry is growing worldwide, and today fully 7% of the world's forests are plantations. One-quarter of these feature non-native tree species.

When a plantation replaces a natural forest, the community undergoes simplification. Indeed, ecologists and foresters view plantations more as crop agriculture than as ecologically functional forests. Because there are few tree species and little variation in tree age, plantations do not offer many forest organisms the habitat they need. For instance, stands of red pine planted near Escanaba, Michigan, host far less biodiversity than the more diverse forests of multiple tree species that surround them. Plantations lack the structural complexity that characterizes a natural forest as seen in Figure 9.3 (p. 188). Plantations are also vulnerable to outbreaks of pest species, as we shall soon see. For all these reasons, some harvesting methods (see Figure 9.11c) aim to maintain *uneven-aged* stands, where a mix of ages (and often a mix of tree species) makes the stand more similar to a natural forest.

All timber-harvesting methods disturb the soil, alter habitat, and affect plants and animals. All methods modify forest structure and composition. Most logging methods speed runoff, promote flooding, and increase soil erosion, which degrades water quality. When steep hillsides are clear-cut, landslides can result. In recent decades, rising awareness of these problems has led citizens to urge that public forests be managed for recreation, wildlife, and ecosystem integrity, as well as for timber.

194

FIGURE 9.10 ▼ Clear-cutting is cost-efficient for timber companies but can have severe ecological consequences, including soil erosion, water pollution, and altered community composition. Public reaction to clear-cutting has driven changes in forestry methods.

Public forests may be managed for recreation and ecosystems

In 1976 the U.S. Congress passed the **National Forest Management Act**, which mandated that every national forest draw up plans for renewable resource management, subject to public input under the National Environmental Policy Act (p. 100). Guidelines specified that these plans assess the ecological impacts of logging. As a result, timber-harvesting methods were integrated with ecosystem-based management goals, and the U.S. Forest Service developed programs to manage wildlife and restore degraded ecosystems.

The Hiawatha National Forest on Michigan's Upper Peninsula provides an example of management under the National Forest Management Act. The most recent revision of its forest management plan seeks a balance of uses, aiming to:

▶ Carefully manage timber harvesting.

▶ Monitor fish and wildlife populations.

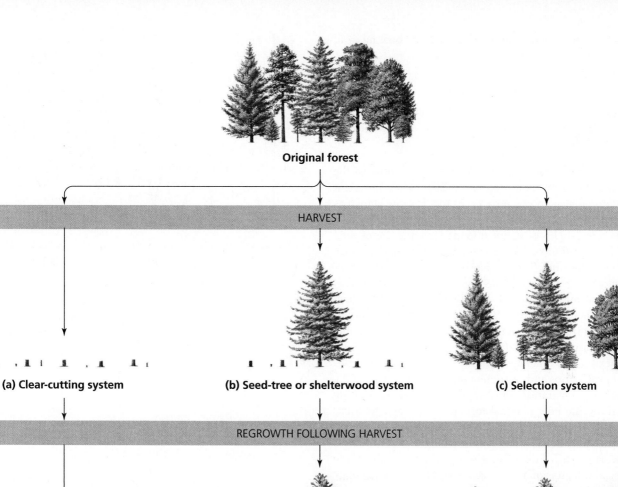

Original forest

HARVEST

(a) Clear-cutting system

(b) Seed-tree or shelterwood system

(c) Selection system

REGROWTH FOLLOWING HARVEST

FIGURE 9.11 ▲ Foresters have devised various methods to harvest timber. In clear-cutting **(a)**, all trees in an area are cut, extracting a great deal of timber inexpensively but leaving a vastly altered landscape. In seed-tree systems and shelterwood systems **(b)**, small numbers of large trees are left in clear-cuts to help reseed the area or provide shelter for growing seedlings. In selection systems **(c)**, a minority of trees is removed at any one time, and most are left standing. These latter methods involve less environmental impact than clear-cutting, but all methods can significantly modify the structure and function of natural forest communities.

FIGURE 9.12 ▶ Even-aged stand management is practiced on tree plantations where all trees are of equal age, as seen in the stand in the foreground, which is regrowing after clear-cutting. In uneven-aged stand management, harvests are designed to maintain a mix of tree ages, as seen in the more mature forest in the background. The greater structural diversity of uneven-aged stands provides superior habitat for most wild species and, if diverse tree species also are intermixed, makes these stands more akin to ecologically functional forests.

- Allow diverse recreation—hiking, fishing, boating, and more—in specified areas.
- Prohibit livestock grazing, protect historic and archaeological sites, and build no new roads.
- Maintain habitats for threatened plants and animals; restore wetlands, streams, forests, and soils; and protect stands of old-growth forest.

Another national policy milestone that accentuated the shift toward conservation occurred in 2001, when President Bill Clinton issued an executive order that became known as the **roadless rule**. The roadless rule put 23.7 million ha (58.5 million acres)—31% of national forest land and 2% of total U.S. land—off-limits to road construction or maintenance (and thus to logging). The roadless rule received strong popular support, including a record 4.2 million public comments.

The administration of President George W. Bush marked a change in policy direction. In 2004, the Bush administration freed forest managers from many requirements of the National Forest Management Act, granting them more flexibility in managing forests but loosening environmental protections and restricting public oversight. In 2005, the Bush administration repealed the roadless rule, inviting states to decide how national forests within their boundaries should be managed. Some states responded favorably, but others sued the administration, asking that the roadless rule be reinstated. Following a series of court rulings, the Obama administration reinstated most of the roadless policy through 2011.

Fire can hurt or help forests

Another area of policy debate involves how to handle wildfire. Smokey Bear, the Forest Service's beloved cartoon bear in a ranger's hat, advises us to fight forest fires—and for over a century, land management agencies suppressed fire wherever it broke out. Yet scientific research shows that many species and ecological communities depend on fire. Some plants have seeds that germinate only in response to fire, and researchers studying tree rings have documented that North America's grasslands and pine woodlands burned frequently. (Burn marks in a tree's growth rings reveal past fires, giving scientists an accurate history of fire events extending back hundreds of years.) Ecosystems dependent on fire are adversely affected when fire is suppressed. Grasslands are invaded by shrubs, whereas pine woodlands become cluttered with hardwood understory. Invasive plants move in, and animal diversity and abundance decline.

In the long term, suppressing frequent fires leads to occasional catastrophic fires that damage forests, destroy property, and threaten human lives. This is because fire suppression allows limbs, logs, sticks, and leaf litter to accumulate on the forest floor, producing kindling for a catastrophic fire. Accordingly, catastrophic fires have become more numerous in recent years (**FIGURE 9.13**). Moreover, residential development on the edges of forested land—in the **wildland-urban interface**—is placing more homes in fire-prone situations (**FIGURE 9.14**).

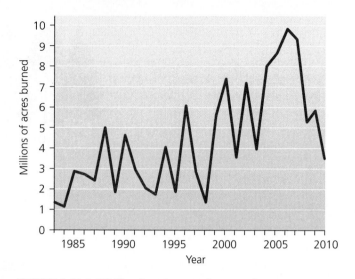

FIGURE 9.13 ▲ Wildfires have become larger and more numerous in the United States, as shown by this graph of acres burned each year throughout the nation. Fuel buildup from decades of fire suppression has contributed to this trend. Data from National Interagency Fire Center.

FAQ

Q: Aren't all forest fires bad?

A: No. Wildfires can alter ecosystems, damage property, and take human life. But ironically, the destructiveness of many recent wildfires results in part from our suppression of fires in past decades. Fire is a natural process that helps to maintain the health of many forests and grasslands. When allowed to occur naturally, fires generally burn moderate amounts of material, return nutrients to the soil, and promote lush growth of new vegetation. By suppressing fire, we have allowed unnaturally large amounts of dead wood, dried grass, and leaf litter to accumulate. This material becomes kindling that eventually can feed a truly damaging fire that is too big and too hot to control. This is why many land managers today intentionally set carefully controlled prescribed burns and also allow some natural fires to run their course. By doing so, they aim to help return our fire-dependent ecosystems to a healthier and safer condition.

To reduce fuel loads, protect property, and improve the condition of forests, land management agencies burn areas of forest intentionally under carefully controlled conditions (see Figure 1.6b, p. 8). These **prescribed burns** clear away fuel loads, nourish the soil with ash, and encourage the vigorous growth of new vegetation.

Once a catastrophic fire burns a forest, it may leave many dead trees. The removal of dead trees, or snags, by timber

FIGURE 9.14 ▲ People's suppression of fire over the past century has led to a buildup of leaf litter, woody debris, and young trees, which act as fuel to intensify fires when they do occur. As a result, catastrophic wildfires damage ecosystems and threaten homes in the wildland-urban interface (such as this fire in southern California in 2007). To prevent these unnaturally severe fires, ecologists suggest allowing natural fires to burn when possible and instituting controlled burns to reduce fuel loads and restore forest ecosystems.

companies following a natural disturbance (such as a fire, windstorm, insect damage, or disease) is called **salvage logging**. From a short-term economic standpoint, salvage logging may seem to make good sense. However, snags are valuable; the insects that decay them provide food for wildlife, and many animals depend on holes in snags for nesting and roosting. Removing timber from recently burned land can also cause erosion and soil damage. Moreover, it may impede forest regeneration and promote further wildfire (see **THE SCIENCE BEHIND THE STORY**, pp. 198–199).

WEIGHING THE ISSUES

How to Handle Fire? A century of fire suppression has left vast areas of North American forests vulnerable to catastrophic wildfires. Prescribed burning helps to alleviate this risk, yet we will never have adequate resources to conduct careful prescribed burning over all these lands. Can you suggest solutions to help protect people's homes in the wildland-urban interface while improving the ecological condition of forests? Do you think people should be allowed to develop homes in fire-prone areas?

Climate change is altering forests

Global climate change (Chapter 14) is now worsening wildfire risk by bringing warmer weather to North America and drier weather to the American West. In addition, milder winters and hotter, drier summers appear to be contributing to

major outbreaks of the mountain pine beetle, an insect whose infestations have killed trees across more than 11 million ha (27 million acres) of western North America since 1990 (**FIGURE 9.15**). Past forest management has resulted in even-aged forests across large regions, and many trees in these forests are at a prime age for beetle infestation. Most affected are plantation forests dominated by single species the beetles prefer. Climate change benefits some species while harming others, and as it interacts with pests, diseases, and management strategies, many forest ecosystems could be altered in profound ways. Some dense, moist forests may be replaced by drier woodlands, shrublands, or grasslands—or by novel types of ecosystems not seen today.

Sustainable forestry is gaining ground

All these challenges can be addressed with sustainable forestry practices. Although the world overall is losing forested land, the FAO's 2010 *Global Forest Resources Assessment* shows that more and more forests are being managed for conservation of soil, water, and biodiversity and that we are increasingly getting our timber efficiently from plantations.

Any company can claim that its timber harvesting practices are sustainable, but how is the purchaser of wood products to know whether they really are? Organizations such as the Forest Stewardship Council (FSC) examine the practices of firms and rate them against criteria for sustainability. The organization then grants **sustainable forest certification** to forests, companies, and products produced using methods it considers sustainable (**FIGURE 9.16**).

Several certification organizations exist, but the Forest Stewardship Council is considered to have the strictest standards. The paper for this textbook is "chain-of-custody certified" by the FSC, meaning that all steps in the life cycle of the paper's production—from timber harvest to transport to pulping to production—have met strict standards. FSC certification rests on 10 general principles and 56 more-detailed criteria.

FIGURE 9.15 ▼ Throughout millions of acres in the West, trees have been killed by the mountain pine beetle (inset), which feeds in tree bark. The beetle thrives in monocultural plantations and with the warm summers and mild winters that climate change is bringing.

Oregon State University graduate student Daniel Donato testifies in a Congressional hearing.

Fighting over Fire and Forests

It's not often that a single scientific paper throws an entire college into turmoil and lands a graduate student in a Congressional hearing to face hostile questioning from federal lawmakers. But such is the political sensitivity of salvage logging.

When a fire burns a forest, should the killed trees be cut and sold for timber? Proponents of salvage logging say yes: We should not let economically valuable wood go to waste. Opponents of salvage logging counter that the burned wood is more valuable left in place—for erosion control, wildlife habitat, and organic material to enhance the soil and nurse the growth of future trees.

Proponents of salvage logging argue that forests regenerate best after a fire if they are logged and replanted with seedlings. Moreover, they maintain, salvage logging reduces fire risk by removing woody debris that could fuel the next fire. When the Biscuit Fire consumed 200,000 ha (500,000 acres) in Oregon in 2002, foresters from the College of Forestry at Oregon State University (OSU) made these arguments in support of plans to log portions of the burned area.

Meanwhile, OSU forestry graduate student Daniel Donato and five other OSU researchers were setting up research plots in areas burned by the Biscuit Fire to test whether salvage logging really does reduce fire risk and help seedlings regenerate. They measured seedling growth and survival and the amount of woody debris in a number of study plots on burned land before (2004) and after (2005) salvage logging took place, and on burned land that was not logged.

The researchers found that conifer seedlings sprouted naturally in the burned areas at densities exceeding what foresters aim for when they replant sites manually. This suggested that manual planting of seedlings may be unnecessary. In contrast, natural seedling densities in logged areas were only 29% as high (**see first figure**). This indicated that salvage logging was hindering seedling survival, presumably because logging disturbs the soil and crushes many seedlings.

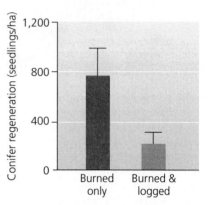

In the wake of the Biscuit Fire, natural growth of conifer seedlings was lower in areas (orange bar) that underwent salvage logging. Data from Donato, D.C., et al., 2006. Post-wildfire logging hinders regeneration and increases fire risk. *Science* 311:352. Reprinted with permission from AAAS.

Donato's team also found that salvage logging more than tripled the amount of woody debris on the ground relative to unlogged sites (**see second figure**). The research team suggested that the best strategy after fire may be to leave the site alone and leave dead trees standing, so that seedlings regenerate safely.

FIGURE 9.16 ◀ Loggers and inspectors in Michigan's Upper Peninsula and many other areas throughout the world work together to confirm that timber was harvested in accord with FSC criteria for sustainable harvesting.

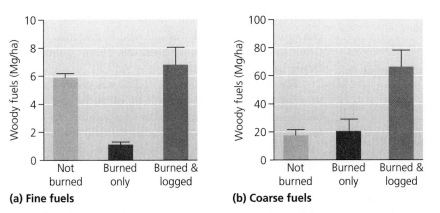

(a) Fine fuels

(b) Coarse fuels

Burned sites that were salvage-logged (orange bar) contained more fine woody debris **(a)** and coarse woody debris **(b)** than did unlogged burned sites (red bar) or sites that did not burn (yellow bar). Data from Donato, D.C., et al., 2006. Post-wildfire logging hinders regeneration and increases fire risk. *Science* 311:352. Reprinted with permission from AAAS.

directly contradicted what some OSU forestry professors had argued following the 2002 fire. When they learned that the prestigious journal *Science* had accepted the Donato team's paper for publication, they took the unusual step of asking the journal's editors to reconsider their decision. Claiming that *Science*'s peer review process had failed, Professor John Sessions and others tried through back channels to stop publication of the paper—actions that were widely condemned as an attempt at censorship.

The paper's publication in 2006 unleashed a torrent of bizarre events. U.S. Congressmen Greg Walden of Oregon and Brian Baird of Washington felt that the paper threatened legislation they had sponsored to accelerate salvage logging. Walden and Baird called Donato and others before a hearing of the House of Representatives' Committee on Resources and grilled them before a packed crowd in Medford, Oregon. There the 29-year-old graduate student stood up for his team's findings under harsh questioning.

The Bureau of Land Management then suspended the team's research grant, in what many viewed as a response to political pressure. This highly unusual action drew media attention, and the BLM quickly reinstated the funding.

A heated debate roiled for months in the OSU College of Forestry. The college receives 12% of its funding from taxes on timber sales, leading many to suggest that the college is open to influence from industry. E-mail correspondence was subpoenaed and showed the college's dean, Hal Salwasser, collaborating with timber industry representatives to refute the paper.

As publicity built, the college's reputation suffered, graduate admissions declined, and a faculty committee on academic freedom criticized Salwasser for "significant failures of leadership." The dean admitted mistakes, survived a no-confidence vote of the faculty, and pledged to make reforms.

In the pages of *Science* and elsewhere, scientific criticisms of the study's methods were largely rebutted, yet many felt that the study's conclusions stretched beyond what its short-term data could support. Scientists on both sides of the debate agreed that long-term research was needed to fully assess the effects of salvage logging on forest regeneration and fire risk.

Two studies by different OSU forestry scientists soon provided the first such long-term data. Jeffrey Shatford and colleagues documented widespread natural regrowth of conifers across areas of Oregon and northern California that had burned 9–19 years earlier. And Jonathan Thompson and colleagues examined satellite data, aerial photography, and government records for regions within the Biscuit Fire area that had burned in a previous fire 15 years earlier. They found that of the regions burned in that 1987 fire, those that were salvage-logged burned more severely in 2002 than regions that were not logged.

As more long-term studies on the impacts of salvage logging are conducted, we should become better able to manage our forests in an age of increasingly frequent wildfire.

The number of FSC-certified forests, companies, and products is growing quickly. Over 5% of the world's forests managed for timber production are now FSC-certified—over 134 million ha (332 million acres) in 81 nations at the start of 2011. Pursuing sustainable forestry practices is often more costly for producers, at least in the short term, but producers recoup these costs when consumers are willing to pay more for certified products. Over the long term, sustainable practices conserve the resource base, thus holding down costs. When we ask businesses if they carry certified wood, we make them aware of our preferences and help drive demand for these products. Consumer demand has already led Home Depot and other major retailers to sell sustainable wood. The purchasing decisions of such retailers, in turn, are influencing timber harvesting practices around the world.

PARKS AND PROTECTED AREAS

As our world fills with more people consuming more resources, the sustainable management of resources from forests and other ecosystems becomes ever-more important. So does our need to preserve functional ecosystems by setting aside tracts of land to remain forever undeveloped. Today 12% of the world's land area is designated for preservation in various types of parks, reserves, and protected areas.

Why create parks and reserves?

People establish parks and protected areas for various reasons:

▶ Enormous or unusual scenic features inspire people to preserve them (**FIGURE 9.17**).

▶ Protected areas offer hiking, fishing, hunting, kayaking, bird-watching, and other recreation.

▶ Parks generate revenue from ecotourism (pp. 60, 174).

▶ Undeveloped land offers us peace of mind, health, exploration, wonder, and spiritual solace.

▶ Protected areas offer utilitarian benefits and ecosystem services. For example, undeveloped watersheds provide cities clean drinking water and a buffer against floods.

▶ Reserves protect biodiversity. These islands of habitat maintain species, communities, and ecosystems.

Federal parks and reserves began in the United States

The striking scenery of the American West impelled the U.S. government to create the world's first **national parks**, public lands protected from development but open to nature

FIGURE 9.17 ▲ The awe-inspiring beauty of places like Yosemite National Park inspires many millions of people to visit America's national parks.

appreciation and recreation. Yellowstone National Park was established in 1872, and was followed by Sequoia, General Grant (now Kings Canyon), Yosemite, Mount Rainier, and Crater Lake National Parks.

The National Park Service (NPS) was created in 1916 to administer the growing system of parks and monuments, which today numbers 393 sites totaling 36 million ha (89 million acres) and includes national historic sites, national recreation areas, national seashores, and other areas (see Figure 9.8).

Because America's national parks are open to everyone and showcase the nation's natural beauty in a democratic way, writer Wallace Stegner famously called them "the best idea we ever had." A mill worker in Escanaba can take his or her family for the weekend to Pictured Rocks National Lakeshore to camp along sandstone cliffs on the shore of Lake Superior. They can head across the lake to Isle Royale National Park to hike and canoe through lands that are home to moose and wolves. Or they can enjoy climbing immense sand dunes at Sleeping Bear Dunes National Lakeshore.

Another type of federal protected area in the United States is the **national wildlife refuge**. The national wildlife refuge system, begun in 1903 by President Theodore Roosevelt, now totals over 550 sites comprising 61 million ha (151 million acres; see Figure 9.8). The U.S. Fish and Wildlife Service administers these refuges, which serve as havens for wildlife and encourage hunting, fishing, birding, wildlife observation, photography, environmental education, and other public uses.

In response to the public's desire for undeveloped areas of land, in 1964 the U.S. Congress passed the Wilderness Act, which allowed some areas of existing federal lands to be designated as **wilderness areas**. These areas are off-limits to development but open to hiking, nature study, and other low-impact recreation. Overall the nation has 756 wilderness areas totaling 44 million ha (109 million acres), covering 5% of U.S. land area (2.7% outside Alaska).

Many agencies and groups help to protect land

Efforts to set aside land at the federal level are paralleled at regional and local levels. Each U.S. state has agencies that manage resources on public lands, as do many counties and municipalities. When Mackinac Island, the nation's second national park, was transferred to the state of Michigan, it became the first officially designated state park in the nation. Today it is one of 3,700 state parks across the United States, along with regional parks, county parks, and others.

Private nonprofit groups also preserve land. **Land trusts** are local or regional organizations that purchase land to preserve in its natural condition. The Nature Conservancy is the world's largest land trust, but nearly 1,700 local land trusts in the United States together own 690,000 ha (1.7 million acres) and have helped preserve an additional 4.1 million ha (10.2 million acres), including scenic areas such as Big Sur on the California coast, Jackson Hole in Wyoming, and Maine's Mount Desert Island. Moreover, thousands of local volunteer groups (often named "Friends of" an area) have organized from the grassroots to help to care for protected lands.

Parks and reserves are increasing internationally

Many nations have established protected areas and are benefiting from ecotourism as a result—from Costa Rica (Chapter 3) to Ecuador to Thailand to Tanzania. The worldwide area in parks and reserves has increased sixfold since 1970, and today 114,000 protected areas cover 12% of the planet's land area. However, parks do not always receive the funding they need to manage resources, provide recreation, and protect wildlife from poaching and timber from logging. Thus many of the world's protected areas are merely *paper parks*—protected on paper but not in reality.

Some types of protected areas are designated or partly managed internationally by the United Nations. **Biosphere reserves** are tracts of land with exceptional biodiversity that couple preservation with sustainable development to benefit local people. Each biosphere reserve consists of (1) a core area that preserves biodiversity, (2) a buffer zone that allows local activities and limited development, and (3) an outer transition zone where agriculture, human settlement, and other land uses can be pursued in a sustainable way (**FIGURE 9.18**).

World heritage sites are another type of international protected area. Over 900 sites in more than 150 countries are listed for their natural or cultural value. One example is a mountain gorilla reserve shared by three African countries. This reserve, which integrates national parklands of Rwanda, Uganda, and the Democratic Republic of Congo, is also an example of a *transboundary park*, an area of protected land overlapping national borders. Some transboundary reserves function as "peace parks," helping to ease tensions by acting as buffers between nations that quarrel over boundary disputes.

Beyond all these efforts on land, the importance of conserving the oceans' natural resources is leading us to establish protected areas and reserves in marine waters (p. 275).

Habitat fragmentation makes preserves still more vital

Protecting large areas of land has taken on new urgency now that scientists understand the risks posed by habitat fragmentation (p. 169). Expanding agriculture, residential development, highways, logging, and other impacts divide large expanses of habitat into small disconnected ones (**FIGURE 9.19A, B**).

When forests are fragmented, many species suffer. Bears, mountain lions, and other animals that need large areas of habitat may disappear. Birds that thrive in the interior of forests may fail to reproduce near the edge of a fragment (**FIGURE 9.19C**). Their nests are attacked by predators and parasites that favor open habitats or that travel along habitat edges. Because of such **edge effects**, avian ecologists judge forest fragmentation to be a main reason why populations of many North American songbirds are declining.

Habitat fragmentation is affecting our national parks, which are, after all, islands of natural habitat surrounded by farms, ranches, roads, and cities. In 1983 conservation biologist William Newmark examined historical records of mammal sightings in national parks of western Ameri-

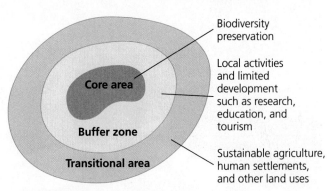

(a) The three zones of a biosphere reserve

(b) Sustainable harvesting by local people

FIGURE 9.18 ▲ Biosphere reserves are international protected areas that couple preservation with sustainable development to benefit local residents. Each reserve **(a)** includes a core area that preserves biodiversity, a buffer zone that allows limited development, and a transition zone that permits various uses. At the Maya Biosphere Reserve in Guatemala **(b)**, local women process and sell Maya nuts harvested from rainforest trees. Here, FSC certification in the transition zone helped prevent illegal deforestation that had been rampant in the core area.

ca. He found that many parks were missing a few species they had held previously. The red fox and river otter had vanished from Sequoia and Kings Canyon National Parks, for example, and the white-tailed jackrabbit and spotted skunk no longer lived in Bryce Canyon National Park. In all, 42 species had disappeared. As ecological theory predicted, smaller parks lost more species than larger parks. Species were disappearing because the parks were too small to sustain their populations, Newmark concluded, and because the parks had become too isolated to be recolonized by new arrivals.

Reserve design has consequences for biodiversity

Because habitat fragmentation is such a central issue in biodiversity conservation, and because there are limits on how much land can be set aside, conservation biologists have debated whether it is better to make reserves large in size and few in number, or many in number but small in size. This

(a) Clear-cuts (snowy patches) in Mount Hood National Forest, Oregon

(c) Wood thrush

FIGURE 9.19 ▲ Forest fragmentation can result from clear-cutting **(a)**, agriculture, or residential development. Shown in **(b)** are historical changes in forested area in a region of Wisconsin from 1831 to 1950. Fragmentation affects forest-dwelling species such as the wood thrush **(c)**, whose nests are parasitized by cowbirds that thrive in the surrounding open country. *Source* (b): Curtis, J.T., 1956. The modification of mid-latitude grasslands and forests by man. In Thomas, W.L. Jr., ed., *Man's role in changing the face of the earth.* Copyright 1956. Used by permission of the publisher, University of Chicago Press.

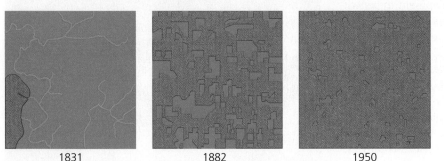

| 1831 | 1882 | 1950 |

(b) Fragmentation of wooded area (green) in Cadiz Township, Wisconsin

so-called **SLOSS dilemma** (for "**s**ingle **l**arge **o**r **s**everal **s**mall") led conservation biologists to establish a gigantic experiment smack in the middle of South America's Amazon rainforest. For over 30 years and across 1,000 km² (386 mi²), researchers with the Biological Dynamics of Forest Fragments Project have studied what happens to animals, plants, and ecosystems when we fragment a forest.

Researchers first worked with Brazilian ranchers who were clearing forest, guiding them to do so in ways that left study plots of various sizes (**FIGURE 9.20**). The scientists then compared these resulting forest fragments to identically sized plots within continuous forest nearby. Seven hundred research papers, graduate theses, and books later, researchers here have found that:

▸ Species diversity of most animal groups declines in fragments.

▸ Small fragments lose more species, and lose them faster, than large fragments.

▸ Even fragments as large as 100 ha (250 acres) lose half their species in less than 15 years.

▸ Fragments distant from continuous forest lose more species than fragments near continuous forest.

▸ Regrowing secondary forest can act as a corridor for some forest species, letting them move between fragments.

▸ Edges and secondary forest also allow new species to invade fragments—generalists adapted to disturbed areas—and these can displace interior-forest species.

▸ Edges receive enough sun, heat, and wind to kill trees adapted to the dark, moist interior, setting in motion many changes as edge effects extend 300 m (1,000 ft) into the fragments.

FIGURE 9.20 ▼ Researchers have studied experimental forest fragments of 1, 10, and 100 ha (2.5, 25, and 250 acres) in the Brazilian Amazon.

Such results are helping scientists learn how large fragments need to be to retain their species, and this helps them work with policymakers to preserve forests in the face of development pressures.

Climate change threatens protected areas

Today global climate change (Chapter 14) threatens our investment in protected areas. As temperatures become warmer, species' ranges are shifting toward the poles and upward in elevation (p. 311). In a landscape of fragmented habitat, organisms that cannot disperse long distances will be unable to move from one fragment to another. Species we had hoped to protect in parks may, in a warming world, become trapped in them. As with the golden toad in Monteverde (Chapter 3), high-elevation species have nowhere to go once a mountaintop becomes too warm or dry, and they may be driven extinct. For this reason, **corridors** of protected land that allow animals to travel between islands of habitat have become more important than ever. Conservation biologists and land management agencies are now trying to use corridors to connect reserves across the landscape as they explore strategies for saving biodiversity that go beyond parks and protected areas.

➤ CONCLUSION

Forests are ecologically vital and economically valuable, yet we are losing forests around the world. In North American forest management, early emphasis on resource extraction evolved into policies on sustainable yield and multiple use as land and resource availability declined. Public forests today are managed not only for timber production, but also for recreation, wildlife habitat, and ecosystem integrity. Sustainable forest certification provides economic incentives for conservation on forested lands.

Meanwhile, public support for the preservation of natural lands has led to the establishment of parks and protected areas in the United States and abroad. As development spreads across the landscape, fragmenting habitats and subdividing populations, scientists trying to conserve species, communities, and ecosystems are thinking and working at the landscape level.

TESTING YOUR COMPREHENSION

1. Name at least two reasons why natural primary forests contain more biodiversity than even-aged single-species forestry plantations.

2. Describe three ecosystem services that forests provide.

3. Name several major causes of deforestation. Where is deforestation most severe today?

4. Compare and contrast maximum sustainable yield, ecosystem-based management, and adaptive management. Why may pursuing maximum sustainable yield sometimes conflict with what is ecologically desirable?

5. Compare and contrast the major methods of timber harvesting.

6. Describe several ecological impacts of logging. How has the U.S. Forest Service responded to public concern over these impacts?

7. Are forest fires a bad thing? Explain your answer.

8. Name at least four reasons that people have created parks and reserves. How do national parks differ from national wildlife refuges? What is a wilderness area?

9. What percentage of Earth's land is protected? Describe one type of protected area that has been established outside North America.

10. Give two examples of how forest fragmentation affects animals. How might research like that of the Biological Dynamics of Forest Fragments Project help us design reserves?

SEEKING SOLUTIONS

1. People in developed nations are fond of warning people in developing nations to stop destroying rainforest. People of developing nations often respond that this is hypocritical, because the developed nations became wealthy by deforesting their land and exploiting its resources in the past. What would you say to the president of a developing nation, such as Indonesia, in which a great deal of forest is being cleared?

2. Do you think maximum sustainable yield represents an appropriate policy for resource managers to follow? Why or why not?

3. Given the effects that climate change may have on species' ranges, if you were trying to preserve an endangered mammal that occurs in a small area and you had unlimited funds to acquire land to help restore its population, how would you design a protected area for it? Would you use corridors? Would you include a diversity of elevations? Would you design few large reserves or many small ones? Explain your answers.

4. **THINK IT THROUGH** You are the supervisor of a national forest. Timber companies are requesting to cut as many trees as you will let them, and environmentalists want no logging at all. Ten percent of your forest is old-growth primary forest, and the remaining 90% is secondary forest. Your forest managers are split among preferring maximum sustainable yield, ecosystem-based

management, and adaptive management. What management approach or approaches will you take? Will you allow logging of all, some, or no old-growth trees? Will you allow logging of secondary forest? If so, what harvesting strategies will you encourage? What would you ask your scientists before deciding on policies on fire management and salvage logging?

5. **THINK IT THROUGH** You run a major nonprofit environmental advocacy organization and are trying to save an ecologically priceless tract of tropical forest in a poor developing nation. You have worked in this region for years and know and care for the local people, who want to save the forest and its animals but also need to make a living and use the forest's resources. The nation's government plans to sell a concession to a foreign multinational timber corporation to log the entire forest unless your group can work out some other solution. Describe what solution(s) you would try to arrange. Consider the range of issues and options discussed in this chapter, including government protected areas, private protected areas, biosphere reserves, forest management techniques, carbon offsets, FSC-certified sustainable forestry, and more. Explain reasons for your choice(s).

CALCULATING ECOLOGICAL FOOTPRINTS

We all rely on forest resources. The average North American consumes 225 kg (500 lb) of paper and paperboard per year. Using the estimates of paper and paperboard consumption for each region of the world, calculate the per capita consumption for each region using the population data in the table. Note that 1 metric ton = 2,205 lb.

	Population (millions)	Total paper consumed (millions of metric tons)	Per capita paper consumed (pounds)
Africa	999	6	13
Asia	4,117	159	
Europe	738	99	
Latin America	580	27	
North America	341	77	
Oceania	36	4	
World	6,810		127
All data are for 2009, from Population Reference Bureau and U.N. Food and Agriculture Organization (FAOSTAT).			

1. How much paper would North Americans save each year if we consumed paper at the rate of Europeans?

2. How much paper would be consumed if everyone in the world used as much paper as the average European? As the average North American?

3. Why do you think people in other regions consume less paper, per capita, than North Americans? Name three things you could do to reduce your paper consumption.

4. Describe three ways in which consuming FSC-certified paper rather than conventional paper can reduce the environmental impacts of paper consumption.

10 Environmental Health and Toxicology

Upon completing this chapter, you will be able to:

➤ Identify major environmental health hazards and explain the goals of environmental health

➤ Describe the types of toxic substances in the environment, the factors that affect their toxicity, and the defenses that organisms possess against them

➤ Explain the movements of toxic substances and how they affect organisms and ecosystems

➤ Discuss the study of chemical hazards, including wildlife toxicology, epidemiology, animal testing, and dose-response analysis

➤ Compare and contrast risk assessment and risk management

➤ Compare philosophical approaches to risk and how they relate to regulatory policy

Is this baby ingesting toxic substances?

Poison in the Bottle: Is Bisphenol A Safe?

"Babies in the U.S. are born pre-polluted with BPA. What more evidence do we need to act?"
—Dr. Janet Gray, Director of the Environmental Risks and Breast Cancer Project, Vassar College

"There is no basis for human health concerns from exposure to BPA."
—The American Chemistry Council

How is it that a chemical found to alter reproductive development in animals gets used in baby bottles? How can it be that a substance linked to breast cancer, prostate cancer, and heart disease is routinely used in food and drink containers? The chemical **bisphenol A** (**BPA** for short) has been associated with everything from neurological effects to miscarriages. Yet it's in hundreds of products we use every day, and there's a 90% chance that it is coursing through your body right now.

To understand how chemicals that may pose health risks come to be widespread in our society, we need to explore how scientists and policymakers study toxic substances and other environmental health risks—and the vexing challenges these pursuits entail.

Bisphenol A is a synthetic organic compound ($C_{15}H_{16}O_2$) used in the resins that line metal food cans and drink cans and water supply pipes, and in dental sealants for our teeth. It's also found in the hard, clear polycarbonate plastic in some water bottles, food containers, eating utensils, eyeglass lenses, CDs and DVDs, electronics, baby bottles, and children's toys.

Unfortunately, bisphenol A leaches out of these products into our food, air, and bodies. The Centers for Disease Control and Prevention (CDC) reports that 93% of Americans carry detectable concentrations in their urine. Because most of the chemical passes through the body within hours of exposure, its widespread presence in urine suggests that most Americans receive continuous exposure to BPA. Babies and children accumulate the most BPA, because they eat more for their body weight and metabolize the chemical less effectively.

Bisphenol A: Worldwide

What, if anything, is BPA doing to us? Over 200 studies with rats, mice, and other animals have shown many apparent effects of BPA, including a wide range of reproductive abnormalities, and a few recent studies suggest human health impacts (see **THE SCIENCE BEHIND THE STORY**, pp. 214–215). Many of these effects are seen at extremely low concentrations. Scientists say this is because BPA mimics the female sex hormone estrogen and can induce some of estrogen's effects in animals. Hormones such as estrogen function at very low concentrations in the body, so a synthetic chemical in the body at similarly low concentrations can fool the body into responding as it would to estrogen.

In reaction to the burgeoning research, a growing number of researchers, doctors, and consumer advocates are calling on governments to regulate bisphenol A and for manufacturers to stop using it. The chemical industry insists that BPA is safe, pointing to industry-sponsored research that finds no health impacts. Expert panels convened to assess the fast-growing body of scientific studies on BPA have struggled with the fact that traditional research methods are not geared to test hormone-mimicking substances that exert effects at low doses. These panels have often arrived at divergent conclusions. For instance, the

FIGURE 10.1 ▲ Researchers for *Consumer Reports* magazine tested these (and more) common packaged foods in 2009; they found that nearly all of them contained bisphenol A that had leached from the linings of their containers.

U.S. Food and Drug Administration (FDA) insisted in 2008 that it saw no reason to regulate BPA, but its own science advisory committee disagreed, and in 2009 the FDA decided to start a testing program.

In 2008, the Canadian government was the first to declare bisphenol A toxic, banning its sale and importation. As of 2011, the use of BPA in certain products for children was banned in China, Malaysia, and in nine U.S. states. Despite the lack of federal regulation of BPA in the United States, grassroots lobbying efforts have led many companies to voluntarily remove BPA from their products, especially those made for children and infants. The six major U.S. manufacturers of plastic baby bottles promised in 2008 to stop using BPA, and the manufacturer Sunoco stopped selling BPA to companies that use it in children's products. Nalgene phased out its BPA-containing polycarbonate water bottles. The retailers Walmart and Toys "R" Us decided to stop carrying children's products with BPA. As a result, concerned parents can now more easily find BPA-free products for their children, but the rest of us remain exposed through thousands of products (**FIGURE 10.1**).

Bisphenol A is by no means one of our greatest environmental health threats. However, it provides a timely example of how we as a society assess health risks and decide how to manage them. As scientists and government regulators assess BPA's potential risks, their efforts give us a window on how hormone-disrupting chemicals are challenging the way we appraise and control the environmental health risks we face. ■

ENVIRONMENTAL HEALTH

Examining the impacts of human-made chemicals such as bisphenol A is just one aspect of the broad field of environmental health. The study and practice of **environmental health** assesses environmental factors that influence our health and quality of life. These factors include wholly natural aspects of the environment over which we have little or no control, as well as anthropogenic (human-caused) factors. Practitioners of environmental health seek to prevent adverse effects on human health and on the ecological systems that are essential to our well-being.

We face four types of environmental hazards

We can categorize environmental health hazards into four main types: physical, chemical, biological, and cultural. Although some amount of risk is unavoidable, much of environmental health focuses on taking steps to minimize the risks of encountering hazards and to mitigate the impacts of the hazards we do encounter.

Physical hazards *Physical hazards* arise from processes that occur naturally in our environment and pose risks to human life or health. Some are ongoing natural phenomena, such as excessive exposure to ultraviolet (UV) radiation from sunlight, which damages DNA and has been tied to skin cancer, cataracts, and immune suppression (**FIGURE 10.2A**). We can reduce these risks by shielding our skin from intense sunlight with clothing and sunscreen.

(a) Physical hazard

(b) Chemical hazard

(c) Biological hazard   **(d) Cultural hazard**

FIGURE 10.2 ▲ Environmental health hazards come in four types. The sun's ultraviolet radiation is an example of a physical hazard **(a)**. Excessive exposure increases the risk of skin cancer. Chemical hazards **(b)** include both synthetic and natural chemicals. Much of our exposure comes from pesticides and household chemical products. Biological hazards **(c)** include diseases and the organisms that transmit them. Some mosquitoes, for example, are vectors for pathogenic microbes, including those that cause malaria. Cultural or lifestyle hazards **(d)** include the behavioral decisions we make, as well as the socioeconomic constraints forced on us. Smoking is a lifestyle choice that raises one's risk of lung cancer and other diseases.

Other physical hazards include discrete events such as earthquakes, volcanic eruptions, fires, floods, landslides, hurricanes, and droughts. We cannot prevent many of these hazards, but we can minimize risk by preparing ourselves with emergency plans and avoiding common practices that make us vulnerable to certain physical hazards. For example, clearing vegetation from hillsides increases the chance of landslides, and channelizing rivers promotes flooding in some areas while preventing it in others (p. 261).

Chemical hazards   *Chemical hazards* include many of the synthetic chemicals that our society manufactures, such as pharmaceuticals, disinfectants, and pesticides (**FIGURE 10.2B**). Some natural substances that we process for our use (such as hydrocarbons, lead, and asbestos) are also harmful to human health. Following our overview of environmental health, much of this chapter will focus on chemical health hazards and the ways we study and regulate them.

Biological hazards   *Biological hazards* result from ecological interactions among organisms (**FIGURE 10.2C**). When we become sick from a virus, bacterial infection, or other pathogen, we are suffering parasitism (pp. 66–67). This is what we call **infectious disease**. Infectious diseases such as malaria, cholera, tuberculosis, and influenza (flu) are major environmental health hazards, especially in developing nations with widespread poverty and few resources for health care. As with physical and chemical hazards, it is impossible for us to avoid risk from biological agents completely, but through monitoring, sanitation, and medical treatment we can reduce the likelihood and impacts of infection.

Cultural hazards   Hazards that result from our place of residence, our socioeconomic status, our occupation, or our behavioral choices can be thought of as *cultural hazards* or *lifestyle hazards*. We can minimize or prevent some of these cultural or lifestyle hazards, but others may be beyond our control. For instance, individuals can choose whether or not to smoke cigarettes (**FIGURE 10.2D**), but exposure to second-hand smoke in the home or workplace may be beyond one's control. Much the same might be said for other cultural hazards such as drug use, diet and nutrition, crime, and mode of transportation. Environmental justice advocates (pp. 14–15) argue that "forced" risks from cultural hazards, such as living near a hazardous waste site, are often higher for people with fewer economic resources or less political clout.

The biological hazard of disease is a focus of environmental health

Despite all our technological advances, we still find ourselves battling disease, which causes the vast majority of human deaths worldwide (**FIGURE 10.3A**). Over half the world's deaths result from noninfectious diseases, such as cancer and heart disease. These diseases are not spread from one person to another, but rather are influenced by genetics, environmental factors, and lifestyle choices. For instance, whether a person develops heart disease depends not only on his or her genes, but also on lifestyle choices such as diet and exercise.

Infectious diseases account for almost one of every four deaths that occur each year—nearly 14 million people worldwide (**FIGURE 10.3B**). Some pathogenic viruses, bacteria, and protists attack us directly; others cause infection through a *vector*, an organism that transfers the pathogen to the host.

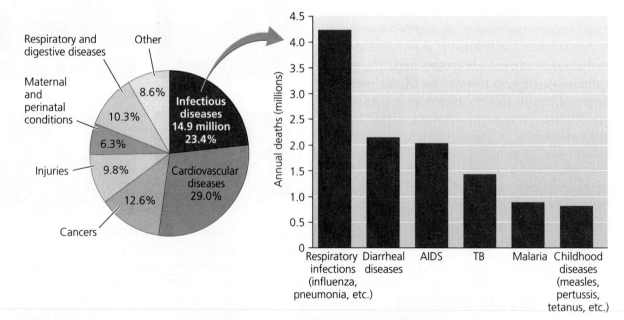

(a) Leading causes of death across the world   **(b) Leading causes of death by infectious disease**

FIGURE 10.3 ▲ Infectious disease is the second-leading cause of death worldwide **(a)**, accounting for nearly one-quarter of all deaths. Six types of diseases **(b)**—respiratory infections, diarrhea, AIDS, tuberculosis (TB), malaria, and childhood diseases—account for 80% of all deaths from infectious disease. Data from World Health Organization, 2009. *World health statistics 2009.* WHO, Geneva, Switzerland.

Infectious disease is a greater problem in developing countries, where it accounts for close to half of all deaths. Infectious disease causes many fewer deaths in developed nations because their wealth allows their citizens better nutrition, sanitation, hygiene, and access to medical care.

Decades of public health efforts have lessened the impacts of infectious disease and even have eradicated some diseases—yet other diseases are posing new challenges. Some, such as acquired immunodeficiency syndrome (AIDS), continue to spread globally despite concerted efforts to stop them. Others, such as tuberculosis and strains of malaria, are evolving resistance to our antibiotics, in the same way that pests evolve resistance to our pesticides (p. 147). Additionally, human-induced global warming of the climate (Chapter 14) is enabling tropical diseases (such as malaria, dengue, and cholera) to gain footholds in temperate regions.

In our world of global mobility and dense human populations, novel infectious diseases (or new strains of old diseases) that emerge in one location are more likely to spread quickly to other locations. Recent examples include severe acute respiratory syndrome (SARS) in 2003, the H5N1 avian flu starting in 2004, and the H1N1 swine flu that spread across the globe in 2009–2010. Diseases like influenza, whose pathogens evolve rapidly, give rise to a variety of strains, making it more likely that one may turn exceedingly dangerous and cause a global pandemic (a widespread outbreak of a disease).

Thousands of dedicated people—from doctors and nurses to policymakers to philanthropists—are dedicating their lives to reducing the incidence of disease and improving human health. They use a diversity of approaches to better the living conditions of those most affected by infectious disease by improving access to clean drinking water, sanitation, medical care, and nutritious foods.

Toxicology is the study of chemical hazards

Although most indicators of human health are improving as the world's wealth increases, our modern society is exposing us to more and more synthetic chemicals. Some of these substances pose threats to human health, but figuring out which of them do—and how, and to what degree—is a complicated scientific endeavor. **Toxicology** is the science that examines the effects of poisonous chemicals on humans and other organisms. Toxicologists assess and compare substances to determine their **toxicity**, the degree of harm a chemical substance can inflict. A toxic substance, or poison, is called a **toxicant**, but any chemical substance may exert negative impacts if we ingest or expose ourselves to enough of it. Conversely, a toxicant in a small enough quantity may pose no health risk at all. These facts are often summarized in the catchphrase, "The dose makes the poison." In other words, a substance's toxicity depends not only on its chemical identity, but also on its quantity.

In recent decades, our ability to produce new chemicals has expanded, concentrations of chemical contaminants in the environment have increased, and public concern for health and the environment has grown. These trends have driven the rise of **environmental toxicology**, which deals specifically with toxic substances that come from or are discharged into the environment. Toxicologists generally focus on human health, using other organisms as models and test subjects. Environmental toxicologists study animals and plants to determine the ecological impacts of toxic substances, and to see if other organisms can serve as indicators of health threats that could soon affect people.

People face environmental health hazards indoors

Modern Americans spend roughly 90% of their lives indoors. Unfortunately, our homes and workplaces, just like the outdoors, can be rife with physical, biological, chemical, and cultural hazards (**TABLE 10.1**; also see Figure 13.21, p. 295).

Cigarette smoke and radon are leading indoor hazards (pp. 294–296) and are the top two causes of lung cancer in developed nations. Homes and offices can have problems with toxic compounds produced by mold, which can flourish in wall spaces when moisture levels are high. Asbestos, used in the past as insulation in walls and other products, can be dangerous when it is inhaled. Lead poisoning from water pipes or old paint can cause damage to the brain, liver, kidney, and stomach; learning problems and behavioral abnormalities; anemia; hearing loss; and even death. Lead poisoning among U.S. children has greatly declined in recent years as a result of education campaigns and the phaseout of lead-based paints and leaded gasoline (p. 5), which was prompted by government regulation to protect public health.

There are also indoor chemical hazards that we have yet to discover. One recently recognized hazard is *polybrominated diphenyl ethers* (PBDEs). These compounds are used as fire retardants in computers, televisions, plastics, and furniture,

TABLE 10.1 Selected Environmental Hazards
Outdoor Air
▶ Chemicals from automotive exhaust
▶ Chemicals from industrial pollution
▶ Photochemical smog (pp. 288–289)
▶ Pesticide drift
▶ Dust and particulate matter
Water
▶ Pesticide and herbicide runoff
▶ Nitrates and fertilizer runoff
▶ Mercury, arsenic, and other heavy metals in groundwater and surface water
Food
▶ Natural toxins
▶ Pesticide and herbicide residues
Indoors
▶ Smoking and secondhand smoke
▶ Radon
▶ Lead in paint and pipes
▶ Asbestos
▶ Toxicants (*e.g.*, PBDEs, phthalates, bisphenol A) in plastics and consumer products
▶ Dust and particulate matter

and they may evaporate at very slow rates throughout the lifetime of the product. Like bisphenol A, PBDEs appear to act as hormone disruptors. The European Union decided in 2003 to ban PBDEs, but in the United States there has so far been little movement to address the issue.

Risks must be balanced against rewards

As we review the impacts of toxic substances throughout this chapter, it is important to keep in mind that artificially produced chemicals have played a crucial role in giving us the standard of living we enjoy today. These chemicals have helped create the industrial agriculture that produces our food, the medical advances that protect our health and prolong our lives, and many of the modern materials and conveniences we use every day. It is appropriate to remember these benefits as we examine some of the unfortunate side effects of these advances and as we search for better alternatives.

TOXIC SUBSTANCES AND THEIR EFFECTS ON ORGANISMS

Our environment contains countless natural substances that may pose health risks. These include oil oozing naturally from the ground; radon gas seeping up from bedrock; and **toxins**, toxic chemicals manufactured in the tissues of living organisms—for example, chemicals that plants use to ward off herbivores or that insects use to defend themselves from predators. In addition, we are exposed to many synthetic (human-made) chemicals.

Synthetic chemicals are all around us—and *in* us

Synthetic chemicals surround us in our daily lives, and each year in the United States we manufacture or import 113 kg (250 lb) of chemical substances for every man, woman, and child. Many of these substances, particularly the pesticides we use to control insects and weeds, find their way into soil, air, and water—and into humans and other organisms (**FIGURE 10.4**).

As a result of all this exposure, every one of us carries traces of hundreds of industrial chemicals in our bodies. The U.S. government's latest National Health and Nutrition Examination Survey gathered data on 148 foreign compounds in Americans' bodies. Among these were several toxic persistent organic pollutants restricted by international treaty (p. 223). Depending on the pollutant, these were detected in 41–100% of the people tested. Smaller-scale surveys have found similar results. Our exposure to synthetic chemicals begins in the womb as substances our mothers ingested while pregnant were transferred to us. A 2009 study by the nonprofit Environmental Working Group found 232 chemicals in the umbilical cords of 10 newborn babies it tested. Nine of the 10 umbilical cords contained BPA, leading researchers to note that we are born "pre-polluted."

All this should not necessarily be cause for alarm. Not all synthetic chemicals pose health risks, and relatively few are

known with certainty to be toxic. However, of the roughly 100,000 synthetic chemicals on the market today, very few have been thoroughly tested. For the vast majority, we simply do not know what effects, if any, they may have on us.

Silent Spring changed public attitudes toward synthetic chemicals

It was not until the 1960s that people began to learn about the risks of exposure to pesticides. The key event was the publication of Rachel Carson's 1962 book *Silent Spring* (pp. 98–100), which brought the insecticide dichlorodiphenyl-trichloroethane (DDT) to the public's attention. The book was written at a time when large amounts of pesticides virtually untested for health effects were indiscriminately sprayed, on the assumption that the chemicals would do no harm to people (**FIGURE 10.5**).

Carson synthesized scientific studies, medical case histories, and other data to contend that DDT in particular, and artificial pesticides in general, were hazardous to people, wildlife, and ecosystems. The book became a best-seller and helped generate significant social change in views and actions toward the environment. The use of DDT was banned in the United States in 1973 and is now illegal in a number of nations. U.S. chemical companies still manufacture and export DDT, however, because developing countries with tropical climates use it to control disease vectors, such as mosquitoes that transmit malaria. In these countries, malaria represents a greater health threat than do the toxic effects of the pesticide.

 WEIGHING THE ISSUES

A Circle of Poison? Although the United States has banned the use of DDT, U.S. companies still manufacture and export the compound to developing nations. Thus, it is possible that pesticide-laden food can be imported back into the United States in what has been called a "circle of poison." How do you feel about this? Is it unethical for one country to sell to others a substance that it has deemed toxic? Or would it be unethical for the United States *not* to sell DDT to African nations if they desire it for controlling malaria?

Not all toxic substances are synthetic, and not all synthetic chemicals are toxic

Although many toxicologists focus on synthetic chemicals, toxic substances also exist naturally in the environment around us and in the foods we eat. Thus, it would be a mistake to assume that all artificial substances are unhealthy and that all natural substances are healthy. In fact, the plants and animals we eat contain many chemicals that can cause us harm. Recall that plants produce toxins to ward off animals that eat them. In domesticating crop plants, we have selected (p. 48) for strains with reduced toxin content, but we have not eliminated these dangers. Furthermore, when we consume animal meat, we ingest toxins the animals obtained from plants or animals they ate. Scientists

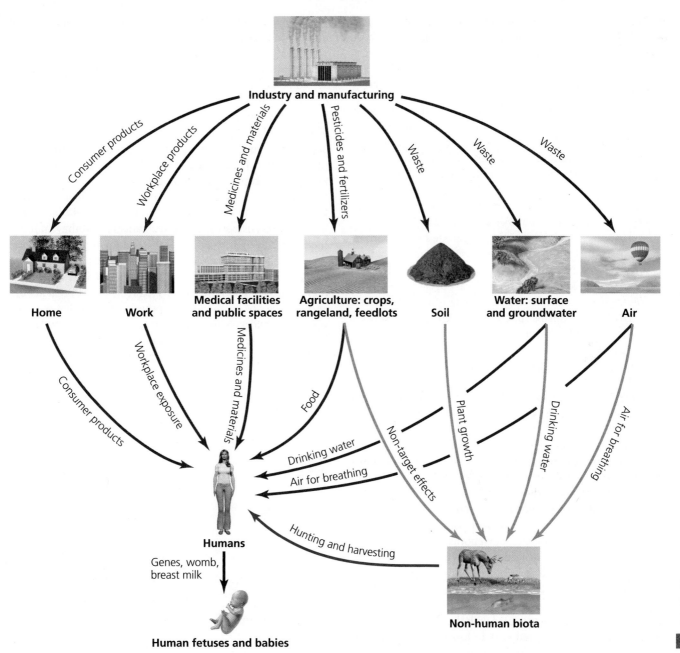

FIGURE 10.4 ▲ Synthetic chemicals take many routes in traveling through the environment. People take in only a tiny proportion of these compounds, and many compounds are harmless. However, people receive small amounts of toxicants from many sources, and developing fetuses and babies are particularly sensitive.

FIGURE 10.5 ▶ Children on a beach in Long Island, New York, are fogged with DDT from a pesticide spray machine being tested in 1945. Before the 1960s, the environmental and health effects of potent pesticides such as DDT were not widely known. Public parks and neighborhoods were regularly sprayed for insect control without safeguards against excessive human exposure.

are actively debating just how much risk natural toxicants pose, and it is clear that more research is required on these questions.

Toxic substances come in different types

Toxic substances can be classified based on their particular effects on health. The best-known toxicants are **carcinogens**, which are substances or types of radiation that cause cancer. In cancer, malignant cells grow uncontrollably, creating tumors, damaging the body, and often leading to death. Cancer frequently has a genetic component, but a wide variety of environmental factors are thought to raise the risk of cancer. In our society today, the greatest number of cancer cases is thought to result from carcinogens contained in cigarette smoke. Carcinogens can be difficult to identify because there may be a long lag time between exposure to the agent and the detectable onset of cancer—up to 15–30 years in the case of cigarette smoke.

Mutagens are substances that cause genetic mutations in the DNA of organisms (p. 28). Although most mutations have little or no effect, some can lead to severe problems, including cancer and other disorders. If mutations occur in an individual's sperm or egg cells, then the individual's offspring suffer the effects.

Chemicals that cause harm to the unborn are called **teratogens**. Teratogens that affect development of human embryos in the womb can cause birth defects. One example involves the drug thalidomide, developed in the 1950s as a sleeping pill and to prevent nausea during pregnancy. Tragically, the drug turned out to be a powerful teratogen. Its use caused birth defects in thousands of babies, and its use by pregnant women was banned in the 1960s.

Other toxicants, known as **neurotoxins**, assault the nervous system. Neurotoxins include venoms produced by animals, heavy metals such as lead and mercury, and some pesticides. A famous case of neurotoxin poisoning occurred in Japan, where a chemical factory dumped mercury waste into Minamata Bay between the 1930s and 1960s. Thousands of people there ate fish contaminated with the mercury and soon began suffering from slurred speech, loss of muscle control, sudden fits of laughter, and in some cases death.

The human immune system protects our bodies from disease. Some toxic substances weaken the immune system, reducing the body's ability to defend itself against bacteria, viruses, allergy-causing agents, and other attackers. Others, called **allergens**, overactivate the immune system, causing an immune response when one is not necessary. One hypothesis for the increase in asthma in recent years is that allergenic synthetic chemicals are more prevalent in our environment. Allergens are not universally considered toxicants, however, because they affect some people but not others and because one's response does not necessarily correlate with the degree of exposure.

Pathway inhibitors are toxicants that interrupt vital biochemical processes in organisms by blocking one or more steps in important biochemical pathways. Rat poisons, for example, cause internal hemorrhaging in rodents by interfering with the biochemical pathways that create blood clotting proteins. Some herbicides, such as atrazine, kill plants by blocking steps in photosynthesis. Cyanide kills by interrupting chemical pathways that produce energy in mitochondria and depriving cells of energy.

Most recently, scientists have recognized **endocrine disruptors**, toxic substances that interfere with the endocrine system. The endocrine system consists of chemical messengers (hormones) that travel through the bloodstream at extremely low concentrations and have many vital functions. They stimulate growth, development, and sexual maturity, and they regulate brain function, appetite, sex drive, and many other aspects of our physiology and behavior. Some hormone-disrupting toxicants affect an animal's endocrine system by blocking the action of hormones or accelerating their breakdown. Others are so similar to certain hormones in their molecular structure and chemistry that they "mimic" the hormone by interacting with receptor molecules just as the actual hormone would (**FIGURE 10.6**). Bisphenol A is one of many chemicals that appear to mimic the female sex hormone estrogen and bind to estrogen receptors. Phthalates are another class of hormone-disrupting chemicals that are used widely in children's toys, perfumes and cosmetics, and other items. Health research on phthalates has linked them to birth defects, breast cancer, reduced sperm counts, and other re-

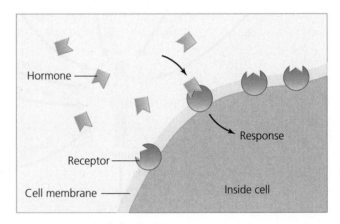

(a) Normal hormone binding

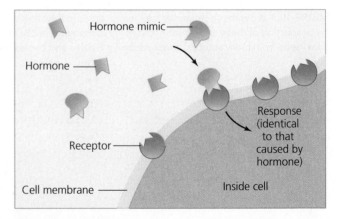

(b) Hormone mimicry

FIGURE 10.6 ▲ Many endocrine-disrupting substances mimic the structure of hormone molecules. Like a key similar enough to fit into another key's lock, the hormone mimic binds to a cellular receptor for the hormone, causing the cell to react as though it had encountered the hormone.

productive effects. Like bisphenol A, phthalates show how a substance can have multiple effects, by being a carcinogen, mutagen, and endocrine disruptor.

Organisms have natural defenses against toxic substances

Although synthetic toxicants are new, organisms have been exposed to natural toxicants for millions of years. Mercury, cadmium, arsenic, and other harmful substances are found naturally in the environment. Some organisms produce biological toxins to avoid predators or capture prey. Examples include venom in poisonous snakes, toxins in sea urchins, and the natural insecticide pyrethrin found in chrysanthemums. These exposures have provided selection pressure (pp. 46–48) for protection from toxins, and over time, organisms able to tolerate these harmful substances have gained an evolutionary advantage.

Barriers such as skin, scales, feathers, and fur are the first line of defense against toxic substances because they help the body to resist uptake from the surrounding environment. However, toxicants can circumvent these barriers and enter the body from vital activities such as eating, drinking, and breathing. Once in the organism, they are distributed widely by the circulatory and lymph systems in animals, and by the vascular system in plants.

Organisms possess biochemical pathways that use enzymes to detoxify harmful chemicals. Some pathways break down, or metabolize, toxic substances to render them inert. Other pathways make toxic substances water soluble so they are easier to excrete through the urinary system. In humans, many of these pathways are found in the liver, so this organ is disproportionately affected by intake of harmful substances such as excessive alcohol.

Some toxic substances cannot be effectively detoxified or made water soluble by detoxification enzymes. These chemicals are sequestered in fatty tissues and cell membranes to keep them away from vital organs. Heavy metals, dioxins, and some insecticides (including DDT) are stored in body tissue in this manner.

These defenses can protect organisms against low levels of some toxicants but can be overwhelmed if exposure exceeds critical levels. For other toxicants, harm occurs with any exposure if organisms have no defense against the substance. Defense mechanisms for natural toxins have evolved over millions of years. Organisms have not had long-term exposure to the synthetic chemicals that are so prevalent in today's environment, so the impacts of these toxic substances can be severe and unpredictable.

Individuals vary in their responses to hazards

Some of the defenses described above have a genetic basis. As a result, individuals may respond quite differently to identical exposures to hazards because they happen to have different combinations of genes. Poorer health also makes an individual more sensitive to biological and chemical hazards. Sensitivity also can vary with sex, age, and weight. Because of their smaller size and rapidly developing organ systems, younger organisms (for example, fetuses, infants, and young children) tend to be much more sensitive to toxicants than are adults. Regulatory agencies such as the U.S. Environmental Protection Agency (EPA) typically set human chemical exposure standards for adults and extrapolate downward for infants and children. However, many scientists contend that these linear extrapolations often do not offer adequate protection to fetuses, infants, and children.

FAQ

Q: Does exposure to a toxic substance cause genetic resistance to the substance?

A: When a population of organisms is exposed to a toxicant, such as a pesticide, a few individuals often survive while the vast majority of the population is killed. These individuals survive because they possess genes (which others in the population do not) that code for enzymes that counteract the toxic properties of the toxicant. Because the effects of these genes are only expressed when the pesticide is applied, many people think the toxicant "creates" detoxification genes by mutating the DNA of a small number of individuals. This is not the case. The genes for detoxifying enzymes were present in the DNA of resistant individuals from birth, but their effects were only seen when pesticide exposure caused selective pressure (pp. 46–48) for resistance to the toxic substance.

The type of exposure can affect the response

The risk posed by a hazard often varies according to whether a person experiences high exposure for short periods of time, known as **acute exposure**, or low exposure over long periods of time, known as **chronic exposure**. Incidences of acute exposure are easier to recognize because they often stem from discrete events, such as accidental ingestion, an oil spill, a chemical spill, or a nuclear accident. Toxicity tests in laboratories generally reflect effects of acute toxicity. However, chronic exposure is more common—and more difficult to detect and diagnose. Chronic exposure often affects organs gradually, as when smoking causes lung cancer, or when alcohol abuse leads to liver or kidney damage. Because of the long time periods involved, relationships between cause and effect may not be readily apparent.

TOXIC SUBSTANCES AND THEIR EFFECTS ON ECOSYSTEMS

When toxicants concentrate in environments and harm the health of many individuals, populations (p. 46) of the affected species become smaller. This decline can then affect other species. For instance, species that are prey of the affected organism could experience population growth due to lower levels of predation. Predators of the poisoned species, however, would decline as their food source became less abundant.

Testing the Safety of Bisphenol A

Dr. Patricia Hunt, Case Western Reserve University

Of the many studies documenting health impacts of bisphenol A on lab animals, one of the first came about because a lab assistant reached for the wrong soap.

At a laboratory at Case Western Reserve University in Ohio in 1998, geneticist Patricia Hunt was making a routine check of her female lab mice. As she extracted and examined developing eggs from the ovaries, she began to wonder what had gone wrong. About 40% of the eggs showed problems with their chromosomes, and 12% had irregular amounts of genetic material, a dangerous condition called aneuploidy, which can lead to miscarriages or birth defects in mice and people alike.

A bit of sleuthing revealed that a lab assistant had mistakenly washed the lab's plastic mouse cages and water bottles with an especially harsh soap. The soap damaged the cages so badly that parts of them seemed to have melted.

The cages were made from polycarbonate plastic, which contains bisphenol A (BPA). Hunt knew at the time that BPA mimics estrogen and that some studies had linked the chemical to reproductive abnormalities in mice, such as low sperm counts and early sexual development. Other research indicated that BPA leaches out of plastic into water and food when the plastic is treated with heat, acidity, or harsh soap.

Hunt wondered whether the chemical might be adversely affecting the mice in her lab. Deciding to re-create the accidental cage-washing incident in a controlled experiment, Hunt instructed researchers in her lab to wash polycarbonate cages and water bottles using varying levels of the harsh soap. They then compared mice kept in damaged cages with plastic water bottles to mice kept in undamaged cages with glass water bottles.

The developing eggs of mice exposed to BPA through the deliberately damaged plastic showed significant problems during meiosis, the division of chromosomes during egg formation— just as they had in the original incident (**first figure**). In contrast, the eggs of mice in the control cages were normal.

In another round of tests, Hunt's team gave sets of female mice daily oral doses of BPA over 3, 5, and 7 days. They observed the same meiotic abnormalities in these mice, although at lower levels (**second figure**). The mice given BPA for 7 days were most severely affected.

Published in 2003 in the journal *Current Biology*, Hunt's findings set off a new wave of concern over the safety of bisphenol A. The findings were disturbing because sex cells of mice and of people divide and function in similar ways. "We have observed meiotic defects in mice at exposure levels close to or even below those considered 'safe' for humans," the research paper stated. "Clearly, the possibility that BPA exposure increases the likelihood of genetically abnormal offspring is too serious to be dismissed without extensive further study."

Since that time, dozens of other studies of BPA at low doses have documented harmful effects in lab animals,

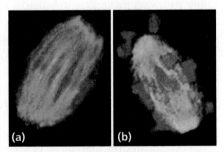

In normal cell division (a), chromosomes (red) align properly. Exposure to bisphenol A causes abnormal cell division (b), whereby chromosomes scatter and are distributed improperly and unevenly between daughter cells.

Cascading impacts can cause changes in the composition of the biological community (p. 68) and threaten ecosystem functioning. There are many ways toxicants can concentrate and persist in ecosystems and affect ecosystem services.

Airborne substances can travel widely

Toxic substances are released around the world from agricultural, industrial, and domestic activities and can sometimes be redistributed far from their emission site. Many chemical substances can be transported by air and exert impacts on ecosystems far from the site of their origin.

Because so many substances are carried by the wind, synthetic chemicals are ubiquitous worldwide, even in seemingly pristine areas. Earth's polar regions are particularly contaminated because natural patterns of global atmospheric circulation (p. 281) tend to move airborne chemicals toward the poles (**FIGURE 10.7**). Thus, although we manufacture and apply synthetic substances mainly in temperate and tropical regions, contaminants are strikingly concentrated in the tissues of Arctic polar bears, Antarctic penguins, and people living in Greenland. Polychlorinated biphenyls (PCBs), which are by-products of chemicals used in transformers and other electrical equipment, are one such example.

214

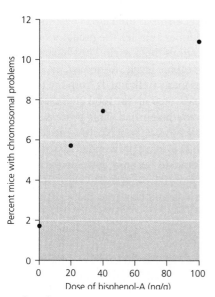

In this dose-response experiment, the percentage of mice showing chromosomal problems during cell division rose with increasing dose of bisphenol A. In the United States and Europe, regulators have set safe intake levels for people at doses of 50 ng/g of body weight per day. Data from Hunt, P.A., et al., 2003. Bisphenol A exposure causes meiotic aneuploidy in the female mouse. *Current Biology* 13: 546–553.

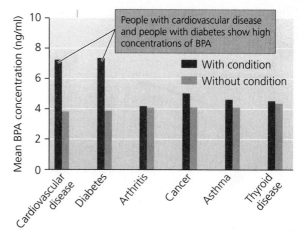

In a 2008 epidemiology study, average bisphenol A concentrations were significantly higher for Americans with diabetes and cardiovascular disease, but not for those with various other conditions. Error bars are 95% confidence intervals. Adapted from Lang, I.A., et al., 2008. Association of urinary bisphenol A concentration with medical disorders and laboratory abnormalities in adults. *JAMA* 300: 1303–1310.

In 2008, the *Journal of the American Medical Association* published research led by Iain Lang and David Melzer of the Peninsula Medical School, Exeter, U.K. Lang and Melzer's team took an epidemiological approach (p. 218) to assess BPA's possible effects on people, using data from the U.S. government's latest National Health and Nutrition Examination Survey. Using data from 1,455 survey participants, they attained a representative sample of adults in the U.S. population. After controlling the data for race/ethnicity, education, income, smoking, body mass, and other variables, they tested for statistical correlations between a series of major health disorders and the concentration of BPA in people's urine.

These researchers' analyses showed that Americans with high urine BPA concentrations showed high rates of diabetes and cardiovascular disease (**third figure**), as well as abnormal concentrations of three liver enzymes. The team found no association with a number of other conditions such as cancer, stroke, arthritis, thyroid disease, and respiratory diseases. The researchers also explored correlations with other estrogenic compounds and found that these did not show the associations that BPA showed.

Previous studies of the mechanisms by which BPA acts in cell cultures and in rodents' bodies helped explain how and why BPA might affect liver enzymes and diabetes. However, the reasons for cardiovascular effects remain unclear.

This first direct indication of human health impacts from BPA was a correlative study that does not establish causation. To demonstrate that BPA actually causes the observed effects, researchers would need to track people with low and high BPA levels for years, predict who would most likely get sick, and test these predictions with future data. It will take many years to complete such long-term studies. In the meantime, more and more scientists are urging regulators to restrict BPA based on the evidence already at hand.

including not only reproductive disorders related to estrogen mimicry, but also other maladies ranging from thyroid problems to liver damage to obesity. Scientist Frederick vom Saal, whose research in 1997 had shown the first evidence for BPA's effects, said in 2007, "This chemical is harming snails, insects, lobsters, fish, frogs, reptiles, birds, and rats, and the chemical industry is telling people that because you're human, unless there's human data, you can feel completely safe."

Vom Saal did not have to wait long for the first human study to appear.

Effects can also occur over relatively shorter distances. Pesticides can be carried by air currents to sites far away from agricultural fields in a process called *pesticide drift*. Frogs in the mountains of the Sierra Nevada, for example, have experienced population declines associated with pesticide drift from agriculture in California's nearby Central Valley region.

Some toxicants persist

Once a toxic substance arrives somewhere, it may degrade quickly and become harmless, or it may remain unaltered and persist for many months, years, or decades. The rate at which a given substance degrades depends on its chemistry and on factors such as temperature, moisture, and sun exposure. The *Bt* toxin (p. 150) used in biocontrol and genetically modified crops has a very short persistence time, whereas chemicals such as DDT and PCBs persist for decades.

Persistent synthetic chemicals exist in our environment today because we have designed them to persist. The synthetic chemicals used in plastics, for instance, are used precisely because they resist breakdown. Sooner or later, however, most toxic substances degrade into simpler compounds called breakdown products. Often these are less harmful than the original substance, but sometimes they are just as toxic as

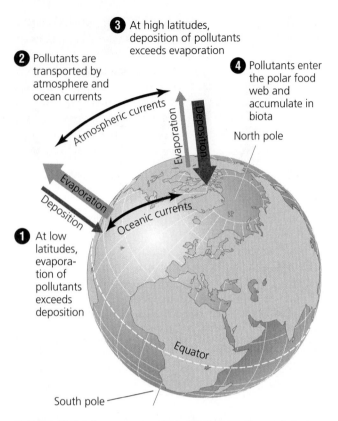

❸ At high latitudes, deposition of pollutants exceeds evaporation

❷ Pollutants are transported by atmosphere and ocean currents

❹ Pollutants enter the polar food web and accumulate in biota

North pole

Atmospheric currents

Evaporation

Deposition

Deposition

Evaporation

Oceanic currents

❶ At low latitudes, evaporation of pollutants exceeds deposition

Equator

South pole

FIGURE 10.7 ▲ In a process called global distillation, pollutants that evaporate and rise high into the atmosphere at lower latitudes, or are deposited in the ocean, are carried toward the poles by atmospheric currents of air and oceanic currents of water. This process concentrates pollutants near the poles.

the original chemical, or more so. For instance, DDT breaks down into DDE, a highly persistent and toxic compound in its own right.

Toxicants may concentrate in water

Toxic substances are not evenly distributed in the environment, and they move about in specific ways. Runoff in watersheds (p. 252) concentrates contaminants in small volumes of surface water. Traces of toxins, pharmaceuticals, and detoxification products excreted by people can enter waters from wastewater treatment plants. Many chemicals are soluble in water and enter organisms' tissues through drinking or absorption. For this reason, aquatic animals such as fish, frogs, and stream invertebrates are effective indicators of pollution. The contaminants that wash into streams and rivers also flow and seep into the water we drink and drift in the air we breathe. Once concentrated in waters, toxicants can move long distances through aquatic systems (pp. 250–251) and affect a diversity of ecosystems.

Toxic substances may accumulate and move up the food chain

Fat- and oil-soluble toxicants accumulate in fatty tissues in a process termed **bioaccumulation**, which results in the animal's tissues having a greater concentration of the substance than exists in the surrounding environment. Toxic substances that bioaccumulate in an organism's tissues may

be transferred to other organisms as predators consume prey. When one organism consumes another, the predator takes in any stored toxicants and stores them itself. Thus bioaccumulation takes place on all trophic levels. Moreover, each individual predator consumes many individuals from the trophic level beneath it, so with each step up the food chain, concentrations of toxicants become magnified. This process, called **biomagnification**, occurred throughout North America with DDT. Top predators, such as birds of prey, ended up with high concentrations of the pesticide because concentrations became magnified as DDT moved from water to algae to plankton to small fish to larger fish and finally to fish-eating birds (**FIGURE 10.8**).

Biomagnification of DDT caused populations of many North American birds of prey to decline precipitously from the 1950s to the 1970s. The peregrine falcon was almost totally wiped out in the eastern United States, and the bald eagle,

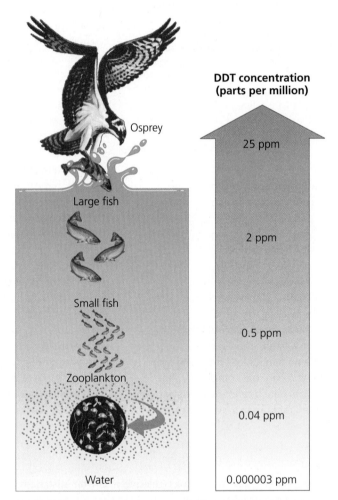

DDT concentration (parts per million)

Osprey

25 ppm

Large fish

2 ppm

Small fish

0.5 ppm

Zooplankton

0.04 ppm

Water

0.000003 ppm

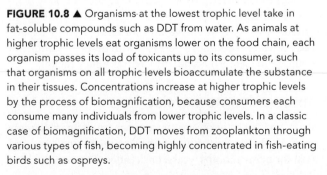

FIGURE 10.8 ▲ Organisms at the lowest trophic level take in fat-soluble compounds such as DDT from water. As animals at higher trophic levels eat organisms lower on the food chain, each organism passes its load of toxicants up to its consumer, such that organisms on all trophic levels bioaccumulate the substance in their tissues. Concentrations increase at higher trophic levels by the process of biomagnification, because consumers each consume many individuals from lower trophic levels. In a classic case of biomagnification, DDT moves from zooplankton through various types of fish, becoming highly concentrated in fish-eating birds such as ospreys.

the U.S. national bird, was virtually eliminated from the lower 48 states. Eventually scientists determined that DDT was causing these birds' eggshells to grow thinner, so that eggs were breaking in the nest and killing the embryos within. In a remarkable environmental success story, populations of all these birds have rebounded (p. 176) since the United States banned DDT.

Effects from biomagnification still persist, though. Mercury bioaccumulates in some commercially important fish species, such as tuna. Polar bears of Svalbard Island in arctic Norway show extremely high levels of PCB contamination from biomagnification. Polar bear cubs suffer immune suppression, hormone disruption, and high mortality—and because the cubs receive PCBs in their mothers' milk, contamination persists and accumulates across generations.

In all these cases, biomagnification affects ecosystem composition and functioning. When populations of top predators such as eagles and polar bears are reduced, species interactions (pp. 65–68) change, and effects cascade through food webs (pp. 70–71).

Toxic substances can threaten ecosystem services

Toxicants can alter the biological composition of ecosystems and the manner in which organisms interact with one another and their environment. In so doing, harmful compounds can threaten the ecosystem services (pp. 2, 36) provided by nature. For example, pesticide exposure has been implicated as a factor in the recent declines in honeybee populations (p. 149). Honeybees pollinate over 100 economically important crops, and reduced pollination by wild bees has increased costs for farmers by forcing them to hire professional beekeepers to pollinate their crops.

Healthy, functioning ecosystems provide the service of nutrient cycling. Decomposers and detritivores in the soil (p. 68) break down organic matter and replenish soils with nutrients for plants to utilize. When soils are exposed to pesticides or antifungal agents, the nutrient cycling rates are altered. This affects the quantity of nutrients available to producers, affects their growth, and produces effects throughout the ecosystem.

STUDYING EFFECTS OF HAZARDS

Determining the effects of particular environmental hazards on individuals and ecosystems is a challenging job, and scientists rely on several different methods to do this, ranging from correlative surveys to manipulative experiments (pp. 9–10).

Wildlife studies integrate work in the field and lab

Scientists study the impacts of environmental hazards on wild animals to help conserve animal populations and also to understand potential risks to people. Often wildlife toxicologists work in the field with animals to take measurements, document patterns, and generate hypotheses before heading

to the laboratory to run controlled manipulative experiments to test their hypotheses. For instance, biologist Louis Guillette and his collaborators discovered that alligators in lakes in Florida receiving agricultural runoff had higher rates of reproductive problems than alligators in less polluted lakes (**FIGURE 10.9A**). Based on these field studies, he hypothesized that chemical contaminants were disrupting the endocrine systems of alligators during their development in the egg. Subsequent laboratory studies showed that contaminants

(a) Louis Guillette taking blood sample from alligator

(b) Tyrone Hayes in lab with frog

FIGURE 10.9 ▲ Researchers Louis Guillette **(a)** and Tyrone Hayes **(b)** found that alligators and frogs, respectively, show reproductive abnormalities that they attribute to endocrine disruption by pesticides.

found in alligator eggs, including the herbicide atrazine, could bind to receptors for estrogen or produce an enzyme that converts testosterone to estrogen.

Building on Guillette's work, researcher Tyrone Hayes (**FIGURE 10.9B**) found in lab experiments that male frogs raised in water containing very low doses of atrazine became feminized and hermaphroditic, developing both testes and ovaries. Field surveys then showed that leopard frogs across North America experienced hormonal problems in areas of heavy atrazine usage. These studies indicated that atrazine, which kills plants by blocking steps in photosynthesis, can also act as an endocrine disruptor in animals.

Human studies rely on case histories, epidemiology, and animal testing

Environmental toxicologists also conduct **epidemiological studies**, large-scale comparisons among groups of people, usually contrasting a group known to have been exposed to some hazard and a group that has not. Epidemiologists track the fate of all people in the study for a long period of time (often years or decades) and measure the rate at which deaths, cancers, or other health problems occur in each group. The epidemiologist then analyzes the data, looking for observable differences between the groups, and statistically tests hypotheses accounting for differences. When a group exposed to a hazard shows a significantly greater degree of harm, it suggests that the hazard may be responsible. The epidemiological process is akin to a natural experiment (p. 10), in which an event creates groups of subjects that researchers can study (for example, people exposed to carcinogenic compounds in their drinking water versus those not similarly exposed).

Epidemiological studies measure a statistical association between a health hazard and an effect, but they do not confirm that the hazard causes the effect. To establish causation, manipulative experiments are needed. However, subjecting people to massive doses of toxic substances in a lab experiment would clearly be unethical. This is why researchers have traditionally used laboratory strains of rats, mice, and other mammals. Experimenting with these creatures elicits fewer ethical objections in society, and because of shared evolutionary history, substances that harm mice and rats are reasonably likely to harm us.

Dose-response analysis is a mainstay of toxicology

The standard method of testing with lab animals in toxicology is **dose-response analysis**. Scientists quantify the toxicity of a substance by measuring the strength of its effects or the number of animals affected at different doses. The **dose** is the amount of substance the test animal receives, and the **response** is the type or magnitude of negative effects the animal exhibits as a result. The response is generally quantified by measuring the proportion of animals exhibiting negative effects. The data are plotted on a graph, with dose on the x axis and response on the y axis (**FIGURE 10.10A**). The resulting curve is called a **dose-response curve**.

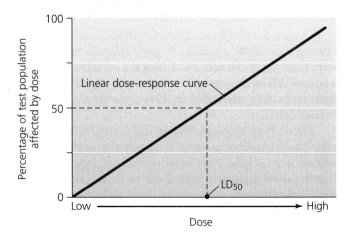

(a) Linear dose-response curve

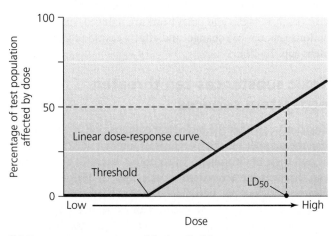

(b) Dose-response curve with threshold

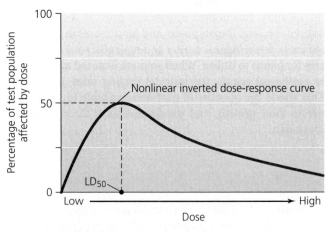

(c) Unconventional dose-response curve

FIGURE 10.10 ▲ In a classic linear dose-response curve **(a)**, the percentage of animals killed or otherwise affected by a substance rises with the dose. The point at which 50% of the animals are killed is labeled the lethal-dose-50, or LD_{50}. For some toxic substances, a threshold dose **(b)** exists, below which doses have no measurable effect. Some substances, in particular endocrine disruptors, show unconventional, nonlinear dose-response curves **(c)** that are U-shaped, J-shaped, or inverted. These curves show that organisms' responses to toxicants may sometimes be complex.

Once they have plotted a dose-response curve, toxicologists can calculate a convenient shorthand gauge of a substance's toxicity: the amount of the substance it takes to kill half the population of study animals used. This lethal dose for 50% of individuals is termed the **LD_{50}**. A high LD_{50} indicates low toxicity for a substance, and a low LD_{50} indicates high toxicity.

If the experimenter is interested in nonlethal health effects, he or she may want to document the level of toxicant at which 50% of a population of test animals is affected in some other way (for instance, what level of toxicant causes 50% of lab mice to lose their hair?). Such a level is called the effective-dose-50%, or **ED_{50}**.

Some substances can elicit effects at any concentration, but for others, responses may occur only above a certain dose, or threshold. Such a **threshold** dose (**FIGURE 10.10B**) might be expected if the body's organs can fully metabolize or excrete a toxicant at low doses but become overwhelmed at higher concentrations. It might also occur if cells can repair damage to their DNA only up to a certain point.

Sometimes a response may *decrease* as a dose increases. Toxicologists are finding that some dose-response curves are U-shaped, J-shaped, or shaped like an inverted U (**FIGURE 10.10C**). Such counterintuitive curves contradict toxicology's traditional assumption that "the dose makes the poison." These unconventional dose-response curves often occur with endocrine disruptors, likely because the hormone system is geared to respond to minute concentrations of substances (normally, hormones in the bloodstream). Because the endocrine system responds to minuscule amounts of chemicals, it may be vulnerable to disruption by contaminants that reach our bodies in very low concentrations. In research with bisphenol A, a number of studies with lab animals have found unconventional dose-response curves.

The shape of dose-response curves is important because estimating effects on humans often requires extrapolation—extending the dose-response curves beyond the doses tested with laboratory animals. Because these extrapolations stretch beyond the actual data obtained, they introduce uncertainty into the interpretation of what doses are safe for people. As a result, to be on the safe side, regulatory agencies set standards for maximum allowable levels of toxic substances that are well below the minimum toxicity levels estimated from lab studies.

Mixes may be more than the sum of their parts

It is difficult enough to determine the impact of a single hazard, but the task becomes astronomically more difficult when multiple hazards interact. Chemical substances, when mixed, may act together in ways that cannot be predicted from the effects of each in isolation. Mixed toxicants may sum each other's effects, cancel out each other's effects, or multiply each other's effects. Interactive impacts that are greater than the simple sum of their constituent effects are called **synergistic effects**.

With Florida's alligators, lab experiments have indicated that the DDT breakdown product DDE can either help cause or inhibit sex reversal, depending on the presence of other chemicals. Mice exposed to a mixture of nitrate, atrazine, and aldicarb have been found to show immune, hormone, and nervous system effects that were not evident from exposure to each of these chemicals alone.

Traditionally, environmental health has tackled effects of single hazards one at a time. In toxicology, the complex experimental designs required to test interactions, and the sheer number of chemical combinations, have meant that single-substance tests have received priority. This approach is changing, but scientists in environmental health and toxicology will never be able to test all possible combinations.

Endocrine disruption poses challenges for toxicology

Unconventional dose-response curves are presenting challenges for scientists studying toxic substances and for policymakers trying to set safety standards for them. Because so many novel synthetic chemicals exist in very low concentrations over wide areas, many scientists suspect that we may have underestimated the dangers of compounds that exert impacts at low concentrations.

Scientists first noted endocrine-disrupting effects decades ago, but the idea that synthetic chemicals might be altering the hormones of animals was not widely appreciated until the 1996 book *Our Stolen Future*, by Theo Colburn, Dianne Dumanoski, and J.P. Myers. Like *Silent Spring*, this book integrated scientific work from various fields and presented a unified picture that shocked many readers—and brought criticism from some scientists and from the chemical industry.

Today, thousands of studies have linked hundreds of substances to effects on reproduction, development, immune function, brain and nervous system function, and other hormone-driven processes. Evidence is strongest so far in nonhuman animals, but many studies suggest impacts on people. Some researchers suggest that endocrine disruptors may account for rising rates of testicular cancer, undescended testicles, and genital birth defects in males. Others argue that the sharp rise in breast cancer rates (one in eight U.S. women today develops breast cancer) may be due to hormone disruption, because an excess of estrogen appears to feed tumor development in older women. Still other scientists attribute male reproductive problems to elevated BPA exposure. For example, a 2009 study determined that workers in Chinese factories that manufactured BPA had four times the rate of erectile dysfunction as workers in factories where BPA wasn't present. A follow-up study in 2010 found that workers with detectable levels of BPA in their urine were 2–4 times more likely to have reduced sperm counts and poorer sperm quality than workers in which no BPA was detected. While the BPA exposure in these workers was far higher than that experienced by the average American, they represent some of the first studies to link bisphenol A exposure to reproductive abnormalities in humans.

Research on hormone disruption has brought about strident debate. This is partly because of the scientific uncertainty inherent in any emerging field of study, but also because of the economic value of the chemicals being tested. The chemical industry has generated research showing that bisphenol A does not produce health effects, and regulatory agencies such as the FDA have relied primarily on this research in vouching for the chemical's safety. However, independent academic scientists unaffiliated with industry are reaching different conclusions. By one count through the end of 2006, 151 of the 178 published studies with lab animals reported harm from low doses of bisphenol A. Almost without exception, the studies reporting harm received public government funding, whereas those reporting no harm were funded by industry.

RISK ASSESSMENT AND RISK MANAGEMENT

Policy decisions on whether to ban chemicals or restrict their use generally follow years of rigorous testing for toxicity. Likewise, strategies for combating disease and other health threats are based on extensive scientific research. However, policy and management decisions also incorporate economics and ethics, and all too often these aspects are influenced by political pressure from powerful interests. The steps between the collection and interpretation of scientific data and the formulation of policy involve assessing and managing risk.

We express risk in terms of probability

Exposure to an environmental health threat does not invariably produce a given effect. Rather, it causes some probability of harm, a statistical chance that damage will result. To understand a health threat, a scientist must know more than just its identity and strength. He or she must also know the chance that one will encounter it, the frequency with which one may encounter it, the amount of substance or degree of threat to which one is exposed, and

one's sensitivity to the threat. Such factors help determine the overall risk posed.

Risk can be measured in terms of *probability*, a quantitative description of the likelihood of a certain outcome. The probability that some harmful outcome (for instance, injury, death, environmental damage, or economic loss) will result from a given action, event, or substance expresses the risk posed by that phenomenon.

Our perception of risk may not match reality

Every action we take and every decision we make involves some element of risk, some (generally small) probability that things will go wrong. We try in everyday life to behave in ways that minimize risk, but our perceptions of risk do not always match statistical reality (**FIGURE 10.11**). People often worry unduly about smaller risks yet readily engage in other activities that pose higher risks. For instance, most people perceive flying in an airplane as a riskier activity than driving a car, but according to a report by the National Safety Council, a person's chance of dying from an automobile accident in 2010 is 69 times higher than from an airplane crash. Psychologists agree that this difference between perception and reality stems from the fact that we feel more at risk when we are not controlling a situation and more safe when we are "at the wheel"—regardless of the actual risk involved.

This psychology may help account for people's anxiety over exposure to BPA, nuclear power, toxic waste, and pesticide residues on foods—environmental hazards that are invisible or little understood and whose presence in our lives is largely outside our personal control. In contrast, people are more ready to accept and ignore the risks of smoking cigarettes, overeating, and not exercising—voluntary activities statistically shown to pose far greater risks to health.

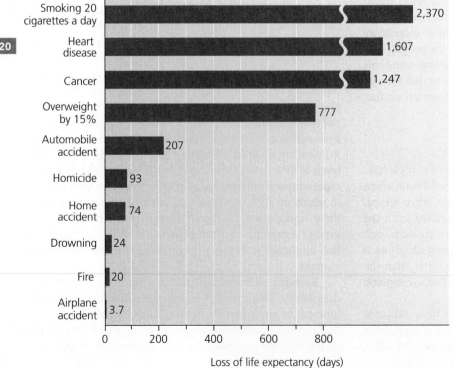

Smoking 20 cigarettes a day — 2,370
Heart disease — 1,607
Cancer — 1,247
Overweight by 15% — 777
Automobile accident — 207
Homicide — 93
Home accident — 74
Drowning — 24
Fire — 20
Airplane accident — 3.7

Loss of life expectancy (days)

FIGURE 10.11 ◀ Our perceptions of risk do not always match the reality of risk. Listed here are several leading causes of death in the United States, along with a measure of the risk each poses. Risk is measured in days of lost life expectancy, that is, the number of days of life lost by people suffering the hazard, spread across the entire population— a measure commonly used by insurance companies. By this measure, one common source of anxiety, airplane accidents, poses 20 times less risk than home accidents, over 50 times less risk than auto accidents, and over 200 times less risk than being overweight. Data from Cohen, B., 1991. Catalog of risks extended and updated. *Health Physics* 61: 317–335.

Risk assessment analyzes risk quantitatively

The quantitative measurement of risk and the comparison of risks involved in different activities or substances together are termed **risk assessment**. Risk assessment is a way to identify and outline problems. In environmental health, it helps ascertain which substances and activities pose health threats to people or wildlife and which are largely safe.

Assessing risk for a chemical substance involves several steps. The first steps involve the scientific study of toxicity we examined above—determining whether a substance has toxic effects and, through dose-response analysis, measuring how effects vary with the degree of exposure. Subsequent steps involve assessing the individual's or population's likely extent of exposure to the substance, including the frequency of contact, the concentrations likely encountered, and the length of encounter. As discussed in the central case, risk assessments for bisphenol A by expert panels have often arrived at divergent conclusions.

Risk management combines science and other social factors

Accurate risk assessment is a vital step toward effective **risk management**, which consists of decisions and strategies to minimize risk (**FIGURE 10.12**). In most nations, risk management is handled largely by federal agencies. In the United States, these agencies include the Environmental Protection Agency (EPA), the Centers for Disease Control and Prevention (CDC), and the Food and Drug Administration (FDA). In risk management, scientific assessments of risk are considered in light of economic, social, and political needs and values. Risk managers assess costs and benefits of addressing risk in various ways, with regard to both scientific and non-scientific concerns, before making decisions on whether and how to reduce or eliminate risk.

In environmental health and toxicology, comparing costs and benefits (p. 90) can be difficult because the benefits are often economic, whereas the costs often pertain to health. Moreover, economic benefits are generally known, easily quantified, and of a discrete and stable amount, whereas health risks are hard-to-measure probabilities, often involving a small percentage of people likely to suffer greatly and a large majority likely to experience little effect. When a government agency bans a pesticide, it may mean considerable economic loss for the manufacturer and potential economic loss for the farmer, whereas the benefits accrue less predictably over the long term through healthier people, lower health care costs, and increased worker productivity. Because of the lack of equivalence in the way costs and benefits are measured, risk management frequently tends to stir up debate.

In the case of bisphenol A, eliminating plastic linings in our food and drink cans could do more harm than good, because the linings help prevent metal corrosion and the contamination of food by pathogens. Alternative substances exist for most of BPA's uses, but replacing BPA with alternatives will entail economic costs to industry, and these costs get passed on to consumers in the prices of products. Such complex considerations can make risk management decisions difficult even if the science of risk assessment is fairly clear. This difficulty may help account for the hesitancy of U.S. regulatory agencies to issue restrictions on BPA so far. As of 2011, both the FDA and the EPA were continuing to review options for managing risk from BPA.

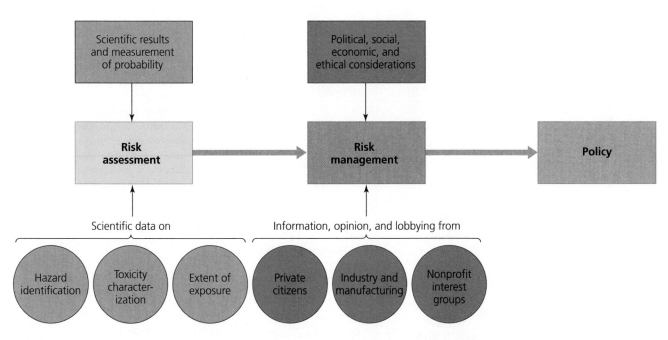

FIGURE 10.12 ▲ The first step in addressing the risk of an environmental hazard is risk assessment, a process of quantifying the risk of the hazard and comparing it to other risks. Once science identifies and measures risks, then risk management can proceed. In this process, economic, political, social, and ethical issues are considered in light of the scientific data from risk assessment. The consideration of all these types of information is intended to result in policy decisions that minimize the risk of the environmental hazard.

Two approaches exist for determining safety

Because we cannot know a substance's toxicity until we measure and test it, and because there are so many untested chemicals and combinations, science will never eliminate the many uncertainties that accompany risk assessment. In such a world of uncertainty, there are two basic philosophical approaches to categorizing substances as safe or dangerous (**FIGURE 10.13**).

One approach is to assume that substances are harmless until shown to be harmful. We might nickname this the "innocent-until-proven-guilty" approach. Because thoroughly testing every existing substance (and combination of substances) for its effects is a hopelessly long, complicated, and expensive pursuit, the innocent-until-proven-guilty approach has the virtue of facilitating technological innovation and economic activity. However, it has the disadvantage of putting into wide use some substances that may later turn out to be dangerous.

The other approach is to assume that substances are harmful until shown to be harmless. This approach follows the *precautionary principle* (p. 152). This more cautious approach should enable us to identify troublesome toxicants before they are released into the environment, but it may also impede the pace of technological and economic advance.

These two approaches are actually two ends of a continuum of possible approaches. The two endpoints differ mainly in where they lay the burden of proof—specifically, whether product manufacturers are required to prove a product is safe or whether government, scientists, or citizens are required to prove a product is dangerous.

WEIGHING THE ISSUES

The Precautionary Principle Industry's critics say chemical manufacturers should bear the burden of proof for the safety of their products before they hit the market. Industry's supporters say that mandating more safety research will hamper the introduction of products that consumers want and increase the price of products. What do you think? Should government follow the precautionary principle and require proof of safety prior to a chemical's introduction into the market?

Philosophical approaches are reflected in policy

The choice of philosophical approach has direct implications for policy, and nations vary in how they blend the two approaches. European nations have recently embarked on a

222

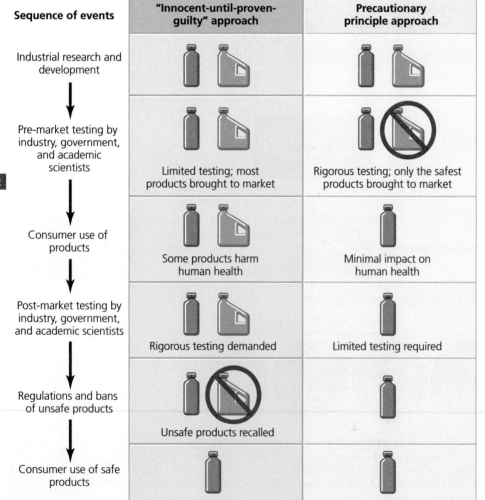

FIGURE 10.13 ◀ Two main approaches can be taken to introduce new substances to the market. In one approach, substances are "innocent until proven guilty"; they are brought to market relatively quickly after limited testing. Products reach consumers more quickly, but some fraction of them (blue bottle in diagram) may cause harm to some fraction of people. The other approach is to adopt the precautionary principle, bringing substances to market cautiously, only after extensive testing. Products that reach the market should be safe, but many perfectly safe products (purple bottle in diagram) will be delayed in reaching consumers.

policy course that incorporates the precautionary principle, whereas the United States largely follows an innocent-until-proven-guilty approach.

In the United States, several federal agencies are assigned responsibility for tracking and regulating synthetic chemicals under various legislative acts. The FDA, under an act first passed in 1938, monitors foods and food additives, cosmetics, drugs, and medical devices. The EPA regulates pesticides under a 1947 act and its amendments. The Occupational Safety and Health Administration (OSHA) regulates workplace hazards under a 1970 act. Several other agencies regulate other substances. Synthetic chemicals not covered by other laws are regulated by the EPA under the 1976 Toxic Substances Control Act (TSCA).

EPA regulation is only partly effective

The **Toxic Substances Control Act (TSCA)** directs the EPA to monitor the roughly 83,000 industrial chemicals manufactured in or imported into the United States, ranging from PCBs to lead to bisphenol A. The act gives the agency power to regulate these substances and ban them if they are found to pose excessive risk.

However, many public health advocates view TSCA as being far too weak. They note that the screening required of industry is minimal and that to mandate more extensive and meaningful testing, the EPA must show proof of the chemical's toxicity. In other words, the agency is trapped in a Catch-22: To push for studies looking for toxicity, it must have proof of toxicity already. The result is that most synthetic chemicals are not thoroughly tested before being brought to market. Of those that fall under TSCA, only 10% have been thoroughly tested for toxicity; only 2% have been screened for carcinogenicity, mutagenicity, or teratogenicity; fewer than 1% are regulated; and almost none have been tested for endocrine, nervous, or immune system damage, according to the U.S. National Academy of Sciences. As a result, products

marketed in the United States are not guaranteed to be free of toxic substances, so consumer choice is especially important.

Toxicants are regulated internationally

The European Union is taking the world's boldest step toward testing and regulating manufactured chemicals. In 2007, the EU's **REACH** program went into effect (*REACH* stands for *R*egistration, *E*valuation, *A*uthorization, and restriction of *CH*emicals). REACH largely shifts the burden of proof for testing chemical safety from national governments to industry and requires that chemical substances produced or imported in amounts of over 1 metric ton per year be registered with a new European Chemicals Agency. It is expected that REACH will require the registration of about 30,000 substances. In an impacts assessment in 2003, EU commissioners estimated that REACH will cost the chemical industry and chemical users 2.8–5.2 billion euros (US $3.8–7.0 billion) over 11 years but that the health benefits to the public would be roughly 50 billion euros (US $67 billion) over 30 years. Changes in the program since then have made the predicted cost-to-benefit ratio even better.

The world's nations have also sought to address chemical pollution with international treaties. The *Stockholm Convention on Persistent Organic Pollutants* (*POPs*) came into force in 2004 and has been ratified by over 150 nations. POPs are toxic chemicals that persist in the environment, bioaccumulate and biomagnify up the food chain, and often can travel long distances. The PCBs and other contaminants found in polar bears are a prime example. Because contaminants often cross international boundaries, an international treaty seemed the best way to deal fairly with such transboundary pollution. The Stockholm Convention aims first to end the use and release of 12 POPs shown to be most dangerous, a group nicknamed the "dirty dozen." It sets guidelines for phasing out these chemicals and encourages transition to safer alternatives.

► CONCLUSION

International agreements such as REACH and the Stockholm Convention represent a sign that governments may act to protect the world's people, wildlife, and ecosystems from toxic substances and other environmental hazards. At the same time, solutions often come more easily when they do not arise from government regulation alone. Consumer choice exercised through the market can often be an effective way to influence industry's decision making. Consumers of products, from plastics to pesticides to cosmetics to kids' toys, can make decisions that influence industry when they have full information from scientific research regarding the risks involved. Once scientific results are in, a society's philosophical approach to risk management will determine what policy decisions are made.

Whether the burden of proof is laid at the door of industry or of government, we will never attain complete scientific knowledge of any risk. Rather, we must make choices based on the information available. Synthetic chemicals have brought us innumerable modern conveniences, a larger food supply, and medical advances that save and extend human lives. Human society would be very different without them. Yet a better future, one that safeguards the well-being of both people and the environment, depends on knowing the risks that some hazards pose and on having in place the means to phase out harmful substances and replace them with safer ones.

TESTING YOUR COMPREHENSION

1. What are the four major types of environmental health hazards?

2. In what way is disease the greatest hazard that people face? What kinds of interrelationships must environmental

health experts study to learn how diseases affect human health?

3. Where does most exposure to lead, asbestos, radon, and PBDEs occur?

4. List and describe the general categories of toxic substances described in this chapter.

5. Explain the mechanisms found in organisms that protect them from damage from toxic substances.

6. How do toxic substances travel through the environment, and where are they most likely to be found? Describe and contrast the processes of bioaccumulation and biomagnification.

7. What are epidemiological studies, and how are they most often conducted?

8. Explain the dose-response curve. Why is a substance with a high LD_{50} considered safer than one with a low LD_{50}?

9. What factors may affect an individual's response to a toxic substance? Why is chronic exposure to toxic agents often more difficult to measure and diagnose than acute exposure?

10. How do scientists identify and assess risks from substances or activities?

SEEKING SOLUTIONS

1. Describe some environmental health hazards that you think you may be living with indoors. How do you think you may have been affected by indoor or outdoor hazards in the past? How could you best deal with these hazards in the future?

2. Do you feel that laboratory animals should be used in experiments in toxicology? Why or why not?

3. Describe differences in the policies of the United States and the European Union toward the study and management of the risks of synthetic chemicals. Which do you believe are better, the policies of the United States or those of the European Union? Why?

4. **THINK IT THROUGH** You are the parent of two young children, and you want to minimize the environmental health risks your kids are exposed to. Name five steps that you could take in your household and in your daily life that would accomplish your goal.

5. **THINK IT THROUGH** You work for a public health organization and have been asked to educate the public about bisphenol A and to suggest ways to minimize exposure to the chemical. You begin by examining your lifestyle and finding ways to use alternatives to BPA-containing products. Create a list of five ways you are exposed daily to bisphenol A, and then list approaches that would avoid or minimize these exposures. Do these steps require more time and/or money? What are some costs of embracing these changes? What would you tell an interested person about bisphenol A as it relates to human health?

CALCULATING ECOLOGICAL FOOTPRINTS

In 2001, the last year the EPA gathered and reported data on pesticide use, Americans used 1.20 billion pounds of pesticide active ingredients, and world pesticide use totaled 5.05 billion pounds of active ingredients. In that same year, the U.S. population was 285 million, and the world's population totaled 6.16 billion. Pesticides include hundreds of chemicals used by farmers, governments, industries, and individuals to control "pest" organisms. In the table, calculate your share of pesticide use as a U.S. citizen in 2001 and the amount used by (or on behalf of) the average citizen of the world.

	Annual pesticide use (pounds of active ingredients)
You	
Your class	
Your state	
United States	1.20 billion
World (total)	5.05 billion
World (per capita)	

1. What is the ratio of your annual pesticide use to the world's per capita average?

2. In 2007, the average U.S. citizen had an ecological footprint of 8.0 hectares and the average world citizen's footprint was 2.7 hectares (Chapter 1). Compare the ratio of pesticide usage with the ratio of the overall ecological footprints. How do these differ, and how would you account for the difference?

3. Does the per capita pesticide use for you as a U.S. citizen seem reasonable for you personally? Why or why not? Do you find this figure alarming, or of little concern? What else would you like to know to assess the risk associated with this level of pesticide use?

Mastering**ENVIRONMENTALSCIENCE**™

Go to **www.masteringenvironmentalscience.com** for homework assignments, practice quizzes, Pearson eText, and more.

11 Geology, Minerals, and Mining

Upon completing this chapter, you will be able to:

➤ Describe Earth's internal structure and explain how plate tectonics shapes its surface

➤ Identify the categories of rocks and explain how the rock cycle shapes the landscape around us and the earth beneath our feet

➤ List the major types of geologic hazards and describe ways to reduce their impacts

➤ Outline types of mineral resources and how they contribute to our products and society

➤ Describe the major methods of mining

➤ Characterize the environmental and social impacts of mining

➤ Assess reclamation efforts and mining policy

➤ Evaluate ways to encourage the sustainable use of mineral resources

Coltan miners in eastern Congo

Mining for . . . Cell Phones?

"The conflict in the Democratic Republic of the Congo has become mainly about access, control, and trade of five key mineral resources: coltan, diamonds, copper, cobalt, and gold."
—REPORT TO THE UNITED NATIONS SECURITY COUNCIL, APRIL 2001

"Coltan . . . is not helping the local people. In fact, it is the curse of the Congo."
—AFRICAN JOURNALIST KOFI AKOSAH-SARPONG

Pulling a cell phone from her pocket, a student on a college campus in the United States dials a friend. Inside her phone is a little-known metal called tantalum—just a tiny amount, but no cell phone could operate without it.

Half a world away, a dirt-poor miner in the heart of Africa toils all day in a jungle streambed, sifting sediment for nuggets of coltan ore, which contain tantalum. At nightfall, rebel soldiers take most of his ore, leaving him to sell what little remains to buy food for his family at the squalid mining camp where they live.

In bedeviling ways, tantalum links our glossy global high-tech economy with one of the most badly wrecked regions on Earth. The Democratic Republic of the Congo has been embroiled in a sprawling conflict that has involved six nations and various rebel militias. Over 5 million people have lost their lives in this war since 1998. It is the latest chapter in the sad history of a nation rich in natural resources—copper, cobalt, gold, diamonds, uranium, and timber—whose impoverished people keep losing control of those resources to others.

At the center of the recent conflict is tantalum (Ta), element number 73 on the Periodic Table (**APPENDIX C**). We rely on this metal for our cell phones, computer chips, DVD players, game consoles, and digital cameras. Tantalum powder is ideal for capacitors (the components that store energy and regulate current in miniature circuit boards) because it is highly heat resistant and readily conducts electricity.

Tantalum comes from a dull blackish mineral called tantalite, which often occurs with a mineral called columbite—so the ore is referred to as

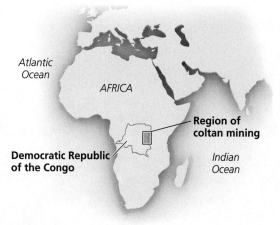

Atlantic Ocean

AFRICA

Region of coltan mining

Democratic Republic of the Congo

Indian Ocean

columbite-tantalite, or *coltan* for short. In eastern Congo, men dig craters in rainforest streambeds, panning for coltan much as early California miners panned for gold.

As information technology boomed in the late 1990s, global demand for tantalum rose, and market prices for the metal shot up to $500/kg ($230/lb) in 2001. High prices led some Congolese men to mine coltan by choice, but many more were forced into it.

In 1998, local militias, supported by forces from neighboring Rwanda and Uganda, overran eastern Congo. Farmers were chased off their land, villages were burned, and civilians were terrorized. Soldiers from each army seized control of mining operations. They forced farmers, refugees, prisoners, and children to work, and the soldiers skimmed profits from the coltan the people mined. Children and teachers abandoned school and worked in the mines. The turmoil also caused ecological havoc as miners and soldiers streamed into national parks, clearing rainforests and killing wildlife for food, including forest elephants, hippopotamuses, endangered gorillas, and the okapi, a rare relative of the giraffe.

Most miners ended up with little, while rebels, soldiers, and bandits enriched themselves by selling coltan to traders, who sold it to processing companies in Europe and the United States. These companies refine and sell tantalum powder to capacitor manufacturers, which in turn sell capacitors to Nokia,

Motorola, Sony, Intel, Compaq, Dell, and other high-tech corporations.

In 2001, an expert panel commissioned by the United Nations Security Council concluded that coltan riches were fueling, financing, and prolonging the war. The panel urged a U.N. embargo on coltan and other minerals smuggled from Congo and exported by neighboring nations. A grass-roots activist movement advanced the slogan, "No blood on my cell phone!"

Sony, Nokia, Ericsson, and other corporations rushed to assure consumers that they were not using tantalum from eastern Congo—and the region was, in fact, producing less than 10% of the world's supply. Meanwhile, some observers felt an embargo could hurt the long-suffering Congolese people, rather than help them. The mining life may be miserable, but it pays better than most alternatives in a land where the average income is 20 cents a day.

Soon, however, the high-tech boom went bust, and global demand for tantalum diminished. This occurred just as Australia and other countries were ramping up industrial-scale tantalite mining. As supply outpaced demand, the market price of tantalum fell, and several major producers quit mining tantalum. But nations began to work through their stockpiles, and by 2010 demand had grown, driving prices up once again.

Today, foreign troops are out of Congo, but internal factions still fight viciously over these resources and thousands of people continue to die or to flee their homes. Western electronics companies avoid knowingly purchasing tantalum from Congo, but as a result, much of it ends up being sold to China. In 2010 the U.S. Congress included in its financial reform bill an amendment requiring all electronics companies to report the origin of the tantalum in the products they sell. Yet the trade has so many middlemen and so little transparency that companies will find it very difficult to determine where their tantalum actually comes from.

In the meantime, some Congolese men are returning to the coltan mines, while others mine for tin, copper, or cobalt. Similar stories are playing out with these and other "conflict minerals" that we in more wealthy nations put to use in our products every day. ■

GEOLOGY: THE PHYSICAL BASIS FOR ENVIRONMENTAL SCIENCE

Coltan provides just one example of how we extract raw materials from beneath our planet's surface and turn them into products we use in our everyday lives. To understand the environmental impacts of extracting resources from the earth, and the many ways we can make mineral extraction less damaging, we first need a working knowledge of some of the physical processes that shape our planet.

Our planet is dynamic, and this is what motivates **geology**, the study of Earth's physical features, processes, and history. A human lifetime is just a blink of an eye in the long course of geologic time, and the Earth we experience is merely a snapshot in our changing planet's long history. We can begin to grasp this long-term dynamism as we consider two processes of fundamental importance to geology—plate tectonics and the rock cycle.

Earth consists of layers

Most geologic processes take place near Earth's surface, but our planet consists of multiple layers (**FIGURE 11.1**). At Earth's center is a dense **core** consisting mostly of iron, solid in the inner core and molten in the outer core. Surrounding the core is a thick layer of dense, elastic rock called the **mantle**. A portion of the upper mantle called the *asthenosphere* contains especially soft rock, melted in some areas. The harder rock above the asthenosphere is what we know as the **lithosphere**. The lithosphere includes both the uppermost mantle and

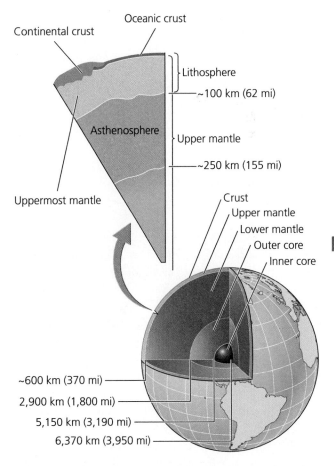

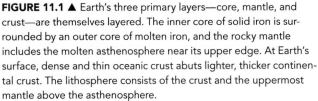

FIGURE 11.1 ▲ Earth's three primary layers—core, mantle, and crust—are themselves layered. The inner core of solid iron is surrounded by an outer core of molten iron, and the rocky mantle includes the molten asthenosphere near its upper edge. At Earth's surface, dense and thin oceanic crust abuts lighter, thicker continental crust. The lithosphere consists of the crust and the uppermost mantle above the asthenosphere.

the entirety of Earth's third major layer, the crust—the thin, brittle, low-density layer of rock that covers Earth's surface. The intense heat in the inner Earth drives convection currents that flow in loops in the mantle, pushing the mantle's soft rock cyclically upward (as it warms) and downward (as it cools), like a gigantic conveyor belt system. As the mantle material moves, it drags large plates of lithosphere along its surface. This movement of lithospheric plates is known as **plate tectonics**, a process of extraordinary importance to our planet.

Plate tectonics shapes Earth's geography

Our planet's surface consists of about 15 major tectonic plates, which fit together like puzzle pieces (**FIGURE 11.2**). Imagine peeling an orange and then placing the pieces of peel back onto the fruit; the ragged pieces of peel are like the tectonic plates riding atop Earth's surface. However, the plates are thinner relative to the planet's size, more like the skin of an apple. These plates move at rates of roughly 2–15 cm (1–6 in.) per year. This slow movement has influenced Earth's climate and life's evolution throughout our planet's history as the continents combined, separated, and recombined in various configurations. By studying ancient rock formations throughout the world, geologists have determined that at least twice, all landmasses were joined together in a "supercontinent." Scientists have dubbed the one that occurred about 225 million years ago *Pangaea* (see Figure 11.2).

There are three types of plate boundaries

The processes that occur at the boundaries between plates have major consequences. We can categorize plate boundaries into three types.

At **divergent plate boundaries**, tectonic plates push apart from one another as **magma** (rock heated to a molten,

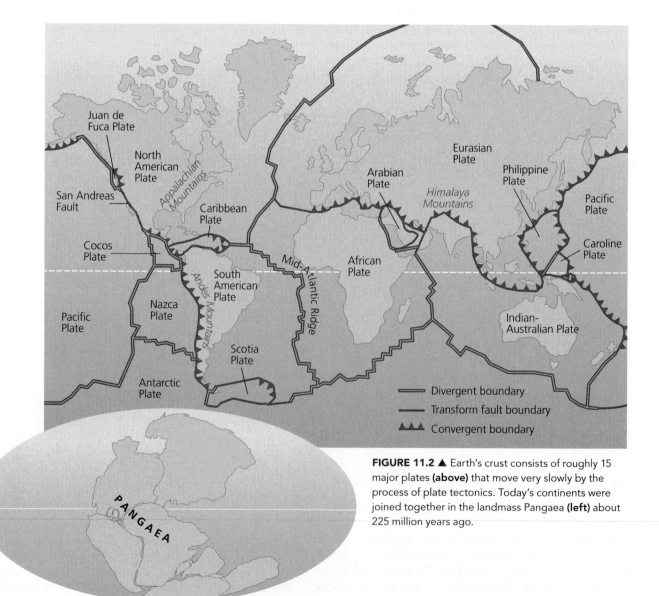

FIGURE 11.2 ▲ Earth's crust consists of roughly 15 major plates **(above)** that move very slowly by the process of plate tectonics. Today's continents were joined together in the landmass Pangaea **(left)** about 225 million years ago.

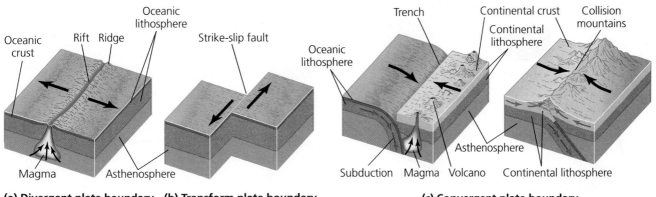

(a) Divergent plate boundary **(b) Transform plate boundary**   **(c) Convergent plate boundary**

FIGURE 11.3 ▲ Different types of boundaries between tectonic plates generate different geologic processes. At a divergent plate boundary, such as a mid-ocean ridge on the seafloor **(a)**, the two plates move gradually away from the boundary in the manner of conveyor belts, and magma from beneath the crust may extrude as lava. At a transform plate boundary **(b)**, two plates slide alongside one another, creating friction that leads to earthquakes. Where plates collide at a convergent plate boundary **(c)**, one plate is subducted beneath another, leading to volcanism. If continental crust from two plates collides, the buckling of rock can form mountain ranges.

liquid state) rises upward to the surface, creating new crust as it cools (**FIGURE 11.3A**). An example is the Mid-Atlantic Ridge, a 16,000-km (10,000-mi) plate boundary that extends from the Arctic Ocean to the southern tip of Africa in the Atlantic Ocean.

Where two plates meet, they may slip and grind alongside one another, forming a **transform plate boundary** (**FIGURE 11.3B**). This movement creates friction that generates earthquakes (p. 231) along strike-slip faults. *Faults* are fractures in Earth's crust, and at strike-slip faults each landmass moves horizontally in opposite directions. The Pacific Plate and the North American Plate rub against one another along California's San Andreas Fault. Southern California is slowly inching its way toward northern California along this fault, and the site of Los Angeles will eventually reach that of San Francisco.

When plates collide at **convergent plate boundaries**, either of two consequences may result (**FIGURE 11.3C**). In the first case, one plate may slide beneath another in a process called **subduction**. The subducted plate is heated and pressurized as it dives into the mantle, and water vapor escapes, helping to melt rock above the sinking plate (by lowering its melting temperature). The molten rock rises and may erupt through the surface at volcanoes (pp. 232–233).

Oceanic crust is denser than continental crust, so at subduction zones, oceanic crust generally slides beneath continental crust, leading to the formation of volcanic mountain ranges that parallel coastlines. The Cascades, where Mount Saint Helens in Washington (p. 283) erupted violently in 1980 and renewed its activity in 2004, are fueled by magma from subduction. South America's Andes Mountains (where the Nazca Plate slides beneath the South American Plate) provide another example. When one plate of oceanic crust is subducted beneath another plate of oceanic crust, the resulting volcanism may form arcs of islands, such as Japan and the Aleutians. This may also create deep trenches, such as the Mariana Trench, our planet's deepest abyss.

Alternatively, when two plates of continental crust collide, this collision may lift material from both plates. The Himalayas, the world's highest mountains, result from the Indian-Australian Plate's collision with the Eurasian Plate beginning 40–50 million years ago, and these mountains are still being uplifted today as these plates converge. The Appalachian Mountains of the eastern United States, once the world's highest mountains themselves, result from a more ancient collision with the edge of what is now Africa.

Tectonics creates Earth's landforms

In these ways, the processes of plate tectonics build mountains; shape the geography of oceans, islands, and continents; and give rise to earthquakes and volcanoes. The coltan mining areas of eastern Congo are situated along the western edge of Africa's Great Rift Valley system, a region where the African plate is slowly pulling itself apart. Some of the world's largest lakes have formed in the immense valley floors, far below towering volcanoes such as Mount Kilimanjaro.

Topography created by tectonic processes, in turn, shapes climate by altering patterns of rainfall, wind, ocean currents, and heating and cooling, all of which affect rates of weathering and erosion and the ability of plants and animals to inhabit different regions. Thus, plate tectonics influences the locations of biomes (pp. 78–84). Moreover, tectonics has affected the history of life's evolution; for instance, the convergence of landmasses into supercontinents is thought to have helped bring about widespread extinctions by limiting the extent of species-rich coastal areas and by creating an arid continental interior with extreme temperature swings.

Only in the last several decades have scientists learned about plate tectonics—this environmental system of such fundamental importance was completely unknown to humanity just half a century ago. Amazingly, our civilization was sending people to the moon by the time our geologists were explaining the movement of land under our very feet.

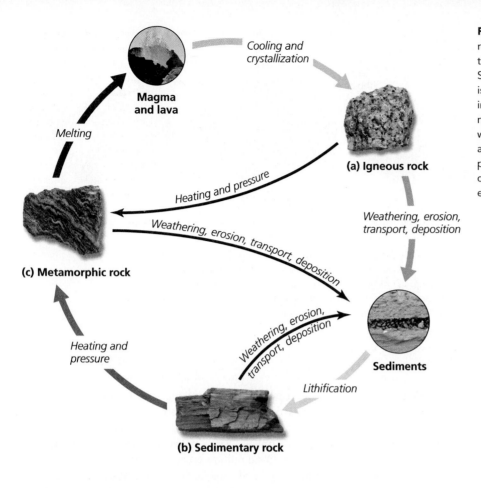

FIGURE 11.4 ◀ In the rock cycle, igneous rock **(a)** is formed when rock melts and the resulting magma or lava then cools. Sedimentary rock **(b)** is formed when rock is weathered and eroded and the resulting sediments are compressed to form new rock. Metamorphic rock **(c)** is formed when rock is subjected to intense heat and pressure underground. Through these processes (shown by arrows of different colors), each type of rock can be converted into either of the other two types.

Figure labels:
- Magma and lava
- Cooling and crystallization
- Melting
- (a) Igneous rock
- Heating and pressure
- Weathering, erosion, transport, deposition
- (c) Metamorphic rock
- Weathering, erosion, transport, deposition
- Heating and pressure
- Weathering, erosion, transport, deposition
- Sediments
- Lithification
- (b) Sedimentary rock

The rock cycle alters rock

We tend to think of rock as pretty solid stuff. Yet in the long run, over geologic time, rocks and the minerals that comprise them are heated, melted, cooled, broken down, and reassembled in a very slow process called the **rock cycle** (**FIGURE 11.4**).

A **rock** is any solid aggregation of minerals. A **mineral**, in turn, is any naturally occurring solid element or inorganic compound with a crystal structure, a specific chemical composition, and distinct physical properties. The type of rock in a given region affects soil characteristics and thereby influences the region's plant community. Understanding the rock cycle enables us to better appreciate the formation and conservation of soils, mineral resources, fossil fuels, and other natural resources.

Igneous rock All rocks can melt. At high enough temperatures, rock will enter a molten, liquid state called **magma**. If magma is released through the lithosphere (as in a volcanic eruption), it may flow or spatter across Earth's surface as **lava**. Rock that forms when magma or lava cools is called **igneous** (from the Latin *ignis,* meaning "fire") **rock** (**FIGURE 11.4A**).

Sedimentary rock All exposed rock weathers away with time. The relentless forces of wind, water, freezing, and thawing eat away at rocks, stripping off one tiny grain (or large chunk) after another. Through weathering (p. 137) and erosion (p. 140), particles of rock blown by wind or washed away by water come to rest downhill, downstream, or downwind from their sources, eventually forming **sediments**. Alternatively, some sediments form not from the physical erosion and accumulation of rock particles, but chemically from the precipitation of substances out of solution. Sediment layers accumulate over time, causing the weight and pressure of overlying layers to increase. **Sedimentary rock** (**FIGURE 11.4B**) is formed as sediments are physically pressed together and as dissolved minerals seep through sediments and act as a kind of glue, binding sediment particles.

Processes of physical compaction and chemical transformation in sedimentary layers also create the fossils of organisms (p. 50) and the fossil fuels we use for energy (pp. 328–336). Because sedimentary layers, or strata, pile up in chronological order, geologists and paleontologists can assign relative dates to fossils they find in sedimentary rock. By studying evidence from sedimentary rock, scientists can thereby make inferences about Earth's history.

Metamorphic rock Geologic forces may bend, uplift, compress, or stretch rock. When rock is subjected to great heat or pressure, it may alter its form, becoming **metamorphic** (from the Greek for "changed form") **rock** (**FIGURE 11.4C**). The forces that metamorphose rock generally occur deep underground, at temperatures lower than the rock's melting point but high enough to change its appearance and physical properties.

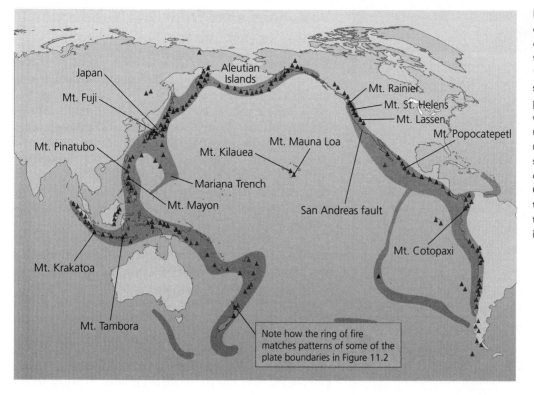

FIGURE 11.5 ◄ Most of our planet's volcanoes and earthquakes occur along the circum-Pacific belt, or "ring of fire," the system of subduction zones and other plate boundaries that encircles the Pacific Ocean. In this map, red symbols indicate major volcanoes, and gray-shaded areas indicate areas of greatest earthquake risk. Compare the distribution of these hazards with the tectonic plate boundaries shown in Figure 11.2.

Note how the ring of fire matches patterns of some of the plate boundaries in Figure 11.2

GEOLOGIC AND NATURAL HAZARDS

Plate tectonics gives rise to creative forces that shape our planet, but some of the consequences of tectonic movement can also pose hazards to us. Earthquakes and volcanic eruptions are examples of such *geologic hazards*. We can see how such hazards relate to tectonic processes by examining a map of the *circum-Pacific belt*, or so-called "ring of fire" (**FIGURE 11.5**). Nine out of 10 earthquakes and over half the world's volcanoes occur along this 40,000-km (25,000-mi) arc of subduction zones and fault systems.

Earthquakes result from movement at plate boundaries and faults

Along tectonic plate boundaries, and in other places where fractures in Earth's crust known as *faults* occur, the earth may relieve built-up pressure in fits and starts. Each release of energy causes what we know as an **earthquake**. Most earthquakes are barely perceptible, but occasionally they are powerful enough to do tremendous damage to human life and property (**FIGURE 11.6, TABLE 11.1**). Damage is generally greatest where soils are loose or saturated with water; areas of cities built atop landfill are particularly susceptible.

To minimize damage from earthquakes, engineers have developed ways to protect buildings from shaking. They do this by strengthening structural components while also designing points at which a structure can move and sway harm-lessly with ground motion. Just as a flexible tree trunk bends in a storm while a brittle one breaks, buildings with built-in flexibility are more likely to withstand an earthquake's violent shaking. Such designs are an important part of new building codes in California, Japan, and other quake-prone regions, and many older structures are being retrofitted to meet these codes.

FIGURE 11.6 ▼ The 2010 earthquake in Haiti devastated the capital city of Port-au-Prince and killed an estimated 230,000 people. One reason for the high number of fatalities was that many of Haiti's buildings were not constructed to withstand earthquakes. The structural damage in Japan from a much larger earthquake in early 2011 was less extensive than that seen in Haiti, due to more stringent building codes for earthquake resistance.

TABLE 11.1 Examples of Large or Recent Earthquakes

Location	Year	Fatalities	Magnitude[1]
Shaanxi Province, China	1556	830,000	~8
Lisbon, Portugal	1755	70,000[2]	8.7
San Francisco, California	1906	3,000	7.8
Kwanto, Japan	1923	143,000	7.9
Anchorage, Alaska	1964	128[2]	9.2
Tangshan, China	1976	255,000+	7.5
Michoacán, Mexico	1985	9,500	8.0
Loma Prieta, California	1989	63	6.9
Northridge, California	1994	60	6.7
Kobe, Japan	1995	5,502	6.9
Northern Sumatra	2004	228,000[2]	9.1
Kashmir, Pakistan	2005	86,000	7.6
Sichuan Province, China	2008	50,000+	7.9
Port-au-Prince, Haiti	2010	236,000	7.0
Maule, Chile	2010	500	8.8
Tohoku, Japan	2011	23,600[2,3]	9.0

[1] *Measured by moment magnitude; each full unit is roughly 32 times as powerful as the preceding full unit.*
[2] *Includes deaths from resulting tsunami.*
[3] *Includes people missing and presumed dead.*

Volcanoes arise from rifts, subduction zones, or hot spots

Where molten rock, hot gas, or ash erupts through Earth's surface, it forms a **volcano**, which can create a mountain over time as cooled lava accumulates. As we have seen, lava can extrude in rift valleys and along mid-ocean ridges, or over subduction zones as one tectonic plate dives beneath another. Lava may also be emitted at "hot spots," localized areas where plugs of molten rock from the mantle erupt through the crust. As a tectonic plate moves across a hot spot, repeated eruptions from this source may create a linear series of volcanoes. The Hawaiian Islands are an example of this process (**FIGURE 11.7A**).

At some volcanoes, lava flows slowly downhill, such as at Mount Kilauea in Hawaii (**FIGURE 11.7B**), which has been erupting continuously since 1983! Other times, a volcano may let loose large amounts of ash and cinder in a sudden explosion, such as during the 1980 eruption of Mount Saint Helens (p. 283). And sometimes a volcano can unleash a *pyroclastic flow*—a fast-moving cloud of toxic gas, ash, and rock fragments that races down the slopes, enveloping everything in its path. Such a flow buried the inhabitants of the ancient Roman cities of Pompeii and Herculaneum in A.D. 79, when Mount Vesuvius erupted.

Volcanic eruptions also exert environmental impacts (**TABLE 11.2**). Ash blocks sunlight, and sulfur emissions lead to a sulfuric acid haze that blocks radiation and cools the atmosphere. Large eruptions—such as that of Mount Pinatubo in

(a) Current and former Hawaiian islands, formed as crust moves over a volcanic hot spot

(b) Mt. Kilauea erupting

FIGURE 11.7 ▲ The Hawaiian Islands **(a)** have been formed by repeated eruptions from a hot spot of magma in the mantle as the Pacific Plate passes over the hot spot. The Big Island of Hawaii is most recently formed, and it is still volcanically active. The other islands are older and have already begun eroding away. To their northwest stretches a long series of former islands, now submerged. In the future, a new island will one day rise above the sea to the southeast of today's Big Island. The active volcano Kilauea **(b)**, on the Big Island's southeast coast, is the youngest of Hawaii's volcanoes, currently located above the edge of the hot spot.

TABLE 11.2 Examples of Notable Volcanic Eruptions

Location	Year	Impacts	Magnitude
Yellowstone Caldera, Wyoming, U.S.	640,000 B.P.[1]	Most recent "mega-eruption" at site of Yellowstone National Park	8
Mount Mazama, Oregon, U.S.	6,870 B.P.	Created Crater Lake	7
Mount Vesuvius, Italy	A.D. 79	Buried Pompeii and Herculaneum	5
Mount Tambora, Indonesia	1815	Created "year without a summer"; killed at least 70,000 people	7
Krakatau, Indonesia	1883	Killed over 36,000 people; heard 5,000 km (3,000 mi) away; affected weather for 5 years	6
Mount Saint Helens, Washington, U.S.	1980	Blew top off mountain; sent ash 19 km (12 mi) into sky and into 11 U.S. states; 57 people killed	5
Kilauea, Hawaii	1983–present	Continuous lava flow	1
Mount Pinatubo, Philippines	1991	Sulfuric aerosols lowered world temperature 0.5 °C (0.9 °F)	6
Eyjafjallajokull, Iceland	2010	Ash cloud disrupted air travel throughout Europe	1

[1]B.P. = years before the present.
[2]Measured by the Volcanic Explosivity Index, which ranges from 0 (least powerful) to 8 (most powerful).

the Philippines in 1991—can depress temperatures throughout the world. When Indonesia's Mount Tambora erupted in 1815, it cooled the planet enough over the following year to cause crop failures worldwide and make 1816 "the year without a summer."

Landslides are a form of mass wasting

At a smaller scale than volcanoes or earthquakes, a **landslide** occurs when large amounts of rock or soil collapse and flow downhill. Landslides are severe, and often sudden, manifestations of the more general phenomenon of **mass wasting**, the downslope movement of soil and rock due to gravity. Mass wasting occurs naturally, but often it is brought about by human land use practices that expose or loosen soil, making slopes more prone to collapse. Heavy rains may saturate soils and trigger *mudslides* of soil, rock, and water.

Most often, mass wasting eats away at unstable hillsides, damaging property one structure at a time (**FIGURE 11.8**). Occasionally mass wasting events can be colossal and deadly; mudslides that followed the torrential rainfall of Hurricane Mitch in Nicaragua and Honduras in 1998 killed over 11,000 people. Mudslides caused when volcanic eruptions melt snow and send huge volumes of destabilized mud racing downhill are called *lahars*, and they are particularly dangerous. A lahar buried the entire town of Armero, Colombia, in 1985 following an eruption, killing 21,000 people.

Tsunamis can follow earthquakes, volcanoes, or landslides

Earthquakes, volcanic eruptions, and large coastal landslides can all displace huge volumes of ocean water instantaneously and trigger a **tsunami**, an immense swell, or wave, of water that can travel thousands of miles across oceans. The world's

attention was drawn to this hazard on December 26, 2004, when a massive tsunami, triggered by an earthquake off Sumatra, devastated coastlines all around the Indian Ocean, from Indonesia to India to Africa. Roughly 230,000 people were killed, 1–2 million were displaced, and whole communities were destroyed.

More recently, a tsunami generated by an offshore earthquake devastated large portions of northeastern Japan on March 11, 2011 (**FIGURE 11.9**). The tsunami and earthquake killed more than 23,000 people, caused over $300 billion in economic impacts, and contributed to the release of radioactive material from the Fukushima nuclear power plant (p. 349).

FIGURE 11.8 ▼ Landslides are a frequent occurrence in sloping areas along the California coast, particularly after heavy winter rains saturate soils. Homes built on unstable slopes, such as these in Laguna Beach, Orange County, can be damaged or destroyed when slopes give way.

FIGURE 11.9 ◄ The tsunami on March 11, 2011, destroyed entire coastal communities in northeastern Japan. Waves up to 10 m (30 ft) high battered the coastline and penetrated up to 9.6 km (6 mi) inland. In this photo, the massive swell crashes over dikes that were thought to be high enough to offer protection.

Those of us who live in the United States and Canada should not consider tsunamis to be something that occurs only in faraway places. The 2011 tsunami from Japan crossed the Pacific and caused some damage and one death in California and Oregon. A large tsunami struck North America's Pacific coast in 1700 following a huge earthquake in the Pacific Northwest, and one following the Alaskan earthquake of 1964 drowned 100 people, some as far south as California. North America's Atlantic coast could be at risk if an unstable portion of a Canary Island volcano that researchers are monitoring were to slump into the sea.

Since the 2004 tsunami, nations and international agencies have stepped up efforts to develop systems to give coastal residents advance warning of approaching tsunamis. In addition, we can lessen the impacts of tsunamis if we leave in place natural vegetation such as mangrove forests (p. 257).

We can worsen or reduce the impacts of natural hazards

Besides the geologic hazards just described, people face other types of natural hazards that result from conditions in the hydrosphere, atmosphere, or biosphere. Heavy rains can lead to *flooding* that ravages low-lying areas near rivers and streams (p. 261). *Coastal erosion* can eat away at beaches (p. 248). *Wildfire* can threaten life and property in fire-prone areas (p. 196). And *tornadoes* and *hurricanes* (p. 282) can cause extensive damage and loss of life.

Although we refer to such phenomena as "natural hazards," the magnitude of their impacts upon us often depends on choices we make. We tend to worsen the impacts of so-called natural hazards in various ways:

▶ As our population grows, more people live in areas susceptible to natural disasters.

▶ Many of us choose to live in areas that we deem attractive but that are also prone to hazards. For instance, coastlines are vulnerable to tsunamis and erosion by storms, and mountainous areas may feature volcanoes and mass-wasting events.

▶ We use and engineer landscapes around us in ways that can increase the frequency or severity of natural hazards. Damming and diking rivers to control floods can sometimes lead to catastrophic flooding, like the flooding along the Mississippi River in 2011 (p. 261). Suppressing natural wildfires puts forests at risk of larger, highly damaging fires. Clear-cutting on slopes (p. 194) and some mining practices (pp. 237–240) can induce mass wasting, speed runoff, compact soil, and change drainage patterns.

▶ As we change Earth's climate by emitting greenhouse gases (Chapter 14), we alter patterns of precipitation, increasing risks of drought, fire, flooding, and mudslides locally and regionally. Rising sea levels induced by global warming increase coastal erosion. And some research suggests that warming ocean temperatures may increase the power and duration of hurricanes.

We can often reduce the impacts of hazards through the thoughtful use of technology, engineering, and policy, informed by a solid understanding of geology and ecology. Examples already noted include building earthquake-resistant structures; designing early warning systems for tsunamis and volcanoes; and conserving coastal forests, reefs, and salt marshes to protect against tsunamis and coastal erosion (pp. 257–259). In addition, better forestry and mining practices can help prevent landslides. Zoning regulations, building codes, and insurance incentives that discourage development in areas prone to landslides, floods, fires, and storm surges can keep us out of harm's way and decrease taxpayer expense when cleaning up after natural disasters. Finally, addressing global climate change may help reduce the frequency of natural hazards in many regions.

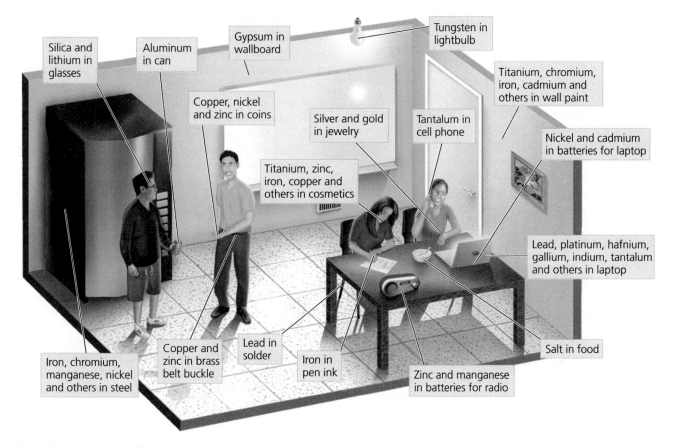

FIGURE 11.10 ▲ Elements from minerals that we mine are everywhere in the products we use in our everyday lives. This scene from a typical college student lounge points out just a few of the many minerals that surround us.

EARTH'S MINERAL RESOURCES

Both gradual geologic processes and catastrophic geologic hazards influence the distribution of rocks and minerals in the lithosphere and their availability to us. We depend on a wide array of mineral resources as raw materials for our products, so we mine and process these resources. Without the resources from beneath the ground that we use to make building materials, wiring, clothing, appliances, fertilizers for crops, and so much more, civilization as we know it could not exist. Consider a typical scene from a student lounge at a college or university (**FIGURE 11.10**), and note how many items are made with elements from the minerals we take from Earth.

We use mined materials extensively

We often don't notice how many mined resources we use every day. Using data from the U.S. governement, the Mineral Information Institute estimated that in 2009 the average American consumed over 17,000 kg (37,000 lb) of new minerals and fuels every year. At current rates of use, a child born in 2009 will use around 1.3 million kg (2.9 million lbs) over his or her lifetime (**FIGURE 11.11**).

More than half (9,600 kg or 21,000 lb) of the annual mineral and fuel use of 17,000 kg is from the coal, oil, and natural gas used to supply our intensive demands for energy. Much of the remaining mineral use is attributable to the sand, gravel, and stone used in constructing our buildings, roads, bridges, and parking lots. Metal use is dwarfed by these other two cat-

egories, but the average American will still use more than 2 tons of aluminum over his or her lifetime. This level of consumption clearly shows the potential of recycling and reuse (such as recycling stone and gravel from old highways into new construction) to make our lifestyle more sustainable.

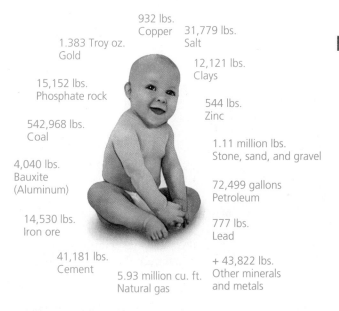

1.383 Troy oz. Gold

932 lbs. Copper

31,779 lbs. Salt

15,152 lbs. Phosphate rock

12,121 lbs. Clays

542,968 lbs. Coal

544 lbs. Zinc

4,040 lbs. Bauxite (Aluminum)

1.11 million lbs. Stone, sand, and gravel

14,530 lbs. Iron ore

72,499 gallons Petroleum

777 lbs. Lead

41,181 lbs. Cement

5.93 million cu. ft. Natural gas

+ 43,822 lbs. Other minerals and metals

FIGURE 11.11 ▲ At current rates of usage, a baby born in 2009 in the United States is predicted to use a stunning 1.3 million kg (2.9 million lb) of mined minerals, metals, and fuels in his or her lifetime. Data from Mineral Information Institute, 2009.

We obtain minerals by mining

We obtain the minerals we use in all these ways through the process of mining. The term *mining* in the broad sense describes the extraction of any resource that is nonrenewable on the timescale of our society. In this sense, we mine fossil fuels and groundwater, as well as minerals. When used specifically in relation to minerals, **mining** refers to the systematic removal of rock, soil, or other material for the purpose of extracting minerals of economic interest. Because most minerals of interest are widely spread but in low concentrations, miners and mining geologists first try to locate concentrated sources of minerals before mining begins.

Metals are extracted from ores

Some minerals can be mined for metals. A **metal** is a type of chemical element that typically is lustrous, opaque, malleable, and can conduct heat and electricity. Most metals are not found in a pure state in Earth's crust, but instead are present within **ore**, a mineral or grouping of minerals from which we extract metals.

Copper, iron, lead, gold, and aluminum are among the many economically valuable metals we extract from mined ore. The tantalum used in the electronic components of computers, cell phones, DVD players, and other devices is a metal that comes from the mineral tantalite (**FIGURE 11.12**). In nature, tantalite is often found with the mineral columbite within the ore called coltan.

We process metals after mining ore

Extracting minerals from the ground is the first step in putting them to use. However, most minerals need to be processed in some way to become useful for our products. For example, after ores are mined, the rock is crushed and pulverized, and the desired metals are isolated by chemical or physical means. The material is then processed to purify the metals we desire. With coltan, processing facilities use acid solvents to separate tantalite from columbite. Other chemicals are then used to

produce metallic tantalum powder. This powder can be consolidated by various melting techniques and can be shaped into wire, sheets, or other forms.

Sometimes we mix, melt, and fuse a metal with another metal or a nonmetal substance to form an *alloy*. For example, steel is an alloy of the metal iron that has been fused with a small quantity of carbon. The strength and malleability of this particular alloy make steel ideal for its many applications in buildings, vehicles, appliances, and more. To make steel, we first mine iron ore, which consists of iron-containing compounds such as iron oxide. Steelmakers then heat the ore and chemically extract the iron with carbon in a process known as **smelting** (heating ore beyond its melting point and combining it with other metals or chemicals). They then melt and reprocess the mixture, removing precise amounts of carbon and shaping the product into rods, sheets, or wires. During this melting process, certain other metals may be added to modify the strength, malleability, or other characteristics of the steel.

Processing minerals exerts environmental impacts. Most methods are water-intensive and energy-intensive. Moreover, many chemical reactions and heating processes used for extracting metals from ore emit air pollution, and smelting plants in particular have long been hot spots of toxic air pollution. In addition, soil and water commonly become polluted by **tailings**, portions of ore left over after metals have been extracted. Tailings may leach heavy metals present in the ore waste as well as chemicals applied in the extraction process. For instance, we use cyanide to extract gold from ore, and we use sulfuric acid to extract copper. Mining operations often store toxic slurries of tailings in large reservoirs called **surface impoundments**. Impoundment walls are designed to prevent leaks and collapse, but accidents can occur if the structural integrity of the impoundment is compromised. In 2000, a breach of a coal tailings impoundment near Inez, Kentucky, released over 1 billion liters (250–300 million gal) of coal slurry, blackening 120 km (75 mi) of streams, killing aquatic wildlife, and affecting drinking water supplies for many communities.

(a) Tantalite ore

(b) Purified tantalum

(c) Capacitor containing tantalum

FIGURE 11.12 ▲ Tantalite ore **(a)** is mined from the ground and then processed to extract the pure metal tantalum **(b)**. This metal is used in capacitors **(c)** and other electronic components in computer chips, cell phones, and many other devices.

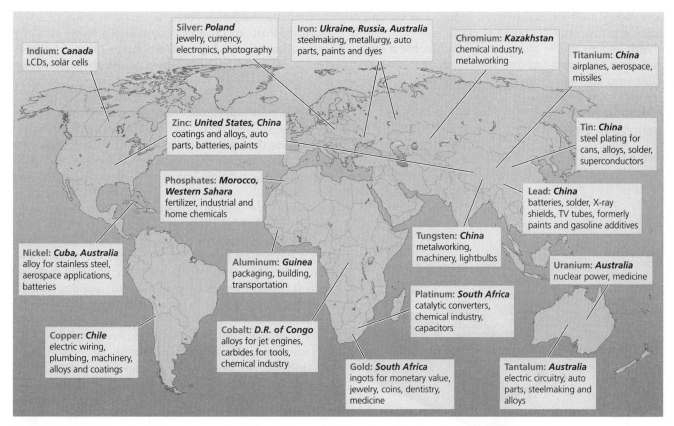

FIGURE 11.13 ▲ The minerals we use come from all over the world. Shown is a selection of economically important minerals together with their major uses and their main nation of origin. Only a minority of minerals, uses, and origins is shown.

We also mine nonmetallic substances

We also mine and use many minerals that do not contain metals. **FIGURE 11.13** illustrates the nation of origin and uses for some economically important mineral resources, both metallic and nonmetallic. As you can see, many geologic resources in the products you use were mined in faraway nations.

Sand and gravel (the most commonly mined mineral resources) provide fill and construction materials. Phosphates provide us with fertilizer. We mine limestone, salt, potash, and other minerals for a number of diverse purposes.

Gemstones are treasured for their rarity and beauty. For instance, diamonds have long been prized—and like coltan, they have fueled resource wars. Besides the conflict in eastern Congo, the diamond trade has acted to fund, prolong, and intensify wars in Angola, Sierra Leone, Liberia, and elsewhere, as armies exploit local people for mine labor and sell the diamonds for profit. This is why you may hear the phrase "blood diamonds," just as coltan has been called a "conflict mineral."

We also mine substances for fuel. Uranium ore is a mineral from which we extract the metal uranium, which we use in nuclear power (pp. 345–350). One of the most common fuels we mine is coal. Coal (p. 329) is not a mineral, because it consists of organic matter, but we consider coal mining in this chapter because it has relevance for many general mining issues. Other fossil fuels—petroleum, natural gas, and alternative fossil fuels such as oil sands, oil shale, and methane hydrates—are also organic and are extracted from the earth (as we will see in Chapter 15).

MINING METHODS AND THEIR IMPACTS

Mining for minerals is an important industry that provides jobs for people and revenue for communities in many regions. Mining supplies us raw materials for countless products we use daily, so it is necessary for the lives we lead. However, mining also exerts a price in environmental and social impacts. Because minerals of interest often make up only a small portion of the rock in a given area, very large amounts of material are removed in order to obtain the desired minerals. This frequently means that mining disturbs large areas of land.

Depending on the nature of the mineral deposit, any of several mining methods may be employed to extract the resource from the ground. Mining companies select which method to use based largely on its economic efficiency.

Strip mining removes surface layers of soil and rock

When a resource occurs in shallow horizontal deposits near the surface, the most effective mining method is often **strip mining**, whereby layers of surface soil and rock are removed from large areas to expose the resource. Heavy machinery removes the overlying soil and rock (termed *overburden*) from a strip of land, and the resource is extracted. This strip is then refilled with the overburden, and miners proceed to an adjacent strip of land and repeat the process. Strip mining is commonly used for coal (**FIGURE 11.14A**) and oil sands (p. 335), and sometimes for sand and gravel.

Strip mining can be economically efficient, but it obliterates natural communities over large areas, and the soil in refilled areas can easily erode away.

In subsurface mining, miners work underground

When a resource occurs in concentrated pockets or seams deep underground, and the earth allows for safe tunneling, then mining companies pursue **subsurface mining**. In this approach, shafts are excavated deep into the ground, and networks of tunnels are dug or blasted out to follow deposits of the mineral (**FIGURE 11.14B**). Miners remove the resource systematically and ship it to the surface.

We use subsurface mining for metals such as zinc, lead, nickel, tin, gold, copper, and uranium, as well as for diamonds, phosphate, salt, and potash. In addition, a great deal of coal is mined using the subsurface technique. The scale of subsurface mining can be mind-boggling; the world's deepest mines (certain gold mines in South Africa) extend nearly 4 km (2.5 mi) underground.

Subsurface mining is the most dangerous form of mining, and fatal accidents are not unusual. In China, coal-mining conditions are so dangerous that in 2010 alone over 2,400 miners lost their lives. Besides risking injury or death from natural gas explosions and collapsing shafts or tunnels, miners inhale toxic fumes and coal dust, which can lead to respiratory diseases, including fatal black lung disease.

Occasionally subsurface mines can affect people long after they are closed. Abandoned mine tunnels can collapse, causing sinkholes at the surface. Both strip mining and subsurface mining can also pollute waterways through the process of **acid drainage**, which occurs when sulfide minerals in newly exposed rock surfaces react with oxygen and rainwater to produce sulfuric acid. As the sulfuric acid runs off, it leaches metals from the rocks, many of which are toxic to organisms. Acid drainage can affect fish and other aquatic organisms when it leaches into streams and it pollutes groundwater supplies people use for drinking water or irrigating crops (**FIGURE 11.15**). Although acid drainage is a natural phenomenon, mining greatly accelerates this process by exposing many new rock surfaces at once.

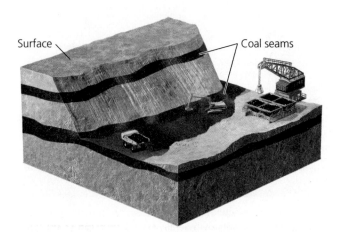

(a) Strip mining

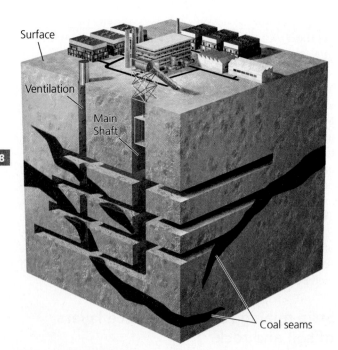

(b) Subsurface mining

FIGURE 11.14 ▲ Coal mining involves two types of mining approaches. In strip mining **(a)**, soil is removed from the surface in strips, exposing seams from which coal is mined. In subsurface mining **(b)**, miners work below ground in shafts and tunnels blasted through the rock. These passageways provide access to underground seams of coal or minerals.

FIGURE 11.15 ▲ Acidic drainage from an underground coal mine streams down a slope in West Virginia. The yellow-orange color is due to iron from the drainage settling out on the soil surface and forming rust.

Q: Why would anyone work in a mine when it's such dangerous work?

A: It seems that mining accidents appear regularly in the headlines. In late 2010, for example, 33 miners in Chile were rescued after being trapped 600 m (2,000 ft) underground for 69 days in a gold and copper mine. Four days later, an explosion at an underground coal mine in China killed 37 miners. Similar explosions claimed the lives of 29 miners in a coal mine in West Virginia earlier in 2010 and 45 miners in Pakistan in 2011. Given these dangers, and the chronic health impacts of working in an underground mine, it's reasonable to wonder why anyone would accept such a job.

Many of the people who work in mines do so because they have few other options. Underground mining often occurs in economically depressed areas, such as Appalachia in the United States, where mining is one of the few well-paying jobs. And for most mining jobs, people can begin working right out of high school. So although the work is dangerous, many miners are willing to accept those risks to provide for themselves and their families because few other career opportunities are available.

Open pit mining creates immense holes in the ground

When a mineral is spread widely and evenly throughout a rock formation, or when the earth is unsuitable for tunneling, the method of choice is **open pit mining**. This essentially involves digging a gigantic hole and removing the desired ore, along with waste rock that surrounds the ore. Some open pit mines are inconceivably enormous. The world's largest, the Bingham Canyon Mine near Salt Lake City, Utah, is 4 km (2.5 mi) across and 1.2 km (0.75 mi) deep (**FIGURE 11.16**). Conveyor systems and immense trucks with tires taller than a person carry out nearly half a million tons of ore and waste rock each day.

Open pit mines are terraced so that men and machinery can move about, and waste rock is left in massive heaps outside the pit. The pit is expanded until the resource runs out or becomes unprofitable to mine. Open pit mining is used for copper, iron, gold, diamonds, and coal, among other resources. We also use this technique to extract clay, gravel, sand, and stone such as limestone, granite, marble, and slate, but we generally call these pits *quarries*.

Open pit mines are so large because huge volumes of waste rock need to be removed in order to extract relatively small amounts of ore, which in turn contain still smaller traces of valuable minerals. The sheer size of these mines means that the degree of habitat loss and aesthetic degradation is considerable.

Once mining is complete, abandoned pits generally fill up with groundwater. If sulfuric minerals are present, acid drainage can form in the pit and percolate into aquifers.

FIGURE 11.16 ▲ The Bingham Canyon open pit mine outside Salt Lake City, Utah, is the world's largest human-made hole in the ground. This immense mine produces mostly copper.

Placer mining uses running water to isolate minerals

Some metals and gems accumulate in riverbed deposits, having been displaced from elsewhere and carried along by flowing water. To search for these metals and gems, miners sift through material in modern or ancient riverbed deposits, generally using running water to separate lightweight mud and gravel from heavier minerals of value (**FIGURE 11.17**). This technique is called **placer mining** (pronounced "plasser").

FIGURE 11.17 ▼ Miners in eastern Congo find coltan by placer mining. Sediment is placed in plastic tubs, and water is run through them. A mixing motion allows the sediment to be poured off while the heavy coltan settles to the bottom.

Placer mining is the method used by Congo's coltan miners, who wade through streambeds, sifting through large amounts of debris by hand with a pan or simple tools, searching for high-density tantalite that settles to the bottom while low-density material washes away. Today's African miners practice small-scale placer mining similar to the method used by American miners who ventured to California in the Gold Rush of 1849, and later to Alaska in the Klondike Gold Rush of 1896–1899. Placer mining for gold is still practiced in areas of Alaska and Canada, although today it uses large dredges and heavy machinery.

Besides the many social impacts of placer mining in places like Congo, placer mining is environmentally destructive because most methods wash large amounts of debris into streams, making them uninhabitable for fish and other life for many miles downstream. Gold mining in northern California's rivers in the decades following the Gold Rush washed so much debris all the way to San Francisco Bay that a U.S. district court ruling in 1884 finally halted this mining practice. Placer mining also disturbs stream banks, causing erosion and harming ecologically important plant communities.

Mountaintop mining reshapes ridges and can fill valleys

When a resource occurs in underground seams near the tops of ridges or mountains, mining companies may practice **mountaintop removal mining**, in which several hundred vertical feet of mountaintop may be removed to allow recovery of entire seams of the resource (**FIGURE 11.18**). This method of mining is used primarily for coal in the Appalachian Mountains of the eastern United States. In mountaintop removal mining, a mountain's forests are clear-cut, the timber is sold, topsoil is removed, and then rock is repeatedly blasted away to expose the coal for extraction. Overburden is placed back onto the mountaintop, but this waste rock is unstable and typically takes up more volume than the original rock, so generally a great deal of waste rock is dumped into adjacent valleys (a practice called "valley filling"). So far, mountaintop removal has blasted away an area the size of Delaware and has buried nearly 3,200 km (2,000 mi) of streams.

FIGURE 11.18 ▼ In the Appalachians, mountaintop mining for coal takes place on massive scales. Mining trucks like those shown are up 15 m (50 ft) long and 7 m (23 ft) tall.

Scientists are finding that dumping tons of debris into valleys degrades or destroys immense areas of habitat, clogs streams and rivers, and pollutes waterways with acid drainage. With slopes deforested and valleys filled with debris, erosion intensifies, mudslides become frequent, and flash floods ravage the lower valleys. Further, the Appalachian forests that are cleared in mountaintop mining are some of the richest forests for biodiversity in the nation.

People living in communities near the sites also experience social and health impacts. Blasts from mines crack house foundations and wells, loose rock tumbles down into yards and homes, and floods tear through properties. Coal dust causes respiratory ailments, and contaminated water unleashes a variety of health problems. In fact, a 2009 study documented that people in mountaintop mining areas show elevated levels of lung cancer, heart disease, kidney disease, pulmonary disorders, hypertension, and mortality. In all these ways, the people of Appalachia—already among the poorest in the United States—suffer substantial external costs (p. 92), while the rest of us benefit from the electricity we produce with their coal.

Critics of mountaintop removal mining argue that valley filling violates the Clean Water Act. In 2010 the Environmental Protection Agency introduced new regulations to limit damage from mountaintop mining and valley filling. Legislation introduced in Congress in 2009 sought to restrict some mountaintop mining practices, but failed to become law.

Solution mining dissolves and extracts resources in place

When a deposit is especially deep and the resource can be dissolved in a liquid, miners may use a technique called *solution mining* or *in-situ recovery*. In this technique, a narrow borehole is drilled deep into the ground to reach the deposit, and water, acid, or another liquid is injected down the borehole to leach the resource from the surrounding rock and dissolve it in the liquid. The resulting solution is then sucked out, and the desired resource is isolated. Sodium chloride (table salt), lithium, boron, bromine, magnesium, potash, copper, and uranium can be mined in this way.

Solution mining generally exerts less environmental impact than other mining techniques, because less area at the surface is disturbed. The primary potential impacts involve accidental leakage of acids into groundwater surrounding the borehole, and the contamination of aquifers with acids, heavy metals, or uranium leached from the rock.

Some mining occurs in the ocean

The oceans hold many minerals useful to our society. We extract some minerals from seawater, such as magnesium from salts held in solution. We extract other minerals from the ocean floor. For example, many minerals are concentrated in manganese nodules, small ball-shaped accretions that are scattered across parts of the ocean floor. Over 1.5 trillion tons of manganese nodules may exist in the Pacific Ocean alone, and their reserves of metal may exceed all terrestrial reserves. As land resources become scarcer and as undersea mining technology develops, mining companies may turn increasingly to the seas. The logistical difficulty of mining offshore resources, however, has kept their extraction limited so far.

Restoration of mined sites is often only partly effective

Because of the environmental impacts of mining, governments of the United States and other developed nations now require that mining companies restore, or reclaim, surface-mined sites following mining. The aim of such restoration, or **reclamation**, is to restore the site to a condition similar to its condition before mining. To restore a site, companies are required to remove buildings and other structures used for mining, replace overburden, fill in shafts, and replant the area with vegetation (**FIGURE 11.19**). In the United States, the 1977 *Surface Mining Control and Reclamation Act* mandates restoration efforts, requiring companies to post bonds to cover reclamation costs before mining can be approved. This ensures that if the company fails to restore the land for any reason, the government will have the money to do so. Most other nations exercise less oversight, and in nations such as the Congo, there is no regulation at all.

FIGURE 11.19 ▲ More mine sites are being restored today, but restoration rarely is able to recreate the natural community present before mining. Here, reclamation workers in Ghana, West Africa, plant trees in an abandoned gold-mining pit.

The mining industry has made great strides in reclaiming mined land and employs many hard-working ecologists and engineers to conduct these efforts. However, even on sites that are restored, impacts from mining (such as soil and water damage from acid drainage) can be severe and long-lasting. Moreover, reclaimed sites do not generally regain the same biotic communities that were naturally present before mining. One reason is that fast-growing grasses are generally used to initiate and anchor restoration efforts. This helps control erosion quickly from the outset, but it can hinder the longer-term establishment of forests, wetlands, or other complex natural communities. Instead, grasses may outcompete slower-growing native plants in the acidic, compacted, nutrient-poor soils that usually result from mining. Moreover, many inconspicuous but vital symbiotic relationships (p. 68) that maintain ecosystems—such as specialized relationships between plants and fungi or plants and insects—are eliminated by mining and are very difficult to restore.

Water polluted by mining and acid drainage can also be reclaimed, if its pH can be moderated and if toxic heavy metals can be removed. Like the reclamation of land, this is a challenging and imperfect process, but researchers and the mining industry are making progress in improving techniques (see **THE SCIENCE BEHIND THE STORY**, pp. 242–243). The need for treatment can be long-lasting. Mines in Spain from the era of the Roman Empire still leach acid drainage into waterways today.

WEIGHING THE ISSUES

Restoring Mined Areas Mining has severe environmental impacts, and restoring a mined site to a condition similar to its state before mining is costly and difficult. How much do you think we should require mining companies to restore after a mine is shut down, and what criteria should we use to guide restoration? Should we require complete restoration? No restoration? What should our priorities be—to minimize water pollution, health impacts, biodiversity loss, soil damage, or other factors? Should the amount of restoration we require depend on how much money the company made from the mine? Explain your recommendations.

An 1872 law still guides U.S. mining policy

The ways that mining companies stake claims and use land in the United States is guided by a law that is well over a century old. The **General Mining Act of 1872** encourages people and companies to prospect for minerals on federally owned land by allowing any U.S. citizen or any company with permission to do business in the United States to stake a claim on any plot of public land open to mining. The person or company owning the claim gains the sole right to take minerals from the area. The claim-holder can also patent the claim (i.e., buy the land) for only about $5 per acre. Regardless of the profits they might make on minerals they extract, the law requires

THE SCIENCE BEHIND THE STORY

Using Bacteria to Clean Mine Water and Recover Metals

Toxic waters of the Berkeley Pit near Butte, Montana

Mining poses two dilemmas: It exerts environmental impacts, and it can deplete nonrenewable minerals. Now, some scientists are seeking to address both these drawbacks in one fell swoop. They're aiming to clean up polluted mine sites and amass valuable minerals at the same time. Their secret weapon? Bacteria.

Let's set the stage by traveling to western Montana, where in 1982 the Atlantic Richfield Company ceased mining at the Berkeley Pit, a massive open pit copper mine covering just over 1 square mile. Once the mine was shut down and pumps were turned off, groundwater started to fill the 540-m-deep (1,780-ft-deep) crater.

As water bathed the rock walls, it mixed with oxygen from the air and reacted with sulfide minerals in the freshly exposed rock, producing acid drainage (p. 238). Sulfuric acid formed, and water accumulating in the Berkeley Pit became as acidic as lemon juice, reaching pH values as low as 2.2. The very low pH caused metals in the rock to leach into the acidic water, including ions of iron, zinc, aluminum,

manganese, nickel, cadmium, cobalt, and arsenic, as well as copper. These metals reacted with the sulfuric acid in the water to form metal sulfates, compounds containing a metal ion bonded to a sulfate ion. For instance, zinc formed zinc sulfate ($ZnSO_4$) and iron formed iron sulfate ($FeSO_4$).

Metal sulfates dissolve in water and high concentrations of these dissolved metals made the water toxic to wildlife. The water here was so unusually acidic and metal-rich that biologists surveying the site discovered new species of microbes in the water.

As groundwater continued to fill the pit at a rate of 13 vertical feet per year, experts estimated that by 2015 the water would overflow into the Clark Fork River drainage, polluting

ecosystems and poisoning people's drinking water. The race was on to find a solution.

Traditionally, mining engineers have tried to treat acid mine drainage by adding a strongly alkaline substance such as lime or sodium hydroxide to raise the water's pH. When the pH is raised, the dissolved heavy metals precipitate and fall out of solution. In recent years, researchers have begun to put sulfate-reducing bacteria to work in converting soluble metal sulfates in acid mine drainage to insoluble metal sulfides.

Sulfate-reducing bacteria are microbes that thrive in the absence of oxygen by chemically reducing sulfates to obtain energy. In so doing, they generate sulfides that they expel as waste. In nature, some such bacteria degrade organic materials in the mud of swamps and produce hydrogen sulfide, the gas that gives swamps and mudflats their distinctive rotten-egg odor.

At the Berkeley Pit, EPA scientist Henry Tabak, University of Cincinnati

no payments of any kind to the public, and until recently no restoration of the land after mining was required.

Supporters of the policy say that it is appropriate and desirable to continue encouraging the domestic mining industry, which must undertake substantial financial risk and investment to locate resources that are vital to our economy. Critics counter that the policy gives valuable public resources away to private interests nearly for free. They also point out that many claims made under this law have eventually led to lucrative land development schemes (such as condominium development) that have nothing to do with mining.

Critics have tried to amend the law many times over the years, mostly without success. The latest effort, the Hardrock Mining and Reclamation Act of 2009, sought to largely end the patenting process, put some public lands off-limits to mining, mandate that mined sites be restored to some semblance of

their former condition, and require miners to pay the government royalties of 4% of profits from new mines and 8% from existing mines. The money would help to fund cleanups of mined sites and reimburse communities affected by mining. The legislation was referred to committee in both houses of Congress but failed to become law.

THE ISSUES

Reforming the 1872 Mining Law You are a legislator in the U.S. Congress, and your colleagues are asking you to help prepare a bill to reform the General Mining Act of 1872. Would you join this effort to reform the law? Why or why not? If you would, then what would you seek to include in the bill?

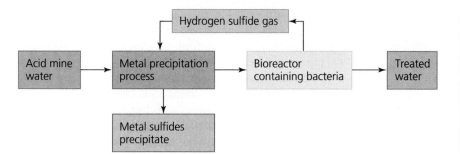

Acidic mine water rich in sulfates is fed into two tanks. Bacteria in the bioreactor produce hydrogen sulfide (H_2S), which circulates to the metal precipitation tanks. In the presence of hydrogen sulfide, metal ions bond to sulfur to form metal sulfides. The sulfides are not soluble, so they precipitate out of the water. The sulfides are collected and may be processed further to harvest minerals of economic value. The treated water is discharged from the system, clean enough to use for agriculture.

Percent Recovery of Metals from Mine Water Using Sulfate-Reducing Bacteria	
Metal	**Percent recovery of metal**
Aluminum	99.8
Cadmium	99.7
Cobalt	99.1
Copper	99.8
Iron	97.1
Manganese	87.4
Nickel	47.8
Zinc	100.0

Adapted from Tabak, H.H., et al., 2003. Advances in biotreatment of acid mine drainage and biorecovery of metals: 1. Metal precipitation for recovery and recycle. Biodegradation 14: 423–436.

chemical engineer Rakesh Govind, and three colleagues examined whether they could use sulfate-reducing bacteria to clean up the water and recover valuable metals.

Highly acidic water harms most bacteria, so Tabak and Govind's team designed a two-step process (**see figure**). They grew their bacteria in water kept at a neutral pH and fed them sulfates from the mine water, along with carbon-containing nutrients, letting them produce hydrogen sulfide. They then funneled this hydrogen sulfide into a series of tanks, where it reacted with the acid mine drainage from the Berkeley Pit. The reactions in these tanks caused the dissolved metals in the mine water to precipitate out as insoluble metal sulfides and settle to the bottom of the tank.

The metal sulfides were then removed from the precipitation tanks. The research team measured its success by calculating the average percent recovery for each of several metals (**see table**). Precipitates of the various metals ranged in purity from 75% to 98%.

Besides recovering high percentages of metals that can be processed and recycled, the process cleaned toxic metals from the mine water and also made it less acidic because the reactions produce bicarbonate ions (HCO_3-), which raise the pH of the water. After full treatment, the researchers determined that the water was pure enough to meet safety standards for irrigating agricultural crops.

This research, published in 2003, was done in the lab at a small scale. More work remains to scale it up to treat large amounts of water from the mine. Today, a wide variety of similar efforts are ongoing around the world, from Montana to South Africa, as scientists develop better ways of removing metals from acid mine drainage and as engineers work to scale these systems up. These approaches are showing promise, and give scientists new tools to reclaim mined sites around the world.

TOWARD SUSTAINABLE MINERAL USE

Mining exerts plenty of environmental impacts, but we also have another concern to keep in mind: Minerals are non-renewable resources (p. 2) in finite supply. As a result, it will benefit us to find ways to conserve the supplies we have left and to make them last.

Minerals are limited in supply

Some minerals we use are abundant and will likely never run out, but others are rare enough that they could soon become unavailable. For instance, geologists in 2010 calculated that the world's known reserves of tantalum will last about 164 more years at today's rate of consumption. If demand for tantalum increases, it could run out faster. And if everyone in the world began consuming tantalum at the rate of U.S. citizens, then it would last for only 18 years! Most pressing may be dwindling supplies of indium. This obscure metal, which is used for LCD screens, might last only another 32 years. Because of these supply concerns and price volatility, industries now are working hard to develop ways of substituting other materials for indium. A lack of indium and gallium would threaten the production of high-efficiency cells for solar power. Platinum is dwindling too, and its unavailability would make it harder to develop fuel cells and catalytic converters for vehicles. However, platinum's high market price encourages recycling, which may keep it available, albeit as an expensive metal.

FIGURE 11.20 shows estimated years remaining for several selected minerals at today's consumption rates. Calculating how long a given mineral resource will be available to us is beset by a great deal of uncertainty, for several reasons:

▶ As we discover new deposits of a mineral, the known reserves—and thus the years this mineral is available to

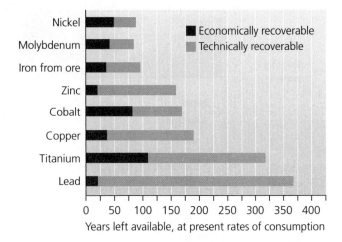

FIGURE 11.20 ▲ Minerals are nonrenewable resources, so supplies of metals are limited. Shown in red are the numbers of remaining years that certain metals are estimated to be economically recoverable at current prices, given known global reserves and assuming current rates of consumption. The entire lengths of the bars (red plus orange) show the numbers of remaining years that certain metals are estimated to be available using current technology on all known deposits, whether economically recoverable or not. All these time periods could increase if more reserves are found, or decrease if consumption rates rise. Data are for 2010, from U.S. Geological Survey, 2011. *Mineral Commodity Summaries 2011.* USGS, Washington, D.C.

us—increase. For this reason, some previously predicted shortages have not come to pass, and we may have access to these minerals for longer than currently estimated.

▸ Rising prices for minerals may favor the development of enhanced mining technologies that can reach more minerals than are currently economically viable to extract.

▸ New societal developments and new technologies can modify demand for minerals. Just as cell phones and computer chips boosted demand for tantalum, fiber-optic cables decreased demand for copper as they replaced copper wiring in communications applications.

▸ Changing consumption patterns alter the speed with which we exploit mineral resources. Economic recession depressed demand and caused a decrease in the production and consumption of most minerals in 2008 and 2009.

▸ Advances in recycling technologies and increased recycling rates can extend the lifetimes of mineral resources by allowing us to reuse them many times.

Despite these sources of uncertainty, we would be wise to be concerned about Earth's finite supplies of mineral resources and to try to use them more sustainably.

We can make our mineral use more sustainable

We can address both major challenges facing us regarding mineral resources—finite supply and environmental damage—by encouraging recycling of these resources. Municipal recycling programs handle used items that we as consumers place in recycling bins and divert from the waste stream. In 2009, fully 35% of metals in the U.S. municipal solid waste stream were diverted for recycling. Car battery recycling programs provide 80% of the lead we need for new products, and 35% of our copper comes from recycled copper sources such as pipes and wires.

In many cases, recycling can decrease energy use substantially. For instance, making steel by remelting recycled iron and steel scrap requires much less energy than producing steel from virgin iron ore. Similarly, over 40% of the aluminum in the United States today is recycled (**FIGURE 11.21**). This is a good thing because it takes over 20 times more energy to extract virgin aluminum from ore (bauxite) than it does to obtain it from recycled sources. This reduced energy use also results in lower emissions of greenhouse gases.

We can recycle metals from e-waste

Electronic waste, or e-waste, from discarded computers, printers, cell phones, handheld devices, and other electronic products is rising fast—and e-waste contains hazardous substances (Chapter 17, pp. 392–394). Recycling old electronic devices helps keep them out of landfills and also helps us conserve valuable minerals such as tantalum.

Tantalum is recycled from scrap by-products generated during the manufacture of electronic components and also from scrap from tantalum-containing alloys and manufactured materials. Currently the industry estimates that recycling accounts for 20–25% of the tantalum available for use in products.

When you turn in your old phone to a recycling and reuse center rather than discarding it, the phone may be refurbished and resold in a developing country. People in African nations in particular readily buy used cell phones because they are inexpensive and land-line phone service does not always exist in poor and rural areas. Alternatively, the phone may be dismantled in a developing country and the various parts refurbished and reused, or recycled for their metals. Either way, you are helping to extend the availability of resources through reuse and recycling and to decrease waste of valuable minerals.

FIGURE 11.21 ▼ When you recycle aluminum cans, you contribute to valuable efforts to save mineral resources, money, and energy.

Today only about 10% percent of old cell phones are recycled. By recycling more, we can reduce demand for virgin ore and decrease pressure on people and ecosystems, such as the miners and natural areas in Africa's coltan-mining regions.

Throughout the world, using recycling to make better use of the mineral resources we have already mined will help minimize the impacts of mining and assure us access to resources further into the future.

➤ CONCLUSION

Physical processes of geology such as plate tectonics and the rock cycle are centrally important because they shape Earth's terrain and form the foundation for living systems. Geologic processes also generate phenomena that can threaten our lives and property, including earthquakes, volcanoes, landslides, and tsunamis. We depend on a diversity of minerals and metals from the Earth and we mine these nonrenewable resources by various methods, according to how the minerals are distributed. Economically efficient mining

methods have greatly contributed to our material wealth, but they have also resulted in extensive environmental impacts, ranging from habitat loss to acid drainage. Restoration efforts and enhanced regulation help to minimize the environmental and social impacts of mining, although to some extent these impacts will always exist. We can prolong our access to mineral resources and make our mineral use more sustainable by maximizing the recovery and recycling of key minerals.

TESTING YOUR COMPREHENSION

1. Name the primary layers that make up our planet. Which portions does the lithosphere include?

2. Describe how plate tectonics accounts for the formation of (a) mountains, (b) volcanoes, and (c) earthquakes. In your answer, explain the roles that divergent plate boundaries, transform plate boundaries, and convergent plate boundaries play in the formation of each.

3. Name the three main types of rocks, and describe how each type may be converted to the others via the rock cycle.

4. Define each of the following and contrast them with one another: mineral, metal, ore, alloy.

5. A mining geologist locates a horizontal seam of coal very near the surface of the land. What type of mining method will the mining company use to extract it? What is one common environmental impact of this type of mining?

6. How does strip mining differ from subsurface mining? How do each of these approaches differ from open pit mining?

7. What is acid drainage, and why can it be toxic to aquatic organisms?

8. Explain why reclamation efforts after mining frequently fail to effectively restore natural communities. Discuss both soil and vegetation in your answer.

9. List five factors that influence how long global supplies of a given mineral will last, and explain how each affects the time span the mineral will be available to us.

10. Name three types of metal that we currently recycle, and identify the products or materials that are recycled to recover these metals.

SEEKING SOLUTIONS

1. For each of the following natural hazards, describe one thing that can be done to minimize or mitigate their impacts on our lives and property:

   ▸ Earthquakes
   ▸ Landslides
   ▸ Flooding

2. List three impacts of mining on the natural environment, and describe how particular mining practices can lead to each of these impacts. How are these impacts being addressed? Can you think of additional solutions to prevent, reduce, or mitigate these impacts?

3. You have won a grant from the U.S. Environmental Protection Agency (EPA) to work with a mining company to develop a more effective way of restoring a mine site that is about to be abandoned. Describe a few preliminary

ideas for carrying out restoration better than it is typically being done. Now describe a field experiment you would like to run to test one of your ideas.

4. **THINK IT THROUGH** The story of coltan in the Congo is just one example of how an abundance of exploitable resources can often worsen or prolong military conflicts in nations that are too poor or ineffectively governed to protect these resources. In such "resource wars," civilians often suffer the most as civil society breaks down. Suppose you are the head of an international aid agency that has earmarked $10 million to help address conflicts related to mining in the Democratic Republic of the Congo. You have access to government and rebel leaders in Congo and neighboring countries, to ambassadors of the world's nations in the United Nations, and to representatives of international mining corporations. Based

on what you know from this chapter, what steps would you consider taking to help improve the situation in the Congo?

5. **THINK IT THROUGH** As you finish your college degree, you learn that the mountains behind your childhood home in the hills of Kentucky are slated to be mined for coal using the mountaintop removal method. Your parents, who still live there, are worried for their health and safety and do not want to lose the beautiful forested creek and ravine behind their property. However, your brother is out of work and could use a mining job. What would you attempt to do in this situation?

CALCULATING ECOLOGICAL FOOTPRINTS

As we saw in Figure 11.20, some metals are in limited enough supply that, at today's prices, they could be available to us for only a few more decades. After that, prices will rise as they become scarcer. The number of years of total availability (at all prices) depends on a number of factors: On the one hand, metals will be available for longer if new deposits are discovered, new mining technologies are developed, or recycling efforts are improved. On the other hand, if our consumption of metals increases, the number of years we have left to use them will decrease.

Currently the United States consumes metals at a much higher per-person rate than the world does as a whole. If one goal of humanity is to lift the rest of the world up to U.S. living standards, then this will sharply increase pressures on mineral supplies.

The chart shows currently known economically recoverable global reserves for several metals, together with the amount used per year (each figure in thousands of metric tons). For each metal, calculate and enter in the fourth column the years of supply left at current prices by dividing the reserves by the amount used annually.

The fifth column shows the amount that the world would use if everyone in the world consumed the metal at the rate that Americans do. Now calculate the years of supply left at current prices for each metal if the world were to consume the metals at the U.S. rate, and enter these values in the sixth column.

Metal	Known economic reserves	Amount used per year	Years of economic supply left	Amount used per year if everyone consumed at U.S. rate	Years of economic supply left if everyone consumed at U.S. rate
Titanium	690,000	6,300		33,390	
Copper	630,000	16,200		38,510	
Nickel	76,000	1,550		5,098	
Tin	5,200	261		846	
Tungsten	2,900	61		312	
Antimony	1,800	135		481	
Silver	510	22.2		130	
Gold	51	2.50		3.34	

Data are for 2010, from U.S. Geological Survey, 2011. Mineral Commodity Summaries 2011. USGS, Washington, D.C. All numbers are in thousands of metric tons. World consumption data are assumed equal to world production data. "Known economic reserves" include extractable amounts under current economic conditions. Additional reserves exist that could be mined at greater cost.

1. Which of these eight metals will last the longest under current economic conditions and at current rates of global consumption? For which of these metals will economic reserves be depleted fastest?

2. If the average citizen of the world consumed metals at the rate that the average U.S. citizen does, which of these eight metals' economic reserves would last the longest? Which would be depleted fastest?

3. In this chart, our calculations of years of supply left do not factor in population growth. All else being equal, how do you think population growth will affect these numbers?

4. Describe two general ways that we could increase the years of supply left for these metals. What do you think it will take to accomplish this?

Go to **www.masteringenvironmentalscience.com** for homework assignments, practice quizzes, Pearson eText, and more.

12 Fresh Water, Oceans, and Coasts

Upon completing this chapter, you will be able to:

➤ Explain water's importance to people and ecosystems, and describe the distribution of fresh water on Earth

➤ Describe the freshwater, marine, and coastal portions of the interconnected aquatic system

➤ Discuss how we use water and how human activities affect aquatic systems

➤ Assess problems of water supply and propose solutions to address depletion of fresh water

➤ Describe the major classes of water pollution and compare and contrast point sources and non-point sources of water pollution

➤ Describe legislation in the United States that addresses water quality

➤ Explain how we treat drinking water and wastewater

➤ Review the state of ocean fisheries and reasons for their decline

➤ Evaluate marine protected areas and reserves as innovative solutions

Louisiana's vanishing coastal wetlands support a diversity of wildlife, such as this Great Egret.

Starving the Louisiana Coast of Sediment

"The Louisiana and Mississippi coastal region is critical to the economic, cultural, and environmental integrity of the nation."
—Nancy Sutley, Chair of the White House Council on Environmental Quality

"What really screwed up the marsh is when they put the levees on the river. They should take the levees out and let the water run; that's what built the land."
—Frank "Blackie" Campo, Resident of Shell Beach, Louisiana

The state of Louisiana is shrinking. Its coastal wetlands straddle the boundary between the land and the ocean, and these wetlands are disappearing beneath the waters of the Gulf of Mexico. Louisiana loses 65 km² (25 mi²) of coastal wetlands each year, and comparisons of wetland area from the mid-1800s to the early 1990s show drastic losses (**FIGURE 12.1A**). Since the 1930s alone, Louisiana has lost nearly 4,900 km² (1,900 mi²) of coastal wetlands—an area roughly the size of Delaware.

Louisiana's coastal wetlands transition from communities of salt-tolerant grasses at the ocean's edge to freshwater bald cypress swamps further inland. These ecosystems support a diversity of animals, including eagles, pelicans, shrimp, oysters, black bears, alligators, and sea turtles. The state's coastal wetlands also protect cities such as New Orleans and Baton Rouge from storms. Vegetation in these wetlands acts as a windbreak on strong winds and as a water break on waves coming inland from the Gulf.

Louisiana's millions of acres of coastal wetlands were created over the past 7,000 years as the Mississippi River fanned out and deposited its sediments at its delta before emptying into the Gulf of Mexico. The Mississippi River accumulates large quantities of sediment from its 3.2-million-km² (1.2-million-mi²) watershed (**FIGURE 12.1B**). Much of this sediment originates from the Missouri River basin that drains America's agricultural heartland.

The salt marshes in the river's delta naturally compact over time. This compaction lowers the level of the marsh bottom and submerges vegetation under increasingly deeper waters. When waters become too deep, the vegetation dies, and soils are then washed away by the ocean. The natural compaction is offset, however, by inputs of sediments from the river and from the deposition of organic matter from marsh grasses. These additions keep soil levels high, water depths relatively stable, and vegetation healthy.

So why are Louisiana's wetlands being swallowed by the sea? It's because people have modified the Mississippi River so extensively that its sediments no longer reach the wetlands that need them. The river's basin contains roughly 2,000 dams, which slow river flow and allow sediments suspended in the water to settle in reservoirs. This not only prevents sediments from reaching the river's delta, but also slowly fills in each dam's reservoir, decreasing its volume and shortening its life span. Therefore, dams in Minnesota and other locations throughout the Mississippi basin affect the Louisiana coastline hundreds of miles downriver.

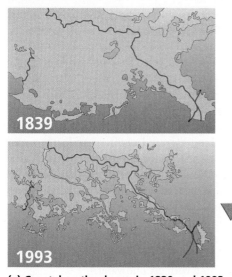

1839

1993

(a) Coastal wetland area in 1839 and 1993

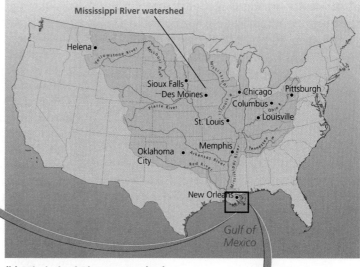

Mississippi River watershed

Helena

Yellowstone River

Missouri River

Sioux Falls
Des Moines

Platte River

St. Louis

Chicago

Columbus

Ohio River

Pittsburgh

Louisville

Oklahoma
City

Arkansas River

Memphis

Tennessee R.

Red River

Mississippi River

New Orleans

Gulf of
Mexico

(b) Mississippi River watershed

FIGURE 12.1 ▲ The size of Louisiana's coastal wetlands shrunk substantially **(a)** from 1839 to 1993 because people modified sediment deposition patterns in the Mississippi River's delta by constructing extensive levees along the river and blocking sediments upriver behind dams. The Mississippi River system **(b)** is the largest in the United States, draining over 40% of the land area of the lower 48 states. A satellite image of south Louisiana **(c)** shows the brown plumes of sediments being released into the deep waters of the Gulf of Mexico from the Mississippi River (plume on the right) and the Atchafalaya River (plume on the left).
(a) Adapted from Environmental Defense Fund.

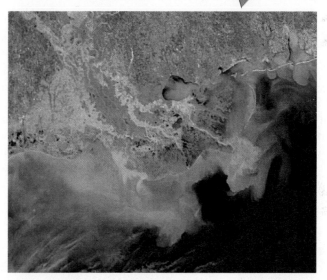

(c) Sediment plumes from Mississippi River entering Gulf

The Mississippi River is also lined with thousands of miles of levees. These structures prevent small-scale flooding, and the mouth of the Mississippi is lined with levees to provide a deep river channel for shipping into the Gulf of Mexico. These levees prevent the river from fanning out into its delta and turn the lower Mississippi into a "barrel" that shoots sediments off the continental shelf into the deep waters of the Gulf (**FIGURE 12.1C**).

Although Louisiana's economy has benefited from oil and gas extraction, these activities have also promoted wetland losses. The extraction of large quantities of oil, natural gas, and saline groundwater associated with oil deposits causes the land to compact, lowering soil levels. Additionally, engineers have cut nearly 13,000 km (8,000 mi) of canals through coastal wetlands to facilitate shipping and oil and gas

exploration. These canals fragment the wetlands and increase erosion rates. They also enable salty ocean water to penetrate inland and damage freshwater marshes. The 2010 *Deepwater Horizon* oil spill in the Gulf of Mexico (pp. 326–327) also affected Louisiana's marshes. Oil from the spill washed into coastal wetlands, coating marsh grasses and impairing their ability to secure oxygen.

Proposed solutions for coastal erosion center on restoring the system to its natural state by diverting large quantities of water from the Mississippi River into coastal wetlands. Proponents of this approach point to the Atchafalaya River, which currently drains one-third of the Mississippi River's volume. The Atchafalaya delta, fed by this water and sediment, is gaining land area of healthy coastal wetlands. In March 2010, the Obama administration announced the creation of a "roadmap" for restoring the Gulf Coast that emphasizes coastal ecosystem restoration, supported by $63 million in funding in 2009–2011. Some residents oppose water diversions from the Mississippi River, however, because they do not wish to see the land they own submerged by floodwaters. Residents also fear contamination of local water supplies from the pollutants carried by the Mississippi River.

Given the conflicting demands we put on waterways for water withdrawal, shipping, and flood control, there are no easy solutions to the problems faced in the Mississippi River and southern Louisiana. But how we tackle problems like those in Louisiana's coastal wetlands will help determine the long-term sustainability of our most precious natural resource—water. ■

FRESHWATER SYSTEMS

"Water, water, everywhere, nor any drop to drink." The well-known line from the poem *The Rime of the Ancient Mariner* describes the situation on our planet well. Water may seem abundant, but water that we can drink is quite rare and limited (**FIGURE 12.2**). About 97.5% of Earth's water resides in the oceans and is too salty to drink or to use to water crops. Only 2.5% is considered **fresh water**, water that is relatively pure with few dissolved salts. Because most fresh water is

tied up in glaciers, icecaps, and underground aquifers, just over 1 part in 10,000 of Earth's water is easily accessible for human use.

Water is renewed and recycled as it moves through the *water cycle* (pp. 37–38). Precipitation falling from the sky either sinks into the ground or acts as runoff to form rivers, which carry water to the oceans or large inland lakes. As they flow, rivers can interact with ponds, wetlands, and coastal aquatic ecosystems. Underground aquifers exchange water with rivers, ponds, and lakes through the sediments on the bottoms of these water bodies. The movement of water in the water cycle creates a web of interconnected aquatic systems (**FIGURE 12.3**) that exchange water, organisms, sediments, pollutants, and other dissolved substances. What happens in one system therefore affects other systems—even those that are far away. Let's examine the components of this interconnected system, beginning with groundwater.

Groundwater plays key roles in the water cycle

Some of the precipitation reaching Earth's land surface infiltrates the surface and percolates downward through the soil to become **groundwater**, water beneath the surface held within pores in soil or rock. Groundwater flows slowly beneath the surface and can remain underground for long periods. Groundwater makes up one-fifth of Earth's fresh water supply and plays a key role in meeting human water needs.

Groundwater is contained within **aquifers**: porous, spongelike formations of rock, sand, or gravel that hold water (p. 38) (**FIGURE 12.4**). An aquifer's upper layer, or *zone of*

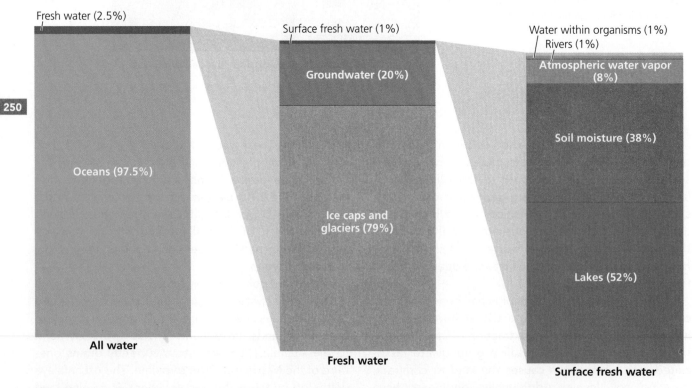

FIGURE 12.2 ▲ Only 2.5% of Earth's water is fresh water. Of that 2.5%, most is tied up in glaciers and ice caps. Of the 1% that is surface water, most is in lakes and soil moisture. Data from United Nations Environment Programme (UNEP) and World Resources Institute.

Groundwater
springs feed
river water

Freshwater wetlands

Salt marsh

Pesticides and
fertilizer enter
groundwater
and surface
water

Agricultural pollutants
and eroded soil

Levees facilitate shipping but
prevent deposition of river
sediments to coastal wetlands

River

Water withdrawals for
irrigation reduce river flow

Dam

Reservoir

Ocean

Urban and
industrial
pollutants

Dam blocks river flows and
traps sediments in reservoir

Groundwater
flowing into ocean

FIGURE 12.3 Water flows through freshwater systems and marine and coastal aquatic systems that interact extensively with one another. People affect the components of the system by constructing dams and levees, withdrawing water for human use, and introducing pollutants. Because the systems are closely connected, these impacts can cascade through the system and cause effects far from where they originated.

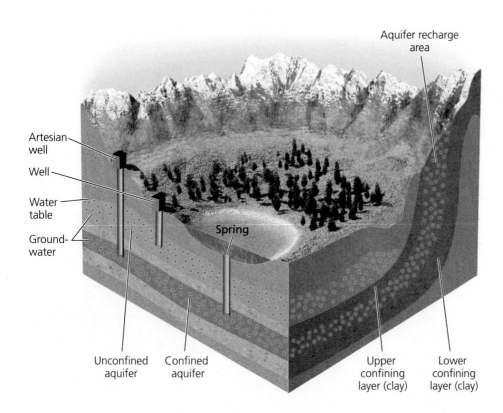

Aquifer recharge
area

Artesian
well

Well

Water
table

Ground-
water

Spring

Unconfined
aquifer

Confined
aquifer

Upper
confining
layer (clay)

Lower
confining
layer (clay)

FIGURE 12.4 ◄ Groundwater may occur in unconfined aquifers above impermeable layers or in confined aquifers under pressure between impermeable layers. Water may rise naturally to the surface at springs and through the wells we dig. Artesian wells tap into confined aquifers to mine water under pressure.

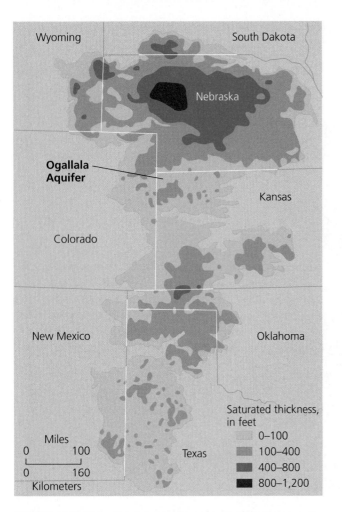

FIGURE 12.5 ▲ The Ogallala Aquifer is the world's largest aquifer, and it held 3,700 km³ (881 mi³) of water before pumping began. This aquifer underlies 453,000 km² (175,000 mi²) of the Great Plains beneath eight U.S. states. Overpumping for irrigation is currently reducing the volume and extent of this aquifer.

Map labels: Wyoming, South Dakota, Nebraska, Ogallala Aquifer, Kansas, Colorado, New Mexico, Oklahoma, Texas

Saturated thickness, in feet
- 0–100
- 100–400
- 400–800
- 800–1,200

Miles: 0   100
Kilometers: 0   160

aeration, contains pore spaces partly filled with water. In the lower layer, or *zone of saturation*, the spaces are completely filled with water. The boundary between these two zones is the **water table**.

The largest known aquifer is the Ogallala Aquifer, which underlies the Great Plains of the United States (**FIGURE 12.5**). Water from this massive aquifer has enabled American farmers to create the most bountiful grain-producing region in the world. However, unsustainable water withdrawals are threatening the long-term use of the aquifer for agriculture.

Surface water converges in river and stream ecosystems

Surface water, liquid fresh water located atop Earth's surface, accounts for just 1% of fresh water, but it is vital for our survival and for the planet's ecological systems. Groundwater and surface water interact, and water can flow from one type of system to the other. Surface water becomes groundwater by infiltration. Groundwater becomes surface water through springs (and human-drilled wells), often keeping streams flowing or wetlands moist when surface conditions are otherwise dry. Each day in the United States, 1.9 trillion L (492 billion gal) of groundwater are released into surface waters—nearly as much as the daily flow of the Mississippi River.

Water that falls from the sky as rain, emerges from springs, or melts from snow or a glacier and then flows over the land surface is called **runoff**. Runoff converges as it flows downhill and forms streams. These small watercourses may merge into rivers, whose water eventually reaches a lake or ocean. A smaller river flowing into a larger one is called a *tributary*. The area of land drained by a river and all its tributaries is the river's **watershed** (p. 21). If you could trace every drop of water in the Mississippi River back to the spot it first fell as precipitation, you would have delineated the river's watershed.

Landscapes determine where rivers flow, but rivers shape the landscapes through which they run. Over thousands or millions of years, a meandering river may shift from one course to another, back and forth over a large area, carving out a flat valley (**FIGURE 12.6**). Areas nearest a river's course that are flooded periodically are said to be within the river's **floodplain**. Frequent deposition of silt from flooding makes floodplain soils especially fertile. As a result, agriculture thrives in floodplains, and *riparian* (riverside) forests are productive and species-rich. A river's meandering is often driven by large-scale flooding events that scour new channels during periods of high flow. However, extensive damming on the Mississippi and other rivers has reduced the rate of river meandering 66–83% from its historic rate. This occurs because floodwaters are trapped by dams and held in reservoirs rather than coursing down the river.

Rivers and streams host diverse biological communities. Algae and detritus (p. 68) support many types of invertebrates, from water beetles to crayfish. Fish and amphibians consume aquatic invertebrates and plants, and birds such as kingfishers, herons, and ospreys dine on fish and amphibians.

FIGURE 12.6 ▲ Rivers, such as the Wood River in Alaska shown here, shape the landscapes through which they flow.

Lakes and ponds are ecologically diverse systems

Lakes and ponds are bodies of standing surface water. The largest lakes, such as North America's Great Lakes, are sometimes known as inland seas. Although lakes and ponds can vary greatly in size, scientists have described several zones common to these waters (**FIGURE 12.7**).

Around the nutrient-rich edges of a water body, the water is shallow enough that aquatic plants grow from the mud and reach above the water's surface. This region, named the *littoral zone*, abounds in invertebrates—such as insect larvae, snails, and crayfish—on which fish, birds, turtles, and amphibians feed. The *benthic zone* extends along the bottom of the lake or pond, from shore to the deepest point, and is home to many invertebrates. In the open portion of a lake or pond, far from shore, sunlight penetrates shallow waters of the *limnetic zone*. Because light enables photosynthesis (p. 30), the limnetic zone supports phytoplankton (algae, protists, and cyanobacteria; p. 21), which in turn support zooplankton (p. 25), both of which are eaten by fish. The open water below the limnetic zone does not receive sunlight, so it has no photosynthetic life and has lower levels of dissolved oxygen than upper waters.

Ponds and lakes change over time as streams and runoff bring them sediment and nutrients. *Oligotrophic* lakes and ponds, which are low in nutrients and high in oxygen, may slowly give way to the high-nutrient, low-oxygen conditions of *eutrophic* water bodies (p. 25). Eventually, water bodies may fill in completely by the process of aquatic succession (p. 74). These changes occur naturally, but eutrophication can also result from human-caused nutrient pollution, as is happening in the Chesapeake Bay (pp. 23–25).

Freshwater wetlands include marshes, swamps, bogs, and vernal pools

Wetlands are systems in which the soil is saturated with water, and they generally feature shallow standing water with ample vegetation. There are many types of freshwater wetlands, and most are enormously rich and productive. In *freshwater marshes*, shallow water allows plants such as cattails and bulrushes to grow above the water surface. *Swamps* also consist of shallow water rich in vegetation, but they occur in

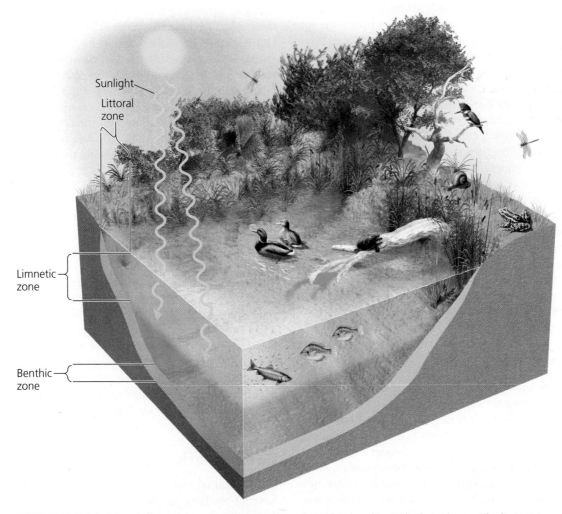

Sunlight

Littoral zone

Limnetic zone

Benthic zone

FIGURE 12.7 ▲ In lakes and ponds, emergent plants grow along the shoreline in the littoral zone. The limnetic zone is the layer of open, sunlit water where photosynthesis takes place. The benthic zone, at the bottom of the water body, often is muddy, rich in detritus and nutrients, and low in oxygen.

FIGURE 12.8 ▲ Freshwater wetlands such as this bald cypress swamp in Louisiana support biologically diverse and productive ecosystems.

forested areas (**FIGURE 12.8**). *Bogs* are ponds covered with thick floating mats of vegetation and can represent a stage in aquatic succession. *Vernal pools* are seasonal wetlands that form in early spring from rain and snowmelt and dry up once weather becomes warmer.

Wetlands are extremely valuable habitat for wildlife. Louisiana's coastal wetlands, for example, provide habitat for approximately 1.8 million migratory waterbirds each year. Wetlands also provide important ecosystem services by slowing runoff, reducing flooding, recharging aquifers, and filtering pollutants. Cypress swamps in coastal Louisiana are home to many rare species that suffer habitat loss when cypress trees are ground up to produce cypress mulch for landscaping. Environmental groups lobbied major home-improvement retailers to cease selling cypress mulch from Louisiana, and in 2007 Walmart announced it would no longer buy or sell cypress mulch from the state. Other major retailers have yet to join Walmart, however, so Louisiana's cypress forests remain threatened.

Despite the vital roles played by wetlands, people have drained and filled them extensively for agriculture. Many wetlands are lost when people divert and withdraw water, channelize rivers, and build dams. Southern Canada and the United States, for example, have lost well over half their wetlands since European colonization.

THE OCEANS

The oceans are an important component of Earth's interconnected aquatic systems. The vast majority of rivers empty into oceans (a small number of rivers empty into inland seas), so the oceans receive most of the inputs of water, sediments, pollutants, and organisms carried by freshwater systems. The oceans touch and are touched by virtually every environmental system and every human endeavor. Even if you live in a landlocked region far from the coast, the oceans affect you. They provide fish for people to eat in Iowa, they supply oil for cars in New Mexico, and they influence the weather in Tennessee.

The physical makeup of the ocean is complex

The world's five major oceans—Pacific, Atlantic, Indian, Arctic, and Antarctic—are all connected, comprising a single vast body of water that covers 71% of Earth's surface. Ocean water contains roughly 96.5% H_2O by mass; most of the remainder consists of ions from dissolved salts. Ocean water is salty primarily because rivers and winds carry sediment and salts from the continents into the ocean. Evaporation then removes pure water, leaving a higher concentration of salts. If we were able to evaporate all the water from the oceans, the world's ocean basins would be covered with a layer of dried salt 63 m (207 ft) thick.

Surface waters of the oceans are warmer than subsurface waters because the sun heats them and because warmer water is less dense. Deep below the surface, water is dense and sluggish, unaffected by winds and storms, sunlight, and daily temperature fluctuations. Ocean water travels in **currents**, vast riverlike flows that move in the upper 400 m (1,300 ft) of water, horizontally and for great distances (**FIGURE 12.9**). Wind, solar heating and cooling, gravity, density differences, and the Coriolis effect (pp. 281–282) drive the global system of ocean currents.

Surface winds and heating also create vertical currents in seawater. **Upwelling** is the rising of cold, dense water toward the surface. Because this water is rich in nutrients from the bottom, upwellings often support high primary productivity (p. 32) and lucrative fisheries. At **downwellings**, warm surface water rich in dissolved gases is displaced downward, providing an influx of oxygen for deep-water life.

Although oceans are depicted on most maps and globes as smooth swaths of blue, parts of the ocean floor are rugged and complex (**FIGURE 12.10**). Underwater volcanoes shoot forth enough magma to build islands above sea level, such as the Hawaiian Islands (see Figure 11.7, p. 232). Steep canyons similar in scale to Arizona's Grand Canyon lie just offshore of some continents. The deepest spot in the oceans—the Mariana Trench in the South Pacific—is deeper than Mount Everest is high, by over a mile. Our planet's longest mountain range is under water—the Mid-Atlantic Ridge (pp. 228–229) runs the length of the Atlantic Ocean.

Some ocean regions support more life than others. The uppermost 10 m (33 ft) of ocean water absorbs 80% of the solar energy that reaches its surface. For this reason, nearly all of the oceans' primary productivity occurs in the well-lit top layer, or *photic zone*. Generally, the warm, shallow waters of continental shelves are most biologically productive and support the greatest species diversity. Habitats and ecosystems occurring between the ocean's surface and floor are classified as **pelagic**, whereas those that occur on the ocean floor are classified as **benthic**.

Currents affect climate

The horizontal and vertical movements of ocean water can have far-reaching effects on climate globally and regionally. The **thermohaline circulation** is a worldwide current system in which warmer, fresher water moves along the surface and colder, saltier water (which is denser) moves deep beneath

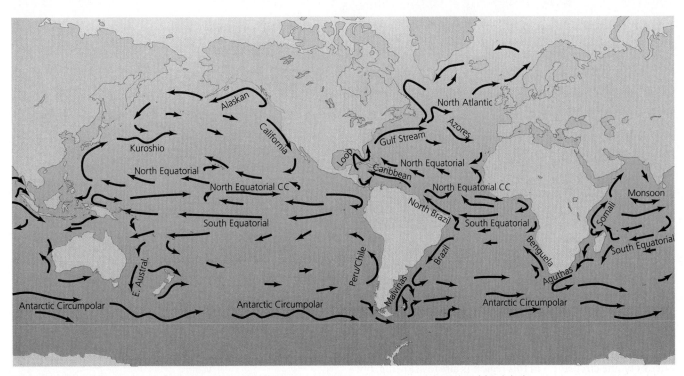

FIGURE 12.9 ▲ The upper waters of the oceans flow in surface currents, long-lasting and predictable global patterns of water movement. Warm- and cold-water currents interact with the planet's climate system, and people have used them for centuries to navigate the oceans. *Source:* Adapted from Rick Lumpkin (NOAA/AOML).

the surface (**FIGURE 12.11**). One segment of this worldwide conveyor-belt system is the warm surface water in the Gulf Stream that flows across the Atlantic Ocean to Europe. As this water releases heat to the air, keeping Europe warmer than it would otherwise be, the water cools, becomes saltier through evaporation, becomes denser, and sinks. This creates a region of downwelling known as the *North Atlantic Deep Water (NADW)*.

Scientists hypothesize that interrupting the thermohaline circulation could trigger rapid climate change. If global warming (Chapter 14) causes much of Greenland's ice sheet to melt, the resulting freshwater runoff into the North Atlantic would make surface waters less dense (because fresh water is less dense than salt water). This could potentially stop the NADW formation and shut down the northward flow of warm water, causing Europe to cool rapidly. Some data suggest that the

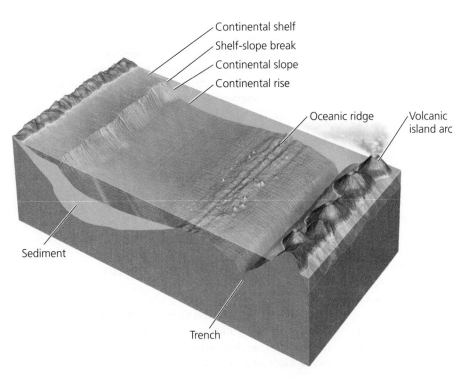

Continental shelf
Shelf-slope break
Continental slope
Continental rise
Oceanic ridge
Volcanic island arc
Sediment
Trench

FIGURE 12.10 ◄ A stylized bathymetric profile shows key geologic features of the submarine environment. Shallow water exists around the edges of continents over the continental shelf, which drops off at the shelf-slope break. The steep continental slope gives way to the more gradual continental rise, all of which are underlain by sediments from the continents. Vast areas of seafloor are flat abyssal plain. Seafloor spreading occurs at oceanic ridges, and oceanic crust is subducted in trenches (p. 229). Volcanic activity along trenches may give rise to island chains such as the Aleutian Islands. Features on the left side of this diagram are more characteristic of the Atlantic Ocean, and features on the right side of the diagram are more characteristic of the Pacific Ocean. Adapted from Thurman, H.V., 1990. *Essentials of oceanography*, 4th ed. New York: Macmillan.

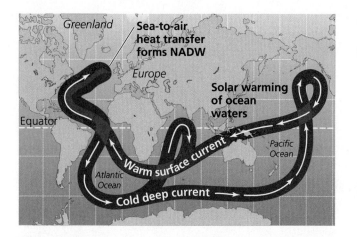

FIGURE 12.11 ▲ As part of the oceans' thermohaline circulation, warm surface currents carry heat from equatorial waters northward toward Europe, where they warm the atmosphere and then cool and sink, forming the North Atlantic Deep Water (NADW). Scientists debate whether rapid melting of Greenland's ice sheet could interrupt this heat flow and cause Europe to cool dramatically.

thermohaline circulation in this region is already slowing, but other researchers maintain that Greenland will not produce enough runoff to cause a shutdown this century.

Another interaction between ocean currents and the atmosphere that influences climate is the **El Niño–Southern Oscillation (ENSO)**, a systematic shift in atmospheric pressure, sea surface temperature, and ocean circulation in the tropical Pacific Ocean. Under normal conditions, prevailing winds blow from east to west along the equator, from a region of high pressure in the eastern Pacific to one of low pressure in the western Pacific, forming a large-scale convective loop in the atmosphere (**FIGURE 12.12A**). The winds push surface waters westward, causing water to "pile up" in the western Pacific. As a result, water near Indonesia can be 50 cm (20 in.) higher and 8 °C warmer than water near South America. The westward-moving surface waters allow cold water to rise up from the deep in a nutrient-rich upwelling along the coast of Peru and Ecuador.

El Niño conditions are triggered when air pressure decreases in the eastern Pacific and increases in the western Pacific, weakening the equatorial winds and allowing the warm water to flow eastward (**FIGURE 12.12B**). This suppresses upwelling along the Pacific coast of the Americas, shutting down the delivery of nutrients that support marine life and fisheries. Coastal industries such as Peru's anchovy fisheries are devastated by El Niño events, and the 1982–1983 El Niño caused over $8 billion in economic losses worldwide. El Niño events alter weather patterns around the world, creating rainstorms and floods in areas that are generally dry (such as southern California) and causing drought and fire in regions that are typically moist (such as Indonesia).

La Niña events are the opposite of El Niño events; in a La Niña event, cold waters rise to the surface and extend westward in the equatorial Pacific when winds blowing to the west strengthen, and weather patterns are affected in opposite ways. ENSO cycles are periodic but irregular, occurring every 2–8 years. Scientists are exploring whether globally warming air and sea temperatures (Chapter 14) may be increasing the frequency and strength of these cycles.

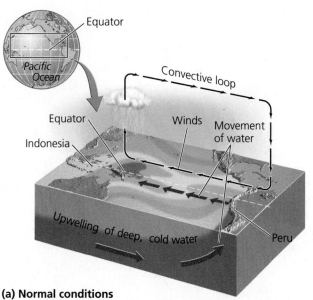

(a) Normal conditions

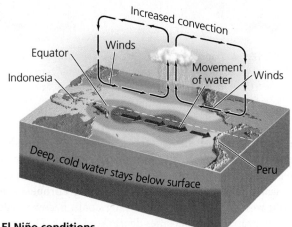

(b) El Niño conditions

FIGURE 12.12 ▲ In these diagrams, red and orange colors denote warmer water, and blue and green colors denote colder water. Under normal conditions **(a)**, prevailing winds push warm surface waters toward the western Pacific. Under El Niño conditions **(b)**, winds weaken, and the warm water flows back across the Pacific toward South America, like water sloshing in a bathtub. This shuts down upwelling along the American coast and alters precipitation patterns regionally and globally. Adapted from National Oceanic and Atmospheric Administration, Tropical Atmospheric Ocean Project.

MARINE AND COASTAL SYSTEMS

With their variation in topography, temperature, salinity, nutrients, and sunlight, marine and coastal environments feature a variety of ecosystems. These systems may not give us the water we need for drinking and growing crops, but they teem with biodiversity and provide many other necessary resources.

Fresh water meets salt water in estuaries

Water bodies where rivers flow into the ocean, mixing fresh water with salt water, are called **estuaries** (**FIGURE 12.13**). Estuaries are biologically productive ecosystems that experience fluctuations in salinity with the daily and seasonal

FIGURE 12.13 ▲ Freshwater rivers mix with salt water to form brackish water in estuaries. Elevated inputs of nutrients from river waters make estuaries one of the most productive ecosystem types on Earth.

FIGURE 12.15 ▲ Mangrove forests line tropical and subtropical coastlines. Mangrove trees, with their unique roots, are adapted for growing in salt water and provide habitat for many fish, birds, crabs, and other animals.

variations in tides and freshwater runoff. The shallow water of estuaries nurtures eelgrass beds and other plant life and provides critical habitat for shorebirds and for many commercially important shellfish species.

Estuaries everywhere have been affected by coastal development, water pollution, habitat alteration, and overfishing. The Chesapeake Bay estuary (profiled in Chapter 2) is one such example. Estuaries and other coastal ecosystems have borne the brunt of human impact because two out of every three people choose to live within 160 km (100 mi) of the ocean.

Salt marshes line temperate shorelines

Along many of the world's coasts at temperate latitudes, **salt marshes** occur where the tides wash over gently sloping sandy or silty substrates. Rising and falling tides flow into and out of channels called *tidal creeks* and at highest tide spill over onto elevated marsh flats, like those in coastal Louisiana (**FIGURE 12.14**). Marsh flats grow thick with salt-tolerant grasses, as well as rushes, shrubs, and other herbaceous plants.

Salt marshes boast very high primary productivity and provide critical habitat for shorebirds, waterfowl, and many

FIGURE 12.14 ▼ Salt marshes occur in temperate intertidal zones where the substrate is muddy. Tidal waters flow in channels called tidal creeks amid flat areas called benches, sometimes partially submerging the salt-adapted grasses.

fish and shellfish species. Salt marshes also filter pollution and stabilize shorelines against storm surges.

Mangrove forests line coasts in the tropics and subtropics

In tropical and subtropical latitudes, mangrove forests replace salt marshes along the coasts. **Mangroves** are salt-tolerant, and they have unique roots that curve upward like snorkels to attain oxygen, or that curve downward like stilts to support the tree in changing water levels (**FIGURE 12.15**). Fish, shellfish, crabs, snakes, and other organisms thrive among the root networks, and birds feed and nest in the dense foliage of these coastal forests. Mangroves protect shorelines from storm surges, filter pollutants, and capture eroded soils, protecting offshore coral reefs. They also provide materials that people use for food, medicine, tools, and construction. Half the world's mangrove forests have been destroyed as people have developed coastal areas, often for tourist resorts and shrimp farms.

Intertidal zones undergo constant change

Where the ocean meets the land, **intertidal**, or **littoral**, ecosystems (**FIGURE 12.16**) spread between the uppermost reach of the high tide and the lowest limit of the low tide. **Tides** are the periodic rising and falling of the ocean's height at a given location, caused by the gravitational pull of the moon and sun. Intertidal organisms spend part of each day submerged in water, part of the day exposed to air and sun, and part of the day being lashed by waves.

Life abounds in the crevices of rocky shorelines, which provide shelter and pools of water (tide pools) during low tides. Sessile (stationary) animals such as anemones, mussels, and barnacles live attached to rocks, filter-feeding on plankton in the water that washes over them. Urchins, sea slugs, chitons, and limpets eat intertidal algae or scrape food from

FIGURE 12.16 ▲ The rocky intertidal zone stretches along rocky shorelines between the lowest and highest reaches of the tides, providing niches for a diversity of organisms. Areas higher on the shoreline are exposed to the air more frequently and for longer periods, so organisms that tolerate exposure best specialize in the upper intertidal zone. The lower intertidal zone is exposed less frequently and for shorter periods, so organisms less tolerant of exposure thrive in this zone.

the rocks. Sea stars (starfish) creep slowly along, preying on the filter-feeders and herbivores. Crabs clamber around the rocks, scavenging detritus. The rocky intertidal zone is so diverse because environmental conditions such as temperature, salinity, and moisture change dramatically from the high to the low reaches.

Kelp forests harbor many organisms

Along many temperate coasts, large brown algae, or **kelp**, grow from the floor of continental shelves, reaching up toward the sunlit surface. Some kelp reaches 60 m (200 ft) in height and can grow 45 cm (18 in.) per day. Dense stands of kelp form underwater "forests" (**FIGURE 12.17**). Kelp forests supply shelter and food for invertebrates and fish, which in turn provide food for predators such as seals and sharks. (Indeed, kelp forests were the setting for our discussion of sea otters as keystone species in Chapter 4, pp. 71–73.) Kelp forests absorb wave energy and protect shorelines from erosion. People eat some types of kelp, and kelp provides compounds that serve as thickeners in cosmetics, paints, ice cream, and other consumer products.

Coral reefs are treasure troves of biodiversity

Shallow subtropical and tropical waters are home to coral reefs. A coral reef is a mass of calcium carbonate composed of the skeletons of tiny invertebrate animals known as *corals*. Corals are related to jellyfish and capture passing food

with stinging tentacles. They also derive nourishment from symbiotic algae known as *zooxanthellae*, which inhabit their bodies and produce food through photosynthesis (and provide the diversity of vibrant colors in reefs). Most corals are colonial, and the colorful surface of a coral reef consists of millions of densely packed individuals. As corals die, their skeletons remain part of the reef while new corals grow atop them, increasing the reef's size.

Coral reefs protect shorelines by absorbing wave energy. They also host tremendous biodiversity (**FIGURE 12.18A**). This is because coral reefs provide complex physical structure (and thus many habitats) in shallow nearshore waters, which are

FIGURE 12.17 ▼ "Forests" of tall brown algae known as kelp grow from the floor of the continental shelf. Numerous fish and other creatures eat kelp or find refuge among its fronds.

(a) Coral reef community

(b) Bleached coral

FIGURE 12.18 ▲ Corals reefs provide food and shelter for a tremendous diversity of fish and other creatures **(a)**. Today these reefs face multiple stresses from human impacts. Many corals have died as a result of coral bleaching **(b)**, in which corals lose their zooxanthellae. Bleaching is evident in the whitened portion of this coral.

regions of high primary productivity. Besides the staggering diversity of anemones, sponges, tubeworms, and other sessile invertebrates, innumerable molluscs, flatworms, and urchins patrol reefs, while thousands of fish species find food and shelter in the reef structure. The beauty and biodiversity of coral reefs makes them valuable ecotourism destinations that provide economic benefit to coastal communities.

Coral reefs are experiencing alarming declines worldwide, however. Nutrient pollution in coastal waters promotes the growth of algae, which smothers reefs. Many reefs have undergone "coral bleaching," a process that occurs when zooxanthellae die or leave the coral, depriving it of nutrition (**FIGURE 12.18B**). Coral bleaching is thought to result from increased sea surface temperatures associated with global climate change, from the influx of pollutants, from unknown natural causes, or from combinations of these factors.

Coral reefs have been affected by climate change in other ways, too. The oceans have soaked up roughly a third of the excess carbon dioxide (CO_2) that people have added to the atmosphere so far, and this has slowed the onset of global climate change. However, this excess CO_2 has lowered the pH of seawater, a phenomenon called **ocean acidification**. This causes a series of chemical reactions that reduce the ocean's concentration of the carbonate ions that coral and other creatures need to build their shells. These reactions also produce molecules that dissolve the calcium carbonate in coral shells. By dissolving shells at an increased rate and decreasing the rate at which new shells are formed, ocean acidification threatens the persistence of coral reefs. Ocean acidification increases along with atmospheric concentrations of CO_2. If atmospheric concentrations of CO_2 reach 500 parts per million (ppm) (up from 391 ppm today), very little ocean area will have carbonate ion concentrations sufficient to support coral reefs.

Open-ocean ecosystems vary in their biodiversity

Biological diversity in pelagic regions of the open ocean is highly variable in its distribution. Primary production (p. 32) and animal life near the surface are concentrated in regions of nutrient-rich upwelling. Microscopic phytoplankton constitute the base of the marine food chain in the pelagic zone and are the prey of zooplankton (p. 25), which in turn become food for fish, jellyfish, whales, and other free-swimming animals (**FIGURE 12.19**). Predators at higher trophic levels include larger fish, sea turtles, and sharks.

In the little-known deep-water ecosystems, animals have adapted to tolerate extreme water pressures and to live in the dark without food from autotrophs (p. 30). Many of these often bizarre-looking creatures scavenge carcasses or organic detritus that falls from above. Others are predators, and still

FIGURE 12.19 ▼ The uppermost reaches of ocean water contain billions upon billions of phytoplankton—tiny photosynthetic algae, protists, and bacteria that form the base of the marine food chain—as well as zooplankton, small animals and protists that dine on phytoplankton and comprise the next trophic level.

others attain food from mutualistic (p. 68) bacteria. Ecosystems also form around hydrothermal vents, where heated water spurts from the seafloor, carrying minerals that precipitate to form rocky structures. Tubeworms, shrimp, and other creatures in these recently discovered deepwater systems use symbiotic bacteria to derive their energy from chemicals in the heated water rather than from sunlight.

Aquatic systems are affected by human activities

Our tour of aquatic systems has shown the ecological and economic value of freshwater and marine ecosystems. We will now see how people affect these systems when we withdraw water for human use, build dams and levees, and introduce pollutants that alter water's chemical, biological, and physical properties.

HUMAN ACTIVITIES AFFECT WATERWAYS

Although water is a limited resource, it is also a renewable resource as long as we manage our use sustainably. Unfortunately, people are withdrawing water at unsustainable levels and are depleting many sources of surface water and groundwater. Already, one-third of the world's people are affected by water shortages.

Additionally, people have intensively engineered freshwater waterways with dams, levees, and diversion canals to satisfy demands for water supplies, transportation, and flood control. As we have seen in our Central Case Study, dams and channelization in the Mississippi River basin have led to adverse impacts at the river's mouth. What we do in one part of the interconnected aquatic system affects many others, sometimes in significant ways.

Fresh water is unevenly distributed across Earth

The availability of fresh water varies widely around the world because different regions possess varying amounts of groundwater, surface water, and precipitation. When we compare human populations with fresh water supplies on a map, we find that people are not distributed across the globe in accordance with the amount of fresh water (**FIGURE 12.20**). Because of the mismatched distribution of water and population, human societies have always struggled to transport fresh water from its source to where people need it.

Fresh water is distributed unevenly in time as well as space. India's monsoon storms can dump half of a region's annual rain in just a few hours, for example. Rivers have seasonal differences in flow. For this reason, dams store water from wetter months that can be used in drier times of the year when river flow is reduced.

As if the existing mismatches between water availability and human need were not enough, global climate change (Chapter 14) may worsen conditions in many regions by altering precipitation patterns, melting glaciers, causing early season runoff, and intensifying droughts and flooding.

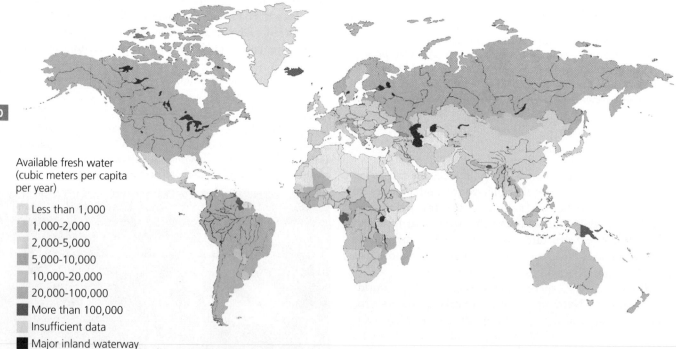

Available fresh water
(cubic meters per capita
per year)

- Less than 1,000
- 1,000–2,000
- 2,000–5,000
- 5,000–10,000
- 10,000–20,000
- 20,000–100,000
- More than 100,000
- Insufficient data
- Major inland waterway

FIGURE 12.20 ▲ Nations vary tremendously in the amount of fresh water per capita available to their citizens. For example, Iceland, Papua New Guinea, Gabon, and Guyana (dark blue in this map) each have over 100 times more water per person than do many Middle Eastern and North African countries. Data from Harrison, P., and F. Pearce, 2000. *AAAS atlas of population and the environment*, edited by the American Association for the Advancement of Science, © 2000 by the American Association for the Advancement of Science. Used by permission of the publisher, University of California Press.

Water supplies households, industry, and especially agriculture

We all use water at home for drinking, cooking, and cleaning. Most mining, industrial, and manufacturing processes require water. Farmers and ranchers use water to irrigate crops and water livestock. Globally, we allot about 70% of our annual fresh water use to agriculture. Industry accounts for roughly 20%, and residential and municipal uses for only 10%.

When we remove water from an aquifer or surface water body and do not return it, this is called **consumptive use**. Our primary consumptive use of water is for agricultural irrigation (p. 145–146). In contrast, **nonconsumptive use** of water does not remove, or only temporarily removes, water from an aquifer or surface water body. Using water to generate electricity at hydroelectric dams is an example of nonconsumptive use; water is taken in, passed through dam machinery to turn turbines, and released downstream.

We use so much water in agriculture because our rapid population growth requires us to feed and clothe more people each year. Overall, we withdraw 70% more water for irrigation today than we did 50 years ago and have doubled the amount of land under irrigation. Irrigation can more than double crop yields; as a result, the 18% of world farmland that we irrigate yields fully 40% of our produce, including 60% of the global grain crop.

Worldwide, roughly 15–35% of water withdrawals for irrigation are thought to be unsustainable. In areas where agriculture is demanding more fresh water than can be sustainably supplied, *water mining*—withdrawing water faster than it can be replenished—is taking place. In these areas, aquifers are being depleted or surface water is being piped in from other regions.

We build dikes and levees to control floods

Flooding is a normal, natural process that occurs when snowmelt or heavy rain swells the volume of water in a river so that water spills over the river's banks. This process naturally spreads nutrient-rich sediments over the floodplain, benefiting natural ecosystems and human agriculture.

As cities and towns grew on the floodplains of the Mississippi River and other waterways, however, people tired of the property damage caused by flooding. Communities and governments constructed *dikes* and *levees* (long raised mounds of earth) along riverbanks to hold water in main channels. These structures prevent flooding most of the time, but they can sometimes worsen flooding because they force water to stay in channels and accumulate, building up enormous energy and leading to occasional catastrophic overflow events (**FIGURE 12.21**). Many of the levees along the Mississippi River were constructed after the lower Mississippi River flooded catastrophically in 1927 and the public demanded greater protection from flooding.

We have erected thousands of dams

A **dam** is any obstruction placed in a river or stream to block its flow. Dams create **reservoirs**, artificial lakes that store water for human use. We build dams to prevent floods,

FIGURE 12.21 ▲ Unusually high water levels in the Mississippi River in May 2011 caused this levee in East Carroll Parish, Louisiana, to collapse, flooding adjacent farmland. Levees upstream in some less-populated areas were intentionally destroyed and the floodplain inundated in order to lower river volumes and protect more-populated areas downriver.

provide drinking water, facilitate irrigation, and generate electricity.

Worldwide, we have erected more than 45,000 large dams (greater than 15 m, or 49 ft, high) across rivers in over 140 nations. We have built tens of thousands of smaller dams. Only a few major rivers in the world remain undammed and free-flowing. These run through the tundra and taiga of Canada, Alaska, and Russia and in remote regions of Latin America and Africa.

Dams produce a mix of benefits and costs, as illustrated in **FIGURE 12.22** and as mentioned in our Central Case Study. As an example of this complex mix, consider the world's largest dam: the Three Gorges Dam on China's Yangtze River, at 186 m (610 ft) high and 2.3 km (1.4 mi) wide. This dam was completed in 2008 (**FIGURE 12.23A**), and its reservoir stretches for 616 km (385 mi; as long as Lake Superior). This dam will help control the large floods that have historically occurred on the banks of the Yangtze. It will also enable boats and barges to travel farther upstream and will generate enough hydroelectric power to replace dozens of large coal or nuclear plants.

However, the Three Gorges Dam cost $39 billion to build, and its reservoir flooded 22 cities and displaced 1.24 million people (**FIGURE 12.23B**). The rising water submerged 10,000-year-old archaeological sites, productive farmlands, and wildlife habitat. Tidal marshes at the Yangtze's mouth are eroding away, deprived of the sediments that now settle in the dam's reservoir as the river water slows. Many scientists also worry that the Yangtze's many pollutants will be trapped in the reservoir, making the water undrinkable.

People who feel that the costs of some dams outweigh their benefits are pushing for such dams to be dismantled. By removing dams and letting rivers flow free, they say, we can restore ecosystems, reestablish economically valuable fisheries, and revive river recreation, such as fly-fishing and rafting. Many aging dams are in need of costly repairs or have outlived their economic usefulness, and these are candidates for removal. Some 400 dams have been removed in the United States

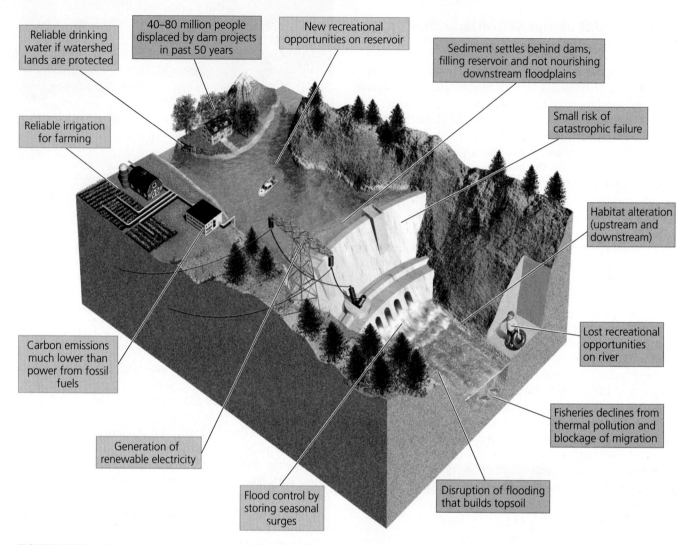

FIGURE 12.22 ▲ Damming rivers has diverse consequences for people and the environment. The generation of clean and renewable electricity is one of several major benefits **(green boxes)** of hydroelectric dams. Habitat alteration is one of several negative impacts **(red boxes)**.

in the past decade, and more will come down in the next 10 years, when the licenses of over 500 dams come up for renewal.

We divert surface water to suit our needs

People have long diverted water from rivers and lakes to farm fields, homes, and cities. The world's largest diversion project is underway in China. There, government leaders are pushing through an ambitious plan to pipe water from the Yangtze River in southern China (where water is plentiful) to northern China's Yellow River, which routinely dries up at its mouth because the climate is drier and people withdraw its water. Three sets of massive aqueducts (human-made river channels), totaling 2,500 km (1,550 mi) in length, are being built to move trillions of gallons of water northward. China's leaders hope the diversions will solve water shortages for northern farms and cities. However, many scientists say the $62 billion project won't transfer enough water to satisfy northern China's water demands, yet it will cause extensive environmental impacts and displace hundreds of thousands of people.

WEIGHING THE ISSUES

Reaching for Water Controversial diversions of fresh water occur in the United States as well. The rapidly growing Las Vegas metropolitan area is exceeding its allotment of water from the Colorado River and has proposed a $3.5-billion project that would divert groundwater from 450 km (280 mi) away to meet the growing demand in Nevada's largest city. Do you think such diversions are ethically justified? If rural communities and wetland ecosystems at the diversion site in eastern Nevada are destroyed by this project, is this an acceptable cost given the economic activity generated in Las Vegas? How else might cities like Las Vegas meet their future water needs?

Excessive water withdrawals are draining rivers and lakes

As a result of our diversions and our consumption, in many places we are withdrawing surface water at unsustainable rates

(a) The Three Gorges Dam in Yichang, China

(b) Displaced people in Sichuan Province, China

FIGURE 12.23 ▲ China's Three Gorges Dam (a), completed in 2008, is the world's largest dam. Over 1.2 million people were displaced and whole cities were leveled for its construction, as shown here (b) in Sichuan Province.

and drastically reducing flows in rivers. Many major rivers—the Colorado River in North America, the Yellow River in China, and the Nile in Africa—regularly run dry before reaching the sea due to excessive water withdrawals. This reduction in flow not only threatens the future of the cities and farms that depend on these rivers, but also drastically alters the ecology of the rivers and their deltas, changing plant communities, wiping out populations of fish and invertebrates, and devastating fisheries.

Nowhere are the effects of surface water depletion so evident as at the Aral Sea, on the border of present-day

Uzbekistan and Kazakhstan. Once the fourth-largest lake on Earth, the Aral Sea lost over four-fifths of its volume in just 45 years (**FIGURE 12.24**), as water was diverted from the two rivers that feed it, to provide irrigation for cotton farming in this arid region. The shrinking of the Aral Sea has caused the loss of 60,000 fishing jobs, but scientists, engineers, and local people are struggling to save the northern portion of the Aral Sea, and may now have begun reversing its decline.

(a) Satellite image of Aral Sea, 1987

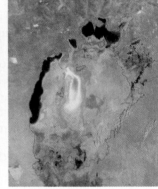

(b) Satellite image of Aral Sea, 2009

(c) Ships stranded by the Aral Sea's fast-receding waters

FIGURE 12.24 ▲ The Aral Sea in central Asia has been shrinking (a, b) because so much water was withdrawn to irrigate cotton crops. Along its former shorelines, ships lie stranded in the sand (c) because the waters receded so far and so quickly. Today restoration efforts are beginning to reverse the decline in the northern portion of the sea, and waters there are slowly rising.

Groundwater depletion affects people and ecosystems

Groundwater is more easily depleted than surface water because most aquifers recharge very slowly. If we compare an aquifer to a bank account, we are making more withdrawals than deposits, and the balance is shrinking. Today we are mining groundwater, extracting 160 km³ (5.65 trillion ft³) more water each year than returns to the ground. This is a problem because one-third of Earth's human population—including 99% of the rural population of the United States—relies on groundwater for its needs.

As aquifers are mined, water tables drop deeper underground. This deprives freshwater wetlands of groundwater inputs, causing them to dry up. Groundwater also becomes more difficult and expensive to extract, and eventually it may run out. In parts of Mexico, India, China, and multiple Asian and Middle Eastern nations, water tables are falling 1–3 m (3–10 ft) per year.

When groundwater is overpumped in coastal areas, salt water can intrude into aquifers from the ocean, making water undrinkable. This has occurred in Florida, the Middle East, and other locations. Moreover, as aquifers lose water, they become less able to support overlying strata, and the land surface above may sink or collapse. Mexico City, Venice, Bangkok, and Beijing are slowly sinking, causing streets to buckle and undergound pipes to rupture. Once the ground sinks, soil and rock becomes compacted, losing the porosity that enabled it to hold water. Recharging a depleted aquifer thereafter becomes more difficult.

Bottled water has ecological costs

These days, our groundwater is being withdrawn for a new purpose—to be packaged in plastic bottles and sold on supermarket shelves. In 2009 the average American drank over 27 gallons of bottled water, and sales topped $10.5 billion in the United States and $60 billion worldwide.

Most people who buy bottled water do so for portability and convenience, or because they believe it is superior to tap water. However, in blind taste tests people think tap water tastes just as good, and chemical analyses show that bottled water is no safer or healthier than tap water (see **THE SCIENCE BEHIND THE STORY**, pp. 268–269).

Bottled water also exerts substantial ecological impact. A 2009 study calculated the energy costs of bottled water to be 1,000–2,000 times greater than the energy costs of tap water. Most energy was used in manufacturing the bottle and transporting the product. Furthermore, since at least three out of four bottles in the United States are thrown away after use, we must dispose of 30–40 billion containers per year.

Will we see a future of water wars?

Population growth, expansion of irrigated agriculture, and industrial development doubled our annual fresh water use between 1960 and 2000. Increased withdrawals of fresh water can lead to shortages, and resource scarcity can lead to conflict. Many predict that water's role in regional conflicts will increase as human population continues to grow and as climate change alters precipitation patterns. A total of 261 major rivers (whose watersheds cover 45% of the world's land area)

cross national borders, and transboundary disagreements are common. Water is already a key element in the disagreements among Israel, the Palestinian people, and neighboring nations.

The United States has its share of conflicts over water. The Colorado River's water allocations have long been a source of conflict between farms and growing cities in the arid West. The states of Georgia, Alabama, and Florida are also currently embroiled in legal disputes over water withdrawals from shared rivers.

Yet on the positive side, many nations have cooperated to resolve water disputes. India has struck agreements to co-manage transboundary rivers with Pakistan, Bangladesh, Bhutan, and Nepal. In Europe, nations along the Rhine and Danube rivers have signed water-sharing treaties. Such progress gives reason to hope that water wars will be few and far between.

SOLUTIONS TO THE DEPLETION OF FRESH WATER

To address the depletion of fresh water, we can aim either to increase supply or to reduce demand. Increasing water supplies by constructing large dams was a common solution to water shortages in the past. However, we have already developed the vast majority of suitable sites for large dams, and most of those that remain are in remote locations in developing regions. Building more dams therefore does not appear to be a viable solution in most locations for meeting people's demands for fresh water.

Another strategy for increasing water supplies is to generate fresh water by desalination, the removal of salt from seawater or other water of marginal quality. This can be done by evaporating the salt water and condensing the freshwater vapor that is produced. Unfortunately, desalination is an energy-intensive process that requires large inputs of fossil fuels or electricity. This makes it expensive and limits its intensive use to arid, oil-rich nations like those in the Persian Gulf region. Hence, its ability to increase water supplies in most nations is limited.

FAQ

Q: Can't we just use desalination to fulfill our demand for water?

A: Given the seemingly endless supply of water in Earth's oceans, many people assume that desalination is the answer to our world's water crises. So why aren't we eagerly utilizing this technology everywhere?

Simply put, we lack the abundant, clean energy sources needed to make the widespread use of desalination economically viable and environmentally sustainable. For example, the United States withdraws over 700 billion liters (185 billion gallons) of fresh water *every day* for use in food production, industry, and public supplies. Diverting the energy necessary to supply even a tiny fraction of this quantity from desalination would cause prices for electricity, gasoline, and other fuels to skyrocket. Using fossil fuels as an energy source for desalination would also drastically increase U.S. emissions of air pollutants and greenhouse gases. Due to these constraints, it is unlikely that desalination will be widely embraced in the United States unless we are able to find abundant, environmentally-friendly energy sources.

We can decrease our demand for water

Because supply-based strategies do not hold great promise for increasing water supplies, people are increasingly turning to demand-based solutions. Strategies for reducing fresh water demand include conservation and efficiency measures. Such strategies require changes in individual behaviors and can therefore be politically difficult, but they offer better economic returns and cause less ecological and social damage. Our existing shift from supply-based to demand-based solutions is already paying dividends. The United States, for example, decreased its water consumption by 5% from 1980 to 2005 thanks to conservation measures, even while its population grew 31%. Let's examine approaches that can conserve water in agriculture, households, industry, and municipalities.

Agriculture   Farmers can improve efficiency by adopting more efficient irrigation methods. "Flood and furrow" irrigation, in which fields are liberally flooded with water, accounts for 90% of irrigation worldwide. However, crop plants end up using only 40% of the water applied, and other methods are far more efficient. Low-pressure spray irrigation squirts water downward toward plants, and drip irrigation systems target individual plants and introduce water directly onto the soil (pp. 145-146). Experts estimate that drip irrigation (in which as little as 10% of water is wasted) could cut water use in half while raising yields by 20–90% and producing $3 billion in extra annual income for farmers of the developing world.

Choosing crops to match the land and climate in which they are farmed can also save huge amounts of water. Eliminating the planting of crops that require a great deal of water (such as cotton, rice, and alfalfa) in arid areas with government-subsidized irrigation could greatly reduce water use. Biotechnology may also play a role by producing crop varieties that require less water through selective breeding (p. 48) and genetic modification (pp. 149–152).

Households   In our residences, we can reduce water use by installing low-flow faucets, showerheads, washing machines, and toilets. High-efficiency toilets and showerheads provide the biggest savings because these are typically the two largest indoor uses of water. Outdoor water use can be minimized by catching and storing rain runoff from your roof in a barrel. Replacing exotic vegetation with native plants adapted to your region's natural precipitation patterns can also reduce watering demand. *Xeriscaping*, landscaping using plants adapted to arid conditions, has become a popular way in the U.S. Southwest to reduce outdoor water use.

Industry and municipalities   Manufacturers are shifting to processes that use less water and in doing so are reducing their costs. Las Vegas is one of many cities that are recycling treated municipal wastewater for irrigation and industrial uses. Finding and patching leaks in pipes has saved some cities and companies large amounts of water—and money. Boston and its suburbs reduced water demand by 31% over 17 years by patching leaks, retrofitting homes with efficient plumbing, auditing industry, and promoting conservation. This program enabled Massachusetts to avoid an unpopular $500 million river diversion scheme.

WATER POLLUTION AND ITS CONTROL

We have seen that people affect aquatic systems by withdrawing too much water and by altering the systems' natural processes by engineering waterways with dams, diversions, and levees. However, people also affect aquatic ecosystems and threaten human health when we introduce toxic substances and disease-causing organisms into surface waters and groundwater.

Developed nations have made admirable advances in cleaning up water pollution over the past few decades. Still, the World Commission on Water recently concluded that over half the world's major rivers remain "seriously depleted and polluted, degrading and poisoning the surrounding ecosystems, threatening the health and livelihood of people who depend on them." The largely invisible pollution of groundwater, meanwhile, has been termed a "covert crisis."

Water pollution comes from point sources and non-point sources

Pollution is the release into the environment of matter or energy that causes undesirable impacts on the health or well-being of people or other organisms. Pollution can be physical, chemical, or biological and can affect water, air, or soil. **Water pollution** comes in many forms and can cause diverse impacts on aquatic ecosystems and human health.

Most forms of water pollution are not conspicuous to the human eye, so scientists and technicians measure water's chemical properties (such as pH, nutrient concentrations, and dissolved oxygen concentration), physical characteristics (such as temperature and turbidity—the density of suspended particles in a water sample), and biological properties (such as the presence of harmful microorganisms or the species diversity in aquatic ecosystems).

Some water pollution is emitted from **point sources**— discrete locations, such as a factory, sewer pipe, or oil tanker. In contrast, **non-point-source** pollution is cumulative, arising from multiple inputs over larger areas, such as farms, city streets, and residential neighborhoods (**FIGURE 12.25**). The U.S. Clean Water Act (pp. 100–101) addressed point-source pollution with some success by targeting industrial discharges. Water quality in the United States today suffers most from non-point-source pollution resulting from countless common activities, such as applying fertilizers and pesticides to lawns, applying salt to roads in winter, and changing automobile oil. To minimize non-point-source pollution of drinking water, governments limit development on watershed land surrounding reservoirs.

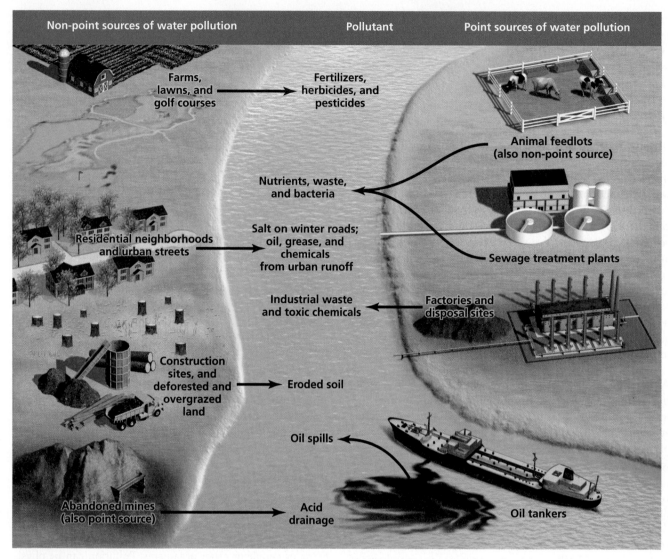

Farms, lawns, and golf courses → Fertilizers, herbicides, and pesticides

Animal feedlots (also non-point source)

Nutrients, waste, and bacteria

Sewage treatment plants

Residential neighborhoods and urban streets → Salt on winter roads; oil, grease, and chemicals from urban runoff

Industrial waste and toxic chemicals ← Factories and disposal sites

Construction sites, and deforested and overgrazed land → Eroded soil

Oil spills

Abandoned mines (also point source) → Acid drainage

Oil tankers

FIGURE 12.25 ▲ Point-source pollution **(on right)** comes from discrete facilities or locations, usually from single outflow pipes. Non-point-source pollution, such as runoff from streets, residential neighborhoods, lawns, and farms **(on left)**, originates from numerous sources spread over large areas.

Water pollution takes many forms

Water pollution comes in many forms that can impair waterways and threaten people and organisms that drink or live in affected waters. Let's survey the major classes of water pollutants affecting waters in the world today.

Toxic chemicals Our waterways and coastal ecosystems have become polluted with toxic organic substances of our own making, including pesticides, petroleum products, and other synthetic chemicals (pp. 146–147, 210). Many of these can poison animals and plants, alter aquatic ecosystems, and cause an array of human health problems, including cancer. In addition, toxic metals such as arsenic, lead, and mercury damage human health and the environment, as do acids from acid precipitation (pp. 291–294) and from acid drainage from mining sites (p. 238). With its massive watershed encompassing rural, urban, and suburban areas, the Mississippi River is one waterway that carries toxic pollutants from sources in agriculture, industry, and homes and businesses.

Issuing and enforcing more stringent regulations of industry can help reduce releases of many toxic chemicals. We can also modify our industrial processes and our purchasing decisions to rely less on these substances.

Pathogens and waterborne diseases Disease-causing organisms (pathogenic viruses, protists, and bacteria) can enter drinking water supplies when they are contaminated with human waste from inadequately treated sewage or animal waste from feedlots (p. 153). Biological pollution by pathogens causes more human health problems than any other type of water pollution, killing around 5 million people a year. This occurs because many people, particularly in rural areas of Asia and Africa, lack access to safe drinking water and sanitation facilities. Although improvements are occurring, the World Health Organization indicates that nearly 1 billon people do not have safe drinking water and that 2.6 billion people—40% of the human population—still lack adequate sewer or sanitation facilities.

We reduce the risks posed by waterborne pathogens by disinfecting drinking water (p. 270) and by treating wastewater (pp. 270–273). Other measures to lessen health risks include public education to encourage personal hygiene and government enforcement of regulations to ensure the cleanliness of food production, processing, and distribution.

Oil pollution Oil pollution in freshwater and marine systems comes from spills of all sizes. Large spills occur infrequently, but their impacts can be staggering near the spill site. The danger of oil spills to fisheries, economies, and ecosystems became clear in 2010 when British Petroleum's *Deepwater Horizon* offshore drilling platform exploded, killing 11 workers and sinking into the ocean off the Louisiana coast (pp. 326–327). Oil gushed from the platform's underwater well at rates of 1,800 gallons per minute, was spread widely by ocean currents, and washed up on coastal areas across the northern Gulf of Mexico (**FIGURE 12.26A**). The economic and ecological impacts of the spill were visited on hundreds of miles of water, sediments, and shoreline along the coasts of Louisiana, Mississippi, Alabama, and Florida.

Yet despite the severity of events like the *Deepwater Horizon* oil spill, nearly half (47%) of the nearly 1.3 million metric tons of petroleum entering the world's oceans each year originates from natural seeps in the ocean bottom. Another 38% accumulates in waters from innumerable, widely spread, small non-point sources. Shipping vessels and recreational boats can leak oil as they ply the waters of rivers, lakes, estuaries, and oceans. Motor oil from vehicles on roads and parking lots is also washed into streams by rains and carried to rivers, coastal areas, and oceans. Spills during petroleum transport account for 12% of oil pollution, and 3% comes from leakage that occurs during the extraction of oil in offshore locations. Although the *Deepwater Horizon* spill was catastrophic, the good news is that the amount of oil spilled from tankers worldwide has decreased over the past three decades, in part

because of an increased emphasis on spill prevention and response (**FIGURE 12.26B**).

The U.S. Oil Pollution Act of 1990 created a $1 billion prevention and cleanup fund. It also required that by 2015 all oil tankers in U.S. waters be equipped with double hulls as a precaution against puncture. In the wake of the *Deepwater Horizon* spill, the U.S. government is considering tighter regulations on offshore drilling operations.

Nutrient pollution The Chesapeake Bay's dead zone shows how nutrient pollution causes eutrophication and hypoxia in surface waters (Chapter 2, pp. 23–25). When excess nitrogen and/or phosphorus enters a water body, it fertilizes algae and aquatic plants, boosting their growth. Algae then spread and cover the water's surface, depriving underwater plants of sunlight. As algae die off, bacteria consume them. Because this decomposition requires oxygen, the increased bacterial activity drives down levels of dissolved oxygen. These levels can drop too low to support fish and shellfish, leading to dramatic changes in freshwater and saltwater ecosystems.

A "dead zone" of very low dissolved oxygen levels appears annually in the northern Gulf of Mexico, fueled by nutrients from Midwest farms carried by the Mississippi and Atchafalaya rivers (**FIGURE 12.27**). The low oxygen conditions have adversely affected marine life and reduced catches of shrimp and fish to half of what they were in the 1980s.

Excessive nutrient concentrations sometimes give rise to population explosions among several species of marine algae that produce powerful toxins. Blooms of these algae are known as **harmful algal blooms** and are sometimes called **red tides** because some toxic algal species produce a red pigment that discolors the water. Harmful algal blooms can cause illness and death in aquatic animals and people and adversely

(a) Oil from *Deepwater Horizon* spill in a Louisiana salt marsh

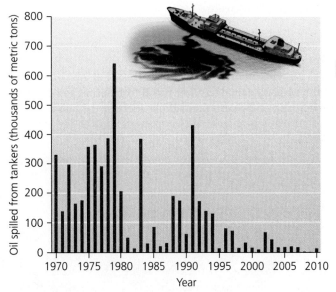

(b) Quantity of petroleum spilled from tankers, 1970–2010

FIGURE 12.26 ▲ Pollution was severe along Louisiana's coastline **(a)** after British Petroleum's *Deepwater Horizon* oil drilling platform exploded and disgorged millions of gallons of crude oil into the Gulf of Mexico in 2010. But less oil is being spilled into ocean waters today in large tanker spills **(b)**, thanks in part to regulations on the oil shipping industry and improved spill response techniques. The bar chart shows cumulative quantities of oil spilled worldwide from nonmilitary spills over 7 metric tons. Data for (b) from International Tanker Owners Pollution Federation Ltd.

Each year, 30–40 billion plastic water bottles are thrown away in the United States.

"Is It Better in a Bottle?"

Which is safer and healthier for you to drink, tap water or bottled water?

It's hard to know the answer, because companies are not required to tell us anything about the quality of the water in their bottles, or even where the water comes from.

Municipalities that provide tap water to their residents need to submit regular reports to the Environmental Protection Agency describing their sources, treatment methods, and contaminants. In contrast, bottled water is regulated much more lightly as a "food" by the Food and Drug Administration (FDA). Bottling companies do not have to inform the public or the government where their water comes from or how it is treated, and they are not required to test samples with certified laboratories or notify the FDA of contamination problems.

So, to find out what's in bottled water, scientists have had to do some detective work. In 2008, research scientists at the Environmental Working Group (EWG), based in Washington, D.C., sent samples of 10 major brands of bottled water to the University of Iowa's Hygienic Laboratory for analysis. The lab's chemists ran a battery of tests and detected 38 chemical pollutants, including traces of heavy metals, radioactive isotopes, caffeine and pharmaceuticals from wastewater pollution, nitrate and ammonia from fertilizer, and various industrial compounds such as solvents and plasticizers (**first figure**).

Each brand contained eight contaminants on average, and two brands had levels of chemicals that exceeded legal limits in California and industry safety guidelines. Two brands showed the chemical composition of standard municipal water treatment. This is not surprising: An estimated 25–44% of bottled water is simply tap water, bottled and sold at elevated prices.

In 2009, researchers Martin Wagner and Jörg Oehlmann of Johann Wolfgang Goethe University in Frankfurt, Germany, tested 20 brands of bottled water for the presence of hormone-disrupting chemicals that mimic estrogen (pp. 206–207). The researchers compared nine brands packaged in glass bottles, nine brands packaged in plastic bottles (PET, or polyethylene terephthalate, the "#1" type plastic), and two brands packaged in "Tetra Pak" paperboard boxes with an inner plastic coating. They placed samples (as well as tap-water samples as controls) in a "yeast estrogen screen," a standard test-tube screening procedure that uses yeast cells engineered with genes to change color when exposed to estrogen-mimicking compounds.

---

affect the economies of communities that rely on beach tourism and fishing.

Eutrophication is a natural process, but nutrient input from wastewater and fertilizer runoff from farms, golf courses, lawns, and sewage can dramatically increase the rate at which it occurs. We can reduce nutrient pollution by specially treating municipal wastewater to remove nutrients, reducing fertilizer applications, using phosphate-free detergents, and planting vegetation and protecting natural areas around streams and rivers to reduce nutrient inputs into waterways.

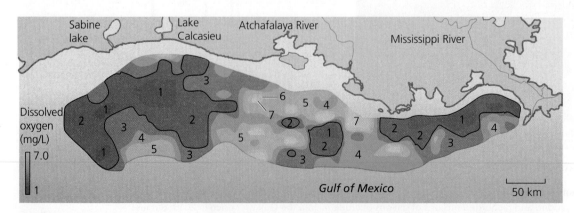

FIGURE 12.27 ▲ Dissolved oxygen concentrations were mapped in the Gulf of Mexico off the Louisiana coast in 2010. Areas in red indicate the lowest oxygen levels. Regions considered hypoxic (< 2 mg/L) are encircled with a black line. Data from N. Rabalais, Louisiana Universities Marine Consortium (LUMCON), and R. E. Turner, Louisiana State University.

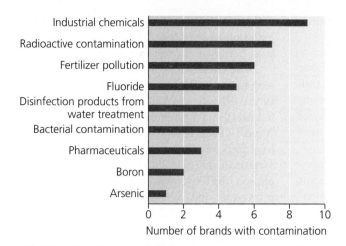

Of 10 leading brands of bottled water tested, most contained industrial chemicals, radioactive isotopes, and fertilizer pollution, as well as other contaminants. *Source:* Naidenko, O., et al., 2008. Bottled water quality investigation: 10 major brands, 38 pollutants. Environmental Working Group, Washington, D.C.

but already these studies and others like them have indicated that bottled water can contain a range of contaminants, some of which may pose health risks. Tap water may also contain plenty of pollutants, but municipalities are required to test and report on their water quality. Based on these findings, more and more people are reassessing their assumptions about the safety of bottled water.

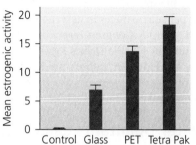

Estrogenic activity (as determined by a yeast cell culture test) was highest in bottled water from Tetra Pak containers, followed by PET plastic containers, and then glass containers. **A negative control showed no appreciable estrogenic potency.** With kind permission from Springer Science and Business Media and the author, from Wagner, M., and J. Oehlmann, 2009. Endocrine disruptors in bottled mineral water: Total estrogenic burden and migration from plastic bottles. *Environmental Science and Pollution Research* 16: 278–286, Fig 3a.

The researchers detected estrogenic contamination in 60% of the samples (**second figure**). Both Tetra Pak brands and seven of nine plastic brands contained hormone-mimicking substances that apparently leached from the packaging. So did three of the glass-bottled brands, presumably from contamination at the bottling plant.

Wagner and Oehlmann then tested whether the chemicals in the water would affect a living animal, a type of snail that is known to increase its reproduction when exposed to an estrogenic substance. They raised some snails in PET plastic containers and others in glass containers. After 56 days, the snails in the PET containers had produced 39–122% more embryos than snails in control conditions, whereas snails in glass containers showed no difference. This suggested that estrogenic compounds were leaching from the plastic. Their results were published in the journal *Environmental Science and Pollution Research*.

Research on the quality and safety of bottled water is just getting started,

Biodegradable Wastes   Introducing large quantities of biodegradable materials into waters decreases dissolved oxygen levels, too. When human wastes, animal manure, paper pulp from paper mills, or yard wastes (grass clippings and leaves) enter waterways, bacterial decomposition escalates as organic material is metabolized. This lowers dissolved oxygen levels in the water, just as in waters receiving elevated inputs of plant nutrients. **Wastewater** is water affected by human activities and is a source of biodegradable wastes. It includes water from toilets, showers, sinks, dishwashers, and washing machines; water used in manufacturing or industrial cleaning processes; and stormwater runoff. The widespread practice of treating wastewater to remove organic matter has greatly reduced impacts from biodegradable wastes in rivers in developed nations. Oxygen depletion remains a major problem in some developing nations, however, where wastewater treatment is less common.

Sediment   As we saw in the Central Case Study, eroded soils are carried to rivers by runoff and transported long distances by river currents. Clear-cutting, mining, clearing land for development, and cultivating farm fields all expose soil to wind and water erosion (pp. 140–144) and increase the amount of soil entering waterways. High sediment concentrations impair aquatic ecosystems by interfering with the respiration of fish and invertebrates and smothering benthic organisms. Sediment also clouds waters, blocking the sunlight needed by rooted aquatic plants. We can reduce sediment pollution by adopting sustainable soil practices, avoiding large-scale disturbance of vegetation, and maintaining riparian vegetation to trap sediments at the water's edge.

Thermal pollution   Water's ability to hold dissolved oxygen decreases as temperature rises, so some aquatic organisms may not survive when human activities raise water temperatures. When we withdraw water from a river and use it to cool an industrial facility, we transfer heat from the facility back into the river where the water is returned. People also raise water temperatures by removing streamside vegetation that shades water.

Too little heat can also cause problems. On the Mississippi River and its tributaries, as in many other dammed rivers, water at the bottoms of reservoirs is colder than water at the surface. When dam operators release water from the depths of a reservoir, downstream water temperatures drop suddenly and affect wildlife.

Nets and plastic debris Discarded fishing nets, plastic bags and bottles, fishing line, buckets, floats, and other trash can harm aquatic organisms. Aquatic mammals, seabirds, fish, and sea turtles may mistake floating plastic debris for food and can die as a result of ingesting material they cannot digest or expel.

In recent years scientists have learned that plastic trash is accumulating in certain regions of the oceans where currents converge. One such area is the "Great Pacific Garbage Patch" in the northern Pacific. The site is often estimated as being twice the size of Texas, and one study documented 3.3 plastic bits per square meter in its waters (see **ENVISIONIT**, p. 271). In 2006, the U.S. Congress responded to such ocean pollution by passing the Marine Debris Research, Prevention, and Reduction Act. However, more is needed. We can all help by reducing our use of unnecessary plastic, reusing the plastic items we do use, and recycling the plastic we discard.

Water pollutants can contaminate groundwater

Many of these pollutants affect groundwater as well as surface waters. Groundwater pollution is more difficult to detect and more problematic to address than surface water pollution. Groundwater flows slowly and may harbor pollutants for long periods. Decomposition is also slower in groundwater than surface water because groundwater is not exposed to sunlight, contains fewer microbes and minerals, and holds less dissolved oxygen and organic matter.

Some chemicals that are toxic at high concentrations, including aluminum, fluoride, nitrates, and sulfates, occur naturally in groundwater. However, groundwater pollution resulting from human activity is widespread. Industrial, agricultural, and urban wastes—from heavy metals to petroleum products to solvents to pesticides—can leach through soil and seep into aquifers. For example, nitrate from fertilizers has leached into aquifers in Canada and in 49 U.S. states. Nitrate in drinking water has been linked to cancers, miscarriages, and "blue-baby" syndrome, which reduces the oxygen-carrying capacity of infants' blood. Other pollutants can enter groundwater through leaky underground tanks, through improperly designed wells, and from the pumping of liquid hazardous waste below ground (pp. 394–395).

Legislative and regulatory efforts have helped to reduce water pollution

As numerous as our freshwater pollution problems may seem, it is important to remember that many were worse a few decades ago (pp. 98–99). Citizen activism and government response during the 1960s and 1970s in the United States resulted in legislation such as the Federal Water Pollution Control Act of 1972 (later amended and renamed the Clean Water Act in 1977). These acts set standards for industrial wastewater, set standards for contaminant levels in surface waters, funded construction of sewage treatment plants, and made it illegal to discharge pollution from a point source

without a permit. Thanks to such legislation, point-source pollution in the United States was reduced, and rivers and lakes became noticeably cleaner.

The Great Lakes of Canada and the United States represent an encouraging success story in fighting water pollution. In the 1970s these lakes were badly polluted with wastewater, fertilizers, and toxic chemicals, and Lake Erie was even declared "dead." Today, efforts of the Canadian and U.S. governments have paid off. According to Environment Canada, releases of seven toxic chemicals into the lakes are down by 71%, municipal phosphorus has decreased by 80%, and chlorinated pollutants from paper mills are down by 82%. Levels of PCBs and DDE are down by 78% and 91%, respectively. Bird populations are rebounding, and Lake Erie is now home to the world's largest walleye fishery. The Great Lakes' troubles are by no means over, however—sediment pollution is still heavy, PCBs and mercury still settle from the air, and fish are not always safe to eat. Even so, the progress made thus far shows how conditions can improve when citizens push their governments to take action.

Such successes require effective enforcement of environmental regulations. A 2009 investigation by the *New York Times* revealed that violations of the Clean Water Act rose in the decade preceding the report and that underfunded and understaffed state and federal regulatory agencies acted on only a tiny percentage of these violations. As a result, 1 in 10 Americans have been exposed to unsafe drinking water. Soon after the report's release, Environmental Protection Agency (EPA) Administrator Lisa Jackson promised to strengthen enforcement to help remedy these issues.

We treat our drinking water

The treatment of drinking water is a widespread practice in developed nations today. Before being sent to your tap, water from a reservoir or aquifer is treated with chemicals to remove particulate matter; passed through filters of sand, gravel, and charcoal; and/or disinfected with small amounts of an agent such as chlorine or ozone. The EPA sets standards for over 90 drinking water contaminants, which local governments and private water suppliers are obligated to meet. The treatment of drinking water treatment has greatly reduced mortality from the waterborne diseases that once claimed many lives, and it is one of the most significant technological advances of modern times.

We treat our wastewater

Prior to the passage of the Clean Water Act, cities and towns regularly released untreated sewage into waterways. This led to oxygen depletion and pathogen contamination in many waters. Wastewater treatment is now a mainstream practice. In rural areas, **septic systems** are the most popular method of treating wastewater. In a septic system, wastewater runs from the house to an underground septic tank, inside which solids and oils separate from water. The clarified water proceeds downhill to a drain field of perforated pipes laid horizontally in gravel-filled trenches underground. Microbes decompose pollutants in the wastewater these pipes emit. Periodically, solid waste from the septic tank is pumped out and taken to a landfill.

In more densely populated areas, municipal sewer systems carry wastewater from homes and businesses to central-

Roadside litter and dumped garbage end up in streams, then rivers, then the ocean.

Booms trap trash headed to the ocean in the Los Angeles River

Countless trillions of pieces of plastic trash are accumulating in the oceans, concentrated by currents.

Charles Moore holds water from the "Great Pacific Garbage Patch"

Dead albatross with stomach full of plastic

Marine debris kills animals that get entangled in nets or swallow plastic.

Endangered Hawaiian Monk Seal caught in fishing tackle

You Can Make a Difference

➤ Clean a beach with the International Coastal Cleanup.

➤ Pick up litter wherever you are; trash in a stream today could be in the ocean tomorrow.

➤ Reduce consumption, and reuse and recycle items. You will generate far less waste.

ized treatment locations, called *wastewater treatment plants* (**FIGURE 12.28**). At the plant, incoming wastewater is first passed through screens to capture large objects. It is then sent to **primary treatment**, where about 60% of the suspended solids in the wastewater settle out when the wastewater is al-lowed to sit in settling tanks. Wastewater then proceeds to **secondary treatment**, in which water is stirred and aerated so that aerobic bacteria consume most of the small particles of organic matter that remain in the wastewater. Roughly 90% of suspended solids may be removed after secondary treat-

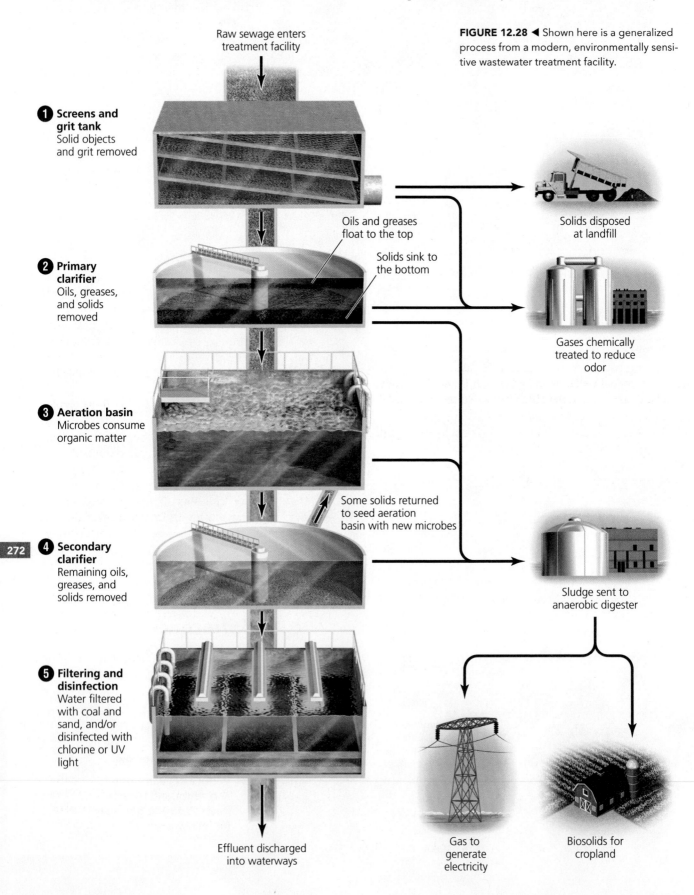

Raw sewage enters treatment facility

FIGURE 12.28 ◄ Shown here is a generalized process from a modern, environmentally sensitive wastewater treatment facility.

❶ **Screens and grit tank**
Solid objects and grit removed

Oils and greases float to the top

Solids sink to the bottom

Solids disposed at landfill

❷ **Primary clarifier**
Oils, greases, and solids removed

Gases chemically treated to reduce odor

❸ **Aeration basin**
Microbes consume organic matter

Some solids returned to seed aeration basin with new microbes

❹ **Secondary clarifier**
Remaining oils, greases, and solids removed

Sludge sent to anaerobic digester

❺ **Filtering and disinfection**
Water filtered with coal and sand, and/or disinfected with chlorine or UV light

Effluent discharged into waterways

Gas to generate electricity

Biosolids for cropland

ment. In some cases, water is subject to additional treatments to remove pollutants of particular concern. Some wastewater treatments plants along the Mississippi River, for example, subject water to additional treatments that remove nitrogen and phosphorus inputs into the northern Gulf of Mexico.

Finally, the clarified water is treated with chlorine, and sometimes ultraviolet light, to kill bacteria. The treated water, called wastewater **effluent**, is then typically released into a river or the ocean. In some situations, effluent is "reclaimed" and used for lawns and golf courses, for irrigation, or for industrial purposes.

The solids collected during the treatment process, called *sludge*, are sent to digesting vats, where microorganisms decompose much of the matter. The resultant "biosolids" are then dried and either landfilled, incinerated, or used as fertilizer on cropland.

Wetlands can aid wastewater treatment

Artificial wetlands have been used to aid wastewater treatment. Most of these sites use microbes, algae, and aquatic plants to "polish" the water released from wastewater treatment plants. Water cleansed in the wetland can then be released into waterways or allowed to percolate underground. One of the first constructed wetlands was established in the 1980s in Arcata, a town on northern California's scenic Redwood Coast. There are currently around 500 artificially constructed or restored wetlands performing this service in the United States.

The release of wastewater effluent even shows promise for preserving coastal wetlands along the Gulf Coast. At a study site in Louisiana, wastewater effluent was released into coastal wetlands, where it elevated growth in marsh grasses due to the effluent's elevated nutrient concentrations. The increased plant growth led to increased deposition of plant organic matter on marsh sediments, offsetting the depth increases caused by natural soil compaction and sustaining the ecosystem.

EMPTYING THE OCEANS

We affect the oceans and their biological resources by engineering waterways with dams and levees and by introducing water pollutants directly into the ocean or into rivers that empty into the sea. In addition, people are putting pressure on oceans by overharvesting marine fish species and threatening the balance and functioning of marine and coastal ecosystems.

Over half the world's marine fish populations are fully exploited, meaning that we cannot harvest them more intensively without depleting them, according to the U.N. Food and Agriculture Organization (FAO). An additional 28% of marine fish populations are overexploited and already being driven toward extinction. If current trends continue, a comprehensive 2006 study in the journal *Science* predicted, populations of *all* ocean species that we fish for today will collapse by the year 2048.

If fisheries collapse as predicted, we will lose the ecosystem services they provide. Productivity will be reduced, ecosystems will become more sensitive to disturbance, and the filtering of water by vegetation and organisms (such as oysters) will decline, making harmful algal blooms, dead zones, fish kills, and beach closures more common. Aquaculture (raising fish in tanks or pens) is booming and is helping to relieve pressure on wild stocks, but fish farming comes with its own set of environmental dilemmas (pp. 153–154). All this makes it vital, many scientists and fisheries managers say, that we turn immediately to more sustainable fishing practices.

Fishing has industrialized

Total global fisheries catch, after decades of increases, leveled off after about 1988 (**FIGURE 12.29**) and has remained fairly constant since then. This seeming stability in catch can be explained by several factors that conceal population declines: Fishing fleets are exploiting increasingly remote fishing areas, are engaging in more intensive fishing, are capturing smaller fish than before, and are targeting less desirable fish species they formerly overlooked.

Today's industrialized commercial fishing fleets employ fossil fuels, huge vessels, and powerful new technologies to capture fish in large numbers using several methods. Some vessels set out long *driftnets* that span large expanses of water (**FIGURE 12.30A**). These chains of transparent nylon mesh nets are arrayed to drift with currents so as to capture passing fish, and they are held vertical by floats at the top and weights at the bottom. *Longline fishing* (**FIGURE 12.30B**) involves setting out extremely long lines (up to 80 km [50 mi] long) with up to several thousand baited hooks spaced along their lengths. *Trawling* entails dragging immense cone-shaped nets through the water, with weights at the bottom and floats at the top. Trawling in open water captures pelagic fish, whereas *bottom-trawling* (**FIGURE 12.30C**) involves dragging weighted nets across the floor of the continental shelf to catch benthic organisms.

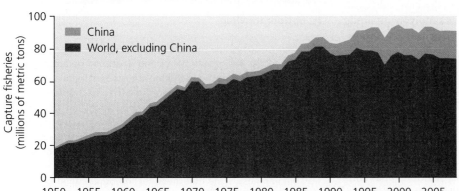

FIGURE 12.29 ◀ The total global fisheries catch has stalled for the past 20 years, and many fear that a global decline is imminent if conservation measures are not taken. The figure shows trends with and without China's data, because research suggests that China's data may be somewhat inflated. Data from the Food and Agriculture Organization of the United Nations, 2010. *The state of world fisheries and aquaculture 2010.* Fig 3. By permission.

Unfortunately, these fishing practices catch more than just the species they target. **Bycatch**, the accidental capture of nontarget animals, accounts for the deaths of millions of animals each year. Driftnetting captures dolphins, seals, and sea turtles, as well as countless nontarget fish. Longline fishing kills turtles, sharks, and seabirds. Bottom-trawling is often likened to clear-cutting (p. 194) and strip-mining (pp. 237–238), and it is especially destructive to structurally complex areas, such as reefs, that provide shelter and habitat for animals.

Industrial fishing fleets have depleted many fisheries

Throughout the world's oceans, today's industrialized fishing fleets are depleting marine populations. In a 2003 study, fisheries biologists Ransom Myers and Boris Worm analyzed fisheries data and concluded that the oceans today contain only one-tenth of the large-bodied fish and sharks they once did.

Many fisheries have collapsed in recent years. Groundfish (species that live in benthic habitats, such as Atlantic cod, haddock, halibut, and flounder) powered the economies of New England and Maritime Canada for close to 400 years. Yet fishing pressure became so intense that most stocks collapsed, bringing fishing economies down with them.

With Canada's cod stocks down by 99% and showing no sign of recovery, the Canadian government in 1992 ordered a complete ban on cod fishing in the Grand Banks region off Newfoundland and Labrador. On the U.S. side of the border, bans are helping to restore depleted fisheries. When the groundfish fisheries of Georges Bank in the Gulf of Maine collapsed in the mid-1990s, the National Marine Fisheries Service (NMFS) closed three prime fishing areas to fishing. The closures worked. Spawning stocks of haddock and yellowtail flounder have risen, and biomass of sea scallops increased 14-fold. Fishers began having better luck, especially just outside the closed regions. Unfortunately, cod have not recovered. Research suggests that once mature cod were eliminated, the species they preyed upon proliferated, and now those species compete with and prey on young cod, preventing the population from rebuilding.

Red snapper stocks have been similarly depleted by overfishing in the Gulf of Mexico. Red snapper were identified as severely overfished in 1989, and current populations are at a mere 3% of their historical abundance. Gulf populations of the species are affected by harvesting and by mortality experienced when they are taken as bycatch on shrimping vessels. The recovery plan for red snapper approved by the NMFS in 2005 was criticized by environmental groups as simply maintaining the "status quo" management approach that had proven unsuccessful for over 15 years. These groups sued in federal court, and the court ruled in their favor in 2007 and ordered NMFS to develop a more comprehensive plan with greater probability of success in restoring populations of red snapper.

Our purchasing choices can influence fishing practices

By exercising careful choice when we buy seafood, we as consumers can encourage more sustainable fisheries practices. Several marine conservation organizations have devised

(a) Driftnetting

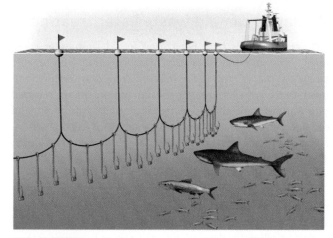

(b) Longlining

(c) Bottom-trawling

FIGURE 12.30 ▲ Commercial fishing fleets use several methods of capture. In drift netting **(a)**, long transparent nylon nets are set out to drift through open water to capture schools of fish. In longlining **(b)**, lines with numerous baited hooks are set out in open water. In bottom-trawling **(c)**, weighted nets are dragged along the floor of the continental shelf. All methods result in large amounts of bycatch, the capture of nontarget animals. The illustrations above are simplified for clarity and do not portray the immense scale that these technologies can attain; for instance, industrial trawling nets can be large enough to engulf several Boeing 747 jumbo jets.

concise guides to help consumers differentiate fish and shellfish that are overfished or whose capture is ecologically damaging from those that are harvested more sustainably. For instance, the Monterey Bay Aquarium provides a wealth of this information on its website.

Marine reserves protect ecosystems

Fisheries managers conduct surveys, study fish population biology, and monitor catches to determine the number of fish of a given species that can be harvested without reducing future catches—a concept called *maximum sustainable yield* (p. 192). Despite the use of this technique, a number of fish and shellfish stocks have plummeted. Thus, many scientists and managers feel it is time to shift the focus away from individual species and toward viewing marine resources as elements of larger ecological systems. One key aspect of such *ecosystem-based management* (p. 192) is to set aside areas of ocean where systems can function without human interference.

Hundreds of **marine protected areas (MPAs)** have been established, most of them along the coastlines of developed countries. However, despite their name, nearly all MPAs allow fishing or other extractive activities and so are not fully protected from impacts from people.

Because of the lack of true refuges from fishing pressure, many scientists want to establish areas where fishing is prohibited. Such "no-take" areas have come to be called **marine reserves**. Designed to preserve ecosystems intact, marine reserves are also intended to improve fisheries. Scientists argue that marine reserves can act as production factories for fish for surrounding areas, because fish larvae produced inside reserves will disperse outside and stock other parts of the ocean. By serving both purposes, proponents maintain, marine reserves are a win-win proposition for environmentalists and fishers alike. However, many commercial and recreational fishers dislike the idea of no-take reserves and have opposed nearly every marine reserve that has been established.

Reserves can work for both fish and fishers

Over the past two decades, data from marine reserves around the world have been indicating that reserves can work as win-win solutions that benefit ecosystems, fish populations, and fishing economies. A comprehensive review of data from marine reserves as of 2001 revealed that just one to two years after their establishment, marine reserves:

▶ Increased densities of organisms on average by 91%.

▶ Increased biomass of organisms on average by 192%.

▶ Increased average size of organisms by 31%.

▶ Increased species diversity by 23%.

If marine reserves work in principle, the question becomes how large reserves need to be, how many there need to be, and where they need to be placed to take best advantage of ocean currents. Of several dozen studies that have estimated how much area of the ocean should be protected in no-take reserves, estimates range from 10% to 65%, with the majority falling between 20% and 50%. Most scientists feel that involving fishers directly in the planning process is crucial for coming up with answers to all these questions. If marine reserves can be made to work and to be accepted, then they may well seed the seas and help lead us toward solutions to one of our most pressing environmental problems.

▶ CONCLUSION

Our planet's diverse aquatic systems comprise an interconnected web of ecosystems that exchange water. The introduction of pollutants and the engineering of waterways therefore cause impacts that cascade through the system. Our expanding population and increasing water use are straining water supplies and affecting surface waters and groundwater around the world. Water pollutants threaten human health and ecosystem stability, and overharvesting of marine fish populations threatens the oceans' biodiversity.

There is plenty of reason for optimism, however. Improvements in water use efficiency show promise for reducing demand for water, even with increasing human populations. Water quality in many freshwater bodies has improved in recent decades, thanks to legislative action from policymakers and the efforts of millions of concerned citizens. In the oceans, marine reserves give hope that we can restore ecosystems and fisheries at the same time.

TESTING YOUR COMPREHENSION

1. Explain why the distribution of water on Earth makes it difficult for many people to access adequate fresh water.

2. Pick one of the aquatic systems profiled in this chapter, and provide three examples of ways it interacts with other aquatic systems.

3. Why are coral reefs biologically valuable? How are they being degraded by human impact? What is causing the disappearance of mangrove forests and salt marshes?

4. Describe three benefits and three costs of damming rivers. What particular environmental, health, and social concerns have China's Three Gorges Dam and its reservoir raised?

5. Why do the Colorado, Rio Grande, Nile, and Yellow rivers now slow to a trickle or run dry before reaching their deltas?

6. Name three major types of water pollutants, and provide an example of each. Explain which classes of pollutants you think are most important in your local area.

7. Define groundwater and list some anthropogenic (human) sources of groundwater pollution. Why do

many scientists consider groundwater pollution a greater problem than surface water pollution?

8. Describe and explain the major steps in the process of wastewater treatment. How can artificially constructed wetlands aid such treatment?

9. Name three industrial fishing practices, and explain how they create by-catch and harm marine life.

10. How does a marine reserve differ from a marine protected area? Why do many fishers oppose marine reserves? Explain why many scientists say no-take reserves will be good for fishers.

SEEKING SOLUTIONS

1. How can we lessen agricultural demand for water? Describe some ways we can reduce household water use. How can industrial uses of water be reduced?

2. Describe three ways in which your own actions contribute to water pollution. Now describe three ways in which you could diminish these impacts.

3. Describe the trends in global fish capture over the past 55 years and over the past 20 years, and explain several factors that account for these trends.

4. **THINK IT THROUGH** Your state's governor has put you in charge of water policy for the state. The aquifer beneath your state has been overpumped, and many wells have run dry. Agricultural production last year decreased for the first time in a generation, and farmers are clamoring for you to do something. Meanwhile, the state's largest city is growing so fast that more water is needed for its burgeoning urban population. What policies would you consider to restore your state's water supply? Would you try to take steps to increase supply, decrease demand, or both? Explain your choices.

5. **THINK IT THROUGH** You are mayor of a coastal town where some residents are employed as commercial fishers and others make a living serving ecotourists who come to snorkel and scuba dive at the nearby coral reef. In recent years, several fish stocks have crashed, and ecotourism is dropping off as fish disappear from the increasingly degraded reef. Scientists are urging you to help establish a marine reserve around portions of the reef, but most commercial and recreational fishers are opposed to this idea. What steps would you take to restore your community's economy and environment?

CALCULATING ECOLOGICAL FOOTPRINTS

One of the single greatest personal uses of water is for showering. Old-style showerheads that were standard in homes and apartments built before 1992 dispense at least 5 gallons of water per minute, but low-flow showerheads produced after that year dispense just 2.5 gallons per minute. Given an average daily shower time of 8 minutes, calculate the amounts of water used and saved over the course of a year with old standard versus low-flow showerheads, and record your results in the table.

	Annual water use with standard showerheads (gallons)	Annual water use with low-flow showerheads (gallons)	Annual water savings with low-flow showerheads (gallons)
You			
Your class			
Your state			
United States			

1. In 2010, under its WaterSense program, the EPA began promoting showerheads that produce still-lower flows of 2 gallons per minute (gpm). Some cities are already requiring these, and some models today go even lower. How much water would you save per year by using a 2-gpm showerhead instead of a 2.5-gpm showerhead?

2. How much water would you be able to save annually by shortening your average shower time from 8 minutes to 6 minutes? Assume you use a 2.5-gpm showerhead.

3. Compare your answers to Questions 1 and 2. Do you save more water by showering 8 minutes with a 2-gpm showerhead or 6 minutes with a 2.5-gpm showerhead?

4. Can you think of any factors that are *not* being considered in this scenario of water savings? Explain.

Mastering**ENVIRONMENTALSCIENCE**™

Go to **www.masteringenvironmentalscience.com** for homework assignments, practice quizzes, Pearson eText, and more.

13 Atmospheric Science and Air Pollution

Upon completing this chapter, you will be able to:

➤ Describe the composition, structure, and function of Earth's atmosphere

➤ Relate weather and climate to atmospheric conditions

➤ Identify major outdoor air pollutants, outline the scope of outdoor air pollution, and assess potential solutions

➤ Explain stratospheric ozone depletion and identify steps taken to address it

➤ Define *acid deposition* and illustrate its consequences

➤ Identify major indoor air pollutants, characterize the scope of indoor air pollution, and assess potential solutions

A bad air day in Tehran, Iran

L.A. and its Sister City, Tehran, Struggle for a Breath of Clean Air

"Poisonous gases emitted from vehicles and industries continue to wreak havoc on Tehran's air."
—YOUSEF RASHIDI, Director of Tehran's Air Quality Control Company

"I left L.A. in 1970, and one of the reasons I left was the horrible smog. And then they cleaned it up. That was one of the greatest things the government has ever done for me. You have beautiful days now. It's a much, much nicer place to live."
—ACTOR AND COMEDIAN STEVE MARTIN, speaking to *Los Angeles Magazine*

Los Angeles has long symbolized air pollution in Americans' minds. Smog blanketed the city from the 1960s through the 1990s, fed by the exhaust of millions of automobiles. But in recent years, policy efforts and technological advances have improved air quality in Los Angeles. Today L.A. still suffers the nation's worst smog, but its skies are clearer than skies in some of its "sister cities" elsewhere in the world.

The Sister Cities program was devised to encourage world peace and understanding by fostering relationships between people of U.S. cities and others around the world. Hundreds of American cities share government visits and cultural ties with their sister cities. One of L.A.'s sister cities is Tehran, the capital of Iran. As it happens, one thing these two cities share is smog.

Tehran's parks and tree-lined streets make it a gorgeous city, but the quality of Tehran's air is so bad that residents commonly wear face masks when they step outside to go to the bank, the laundromat, or the grocery store. On especially bad days the government closes schools and advises people to stay indoors.

Mehdi Gharakhanlou, a 38-year-old businessman, says he returns home from the central city each winter day "fatigued, nauseated, with itchy eyes, a sore throat, a bad headache or even with all of these symptoms. . . . I wear a mask when I am out, but even that doesn't help much."

"People suffer from eye allergies, nose irritation, . . . headaches, burning eyes," adds Tehran journalist Alborz Maleki. "Children and elderly people are facing many [health] problems."

Indeed, each day people die in Iran's capital from outdoor air pollution. Health authorities blame several thousand premature deaths per year in Tehran on respiratory diseases resulting from air pollution. In 2006, 3,600 people succumbed in just a month.

As in Los Angeles, automobile traffic generates most of the pollution in Tehran. An estimated 80% of Tehran's pollutants come from the exhaust of its 3 million vehicles, many of which are over 20 years old and lack basic pollutant-filtering technology such as catalytic converters. Tehran's limited public transportation system forces most people to rely on cars, motorcycles, and taxis to get around. Moreover, Iran's government spends billions of dollars in subsidies to give its citizens some of the cheapest gasoline in the world. With 40-cent-per-gallon gas, people have little financial incentive to conserve, and Iranian vehicles guzzle gas at four times the rate of European vehicles.

In both Los Angeles and Tehran, topography worsens air quality problems. Each city lies in a valley, surrounded by mountains that trap pollution. In Tehran, wintertime is the worst, when thermal inversions confine pollutants over the city and there is little wind to blow them away. And as with Los Angeles in recent decades, people are streaming into Tehran from elsewhere, so efforts to rein in pollution are being overwhelmed by population growth.

All in all, modern-day Tehran is dangerously polluted for the same reasons that Los Angeles was a few decades ago. L.A. and Tehran today typify a difference between cities of developed nations and cities of developing nations. Nations that are industrializing as they try to build wealth for their citizens are running into the same urban and industrial air pollution problems that plagued the United States a generation and more ago.

Los Angeles has 24 other sister cities. Many of them also struggle with severe air pollution: Athens, Greece; Jakarta, Indonesia; Mumbai, India; Guangzhou, China; Taipei, Taiwan; and San Salvador, El Salvador all experience poor air quality on a regular basis. Best known among L.A.'s sister cities is Mexico City, Mexico, which may have the most polluted air in the world.

Most of these cities are taking steps to improve their air quality, just as Los Angeles and other American cities have done. We will examine the solutions sought in Los Angeles, Tehran, and L.A.'s other sister cities as we learn about Earth's atmosphere and how to reduce the pollutants we release into it. ∎

THE ATMOSPHERE

Every breath we take reaffirms our connection to the **atmosphere**, the layer of gases that surrounds Earth. The atmosphere provides oxygen, absorbs hazardous solar radiation, burns up incoming meteors, transports and recycles water and nutrients, and moderates climate.

Earth's atmosphere consists of 78% nitrogen (N_2) and 21% oxygen (O_2). The remaining 1% is composed of argon (Ar) and minute concentrations of water vapor and several other gases (**FIGURE 13.1**). Over our planet's long history, the atmosphere's composition has changed. Oxygen began to build up about 2.7 billion years ago, with the emergence of microbes that emitted oxygen by photosynthesis (p. 30). Today, human activity is altering the concentrations of some atmospheric gases, such as carbon dioxide (CO_2), methane (CH_4), and ozone (O_3).

The atmosphere is layered

The atmosphere that seems to stretch so high above us is actually just 1/100th of Earth's diameter, like the fuzzy skin of

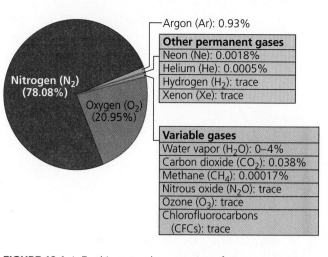

FIGURE 13.1 ▲ Earth's atmosphere consists of nitrogen, oxygen, argon, and several gases at dilute concentrations. Permanent gases are fixed in concentration. Variable gases vary in concentration as a result either of natural processes or of human activities. Data from Ahrens, C.D., 2007. *Meteorology today*, 8th ed. Belmont, CA: Brooks/Cole.

a peach. It consists of four layers that differ in temperature, density, and composition (**FIGURE 13.2**).

Movement of air within the bottommost layer, the **troposphere**, is largely responsible for the planet's weather. Although it is thin (averaging 11 km [7 mi] high) relative to the atmosphere's other layers, the troposphere contains three-quarters of the atmosphere's mass, because gravity pulls mass downward, making air denser near Earth's surface. Tropospheric air gets colder with altitude, dropping to roughly −52 °C (−62 °F) at the top of the troposphere. At this point, temperatures cease to decline with altitude, marking a boundary called the *tropopause*. The tropopause acts as a cap, limiting mixing between the troposphere and the atmospheric layer above it, the stratosphere.

The **stratosphere** extends 11–50 km (7–31 mi) above sea level. Similar in composition to the troposphere, the stratosphere is 1,000 times drier and less dense. Its gases experience little vertical mixing, so once substances (including pollutants) enter it, they tend to remain for a long time. The stratosphere warms with altitude because its ozone and oxygen absorb and scatter the sun's ultraviolet (UV) radiation (p. 30). Most of the atmosphere's ozone concentrates in a portion of the stratosphere roughly 17–30 km (10–19 mi) above sea level, a region called the **ozone layer**. The ozone layer greatly reduces the amount of UV radiation that reaches Earth's surface. Because UV light can damage living tissue and induce genetic mutations, the ozone layer's protective effects are vital for life on Earth.

Above the stratosphere lies the mesosphere, where air pressure is extremely low and temperatures decrease with altitude. The thermosphere, our atmosphere's top layer, extends to an altitude of 500 km (300 mi).

The sun and the atmosphere drive weather and climate

An enormous amount of energy from the sun continuously bombards the upper atmosphere—over 1,000 watts/m², thousands of times more than all the electricity generated by human society. Of that solar energy, about 70% is absorbed by

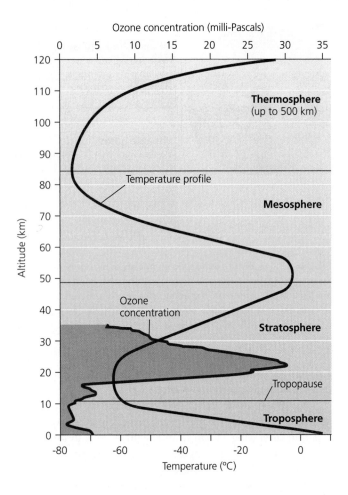

FIGURE 13.2 ▲ Temperature (red line) drops with altitude in the troposphere, rises with altitude in the stratosphere, drops in the mesosphere, then rises again in the thermosphere. The tropopause separates the troposphere from the stratosphere. Ozone (blue shaded area) reaches a peak in a portion of the stratosphere, giving rise to the term *ozone layer*. Adapted from Jacobson, M.Z., 2002. *Atmospheric pollution: History, science, and regulation*. Cambridge: Cambridge University Press; and Parson, E.A., 2003. *Protecting the ozone layer: Science and strategy*. Oxford: Oxford University Press.

the atmosphere and planetary surface, and the rest is reflected back into space (see Figure 14.1, p. 301).

Land and surface water absorb solar energy and then radiate heat, causing some water to evaporate. Air near Earth's surface therefore tends to be warmer and moister than air at higher altitudes. These differences set into motion a process of **convective circulation**. Warm air, being less dense, rises and creates vertical currents. As air rises into regions of lesser atmospheric pressure, it expands and cools. Once the air cools, it becomes denser and descends, replacing warm air that is rising. The descending air picks up heat and moisture near ground level and prepares to rise again, continuing the process. Convective circulation influences both weather and climate.

Weather and climate both involve physical properties of the troposphere, such as temperature, pressure, humidity, cloudiness, and wind. **Weather** specifies atmospheric conditions within small geographic areas over minutes, hours, or days. In contrast, **climate** describes patterns of atmospheric conditions found across large geographic regions over seasons, years, or millennia. Mark Twain once noted the distinction by saying, "Cli-

mate is what we expect; weather is what we get." For example, Los Angeles has a climate characterized by warm dry summers and mild rainy winters, yet on occasional autumn days, dry Santa Ana winds blow in from the desert and bring extremely hot weather.

Under most conditions, air in the troposphere decreases in temperature as altitude increases. Because warm air rises, vertical mixing results (**FIGURE 13.3A**). Occasionally, however, a layer of cool air may form beneath a layer of warmer air. This departure from the normal temperature profile is known as a **temperature inversion**, or **thermal inversion** (**FIGURE 13.3B**). The band of air in which temperature rises with altitude is called an **inversion layer** (because the normal direction of temperature change is inverted). Thermal inversions can occur in different ways, sometimes involving cool air at ground level and sometimes producing an inversion layer higher above the ground. One common type of inversion (shown in Figure 13.3B) occurs in mountain valleys where slopes block morning sunlight, keeping ground-level air within the valley shaded and cool.

Because the cooler air at the bottom of an inversion layer is denser than the warmer air above it, it resists vertical mixing and remains stable. Vertical mixing allows air pollution to be carried

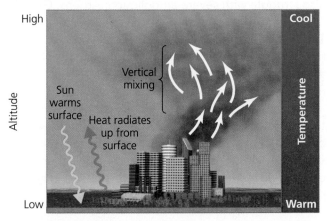

(a) Normal conditions

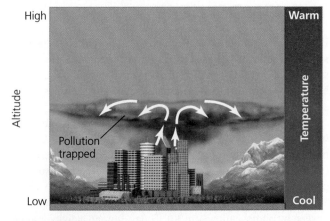

(b) Thermal inversion

FIGURE 13.3 ▲ Under normal conditions **(a)**, air becomes cooler with altitude, and air of different altitudes mixes, dispersing pollutants upward and outward. During a thermal inversion **(b)**, dense cool air remains near the ground, and air warms with altitude within the inversion layer. Little mixing occurs, and pollutants are trapped near the surface.

upward and diluted, but thermal inversions trap pollutants near the ground. An inversion persisting for several days sparked a "killer smog" crisis in London, England, in 1952. The inversion trapped pollutants from factories and coal-burning stoves, creating foul conditions that killed 4,000 people—and by some estimates up to 12,000. Both Los Angeles and Tehran suffer their worst pollution when thermal inversions prevent pollutants from being dispersed. Both cities are encircled by mountains, which trap pollutants by promoting inversion layers and by interrupting air flow. Tehran experiences thermal inversions on more than 250 days each year. Inversions regularly concentrate pollution over large metropolitan areas in valleys ringed by mountains, from Mexico City to Seoul, Korea, to São Paulo, Brazil.

Large-scale circulation systems produce global climate patterns

At large geographic scales, convective air currents contribute to long-term climate patterns (**FIGURE 13.4A**). Near the equator, solar radiation sets in motion a pair of convective cells known as *Hadley cells*. Here, where sunlight is most intense, surface air warms, rises, and expands. As it does so, it releases moisture, producing the heavy rainfall that gives

rise to tropical rainforests near the equator. After releasing much of its moisture, this air diverges and moves in currents heading north and south. The air in these currents cools and descends back to Earth at about 30 degrees latitude north and south. Because the descending air is now dry, the regions around 30 degrees latitude are quite arid, giving rise to deserts. Two further pairs of convective cells, called *Ferrel cells* and *polar cells*, lift air and create precipitation around 60 degrees latitude north and south and cause drier air to descend at around 30 degrees latitude and in the polar regions.

These three pairs of convective cells create wet climates near the equator, arid climates near 30 degrees latitude, moist regions near 60 degrees latitude, and dry conditions near the poles. These patterns, combined with temperature variation, help explain why biomes tend to be arrayed in latitudinal bands (Figure 4.15, p. 78).

The Hadley, Ferrel, and polar cells also interact with Earth's rotation to produce global wind patterns (**FIGURE 13.4B**). As Earth rotates on its axis, locations on the equator spin faster than locations near the poles. This means that as air currents of the convective cells flow north to south, some regions of the planet's surface move west to east beneath them more quickly than others. As a result, from the perspective of an Earth-bound observer, these air currents appear to be deflected from

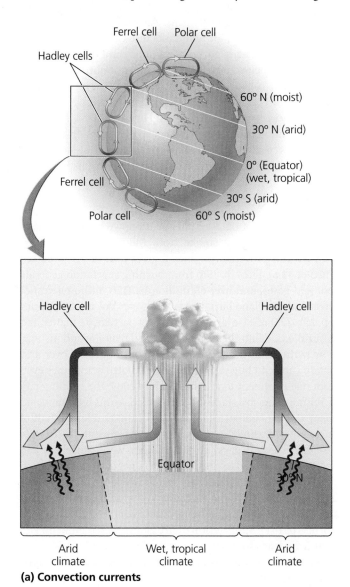

(a) **Convection currents**

FIGURE 13.4 ◀ A series of large-scale convective cells **(a)** helps determine global patterns of humidity and aridity. Warm air near the equator rises, expands, and cools; and moisture condenses, giving rise to a wet climate in tropical regions. Air travels toward the poles and descends around 30 degrees latitude. This air, which lost its moisture in the tropics, causes regions around 30 degrees latitude to be arid. This convective circulation, a Hadley cell, occurs on both sides of the equator. Between roughly 30 and 60 degrees latitude north and south, Ferrel cells occur; and between 60 and 90 degrees latitude, polar cells occur. Air rises around 60 degrees latitude, creating a moist climate, and falls around 90 degrees, creating a dry climate. Global wind currents **(b)** show latitudinal patterns as well. Trade winds between the equator and 30 degrees latitude blow westward, whereas westerlies between 30 and 60 degrees latitude blow eastward.

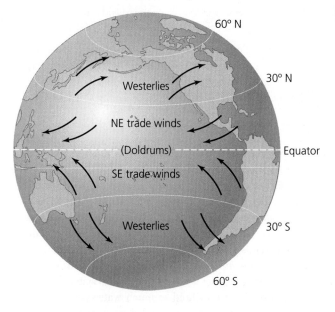

(b) **Global wind patterns**

(a) Satellite image of a hurricane

(b) Photograph of a tornado

FIGURE 13.5 ▲ Hurricanes **(a)** and tornadoes **(b)** are two types of cyclonic storms that pose hazards to our life and property.

a straight path. This deflection is called the *Coriolis effect*, and it results in the curving global wind patterns displayed in Figure 13.4B. For centuries, people used these global wind patterns to facilitate ocean travel by wind-powered sailing ships.

Storms pose hazards

Atmospheric conditions can sometimes create storms that threaten life and property. **Hurricanes** (**FIGURE 13.5A**) can form when winds rush into areas of low pressure where warm moisture-laden air over tropical oceans is rising. In the Northern Hemisphere, these winds turn counterclockwise because of the Coriolis effect. In other regions, such cyclonic storms are called *cyclones* or *typhoons*. The powerful convective currents of these storms draw up immense amounts of water vapor. As the warm moist air rises and cools, water condenses (because cool air cannot hold as much water vapor as warm air) and falls heavily as rain. In North America, the Gulf Coast and Atlantic Coast are most susceptible to hurricanes.

Tornadoes (**FIGURE 13.5B**) can form when a mass of warm air meets a mass of cold air and the warm air rises quickly, setting a powerful convective current in motion. If high-altitude winds are blowing faster and in a different direction from low-altitude winds, the rising column of air may begin to rotate. Eventually the spinning funnel of rising air may lift up soil and objects in its path, with winds up to 500 km per hour (310 mph). In North America, tornadoes are most apt to form in the Great Plains and the Southeast, where cold air from Canada and warm air from the Gulf of Mexico frequently meet.

Understanding how the atmosphere functions can help us predict violent storms and warn people of their approach. Such knowledge can also help us comprehend how our pollution of the atmosphere affects climate, ecological systems, economies, and human health.

OUTDOOR AIR POLLUTION

Throughout human history, we have made the atmosphere a dumping ground for our airborne wastes. Whether from simple wood fires or modern coal-burning power plants, people have generated **air pollutants**, gases and particulate material added to the atmosphere that can affect climate or harm people or other organisms. **Air pollution** refers to the emission or release of air pollutants into the atmosphere.

In recent decades, government policy and improved technologies have helped us reduce most types of **outdoor air pollution** (often called *ambient air pollution*) in industrialized nations. However, outdoor air pollution remains a problem, particularly in developing nations and in urban areas. The greatest air pollution problem today may be our emission of greenhouse gases (p. 300), which contribute to global climate change. (We discuss this issue separately and in depth in Chapter 14.)

Natural sources can pollute

When we think of outdoor air pollution, we tend to envision smokestacks belching smoke from industrial plants. However, natural processes produce a great deal of air pollution (**FIGURE 13.6**). Fires (p. 196) from burning vegetation generate soot and gases, and over 60 million ha (150 million acres) of forest and grassland burn in a typical year. Volcanic eruptions (pp. 231–233) release large quantities of particulate matter and sulfur dioxide into the troposphere, and major eruptions may blow matter into the stratosphere. Winds sweeping over arid terrain can send huge amounts of dust aloft—sometimes even from one continent to another. In July 2009, the bustling city of Tehran came to a standstill when windstorms blew sand and dust from drought-stricken Iraq into Iran, enveloping half of the country. Businesses, schools, and government offices were closed for several days, airplane flights were cancelled, and people were warned to stay indoors to safeguard their health.

Some natural impacts are made worse by human activity and land use policies. Farming and grazing practices that strip vegetation from the soil promote wind erosion and dust storms (p. 141). Suppression of fire allows fuel to build up and eventually leads to more-destructive fires (pp. 196–197). And in the tropics, many farmers set fires to clear forest for agriculture (pp. 139–140).

  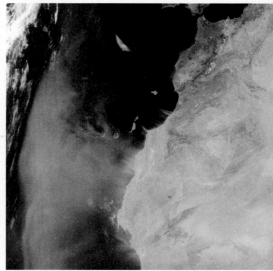

(a) Natural fire in California **(b) Mount Saint Helens eruption, 1980** **(c) Dust storm blowing dust from Africa to the Americas**

FIGURE 13.6 ▲ Fires **(a)** in forests and grasslands are one source of natural air pollution. Volcanoes are another, as shown by Mount Saint Helens **(b)**, which erupted in the state of Washington in 1980. Dust storms are a third source. Trade winds blowing soil across the Atlantic Ocean from Africa to the Americas **(c)** carry fungal and bacterial spores linked to die-offs in Caribbean coral reef systems, although they also bring nutrients to the Amazon rainforest.

We create outdoor air pollution

Human activity introduces many sources of air pollution. As with water pollution, air pollution can emanate from *point sources* or *non-point sources* (pp. 265–266). A point source describes a specific location from which large quantities of pollutants are discharged, such as a coal-fired power plant. Non-point sources are more diffuse, consisting of many small, widely spread sources (such as thousands of automobiles).

Primary pollutants, such as soot and carbon monoxide, are pollutants emitted into the troposphere in a form that can cause harm or can react to form harmful substances. Harmful substances produced once primary pollutants react with constituents of the atmosphere are called **secondary pollutants**.

Pollutants differ in the amount of time they spend in the atmosphere—called their **residence time**—because substances differ in how readily they react in air and in how quickly they settle to the ground. Pollutants with brief residence times exert localized impacts over short time periods. Most particulate matter and most pollutants from automobile exhaust stay aloft only hours or days, which is why air quality in a city like Tehran or Los Angeles can change from day to day. In contrast, pollutants with long residence times can exert impacts regionally or globally for long periods, even centuries. The pollutants that drive global climate change and those that deplete Earth's ozone layer (two separate phenomena!—see FAQ, p. 290) are each able to cause these global and long-lasting impacts because they persist in the atmosphere for so long. **FIGURE 13.7** shows this relationship, with examples.

Clean Air Act legislation addresses pollution in the United States

To address air pollution in the United States, Congress has passed a series of laws, notably the **Clean Air Act**, first enacted in 1963 and amended multiple times since, particularly in 1970 and 1990. This body of legislation funds research on pollution control, sets standards for air quality, imposes limits on emissions from new sources, and enables citizens to sue parties violating the standards. It also introduced an emissions trading program (p. 108) for sulfur dioxide. Beginning in 1995, businesses and utilities were allocated permits for emitting this pollutant and could buy, sell, or trade these allowances. Each year the overall amount of allowed pollution was lowered. This market-based incentive program has helped reduce sulfur dioxide emissions nationally (see Figure 5.18, p. 109). The Los Angeles region adopted its own cap-and-trade program in 1994. The Regional Clean Air Incentives Market (RECLAIM) helped the L.A. basin decrease sulfur dioxide emissions by 47% and nitrogen oxide emissions by 61% by 2003, and further cuts are being achieved as the program continues.

Under the Clean Air Act, the U.S. Environmental Protection Agency (EPA) sets nationwide standards for emissions and for concentrations of pollutants in ambient air throughout the nation. It is largely up to the states to monitor air quality and develop, implement, and enforce regulations within their borders. States submit plans to the EPA for approval, and if a state's plans are not adequate, the EPA can take over enforcement. When a region fails to clean up its air, the EPA can prevent it from receiving federal money for transportation projects.

The EPA sets standards for "criteria pollutants"

The EPA and the states focus on six **criteria pollutants**, pollutants judged to pose especially great threats to human health—carbon monoxide (CO), sulfur dioxide (SO_2), nitrogen dioxide (NO_2), tropospheric ozone (O_3), particulate matter, and

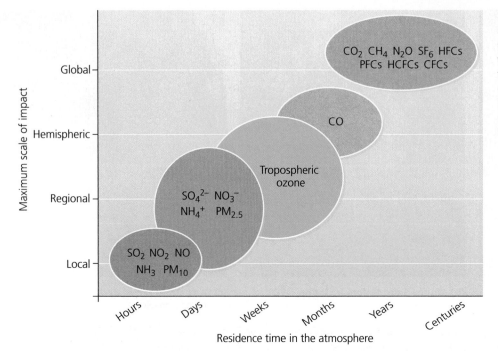

FIGURE 13.7 ◀ Substances with short residence times in the atmosphere affect air quality only locally, whereas those with long residence times affect air quality regionally or globally. Data from United Nations Environment Programme, 2007. *Global environmental outlook (GEO-4)*, Nairobi, Kenya.

lead (Pb). For these, the EPA has established maximum concentrations allowable in outdoor air.

Carbon monoxide **Carbon monoxide** is a colorless, odorless gas produced primarily by the incomplete combustion of fuel. Vehicles and engines account for 78% of CO emissions in the United States. Other sources include industrial processes, fires, waste combustion, and residential wood burning. Carbon monoxide can deprive us of oxygen because it can bind to hemoglobin in red blood cells, preventing the hemoglobin from binding with oxygen.

Sulfur dioxide **Sulfur dioxide** is a colorless gas with a pungent odor. Most SO_2 pollution results from the combustion of coal for electricity generation and industry. During combustion, elemental sulfur (S) in coal reacts with oxygen (O_2) to form SO_2. Once in the atmosphere, SO_2 may react to form sulfur trioxide (SO_3) and sulfuric acid (H_2SO_4), which may return to Earth in acid deposition (pp. 291–294).

Nitrogen dioxide **Nitrogen dioxide** is a foul-smelling reddish-brown gas that contributes to smog and acid deposition. Along with nitric oxide (NO), NO_2 belongs to a family of compounds called **nitrogen oxides** (NO_x). Nitrogen oxides result when atmospheric nitrogen and oxygen react at the high temperatures created by combustion engines. Most U.S. NO_x emissions result from combustion in vehicle engines. Electrical utility and industrial combustion account for most of the remainder.

Tropospheric ozone Although ozone in the stratosphere protects us by filtering UV radiation, ozone from human activity accumulates low in the troposphere. Here, this colorless gas is a secondary pollutant, created by the interaction of sunlight, heat, nitrogen oxides, and volatile carbon-containing chemicals. A major component of smog, O_3 poses health risks as a result of its instability as a molecule; this triplet of oxygen atoms will readily split into a molecule of oxygen gas and a free oxygen atom. The oxygen atom may then participate in reactions that can damage living tissues and cause respiratory problems. **Tropospheric ozone** is the pollutant that most frequently exceeds its EPA standard.

Particulate matter **Particulate matter** is composed of solid or liquid particles small enough to be suspended in the atmosphere and able to damage respiratory tissues when inhaled. Particulate matter includes primary pollutants such as dust and soot, as well as secondary pollutants such as sulfates and nitrates. The EPA classifies particulate pollution by the size of the particles. PM_{10} pollution consists of particles less than 10 microns in diameter (one-seventh the width of a human hair), whereas $PM_{2.5}$ pollution consists of still-finer particles less than 2.5 microns in diameter. Most PM_{10} pollution is from road dust, whereas most $PM_{2.5}$ pollution results from combustion processes.

Lead **Lead** is a heavy metal that enters the atmosphere as a particulate pollutant. The lead-containing compounds tetraethyl lead and tetramethyl lead, when added to gasoline, improve engine performance. However, exhaust from the combustion of leaded gasoline emits airborne lead, which can be inhaled or can settle on land and water. Lead can enter the food chain, accumulate in body tissues, and cause central nervous system malfunction, developmental problems in children, and other ailments (p. 209). Since the 1980s, leaded gasoline has been phased out in most industrialized nations (p. 5), and lead pollution has plummeted. However, auto exhaust still generates lead pollution in developing nations where leaded gasoline has not been banned.

EPA monitoring finds that many Americans live where concentrations of criteria pollutants regularly reach unhealthy levels. Residents of Los Angeles County, for instance, breathe air that violates safety standards for four of the six criteria pollutants. All together, as of 2008, 127 million Americans lived in counties that violated the national ambient air quality standards for at least one of the six criteria pollutants.

Agencies monitor emissions

Besides measuring concentrations of the six criteria pollutants in ambient air, state and local agencies also monitor and report to the EPA emissions of pollutants that affect ambient concentrations of the criteria pollutants. Emissions are monitored for the four criteria pollutants that are primary pollutants (carbon monoxide, sulfur dioxide, particulate matter, and lead), as well as for all nitrogen oxides (because NO reacts in the atmosphere to form NO_2, which is both a primary and secondary pollutant). Tropospheric ozone is a secondary pollutant only; we do not emit it. Instead, agencies monitor emissions of **volatile organic compounds (VOCs)**, carbon-containing chemicals (such as hydrocarbons; p. 28) that can react to produce ozone and other secondary pollutants. The largest sources of anthropogenic VOC emissions include industrial use of solvents and vehicle emissions. In the United States in 2008, human activity polluted the air with 123 million tons of the six monitored pollutants (**FIGURE 13.8**).

We have reduced U.S. air pollution

Since the Clean Air Act of 1970, we have reduced emissions of each of the six monitored pollutants (**FIGURE 13.9A**), and total emissions of the six together have declined by 60%. These dramatic reductions in emissions have occurred despite substantial increases in the nation's population, energy consumption, miles traveled by vehicle, and gross domestic product (**FIGURE 13.9B**).

We have achieved this success as a result of policy steps and technological developments, each motivated by grassroots social demand for cleaner air. Cleaner-burning motor vehicle engines and automotive technologies such as catalytic converters have played a large part. In a catalytic converter, engine exhaust reacts with several metals that convert hydrocarbons, CO_2, and NO_x into carbon dioxide, water vapor, and nitrogen gas. The sulfur dioxide permit-trading program

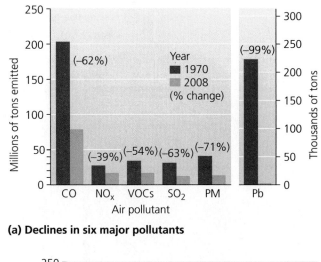

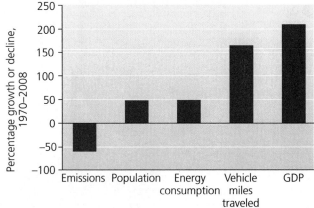

(a) Declines in six major pollutants

(b) Trends in major indicators

FIGURE 13.9 ▲ The EPA tracks emissions of several major pollutants into U.S. air. Each of these pollutants has shown substantial declines since 1970 **(a)**, and emissions from all six together have declined by 60%. We have achieved these reductions in emissions despite increases **(b)** in U.S. population, energy consumption, vehicle miles traveled, and gross domestic product (GDP). Data for particulate matter is for PM_{10} since 1985. Data from U.S. EPA.

(pp. 108–109) and clean coal technologies (p. 336) have reduced SO_2 emissions. Technologies such as baghouse filters, electrostatic precipitators, and **scrubbers** (**FIGURE 13.10**) that chemically convert or physically remove airborne pollutants before they are emitted from smokestacks have allowed factories, power plants, and refineries to reduce emissions of several pollutants. And the phaseout of leaded gasoline caused U.S. lead emissions to plummet by 93% in the 1980s alone.

The reduction of outdoor air pollution since 1970 represents one of the greatest accomplishments achieved by the United States in safeguarding human health and environmental quality. The EPA estimates that between 1970 and 1990, clean air regulations saved the lives of 200,000 Americans.

Plenty of room for improvement remains, however, because concerns over new pollutants are emerging and because greenhouse gas emissions (p. 302) continue to rise. U.S. carbon dioxide emissions rose 44% from 1970 to 2008. The U.S. Supreme Court in 2007 ruled that the EPA has the legal authority to regulate carbon dioxide as a pollutant. The EPA took the first steps toward doing so in 2011, but faces logistical challenges and formidable political opposition. Yet because we

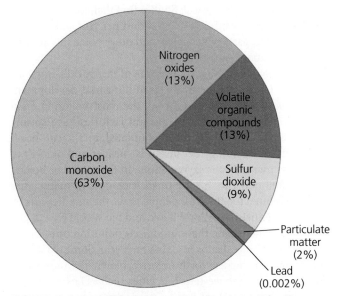

FIGURE 13.8 ▲ In 2008, the United States emitted 123 million tons of the six major pollutants whose emissions are monitored by the EPA and state and local agencies. Carbon monoxide accounted for most of these emissions, by mass. Data from U.S. EPA.

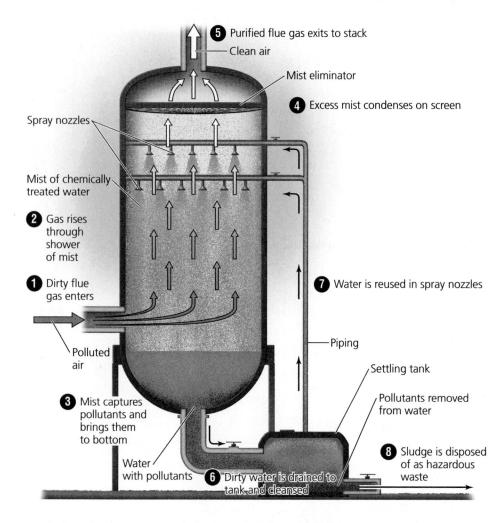

⑤ Purified flue gas exits to stack
— Clean air
— Mist eliminator
④ Excess mist condenses on screen

Spray nozzles

Mist of chemically treated water

② Gas rises through shower of mist

① Dirty flue gas enters

Polluted air

③ Mist captures pollutants and brings them to bottom

Water with pollutants

⑥ Dirty water is drained to tank and cleansed

⑦ Water is reused in spray nozzles

Piping

Settling tank

Pollutants removed from water

⑧ Sludge is disposed of as hazardous waste

FIGURE 13.10 ◄ In this spray-tower wet scrubber, polluted air ① rises through a chamber while arrays of nozzles spray a mist of water mixed with lime or other active chemicals ②. The falling mist captures pollutants and carries them to the bottom of the chamber ③, essentially washing them out of the air. Excess mist is captured on a screen ④, and air emitted from the scrubber has largely been cleansed ⑤. Periodically, the dirty water is drained from the chamber ⑥, cleansed in a settling tank, and recirculated ⑦ through the spray nozzles. The resulting sludge must be disposed of ⑧ as hazardous waste (pp. 392–395). Scrubbers and other pollution control devices come in many designs; the type shown here typically removes at least 90% of particulate matter and gases such as sulfur dioxide.

were able to reduce emissions of several major pollutants by 60% while expanding our economy, we can hope that similar success might soon be achieved with greenhouse gas emissions.

Toxic pollutants pose health risks

We are also reducing emissions of **toxic air pollutants**, substances known to cause cancer, reproductive defects, or neurological, developmental, immune system, or respiratory problems. Under the 1990 Clean Air Act, the EPA regulates 188 toxic air pollutants, ranging from mercury (from coal-burning power plant emissions and other sources) to VOCs such as benzene (a component of gasoline) and methylene chloride (found in paint stripper).

Based on monitoring at 300 sites across the United States, experts estimate that toxic air pollutants cause cancer in 1 out of every 28,000 Americans (36 cancer cases per 1 million people). Although residents of areas such as Los Angeles still experience high health risks, the EPA estimates that Clean Air Act regulations have helped to reduce emissions of toxic air pollutants since 1990 by more than 35%.

Industrializing nations are suffering increasing air pollution

Although the United States and other industrialized nations have improved their air quality, outdoor air pollution is growing worse in many industrializing countries. In these societies, proliferating factories and power plants are emitting more pollutants as governments encourage economic growth. Additionally, more citizens own and drive automobiles (**FIGURE 13.11**). At the same time, most people continue to burn traditional sources of fuel such as wood, charcoal, and coal for cooking and home heating. Iran is typical: Studies find that each resident of Tehran inhales 7–9 kg (15–20 lb) of dust per year and that levels of CO, SO_2, particulate matter, and other pollutants are well above international safety standards.

The people of China suffer some of the world's worst air pollution. China has fueled its rapid industrial development with its abundant reserves of coal, the most-polluting fossil fuel (pp. 329–330). Power plants and factories have sprung up across the nation, often using outdated, inefficient, heavily polluting technology because it is cheaper and quicker to build. Car ownership is skyrocketing; in the capital of Beijing alone, 1,500 new cars hit the roads each day. In many cities the haze is often too thick for people to see the sun. Reports by Chinese scientists, the World Bank, and the World Health Organization all estimate that outdoor air pollution causes over 300,000 premature deaths in China each year. Some of China's pollution even reaches North America, blown by winds across the Pacific Ocean to western U.S. cities such as Los Angeles.

China's government is now striving to reduce pollution. The government is closing down some heavily polluting factories and mines, phasing out some subsidies for polluting industries, and installing pollution controls in power plants. It

FIGURE 13.11 ▲ Automobile traffic in Tehran illustrates a major cause of air pollution in today's industrializing nations. Officials estimate that Tehran's vehicles emit 5,000 tons of pollutants each day as traffic creeps along at an average of 18 kph (11 mph).

is subsidizing people to buy efficient electric heaters for their homes, mandating cleaner formulations for gasoline and diesel, raising standards for fuel efficiency, and tightening regulations on automotive emissions. In Beijing, mass transit is being expanded, many buses run on natural gas, and heavily pollut-

ing vehicles are restricted in the central city. China is also aggressively developing cleaner wind, solar, and nuclear power to substitute for power produced by burning coal.

Smog is our most common air quality problem

Let's now take a closer look at our most widespread type of air pollution: smog. *Smog* is an unhealthy mixture of air pollutants that often forms over urban areas as a result of fossil fuel combustion. Since the onset of the industrial revolution, cities have suffered a type of smog known as **industrial smog**. When coal or oil is burned, some portion is completely combusted, forming CO_2; some partially combusts, producing CO; and some remains unburned and is released as soot (particles of carbon). Moreover, coal contains contaminants such as mercury and sulfur. Sulfur reacts with oxygen to form sulfur dioxide, which can undergo a series of reactions to form sulfuric acid and ammonium sulfate (**FIGURE 13.12A**). These chemicals and others produced by further reactions, along with soot, are the main components of industrial smog.

In the wake of London's 1952 "killer smog" and other fatal pollution episodes, governments of developed nations began regulating industrial emissions and have greatly reduced

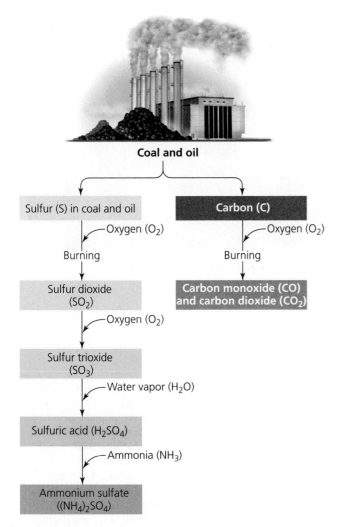

(a) Formation of industrial smog

(b) Donora, Pennsylvania, at midday in the 1948 smog event

FIGURE 13.12 ◄ Emissions from the combustion of coal and oil in manufacturing plants and utilities without pollution control technologies can create industrial smog. Industrial smog consists primarily of sulfur dioxide and particulate matter, as well as carbon monoxide and carbon dioxide from the carbon component of fossil fuels. When fossil fuels are combusted, sulfur contaminants give rise to sulfur dioxide, which in the presence of other chemicals in the atmosphere can produce several other sulfur compounds **(a)**. Under certain weather conditions, industrial smog can blanket whole towns or regions, as it did in Donora, Pennsylvania, shown here **(b)** in the daytime during its deadly 1948 smog episode.

industrial smog. However, in industrializing regions such as China, India, and eastern Europe, coal burning and lax pollution controls result in industrial smog that poses significant health risks.

As we've seen, weather and topography play roles in smog formation. Four years before London's killer smog, a similar event occurred in Pennsylvania in the small town of Donora (**FIGURE 13.12B**). Air near the ground cooled in the night, and because Donora is in hilly terrain, morning sun did not reach the valley floor to warm and disperse the air. The resulting thermal inversion trapped smog from a steel and wire factory. Twenty-one people died, and over 6,000 people—nearly half the town—became ill.

Photochemical smog results from a series of reactions

In most cities today, including Los Angeles and Tehran, pollution from automobile exhaust generates a different type of smog: photochemical smog. **Photochemical smog** forms when sunlight drives chemical reactions between primary pollutants and normal atmospheric compounds, producing a mix of over 100 different chemicals, with tropospheric ozone often the most abundant (**FIGURE 13.13A**). Because it

also includes NO_2, photochemical smog generally appears as a brownish haze (**FIGURE 13.13B**).

Hot, sunny, windless days in urban areas provide perfect conditions for the formation of photochemical smog. Exhaust from morning traffic releases NO and VOCs into a city's air, and sunlight then promotes the production of ozone and other pollutants. Photochemical smog in urban areas typically peaks in midafternoon. Cities like Los Angeles and Tehran are prone to photochemical smog because they have sunny climates and because nearby mountains trap air and promote thermal inversions. L.A.'s sister city of Mexico City, also ringed by mountains and with a sunny climate, suffers some of the world's worst photochemical smog. In 1996, a 5-day crisis there killed at least 300 people and sent 400,000 to hospitals with eye, nose, throat, and respiratory problems.

We can take steps to reduce smog

Los Angeles's struggle with air pollution began in 1943, when the city's first major smog episode cut visibility to three blocks. Since then, L.A. residents have dealt with headaches, eye irritation, asthma, lung damage, and related illnesses. However, Los Angeles confronted its problem and has made great progress in clearing the air since the 1970s.

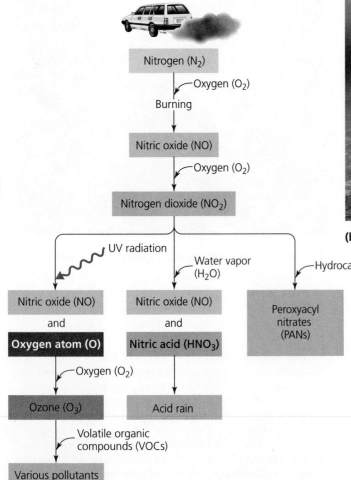

(a) Formation of photochemical smog

(b) Photochemical smog over Mexico City

FIGURE 13.13 ◄ Nitric oxide, a key element of photochemical smog, can start a chemical chain reaction **(a)** that results in the production of other compounds, including nitrogen dioxide, nitric acid, ozone, and peroxyacyl nitrates (PANs). PANs can induce further reactions that damage living tissues in animals and plants. Nitric acid contributes to acid deposition as well as photochemical smog. Photochemical smog is common today over many urban areas, especially those with hilly topography or frequent inversion layers. Mexico City **(b)** is one city that frequently experiences photochemical smog.

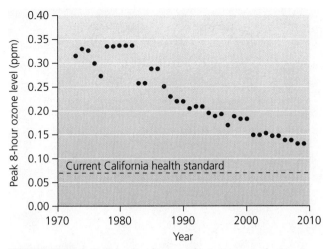

FIGURE 13.14 ▲ Peak levels of tropospheric ozone in the Los Angeles region have been reduced since the 1970s, thanks to government policy and improved automotive technology. Ozone pollution still violates the state health standard, however. Data from Environment California, 2010. *Clean cars in California: Four decades of progress in the unfinished battle to clean up our air.*

California took the lead among U.S. states in adopting pollution control technology, setting emissions standards for vehicles, and pushing the federal government to do the same. California's demands helped lead the auto industry to develop less-polluting cars. A 2010 study by the nonprofit group Environment California concluded that a new car today generates just 1% of the smog-forming emissions of a 1960s-era car. Because today's cars are 99% cleaner, the air is cleaner, even with more drivers on the road. In Los Angeles, peak smog levels have decreased 60–70% since 1980 (**FIGURE 13.14**).

Today in California and 33 other U.S. states, drivers are required to have their vehicle exhaust inspected periodically. Vehicle inspection programs have cut emissions leading to photochemical smog by 30% in these states.

Despite its progress, Los Angeles still suffers the worst tropospheric ozone pollution of any U.S. metropolitan area, according to a 2010 study by the American Lung Association. L.A. residents breathe air exceeding California's health standard for ozone on more than 130 days per year. A 2008 study calculated that air pollution in the L.A. basin and the nearby San Joaquin Valley each year caused nearly 3,900 premature deaths and cost society $28 billion (due to hospital admissions, lost work days, etc.).

Tehran is estimated to lose $130 million for each "smog holiday" its government declares, when pollution forces it to close schools and offices and advise people to stay home. City officials there have taken steps to combat the problem. Vehicle inspections are now required in Tehran, traffic into the city center is restricted, and people are paid to turn in old cars for newer, cleaner ones. Sulfur was reduced in diesel fuel, lead was removed from gasoline, and buses began running on (cleaner-burning) natural gas. Citywide pollution maps were made available online, and 22 electronic indicator boards were installed around the city, displaying current pollutant levels. All these efforts helped reduce pollution. Yet people keep streaming into the city, and residents buy cars even faster. Together these trends have overtaken the government's efforts, and pollution worsened again after 2006. In response, officials lowered gasoline subsidies, rationed fuel, and began expanding the subway system.

Of all the world's cities, Mexico City is gaining attention today for its success in reducing its smog problem—once the world's worst—even as its population, cars, and economic activity have grown. Regulations now require cars to have catalytic converters, get emissions tests, and stay off the roads one day per week. Some industrial facilities cleaned up their processes, and others were forced out. The national oil company Pemex removed lead from gasoline, improved its refineries, imported cleaner gasoline, and removed pollutants from the fuel that city residents use for cooking and heating. The subway system and a fleet of low-emission buses are being expanded, and a bike-hiring system has been introduced. As a result, smog is down, along with other pollutants such as lead and sulfur dioxide.

THE ISSUES

Your Region's Air Quality What outdoor air pollution challenges exist in your region? Explore one of the EPA websites that let you browse information on the air you breathe: www.airnow.gov, www.epa.gov/aircompare, or www.epa.gov/air/emissions/where.htm. How does your region's air quality compare to the rest of the nation? What factors do you think influence the quality of your region's air? Propose three steps for reducing air pollution in your region. What benefits might your region enjoy after taking such steps?

Synthetic chemicals deplete stratospheric ozone

Although ozone in the troposphere is a pollutant in photochemical smog, ozone in the stratosphere protects life on Earth by absorbing the sun's ultraviolet radiation, which can damage tissues and DNA. One generation ago, scientists discovered that our planet's stratospheric ozone was being depleted, posing a threat to human health and the environment. Years of research by hundreds of scientists (see **THE SCIENCE BEHIND THE STORY**, pp. 292–293) revealed that certain airborne chemicals destroy ozone, and that most of these **ozone-depleting substances** are human-made.

In particular, researchers pinpointed **halocarbons**—human-made compounds derived from simple hydrocarbons (p. 28) in which hydrogen atoms are replaced by halogen atoms such as chlorine, bromine, or fluorine. Industry was mass-producing one class of halocarbon, **chlorofluorocarbons (CFCs)**, at 1 million tons per year in the 1970s. CFCs were useful as refrigerants, as fire extinguishers, as propellants for aerosol spray cans, as cleaners for electronics, and for making polystyrene foam. Because CFCs rarely reacted with other chemicals, scientists surmised that they would be harmless. Alas, because they are nonreactive, CFCs reach the

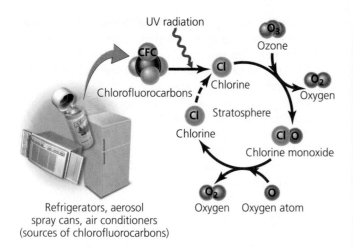

FIGURE 13.15 ▲ A chlorine atom released from a CFC molecule in the presence of UV radiation reacts with an ozone molecule, forming one molecule of oxygen gas and one chlorine monoxide (ClO) molecule. The oxygen atom of the ClO molecule will then bind with a stray oxygen atom to form oxygen gas, leaving the chlorine atom to begin the destructive cycle anew. In this way, a chlorine atom can destroy up to 100,000 ozone molecules.

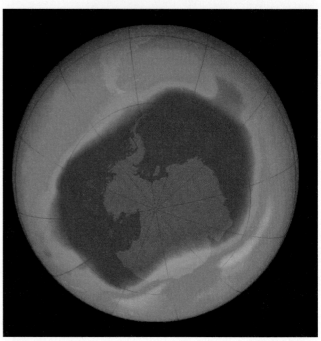

FIGURE 13.16 ▲ The "ozone hole" consists of a region of thinned ozone density in the stratosphere over Antarctica and the southernmost ocean regions. It has reappeared seasonally each September in recent decades. This colorized satellite imagery of Earth's Southern Hemisphere from September 24, 2006, shows the ozone hole (purple and blue colors) at its maximal recorded extent to date.

stratosphere unchanged and can linger there for a century or more. In the stratosphere, intense solar radiation breaks bonds in CFC molecules, releasing their constituent chlorine atoms. In a two-step chemical reaction (**FIGURE 13.15**), each newly freed chlorine atom can split an ozone molecule and then ready itself to split another one. During its long residence time in the stratosphere, each free chlorine atom can destroy as many as 100,000 ozone molecules!

In 1985, researchers shocked the world by announcing that stratospheric ozone levels over Antarctica in springtime had declined by nearly half in just the previous decade, leaving a thinned ozone concentration that was soon dubbed the **ozone hole** (**FIGURE 13.16**). During each Southern Hemisphere spring since then, ozone concentrations over this immense region have dipped to roughly half their historic levels.

FAQ

Q: Is the ozone hole related to global warming?

A: This is a common misconception held by the public. Some people believe that the depletion of stratospheric ozone helps to prevent global warming by letting heat or greenhouse gases out of the atmosphere. Other people believe that ozone depletion worsens warming by letting heat into the atmosphere. Neither is true. Ozone depletion lets in excess ultraviolet radiation from the sun, but this does not appreciably warm or cool the atmosphere. However, research published in 2011 suggested that the ozone hole does apparently affect atmospheric circulation and rainfall in the Southern Hemisphere. We still have more to learn, and perhaps researchers will discover further connections between aspects of climate change and ozone depletion.

The Montreal Protocol addressed ozone depletion

International policy efforts to restrict production of CFCs bore fruit in 1987 with the **Montreal Protocol**. In this treaty, signatory nations (eventually numbering 196) agreed to cut CFC production in half by 1998. Five follow-up agreements deepened the cuts, advanced timetables for compliance, and addressed additional ozone-depleting substances. The substances covered by these agreements have now been mostly phased out, and industry has been able to shift to safer alternative chemicals.

As a result, we have evidently stopped the Antarctic ozone hole from growing worse (**FIGURE 13.17**). However, the ozone layer is not expected to recover completely until 2060–2075. Much of the 5 billion kg (11 billion lb) of CFCs emitted into the troposphere has yet to diffuse up into the stratosphere, so concentrations may not peak there until 2020. Because of this and the long residence times of many halocarbons, we can expect a long time lag between implementation of policy and the desired environmental effect.

Because of its success in addressing ozone depletion, the Montreal Protocol is widely viewed as a model for international cooperation on other global problems, from biodiversity loss (p. 177) to persistent organic pollutants (p. 223) to climate change (p. 319).

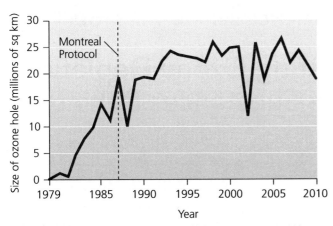

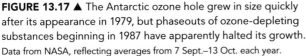

FIGURE 13.17 ▲ The Antarctic ozone hole grew in size quickly after its appearance in 1979, but phaseouts of ozone-depleting substances beginning in 1987 have apparently halted its growth. Data from NASA, reflecting averages from 7 Sept.–13 Oct. each year.

Acid deposition is another transboundary pollution problem

Just as stratospheric ozone depletion crosses political boundaries, so does **acid deposition**, the deposition of acidic (p. 27) or acid-forming pollutants from the atmosphere onto Earth's surface. This can take place by precipitation (commonly referred to as **acid rain**, but also including acid snow, sleet, and hail), by fog, by gases, or by the deposition of dry particles. Acid deposition is one type of **atmospheric deposition**, which refers more broadly to the wet or dry deposition of a variety of pollutants, including mercury, nitrates, organochlorines, and others.

Acid deposition originates primarily with the emission of sulfur dioxide and nitrogen oxides, largely through fossil fuel combustion by automobiles, electric utilities, and industrial facilities. Once airborne, these pollutants can react with water, oxygen, and oxidants to produce compounds of low pH (pp. 27–28), primarily sulfuric acid and nitric acid. Suspended in the troposphere, droplets of these acids may travel days or weeks for hundreds of kilometers (**FIGURE 13.18**).

Acid deposition has many impacts

Acid deposition has wide-ranging detrimental effects on ecosystems and on our infrastructure (**TABLE 13.1**). Acids leach nutrients such as calcium, magnesium, and potassium ions out of the topsoil, altering soil chemistry and harming plants and soil organisms. Acid precipitation also "mobilizes" toxic metal ions such as aluminum, zinc, mercury, and copper by chemically converting them from insoluble forms to soluble forms. Elevated soil concentrations of

TABLE 13.1 Effects of Acidic Deposition

Acidic deposition in northeastern U.S. forests has

- ▶ Accelerated leaching of base cations (ions such as Ca²⁺, Mg²⁺, Na⁺, and K⁺, which counteract acid deposition) from soil
- ▶ Allowed sulfur and nitrogen to accumulate in soil (Excess N may overfertilize native plants and encourage weeds.)
- ▶ Increased dissolved inorganic aluminum in soil, hindering plant uptake of water and nutrients
- ▶ Leached calcium from needles of red spruce, leading to tree mortality from wintertime freezing
- ▶ Increased mortality of sugar maples due to leaching of base cations from soil and leaves
- ▶ Acidified 41% of Adirondack, New York, lakes and 15% of New England lakes
- ▶ Lowered lakes' capacity to neutralize further acids
- ▶ Elevated aluminum levels in surface waters
- ▶ Reduced species diversity and abundance of aquatic life, affecting entire food webs

Source: Adapted from Driscoll, C.T., et al., 2001. *Acid rain revisited.* Hubbard Brook Research Foundation. Copyright 2001 C.T. Driscoll. Used with permission.

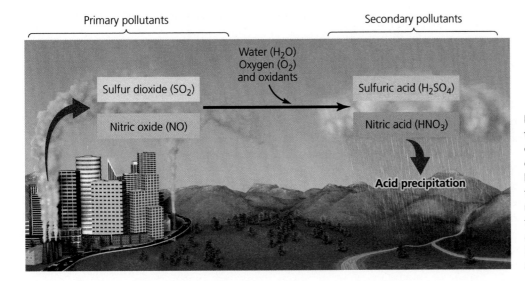

Primary pollutants

Secondary pollutants

Sulfur dioxide (SO₂)

Nitric oxide (NO)

Water (H₂O) Oxygen (O₂) and oxidants

Sulfuric acid (H₂SO₄)

Nitric acid (HNO₃)

Acid precipitation

FIGURE 13.18 ◀ Acid deposition can have consequences far downwind from its source. Sulfur dioxide and nitric oxide emitted by industries and utilities are transformed into sulfuric acid and nitric acid through chemical reactions in the atmosphere. These acidic compounds descend to Earth's surface in rain, snow, fog, and dry deposition.

Drs. F. Sherwood Rowland (left), Mario Molina (center), and Paul Crutzen (right) at a press conference after receiving the Nobel Prize

Discovering Ozone Depletion and the Substances Behind It

In discovering and coming to understand the depletion of stratospheric ozone, scientists have used historical records, field observations, laboratory experiments, computer models, and satellite technology.

The story starts back in 1924, when British scientist G.M.B. Dobson built an instrument that measured atmospheric ozone concentrations by sampling sunlight at ground level and comparing the intensities of wavelengths that ozone does and does not absorb. By the 1970s, the Dobson ozone spectrophotometer was being used to monitor ozone by a global network of observation stations.

Meanwhile, atmospheric chemists were learning how stratospheric ozone is created and destroyed. Ozone and oxygen exist in a natural balance, with one occasionally reacting to form the other, and oxygen being far more abundant. Researchers found that certain chemicals naturally present in the atmosphere, such as nitric oxide (NO), destroy ozone. Dutch meteorologist Paul Crutzen reported in 1970 that nitrous oxide (N_2O) produced by soil bacteria can make its way to the stratosphere and produce NO. And some human activities, such as fertilizer application, were increasing emissions of N_2O.

Following Crutzen's report, American scientists Richard Stolarski and Ralph Cicerone showed in 1973 that chlorine atoms can destroy ozone even more effectively than N_2O can. And two years earlier, British scientist

James Lovelock had developed an instrument to measure trace amounts of atmospheric gases and found that virtually all the chlorofluorocarbons (CFCs) humanity had produced in the past four decades were still aloft, accumulating in the stratosphere.

This set the stage for the key insight. In 1974, American chemist F. Sherwood Rowland and his Mexican postdoctoral associate, Mario Molina, took note of all the preceding research and realized that CFCs were rising into the stratosphere, being broken down by UV radiation, and releasing chlorine atoms that ravaged the ozone layer (see Figure 13.15, p. 290). Molina and Rowland's analysis, published in the journal *Nature*, eventually earned them the Nobel Prize in chemistry jointly with Crutzen.

The paper also sparked discussion about setting limits on CFC emissions.

Industry leaders attacked the research; DuPont's chairman of the board reportedly called it "a science fiction tale . . . a load of rubbish . . . utter nonsense." But measurements by numerous researchers soon confirmed that CFCs and other halocarbons were indeed depleting ozone. As a result, the United States and several other nations banned the use of CFCs in aerosol spray cans in 1978. Other uses continued, however, and by the early 1980s global production of CFCs was on the rise.

Then, a shocking new finding spurred the international community to take action. In 1985, Joseph Farman and colleagues analyzed data from a British research station in Antarctica that had been recording ozone concentrations since the 1950s. Farman's team reported in *Nature* that springtime Antarctic ozone concentrations had plummeted by 40–60% just since the 1970s (**see figure, part (a)**).

Farman's team had beaten a group of NASA scientists to the punch. The NASA scientists were sitting on reams of data from satellites showing a global drop in ozone levels (**see figure, part (b)**), but they had not yet submitted their analysis for publication.

---

metal ions such as aluminum weaken plants by damaging root tissue, hindering their uptake of water and nutrients. In some areas, acid fog with a pH of 2.3 (equivalent to vinegar, and over 1,000 times more acidic than normal rainwater) has enveloped forests and killed trees.

When acidic water runs off from land, it affects streams, rivers, and lakes. Thousands of lakes in Canada, Scandinavia, the United States, and elsewhere have lost their fish because acid precipitation leaches aluminum ions from soil and rock into waterways, where they damage the gills of fish and disrupt their salt balance, water balance, breathing, and circulation. Terrestrial animals are affected, too; populations of snails and

other invertebrates typically decline, and this reduces the food supply for birds.

Acid deposition damages crops, erodes stone buildings, corrodes cars, and erases the writing on tombstones. Ancient cathedrals in Europe, temples in Asia, and monuments in Washington, D.C., are experiencing billions of dollars of damage as their features dissolve away.

Because the pollutants leading to acid deposition can travel long distances, their effects may be felt far from their sources. Regions of greatest acidification, shown in **FIGURE 13.19,** tend to be downwind from heavily industrialized source areas of pollution.

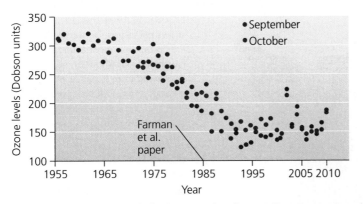

(a) Monthly mean stratospheric ozone levels at Halley, Antarctica

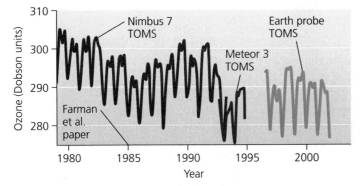

(b) Global stratospheric ozone readings from 3 satellites

Data from Halley, Antarctica (a), show a decrease in stratospheric ozone concentrations from the 1960s to 1990. Once ozone-depleting substances began to be phased out under the 1987 Montreal Protocol, ozone concentrations stopped declining. The paper by Joseph Farman et al. describing the ozone hole and using these data was published in 1985. Ozone decline and stabilization are also evident globally, as seen in these data (b) from three NASA satellites (averages of all regions between 65°N and 65°S latitude). Data in (a) from British Antarctic Survey. Data in (b) from NASA.

But why should ozone loss be localized over Antarctica in the southern spring? To answer this question, atmospheric chemists Susan Solomon, James Anderson, Crutzen, and others mounted expeditions in 1986 and 1987 to analyze atmospheric gases using ground stations and high-altitude balloons and aircraft. They learned that in the frigid Antarctic winter, *polar stratospheric clouds* form at high-altitudes. These icy clouds contain nitric acid, which splits chlorine atoms off from CFCs. The freed chlorine atoms accumulate in the clouds, trapped over Antarctica by wind currents that swirl around in a circular *polar vortex*. In the Antarctic spring (starting in September), sunshine returns and speeds the chlorine atoms' destruction of ozone. The ozone hole lingers over Antarctica until December, when warmth shuts down the polar vortex, allowing air to mix with air from elsewhere. The ozone hole vanishes until the following spring.

By 1987, the mass of scientific evidence helped convince the world's nations to agree on the Montreal Protocol. Within two years, further scientific evidence and computer modeling showed that more drastic measures were needed. In 1990, the Montreal Protocol was strengthened to include a complete phaseout of CFCs by 2000, in the first of several follow-up agreements. Today, amounts of ozone-depleting substances in the stratosphere are beginning to level off.

As the ozone layer begins a long-term recovery, scientists continue their research. In 2009, a team led by A.R. Ravishankara of the National Oceanic and Atmospheric Administration determined that nitrous oxide (N_2O) had now become the leading cause of ozone depletion. Its emissions are not regulated, so its impacts now surpass those currently exerted by the remaining halocarbons. Ravishankara's team points out that regulating nitrous oxide, which is also a potent greenhouse gas, would help mitigate climate change as well as speed ozone recovery.

We have begun to address acid deposition

Policy has helped us address acid deposition. The emissions trading program for sulfur dioxide established by the Clean Air Act of 1990 has reduced SO_2 emissions (see Figure 5.18, p. 109) by 64% so far. A 2005 study calculated that the program's economic benefits outweighed its costs by 40 to 1.

The economic incentives created by this cap-and-trade program encourage polluters to invest in technologies such as scrubbers (p. 286) and to devise other ways to become cleaner and more efficient. As a result of declining SO_2 emissions, av-erage sulfate precipitation in 2007–2009 was 43% lower than in 1989–1991 in the eastern United States. Emissions of NO_x also have been lowered by EPA regulation and by technological advances, and wet nitrogen deposition declined between these periods as well. As with ozone depletion, however, it will take time for ecosystems to recover—and scientists advise that further pollution reductions are needed.

Meanwhile, in the industrializing world, acid deposition is becoming worse. Coal-dependent China emits the most sulfur dioxide of any nation and has the world's worst acid rain problem. Overall, data on acid deposition show that although we have made advances in the control

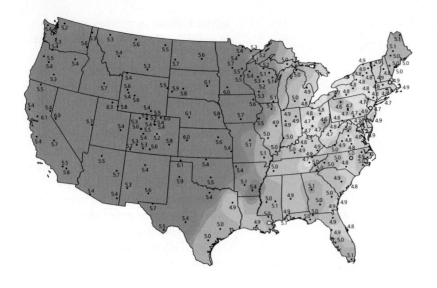

FIGURE 13.19 ◀ This U.S. map shows pH values for precipitation. Precipitation is most acidic in the Northeast and Midwest, near and downwind from (roughly east of) areas of heavy industry. Data are for 2009, from the National Atmospheric Deposition Program.

Lab pH

≥ 5.3	4.7 – 4.8
5.2 – 5.3	4.6 – 4.7
5.1 – 5.2	4.5 – 4.6
5.0 – 5.1	4.4 – 4.5
4.9 – 5.0	4.3 – 4.4
4.8 – 4.9	< 4.3

of outdoor air pollution, more can be done. The same can be said for indoor air pollution, a source of human health threats that is less familiar to most of us, but statistically more dangerous.

INDOOR AIR POLLUTION

Indoor air generally contains higher concentrations of pollutants than does outdoor air. As a result, the health effects from **indoor air pollution** in workplaces, schools, and homes outweigh those from outdoor air pollution. The World Health Organization (WHO) and other international agencies attribute the majority of the world's 2–3 million annual deaths from air pollution to indoor air pollution. This means that indoor air pollution takes several thousand lives each day.

Risks differ in developing and developed nations

Indoor air pollution exerts most impact in the developing world, where poverty forces millions of people to burn wood, charcoal, animal dung, or crop waste inside their homes for cooking and heating with little or no ventilation (**FIGURE 13.20**). In the process, people inhale dangerous amounts of soot and carbon monoxide. The WHO estimates that indoor air pollution from fuelwood burning kills 1.6 million people each year, comprising over 5% of all deaths in some developing nations and 2.7% of the entire global disease burden.

In developed nations, by contrast, the primary indoor air health risks are cigarette smoke and radon. Smoking cigarettes irritates the eyes, nose, and throat; worsens asthma and other respiratory ailments; and greatly increases the risk of lung cancer. Inhaling secondhand smoke causes many of the same problems. Tobacco smoke is a brew of over 4,000 chemical compounds, some of which are known or suspected to be toxic or carcinogenic. Although smoking has become

less popular in developed nations in recent years, it is still estimated to cause over 160,000 lung cancer deaths per year in the United States alone.

Radon gas is the second-leading cause of lung cancer in the developed world, responsible for an estimated 21,000 deaths per year in the United States and for 15% of lung cancer cases worldwide. Radon (p. 209) is a radioactive gas resulting from the natural decay of uranium in soil, rock, or water. It seeps up from the ground and can infiltrate buildings. Colorless and odorless, radon's presence can be impossible to predict without knowing an area's underlying geology. The only way to determine whether radon is entering a building is to sample air with a test kit. The EPA estimates that 6% of U.S. homes exceed its safety standard for radon. Since the 1980s, millions of U.S. homes have been tested for radon, close to a

FIGURE 13.20 ▼ In the developing world, many people build fires inside their homes for cooking and heating, as seen here in a Maasai home in Kenya. Indoor fires expose family members to severe particulate matter and carbon monoxide pollution.

million have undergone radon mitigation, and new homes are being built with radon-resistant features.

Many substances pollute indoor air

In our daily lives at home, we are exposed to many indoor air pollutants (**FIGURE 13.21**). The most diverse are volatile organic compounds. These airborne carbon-containing compounds are released by plastics, oils, perfumes, paints, cleaning fluids, adhesives, and pesticides. VOCs evaporate from furnishings, building materials, color film, carpets, laser printers, fax machines, and sheets of paper. Some products, such as chemically treated furniture, release large amounts of VOCs when new and progressively less as they age. Other items, such as photocopying machines, emit VOCs each time they are used. Formaldehyde—a VOC widely used in pressed wood, insulation, and other products—irritates mucous membranes, induces skin allergies, and causes other ailments. The "new car smell" that fills the interiors of new automobiles comes from a complex mix of dozens of VOCs as they outgas from the newly manufactured plastic, metal, and leather components of the car. Some scientific studies warn of health risks from this brew and recommend that you keep a new car well-ventilated.

VOCs are often held responsible for *sick-building syndrome*, an illness resulting from indoor pollution in which the specific cause is not identified. Microorganisms such as bacteria, fungi, and mold can also induce allergic responses and cause building-related illness. Heating and cooling systems in buildings make ideal breeding grounds for microbes,

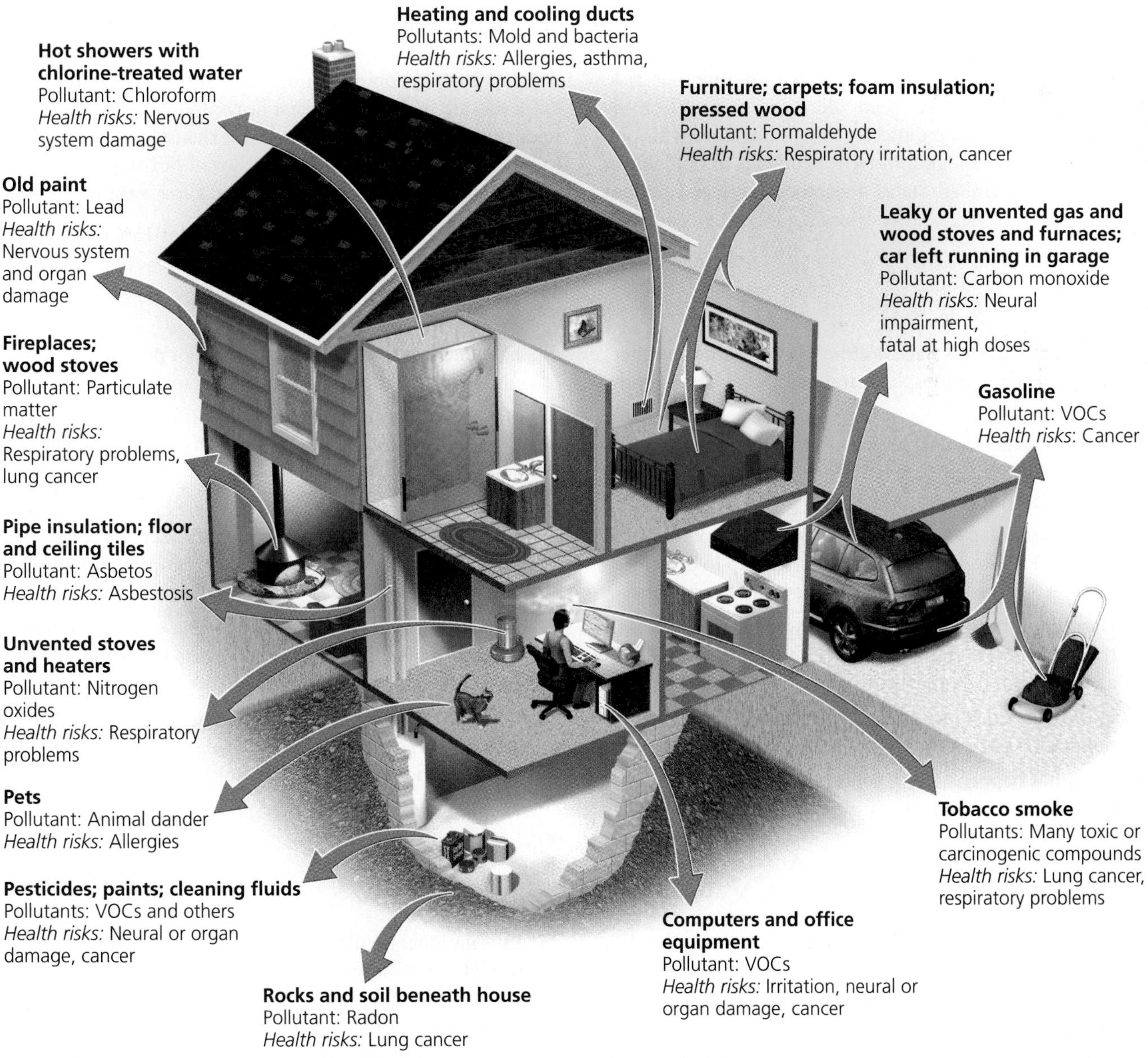

Hot showers with chlorine-treated water
Pollutant: Chloroform
Health risks: Nervous system damage

Old paint
Pollutant: Lead
Health risks: Nervous system and organ damage

Fireplaces; wood stoves
Pollutant: Particulate matter
Health risks: Respiratory problems, lung cancer

Pipe insulation; floor and ceiling tiles
Pollutant: Asbetos
Health risks: Asbestosis

Unvented stoves and heaters
Pollutant: Nitrogen oxides
Health risks: Respiratory problems

Pets
Pollutant: Animal dander
Health risks: Allergies

Pesticides; paints; cleaning fluids
Pollutants: VOCs and others
Health risks: Neural or organ damage, cancer

Rocks and soil beneath house
Pollutant: Radon
Health risks: Lung cancer

Heating and cooling ducts
Pollutants: Mold and bacteria
Health risks: Allergies, asthma, respiratory problems

Furniture; carpets; foam insulation; pressed wood
Pollutant: Formaldehyde
Health risks: Respiratory irritation, cancer

Leaky or unvented gas and wood stoves and furnaces; car left running in garage
Pollutant: Carbon monoxide
Health risks: Neural impairment, fatal at high doses

Gasoline
Pollutant: VOCs
Health risks: Cancer

Tobacco smoke
Pollutants: Many toxic or carcinogenic compounds
Health risks: Lung cancer, respiratory problems

Computers and office equipment
Pollutant: VOCs
Health risks: Irritation, neural or organ damage, cancer

FIGURE 13.21 ▲ The typical U.S. home contains many sources of indoor air pollution. Shown are common sources, the major pollutants they emit, and some of the health risks they pose.

providing moisture, dust, and foam insulation as substrates, as well as air currents that blow the organisms through the air. The U.S. Occupational Safety and Health Administration (OSHA) has estimated that 30–70 million Americans have suffered ailments related to the environment of the building in which they live.

WEIGHING THE ISSUES

How Safe Is Your Indoor Environment? Think about the amount of time you spend indoors. Name some potential indoor air quality hazards in your home, work, or school environment. Are these spaces well ventilated? What could you do to improve the safety of the indoor spaces you use?

We can reduce indoor air pollution

Using low-toxicity materials, monitoring air quality, keeping rooms clean, and providing adequate ventilation are the keys to alleviating indoor air pollution in most situations. In the developed world, we can limit our use of plastics and treated wood when possible and limit our exposure to pesticides, cleaning fluids, and other toxic substances by keeping them in a garage or outdoor shed. The EPA recommends that we test our homes and offices for radon and mold and install detectors for carbon monoxide. Keeping rooms and air ducts clean and free of mildew and other biological pollutants will reduce potential irritants and allergens. Most of all, keeping our indoor spaces well ventilated will minimize concentrations of pollutants.

➤ CONCLUSION

Indoor air pollution poses potentially serious health hazards, but by keeping informed of the latest scientific findings and taking appropriate precautions we can minimize our risk. Outdoor air pollution has been addressed more effectively by government legislation and regulation, together with pollution-control technologies. In fact, reductions in outdoor air pollution in the United States and other developed nations represent some of the greatest strides made in environmental

protection to date. Room for improvement remains, however, particularly in reducing acid deposition and photochemical smog. In the developing world, indoor and outdoor air pollutant levels are higher and take a heavy toll on people's health. Reducing pollution from automobile exhaust, coal combustion in outmoded facilities, indoor fuelwood burning, and other sources pose challenges as the world's less-wealthy nations industrialize.

TESTING YOUR COMPREHENSION

1. About how thick is Earth's atmosphere? Name one characteristic of the troposphere and one characteristic of the stratosphere.

2. Where is the "ozone layer" located? How and why is stratospheric ozone beneficial for people, whereas tropospheric ozone is harmful?

3. How does solar energy influence weather and climate? How do Hadley, Ferrel, and polar cells help to determine long-term climatic patterns and the location of biomes?

4. Describe a thermal inversion. How do inversions contribute to severe smog episodes like the ones in London and in Donora, Pennsylvania?

5. How does a primary pollutant differ from a secondary pollutant? Give an example of each.

6. What has happened with concentrations of "criteria pollutants" in U.S. ambient air in recent decades? What has

happened with our emissions of major pollutants? Name one health risk from toxic air pollutants.

7. How does photochemical smog differ from industrial smog? How do the weather and topography influence smog formation?

8. How do chlorofluorocarbons (CFCs) deplete stratospheric ozone? Why is this depletion considered a long-term international problem? What has been done to address this problem?

9. Why are the effects of acid deposition often felt in areas far from where the primary pollutants are produced? List three impacts of acid deposition.

10. Name three sources of indoor pollution and their associated health risks. For each pollution source, describe one way to reduce exposure to the source.

SEEKING SOLUTIONS

1. Consider responses to the photochemical smog that has plagued Los Angeles, Tehran, Mexico City, and other metropolitan areas. Describe several ways in which major cities have tried to improve their air quality.

2. Describe how and why emissions of major pollutants have been reduced by over 50% in the United States since 1970, despite increases in population, energy use, and economic activity.

3. International regulatory action has produced reductions in CFCs, but other transboundary pollution issues, including acid deposition, have not yet been addressed as effectively. What types of actions do you feel are appropriate for pollutants that cross political boundaries?

4. **THINK IT THROUGH** You have become the head of your county health department, and the EPA informs you that your county has failed to meet the national

ambient air quality standards for ozone, sulfur dioxide, and nitrogen dioxide. Your county is partly rural but is home to a city of 200,000 people and 10 sprawling suburbs. There are several large and aging coal-fired power plants, a number of factories with advanced pollution control technology, and no public transportation system. What steps would you urge the county government to take to meet the air quality standards? Explain how you would prioritize these steps.

5. **THINK IT THROUGH** You have been elected mayor of the largest city in your state. Your city's residents are complaining about photochemical smog and traffic congestion. Traffic engineers and city planners project that population and traffic will grow by 20% in the next decade. Some experts are urging you to restrict traffic into the city, allowing only cars with odd-numbered license plates on odd-numbered days, and those with even-numbered plates on even-numbered days. However, business-owners fear losing money if shoppers are discouraged from visiting. Consider the particulars of your city, and then decide whether you will pursue an odd-day/even-day driving program, and explain why or why not. What other steps would you take to address your city's smog problem?

CALCULATING ECOLOGICAL FOOTPRINTS

"While only some motorists contribute to traffic fatalities, all motorists contribute to air pollution fatalities." So stated a writer for the Earth Policy Institute, in pointing out that air pollution kills far more people than vehicle accidents. According to EPA data, emissions of nitrogen oxides in the United States in 2008 totaled 16.3 million tons. Nitrogen oxides come from fuel combustion in motor vehicles, power plants, and other industrial, commercial, and residential sources, but fully 9.5 million tons of the 2008 total came from vehicles. The U.S. Census Bureau estimates the nation's population to have been 304.1 million in 2008 and projects that it will reach 334.1 million in 2020. Considering these data, calculate the missing values for 2008 in the table below (1 ton = 2,000 lb).

	Total NO_x emissions (lb)	NO_x emissions from vehicles (lb)
You		
Your class		
Your state		
United States		

Data from U.S. EPA.

1. By what percentage is the U.S. population projected to increase between 2008 and 2020? Do you think that NO_x emissions will increase, decrease, or remain the same over that period of time? Why? (You may want to refer to Figure 13.9.)

2. Assume you are an average American driver. Using the 2008 emissions totals, how many pounds of NO_x emissions are you responsible for creating? How many pounds would you prevent if you were to reduce by half the vehicle miles you travel? What percentage of your total NO_x emissions would that be?

3. How might you reduce your vehicle miles traveled by 50%? What other steps could you take to reduce the NO_x emissions for which you are responsible?

Mastering ENVIRONMENTALSCIENCE™

Go to **www.masteringenvironmentalscience.com** for homework assignments, practice quizzes, Pearson eText, and more.

14 Global Climate Change

Upon completing this chapter, you will be able to:

- ➤ Describe Earth's climate system and explain factors influencing global climate
- ➤ Characterize human influences on the atmosphere and on climate
- ➤ Summarize how researchers study climate
- ➤ Outline current trends and impacts of global climate change
- ➤ Describe predicted future trends and impacts of global climate change
- ➤ Suggest and assess ways we may respond to climate change

The Maldives' underwater cabinet meeting

Hassan Latheef

Dr. Ibrahim Didi
Minister of Fisheries and Agri

Rising Seas May Flood the Maldives

"Global warming and climate change can effectively kill us off, make us refugees. . . ."
—Ismail Shafeeu, Minister of Environment, Maldives

"If we can't save the Maldives today, we can't save London, New York, or Hong Kong tomorrow."
—Mohamed Nasheed, President, Maldives

On October 17, 2009, President Mohamed Nasheed of the Maldives, a nation of low-lying islands in the Indian Ocean, donned scuba gear and dove into the blue waters of Girifushi Island lagoon. He was followed by his entire cabinet.

These officials proceeded to hold a cabinet meeting underwater—no doubt the world's first such meeting ever. Sitting at a table beneath the waves, they signed a declaration reading:

> *SOS from the front line:* Climate change is happening and it threatens the rights and security of everyone on Earth. With less than one degree of global warming, the glaciers are melting, the ice sheets collapsing, and low-lying areas are in danger of being swamped. We must unite in a global effort to halt further temperature rises, by slashing carbon dioxide emissions to a safe level of 350 parts per million.

Known for its spectacular tropical setting, colorful coral reefs, and sun-drenched beaches, the Maldives seems a paradise to its many visiting tourists—while for 370,000 Maldives residents, the islands are home. But residents and tourists alike now fear that the Maldives could be submerged by rising seas brought by global climate change.

Nearly 80% of the Maldives' land area lies less than 1 m (39 in.) above sea level. In a nation of 1,200 islands whose highest point is just 2.4 m (8 ft) above sea level, rising seas are a matter of life or death. The world's oceans rose 10–20 cm (4–8 in.) during the 20th century as warming temperatures expanded ocean water and as melting polar ice discharged water into the ocean. According to current projections,

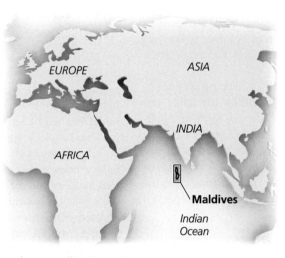

sea level will rise another 18–59 cm (7–23 in.) by the year 2100.

Higher seas would flood large areas of the Maldives and cause salt water to contaminate drinking water supplies. Storms intensified by warmer water temperatures are expected to erode beaches and damage the coral reefs that are so vital to the tourism and fishing industries that drive the nation's economy. The Maldives government has evacuated residents from several of the lowest-lying islands, and residents of other islands are considering moving.

President Nasheed is making sure the world knows of his country's situation. He points out that small island nations like his are not responsible for the carbon emissions driving global climate change, yet they are the ones bearing the brunt of the consequences. "If things go business as usual," he has said, "we will not live; we will die. Our country will not exist."

The underwater cabinet meeting kicked off a global campaign to draw attention to the impacts of climate change, sponsored by the nonprofit group 350.org. This campaign culminated in an International Day of Climate Action on October 24, 2009, when 5,200 events took place in 181 nations.

Residents of the Maldives are not alone in their predicament. Other island nations, from the Galápagos to Fiji to the Seychelles, also fear encroaching seawater.

These island nations have organized to make their position on climate change known to the world through AOSIS, the Alliance of Small Island States.

Mainland coastal areas of the world, from the hurricane-battered coasts of Florida, Louisiana, Texas, the Carolinas, and other states, to coastal cities such as New York and San Francisco, will face similar challenges from sea level rise—and this is just one of the many consequences of global climate change. In one way or another, climate change will affect each and every one of us for the remainder of our lifetimes. ■

OUR DYNAMIC CLIMATE

Climate influences virtually everything around us, from the day's weather to major storms, from crop success to human health, and from national security to the ecosystems that support our economies. If you are a student in your teens or twenties, the accelerating changes in our climate today may well be *the* major event of your lifetime and the phenomenon that most shapes your future.

Climate change is also the fastest-developing area of environmental science. New scientific studies that refine our understanding of climate are published every week, and policymakers and businesspeople make decisions and announcements just as quickly. By the time you read this chapter, some of its information will already be out of date. We urge you to explore further, with your instructor and on your own, the most recent information on climate change and the impacts it will have on your future.

What is climate change?

Climate describes an area's long-term atmospheric conditions, including temperature, precipitation, wind, humidity, barometric pressure, solar radiation, and other characteristics. *Climate* differs from *weather* (p. 281) in that weather specifies conditions at localized sites over hours or days, whereas climate describes conditions across broader regions over seasons, years, or centuries. **Global climate change** describes modifications in aspects of Earth's climate, such as temperature, precipitation, and storm frequency and intensity. People often use the term *global warming* synonymously in casual conversation, but **global warming** refers specifically to an increase in Earth's average surface temperature. Global warming is only one aspect of global climate change, although warming does in turn drive other components of climate change.

Over the long term, our planet's climate varies naturally through time. However, today's climatic changes are unfolding at an exceedingly rapid rate and are creating conditions humanity has never before experienced. Scientists agree that human activities, notably fossil fuel combustion and deforestation, are largely responsible. Understanding why today's climate is changing requires understanding

how our planet's climate functions. Thus, we first will survey the fundamentals of Earth's climate system—a complex and finely tuned system that has nurtured life for billions of years.

Three factors influence climate

Three natural factors exert the most influence on Earth's climate. The first is the sun. Without it, Earth would be dark and frozen. The second is the atmosphere. Without it, Earth would be on average 33 °C (59 °F) colder, and temperature differences between night and day would be far greater than they are. The third is the oceans, which store and transport heat and moisture.

The sun supplies most of our planet's energy. Earth's atmosphere, clouds, land, ice, and water together absorb about 70% of incoming solar radiation, and reflect the remaining 30% back into space (**FIGURE 14.1**).

Greenhouse gases warm the lower atmosphere

As Earth's surface absorbs solar radiation, the surface increases in temperature and emits infrared radiation (p. 30), radiation with wavelengths longer than those of visible light. Atmospheric gases with three or more atoms in their molecules tend to absorb infrared radiation very effectively. These include water vapor, ozone (O_3), carbon dioxide (CO_2), nitrous oxide (N_2O), and methane (CH_4), as well as halocarbons, a diverse group of mostly human-made gases that includes chlorofluorocarbons (CFCs; p. 289). Such gases are known as **greenhouse gases**. After absorbing radiation emitted from the surface, greenhouse gases subsequently re-emit infrared radiation in all directions. Some of this re-emitted energy is lost to space, but some travels back downward, warming the atmosphere (specifically the *troposphere*; p. 279) and the planet's surface in a phenomenon known as the **greenhouse effect**.

Q: The greenhouse effect works just like a greenhouse, right?

A: Actually, not quite. A greenhouse helps plants grow because its glass walls trap heat. In contrast, greenhouse gas molecules in our atmosphere absorb particular wavelengths of light reflected up from the surface, then re-emit radiation at different wavelengths. Some of this radiation travels back toward the surface, keeping the surface and lower atmosphere warmer than they would otherwise be. This phenomenon differs from what happens in a greenhouse, but it was called the "greenhouse effect" in the past, and the name has stuck.

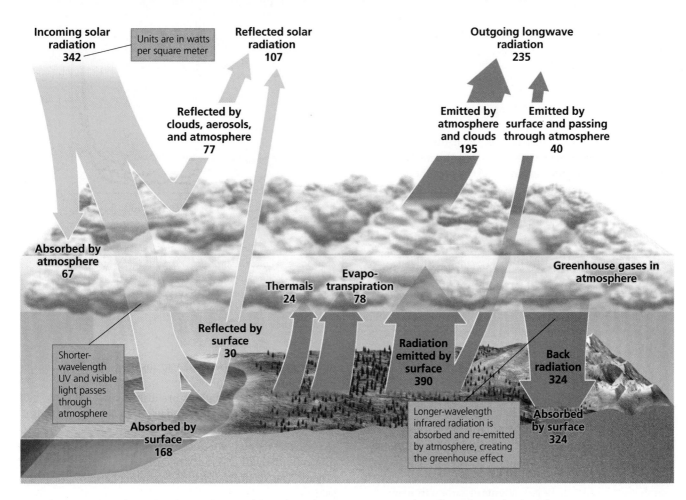

Incoming solar radiation 342

Units are in watts per square meter

Reflected solar radiation 107

Outgoing longwave radiation 235

Reflected by clouds, aerosols, and atmosphere 77

Emitted by atmosphere and clouds 195

Emitted by surface and passing through atmosphere 40

Absorbed by atmosphere 67

Greenhouse gases in atmosphere

Thermals 24

Evapo-transpiration 78

Shorter-wavelength UV and visible light passes through atmosphere

Reflected by surface 30

Radiation emitted by surface 390

Back radiation 324

Absorbed by surface 168

Longer-wavelength infrared radiation is absorbed and re-emitted by atmosphere, creating the greenhouse effect

Absorbed by surface 324

FIGURE 14.1 ▲ Our planet absorbs nearly 70% of the solar radiation it receives, and it reflects the rest back into space (yellow arrows). Most visible and ultraviolet radiation from the sun readily passes through the atmosphere and reaches the surface, and this radiation is absorbed and then re-emitted (orange arrows) as infrared radiation, which has longer wavelengths. Greenhouse gases absorb some of this long-wavelength radiation and then re-emit it, sending some downward to warm the atmosphere and surface by the greenhouse effect. This illustration of major pathways of energy flow shows that our planet naturally emits and reflects 342 watts per square meter, the same amount it receives from the sun. Arrow thicknesses are proportional to flows of energy in each pathway. Data from Kiehl, J.T., and K.E. Trenberth, 1997. Earth's annual global mean energy budget. *Bulletin of the American Meteorological Society* 78: 197–208.

Greenhouse gases differ in their ability to warm the troposphere and surface. *Global warming potential* refers to the relative ability of one molecule of a given greenhouse gas to contribute to warming. **TABLE 14.1** shows global warming potentials for several greenhouse gases. Values are expressed in relation to carbon dioxide, which is assigned a global warming potential of 1. Thus, a molecule of methane is 25 times as potent as a molecule of carbon dioxide, and a molecule of nitrous oxide is 298 times as potent as a CO_2 molecule.

Although carbon dioxide is less potent on a per-molecule basis than methane or nitrous oxide, it is far more abundant in the atmosphere, so it contributes more to the greenhouse effect. Moreover, greenhouse gas emissions from human activity consist mostly of carbon dioxide. According to the latest data, CO_2 is exerting nearly six times more impact than methane, nitrous oxide, and halocarbons combined.

TABLE 14.1 Global Warming Potentials of Four Greenhouse Gases

Greenhouse gas	Relative heat-trapping ability (in CO_2 equivalents)
Carbon dioxide	1
Methane	25
Nitrous oxide	298
Hydrochlorofluorocarbon HFC-23	14,800

Data are for a 100-year time horizon, from Intergovernmental Panel on Climate Change, 2007. Fourth assessment report. Climate change 2007: The physical science basis.

Greenhouse gas concentrations are rising fast

The greenhouse effect is a natural phenomenon, and greenhouse gases have been present in our atmosphere for all of Earth's history. That's a good thing, because without the natural greenhouse effect, our planet would be too cold to support life as we know it. Thus, it is not the natural greenhouse effect that concerns scientists today, but the *anthropogenic* (human-generated) intensification of the greenhouse effect. By adding novel greenhouse gases (certain halocarbons) to the atmosphere, and by increasing the concentrations of several natural greenhouse gases over the past 250 years (**FIGURE 14.2**), we are intensifying our planet's greenhouse effect beyond what our species has ever experienced. For example, we have boosted Earth's atmospheric concentration of carbon dioxide from 280 parts per million (ppm) in the late 1700s to over 392 ppm in 2011 (see Figure 14.2). Today the atmospheric CO_2 concentration is at its highest level by far in over 800,000 years, and likely the highest in the last 20 million years.

Why have atmospheric carbon dioxide levels risen so rapidly? Most carbon is stored for long periods in the upper layers of the lithosphere (pp. 38–39, 329). The deposition, partial decay, and compression of organic matter (mostly plants) in wetland or marine areas hundreds of millions of years ago led to the formation of coal, oil, and natural gas in buried sediments. Over the past two centuries we have extracted these fossil fuels and burned them in our homes, factories, and automobiles, transferring large amounts of carbon from one reservoir (the underground deposits that stored the carbon for millions of years) to another (the atmosphere). This sudden flux of carbon is the main reason atmospheric CO_2 concentrations have increased so dramatically.

At the same time, people have cleared and burned forests to make room for crops, pastures, villages, and cities. Forests serve as a reservoir for carbon as plants conduct photosynthesis and then store carbon in their tissues. Thus, when we clear forests it reduces the biosphere's ability to remove carbon dioxide from the atmosphere. In this way, deforestation (pp. 188–191) has contributed to rising atmospheric CO_2 concentrations. **FIGURE 14.3** summarizes scientists' current understanding of the fluxes (both natural and anthropogenic)

of carbon dioxide between the atmosphere and reservoirs on Earth's surface.

Methane concentrations are also rising—2.5-fold since 1750 (see Figure 14.2)—and today's atmospheric concentration is the highest by far in over 800,000 years. We release methane by tapping into fossil fuel deposits, raising livestock that emit methane as a metabolic waste product, disposing of organic matter in landfills, and growing certain crops such as rice.

Human activities have also enhanced atmospheric concentrations of nitrous oxide. This greenhouse gas, a by-product of feedlots, chemical manufacturing plants, auto emissions, and synthetic nitrogen fertilizers, has risen by nearly 20% since 1750 (see Figure 14.2).

Among other greenhouse gases, ozone concentrations in the troposphere have risen roughly 36% since 1750 because of photochemical smog (p. 288). The contribution of halocarbon gases to global warming has begun to slow because of the Montreal Protocol and subsequent controls on their production and use (pp. 290–291).

Water vapor is the most abundant greenhouse gas in our atmosphere and contributes most to the natural greenhouse

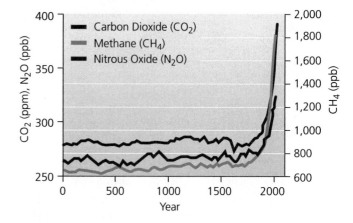

FIGURE 14.2 ▲ Since the start of the industrial revolution, global atmospheric concentrations of carbon dioxide, methane, and nitrous oxide have increased markedly. Data from Intergovernmental Panel on Climate Change, 2007. *Fourth assessment report;* NOAA; and Carbon Dioxide Information Analysis Center.

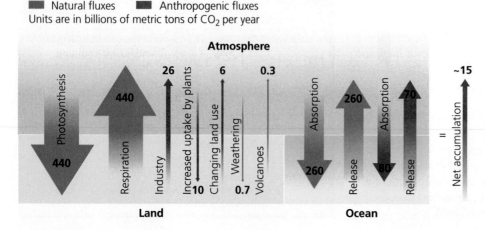

Natural fluxes   Anthropogenic fluxes
Units are in billions of metric tons of CO_2 per year

FIGURE 14.3 ◄ Human activities since the industrial revolution have sent more carbon dioxide from the Earth to its atmosphere than is moving back from the atmosphere to the Earth. Shown here are all current fluxes of carbon dioxide, with arrows sized according to their mass of CO_2. Green arrows indicate natural fluxes, and red arrows indicate anthropogenic fluxes. Adapted from Intergovernmental Panel on Climate Change, 2007. *Fourth assessment report.*

effect. Its concentrations vary locally, but its global concentration has not changed over recent centuries, so it is not viewed as a driver of industrial-age climate change.

Other factors warm or cool the surface

Whereas greenhouse gases exert a warming effect on the atmosphere, **aerosols**, microscopic droplets and particles, can have either a warming or cooling effect. Soot particles, or black carbon aerosols, generally cause warming by absorbing solar energy, but most other tropospheric aerosols cool the atmosphere by reflecting the sun's rays. Sulfate aerosols produced by fossil fuel combustion may slow global warming, at least in the short term. When sulfur dioxide enters the atmosphere, it undergoes various reactions, some of which lead to acid precipitation (pp. 291–294). These reactions can form a sulfur-rich aerosol haze in the upper atmosphere that reduces the sunlight reaching Earth's surface. Aerosols released by major volcanic eruptions can cool Earth's climate for up to several years. This occurred in 1991 with the eruption of Mount Pinatubo in the Philippines.

To measure the degree of impact that any given factor exerts on Earth's temperature, scientists calculate its **radiative forcing**, the amount of change in thermal energy that a given factor causes. Positive forcing warms the surface, whereas negative forcing cools it. **FIGURE 14.4** shows researchers' best calculations of the radiative forcing that our planet is experiencing today from aerosols, greenhouse gases, and other factors. When scientists sum up the effects of all factors, they find that Earth today is experiencing overall radiative forcing of about 1.6 watts/m^2. This means that compared with the pre-industrial Earth of 1750, today's planet is receiving and retaining 1.6 watts/m^2 more thermal energy than it is emitting into space. Look back at Figure 14.1 and note that Earth is estimated naturally to receive and give off 342 watts/m^2 of energy. Although 1.6 may seem like a small proportion of 342, it is enough to alter climate significantly.

Climate varies naturally for several reasons

Besides atmospheric composition, our climate is influenced by cyclic changes in Earth's rotation and orbit, variation in energy released by the sun, absorption of carbon dioxide by the oceans, and ocean circulation patterns.

Milankovitch cycles In the 1920s, Serbian mathematician Milutin Milankovitch described three types of periodic changes in Earth's rotation and orbit around the sun. Over thousands of years, our planet wobbles on its axis, varies in the tilt of the axis, and experiences change in the shape of its orbit, all in regular long-term cycles of different lengths. These variations, now known as **Milankovitch cycles**, alter the way solar radiation is distributed over Earth's surface (**FIGURE 14.5**). By modifying patterns of atmospheric heating,

these cycles trigger long-term climate variation. This includes periodic episodes of *glaciation* during which global surface temperatures drop and ice sheets advance from the poles toward the midlatitudes.

Solar output The sun varies in the amount of radiation it emits. For example, at each peak of its 11–year sunspot cycle the sun may emit solar flares, bursts of energy strong enough to disrupt satellite communications. However, scientists are concluding that the variation in solar energy reaching our planet in recent centuries has simply not been great enough to drive significant temperature change on Earth's surface. Estimates place the radiative forcing of natural changes in solar output at only about 0.12 watts/m^2—less than any of the anthropogenic causes shown in Figure 14.4.

Ocean absorption The oceans hold 50 times more carbon than the atmosphere holds. They absorb carbon dioxide from the atmosphere when this gas dissolves directly in water and when marine phytoplankton use it for photosynthesis. However, the oceans are absorbing less CO_2 than we are adding to the atmosphere (see Figure 2.20, p. 39). Thus, carbon absorption by the oceans is slowing global warming but is not preventing it. Moreover, recent evidence indicates

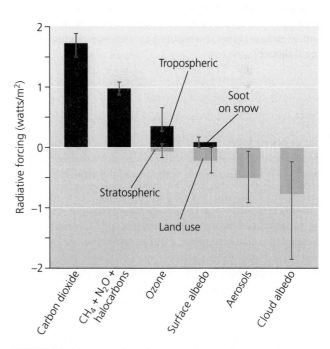

FIGURE 14.4 ▲ For each emitted gas or other human impact on the atmosphere since the industrial revolution, we can estimate the warming or cooling effect this has had on Earth's climate. We express this as *radiative forcing*, which in this graph is shown as the amount of influence on climate today relative to 1750, in watts per square meter. Red bars indicate positive forcing (warming), and blue bars indicate negative forcing (cooling). *Albedo* (p. 309) refers to the reflectivity of a surface. A number of more minor influences are not shown. In total, scientists estimate that human impacts on the atmosphere exert a cumulative radiative forcing of 1.6 watts/m^2. Data from Intergovernmental Panel on Climate Change, 2007. *Fourth assessment report.*

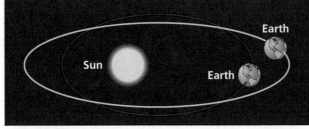

(a) Axial wobble   **(b) Variation of tilt**

25°
22°
Orbital plane   Equator

Sun

Earth

Earth

(c) Variation of orbit

FIGURE 14.5 ▲ There are three types of Milankovitch cycles. The first is an axial wobble **(a)** that occurs on a 19,000- to 23,000-year cycle. The second is a 3-degree shift in the tilt of Earth's axis **(b)** that occurs on a 41,000-year cycle. The third is a variation in Earth's orbit from almost circular to more elliptical **(c)**, which repeats itself every 100,000 years. These variations affect the intensity of solar radiation that reaches portions of Earth at different times, contributing to long-term changes in global climate.

Q: The climate changes naturally, so why worry about climate change?

A: Earth's climate does indeed change naturally across very long periods of time. However, no known natural factors can account for the change we are experiencing today, and our civilization has never before experienced the degree of change predicted for this century. One challenge is that today's climate is changing unusually fast. Another is the sheer amount of change: The quantity by which the world's temperature is forecast to rise is greater than the amount of cooling needed to bring on an ice age. Greenhouse gas concentrations are already higher than they've been in over 800,000 years, and are still rising. Our entire civilization arose only in the last few thousand years during an exceptionally stable period in Earth's climate history. Unless we reduce our emissions, we will soon be challenged by a climate that the human species has never lived through before.

STUDYING CLIMATE CHANGE

To comprehend any phenomenon that is changing, we must study its past, present, and future. Scientists monitor present-day climate, but they also have devised clever means of inferring past change and sophisticated methods to predict future conditions.

Proxy indicators tell us about the past

To understand past climate, scientists have developed techniques to decipher clues from thousands or millions of years ago. **Proxy indicators** are types of indirect evidence that serve as proxies, or substitutes, for direct measurement and that shed light on past climate.

For example, Earth's ice caps, ice sheets, and glaciers hold clues to climate history. In frigid areas near the poles and atop high mountains, snow falling year after year for millennia compresses into ice. Over the ages, this ice accumulates to great depths, preserving within its layers tiny bubbles of the ancient atmosphere (**FIGURE 14.6**). Scientists examine the trapped air bubbles by drilling into the ice and extracting long columns, or cores. The layered ice, accumulating season after season over thousands of years, provides a timescale. By studying the chemistry of the ice and the bubbles in each layer, scientists can determine atmospheric composition, greenhouse gas concentrations, temperature, snowfall, solar activity, and frequency of forest fires and volcanic eruptions during each time period.

Recently, researchers drilled and analyzed the deepest ice core ever. At a remote and pristine site in Antarctica, they drilled down 3,270 m (10,728 ft) to bedrock and pulled out more than 800,000 years' worth of ice! This core chronicles Earth's history across eight glacial cycles. By analyzing air bubbles trapped in the ice, researchers discovered that over the past 800,000 years, atmospheric concentrations of carbon dioxide, methane, and nitrous oxide have never been as high

that as ocean water warms, it absorbs less CO_2 because gases are less soluble in warmer water—a positive feedback effect (pp. 22–23) that accelerates warming of the atmosphere.

Ocean circulation Ocean water exchanges heat with the atmosphere, and ocean currents (pp. 254–255) move energy from place to place. For example, the oceans' thermohaline circulation system (pp. 254–256) moves warm tropical water northward toward Europe, providing that continent a far milder climate than it would otherwise have. Scientists are studying whether freshwater input from Greenland's melting ice sheet might shut down this warmwater flow—an occurrence that could have devastating impacts on European society.

Multiyear climate variability results from the El Niño–Southern Oscillation (p. 256), which involves systematic shifts in atmospheric pressure, sea surface temperature, and ocean circulation in the tropical Pacific Ocean. These shifts overlie longer-term variability from a phenomenon known as the Pacific Decadal Oscillation. El Niño and La Niña events alter weather patterns from region to region in diverse ways, often leading to rainstorms and floods in dry areas and drought and fire in moist areas. This leads to impacts on wildlife, agriculture, and fisheries.

(a) Ice core

(b) Micrograph of ice core

FIGURE 14.6 ▲ In Greenland and Antarctica, scientists have drilled deep into ancient ice sheets and removed cores of ice like this one **(a)**, held by Dr. Gerald Holdsworth of the University of Calgary, to extract information about past climates. Bubbles (black shapes) trapped in the ice **(b)** contain small samples of the ancient atmosphere.

as they are today (**FIGURE 14.7**). These data demonstrate that by emitting greenhouse gases since the industrial revolution, we have brought ourselves deep into uncharted territory.

The ice core results also confirm that temperature swings in the past were tightly correlated with greenhouse gas concentrations (compare the top two datasets in Figure 14.7 with the temperature dataset at bottom). This finding bolsters the scientific consensus that greenhouse gas emissions are causing our planet to warm today.

Researchers also drill cores into beds of sediment beneath bodies of water. Sediments often preserve pollen grains and other remnants from plants that grew in the past, as we saw with the study of Easter Island (pp. 6–7). Because climate influences the types of plants that grow in an area, knowing what plants occurred can tell us a great deal about the climate at that place and time. Other types of proxy indicators include tree rings (which reveal year-by-year precipitation history and fire occurrence), pack-rat middens (which preserve plant parts for centuries), and coral reefs (which reveal aspects of ocean chemistry).

Direct measurements tell us about the present

Today we measure temperature with thermometers, rainfall with rain gauges, wind speed with anemometers, and air pressure with barometers, using computer programs to integrate and analyze this information in real time. With these technologies, we document the fluctuations in weather day by day and hour by hour across the globe. As a result, we have gained an understanding of present-day climate in every region of our planet.

We also measure the chemistry of the atmosphere and the oceans. Direct measurements of carbon dioxide concentrations in the atmosphere reach back to 1958, when scientist Charles

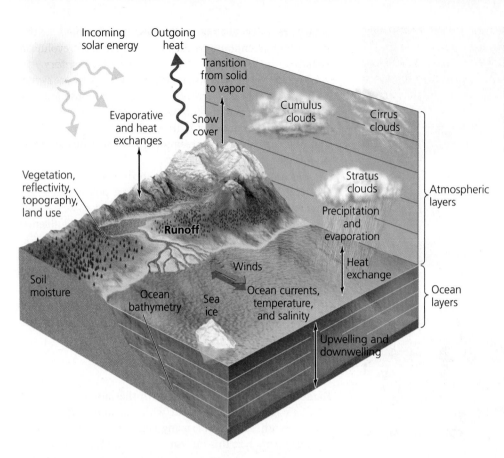

Keeling began analyzing hourly air samples from a monitoring station at the Mauna Loa Observatory in Hawaii. These data show that atmospheric CO_2 concentrations have increased from 315 ppm in 1958 to 392 ppm in 2011.

Models help us predict the future

To understand how climate systems function and to predict future climate change, scientists simulate climate processes with sophisticated computer programs. **Climate models** are programs that combine what is known about atmospheric circulation, ocean circulation, atmosphere-ocean interactions, and feedback cycles to simulate climate processes (**FIGURE 14.8**). This requires manipulating vast amounts of data with complex mathematical equations—a task not possible until the advent of modern computers.

Climate modelers essentially provide starting information to the model, set up rules for the simulation, and then let it run. Researchers test the efficacy of a model by entering past climate data and running the model toward the present. If a model accurately reconstructs current climate, based on well-established data from the past, then we have reason to believe that it simulates climate mechanisms realistically and that it may accurately predict future climate.

Plenty of challenges remain for climate modelers because Earth's climate system is so complex and because many uncertainties remain in our understanding of feedback processes (pp. 22–23). Yet as scientific knowledge of climate processes improves, as computing power intensifies, and as we glean enhanced data from proxy indicators, climate models are becoming better and better at predicting climate change region by region across the world.

CURRENT AND FUTURE TRENDS AND IMPACTS

It seems that virtually everyone is noticing changes in the climate these days. Maldives fishermen note the seas encroaching on their home island. Texas ranchers suffer a multiyear drought. Florida homeowners find it difficult to obtain insurance against hurricanes and storm surges. New Yorkers, Bostonians, Chicagoans, and Los Angelenos face one unprecedented weather event after another.

We cannot blame any single unusual heat wave, flood, or blizzard on climate change, but extreme weather events are indeed part of a real pattern backed by a tremendous volume of scientific evidence. Climate change has already had numerous impacts on the physical properties of our planet, on organisms and ecosystems, and on human well-being. If we continue to emit greenhouse gases into the atmosphere, the consequences of climate change will grow more severe.

The IPCC summarizes evidence and predicts impacts

The immense amount of scientific information on climate change is reviewed and summarized in periodic reports issued by the **Intergovernmental Panel on Climate Change (IPCC)**. This international body consists of many hundreds of scientists and governmental representatives. Established in 1988 by the United Nations Environment Programme (UNEP) and the World Meteorological Organization (WMO), the IPCC was awarded the Nobel Peace Prize in 2007 for its work in informing the world of the trends and impacts of climate change.

In that year the IPCC released its **Fourth Assessment Report**. This report summarized many thousands of scientific studies, and it documented observed trends in surface temperature, precipitation patterns, snow and ice cover, sea levels, storm intensity, and other factors. It also predicted future changes in these phenomena after considering a range of potential scenarios for future greenhouse gas emissions. The report addressed impacts of current and future climate change on wildlife, ecosystems, and society. Finally, it discussed strategies we might pursue in response to climate change. **FIGURE 14.9** summarizes some of the IPCC report's major observed and predicted trends and impacts.

Like all science, the IPCC report deals in uncertainties. Its authors therefore took great care to assign statistical probabilities to its conclusions and predictions. Its estimates regarding impacts on society are conservative, because its scientific conclusions had to be approved by representatives of the world's national governments, some of which are reluctant to move away from a fossil-fuel-based economy.

Temperatures continue to rise

The IPCC's 2007 report concluded that average surface temperatures on Earth rose by an estimated 0.74 °C (1.33 °F) in the century from 1906 to 2005, with most of this increase occurring in the last few decades (**FIGURE 14.10**). According to the WMO, the 17 warmest years on record since global measurements began 150 years ago have all been since 1990. The decade from 2001–2010 was the hottest ever, and since the 1960s each decade has been warmer than the last.

Major Observed Impacts of Climate Change, from IPCC Fourth Assessment Report, 2007	
Global physical indicators	**Social indicators**
Earth's average surface temperature increased 0.74 °C (1.33 °F) in the past 100 years, *and will rise 1.8–4.0 °C (3.2–7.2 °F) in the 21st century.*	Farmers and foresters have had to adapt to altered growing seasons and disturbance regimes.
Eleven of the years from 1995 to 2006 were among the 12 warmest on record.	*Temperate-zone crop yields will rise until temperature warms beyond 3 °C (5.4 °F), but in the dry tropics and subtropics, crop productivity will fall and lead to hunger.*[5]
Oceans absorbed >80% of heat added to the climate system, and warmed to depths of at least 3,000 m (9,800 ft).	*Impacts on biodiversity will cause losses of food, water, and other ecosystem goods and services.*[2]
Glaciers, snow cover, ice caps, ice sheets, and sea ice will continue melting, contributing to sea-level rise.	*Sea level rise will displace people from islands and coasts.*[3]
Sea level rose by an average of 17 cm (7 in.) in the 20th century, *and will rise 18–59 cm (7–23 in.) in the 21st century.*	*Melting of mountain glaciers will reduce water supplies to millions of people.*[2]
Ocean water became more acidic by about 0.1 pH unit, *and will decrease in pH by 0.14–0.35 units more by century's end.*	*Economic costs will outweigh benefits as climate change worsens;*[2] *costs could average 1–5% of GDP globally for 4 °C (7.2 °F) of warming.*
Storm surges increased, *and will increase further.*[1]	Poorer nations and communities suffer more from climate change, because they rely more on climate-sensitive resources and have less capacity to adapt.[2]
Carbon uptake by terrestrial ecosystems will peak by mid-21st century and then weaken or reverse, amplifying climate change.[2]	*Human health will suffer as increased warm-weather health hazards outweigh decreased cold-weather health hazards.*[2]
Regional physical indicators	**Biological indicators**
Arctic areas warmed fastest. *Future warming will be greatest in the Arctic and greater over land than over water.*	Species ranges are shifting toward the poles and upward in elevation, *and will continue to shift.*
Summer Arctic sea ice thinned by 7.4% per decade since 1978.	The timing of seasonal phenomena (such as migration and breeding) is shifting, *and will continue to shift.*
Precipitation will increase at high latitudes and decrease at subtropical latitudes, making wet areas wetter and dry ones drier.[1]	*About 20–30% of species studed so far will face extinction risk if temperature rises more than 1.5–2.5 °C (2.7–4.5 °F).*[5]
Droughts became longer, more intense, and more widespread since the 1970s, especially in the tropics and subtropics.[1]	*Species interactions and ecosystem structure and function could change greatly, resulting in biodiversity loss.*
Droughts and flooding will increase, leading to agricultural losses.[2]	*Corals will experience further mortality from bleaching and ocean acidification.*[5,4]
Hurricanes intensified in the North Atlantic since 1970[1], *and will continue to intensify.*[1]	
The thermohaline circulation will slow, *but will not shut down and chill Europe in the 21st century.*[3]	

FIGURE 14.9 ▲ Listed here are some of the main observed and predicted trends and impacts described in the Intergovernmental Panel on Climate Change's *Fourth Assessment Report.* Observed phenomena are in plain text, whereas predicted future phenomena are in italicized text. For simplicity, this table expresses mean estimates only. The IPCC report provides ranges of estimates as well. Certainty levels are as follows: [1]66–90% probability of being correct; [2]~80% probability of being correct; [3]90–99% probability of being correct; [4]>99% probability of being correct; [5]~50% probability of being correct. Data from the Intergovernmental Panel on Climate Change, 2007. *Fourth assessment report.*

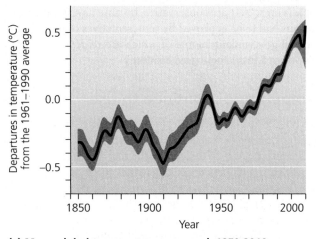

(a) Mean global temperature measured, 1850-2010

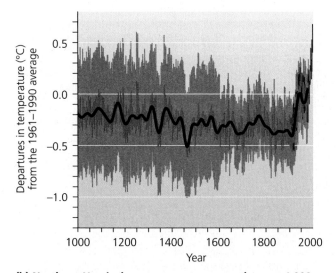

(b) Northern Hemisphere temperature over the past 1,000 years

FIGURE 14.10 ▲ Data from thermometers **(a)** show changes in Earth's average surface temperature from 1850 to 2010. Gray shaded area indicates range of uncertainty. In **(b)**, proxy indicators (blue line) and thermometer data (red line) together show average temperature changes in the Northern Hemisphere over the past 1,000 years. The gray shaded zone represents the 95% confidence range. Data (a) from the Intergovernmental Panel on Climate Change (IPCC), 2007, *Fourth assessment report*, and the National Oceanic and Atmospheric Administration (NOAA); and (b) IPCC 2001, *Third assessment report*.

In the next 20 years, we can expect average surface temperatures on Earth to rise roughly 0.4 °C (0.7 °F), according to IPCC analysis. At the end of the 21st century, the IPCC predicts global temperatures will be 1.8–4.0 °C (3.2–7.2 °F) higher than today's, depending on the emission scenario. Temperature changes are predicted to vary from region to region in ways that intensify regional differences already apparent (**FIGURE 14.11**). For example, polar regions will continue to experience the most intense warming.

Sea surface temperatures are also increasing as the oceans absorb heat from the atmosphere. The record number of hurricanes and tropical storms in 2005—Hurricane Katrina and 27 others—left many people wondering whether global warming was to blame. Recent scientific analyses suggest that warmer seas may not be increasing the number of tropical storms but may be increasing their power, and possibly their duration.

Precipitation is changing, too

A warmer atmosphere holds more water vapor, but changes in precipitation patterns have been complex, with some regions of the world receiving more rain and snow than usual and others receiving less. In regions such as the southwestern United States, droughts have become more frequent and severe, harming agriculture, worsening soil erosion, reducing water supplies, and triggering wildfire. Meanwhile, in dry and humid regions alike, heavy rain events have increased, contributing to flooding, such as the 2008 floods in Iowa and other parts of the Midwest and the 2011 floods along the Mississippi River that killed dozens of people, left thousands homeless, and inflicted billions of dollars in damage.

Future changes in precipitation are predicted to intensify regional changes seen over the past century (**FIGURE 14.12**). In general, precipitation will increase at high latitudes and decrease at low and middle latitudes, magnifying differences that already exist and worsening water shortages in many developing countries of the arid subtropics.

Melting ice and snow have far-reaching effects

As the world warms, mountaintop glaciers are disappearing (**FIGURE 14.13**). Between 1980 and 2009, the world's glaciers on average have each lost mass equivalent to 14 m (46 ft) vertical thickness of water, according to the World Glacier

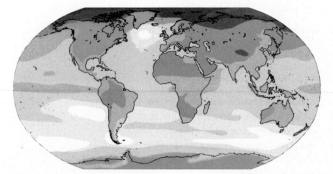

Percent increase in temperature (°C)

- ☐ 0.5–1.5
- 1.5–2.5
- 2.5–3.5
- 3.5–4.5
- 4.5–5.5
- 5.5–6.5
- 6.5–7.5

FIGURE 14.11 ◄ This map shows projected increases in surface temperature for the decade 2090–2099, relative to temperatures in 1980–1999. Landmasses are expected to warm more than oceans, and the Arctic will warm the most. The Intergovernmental Panel on Climate Change uses multiple emission scenarios, and this map was generated using an emission scenario that is intermediate in its assumptions, involving an average global temperature rise of 2.8 °C (5.0 °F) by 2100. Data from Intergovernmental Panel on Climate Change, 2007. *Fourth assessment report.*

FIGURE 14.12 ▶ This map shows projected changes in June–August precipitation for the decade 2090–2099, relative to precipitation in 1980–1999. Browner shades indicate less precipitation, and bluer shades indicate more precipitation. White indicates areas for which models could not agree. This map was generated using an emission scenario that is intermediate in its assumptions, involving an average global temperature rise of 2.8 °C (5.0 °F) by 2100. Data from Intergovernmental Panel on Climate Change, 2007. *Fourth assessment report.*

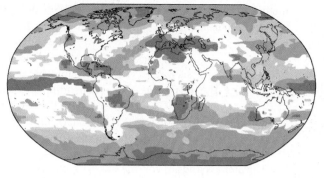

Percent change in precipitation

▪ >20% decrease
▪ 10–20% decrease
▪ 5–10% decrease
▪ 5% decrease to 5% increase
▪ 5–10% increase
▪ 10–20% increase
▪ >20% increase

Monitoring Service. Many glaciers on tropical mountaintops have disappeared already. In Glacier National Park in Montana, only 25 of 150 glaciers present at the park's inception remain, and scientists estimate that by 2030 even these will be gone.

Mountains accumulate snow in winter and release meltwater gradually during summer. Over one-sixth of the world's people live in regions that depend on mountain meltwater. As warming temperatures diminish mountain glaciers, this will reduce summertime water supplies to millions of people, likely forcing whole communities to look elsewhere for water, or to move.

Warming temperatures are also melting vast amounts of ice in the Arctic. Recent research reveals that the immense ice sheet that covers Greenland is melting faster and faster. At the other end of the world, in Antarctica, coastal ice shelves the size of Rhode Island have disintegrated as a result of contact with warmer ocean water.

One reason warming is accelerating in the Arctic is that as snow and ice melt, darker, less-reflective surfaces (such as bare ground and pools of meltwater) are exposed, and Earth's *albedo*, or capacity to reflect light, decreases. As a result, more of the sun's rays are absorbed at the surface, fewer reflect back into space, and the surface warms. In a process of positive feedback, this warming causes more ice and snow to melt, which in turn causes more absorption of radiation and more warming (see Figure 2.1b, p. 23).

Scientists predict that snow cover, ice sheets, and sea ice will continue to diminish near the poles. As Arctic sea ice disappears, new shipping lanes are opening up for commerce, and governments and companies are rushing to exploit newly accessible underwater oil and mineral reserves. Russia, Canada, the United States, and other nations are jockeying for position, using new survey data to try to lay claim to regions of the Arctic as the ice melts.

Warmer temperatures in the Arctic are also causing *permafrost* (permanently frozen ground) to thaw. As ice crystals within permafrost melt, the thawing soil settles, destabilizing buildings, pipelines, and other infrastructure. When permafrost thaws, it also can release methane that it has stored for

(a) Grinnell Glacier in 1938

(b) Grinnell Glacier in 2005

FIGURE 14.13 ◀ Glaciers are melting rapidly around the world as global warming proceeds. The Grinnell Glacier in Glacier National Park, Montana, retreated substantially between **(a)** 1938 and **(b)** 2005.

thousands of years. Because methane is a potent greenhouse gas, this acts as a positive feedback mechanism (p. 22) that intensifies climate change.

Rising sea levels may affect hundreds of millions of people

As glaciers and ice sheets melt, increased runoff into the oceans causes sea levels to rise. Sea levels also are rising because ocean water is warming, and water expands in volume as it warms. Worldwide, average sea levels rose an estimated 17 cm (6.7 in.) during the 20th century (**FIGURE 14.14**), reaching a rate of 3.2 mm/year since 1993. These numbers represent vertical rises in water level, and on most coastlines a vertical rise of a few inches means many feet of incursion inland.

Higher sea levels lead to beach erosion, coastal flooding, intrusion of salt water into aquifers, and storm surges. A *storm surge* is a temporary and localized rise in sea level brought on by the high tides and winds associated with storms. The higher that sea level is to begin with, the further inland a storm surge can reach. In 1987, unusually high waves struck the Maldives and triggered a campaign to build a seawall around Malé, the nation's capital. "The Great Wall of Malé" is intended to protect buildings and roads by dissipating the energy of incoming waves during storm surges.

On December 26, 2004, the Maldives got a taste of what could be in store in the future when a massive *tsunami* (pp. 233–234) devastated coastal areas throughout the Indian Ocean. The tsunami killed 100 Maldives residents and left 20,000 homeless. Property damage in the Maldives was estimated at $470 million, an astounding 62% of the nation's gross domestic product (GDP). Indirect damage from soil erosion, saltwater contamination of aquifers, and other impacts continues to cause further economic losses. The tsunami was caused by an earthquake, not by climate change. Yet as sea levels rise, the damage that natural events can inflict increases considerably.

The Maldives has fared better than many other island nations. It saw sea level rise about 2.5 mm per year throughout the 1990s, but most Pacific islands are experiencing greater sea level rise. Regions experience differing amounts of sea level change because land may be rising or subsiding naturally, depending on local geologic conditions.

In the United States, 53% of the population lives in coastal counties. Vulnerability to storm surges became tragically apparent when Hurricane Katrina struck the Gulf Coast. Outside New Orleans today, marshes of the Mississippi River delta are being lost as rising seas eat away at coastal vegetation. These coastal wetlands are also being lost because dams upriver hold back silt that once maintained the delta, because petroleum extraction has caused land to subside, and because salt water is encroaching up channelized waterways, killing freshwater plants. All told, more than 2.5 million ha (1 million acres) of Louisiana's coastal wetlands have vanished since 1940. Continued wetland loss will deprive New Orleans of protection against future storm surges.

At the end of the 21st century, the IPCC predicts mean sea level will be 18–59 cm (7–23 in.) higher than today's, depending on our level of emissions. However, these estimates do not take into account recent findings on accelerated ice melting in Greenland (and apparently Antarctica as well), because that research was so new that it had not yet been incorporated into climate models when the latest IPCC report was being prepared. If polar melting continues to accelerate, then sea levels will rise more quickly.

If sea levels rise as predicted, hundreds of millions of people will be displaced or will need to invest in costly efforts to protect against high tides and storm surges. Densely populated regions on low-lying river deltas, such as Bangladesh, would be most affected. So would storm-prone regions such as Florida, coastal cities such as Houston and Charleston, and areas where land is subsiding, such as the U.S. Gulf Coast. Many Pacific islands would need to be evacuated. In the meantime, island nations such as the Maldives are likely to suffer from shortages of fresh water as rising seas bring salt water into aquifers. The contamination of groundwater and soils by seawater also threatens coastal areas such as Tampa, Florida, which depend on small lenses of fresh water that float atop saline groundwater.

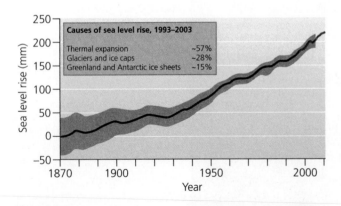

FIGURE 14.14 ▲ Data from tide gauges (black line) and satellite observations (red line) show that global average sea level has risen over 200 mm (7.9 in.) since 1870. Gray shaded area indicates range of uncertainty. Thermal expansion of water accounts for most sea level rise. Data from Intergovernmental Panel on Climate Change, 2007. *Fourth assessment report*, and CSIRO.

WEIGHING THE ISSUES

Environmental Refugees Citizens of the Maldives see an omen of their future in the Pacific island nation of Tuvalu, which has been losing 9 cm (3.5 in.) of elevation per decade to rising seas. Appeals from Tuvalu's 11,000 citizens were heard by New Zealand, which began accepting "environmental refugees" from Tuvalu in 2003. Do you think the rest of the world should grant such environmental refugees international status and assume some responsibility for taking care of them? Do you think a national culture can survive if its entire population is relocated? Think of the tens of thousands of refugees from Hurricane Katrina. How did their lives and culture fare in the wake of that tragedy?

Climate change threatens coral reefs

Around the world, rising seas are eroding the salt marshes and mangrove forests that protect our coasts (p. 257). However, scientists are most concerned about coral reefs (pp. 258–259), which provide habitat for marine species, enhance fisheries, offer snorkeling and scuba diving sites for tourism, and shield coastlines from destructive waves.

Climate change poses two major threats to coral reefs. First, warmer waters contribute to coral bleaching (p. 259), which kills corals. Second, enhanced CO_2 concentrations in the atmosphere alter ocean chemistry, leading to **ocean acidification** (p. 259). As ocean water absorbs atmospheric CO_2, it becomes more acidic, and this impairs the ability of coral and other organisms to build exoskeletons of calcium carbonate. The oceans have already decreased by 0.1 pH unit, and are predicted to decline in pH by 0.14–0.35 more units over the next 100 years. This could easily be enough to destroy most or all of our planet's living coral reefs. Such destruction could be catastrophic for marine biodiversity and fisheries, because so many organisms depend on living coral reefs for food and shelter.

Climate change affects organisms and ecosystems

As the coral reef crisis shows, changes in Earth's physical systems have consequences for living things. Organisms are adapted to their environments, so they are affected when we alter those environments. As global warming proceeds, it is modifying biological phenomena that rely on temperature. In the spring, plants are now leafing out earlier, insects are hatching earlier, birds are migrating earlier, and animals are breeding earlier. These shifts can create mismatches in seasonal timing. For example, European birds known as great tits had evolved to time their breeding to raise their young when caterpillars peak in abundance. Now caterpillars are peaking earlier, but the birds have been unable to adjust, and fewer young birds are surviving.

Biologists are also recording spatial shifts in the ranges of organisms, as plants and animals move toward the poles or upward in elevation (i.e., toward cooler regions) as temperatures warm (**FIGURE 14.15A**). Some organisms will not be able to cope, and the IPCC estimates that up to 20–30% of all plant and animal species could be threatened with extinction. Trees may not be able to shift their distributions fast enough. Rare species may be forced out of preserves and into developed areas where they cannot survive. Animals and plants adapted to mountainous environments may be forced uphill until there is nowhere left to go (**FIGURE 14.15B**; and p. 57).

Effects on plant communities comprise an important component of climate change, because by drawing in CO_2 for photosynthesis, plants act as reservoirs for carbon. If higher CO_2 concentrations enhance plant growth, then more CO_2 might be removed from the air, helping to mitigate carbon emissions. However, if climate change decreases plant growth (through drought, fire, or disease, for instance), then carbon flux to the atmosphere could increase. Today large-scale experiments are revealing complex answers, showing that extra

(a) Birds are moving north

> Center of Purple Finch range shifted 700 km north in 40 years

(b) Pikas are being forced upslope

FIGURE 14.15 ▲ Fully 177 out of 305 North American bird species have shifted their winter ranges significantly northward in the past 40 years, according to a 2009 analysis of Christmas Bird Count data by National Audubon Society researchers. The purple finch **(a)** has shown the greatest shift. Its center of abundance moved 697 km (433 mi) north, from southeastern Missouri to northern Iowa. Montane animals such as the pika **(b)**, a unique mammal that lives at high elevations in western North America, are being forced upslope (into more limited habitat) as temperatures warm. Many pika populations in the Great Basin have disappeared from mountains already.

> Pikas are disappearing from mountains after being forced upwards

carbon dioxide can bring both positive and negative results for plant growth (see **THE SCIENCE BEHIND THE STORY**, pp. 312–313).

In regions where precipitation and stream flow increase, erosion and flooding will pollute and alter aquatic systems. In regions where precipitation decreases, lakes, ponds, wetlands, and streams will shrink. The many impacts of climate change on ecological systems will diminish the ecosystem goods and services we receive from nature and that our societies depend on, from food to clean air to drinking water.

Climate change affects society

Drought, flooding, storm surges, and sea level rise have already taken a toll on the lives and livelihoods of millions of

Aspen FACE site researcher Dr. Mark Kubiske of the U.S. Forest Service

THE SCIENCE BEHIND THE STORY

FACE-ing a High-CO$_2$ Future

Can fumigating trees with carbon dioxide tell us what to expect from global climate change? Hundreds of scientists think so, and they are testing plants' responses to atmospheric change at unique outdoor Free-Air CO$_2$ Enrichment ("FACE") facilities.

Plants remove carbon dioxide (CO$_2$) from the atmosphere to use in photosynthesis, and all organisms return CO$_2$ to the atmosphere by cellular respiration (pp. 30–31). As we burn fossil fuels and clear forests, we add carbon dioxide to the atmosphere, which now contains 35% more CO$_2$ than it did just two centuries ago. Will more CO$_2$ mean more plant growth, and will plants be able to absorb and store the extra CO$_2$? Perhaps, but before we rely on forests and phytoplankton to save us from our emissions, we'd better be sure they can do so.

Historically, if a researcher wanted to measure how plants respond to increased carbon dioxide, he or she would adjust gas levels in a small enclosure, such as a lab or a greenhouse. But can results from such small indoor experiments indicate how entire forests will behave? Many scientists thought not, and eventually some pioneered Free-Air CO$_2$ Enrichment—"FACE" for short. In FACE experiments, researchers precisely control ambient levels of CO$_2$ encompassing areas of forest (or other vegetation) outdoors. With their large scale and open-air conditions, FACE experiments include most factors that influence a plant community in the wild, such as variation in temperature, sunlight, precipitation, herbivorous insects, disease pathogens, and competition among plants. By measuring how plants respond to changing gas compositions in such real-world conditions, we can better learn how ecosystems may change in the carbon dioxide–soaked world that awaits us.

Dozens of organizations have sponsored FACE facilities—36 sites in 17 nations so far, including U.S. sites in Arizona, California, Illinois, Minnesota, North Carolina, Tennessee, Nevada, Wisconsin, and Wyoming. The sites cover a variety of ecosystems, from forests to grasslands to rice paddies, and the plots range in size from 1 m to 30 m (from 3 to 98 ft) in diameter.

To understand how a typical FACE study works, let's visit the Aspen FACE Experiment at the Harshaw Experimental Forest (where aspen trees are common) near Rhinelander, Wisconsin. Here, tall steel and plastic towers and pipes ring 12 circular plots of forest 30 m (98 ft) in diameter (**see photo**). The pipes release CO$_2$, bathing the plants in an atmosphere 50% richer in CO$_2$ than today's (equal to what is expected worldwide for the year 2050). Sensors monitor wind conditions, and computers control for the influence of wind by adjusting CO$_2$ releases, keeping ambient concentrations stable within each plot.

The pipes at the Aspen plots also release tropospheric ozone (O$_3$, a major pollutant in urban smog; p. 288), and researchers study how this gas and CO$_2$ affect plant growth, leaf and root conditions, soil carbon content, and much else. Pipes at some plots release normal air, serving as controls for the treatment plots.

Researchers using the Aspen FACE facility have learned a number of things so far, among them:

- Elevated CO$_2$ levels increase photosynthesis and tree growth—but moderate levels of ozone offset this increased growth (**see graph**). Because many modelers have not taken ozone into account when estimating how much carbon trees can sequester, the Aspen FACE data

people. However, climate change will have still more consequences. These include impacts on agriculture, forestry, health, and economics.

Agriculture For some crops in the temperate zones, moderate warming may slightly increase production because growing seasons become longer. The availability of additional carbon dioxide to plants for photosynthesis may also increase yields, but as mentioned above, elevated CO$_2$ can have mixed results. Moreover, some research shows that crops become less nutritious when supplied with more carbon dioxide. If rainfall shifts in space and time, intensified droughts and floods will likely cut into agricultural productivity. Considering all factors together, the IPCC predicts global crop yields to increase somewhat, but beyond a rise of 3 °C (5.4 °F), it expects crop yields to decline. In seasonally dry tropical and subtropical regions, growing seasons may be shortened, and harvests may be more susceptible to drought. Thus, scientists predict that crop production will fall in these regions even with minor warming. This

At the Aspen FACE facility in Wisconsin, tall towers and pipes control the atmospheric composition around selected patches of trees.

content. This would provide urgently needed data on carbon sequestration, the DOE said, and then millions of dollars could be shifted toward a new and improved generation of FACE experiments.

Many researchers were aghast, however, and argued that the precious and unique long-term sites still had much to teach us. How else can we know how forests and climate will interact after 25 years, or 50, they asked? . . . Except, of course, to wait and let Earth show us—by which time it may be too late to do anything about it.

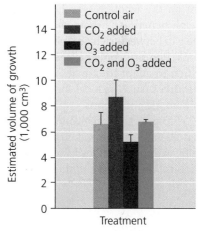

Data from clusters of aspens show that trees supplied with carbon dioxide grew more than control trees during the study period, while those supplied with ozone grew less. Trees supplied with both gases did not grow differently from the controls. Shown are data for one typical cluster. Adapted from Isebrands, J.G., et al., 2001. Growth responses of *Populus tremuloides* clones to interacting elevated carbon dioxide and tropospheric ozone. *Environmental Pollution* 115: 359–371. Fig. 2, with permission from Elsevier.

suggest that existing models may overestimate the amount of CO_2 that trees will pull out of the air.

► As atmospheric levels of ozone and CO_2 rise, insects and diseases that attack aspen and birch trees increase.

► High CO_2 concentrations delay aspen leaf death, which can make aspens vulnerable to frost damage.

Together, such results indicate that rising carbon dioxide levels could have a variety of negative impacts on trees and forests—belying the old expectation that more CO_2 makes for happier plants. Indeed, research from other FACE sites is showing that increased growth from enhanced CO_2 is often temporary and that growth rates later flatten out or decline.

Obtaining solid answers to questions like these takes years or decades, and FACE experiments are designed to monitor plots for the long term as the plants mature. Some FACE sites have been operating for 20 years and are now producing data that cannot be gathered in any other way.

Thus, researchers were shocked in 2008 when the U.S. Department of Energy (DOE), which funds Aspen and other major sites, announced it would cease funding. The DOE advised scientists to cut the trees down and dig up the soil to analyze carbon

would worsen hunger in many of the world's developing nations.

Forestry In the forests that provide our timber and paper products, enriched atmospheric CO_2 may spur greater growth in the near term, but other climatic effects such as drought, fire, and disease may eliminate these gains. Forest managers increasingly find themselves battling catastrophic fires, invasive species, and insect and disease outbreaks. Catastrophic fires are caused in part by decades of fire suppression (p. 196) but are also promoted by longer, warmer, drier fire seasons. Milder winters and hotter, drier summers are promoting outbreaks of pine beetles, pest insects that are destroying millions of acres of trees (see Figure 9.15, p. 197).

Health As climate change proceeds, we will face more heat waves—and heat stress can cause death, especially among older adults. A 1995 heat wave in Chicago killed at least 485 people, and a 2003 heat wave in Europe killed 35,000 people. A warmer climate also exposes us to other health problems:

- ▶ Respiratory ailments from air pollution, as hotter temperatures promote photochemical smog (p. 288)
- ▶ Expansion of tropical diseases, such as dengue fever, into temperate regions as vectors of infectious disease (such as mosquitoes) move toward the poles
- ▶ Disease and sanitation problems when floods overcome sewage treatment systems
- ▶ Injuries and drowning if storms become more frequent or intense

Health hazards from cold weather will decrease, but most researchers feel that the increase in warm-weather hazards will more than offset these gains.

Economics   People will experience a variety of economic costs and benefits from the impacts of climate change, but on the whole researchers predict that costs will outweigh benefits. Climate change is also expected to widen the gap between rich and poor, both within and among nations. Poorer people have less wealth and technology with which to adapt to climate change, and they rely more on resources (such as local food and water) that are sensitive to climatic conditions.

From a variety of economic studies, the IPCC estimated that climate change will cost 1–5% of GDP on average globally, with poor nations losing proportionally more than rich nations. The highest-profile economic study to date has been the *Stern Review on the Economics of Climate Change* headed by economist Nicholas Stern and commissioned by the British government. This exhaustive review concluded that climate change could cost us roughly 5–20% of GDP by the year 2200, but that investing just 1% of GDP starting now could enable us to avoid these future costs.

Impacts will vary regionally

The impacts of climate change are subject to regional variation, so the way each of us experiences these impacts will depend on where we live. Temperature changes have been greatest in the Arctic (**FIGURE 14.16**). Here, ice sheets are melting, sea ice is thinning, storms are increasing, and altered conditions are posing challenges for people and wildlife. As sea ice melts earlier, freezes later, and recedes from shore, it becomes harder for Inuit people and for polar bears alike to hunt the seals they each rely on for food. Permafrost is thawing, destabilizing countless buildings. As the strong Arctic warming melts ice caps and ice sheets, this contributes to sea level rise.

For the United States, potential impacts are analyzed by the U.S. Global Change Research Program, which Congress created in 1990 to coordinate federal climate research. In a 2009 report, scientists for this program reviewed current research, summarized the effects of climate change on the United States, and predicted future impacts (**TABLE 14.2**). Many impacts will vary by region, yet the report predicted that some would be felt across the nation. Average temperatures

ALASKA   2000   2002   2010–2030   2040–2060   2070–2090   RUSSIA   CANADA   GREENLAND   ICELAND   FINLAND   NORWAY   SWEDEN

FIGURE 14.16 ◀ The Arctic has borne the brunt of climate change's impacts so far. As Arctic sea ice melts, it recedes from large areas, as shown by the map indicating the mean minimum summertime extent of sea ice for the recent past, present, and future. Inuit people find it difficult to hunt and travel in their traditional ways, and polar bears starve because they are less able to hunt seals. Human-made structures are damaged as permafrost thaws beneath them: Buildings can lean, buckle, crack, and fall. Map data from National Center for Atmospheric Research and National Snow and Ice Data Center.

TABLE 14.2 Some Predicted Impacts of Climate Change in the United States

▶ Average temperatures will rise 2.2–6.1 °C (4–11 °F) further by the end of this century.

▶ Droughts and flooding will worsen.

▶ Longer growing seasons and enhanced CO_2 will favor crops, but more drought, heat stress, pests, and diseases will decrease most yields.

▶ Snowpack will decrease in the West; water shortages will worsen.

▶ Cold-weather illness will decline, but health problems due to heat stress, disease, and pollution will rise. Some tropical diseases will spread north.

▶ Sea level rise and storm surges will erode beaches and destroy coastal wetlands and real estate.

▶ Alpine ecosystems and barrier islands will begin to vanish.

▶ Drought, fire, and pest outbreaks will continue to alter forests.

▶ Northeast forests will lose sugar maples; Southeast forests will be invaded by grassland; Southwest ecosystems will turn more desertlike.

▶ Melting permafrost will undermine Alaskan buildings and roads.

Adapted from Karl, T.R., et al., eds., 2009. Global climate change impacts in the United States. *U.S. Global Change Research Program and Cambridge University Press.*

in most of the United States have already increased by 0.6–1.1 °C (1–2 °F) since the 1960s and 1970s, and are predicted to rise by another 2.2–6.1 °C (4–11 °F) by the end of this century (**FIGURE 14.17**). Plant communities will likely shift northward

and upward in elevation. Extreme weather events are projected to become more frequent.

We are bound to experience a variety of consequences from anthropogenic climate change in coming years (**FIGURE 14.18**). Yet by addressing its root causes now, we may still be able to prevent the most severe future impacts.

Are we responsible for climate change?

Scientists agree that most or all of today's global warming is due to the well-documented recent increase in greenhouse gas concentrations in our atmosphere. They also agree that this rise in greenhouse gases results primarily from our combustion of fossil fuels for energy and secondarily from the loss of carbon-absorbing vegetation due to deforestation.

Yet despite the overwhelming scientific evidence for climate change and its impacts, many people, especially in the United States, have long tried to deny that it is happening. Indeed, while most nations moved forward to confront climate change through international dialogue, in the United States public discussion of climate change remained mired in debates over whether the phenomenon was real and whether humans were to blame. These debates have been fanned by spokespeople from think tanks and a handful of scientists, many funded by fossil fuel industries. These individuals have aimed to cast doubt on the scientific consensus, and their views are amplified by the American news media, which seeks to present two sides to every issue, even when the sides' arguments are not equally supported by evidence.

In 2009, a hacker illegally broke into computers at the University of East Anglia, U.K., and made public several thousand documents, including over 1,000 private e-mails among a handful of climate scientists. A few of these messages appeared to show questionable behavior in the use of

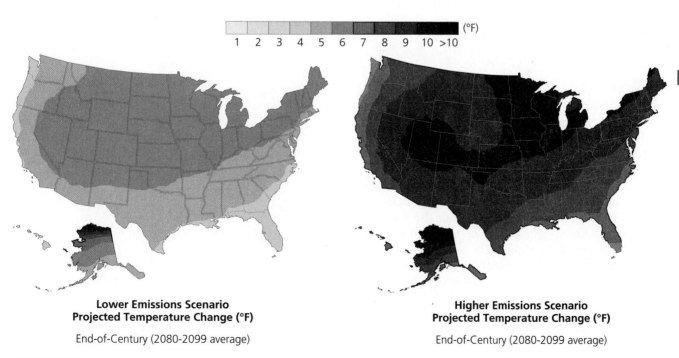

Lower Emissions Scenario
Projected Temperature Change (°F)

End-of-Century (2080-2099 average)

Higher Emissions Scenario
Projected Temperature Change (°F)

End-of-Century (2080-2099 average)

FIGURE 14.17 ▲ Average temperatures across the United States are predicted to rise by 4–6 °F by the end of this century under a low-emissions scenario, and 7–11 °F under a high-emissions scenario. Data from Karl, T.R., et al., eds., 2009. *Global climate change impacts in the United States.* U.S. Global Change Research Program and Cambridge University Press.

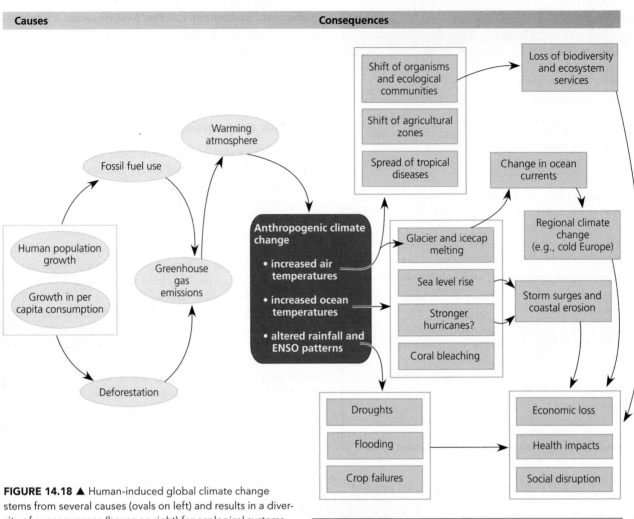

FIGURE 14.18 ▲ Human-induced global climate change stems from several causes (ovals on left) and results in a diversity of consequences (boxes on right) for ecological systems and human well-being. Arrows in this concept map lead from causes to consequences. Note that items grouped within outlined boxes do not necessarily share any special relationship; the outlined boxes are intended merely to streamline the figure.

Solutions

As you progress through this chapter, try to identify as many solutions to anthropogenic climate change as you can. What could you personally do to help address this issue? Consider how each action or solution might affect items in the concept map above.

data and the treatment of other researchers. Climate-change deniers named the incident "Climategate" and used it to accuse the entire scientific establishment of wrongdoing and conspiracy. The news media disseminated the story widely. However, subsequent investigations into the affair by several independent panels cleared the climate scientists, finding no evidence of wrongdoing. Each panel concluded that some individuals may have exercised poor taste or judgment and that some practices could be improved, but they also found that many media accounts trumpeting the news had misrepresented the content of the e-mails. The panels agreed that the hacked messages among a few individuals in no way called into question the vast array of research results compiled by thousands of hard-working independent climate scientists over several decades.

Questions were raised the same year about some of the IPCC report's conclusions, and subsequent inquiry revealed that several statements were inadequately backed by evidence or otherwise misstated or misleading. It is hardly surprising

that a few statements out of thousands in such a vast collaborative project would be in error. Yet scientists wanted to assure that the IPCC's reputation for reliability not be tarnished, and so reforms are underway that should strengthen the IPCC's process during the preparation of its *Fifth Assessment*, due out in 2014.

RESPONDING TO CLIMATE CHANGE

Today, most of the world's people recognize that our fossil fuel consumption is altering the planet that our children will inherit. From this point onward, our society will be focusing on the difficult question of how best to respond to the challenges of climate change. The good news is that everyone—not just leaders in government and business, but everyday people, and especially today's youth—can play a part in this all-important search for solutions.

Shall we pursue mitigation or adaptation?

We can respond to climate change in two fundamental ways. One is to pursue actions that reduce greenhouse gas emissions, so as to lessen the severity of climate change. This strategy is called **mitigation** because the aim is to mitigate, or alleviate, the problem. Examples include improving energy efficiency, switching to clean and renewable energy sources, preventing deforestation, recovering landfill gas, and encouraging farm practices that protect soil quality.

Alternatively, we can pursue strategies to cushion ourselves from the impacts of climate change. This strategy is called **adaptation** because the goal is to adapt to change. Erecting a seawall like the Maldives' Great Wall of Malé is one example of adaptation. Other examples include restricting coastal development; adjusting farming practices to cope with drought; and modifying water management practices to deal with reduced river flows, glacial outburst floods, or salt contamination of groundwater.

We need to pursue adaptation because even if we could halt all our emissions right now, the greenhouse gas pollution already in the atmosphere would continue driving global warming until the planet's systems reach a new equilibrium, with temperature rising an estimated 0.6 °C (1.0 °F) more by the end of the century. Because we will face this change no matter what we do, it will be wise to develop ways to minimize its impacts.

We also need to pursue mitigation, because if we do nothing to diminish climate change, it will eventually overwhelm any efforts we might make to adapt. The sooner we begin reducing our emissions, the lower the level at which they will peak, and the less we will alter climate. We will spend the remainder of our chapter examining approaches for the mitigation of climate change.

We are developing solutions in electricity generation and transportation

The generation of electricity produces the largest portion (40%) of U.S. carbon dioxide emissions, and transportation is not far behind.

Electricity generation From cooking to heating to lighting, much of what we do each day depends on electricity. Fossil fuel combustion generates 70% of U.S. electricity, and coal accounts for most of the resulting emissions.

We can reduce our use of fossil fuels by encouraging conservation and efficiency (pp. 343–345). Power producers can use approaches such as cogeneration (p. 344) to produce fewer emissions per unit energy generated. Manufacturers can produce and consumers can use technologies such as compact fluorescent lightbulbs; high-efficiency appliances; and energy-efficient windows, ducts, insulation, and heating and cooling systems. In addition, each of us can make lifestyle choices to reduce electricity consumption.

We can also reduce greenhouse gas emissions by switching to cleaner energy sources. Natural gas generates the same amount of energy as coal, with half the emissions. Cleaner still are alternatives to fossil fuels, including nuclear power (pp. 345–351), bioenergy, hydroelectric power, geothermal power, solar photovoltaic cells, and wind power (Chapter 16).

While our society transitions to clean and renewable energy alternatives, we are also attempting to capture emissions before they leak to the atmosphere. **Carbon capture** refers to technologies or approaches that remove carbon dioxide from power plant emissions. The next step is **carbon sequestration**, or **carbon storage**, in which the carbon is sequestered, or stored, under pressure in deep salt mines, depleted oil or gas deposits, or other underground reservoirs (see Figure 15.12, p. 337). However, we are a long way from developing adequate technology and secure storage space to accomplish this without leakage—and it is questionable whether we would ever be able to sequester enough carbon to make a sizeable dent in our emissions.

Transportation The average American family makes 10 trips by car each day, and U.S. taxpayers spend over $200 million per day on road construction and repairs for the nation's 250 million registered automobiles. Unfortunately, the typical automobile is highly inefficient. Over 85% of the fuel you pump into your gas tank does something other than move your car down the road (**FIGURE 14.19**).

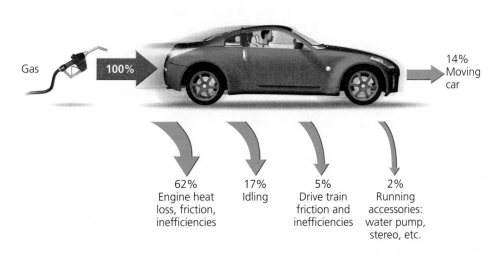

Gas   100%   14% Moving car

62% Engine heat loss, friction, inefficiencies   17% Idling   5% Drive train friction and inefficiencies   2% Running accessories: water pump, stereo, etc.

FIGURE 14.19 ◀ Conventional automobiles are inefficient. Only about 13–14% of the energy from a tank of gas actually moves the typical car down the road, while almost 85% of useful energy is lost, primarily as heat, according to the U.S. Department of Energy.

More aerodynamic designs, increased engine efficiency, and improved tire design can help make our vehicles more fuel-efficient (p. 344). Indeed, the vehicles of many nations are more fuel-efficient than those of the United States. As gasoline prices rise, consumer demand for fuel-efficient automobiles will intensify. Advancing technology is also bringing us alternatives to the traditional combustion-engine automobile. These include gasoline-electric hybrid vehicles (p. 344); fully electric vehicles; alternative fuels such as compressed natural gas and biodiesel (p. 361); and hydrogen fuel cells (p. 375).

We can also make lifestyle choices that reduce our reliance on cars. Some people are choosing to live nearer to their workplaces. Others use mass transit such as buses, subway trains, and light rail. Still others bike or walk to work or on errands. Making automobile-oriented cities and suburbs more friendly to pedestrian and bicycle traffic and improving people's access to public transportation stand as central challenges for city and regional planners (pp. 404–407).

We will need multiple strategies

Advances in agriculture, forestry, and waste management can help us mitigate climate change. In agriculture, sustainable management of cropland and rangeland enables soil to store more carbon. New techniques reduce the emission of methane from rice cultivation and from cattle and their manure and lessen nitrous oxide emissions from fertilizer. We can also grow renewable biofuel crops, although whether these decrease or increase emissions is an active area of research (pp. 358–361).

In forest management, preserving existing forests, reforesting cleared areas, and pursuing sustainable forestry practices (Chapter 9) all help to absorb carbon dioxide from the air. Waste managers are doing their part to cut emissions by treating wastewater (pp. 270, 272–273), generating energy from waste in incinerators (p. 385), and recovering methane seeping from landfills (pp. 385–386). Individuals, communities, and waste haulers can encourage waste reduction, recycling, composting, and the reuse of materials and products (pp. 386–390).

We should not expect to find a single "magic bullet" for mitigating climate change. Reducing emissions will require many steps by people and institutions across many sectors of our economy. However, most reductions can be achieved using current technology—and we can begin implementing these changes right away. Environmental scientists Stephen Pacala and Robert Socolow advise that we follow some age-old wisdom: When the job is big, break it into small parts. Pacala and Socolow identify 15 strategies (**TABLE 14.3**) that could each eliminate 1 billion tons of carbon per year by 2050 if deployed at a large scale. Achieving just 7 of these 15 aims would stabilize our CO_2 emissions at current levels. If we achieve more, then we reduce emissions.

What role should government play?

Even if people agree on strategies and technologies to reduce emissions, they may disagree on how to encourage those strategies and technologies. A major debate is what role government should play: Should it mandate change through laws

TABLE 14.3 Fifteen Ways to Eliminate 1 Billion Tons of Carbon Per Year by 2050
▶ Double the fuel economy of cars.
▶ Halve the miles driven by cars.
▶ Maximize efficiency in all buildings.
▶ Double the efficiency of coal-powered plants.
▶ Switch from coal to natural gas at 1,400 plants.
▶ Capture and store carbon from 800 coal plants.
▶ Capture and store carbon at hydrogen plants.
▶ Capture and store carbon from 180 "synfuels" plants.
▶ Increase hydrogen fuel production by 10 times.
▶ Triple the world's nuclear capacity.
▶ Increase wind power capacity by 50 times.
▶ Increase solar power capacity by 700 times.
▶ Increase ethanol production by 50 times.
▶ Halt tropical deforestation, and double reforestation.
▶ Adopt conservation tillage on all croplands.
Adapted from Pacala, S., and R. Socolow, 2004. Stabilization wedges: Solving the climate problem for the next 50 years with current technologies. Science 305: 968–972.

and regulations? Should it impose no policies and hope that private enterprise will develop solutions on its own? Should it take the middle ground and design programs that give private entities financial incentives to reduce emissions? This debate has been vigorous in the United States, where many business leaders and politicians have opposed all government action to address climate change, fearing that emissions reductions will impose economic costs on industry and consumers.

In 2007, the U.S. Supreme Court ruled that carbon dioxide was a pollutant that the Environmental Protection Agency (EPA) should regulate under the Clean Air Act (p. 283). When Barack Obama became president, he instead urged that Congress craft legislation to address emissions. In 2009, the House of Representatives passed legislation to create a cap-and-trade system (pp. 108–109) in which industries and utilities would compete to reduce emissions for financial gain, and under which emissions were mandated to decrease 17% by 2020. However, similar legislation did not pass in the Senate. As a result, responsibility for addressing emissions passed to the EPA, which embarked on the complex process of developing regulations beginning in 2011. The EPA plans to phase in emissions limits on industry and utilities gradually over many years, hoping to spur energy efficiency retrofits and renewable energy use at a pace that will minimize economic impacts and political opposition.

The Kyoto Protocol sought to limit emissions

Climate change is a global problem, so global cooperation is needed to forge effective solutions. This is why the world's policymakers have tried to tackle climate change with international treaties. In 1992, most of the world's nations signed the

TABLE 14.4 Emissions Reductions Required and Achieved

Nation	Required change, 1990–2008/2012[1]	Observed change, 1990–2009[2]
Russia	0.0%	−35.6%[3]
Germany	−21.0%	−26.3%
United Kingdom	−12.5%	−26.9%
France	0.0%	−7.7%
Italy	−6.5%	−5.4%
Japan	−6.0%	−4.5%
United States[4]	−7.0%	+7.2%
Canada	−6.0%	+16.9%

[1]Percentage decrease in emissions (carbon-equivalents of six greenhouse gases) from 1990 to period 2008–2012, as mandated under the Kyoto Protocol.

[2]Actual percentage change in emissions (carbon-equivalents of six greenhouse gases) from 1990 to 2009. Negative values indicate decreases; positive values indicate increases. Values do not include influences of land use and forest cover.

[3]Russia's decrease was due mainly to economic contraction following the breakup of the Soviet Union.

[4]The United States has not ratified the Kyoto Protocol but was assigned a reduction requirement and reports its emissions, which are included here for comparison.

Data from U.N. Framework Convention on Climate Change, National Greenhouse Gas Inventory Reports, 2011.

U.N. Framework Convention on Climate Change (FCCC), which outlined a plan to reduce greenhouse gas emissions to 1990 levels by the year 2000 through a voluntary approach. Emissions kept rising, however, so nations came together to forge a binding treaty to *require* emissions reductions. Drafted in 1997 in Kyoto, Japan, the **Kyoto Protocol** mandated signatory nations, by the period 2008–2012, to reduce emissions of six greenhouse gases to levels below those of 1990 (**TABLE 14.4**). This treaty took effect in 2005 after Russia became the 127th nation to ratify it.

The United States was the only developed nation not to ratify the Kyoto Protocol. Because the United States emits one-fifth of the world's greenhouse gases, its refusal to join this global effort generated widespread resentment and undermined the treaty's effectiveness. U.S. leaders called the treaty unfair because it required industrialized nations to reduce emissions but did not require the same of rapidly industrializing nations such as China and India. Proponents of the Kyoto Protocol countered that the differential requirements were justified because industrialized nations created the current problem and thus should take the lead in resolving it.

As of 2009 (the most recent year with full international data), nations that signed the Kyoto Protocol had reduced their emissions by 11.3% from 1990 levels. However, much of this reduction was due to economic contraction in Russia and former Soviet-Bloc nations following the breakup of the Soviet Union. When these nations are factored out, the remaining signatories showed a 2.1% *increase* in emissions from 1990 to 2009. Moreover, emissions continued rising from 2009 to 2011 as nations emerged from economic recession.

Climate negotiations have progressed from Copenhagen to Cancun to South Africa

In recent years, representatives of the world's nations have met at a series of conferences, trying to design a treaty to take effect once the Kyoto Protocol ends in 2012. These climate negotiators could not reach consensus at their 2009 meeting in Copenhagen, Denmark. Here, the world's two biggest greenhouse gas emitters, China and the United States, did not offer enough to satisfy other nations (**FIGURE 14.20**). China promised steep emissions cuts but proved unwilling to allow international monitoring to confirm them. U.S. President Obama chose not to promise more than the U.S. Congress had already agreed to. The conference ended amid discord, without specific targets or solid commitments.

The process got back on track in Cancun, Mexico, in 2010, where nations held productive discussion and fleshed out proposals from the Copenhagen conference. Developed nations promised to pay developing nations to help with their mitigation and adaptation efforts—up to $100 billion per year by 2020—through a fund overseen by the World Bank. Nations broadly agreed upon a plan, nicknamed "REDD" (p. 192), to help tropical nations reduce forest loss. Developed nations agreed to transfer clean energy technology to developing nations. And rapidly industrializing nations such as China and India agreed in principle to emission targets and international monitoring. Participants also shared examples of ways they are reducing greenhouse gas pollution. For example, China is accelerating its renewable energy efforts and Brazil announced plans to limit deforestation and encourage no-till agriculture (pp. 133, 143).

Climate negotiators have one last chance to create a successor treaty to the Kyoto Protocol in December 2011, in Durban, South Africa. However, reaching consensus among 200 nations through the treaty process is a daunting challenge, and the political steps outlined so far will not be adequate to halt climate change. As a result, experts predict that most success in mitigating climate change will come from

FIGURE 14.20 ▼ At the Copenhagen climate conference in 2009, these activists showed the negotiating delegates their support for island nations such as Tuvalu and the Maldives and for the goal of bringing the atmospheric carbon dioxide level down to 350 parts per million.

technological advances, carbon trading markets, and national, regional, and local initiatives. Business and industry are accelerating renewable energy and energy efficiency efforts, and policymakers are looking for ways to help the private sector generate productive solutions.

Will emissions cuts hurt the economy?

The U.S. Senate has opposed emissions reductions out of fear that they will dampen the U.S. economy. China, India, and other industrializing nations have so far resisted emissions cuts under the same assumption. This assumption is understandable, given that so much of our economy depends on fossil fuels. Yet nations such as Germany, England, and France have reduced their emissions since 1990 while enhancing their technologically advanced economies and providing their citizens standards of living comparable to those of U.S. citizens.

Because resource use and per capita emissions are high in the United States and other industrialized nations, governments and industries there often feel they have more to lose economically from limiting emissions than developing nations do. However, industrialized nations are also the ones most likely to *gain* economically from major energy transitions, because they are best positioned to invent, develop, and market new technologies to power the world in a post-fossil-fuel era. Germany, Japan, and China have realized this, and are now leading the world in production, deployment, and sales of solar energy technology (**FIGURE 14.21**). If the United States does not act quickly to develop energy technologies for the future, then the future could belong to nations like China, Germany, and Japan.

States and cities are advancing climate change policy

In the absence of action by the U.S. federal government to address climate change, state and local governments across the country are advancing policies to limit emissions. By 2010, mayors from over 1,000 cities from all 50 U.S.

320

FIGURE 14.21 ▼ Workers at a Chinese factory produce photovoltaic solar panels (p. 365). China is racing to develop renewable energy technology and is on track to surpass the United States in becoming a leader in green energy technology.

(a) Greg Nickels   **(b) Arnold Schwarzenegger**

FIGURE 14.22 ▲ In the absence of leadership at the federal level, elected officials at the state and local levels have taken charge. Seattle mayor Greg Nickels (a) convinced over 1,000 U.S. mayors to commit to fighting climate change, while California governor Arnold Schwarzenegger (b) promoted ambitious steps to lower his state's greenhouse emissions.

states had signed on to the U.S. Mayors Climate Protection Agreement, initiated by Mayor Greg Nickels of Seattle (**FIGURE 14.22A**). Under this agreement, mayors commit their cities to pursue policies to "meet or beat" Kyoto Protocol guidelines.

At the state level, the boldest action so far has come in California. In 2006 that state's legislature worked with Governor Arnold Schwarzenegger (**FIGURE 14.22B**) to pass the Global Warming Solutions Act, which aims to cut California's greenhouse gas emissions 25% by the year 2020. This law was the first state legislation with penalties for noncompliance, and it followed earlier efforts in California to mandate higher fuel efficiency for automobiles.

Action was also taken by 10 northeastern states that launched the Regional Greenhouse Gas Initiative (RGGI) in 2007. In this effort, Connecticut, Delaware, Maine, Maryland, Massachusetts, New Hampshire, New Jersey, New York, Rhode Island, and Vermont set up a cap-and-trade program for carbon emissions from power plants. A similar effort, the Western Climate Initiative, involves Arizona, British Columbia, California, Manitoba, Montana, New Mexico, Ontario, Oregon, Quebec, Utah, and Washington. These emissions trading programs (pp. 108–109) show how government can engage the market economy to pursue public policy goals.

Market mechanisms are being used to address climate change

Permit trading programs (pp. 108–109) aim to harness the economic efficiency of market capitalism to achieve public policy goals while allowing business, industry, or utilities flexibility in how they meet those goals. Supporters of permit trading programs argue that they provide the fairest, least expensive, and most effective method of reducing emissions. Polluters choose how to cut their emissions and have financial incentives to reduce emissions below the legally required amount

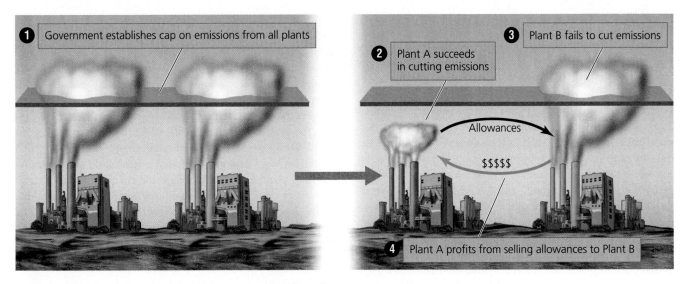

FIGURE 14.23 ▲ In a cap-and-trade emissions trading system, ❶ government first sets an overall cap on emissions. As polluting facilities respond, some will have better success reducing emissions than others. In this figure, ❷ Plant A succeeds in cutting its emissions well below the cap, whereas ❸ Plant B fails to cut its emissions at all. As a result, ❹ Plant B must pay money to Plant A to purchase allowances that Plant A is no longer using. Plant A profits from this sale, and the government cap is met, reducing pollution overall. Over time, the cap can be lowered to achieve further emissions cuts.

Labels in figure:
❶ Government establishes cap on emissions from all plants
❷ Plant A succeeds in cutting emissions
❸ Plant B fails to cut emissions
Allowances
$$$$$
❹ Plant A profits from selling allowances to Plant B

(**FIGURE 14.23**). As an example of how a **cap-and-trade** program for carbon emissions can work, consider the Regional Greenhouse Gas Initiative:

1. Each state decided what polluting sources it would require to participate.

2. Each state set a cap on the total CO_2 emissions it would allow, equal to its 2009 level.

3. Each state distributed to each polluter one permit for each ton it emits, up to the amount of the cap.

4. Each state will lower its cap gradually—10% by 2018.

5. Sources with too few permits to cover their pollution must find ways to reduce their emissions, buy permits from other sources, or pay for credits through a carbon offset project (p. 322). Sources with excess permits may sell them.

6. Any source emitting more than its permitted amount will face penalties.

Once up and running, it is hoped that the system will be self-sustaining. The price of a permit is meant to fluctuate freely in the market, creating the same kinds of financial incentives as any other commodity that is bought and sold.

The world's largest cap-and-trade program is the European Union Emission Trading Scheme. This market got off to a successful start in 2005—until investors discovered that national governments had allocated too many permits to their industries. The overallocation gave companies little incentive to reduce emissions, so permits lost their value, and prices in the market fell to 1/100th of their high value. Europeans tried to correct these problems by making emitters pay for permits and setting emissions caps across the entire European Union while expanding the program to include more greenhouse gases, more emissions sources, and additional members. In the long run, permits will be valuable and the market will work only if policies are in place to limit emissions.

Carbon taxes are another option

As the world's carbon trading markets show mixed results, a number of economists, scientists, and policymakers are saying that cap-and-trade systems are not effective enough, don't work quickly enough, or leave too much to chance. Many of these critics would prefer that governments enact a **carbon tax** instead. In this approach, governments charge polluters a fee for each unit of greenhouse gases they emit. This gives polluters a clear financial incentive to reduce emissions: If they can find ways to reduce their emissions, they reduce their tax and save money. Carbon taxes have so far been established in several European nations, in the Canadian province of British Columbia, and in Boulder, Colorado.

The downside of a carbon tax is that most polluters simply pass the cost along to consumers by charging higher prices for the products or services they sell. Proponents of carbon taxes have responded by proposing an approach called **fee-and-dividend**. In this approach, funds from the carbon tax, or "fee," paid to government by polluters are transferred as a tax refund, or "dividend," to taxpayers. This way, if polluters pass their costs along to consumers, those consumers will be reimbursed by the tax refund they receive. In theory, the system should provide polluters a financial incentive to reduce emissions while imposing no financial burden on taxpayers.

WEIGHING THE ISSUES

Cap-and-Trade or a Carbon Tax? What advantages and disadvantages do you see in using a cap-and-trade system to reduce greenhouse gas emissions? What pros and cons do you see in using carbon taxes to achieve this goal? What do you think of the idea of a "fee-and-dividend" program? If you were a U.S. senator, what type of policy would you support in order to address emissions in the United States, and why?

Carbon offsets are popular

Emissions trading programs generally allow participants to buy **carbon offsets**, voluntary payments intended to enable another entity to reduce emissions that one is unable to reduce oneself. The payment thus offsets one's own emissions. For example, a coal-burning power plant could pay a reforestation project to plant trees that will soak up as much carbon dioxide as the coal plant emits. Or a university could fund the development of clean renewable energy projects to make up for fossil fuel energy the university uses. Carbon offsets have fast become popular among utilities, businesses, universities, governments, and individuals trying to achieve **carbon-neutrality**, a state in which no net carbon is emitted.

In principle, carbon offsets seem a great idea, but rigorous oversight is needed to make sure that the offset money actually accomplishes what it is intended for—and that offsets fund only emissions reductions that would not occur otherwise. Efforts to create a transparent and enforceable system to assess offsets are ongoing.

Corporations are going carbon-neutral

Carbon offsets are a major route toward carbon-neutrality among businesses and corporations seeking to make their practices more sustainable (pp. 95–96), but corporations also can find ways to reduce their carbon footprints directly. An excellent example is Pearson Education, the publisher of your textbook. In 2009 Pearson achieved carbon-neutrality after a concerted two-year effort. Pearson reduced its energy consumption and carbon footprint by 12% by upgrading buildings for energy efficiency, designing more efficient computer servers, reducing the number of vehicles in its fleets, increasing the proportion of hybrid vehicles, and cutting back on employee business travel while enhancing the use of video conferencing. Pearson eliminated a further 47% of its emissions by purchasing clean renewable energy instead of fossil fuel energy and by installing a large solar panel array at one of its sites in New Jersey. To offset the remaining 41% of its emissions, the company is funding a number of programs to preserve forest and replant trees in various areas of the world, from England to Costa Rica.

Should we engineer the climate?

What if all our efforts to reduce emissions are not adequate to rein in climate change? As severe climate change begins looking more likely, some scientists and engineers are reluctantly considering drastic, assertive steps to alter Earth's climate in a last-ditch attempt to reverse global warming—an approach called **geoengineering** (**FIGURE 14.24**).

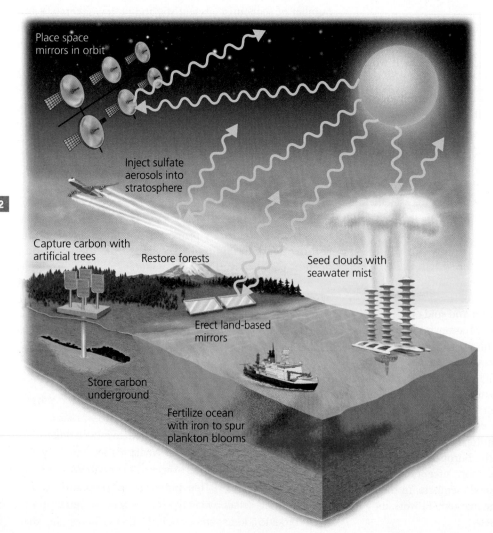

Place space mirrors in orbit

Inject sulfate aerosols into stratosphere

Capture carbon with artificial trees

Restore forests

Seed clouds with seawater mist

Erect land-based mirrors

Store carbon underground

Fertilize ocean with iron to spur plankton blooms

FIGURE 14.24 ◀ Geoengineering proposals seek to cool the planet's climate by removing carbon dioxide from the air or reflecting sunlight away from Earth. Long considered fringe science or science fiction, geoengineering is now being considered by researchers who fear our efforts to reduce emissions will be too little and too late. However, most geoengineering ideas are untested, would take years to develop, might not work well enough, or might lead to other environmental impacts—so they are not a substitute for reducing emissions.

One main geoengineering approach would be to suck carbon dioxide out of the air. We might enhance photosynthesis in natural systems by planting trees or by fertilizing ocean phytoplankton with nutrients such as iron. A more high-tech method would be to design "artificial trees," structures that chemically filter CO_2 from the air.

The second main geoengineering approach would be to block sunlight before it reaches Earth, thus cooling the planet. People have proposed deflecting sunlight by injecting sulfates or other fine dust particles into the stratosphere, by seeding clouds with seawater, or by deploying fleets of reflecting mirrors on land, at sea, or in orbit in space.

Scientists have long been reluctant even to discuss the notion of geoengineering. The potential methods are technically daunting and would take years or decades to develop, and they could pose unforeseen environmental risks. Moreover, blocking sunlight does not reduce greenhouse gas concentrations, so problems such as ocean acidification would continue. In addition, many experts are wary of promulgating hope for easy technological fixes, lest politicians lose their incentive to try to reduce emissions.

However, as climate change intensifies, more scientists are becoming willing to contemplate geoengineering as a back-up plan in case our efforts to reduce emissions fall short. As a result, some researchers and scientific institutions are beginning to assess the risks and benefits of geoengineering options, so that we can be ready to take action in future years if climate change becomes severe enough to justify it.

You can reduce your carbon footprint

Government policies, corporate actions, international treaties, technological innovations—and perhaps even geoengineering—will all play roles in mitigating climate change. But in the end the most influential factor may be the collective actions of millions of regular people. Just as we each have an ecological footprint (p. 4), we each have a **carbon footprint** that expresses the amount of carbon we are responsible for emitting. To help reduce emissions, each of us can take steps in our everyday lives—from turning off lights and choosing appliances to deciding where to live and how to get to work.

College students are vital to driving the personal and societal changes needed to reduce carbon footprints and address climate change. Today a groundswell of interest is sweeping across campuses (pp. 414–415). This was evident on January 31, 2008, when over 1,900 schools participated in the Focus the Nation teach-in on global warming. Young people have played a large role in subsequent grassroots events, including 350.org's International Day of Climate Action—kicked off by the Maldives' underwater cabinet meeting—which featured 5,200 events in 181 nations and was called "the most widespread day of political action in the planet's history." Global climate change may be the biggest challenge we face, but with concerted action we can still avert the most severe impacts. Through outreach, education, innovation, and lifestyle choices, today's youth have the power to turn the tables on climate change and help bring about a bright future for humanity and our planet.

➤ CONCLUSION

Many factors influence Earth's climate, and human activities have come to play a major role. Climate change is well underway, and further greenhouse gas emissions will intensify global warming and cause increasingly severe and diverse impacts. Sea level rise and other consequences of global climate change will affect locations worldwide from the Maldives to Bangladesh to Alaska to Florida. As scientists and policymakers come to better understand anthropogenic climate change and its environmental, economic, and social consequences, more and more of them are urging immediate action. Reducing greenhouse gas emissions and taking other steps to mitigate and adapt to climate change represents our society's foremost challenge in the coming years.

TESTING YOUR COMPREHENSION

1. What happens to solar radiation after it reaches Earth? How do greenhouse gases warm the lower atmosphere?

2. Why is carbon dioxide considered the main greenhouse gas? Why are carbon dioxide concentrations increasing in the atmosphere?

3. How do scientists study the ancient atmosphere? Describe what a *proxy indicator* is, and give two examples.

4. List five major trends in climate that scientists have documented so far.

5. Now list five future trends that researchers are predicting, along with their potential impacts.

6. Describe how rising sea levels, caused by global warming, can create problems for people. How may climate change affect marine ecosystems?

7. How might a warmer climate affect agriculture? How is it affecting distributions of plants and animals? How might it affect human health?

8. What are the two largest sources of greenhouse gas emissions in the United States? How can we reduce these emissions?

9. What roles have international treaties played in addressing climate change? Give two specific examples.

10. Describe one market-based approach for reducing greenhouse gas emissions. Explain one reason it may work well and one reason it may not work well.

SEEKING SOLUTIONS

1. Some people argue that we need "more proof" or "better science" before we commit to substantial changes in our energy economy. How much certainty do you think we need before we should take action regarding climate change? How much certainty do you need in your own life before you make a major decision? Should nations and elected officials follow a different standard? Do you believe that the precautionary principle (pp. 152, 222) is an appropriate standard in the case of global climate change? Why or why not?

2. Describe several ways in which we can reduce greenhouse gas emissions from transportation. Which approach do you think is most realistic, which approach do you think is least realistic, and why?

3. Suppose that you would like to make your own lifestyle carbon-neutral and that you aim to begin by reducing the emissions you are responsible for by 25%. What three actions would you take first to achieve this reduction?

4. **THINK IT THROUGH** You have been appointed as the United States representative to an international conference to negotiate terms of a treaty to take hold after the Kyoto Protocol ends. All nations recognize that the Kyoto Protocol was not fully effective, and most are committed to creating a stronger agreement. The U.S. government has instructed you to take a leading role in designing the new treaty and to engage constructively with other nations' representatives while protecting your nation's economic and political interests. What type of agreement will you try to shape? Describe at least three components that you would propose or agree to, and at least one that you would oppose.

5. **THINK IT THROUGH** You have just been elected governor of a medium-sized U.S. state. Polls show that the public wants you to take bold action to reduce greenhouse gas emissions. However, polls also show that the public does not want gasoline or electricity prices to rise. Carbon-emitting industries in your state are wary of emissions reductions being required of them but are willing to explore ideas with you. Your state legislature will support you in your efforts as long as you remain popular with voters. The state to your west has just passed ambitious legislation mandating steep greenhouse gas emissions reductions. The state to your east has just joined a new regional emissions trading consortium. What actions will you take in your first year as governor?

CALCULATING ECOLOGICAL FOOTPRINTS

Global climate change is something to which we all contribute, because fossil fuel combustion plays such a large role in supporting the lifestyles we lead. Conversely, as individuals, each one of us can help to mitigate climate change through personal decisions and actions in how we live our lives. Several online calculators enable you to calculate your own personal *carbon footprint*, the amount of carbon emissions for which you are responsible. Go to www.nature.org/initiatives/climatechange/calculator, take the quiz, and enter the relevant data in the table.

	Carbon footprint (tons per person per year)
World average	
U.S. average	
Your footprint	
Your footprint with three changes	

1. How does your personal carbon footprint compare to that of the average U.S. resident? How does it compare to that of the average person in the world? Why do you think your footprint differs from these in the ways it does?

2. As you took the quiz and noted the impacts of various choices and activities, which one surprised you the most?

3. Think of three changes you could make in your lifestyle that would lower your carbon footprint. Now take the footprint quiz again, incorporating these three changes. Enter your resulting footprint in the table. By how much did you reduce your yearly emissions?

4. What do you think would be an admirable yet realistic goal for you to set as a target value for your own footprint? Would you choose to purchase carbon offsets to help reduce your impact? Why or why not?

15 Nonrenewable Energy Sources, Their Impacts, and Energy Conservation

Upon completing this chapter, you will be able to:

➤ Identify the energy sources that we use

➤ Describe the nature and origin of coal, natural gas, and crude oil, and evaluate their extraction and use

➤ Assess concerns over the future depletion of global oil supplies

➤ Describe the nature and potential of alternative fossil fuels

➤ Outline and assess environmental, political, social, and economic impacts of fossil fuel use, and explore potential solutions

➤ Specify strategies for conserving energy and enhancing efficiency

➤ Describe nuclear energy and how we harness it

➤ Assess the benefits and drawbacks of nuclear power, and outline the societal debate over this energy source

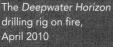

The *Deepwater Horizon* drilling rig on fire, April 2010

Offshore Drilling and the *Deepwater Horizon* Blowout

"This oil spill is the worst environmental disaster America has ever faced."
—U.S. PRESIDENT BARACK OBAMA, 2010

"The Deepwater Horizon *incident is a direct consequence of our global addiction to oil.
. . . If this isn't a call to green power, I don't know what is."*
—UNIVERSITY OF GEORGIA RESEARCHER DR. MANDY JOYE, 2010

It began with a spectacular and deadly explosion that killed 11 people far out to sea. It captivated a horrified nation for three months. And its consequences will stretch on for years. The collapse of British Petroleum's *Deepwater Horizon* drilling rig and the resulting oil spill from its Macondo well in the Gulf of Mexico polluted water, beaches, and marshes; shut down fisheries; ruined tourism; and killed countless animals. The oil contaminated over 1,050 km (650 mi) of coastline in Louisiana, Mississippi, Alabama, and Florida (**FIGURE 15.1**). Ultimately, it raised the question of what costs we are prepared to accept in order to continue relying on fossil fuel energy.

The catastrophe in the Gulf began on April 20, 2010, when a large bubble of natural gas rose through the drill pipe at the Macondo well being drilled by British Petroleum (BP) a mile underwater. The gas bubble shot past a malfunctioning blowout preventer and set off a fiery explosion atop the *Deepwater Horizon* platform, which sank two days later. The stage had been set by a series of setbacks that put the drilling behind schedule and led BP and its contractors to cut corners while government regulators looked the other way.

As oil spewed from the seafloor at a rate of 2,000 gallons every minute, response efforts swung into action. Dozens of ships and boats tried to corral the rising oil at the surface and burn off what they could. Planes and helicopters dumped chemical dispersants from the air. Thousands of people in protective Tyvek suits walked the beaches and spread booms to soak up oil. Teams surveyed marshes for contamination

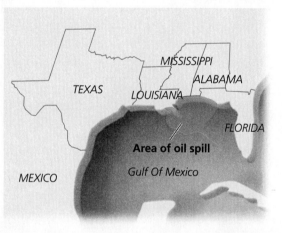

and captured oiled birds and wildlife to clean and release. The work was hot, dirty, and difficult, and the scale of the job seemed overwhelming.

By the time BP engineers finally got the well sealed 86 days later, roughly 4.9 million barrels (230 million gallons) of crude oil had entered the Gulf, creating the largest accidental oil spill in history. As oil washed ashore, it coated beaches and salt marshes, killing birds, turtles, crabs, fish, and plants, and spoiling tourism for an entire summer. Thousands of fishermen were thrown out of work as some of the nation's most productive fisheries were shut down.

Many Americans who watched news coverage of the spill day after day felt shock and outrage. Indeed, the Gulf oil spill resulted from careless missteps by a corporation and its contractors under weak oversight from the federal government. However, the spill is perhaps best viewed not as a single isolated instance of bad practice or misfortune, but as a by-product of

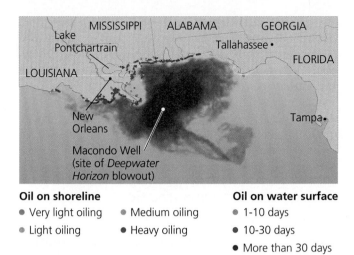

Oil on shoreline
- Very light oiling
- Light oiling
- Medium oiling
- Heavy oiling

Oil on water surface
- 1-10 days
- 10-30 days
- More than 30 days

(a) Extent of the oil spill

(b) Workers scrub oil from a Louisiana beach

FIGURE 15.1 ▲ Oil from the Macondo well blowout spread over thousands of square miles of the Gulf of Mexico **(a)** in the spring and summer of 2010. Darker areas indicate more days with signs of oil at the surface. Thousands of volunteers, government officials, and citizens paid by British Petroleum assisted **(b)** in the vast cleanup effort. Source (a): National Geographic and NOAA.

our society's insatiable appetite for petroleum, driven largely by our reliance on automobiles. Our thirst for fossil fuels has led the oil industry to drill farther and farther out to sea, in search of larger and more profitable untapped deposits. In many cases it has found them, but the farther it moves offshore, the more risks build for major accidents that are hard to control.

Until we reduce our dependence on oil and shift to clean and renewable energy sources, we will suffer pollution in the sea and in the air, climate change and health impacts from fossil fuel combustion, and economic uncertainty from reliance on foreign sources of oil. Every once in a while, some drastic event makes these costs painfully apparent. The *Deepwater Horizon* spill was not the first such event, and it will likely not be the last. ■

SOURCES OF ENERGY

Humanity has devised many ways to harness the renewable and nonrenewable forms of energy available on our planet (**TABLE 15.1**). We use these energy sources to heat and light our homes; power our machinery; fuel our vehicles; produce plastics, pharmaceuticals, and synthetic fibers; and provide the comforts and conveniences to which we've grown accustomed in the industrial age.

We use a variety of energy sources

Most of Earth's energy comes from the sun. We can harness energy from the sun's radiation directly by using solar power technologies. Solar radiation also helps drive wind and the water cycle, enabling us to harness wind power and hydroelectric power. And of course, sunlight drives photosynthesis (p. 30) and the growth of plants, from which we take wood and other biomass as a fuel source. Finally, when plants die, some may impart their stored chemical energy to **fossil fuels**, highly combustible substances formed from the remains of organisms from past geologic ages. The three fossil fuels we use widely today are oil, coal, and natural gas.

Fossil fuels provide most of the energy that our economy buys, sells, and consumes, because their high energy content makes them efficient to ship, store, and burn. We use these fuels for transportation, heating, and cooking, and also to generate **electricity**, a secondary form of energy that is easier to transfer over long distances and apply to a variety of uses. Global consumption of the three main fossil fuels

TABLE 15.1 Energy Sources We Use Today

Energy source	Description	Type of energy
Crude oil	Fossil fuel extracted from ground (liquid)	Nonrenewable
Natural gas	Fossil fuel extracted from ground (gas)	Nonrenewable
Coal	Fossil fuel extracted from ground (solid)	Nonrenewable
Nuclear energy	Energy from atomic nuclei of uranium	Nonrenewable
Biomass energy	Energy stored in plant matter from photosynthesis	Renewable
Hydropower	Energy from running water	Renewable
Solar energy	Energy from sunlight directly	Renewable
Wind energy	Energy from wind	Renewable
Geothermal energy	Earth's internal heat rising from core	Renewable
Tidal and wave energy	Energy from tidal forces and ocean waves	Renewable

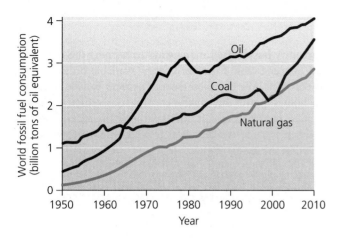

FIGURE 15.2 ▲ Global consumption of fossil fuels has risen greatly over the past half century. Oil use rose steeply during the 1960s to overtake coal, and today it remains our leading energy source. Data from U.S. Energy Information Administration, International Energy Agency, and BP plc. 2011. *Statistical review of world energy 2011.*

has risen steadily for years and is now at its highest level ever (**FIGURE 15.2**).

Energy sources such as sunlight, geothermal energy, and tidal energy are considered perpetually renewable because they are readily replenished, and so we can keep using them without depleting them (pp. 2–3). In contrast, energy sources such as oil, coal, and natural gas are considered nonrenewable. These nonrenewable fuels result from ongoing natural processes, but it takes so long for fossil fuels to form that, once depleted, they cannot be replaced within any time span useful to our civilization. It takes a thousand years for the biosphere to generate the amount of organic matter that must be buried to produce a single day's worth of fossil fuels for our society. At our current rate of consumption, we will use up Earth's accessible store of fossil fuels in just decades to centuries.

Nuclear power as currently harnessed through the fission of uranium (p. 346) is nonrenewable to the extent that uranium ore is in limited supply. However, we can also reprocess some uranium and reuse it.

It takes energy to make energy

We do not simply get energy for free. To harness, extract, process, and deliver the energy we use, we need to invest substantial inputs of energy. For instance, drilling for oil offshore in the Gulf of Mexico requires the construction of immense drilling platforms (the *Deepwater Horizon* cost $560 million) and extensive infrastructure to extract and transport oil—all requiring the use of huge amounts of energy. Thus, when evaluating how much energy a source gives us, it is important to subtract costs in energy invested from benefits in energy received. **Net energy** expresses the difference between energy returned and energy invested:

$$\text{Net energy} = \text{Energy returned} - \text{Energy invested}$$

When assessing energy sources, it is useful to use a ratio often denoted as **EROI**, which stands for **energy returned on investment**. EROI ratios are calculated as follows:

$$\text{EROI} = \text{Energy returned} / \text{Energy invested}$$

Higher EROI ratios mean that we receive more energy from each unit of energy that we invest. Fossil fuels are widely used because their EROI ratios have historically been high. However, EROI ratios can change over time. Those for U.S. oil and natural gas have declined from over 100:1 in the 1940s to about 5:1 today. This means that we used to be able to gain 100 units of energy for every unit of energy expended, but now we can gain only five. The EROI ratios for oil and gas declined because we extracted the easiest deposits first and now must work harder and harder to extract the remaining amounts.

Energy and its consumption are unevenly distributed

Most energy sources are localized and unevenly distributed over Earth's surface. This is true of oil, coal, and natural gas, and as a result, some regions have substantial reserves of fossil fuels whereas others have very few. Nearly two-thirds of the world's proven reserves of crude oil lie in the Middle East. The Middle East is also rich in natural gas, but Russia holds more natural gas than any other country. Russia is also rich in coal, as is China, but the United States possesses the most coal of any nation (**TABLE 15.2**).

TABLE 15.2 Nations with the Largest Proven Reserves of Fossil Fuels		
Oil (% world reserves)	**Natural gas** (% world reserves)	**Coal** (% world reserves)
Saudi Arabia, 17.3	Russia, 23.9	United States, 27.6
Venezuela, 13.8*	Iran, 15.8	Russia, 18.2
Canada, 11.5*	Qatar, 13.5	China, 13.3
Iran, 9.0	Turkmenistan, 4.3	Australia, 8.9
Iraq, 7.5	Saudi Arabia, 4.3	India, 7.0
Most of Canada's and Venezuela's oil reserves occur as oil sands (p. 335), which are included in these figures. *Data are for 2010, from BP plc. 2011. Statistical review of world energy 2011.*		

Consumption rates across the world are also uneven. Citizens of developed regions generally consume far more energy than do those of developing regions. The United States has only 4.5% of the world's population, but it consumes over 20% of the world's energy. Nations also differ in how they use energy. Developing nations devote a greater proportion of energy to subsistence activities, such as growing and preparing food and heating homes, whereas industrialized countries use a greater proportion for transportation and industry. Because industrialized nations rely more on mechanized equipment and technology, they use more fossil fuels. In the United States, fossil fuels supply 83% of energy needs.

COAL, NATURAL GAS, AND OIL

The three major fossil fuels on which we rely today are coal, natural gas, and oil. We will first consider how these fossil fuels are formed, how we locate deposits, how we extract these resources, and how our society puts them to use. We will then examine some environmental and social impacts of their use.

Fossil fuels are indeed fuels created from "fossils"

Fossil fuels form only after organic material is broken down over millions of years in an **anaerobic** environment, one with little or no oxygen. Such environments include the bottoms of lakes, swamps, and shallow seas. The fossil fuels we burn today in our vehicles, homes, industries, and power plants were formed from the tissues of organisms that lived 100–500 million years ago. When organisms were buried quickly in anaerobic sediments after death, chemical energy in their tissues became concentrated as the tissues decomposed and their hydrocarbon compounds (p. 28) were chemically altered amid heat and compression (**FIGURE 15.3**).

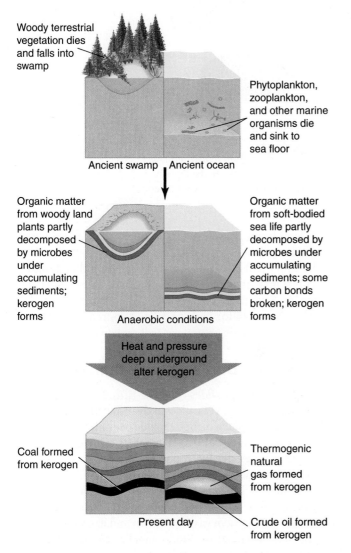

FIGURE 15.3 ▲ Fossil fuels begin to form when organisms die and end up in oxygen-poor conditions, such as when trees fall into lakes and are buried by sediment, or when phytoplankton and zooplankton drift to the seafloor and are buried (**top diagram**). Organic matter that undergoes slow anaerobic decomposition deep under sediments forms kerogen (**middle diagram**). Geothermal heating then acts on kerogen to create crude oil and natural gas (**bottom diagram**). Oil and gas come to reside in porous rock layers beneath dense, impervious layers. Coal is formed when plant matter is compacted so tightly that there is little decomposition.

Coal is a hard blackish substance formed from organic matter (generally woody plant material) that was compressed under very high pressure, creating dense, solid carbon structures. Coal typically results when water is squeezed out of the material as pressure and heat increase and time passes, and when little decomposition takes place because the material cannot be digested or appropriate decomposers are not present. The proliferation 300–400 million years ago of swampy environments where organic material was buried has created coal deposits throughout the world.

Natural gas is a gas consisting primarily of methane (CH_4) and including varying amounts of other volatile hydrocarbons. **Oil**, or **crude oil**, is a sludge-like liquid containing a mix of various hydrocarbon molecules. Oil is also known as **petroleum**, although this term is commonly used to refer to oil and natural gas collectively. Both natural gas and oil have formed from organic material (especially dead plankton) that drifted down through coastal marine waters millions of years ago and was buried in sediments on the ocean floor. This organic material was transformed by time, heat, and pressure into today's natural gas and crude oil.

Two processes give rise to natural gas. *Biogenic* gas is created at shallow depths by the anaerobic decomposition of organic matter by bacteria. An example is the "swamp gas" you may smell when stepping into the muck of a swamp. One source of biogenic natural gas is the decay process in landfills, and many landfill operators are now capturing this gas to sell as fuel (p. 385). *Thermogenic* gas results from compression and heat deep underground. Thermogenic gas may form directly, along with coal or crude oil, or from coal or oil that is altered by heating. Most gas that we extract commercially is thermogenic and is found above deposits of crude oil or seams of coal, so it is often extracted along with those fossil fuels. Indeed, the *Deepwater Horizon* blowout occurred because natural gas accompanying the oil deposit shot up the well shaft once drilling relieved the pressure, and ignited atop the platform.

Because fossil fuels form only under certain conditions, they occur in isolated deposits. For instance, oil and natural gas tend to rise upward through cracks and fissures in porous rock until meeting a dense impermeable rock layer that traps them. Geologists searching for fossil fuels drill cores and conduct ground, air, and seismic surveys to map underground rock formations and predict where fossil fuel deposits might lie.

We mine coal and use it to generate electricity

Coal is the world's most abundant fossil fuel, and it provides 27% of our global primary energy consumption. Once a coal seam is located, we extract coal from the ground using several methods. For deposits near the surface, we use *strip mining*, whereas for deposits deep underground, we use *subsurface mining* (see Figure 11.14, p. 238). Recently, we have begun mining coal on immense scales in the Appalachian Mountains, essentially scraping off entire mountaintops in a process called *mountaintop removal mining* (p. 240). (We explored mining practices and their impacts more fully in Chapter 11.)

People have burned coal to cook food, heat homes, and fire pottery for thousands of years. Coal-fired steam engines helped drive the industrial revolution, powering factories,

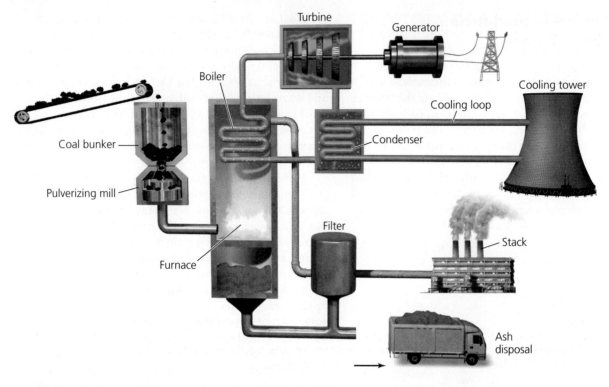

FIGURE 15.4 ▲ At a coal-fired power plant, coal is pulverized and blown into a high-temperature furnace. Heat from the combustion boils water, and the resulting steam turns a turbine, generating electricity by passing magnets past copper coils. The steam is then cooled and condensed in a cooling loop and returned to the furnace. "Clean coal" technologies (pp. 336–337) help filter out pollutants from the combustion process, and toxic ash residue is taken to hazardous waste disposal sites.

agriculture, trains, and ships. Today we burn coal largely to generate electricity. In coal-fired power plants, coal combustion converts water to steam, which turns a turbine to create electricity (**FIGURE 15.4**). Coal provides half the electrical generating capacity of the United States, and it powers China's surging economy. China and the United States are the primary producers and consumers of coal (**TABLE 15.3**).

Coal varies from deposit to deposit in its water content, carbon content, and potential energy. Coal deposits also vary in the amount of impurities they contain, including sulfur, mercury, arsenic, and other trace metals. Coal from the eastern United States tends to be high in sulfur because it was formed in marine sediments, where sulfur from seawater was present. The impurities in coal are emitted during its combustion unless pollution control measures are in place. We will examine the many health and environmental impacts from

coal combustion, along with solutions to these problems, later in this chapter (pp. 336, 341). Reducing pollution from coal is important because society's demand for this abundant fossil fuel may rise once supplies of oil and natural gas begin to decline. Scientists estimate that Earth holds enough coal to supply our society for perhaps a few hundred years more—far longer than oil or natural gas will remain available.

Natural gas burns cleaner than coal

Natural gas today provides over one-fifth of global primary energy consumption. Versatile and clean-burning, natural gas emits just half as much carbon dioxide per unit of energy produced as coal and two-thirds as much as oil. We use natural gas to generate electricity in power plants, to heat and cook in our homes, and for much else. Converted to a liquid at low temperatures (*liquefied natural gas*, or *LNG*), it can be shipped long distances in refrigerated tankers. Russia and the United States lead the world in gas production and gas consumption, respectively (**TABLE 15.4**). World supplies of natural gas are projected to last perhaps 60 more years.

Oil is the world's most-used fuel

Oil today accounts for one-third of the world's primary energy consumption. Global oil consumption has risen 15% in the past decade, and today our society produces and consumes over 750 L (200 gal) of oil annually for every man, woman, and child. **TABLE 15.5** shows the top oil-producing and oil-consuming nations.

TABLE 15.3 Top Producers and Consumers of Coal	
Production (% world production)	**Consumption (% world consumption)**
China, 43.8	China, 45.9
United States, 14.0	United States, 13.2
India, 8.0	India, 9.0
Australia, 5.7	Germany, 3.3
Indonesia, 4.3	Russia, 2.9
Data are for 2009, from U.S. Energy Information Administration, 2011.	

TABLE 15.4 Top Producers and Consumers of Natural Gas	
Production (% world production)	Consumption (% world consumption)
United States, 19.7	United States, 21.4
Russia, 19.4	Russia, 14.5
Canada, 5.3	Iran, 4.4
Iran, 4.4	Japan, 3.3
Norway, 3.4	Germany, 3.1

Data are for 2009, from U.S. Energy Information Administration, 2011.

TABLE 15.5 Top Producers and Consumers of Oil	
Production (% world production)	Consumption (% world consumption)
Russia, 11.8	United States, 22.3
Saudi Arabia, 11.6	China, 9.9
United States, 10.8	Japan, 5.3
Iran, 4.9	India, 3.7
China, 4.7	Russia, 3.2

Data are for 2009, from U.S. Energy Information Administration, 2011.

We drill to extract oil and gas

Once geologists have identified a promising location for an oil or natural gas deposit, a company will typically conduct *exploratory drilling*, drilling small holes that descend to great depths. If enough oil or gas is encountered, extraction begins. Because oil and gas are generally under pressure while in the ground, they will rise to the surface of their own accord when a deposit is tapped. Once pressure is relieved and some oil or gas has risen to the surface, however, the remainder becomes more difficult to extract and may need to be pumped out. As much as two-thirds of a deposit may remain in the ground following **primary extraction**, the initial extraction of available oil or gas. Companies may then begin **secondary extraction**. In secondary extraction for oil, solvents are used or underground rocks are flushed with water or steam (**FIGURE 15.5**). For gas, we use "fracturing techniques" to break into rock formations and pump gas upward. One such technique is to pump salt water under high pressure into rocks to crack them. Sand or small glass beads are injected to hold the cracks open once the water is withdrawn. Even after secondary extraction, quite a bit of oil or gas can remain; we lack technology to remove the entire amounts.

While technology sets a limit on how much *can* be extracted, economics determines how much *will* be extracted. This is because extraction becomes increasingly difficult and costly as oil or gas is removed, so companies will not find it profitable to extract the entire amount. Instead, a company will consider the costs of extraction (and other expenses), and balance them against the current price of the fuel on the world market. Because fuel prices fluctuate, the portion of oil or gas from a given deposit that is "economically recoverable" fluctuates as well. At higher prices, economically recoverable amounts approach technically recoverable amounts.

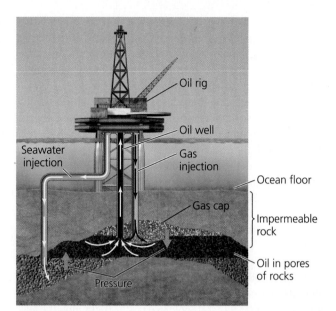

FIGURE 15.5 ▲ Once pressure on an oil deposit drops, material must be injected to increase the pressure. Secondary extraction involves injecting seawater beneath the oil and/or injecting gases just above the oil to force more oil up and out of the deposit.

In the United States, many oil fields did not undergo secondary extraction because the price of oil was too low to make it economical. Once oil prices rose in the 1970s, companies reopened those drilling sites for secondary extraction. More are being reopened today. The amount of a fossil fuel that is technologically and economically feasible to remove under current conditions is termed its **proven recoverable reserve**.

Some drilling occurs offshore

We drill for oil and natural gas not only on land but also below the seafloor on the continental shelves (**FIGURE 15.6**). Offshore drilling has required us to develop technology that can withstand wind, waves, and ocean currents. Some drilling rigs are fixed, standing platforms built with unusual strength. Others are resilient floating platforms anchored in

FIGURE 15.6 ▼ Offshore drilling platforms allow the oil industry to drill for petroleum in the seafloor on the continental shelves.

place above the drilling site. Roughly 35% of the oil and 10% of the natural gas extracted in the United States today comes from offshore sites, primarily in the Gulf of Mexico and secondarily off southern California. The Gulf today is home to 90 drilling rigs and 3,500 production platforms.

Geologists estimate that most U.S. oil and gas remaining to be extracted occurs offshore, and that deepwater sites in the Gulf of Mexico alone may hold 59 billion barrels of oil. We have been drilling in shallow water for several decades, but as oil and gas are depleted at shallow-water sites and as drilling technology improves, the industry is moving into deeper and deeper water. BP's Macondo well lay beneath 1,500 m (5,000 ft) of water, but the deepest wells in the Gulf of Mexico are now twice that depth. Globally, recent discoveries off the coasts of Brazil, Angola, Nigeria, and other nations suggest that a great deal of oil and gas could lie well offshore, and companies are racing one another to get there. Unfortunately, our ability to drill in deep water has outpaced our capacity to deal with accidents there. The fact that it took 86 days for BP to plug the leak in its Macondo well demonstrates the challenge of addressing

an emergency situation a mile or more beneath the surface of the sea.

In 2008 the U.S. Congress lifted a long-standing moratorium on offshore drilling along much of the nation's coastline. The administration of President Barack Obama in 2010 then designated vast areas open for drilling. These included most waters along the Atlantic coast from Delaware south to central Florida, a region of the eastern Gulf of Mexico, and most waters off Alaska's North Slope. However, just weeks after this announcement, the *Deepwater Horizon* spill occurred. Public reaction forced the Obama administration to backtrack, canceling offshore drilling projects it had approved and putting a hold on further approvals until new safety measures could be devised.

Petroleum products have many uses

Once we extract crude oil, we refine it (**FIGURE 15.7**). Crude oil is a mixture of hundreds of types of hydrocarbon molecules characterized by carbon chains of different lengths (p. 28). A

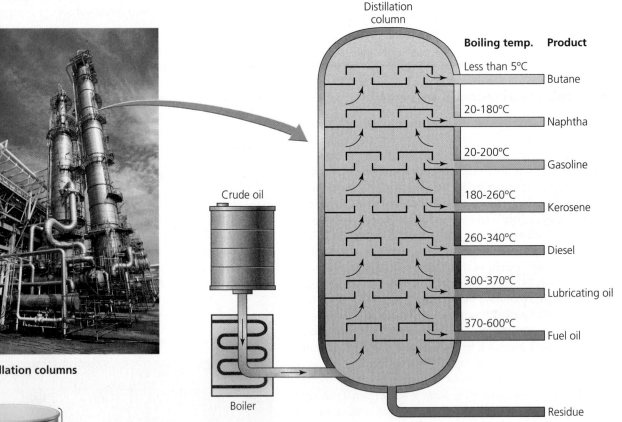

(a) Distillation columns

(b) Distillation process

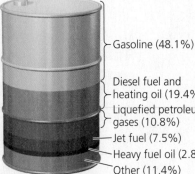

- Gasoline (48.1%)
- Diesel fuel and heating oil (19.4%)
- Liquefied petroleum gases (10.8%)
- Jet fuel (7.5%)
- Heavy fuel oil (2.8%)
- Other (11.4%)

(c) Typical composition of refined oil

FIGURE 15.7 ◀ At oil refineries **(a)**, crude oil is boiled, causing its many hydrocarbon constituents to volatilize and proceed upward through a distillation column **(b)**. Constituents that boil at the hottest temperatures and condense readily once the temperature cools will condense at low levels in the column. Constituents that volatilize at cooler temperatures will continue rising through the column and condense at higher levels, where temperatures are cooler. In this way, heavy oils (generally those with hydrocarbon molecules with long carbon chains) are separated from lighter oils (generally those with short-chain hydrocarbon molecules). The refining process produces a range of petroleum products. Shown in **(c)** are percentages of each major category of product typically generated from a barrel of crude oil. Data for (c) from U.S. Energy Information Administration.

chain's length affects its chemical properties, and these have consequences for human use, such as whether a given fuel burns cleanly in a car engine. Oil refineries sort the various hydrocarbons of crude oil, separating those intended for use in gasoline engines from those, such as tar and asphalt, used for other purposes.

Since the 1920s, refining techniques and chemical manufacturing have greatly expanded our uses of petroleum to include a wide array of products and applications, from lubricants to plastics to fabrics to pharmaceuticals. Today, petroleum-based products are all around us in our everyday lives (**FIGURE 15.8**). Because petroleum products have become so central to our lifestyles, many fossil fuel experts today are voicing concern that oil production may soon decline as we continue to deplete the world's recoverable oil reserves.

We may already have depleted half our oil reserves

Some scientists and oil industry analysts calculate that we have already extracted half the world's oil reserves. So far we have used up about 1.1 trillion barrels of oil, and most estimates hold that somewhat more than 1 trillion barrels remain. To estimate how long this remaining oil will last, analysts calculate the **reserves-to-production ratio**, or **R/P ratio**, by dividing the amount of total remaining reserves by the annual rate of production (i.e., extraction and processing). At current levels of production (30 billion barrels globally per year), 1.2 trillion barrels would last about 40 more years.

Unfortunately, this does not mean that we have 40 years to figure out what to do once the oil runs out. A growing number of scientists and analysts insist that we will face a crisis as soon as the rate of production comes to a peak and then begins to decline—a point in time nicknamed **peak oil**. They point out that if demand continues to increase (because of rising global population and consumption) while production declines, an oil shortage will result. Because production tends to decline once reserves are depleted halfway, many of these experts calculate that a peak oil crisis will likely begin in the very near future.

To understand the basis of these concerns, we must turn back the clock to 1956. In that year, Shell Oil geologist M. King Hubbert calculated that U.S. oil production would peak around 1970. His prediction was ridiculed at the time, but it proved to be accurate; U.S. production peaked in that very year and has continued to fall since then (**FIGURE 15.9A**). The peak in production came to be known as **Hubbert's peak**.

In 1974, Hubbert analyzed data on technology, economics, and geology, predicting that global oil production would peak in 1995. It grew past 1995, but many scientists using newer, better data today predict that at some point in the coming decade, production will begin to decline (**FIGURE 15.9B**). Discoveries of new oil fields peaked 30 years ago, and since then we have been extracting and consuming more oil than we have been discovering.

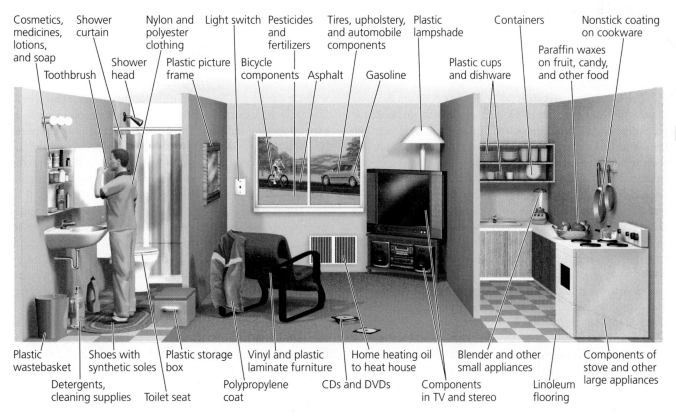

FIGURE 15.8 ▲ Petroleum products are everywhere in our daily lives. Besides the fuels we use for transportation and heating, petroleum products include many of the fabrics we wear and the plastics in countless items we use every day.

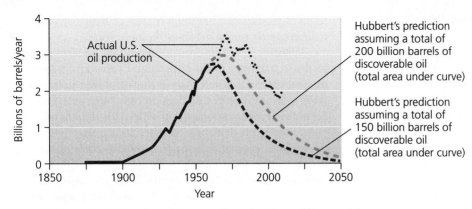

(a) Hubbert's prediction of peak in U.S. oil production, with actual data

In figure (a):
- y-axis: Billions of barrels/year (0 to 4)
- x-axis: Year (1850 to 2050)
- Labels: "Actual U.S. oil production", "Hubbert's prediction assuming a total of 200 billion barrels of discoverable oil (total area under curve)", "Hubbert's prediction assuming a total of 150 billion barrels of discoverable oil (total area under curve)"

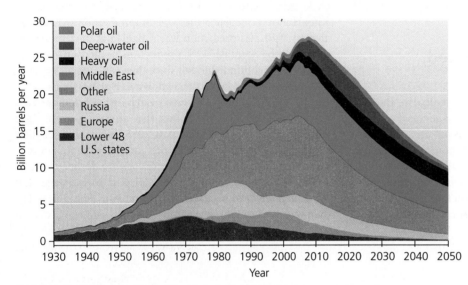

(b) Modern prediction of peak in global oil production

In figure (b):
- y-axis: Billion barrels per year (0 to 30)
- x-axis: Year (1930 to 2050)
- Legend: Polar oil, Deep-water oil, Heavy oil, Middle East, Other, Russia, Europe, Lower 48 U.S. states

FIGURE 15.9 ◄ Because fossil fuels are nonrenewable resources, supplies at some point pass the midway point of their depletion, and annual production begins to decline. U.S. oil production peaked in 1970 **(a)**, just as geologist M. King Hubbert had predicted. Success in Alaska, the Gulf of Mexico, and with secondary extraction increased production above his prediction during the decline. Today many analysts believe global oil production is about to peak. Shown **(b)** is a recent projection, from a 2009 analysis by scientists at the Association for the Study of Peak Oil. Data for (a) from Hubbert, M.K., 1956. *Nuclear energy and the fossil fuels.* Shell Development Co. Publ. No. 95, Houston, TX; and U.S. Energy Information Administration; and for (b) from Campbell, C.J., and Association for the Study of Peak Oil.

Predicting an exact date for peak oil is difficult. Because of year-to-year variability in production, we will not be able to recognize that we have passed the peak until several years after the fact. Many companies and governments do not reveal their true data on oil reserves, and estimates differ as to how much oil we can extract secondarily from existing deposits. Indeed, a recent U.S. Geological Survey report estimated 2 trillion barrels remaining in the world, rather than 1 trillion, and some estimates predict still greater amounts. A 2007 report by the U.S. General Accounting Office reviewed 21 studies and found that most estimates for the timing of the oil production peak ranged from now through 2040.

Whenever it occurs, the coming divergence of demand and supply will likely have momentous economic, social, and political consequences that will profoundly affect the lives of each and every one of us. One prophet of peak oil, writer James Howard Kunstler, has sketched a frightening scenario of our post-peak world during what he calls "the long emergency": Lacking cheap oil with which to transport goods long distances, today's globalized economy would collapse, and our economies would become intensely localized. Large cities could no longer be supported without urban agriculture, and

without petroleum-based fertilizers and pesticides we could feed only a fraction of the world's 7 billion people, even if we expand agricultural land. The American suburbs would be hit particularly hard because of their utter dependence on the automobile. Kunstler argues that the suburbs will become the slums of the future, a bleak and crime-ridden landscape littered with the hulls of rusted-out SUVs.

More optimistic observers argue that as oil supplies dwindle, rising prices will create powerful incentives for businesses, governments, and individuals to conserve energy, use the more expensive types of fossil fuels like those described on the next page, and develop alternative energy sources (Chapter 16)—and that these developments will save us from major disruptions caused by the coming oil peak.

Indeed, to achieve a sustainable society, we will need to switch to renewable energy sources. Investments in energy efficiency and conservation (pp. 343–345) can extend the time we have to make this transition. However, the research and development needed to construct the infrastructure for a new energy economy depend on having cheap oil, and the time we will have to make this enormous transition will be quite limited.

Q: Why should I worry about "peak oil" if there are still years of oil left in the ground?

A: The first thing to bear in mind is that the term "peak oil" doesn't refer to running out of oil. It refers to the point at which our production of oil comes to a peak. Once we pass this peak and production begins to decline, the economics of supply and demand take over. Supply will fall, with some estimates putting the decline at 5% per year. Demand, meanwhile, is forecast to continue rising, especially as nations like China and India put millions of new vehicles on the road. The divergence of demand and supply will drive up oil prices, causing substantial economic ripple effects. Although high oil prices will provide financial incentive to develop alternative energy sources, we may be challenged in a depressed economy to find adequate time and resources to develop new renewable sources.

Other fossil fuels exist

As oil production declines, we will rely more on natural gas and coal—yet these in turn will also eventually peak and decline. At least three further types of fossil fuels exist in large amounts: oil sands, oil shale, and methane hydrate.

Oil sands   **Oil sands** (also called **tar sands**) are deposits of moist sand and clay containing 1–20% *bitumen*, a thick and heavy form of petroleum that is rich in carbon and poor in hydrogen. Oil sands represent crude oil deposits that have been degraded and chemically altered by water erosion and bacterial decomposition. Bitumen is too thick to extract by conventional oil drilling, so oil sands are generally removed by strip mining (**FIGURE 15.10**). After extraction, bitumen may be sent to specialized refineries, where chemical reactions that add hydrogen or remove carbon can upgrade it into more valuable synthetic crude oil.

Three-quarters of the world's oil sands occur in Venezuela and in Alberta, Canada. In Alberta, strip mining began in 1967. Rising crude oil prices have made oil sands more profitable, and dozens of companies are now angling to begin mining projects in the region. Canadian oil sands are now producing well over 1 million barrels of oil per day, contributing half of Canada's petroleum production.

Oil shale   **Oil shale** is sedimentary rock filled with kerogen (organic matter) and can be processed to produce liquid petroleum. Oil shale is formed by the same processes that form crude oil but results when kerogen was not buried deeply enough or subjected to enough heat and pressure to form oil.

We mine oil shale using strip mines or subsurface mines. Once mined, oil shale can be burned directly like coal, or can be baked in the presence of hydrogen and in the absence of air to extract liquid petroleum. The world's known deposits of oil shale may be able to produce over 600 billion barrels of oil (roughly half as much as the conventional crude oil remaining in the world). About 40% of global oil shale reserves are in the United States, mostly on federally owned land in Colorado, Wyoming, and Utah.

Methane hydrate   Another novel potential source of fossil fuel energy occurs in sediments on the ocean floor. **Methane hydrate** (also called *methane clathrate* or *methane ice*) is an ice-like solid consisting of molecules of methane (CH_4, the main component of natural gas) embedded in a crystal lattice of water molecules. Methane hydrate is stable at temperature and pressure conditions found in many sediments on the Arctic seafloor and the continental shelves.

Scientists believe there to be immense amounts of methane hydrate on Earth, holding perhaps twice as much carbon as all known deposits of oil, coal, and natural gas combined. However, we do not yet know how to extract these energy sources safely. Destabilizing a methane hydrate deposit during extraction could lead to a catastrophic release of gas. This could cause a massive landslide and tsunami and would also release huge amounts of methane, a potent greenhouse gas, into the atmosphere, worsening global climate change.

FIGURE 15.10 ▶ In Alberta, companies strip-mine oil sands with the world's largest dump trucks and power shovels. On average, 2 metric tons of oil sands are required to produce 1 barrel of synthetic crude oil.

Alternative fossil fuels have drawbacks

Oil sands, oil shale, and methane hydrate are abundant, but they are no panacea for our energy challenges. For one thing, their net energy values are low, because they are expensive to extract and process. Thus the ratio of energy returned on energy invested (EROI) is low. For instance, much of the energy content of oil shale is consumed in its production, and oil shale's EROI is only about 2:1 or 3:1, compared to a 5:1 or greater ratio for conventional crude oil.

Second, these fuels exert severe environmental impacts. Oil sands and oil shale require extensive strip mining, which devastates landscapes and pollutes waterways. Mining these resources also requires an immense amount of water (which is often scarce in mining regions). At Alberta's oil sands mines, polluted wastewater is left to sit in gigantic reservoirs where waterfowl can die once they land. Besides impacts from their extraction, our combustion of alternative fossil fuels would emit at least as much carbon dioxide, methane, and other air pollutants as our use of coal, oil, and gas. This would worsen the impacts that fossil fuels are already causing, including air pollution and global climate change.

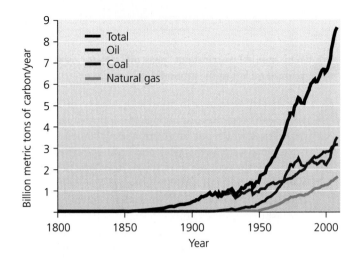

FIGURE 15.11 ▲ Emissions from fossil fuel combustion have risen dramatically as industrialization has proceeded and as population and consumption have grown. Here, global emissions of carbon from carbon dioxide are subdivided by their source (oil, coal, or natural gas). Other minor sources (such as cement production) are also included in the graphed total. Data from Carbon Dioxide Information Analysis Center, Oak Ridge National Laboratory, U.S. Department of Energy, Oak Ridge, TN.

ADDRESSING IMPACTS OF FOSSIL FUEL USE

Our society's love affair with fossil fuels and the many petrochemical products we develop from them has eased constraints on travel, helped lengthen our life spans, and boosted our material standard of living beyond what our ancestors could have dreamed. However, it also causes harm to the environment and human health, and it can lead to political and economic instability.

Fossil fuel emissions pollute air and drive climate change

When we burn fossil fuels, we alter fluxes in Earth's carbon cycle (pp. 38–39). We essentially take carbon that had been retired into a long-term reservoir underground and release it into the air. This occurs as carbon from the hydrocarbon molecules of fossil fuels unites with oxygen from the atmosphere during combustion, producing carbon dioxide (CO_2). Carbon dioxide is a greenhouse gas (p. 300), and CO_2 released from fossil fuel combustion warms our planet and drives changes in global climate (Chapter 14). Because climate change may have diverse, severe, and widespread ecological and socioeconomic impacts, carbon dioxide pollution (**FIGURE 15.11**) is becoming recognized as the greatest environmental impact of fossil fuel use. Moreover, methane is a potent greenhouse gas that drives climate warming. Across the world today, many avenues are being considered to address climate change (pp. 316–323).

Besides modifying our climate, fossil fuel emissions affect human health. Gasoline combustion in automobiles releases pollutants that irritate the nose, throat, and lungs, as well as hydrocarbons (such as benzene and toluene) and impurities (such as lead and arsenic) known to cause cancer or other serious health risks. The combustion of oil in our vehicles and coal in our power plants releases sulfur dioxide and nitrogen oxides, which contribute to smog (pp. 287–289) and acid deposition (pp. 291–294). Combustion of coal may emit mercury that can bioaccumulate in organisms' tissues, poisoning animals as it moves up food chains (pp. 216–217) and presenting health risks to people.

Air pollution from fossil fuel combustion is increasing in developing nations that are industrializing rapidly, but it has been reduced in developed nations as a result of laws such as the U.S. Clean Air Act and government regulations to protect public health (Chapter 13). Technologies such as catalytic converters that cut down on vehicle exhaust pollution have helped a great deal, and their wider adoption in the developing world would reduce pollution further. Regarding pollution from coal-fired power plants, scientists and engineers are seeking ways to cleanse coal of its impurities so that it can continue to be used as an energy source while minimizing impacts on health and the environment.

Clean coal technologies aim to reduce pollution from coal

Clean coal technologies refer to a wide array of techniques, equipment, and approaches that aim to remove chemical contaminants during the process of generating electricity from coal. Among these technologies are various types of scrubbers (pp. 285–286), devices that chemically convert or physically remove pollutants. Another approach is to dry coal that has high water content, making it cleaner-burning. We can also gain more power from coal with less pollution through a process called *gasification*, in which coal is converted into a cleaner synthesis gas, or *syngas*, by reacting it with oxygen and steam at a high temperature. Syngas from coal can be used to turn a gas turbine or to heat water to turn a steam turbine.

The U.S. government and the coal industry have each invested billions of dollars in clean coal technologies, and these have helped to reduce air pollution from sulfates, nitrogen oxides, mercury, and particulate matter (p. 285). At the same time, the coal industry spends a great deal of money fighting regulations and mandates on its practices. As a result, many power plants are built with little in the way of clean coal technologies, and these plants will continue polluting our air for decades. Moreover, many energy analysts and environmental advocates emphasize that these technologies will never result in energy production that is completely clean. They argue that coal is an inherently dirty way of generating power and that it should be replaced outright with cleaner energy sources.

Can we capture and store carbon?

Even if our clean coal technologies were able to remove every last chemical contaminant from power plant emissions, coal combustion would still pump huge amounts of carbon dioxide into the air, intensifying the greenhouse effect and worsening global climate change. This is why many current efforts focus on **carbon capture** and **carbon storage** or **sequestration** (p. 317). This approach consists of capturing carbon dioxide emissions, converting the gas to a liquid form, and then sequestering (storing) it in the ocean or underground in a geologically stable rock formation (**FIGURE 15.12**).

Carbon capture and storage (abbreviated as CCS) is being attempted at a variety of new and retrofitted facilities. The world's first coal-fired power plant to approach zero emissions opened in 2008 in Germany. This plant captures its sulfates and carbon dioxide, compresses the CO_2 into liquid form, trucks it 160 km (100 mi) away, and injects it 900 m (3,000 ft) underground into a depleted natural gas field. In North Dakota, the Great Plains Synfuels Plant gasifies its coal and then sends half the CO_2 through a pipeline into Canada,

where a Canadian oil company buys it to inject into an oilfield to help it pump out oil.

Currently the U.S. Department of Energy is teaming up with nine energy companies to build a prototype of a near-zero-emissions coal-fired power plant. The $1.5 billion *FutureGen* project, located in Mattoon, Illinois, aims to design, construct, and operate a power plant that burns coal using gasification and combined cycle generation, produces electricity and hydrogen, then captures its carbon dioxide emissions and sequesters the CO_2 underground. If this showcase project succeeds, it could be a model for a new generation of power plants.

Energy experts are not ready to rely on the unproven technology of carbon capture and storage quite yet, however. We do not know how to ensure that CO_2 will stay underground once injected there, and we do not know whether these attempts might trigger earthquakes. Injection could in some cases contaminate groundwater supplies, and injecting carbon dioxide into the ocean would acidify its waters (pp. 259, 311). Moreover, CCS is energy-intensive and decreases the EROI of coal, adding to its cost and the amount we need to use. Finally, many renewable energy advocates fear the CCS approach takes the burden off emitters and prolongs our dependence on fossil fuels rather than facilitating a shift to renewables.

WEIGHING THE ISSUES

Clean Coal and Carbon Capture Do you think we should be spending billions of dollars to try to find ways to burn coal cleanly and to sequester carbon emissions from fossil fuels? Or is our money better spent on developing new clean and renewable energy sources, even though they do not yet have enough infrastructure to produce power at the scale that coal can? What pros and cons do you see in each approach?

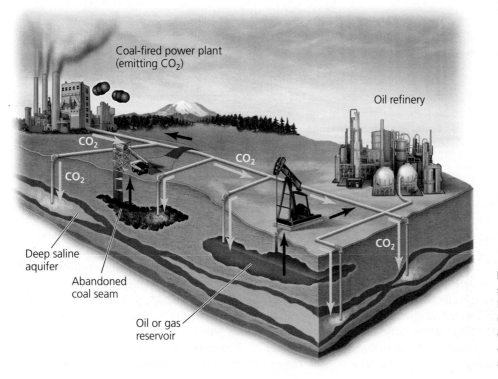

FIGURE 15.12 ◄ Carbon capture and storage schemes propose to inject liquefied carbon dioxide emissions underground into depleted fossil fuel deposits, deep saline aquifers, or oil or gas deposits undergoing secondary extraction.

An EPA scientist takes samples to assess impacts of oil on beaches.

Discovering the Impacts of the Gulf Oil Spill

President Barack Obama echoed the perceptions of many Americans when he called the *Deepwater Horizon* oil spill "the worst environmental disaster America has ever faced." But what has scientific research told us about the actual impacts of the Gulf oil spill?

We don't have all the answers, because the deep-water nature of the spill has made it difficult for scientists to study. A great deal will remain unknown. Yet the intense and focused scientific response to the spill demonstrates the dynamic way in which science can assist society today.

Scientists' first order of business as the spill proceeded was to determine how much oil was leaking and where it was going. Early estimates of the flow rate from BP and the U.S. government proved to be too low, and researchers eventually determined the rate as reaching 62,000 barrels per day. Using underwater imaging, aerial surveys, and shipboard water samples, researchers tracked the movement of oil up through the water column and across the Gulf. These data helped predict

when and where oil might reach shore, thereby helping to direct prevention and cleanup efforts.

Tracking movement of the oil underwater was trickier than monitoring its spread on the surface. University of Georgia biochemist Mandy Joye, who had studied natural seeps in the Gulf for years, documented that the leaking wellhead was creating a plume of oil the size of Manhattan. She also found evidence of low oxygen concentrations, or hypoxia (pp. 22, 25, 267), resulting from the fact that some bacteria consume oil and gas, depleting the water of oxygen as they proliferate. Hypoxia can make underwater regions uninhabitable for fish and other creatures.

Joye and others worried about impacts on marine life in the open ocean and deep beneath the surface. Some researchers feared that the thinly dispersed oil could prove devastating to plankton (the base of the marine food chain) and to the tiny larvae of shrimp,

fish, and oysters (the pillars of the fishing industry). Scientists taking water samples documented sharp drops in plankton during the spill, but it will take a few years to learn whether so many larvae were lost as to diminish populations of adult fish and shellfish.

Other questions revolve around impacts of the chemical dispersant that BP used to break up the oil, a compound called Corexit 9500. Work by biologist Philippe Bodin following the *Amoco Cadiz* oil spill in France in 1978 had found that Corexit 9500 appeared more toxic to marine life than the oil itself. BP applied an unprecedented amount of the chemical to the *Deepwater Horizon* spill, injecting a great deal directly into the path of the oil at the wellhead. This caused the oil to dissociate into trillions of tiny droplets that dispersed across large regions. Many scientists worried that this expanded the oil's reach, affecting more plankton, larvae, and fish.

Impacts of the oil on birds, sea turtles, and marine mammals were somewhat easier to assess. Officially confirmed deaths numbered 6,104 birds, 605 turtles, and 97 mammals—and hundreds of animals were cleaned and saved by wildlife rescue teams—but a much larger, unknown, number succumbed to the oil. What effects this mortality may have on populations in coming years is unclear. Following the *Exxon Valdez* spill in Alaska in 1989, populations of some species rebounded after several years, but populations of others have never recovered. Researchers are following the movements of some marine animals with radio transmitters to try to learn what effects the oil may have had.

As images of oil-coated marshes saturated the media, researchers worried that widespread death of marsh grass would leave the shoreline vulnerable to severe erosion by waves. Louisiana has already lost many of its coastal wetlands to subsidence, dredging, sea

Oil spills pollute marine and coastal environments

Even if we can clean up air pollution from power plants, fossil fuels pollute water in many ways. What comes most readily to mind is the pollution that occurs when massive oil spills from tanker ships or drilling platforms foul coastal waters and beaches.

The *Deepwater Horizon* spill proved so difficult to control because we had never had to deal with a spill so deep

underwater. It revealed that offshore drilling presents serious risks of environmental impact that may be difficult to address, even with our best engineering. As the oil spread through the Gulf of Mexico and washed ashore, the region suffered a wide array of impacts (see **ENVISIONIT**, p. 340). Of the countless animals killed, most conspicuous were birds, which cannot regulate their body temperature once their feathers become coated with oil. However, the underwater nature of the BP spill meant that unknown numbers of fish, shrimp, corals, level rise, and silt capture by dams on the Mississippi River (p. 249). Fortunately, researchers found that oil did not penetrate to the roots of most plants, and that many oiled grasses were sending up new growth. Indeed, Louisiana State University researcher Eugene Turner said that loss of marshland in 2010 from the oil "pales in comparison" with marshland already lost each year due to other factors.

The ecological impacts of the spill had measurable impacts on people. The region's mighty fisheries were shut down, forcing thousands of fishermen out of work. The government tested fish and shellfish for contamination and reopened fishing once they were found to be safe, but consumers did not want to buy Gulf seafood. Beach tourism remained low all summer as visitors avoided the region. Together, the losses in fishing and tourism totaled billions of dollars.

Stress and anxiety over economic losses affected people's health, studies showed. Over one-third of parents told Columbia University researchers in a survey that their children had suffered physical or mental health effects as a result of the spill—and this figure increased to one-half for low-income residents. Other studies found rises in depression, headaches, respiratory problems, and domestic violence.

Scientists expect some impacts from the Gulf spill to be long-lasting. Oil from the similar *Ixtoc* blowout off Mexico's coast in 1979 still lies in coastal mangrove forests and in sediments near dead coral reefs. Fishermen there say it took years for catches to return to normal, and oysters have never come back. After the *Amoco Cadiz* tanker spill, it took seven years for oysters and other marine species to recover. In Alaska, oil from the *Exxon Valdez* spill remains embedded in beach sand, and researchers debate whether it is

SHORELINES
• Air and ground surveys
• Habitat assessment
• Measurements of subsurface oil

WATER COLUMN AND SEDIMENTS
• Water quality surveys
• Sediment sampling
• Transect surveys to detect oil
• Oil plume modeling

AQUATIC VEGETATION
• Air and coastal surveys

HUMAN USE
• Air and ground surveys

Wellhead

FISH, SHELLFISH, AND CORALS
• Population monitoring of adults and larvae
• Surveys of food supply (plankton and invertebrates)
• Tissue collection and sediment sampling
• Testing for contaminants

BIRDS, TURTLES, MARINE MAMMALS
• Air, land, and boat surveys
• Radiotelemetry, satellite tagging, and acoustic monitoring
• Tissue sampling
• Habitat assessment

The effort to assess damage to natural resources from the *Deepwater Horizon* oil spill is the largest-ever undertaking of its kind. In this multi-pronged endeavor, thousands of researchers are surveying habitats, collecting samples and testing them in the lab, tracking wildlife, monitoring populations, and more.

best to try to remove it or to leave it undisturbed.

However, researchers agree on reasons to be hopeful about the Gulf's recovery. One is that the Gulf's warm and sunny climate speeds the natural breakdown of oil. In hot sunlight, volatile components of oil evaporate from the surface and degrade in the water, so that fewer toxic compounds such as benzene, naphthalene, and toluene reach marine life. In addition, bacteria that consume hydrocarbons live in the Gulf's waters, sediments, and marshes, because some oil seeps naturally from the seafloor, and leakage from platforms, tankers, and pipelines are common. Thus, whereas for other major spills, responders tried to apply

oil-eating bacteria or fertilize beaches to encourage bacterial growth, in the Gulf these microbes are already thriving, giving the region a natural self-cleaning capacity.

Researchers are now conducting a wide range of scientific studies (**see figure**). New funding should help: BP has promised to provide half a billion dollars for research over the next 10 years, which is 10 times what the federal government had been providing before the spill. Answers to our many questions will come in gradually as long-term impacts become clear. Scientists can only hope that many findings will be happy ones and that the Gulf's systems will recover more fully than expected.

and other marine animals also died, affecting coastal and ocean ecosystems in complex ways. Marsh plants were also killed, and any resulting erosion of marshes puts New Orleans and other coastal cities at greater risk from storm surges and flooding. Gulf Coast fisheries, which supply much of the nation's seafood, were hit hard by the spill, with thousands of fishermen and shrimpers put out of work. Beach tourism suffered, and economic and social impacts were expected to last for years. Throughout this process, scientists have been

studying aspects of the spill and its impact on the region's people and natural systems (see **THE SCIENCE BEHIND THE STORY**, above).

As climate change melts sea ice in the Arctic, new shipping lanes are opening and nations are jockeying for position, hoping to stake claim to oil and gas deposits that lie beneath the seafloor in this region. Offshore drilling in Arctic waters, however, would pose severe pollution risks, because if a spill were to occur, icebergs, pack ice, storms, cold temperatures,

The *Deepwater Horizon* spill disgorged tens of millions of gallons of oil into the Gulf of Mexico ...

Oil slicks on surface

... where it fouled beaches, killed countless animals, and devastated fisheries and tourism.

Oiled brown pelican

Thousands of people threw themselves into the cleanup effort, but the spill's impacts will last for years.

Cleaning oiled beaches

You Can Make a Difference

➤ Volunteer for cleanups of oil spills in your region.

➤ Urge policymakers to strengthen regulations on offshore drilling.

➤ Limit your own oil consumption by driving less, driving a fuel-efficient car, eating local foods, reusing and recycling products, and supporting renewable energy.

and wintertime darkness would hamper response efforts. Frigid water temperatures would also slow the natural breakdown of oil.

Fossil fuel use and extraction pollute in various ways

Although large catastrophic oil spills have significant impacts on the marine environment, most water pollution from oil results from numerous non-point sources (pp. 265–266) to which all of our actions contribute (p. 267). Oil from automobiles, homes, industries, gas stations, and businesses runs off roadways and enters rivers and wastewater facilities, being discharged eventually into the ocean. Oil can also contaminate groundwater supplies when underground storage tanks leak. In addition, atmospheric deposition of pollutants from the combustion of fossil fuels exerts many impacts on freshwater ecosystems (p. 291).

Extracting fossil fuels on land exerts environmental impacts as well. Mining coal causes water pollution through acid drainage (p. 238), habitat destruction, and a number of other environmental and social impacts (pp. 238, 240). At extraction sites for *coalbed methane* (methane extracted from coal seams), groundwater is pumped out to free gas to rise, but salty groundwater dumped on the surface can contaminate soil and kill vegetation over large areas.

At many natural gas wells, gas is extracted by hydraulic fracturing, or "hydrofracking," in which drillers fracture rock formations by injecting pressurized water mixed with sand and chemicals. The immense volumes of wastewater returned to the surface in this process are often laced with salts, radioactive elements such as radium, and toxic chemicals such as benzene picked up from deep underground. This wastewater is often sent to sewage treatment plants that are not designed to handle all the contaminants and that do not regularly test for radioactivity. This is currently causing concern in Pennsylvania, where a boom in natural gas extraction from the vast Marcellus Shale deposit is sending millions of gallons of drilling waste to treatment plants, which then release their treated water into rivers that supply drinking water for people in Pittsburgh, Philadelphia, Harrisburg, and other cities.

To drill for oil or gas on land, road networks must be constructed, and many sites may be explored in the course of prospecting. The extensive infrastructure needed to support a full-scale drilling operation typically includes housing for workers, access roads, transport pipelines, waste piles for removed soil, and ponds to collect the toxic sludge that remains after oil is removed. These activities can pollute soil and water, fragment habitats, and disturb wildlife. All these impacts have been documented on the tundra of Alaska's North Slope, where policymakers continue to debate whether to open the Arctic National Wildlife Refuge to drilling.

Fortunately, drilling technology is more environmentally sensitive than in the past. **Directional drilling** involves drilling wells in directions outward from a drilling pad, as drillers bore down vertically and then curve to drill horizontally. This allows extraction companies to follow horizontal layered deposits to extract the most they can from them. It also allows drilling to reach a large underground area (up to several thou-

sand meters in radius) around a drill pad. As a result, fewer drill pads are needed, and the surface footprint of drilling is smaller.

The costs of alleviating the health and environmental impacts of fossil fuel extraction are generally not internalized in the market prices of fossil fuels. Instead, we all pay these external costs (p. 92) through medical expenses, costs of environmental cleanup, and impacts on our quality of life. Moreover, the prices we pay at the gas pump or on our monthly utility bill do not even cover the financial costs of fossil fuel production. Rather, fossil fuel prices have been kept inexpensive as a result of government subsidies to extraction companies (p. 108). Thus, we all pay extra for our fossil fuel energy through our taxes, generally without even realizing it.

Fossil fuel extraction has mixed consequences for local people

For people who live in fossil-fuel-bearing regions, development and extraction can yield jobs and economic benefits, but can also result in pollution. In the Gulf of Mexico region, oil and gas industries employ 107,000 people and contribute money toward local economies. However, far more people are employed in tourism, service industries, and fishing industries, which all were negatively affected by the *Deepwater Horizon* spill. Thus, single incidents can sometimes overwhelm the economic benefits of fossil fuel development. Similar tradeoffs exist between jobs and environmental impact in mountaintop-mining areas of Appalachia (p. 240).

In most parts of the world where fossil fuels have been extracted, local residents have suffered pollution without the economic gains to compensate. When multinational corporations extract oil or gas in developing countries, paying those countries' governments for access, the money often does not trickle down to the people who live where the extraction takes place. Moreover, oil-rich developing nations such as Ecuador, Venezuela, and Nigeria tend to have few environmental regulations, and existing regulations may go unenforced if a government does not want to risk losing the large sums of money associated with oil development.

In Ecuador, local people brought suit against Chevron for environmental and health impacts from years of oil extraction in the nation's rainforests. An Ecuadorian court in 2011 found the oil company guilty and ordered it to pay $9.5 billion for cleanup—the largest-ever such judgment. However, Chevron succeeded in getting a U.S. court to issue an injunction, and the two sides are now tussling over jurisdiction. The complex legal battle is being watched internationally, as its outcome could set an influential precedent.

In Nigeria, the Shell Oil Company extracted $30 billion of oil from land of the native Ogoni people, yet the Ogoni still live in poverty, with no running water or electricity. Profits from the oil extraction went to Shell and to the military dictatorships of Nigeria. The development resulted in oil spills, noise, and constantly burning gas flares, all of which caused illness among people living nearby. Starting in 1962, Ogoni activist and leader Ken Saro-Wiwa fought for fair compensation to the Ogoni. After years of persecution by the Nigerian government, Saro-Wiwa was arrested in 1994, given a trial universally regarded as a sham, and put to death by military tribunal.

Dependence on foreign energy affects the economies of nations

Putting all of one's eggs in one basket is always a risky strategy. Because virtually all our modern technologies and services depend in some way on fossil fuels, we are vulnerable to supplies' becoming unavailable or costly. Nations that lack adequate fossil fuel reserves of their own are especially vulnerable (**FIGURE 15.13**). Since its 1970 oil production peak, the United States has relied more and more on foreign energy, and today the nation imports two-thirds of its crude oil.

Such reliance means that seller nations can control energy prices, forcing buyer nations to pay more as supplies dwindle. This became clear in 1973, when the *Organization of Petroleum Exporting Countries (OPEC)* resolved to stop selling oil to the United States. The predominantly Arab nations of OPEC opposed U.S. support of Israel in the Arab-Israeli Yom Kippur War and sought to raise prices by restricting supply. The embargo created panic in the West and caused oil prices to skyrocket (**FIGURE 15.14**), spurring inflation. Fear of oil shortages drove American consumers to wait in long lines at gas pumps. More recently, when Hurricanes Katrina and Rita slammed into the Gulf Coast in 2005, they damaged refineries and offshore platforms, causing oil and gas prices to spike.

With the majority of global oil reserves located in the politically volatile Middle East, crises in this region of the world are a constant concern for U.S. policymakers. The democratic street uprisings of 2011 that began in Tunisia and Egypt and spread across the region put leaders of the United States and other Western nations in an awkward po-

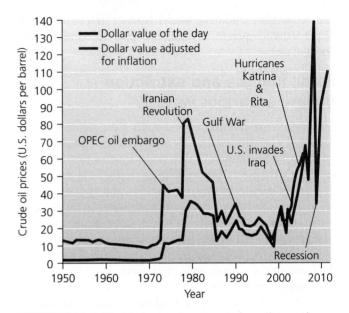

FIGURE 15.14 ▲ World oil prices have gyrated greatly over the decades, often because of political and economic events in oil-producing countries. The greatest price hikes in recent times have resulted from wars and unrest in the oil-rich Middle East. Data from U.S. Energy Information Administration.

sition, because they had long supported many of the region's autocratic rulers. These rulers had facilitated Western access to oil, even as they suppressed democracy in their own societies. The 2011 uprisings were only the most recent in a long history of events that have affected oil prices and global access to oil, stretching back through the U.S.-led wars in Iraq and the Iran-Iraq War of the 1980s to the 1973 OPEC embargo.

In response to the 1973 embargo, the U.S. government enacted a series of policies to reduce reliance on foreign oil. It urged oil companies to pursue secondary extraction at old oil wells. It established an emergency stockpile (which today stores one month's worth of oil). It capped the price that domestic producers could charge for oil, funded research into renewable energy sources, and enacted conservation measures we will discuss below. The new U.S. policies also called for developing additional domestic sources, including offshore oil from the Gulf of Mexico. Since then, the desire to reduce reliance on foreign oil by boosting domestic production has driven the expansion of offshore drilling into deeper and deeper water. It has also repeatedly driven a proposal to open the Arctic National Wildlife Refuge on Alaska's North Slope to drilling, despite critics' charges that drilling there would spoil America's last true wilderness while doing little to boost the nation's oil supply.

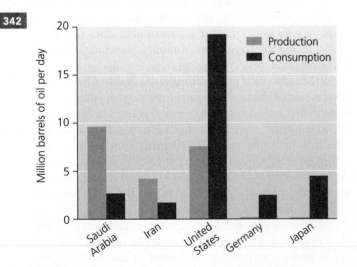

FIGURE 15.13 ▲ Japan, Germany, and the United States are among nations that consume far more oil than they produce. Iran and Saudi Arabia produce more oil than they consume and are able to export oil to high-consumption countries. Data are for 2010, from U.S. Energy Information Administration and British Petroleum.

WEIGHING THE ISSUES

Drill, Baby, Drill? Do you think the United States should open more of its offshore waters to oil extraction? Would the benefits exceed the potential costs? Should we place limits on how far from shore drilling should take place? Should the government regulate offshore drilling more strongly? Give reasons for your answers.

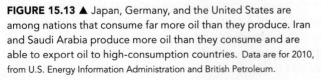

Despite all these policies, U.S. demand for oil has remained high enough that the nation has needed to import more and more each year. The United States has diversified its foreign sources, however, and today receives most oil from non–Middle Eastern nations, including Canada, Mexico, Venezuela, and Nigeria.

How will we convert to renewable energy?

Fossil fuels are not a sustainable long-term solution to our energy needs. Fossil fuels are limited in supply, and their use has health, environmental, political, and socioeconomic consequences (**FIGURE 15.15**). Concern over these issues is a prime reason many scientists, environmental advocates, businesspeople, and policymakers are looking to shift to clean and renewable sources of energy that exert less impact on natural systems and human health. Many nations are moving far faster than the United States. France relies on nuclear power for its electricity needs, Germany is investing in solar power (pp. 354–355), and China is forging ahead and developing multiple renewable energy technologies.

As we make the transition to renewable energy sources, it will benefit us to prolong the availability of fossil fuels. We can prolong our access to fossil fuels by instituting measures to con-serve energy, through lifestyle changes that reduce energy use, and through technological advances that improve efficiency.

ENERGY EFFICIENCY AND CONSERVATION

Until our society makes the transition to renewable energy sources, we will need to find ways to minimize use of our dwindling fossil fuel resources. **Energy efficiency** describes the capacity to obtain a given result or amount of output while using less energy input. **Energy conservation** describes the practice of reducing energy use. Because greater efficiency allows us to reduce energy use, efficiency is one primary means toward conservation. Efficiency and conservation allow us to extend the lifetimes of our nonrenewable energy supplies, to be less wasteful, and to reduce our environmental impact.

Personal choice and efficient technologies are two routes to conservation

As individuals, we can make conscious choices to reduce our own energy consumption by driving less, turning off lights when rooms are not being used, dialing down thermostats,

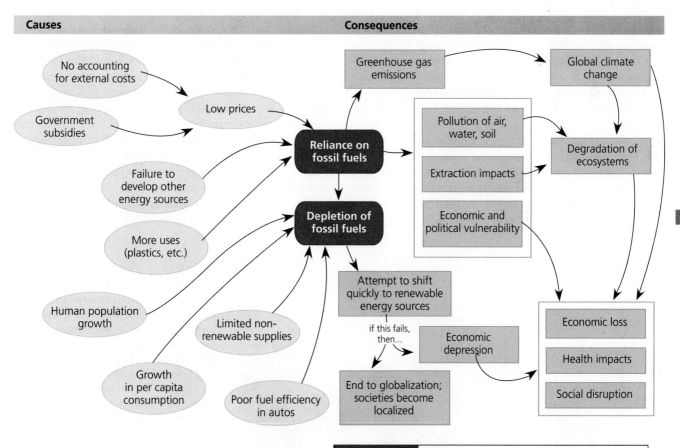

FIGURE 15.15 ▲ Our reliance on, and depletion of, fossil fuels has many causes **(ovals on left)** and many consequences **(boxes on right)**. Arrows in this concept map lead from causes to consequences. Note that items grouped within outlined boxes do not necessarily share any special relationship; the outlined boxes are merely intended to streamline the figure.

Solutions

As you progress through this chapter, try to identify as many solutions to our reliance on and depletion of fossil fuels as you can. What could you personally do to help address this issue? Consider how each action or solution might affect items in the concept map above.

and cutting back on the use of energy-intensive machines and appliances. Many European nations use less energy per capita than the United States, yet enjoy equivalent standards of living. This indicates that U.S. citizens could reduce their energy consumption without diminishing their quality of life. Moreover, for any given individual or business, reducing energy consumption can save money while helping to conserve resources.

As a society, we can conserve energy by developing technologies and strategies to make our energy-consuming devices and processes more efficient. Currently, more than two-thirds of the fossil fuel energy we use is simply lost, as waste heat, in automobiles and power plants. The United States burns through twice as much energy per dollar of Gross Domestic Product (GDP) as do most other industrialized nations. However, the good news is that over the past three decades the United States has decreased its energy use per dollar of GDP by about 50%. Given such tremendous gains in efficiency, we should be able to make still-greater progress in the future.

We can improve the efficiency of power plants through **cogeneration**, in which excess heat produced during electricity generation is captured and used to heat nearby workplaces and homes and to produce other kinds of power. Cogeneration can almost double the efficiency of a power plant.

In homes, offices, and public buildings, a significant amount of heat is needlessly lost in winter and gained in summer because of poor design and inadequate insulation. Improvements in design can reduce the energy required to heat and cool buildings (p. 408). Such improvements may involve passive solar design (p. 364), better insulation, a building's location, the vegetation around it, and even the color of its roof (light colors keep buildings cooler by reflecting the sun's rays).

Consumer products such as lightbulbs and appliances have been reengineered through the years to enhance efficiency. Compact fluorescent bulbs are far more efficient than incandescent light bulbs, and many governments are phasing out incandescent bulbs for this reason; the U.S. phase-out is scheduled to be complete in 2014. The U.S. EPA's Energy Star program, which labels refrigerators, dishwashers, and other appliances for their energy efficiency, has helped to reduce U.S. per-person home electricity use below what it was in the 1970s. For the consumer, studies show that savings on utility bills readily offset the higher costs of energy-efficient appliances.

Automotive technology represents perhaps our best opportunity to save large amounts of fossil fuels fairly easily. We can accomplish this with alternative-technology vehicles such as electric cars, electric/gasoline hybrids (**FIGURE 15.16**), or vehicles that use hydrogen fuel cells (pp. 375–376). Among electric/gasoline hybrids, current U.S. models of the Toyota Prius and the Chevrolet Volt average fuel economy ratings of 50 miles per gallon (mpg) and 60 mpg, respectively—two to three times better than the average American car. Even without alternative vehicles, however, we already possess the means to increase fuel efficiency for gasoline-powered vehicles by using lightweight materials, continuously variable transmissions, and more-efficient gasoline engines.

Automobile fuel efficiency is a key to conservation

Among the measures enacted by the U.S. government in response to the OPEC embargo of 1973–1974 were a mandated increase in the mile-per-gallon (mpg) fuel efficiency of automobiles and a reduction in the national speed limit to 55 miles per hour. Over the next three decades, however, many of the conservation initiatives that followed the 1973–1974 oil crisis were abandoned. Without high market prices and an immediate threat of shortages, people lacked economic motivation to conserve. Government funding for research into alternative energy sources dwindled, speed limits rose, and U.S. policymakers repeatedly failed to raise the *corporate average fuel efficiency (CAFE) standards*, which set benchmarks for auto manufacturers to meet. The average fuel efficiency of new vehicles fell from 22.0 mpg in 1987 to 19.3 mpg in 2004 (as sales of sport-utility vehicles increased relative to sales of cars).

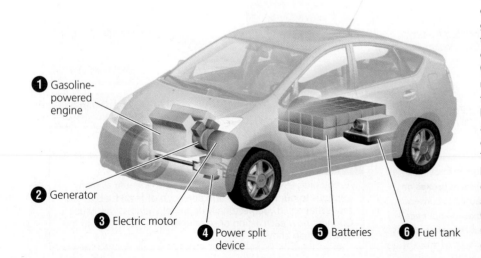

1 Gasoline-powered engine

2 Generator

3 Electric motor

4 Power split device

5 Batteries

6 Fuel tank

FIGURE 15.16 ◄ A hybrid car, such as the Toyota Prius diagrammed here, uses a small, clean, and efficient gasoline-powered engine **1** to produce power that the generator **2** can convert to electricity to drive the electric motor **3**. The power split device **4** integrates the engine, generator, and motor, serving as a continuously variable transmission. The car automatically switches between all-electrical power, all-gas power, and a mix of the two, depending on the demands being placed on the engine. Typically, the motor provides power for low-speed city driving and adds extra power on hills. The motor and generator charge a pack of nickel-metal-hydride batteries **5**, which can in turn supply power to the motor. Energy for the engine comes from gasoline carried in a typical fuel tank **6**.

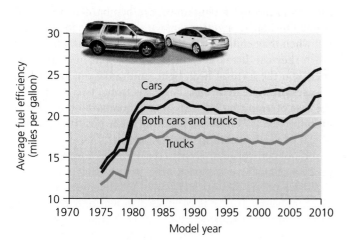

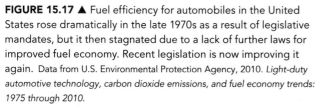

FIGURE 15.17 ▲ Fuel efficiency for automobiles in the United States rose dramatically in the late 1970s as a result of legislative mandates, but it then stagnated due to a lack of further laws for improved fuel economy. Recent legislation is now improving it again. Data from U.S. Environmental Protection Agency, 2010. *Light-duty automotive technology, carbon dioxide emissions, and fuel economy trends: 1975 through 2010.*

Since then, however, fuel economy has climbed up to 22.5 mpg in 2010 (**FIGURE 15.17**). Much of this recent rise in fuel efficiency occurred after Congress passed legislation in 2007 mandating that automakers raise average fuel efficiency to 35 mpg by the year 2020. This was a substantial advance, yet even after this boost, American automobiles will still lag behind the vehicles of most other developed nations. The fuel efficiency of European and Japanese cars is nearly twice that of U.S. cars and is slated to keep improving.

The United States has also kept its taxes on gasoline extremely low, relative to other nations. Americans pay two to three times *less* per gallon of gas than drivers in many European countries, for example. As a result, U.S. gasoline prices do not account for the substantial external costs (p. 92) that oil production and consumption impose on society. The low prices also diminish our economic incentive to conserve.

WEIGHING THE ISSUES

More Miles, Less Gas If you drive an automobile, what gas mileage does it get? How does it compare to the vehicle averages in Figure 15.17? If your vehicle's fuel efficiency were 10 mpg greater, and if you drove the same amount, how many gallons of gasoline would you no longer need to purchase each year? How much money would you save?

Do you think the U.S. government should mandate further increases in the CAFE standards? Should the government raise taxes on gasoline sales to encourage consumers to conserve energy? What effects (on economics, on health, and on environmental quality, for instance) might these steps have?

In 2009, Congress and the Obama administration sought to improve automobile fuel efficiency while stimulating economic activity and saving jobs during a severe recession. The popular "Cash for Clunkers" program—formally named the Consumer Assistance to Recycle and Save (CARS) Act—paid Americans $3,500 or $4,500 each to turn in old vehicles and purchase newer, more fuel-efficient ones. The $3 billion program subsidized the sale or lease of 678,000 vehicles averaging 24.9 mpg that replaced vehicles averaging 15.8 mpg. It is estimated that 824 million gallons of gasoline will be saved as a result, preventing 9 million metric tons of greenhouse gas emissions and creating social benefits worth $278 million.

The rebound effect cuts into efficiency gains

Energy efficiency is a vital pursuit, but it may not always save as much energy as we expect. This is because gains in efficiency from better technology can be partly offset if people engage in more energy-consuming behavior as a result. For instance, a person who buys a fuel-efficient car may choose to drive more because he or she feels it's OK to do so now that less gas is being used per mile. This phenomenon is called the "rebound effect," and studies indicate that it is widespread. In some instances, the rebound effect may completely erase efficiency gains, and attempts at energy efficiency may end up actually causing greater energy consumption! As our society pursues energy efficiency in more and more ways, this will be an important factor to consider.

Nonetheless, efficiency will play a necessary role in the conservation efforts we make toward reducing energy use. It is often said that reducing our energy use is equivalent to finding a new oil reserve. Some estimates hold that effective conservation and efficiency in the United States could save 6 million barrels of oil a day—nearly the amount we gain from all offshore drilling—while also reducing the negative impacts of fossil fuel extraction and use. Indeed, conserving energy is better than finding a new reserve, because it lessens health and environmental impacts while extending our access to fossil fuels. However, regardless of how much we conserve, we will still need energy. Among the alternatives to fossil fuels for our energy economy is nuclear power.

NUCLEAR POWER

Nuclear power occupies an odd and conflicted position in our modern debate over energy. Free of the air pollution produced by fossil fuel combustion, it has long been put forth as an environmentally friendly alternative to fossil fuels. Yet nuclear power's great promise has been clouded by nuclear weaponry, the dilemma of radioactive waste disposal, and the long shadow of Chernobyl and other power plant accidents. As a result, public safety concerns and the costs of addressing them have constrained nuclear power's spread.

First developed commercially in the 1950s, nuclear power experienced most of its growth during the 1970s and 1980s. The United States generates the most electricity from nuclear power—over a quarter of the world's production—yet only 20% of U.S. electricity comes from nuclear power. A

TABLE 15.6 Top Producers of Nuclear Power			
Nation	Nuclear power produced*	Number of reactors†	Percentage of electricity from nuclear power†
United States	807.1	104	19.6
France	410.1	58	74.1
Japan	280.3	50	29.2
Russia	159.4	32	17.1
South Korea	141.9	21	32.2
Germany	133.0	17	28.4
Canada	85.5	18	15.1
Ukraine	84.0	15	48.1
China	71.0	14	1.8
Spain	59.3	8	20.1

*In gigawatt-hours, 2011 data, from the World Nuclear Association.
†2010 data, from the International Atomic Energy Agency.

number of other nations rely more heavily on nuclear power (**TABLE 15.6**).

Fission releases nuclear energy in reactors to generate electricity

Strictly defined, **nuclear energy** is the energy that holds together protons and neutrons (p. 26) within the nucleus of an atom. We harness this energy by converting it to thermal energy, which can then be used to generate electricity. The reaction that drives the release of nuclear energy in power plants is **nuclear fission**, the splitting apart of atomic nuclei (**FIGURE 15.18**). In fission, the nuclei of large, heavy atoms,

such as uranium or plutonium, are bombarded with neutrons. Ordinarily, neutrons move too quickly to split nuclei when they collide with them, but if neutrons are slowed down they can break apart nuclei. Each split nucleus emits energy in the form of heat, light, and radiation, and it also releases multiple neutrons. These neutrons (two to three in the case of uranium-235) can in turn bombard other nearby uranium-235 (^{235}U) atoms, resulting in the positive feedback (p. 22) of a self-sustaining chain reaction.

If not controlled, this chain reaction becomes a runaway process of positive feedback—the process that creates the explosive power of a nuclear bomb. Inside a nuclear power plant, however, fission is controlled so that only one of the two or three neutrons emitted with each fission event goes on to induce another fission event. In this way, the chain reaction maintains a constant output of energy at a controlled rate.

For fission to begin in a nuclear reactor, the neutrons bombarding uranium are slowed down with a substance called a *moderator*, most often water or graphite. As fission proceeds, it becomes necessary to soak up the excess neutrons produced when uranium nuclei divide, so that on average only a single neutron from each nucleus goes on to split another nucleus. For this purpose, *control rods*, made of a metallic alloy that absorbs neutrons, are placed into the reactor among the water-bathed fuel rods of uranium. Engineers move these control rods into and out of the water to maintain the fission reaction at the desired rate. All this takes place within the reactor core and is the first step in the electricity-generating process of a nuclear power plant (**FIGURE 15.19**).

Nuclear energy comes from processed and enriched uranium

We use the element uranium for nuclear power because its atoms are radioactive, emitting subatomic particles and high-energy radiation as they decay into a series of daughter isotopes (p. 26). We obtain uranium from various minerals in naturally occurring uranium ore (*ore* is rock that contains minerals of economic interest [p. 236]). Uranium-containing minerals are uncommon, and uranium ore is in finite supply, so nuclear power is generally considered a nonrenewable energy source.

In the uranium ore we mine from the ground, over 99% of the uranium occurs as the isotope uranium-238, whereas less than 1% is uranium-235 (which has three fewer neutrons). Because ^{238}U does not emit enough neutrons to maintain a chain reaction when fissioned, we use ^{235}U for commercial nuclear power. As a result, we must process the ore we mine to enrich the concentration of ^{235}U to at least 3%. The enriched uranium is formed into pellets of uranium dioxide (UO_2), which are incorporated into the fuel rods used in reactors.

After several years in a reactor, enough uranium has decayed so that the fuel no longer generates adequate energy, and it must be replaced with new fuel. In some countries, the spent fuel is reprocessed to recover the remaining usable energy. However, this is costly relative to the low prices of uranium on the world market in recent years, so

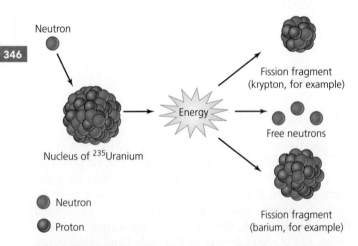

Neutron

Fission fragment (krypton, for example)

Energy

Free neutrons

Nucleus of ^{235}Uranium

● Neutron

● Proton

Fission fragment (barium, for example)

FIGURE 15.18 ▲ In nuclear fission, the nucleus of an atom of uranium-235 is bombarded with a neutron. The collision splits the uranium atom into smaller atoms and releases two or three neutrons, along with energy in the form of heat, light, and radiation. The neutrons can continue to split uranium atoms and set in motion a runaway chain reaction, so engineers at nuclear plants must absorb excess neutrons with control rods to regulate the rate of the reaction.

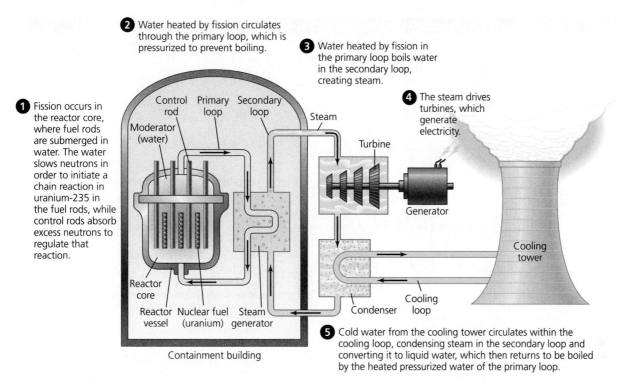

❷ Water heated by fission circulates through the primary loop, which is pressurized to prevent boiling.

❸ Water heated by fission in the primary loop boils water in the secondary loop, creating steam.

❹ The steam drives turbines, which generate electricity.

❶ Fission occurs in the reactor core, where fuel rods are submerged in water. The water slows neutrons in order to initiate a chain reaction in uranium-235 in the fuel rods, while control rods absorb excess neutrons to regulate that reaction.

Control rod   Primary loop   Secondary loop
Moderator (water)
Steam
Turbine
Generator
Cooling tower

Reactor core
Reactor vessel   Nuclear fuel (uranium)   Steam generator
Condenser   Cooling loop

Containment building

❺ Cold water from the cooling tower circulates within the cooling loop, condensing steam in the secondary loop and converting it to liquid water, which then returns to be boiled by the heated pressurized water of the primary loop.

FIGURE 15.19 ▲ In a pressurized light water reactor, the most common type of nuclear reactor, uranium fuel rods are placed in water, which slows neutrons so that fission can occur ❶. Control rods that can be moved into and out of the reactor core absorb excess neutrons to regulate the chain reaction. Water heated by fission circulates through the primary loop ❷ and warms water in the secondary loop, which turns to steam ❸. Steam drives turbines, which generate electricity ❹. The steam is then cooled in the cooling tower by water from an adjacent river or lake and returns to the containment building ❺, to be heated again by heat from the primary loop.

most spent fuel has been disposed of as radioactive waste (p. 392).

Nuclear power delivers energy more cleanly than fossil fuels

Using fission, nuclear power plants generate electricity without creating the air pollution from stack emissions that fossil fuels do. After considering all the steps involved in building plants and generating power, researchers with the International Atomic Energy Agency (IAEA) have calculated that nuclear power releases 4–150 times fewer emissions than fossil fuel combustion. IAEA scientists estimate that at current global levels of use, nuclear power helps us avoid emitting 600 million metric tons of carbon each year, equivalent to 7% of global greenhouse gas emissions.

Nuclear power has additional advantages over fossil fuels—coal, in particular. For residents living downwind from power plants, scientists calculate that nuclear power poses far fewer chronic health risks from pollution than does fossil fuel combustion. And because uranium generates far more power than coal by weight or volume, less of it needs to be mined, so uranium mining causes less damage to landscapes and generates less solid waste than coal mining. Moreover, in the course of normal operation, nuclear power plants are safer for workers than coal-fired plants.

Nuclear power also has drawbacks. One is that the waste it produces is radioactive, and arranging for safe disposal of

this waste is challenging. The second main drawback is that if an accident occurs at a power plant, or if a plant is sabotaged, the consequences can potentially be catastrophic.

Given this mix of advantages and disadvantages (**FIGURE 15.20**), most governments (although not necessarily most citizens) have judged the good to outweigh the bad, and today the world has 441 operating nuclear plants in 30 nations.

WEIGHING THE ISSUES

Choose Your Risk Consult Figure 15.20 on the next page. Given the choice of living next to a nuclear power plant or living next to a coal-fired power plant, which would you choose? What would concern you most about each option?

Nuclear power poses small risks of large accidents

Although nuclear power delivers energy more cleanly than fossil fuels, the possibility of catastrophic accidents has spawned a great deal of public anxiety over nuclear power. Three events have been most influential in shaping public opinion about nuclear energy: Three Mile Island in the United States; Chernobyl, the world's most severe accident; and most recently, Fukushima Daiichi following the 2011 Japanese earthquake and tsunami.

Environmental Impacts of Coal-fired and Nuclear Power		
Type of Impact	**Coal**	**Nuclear**
Land and ecosystem disturbance from mining	Extensive, on surface or underground	Less extensive
Greenhouse gas emissions	Considerable emissions	None from plant operation; much less than coal over the entire life cycle
Other air pollutants	Sulfur dioxide, nitrogen oxides, particulate matter, and other pollutants	No pollutant emissions
Radioactive emissions	No appreciable emissions	Possibility of dangerous emissions if severe accident occurs
Occupational health among workers	More known health problems and fatalities	Fewer known health problems and fatalities
Health impacts on nearby residents	Air pollution impairs health	No appreciable known health impacts under normal operation
Effects of accident or sabotage	No widespread effects	Potentially catastrophic widespread effects
Solid waste	More generated	Less generated
Radioactive waste	None	Radioactive waste generated
Fuel supplies remaining	Should last several hundred more years	Uncertain; supplies could last longer or shorter than coal supplies

FIGURE 15.20 ▲ Coal-fired power plants and nuclear power plants pose very different risks and impacts to human health and the environment. This chart compares the major impacts of each mode of electricity generation. For each type of impact, a red box indicates the more severe impact.

Three Mile Island The first event took place at the **Three Mile Island** plant in Pennsylvania in 1979. Through a combination of mechanical failure and human error, coolant water drained from the reactor vessel, temperatures rose inside the reactor core, and metal surrounding the uranium fuel rods began to melt, releasing radiation. This process is termed a **meltdown**, and at Three Mile Island it proceeded through half of one reactor core. Area residents stood ready to be evacuated as the nation held its breath, but fortunately most radiation remained trapped inside the containment building.

Once this accident was brought under control, the damaged reactor was shut down, and multi-billion-dollar cleanup efforts stretched on for years. Three Mile Island is best regarded as a near miss; the emergency could have been far worse had the meltdown proceeded through the entire stock of uranium fuel or had the containment building not contained the radiation.

Chernobyl In 1986, an explosion at the **Chernobyl** plant in Ukraine (part of the Soviet Union at the time) caused the most severe nuclear power plant accident the world has seen. Engineers had turned off safety systems to conduct tests, and human error, combined with unsafe reactor design, led to explosions that destroyed the reactor and sent clouds of radioactive debris billowing into the atmosphere. Atmospheric currents carried radioactive fallout across much of

the Northern Hemisphere, particularly Ukraine, Belarus, and parts of Russia and Europe. For 10 days, radiation escaped from the plant while emergency crews risked their lives putting out fires. Most residents of the surrounding countryside remained at home for these 10 days, exposed to radiation, before the Soviet government belatedly began evacuating more than 100,000 people.

In the months and years afterwards, workers removed irradiated materials, scrubbed buildings and roads, and erected a gigantic concrete sarcophagus around the demolished reactor (**FIGURE 15.21**). However, the landscape for at least 30 km (19 mi) around the plant remains contaminated, the demolished reactor is still full of dangerous fuel and debris, and radioactivity leaks from the hastily built and quickly deteriorating sarcophagus. Today an international team is trying to build a larger sarcophagus around the original one to prevent a re-release of radiation.

The accident at Chernobyl killed 31 people directly and sickened or caused cancer in thousands more. Exact numbers are uncertain because of inadequate data and the difficulty of determining long-term radiation effects. Health authorities estimate that most of the over 5,000 cases of thyroid cancer diagnosed in people who were children at the time resulted from radioactive iodine spread by the accident. Estimates for the total number of cancer cases attributable to Chernobyl, past and future, vary widely, but an international consensus effort 20 years after the event estimated that radiation raised

FIGURE 15.21 ▲ The world's worst nuclear power plant accident unfolded in 1986 at Chernobyl, in present-day Ukraine (then part of the Soviet Union). As part of the extensive cleanup operation, the destroyed reactor was encased in a massive concrete sarcophagus to contain further radiation leakage.

the cancer rate among exposed people by up to a few percent, resulting in up to several thousand fatal cancer cases.

Fukushima Daiichi On March 11, 2011, a magnitude 9.0 earthquake struck eastern Japan and sent an immense tsunami roaring onshore (pp. 232, 234). Over 23,000 people were killed and many thousands of buildings were destroyed. This natural disaster affected the operation of several of Japan's nuclear plants, most notably the Fukushima Daiichi nuclear power plant. Here, the earthquake shut down power and the tsunami flooded the plant's emergency power generators. Without electricity, workers could not use moderators and control rods to cool the uranium fuel, and the fuel began to overheat as fission proceeded, uncontrolled.

Amid the damage and chaos across the region, help was slow to arrive, and workers had to begin flooding the reactors with seawater in a desperate effort to prevent meltdowns. Several explosions and fires occurred over the next few days, and eventually three reactors experienced full meltdowns, while the plant's other three reactors were seriously damaged. Parts of the plant remained inaccessible for months because of radioactive water, and it is estimated that it will require years or decades to fully clean up the site.

Radioactivity was released during and after these events at levels lower than but comparable to those from Chernobyl. Much of it spread by air or water into the Pacific Ocean, and trace amounts were detected around the world. Thousands of residents of areas near the plant were evacuated and screened for radiation (**FIGURE 15.22**), while restrictions were placed on food and water from the region. At the time of this writing,

releases of radioactivity continued, and long-term health effects on the area's people remain uncertain.

The disaster at Fukushima Daiichi could probably have been avoided had the emergency generators not been located in the basement where a tsunami could flood them. And the design of most modern reactors is safer than Chernobyl's. Yet natural disasters and human error will always pose risks—and as plants age, they require more maintenance and become less safe. Moreover, radioactive material could be stolen from plants and used in terrorist attacks. This possibility is especially worrisome in the cash-strapped nations of the former Soviet Union, where hundreds of former nuclear sites have gone without adequate security for years. In a cooperative international agreement, the U.S. government has been buying up some of this material and diverting it to peaceful use in power generation.

Waste disposal remains a problem

Even if nuclear power generation could be made completely safe, we would still be left with the conundrum of what to do with spent fuel rods and other radioactive waste, which will continue emitting radiation for thousands of years. Currently, such waste is held in temporary storage at nuclear power plants. Spent fuel rods are sunken in pools of cooling water or encased in thick casks of steel, lead, and concrete to minimize radiation leakage.

In total, U.S. power plants are storing over 60,000 metric tons of high-level radioactive waste—enough to fill a football field to the depth of 6 m (20 ft)—as well as much more low-level radioactive waste. This waste is held at more than 120 sites spread across 39 states (**FIGURE 15.23**). A 2005 National Academy of Sciences report judged that most of these sites were vulnerable to terrorist attacks. Over 161 million U.S. citizens live within 125 km (75 mi) of temporarily stored waste.

FIGURE 15.22 ▼ A Japanese child evacuated from the Fukushima area is screened for radiation two weeks after the nuclear disaster that followed Japan's devastating earthquake and tsunami. Radiation exposure can have serious long-term health effects.

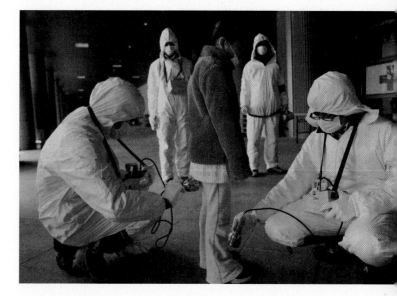

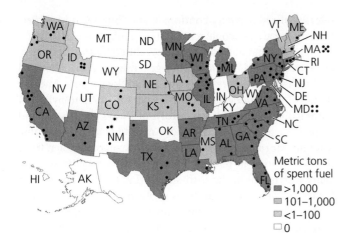

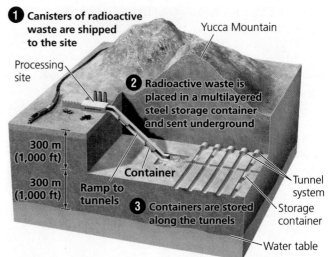

FIGURE 15.23 ▲ High-level radioactive waste from civilian reactors is currently stored at over 120 sites in 39 states across the United States. In this map, dots indicate storage sites, and the four shades of color indicate the total amount of waste stored in each state. Slightly different classifications of waste mean that some states shaded white show storage sites for certain types of waste. Data from Office of Civilian Radioactive Waste Management, U.S. Department of Energy; and Nuclear Energy Institute, Washington, D.C.

FIGURE 15.24 ▲ Yucca Mountain, in a remote part of Nevada, was being developed as the central repository site for all commercial nuclear waste in the United States until support was withdrawn in 2010. Waste was to be buried in a network of tunnels deep underground, yet still high above the water table.

Because storing waste at many dispersed sites creates a large number of potential hazards, nuclear waste managers would prefer to send all waste to a central repository that can be heavily guarded. In the United States, the multiyear search homed in on Yucca Mountain, a remote site in the desert of southern Nevada, 160 km (100 mi) from Las Vegas. Choice of this site followed extensive study by government scientists, and $13 billion was spent on its development, although Nevadans were not happy about the choice. In 2010 as the site was awaiting approval from the Nuclear Regulatory Commission, President Barack Obama's administration ended support for the project. However, some Congresspeople, agencies, and state governments are challenging this decision, and as of 2011 the issue remains unresolved. Without Yucca Mountain, the United States has no place designated to dispose of its radioactive waste from commercial nuclear power plants, so this waste will remain at its numerous current locations across the country.

At Yucca Mountain, waste would be stored in a network of tunnels 300 m (1,000 ft) underground, yet 300 m (1,000 ft) above the water table (**FIGURE 15.24**). Scientists and policymakers chose the Yucca Mountain site because they determined that it is remote and unpopulated, has minimal risk of earthquakes, receives little rain that could cause radioactivity to percolate down into the groundwater, has a deep water table atop an isolated aquifer, and is on federal land that can be protected from sabotage. However, some scientists, antinuclear activists, and concerned Nevadans have challenged these conclusions.

Another concern with any centralized repository is that waste would need to be transported there from the 120-plus current storage areas and from current and future nuclear plants and military installations. Because this would involve many thousands of shipments by rail and truck across hundreds of public highways through almost every state of the union, some people worry that the risk of an accident or of sabotage is unacceptably high.

THE ISSUES

How to Store Waste? Which do you think is a better option—to transport nuclear waste cross-country to a single repository or to store it permanently at numerous power plants and military bases scattered across the nation? Would your opinion be affected if you lived near the repository site? Near a power plant? On a highway route along which waste is transported?

Multiple dilemmas have slowed nuclear power's growth

Dogged by concerns over waste disposal, safety, and expensive cost overruns, nuclear power's growth has slowed. Since the late 1980s, nuclear power has grown by 2.5% per year worldwide, about the same rate as electricity generation overall. Public anxiety in the wake of Chernobyl made utilities less willing to invest in new plants. So did the enormous expense of building, maintaining, operating, and ensuring the safety of nuclear facilities. Almost every nuclear plant has turned out to be more expensive than expected. In addition, plants have aged more quickly than expected because of problems that were underestimated, such as corrosion in coolant pipes. The plants that have been shut down—well over 100 around the world to date—have served on average less than half their expected lifetimes. Moreover, shutting down, or decommissioning, a plant can sometimes be more expensive than the original construction.

As a result of these economic issues, electricity from nuclear power today remains more expensive than electricity from coal and other sources. Governments are still subsidizing nuclear power to keep electricity costs to ratepayers down, but many private investors lost interest long ago. Nonetheless, nuclear power remains one of the few currently viable alternatives to fossil fuels with which we can generate large amounts of electricity in short order. This is why an increasing number of environmental advocates propose expanding U.S. nuclear

capacity using a new generation of reactors designed to be safer and less expensive. Indeed, nuclear power was beginning to experience a bit of a renaissance before the Fukushima tragedy raised new concerns.

With slow growth expected for nuclear power, fossil fuels in limited supply, an oil production peak looming, and climate change worsening, where will our growing human population turn for clean and sustainable energy? People increasingly are turning to renewable sources of energy (Chapter 16): energy sources that cannot be depleted by our use.

➤ CONCLUSION

Over the past 200 years, fossil fuels have helped us build the complex industrialized societies we enjoy today. However, we are now approaching a turning point in history: Our production of fossil fuels will begin to decline. We can respond to this new challenge by encouraging conservation and developing alternative energy sources. Or we can continue our current dependence on fossil fuels and wait until they near depletion before we try to develop new technologies and ways of life. The path we choose will have far-reaching consequences

for human health and well-being, for Earth's climate, for our environment, and for the stability and progress of our civilization.

Nuclear power showed promise to be a pollution-free and highly efficient form of energy. However, high costs and public fears over safety in the wake of accidents have stalled its growth. Nuclear power will likely be part of our future energy economy, but we will need to turn to renewable energy sources as well.

TESTING YOUR COMPREHENSION

1. Why are fossil fuels our most prevalent source of energy today? Why are they considered nonrenewable sources of energy? How are fossil fuels formed?

2. Describe how *net energy* differs from *energy returned on investment (EROI)*. Why are these concepts important when evaluating energy sources?

3. Describe how coal is used to generate electricity. Now, describe how we create petroleum products. Provide examples of several of these products.

4. Why do many experts think we are about to pass the global production peak for oil? What consequences might there be for our society if we do not shift soon to renewable energy sources?

5. Describe three environmental impacts of fossil fuel production and consumption. What impacts have resulted from drilling for oil offshore in the Gulf of Mexico?

6. Give an example of clean coal technology. Now describe how carbon capture and storage is intended to work.

7. Describe one specific example of how technological advances can improve energy efficiency. Now describe one specific action you could take to conserve energy.

8. Describe how nuclear fission works. How do nuclear plant engineers control fission and prevent a runaway chain reaction?

9. In terms of greenhouse gas emissions, how does nuclear power compare to coal, oil, and natural gas?

10. In what ways did the events at Three Mile Island, Chernobyl, and Fukushima Daiichi differ from one another? What consequences resulted from each of these incidents? Now list several concerns about the disposal of radioactive waste. What has been done so far about disposing of radioactive waste?

SEEKING SOLUTIONS

1. What impacts might you expect on your lifestyle once our society arrives at peak oil? What lessons do you think we can take from the conservation methods adopted by the United States in response to the "energy crisis" of 1973–1974? What steps do you think we should take to avoid energy shortages in a post-peak-oil future?

2. Describe and compare the environmental and social impacts of coal and oil extraction and consumption. What

steps could governments, industries, and individuals take to reduce these impacts?

3. Nuclear power has by now been widely used for over four decades, and the world has experienced only two major accidents (Chernobyl and Fukushima Daiichi) responsible for any significant number of injuries or deaths. Would you call this a good safety record? Should we maintain, decrease, or increase our reliance on nuclear

power? Why might safety at nuclear power plants be better in the future? Why might it be worse?

4. **THINK IT THROUGH** You have been elected governor of the state of Florida as the federal government is debating what waters to open to offshore drilling for oil and natural gas. Drilling in Florida waters would create jobs for Florida citizens as well as revenue for the state in the form of royalty payments from oil companies. However, there is always the risk of a catastrophic oil spill, with its ecological, social, and economic impacts. Would you support or oppose offshore drilling off the Florida coastline? Why? What questions would you ask of scientists before making your decision? What factors would you consider in making your decision?

5. **THINK IT THROUGH** You are the head of the national department of energy in a country that has just experienced a minor accident at one of its nuclear plants. A partial meltdown released radiation, but the radiation was fully contained inside the containment building, and there were no health impacts on area residents. However, citizens are terrified, and the media is stressing the dangers of nuclear power. Your country relies on its five nuclear plants for 25% of its energy and 50% of its electricity needs. It has no fossil fuel deposits and recently began a promising but still-young program to develop renewable energy options. What will you tell the public at your next press conference, and what policy steps will you recommend taking to ensure a safe and reliable national energy supply?

CALCULATING ECOLOGICAL FOOTPRINTS

Scientists at the Global Footprint Network calculate the energy component of our ecological footprint by estimating the amount of ecologically productive land and sea required to absorb the carbon released from fossil fuel combustion. This translates into nearly 5.6 ha of the average American's 8.0-ha ecological footprint. Another way to think about our footprint, however, is to estimate how much land would be needed to grow biomass with an energy content equal to that of the fossil fuel we burn.

Assume that you are an average American who burns about 6.7 metric tons of oil-equivalent in fossil fuels each year,

and that average terrestrial net primary productivity (p. 32) can be expressed as 0.0037 metric tons/ha/year. Calculate how many hectares of land it would take to supply our fuel use by present-day photosynthetic production.

	Hectares of land for fuel production
You	1,811
Your class	
Your state	
United States	

1. Compare the energy component of your ecological footprint calculated in this way with the 5.6 ha calculated using the method of the Global Footprint Network. Explain why results from the two methods may differ.

2. Earth's total land area is approximately 15 billion hectares. Compare this to the hectares of land for fuel production from the table.

3. In the absence of stored energy from fossil fuels, how large a human population could Earth support at the level of consumption of the average American, if all of Earth's area were devoted to fuel production? Do you consider this realistic? Provide two reasons why or why not.

Mastering**ENVIRONMENTALSCIENCE**™

Go to **www.masteringenvironmentalscience.com** for homework assignments, practice quizzes, Pearson eText, and more.

16 Renewable Energy Alternatives

Upon completing this chapter, you will be able to:

➤ Discuss the reasons for seeking alternatives to fossil fuels

➤ Outline the major sources of renewable energy and assess their potential for growth

➤ Describe the major sources, scale, and impacts of bioenergy

➤ Describe the scale, methods, and impacts of hydroelectric power

➤ Describe solar energy and the ways it is harnessed, and evaluate its advantages and disadvantages

➤ Describe wind power and how we harness it, and evaluate its benefits and drawbacks

➤ Describe geothermal energy and the ways we make use of it, and assess its advantages and disadvantages

➤ Describe ocean energy sources and how we could harness them

➤ Explain hydrogen fuel cells and weigh options for energy storage and transportation

Homes in the Vauban neighborhood of Freiburg, Germany, which produce more solar power than they use, and sell it to the grid

Germany Goes Solar

"Someday we will harness the rise and fall of the tides and imprison the rays of the sun."
—Thomas A. Edison, 1921

"[Renewable energy] will provide millions of new jobs. It will halt global warming. It will create a more fair and just world. It will clean our environment and make our lives healthier."
—Hermann Scheer, energy expert and German parliament member, 2009

When we think of solar energy, most of us envision a warm sunny place like Arizona or southern California. Yet the country that produces the most solar power is Germany, a European nation as far north as Canada with a climate like Maine's. Germany is the world's top user of photovoltaic (PV) solar technology, which produces electricity from sunshine. In recent years Germany has installed half the world's total of this technology. Germany now obtains more of its energy from solar power than any other nation, and the amount grows each year.

How is this happening in such a cool and cloudy country? A bold federal policy is offering economic incentives to businesses and homeowners to promote solar power and other forms of renewable energy. Germany has a **feed-in tariff** system whereby utilities are mandated to buy power from anyone who can generate power from renewable energy sources and feed it into the electric grid. Under this

system, utilities must pay guaranteed premium prices for this power under long-term contract. As a result, German homeowners and businesses have rushed to install more and more PV panels each year, and are selling their extra solar power to the utilities at a profit.

The feed-in tariffs apply to all forms of renewable energy. As a result, Germany ranks third in the world in electric power capacity from renewable sources, trailing only China and the United States, which have far more people and businesses. Germany aims to obtain 30% of its electricity and 14% of its heating energy from renewable sources by 2020. To make this happen, the German government has been allotting more public money to renewable energy than any

other nation—over $25 billion annually in recent years.

Boosted by domestic demand, German industries have become global leaders in "green tech," designing and selling renewable energy technologies around the world. Germany is second in PV production behind China, leads the world in production of biodiesel, and has recently developed several cellulosic ethanol facilities. Renewable energy industries in Germany today employ over 300,000 citizens.

Germany's push for renewable energy dates back to 1990. The government had decided to phase out its nuclear power plants because of safety concerns, yet by shutting these down, the nation would lose virtually all its clean energy. With few domestic fossil fuel supplies, Germans would find their economy utterly dependent on oil, gas, and coal imported from Russia and the Middle East.

Enter Hermann Scheer, a member of the German parliament and an expert on renewable energy. While everyone else assumed that technologies for harnessing solar, wind, and geothermal energy were costly, risky, and not ready for prime time, Scheer saw them as a great economic opportunity—and as the only long-term answer. In 1990, Scheer helped push through a landmark law establishing feed-in tariffs. Ten years later, the law was revised and strengthened: The Renewable Energy Sources Act of 2000 aimed to promote renewable energy production and use, enhance the security of the energy supply, reduce carbon emissions, and lessen the many external costs (p. 92) of fossil fuel use.

Under the law, each renewable source is assigned its own payment rate according to market considerations, and most rates are reduced year by year in order to encourage increasingly efficient means of producing power. In 2004 and in 2009, the government adjusted the amounts utilities were required to pay homeowners and businesses for their energy production. Then in 2010, the German government slashed PV solar tariff rates by 16% in order to reduce the cost of the subsidies to taxpayers and because PV market prices had already fallen by half. In response, sales of PV modules skyrocketed as Germans rushed to lock in the old rates. In 2010 alone, Germans installed 7 gigawatts of PV solar capacity—over 2.5 times the total cumulative capacity of the United States.

By replacing some of its fossil fuel use with renewable energy, Germany has reduced its emissions of carbon dioxide by 140 million tons per year—equal to taking 24 million cars off the road. Half of this total is due to energy paid for under the feed-in tariff system. Since 1990, carbon dioxide emissions from German energy sources have fallen by over 20%, and emissions of seven other major pollutants (CH_4, N_2O, SO_2, NO_X, CO, VOCs, and dust) have been reduced by 12–95%.

Germany's success is serving as a model for other nations. As of 2011, more than 60 nations had implemented some sort of feed-in tariffs. Spain and Italy ignited their wind and solar development as a result. In North America, Vermont and Ontario established feed-in tariff systems similar to Germany's, while California, Hawaii, Oregon, and Washington conduct more-limited programs. In 2010, Gainesville, Florida, became the first U.S. city to establish feed-in tariffs. Moreover, utilities in 46 U.S. states now offer **net metering**, in which utilities credit customers who produce renewable power and feed it into the grid. As more nations, states, and cities develop policies to encourage renewable energy, we may soon experience a historic transition in the way we meet our energy demands. ■

RENEWABLE ENERGY SOURCES

Germany's bold federal policy is just one facet of a global shift toward renewable energy. Across the world, nations are searching for ways to move away from fossil fuels while ensuring a reliable and affordable supply of energy for their economies.

Renewable sources are growing fast

Today's economies are powered largely by fossil fuels; 81% of our energy comes from oil, coal, and natural gas (**FIGURE 16.1A**). These three fuels also power two-thirds of the world's electricity generation (**FIGURE 16.1B**). Fossil fuels helped to drive the industrial revolution, increase our material prosperity, and create the society we enjoy today. However, these nonrenewable energy sources will not last forever. Easily extractable supplies of oil and natural gas will likely soon dwindle (Chapter 15). Moreover, our use of coal, oil, and natural gas imposes health and environmental impacts, social costs, and security risks (Chapters 14 and 15). For these reasons, most energy experts accept that the world's economies

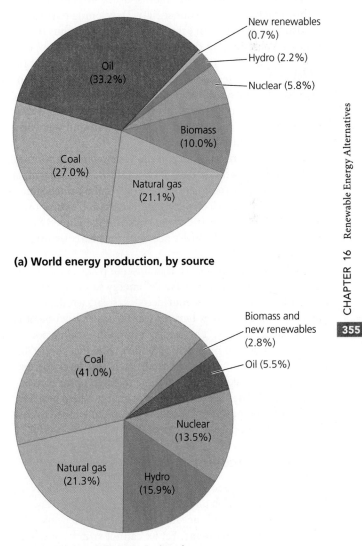

(a) World energy production, by source

(b) World electricity generation, by source

FIGURE 16.1 ▲ Fossil fuels account for 81% **(a)** of the world's energy production. Nuclear power and hydroelectric power contribute substantially to global electricity generation **(b)**, but fossil fuels still power two-thirds of our electricity. Data are for 2008, from International Energy Agency, 2010. *Key world energy statistics 2010.* Paris: IEA.

will need to shift from fossil fuels to energy sources that are less easily depleted and are gentler on our environment and health.

Our scientists, engineers, and entrepreneurs have developed a range of alternatives to fossil fuels. The main non-renewable alternative is nuclear energy (pp. 345–350). Renewable alternatives include biomass, hydropower, solar, wind, geothermal, and ocean energy sources. Biomass and hydropower are well-established and widely used sources. The other renewable sources are often termed "new renewables" because they are not yet widely used and they are harnessed using technologies still in a rapid phase of development. These sources can provide energy for three types of applications: (1) power for electricity, (2) heating of air or water, and (3) fuel for vehicles.

As renewable energy sources replace fossil fuels, they help alleviate air pollution (Chapter 13) and the greenhouse gas emissions that drive global climate change (Chapter 14). Unlike fossil fuels, many renewable sources are inexhaustible on time scales relevant to our society. Developing renewables can also help diversify an economy's energy mix, thus reducing price volatility and dependence on foreign fuel imports (p. 342). Finally, the design, installation, and management required to develop technologies and rebuild our society's energy infrastructure will be a major source of employment for young people today, through **green-collar jobs**. Over 3 million people work in renewable energy jobs around the world already, and the number is rising.

Nations and regions vary in the renewable sources they use. Developing nations account for most use of combustible renewables, or biomass, such as fuelwood. In the United States, most renewable energy comes from biomass and hydropower (**FIGURE 16.2A**). Of electricity generated in the United States from renewables, hydropower accounts for nearly two-thirds (**FIGURE 16.2B**).

Although they comprise a minuscule proportion of our energy budget, the "new renewable" energy sources are growing quickly. Over the past four decades, solar, wind, and geothermal energy sources have grown far faster than has the overall energy supply. The leader in growth is wind power, which has expanded by nearly 50% *each year* since the 1970s. Because these sources started from such low levels of use, however, it will take them some time to catch up to conventional sources. The absolute amount of energy added by a 50% increase in wind power today equals the amount added by just a 1% increase in oil, coal, or natural gas!

Policy can accelerate our transition

Rapid growth in renewable energy sectors seems likely to continue as population and consumption grow, global energy demand expands, fossil fuel supplies decline, and people demand cleaner environments. Yet we cannot switch completely to renewable energy sources overnight, because there are technological and economic barriers. Currently, most renewables lack adequate technology and infrastructure to transfer power on the required scale.

Rapid technological advances in recent years, however, suggest that most remaining barriers are political. Renew-

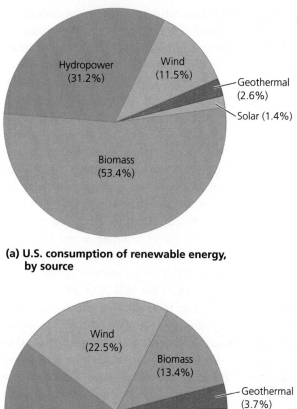

(a) U.S. consumption of renewable energy, by source

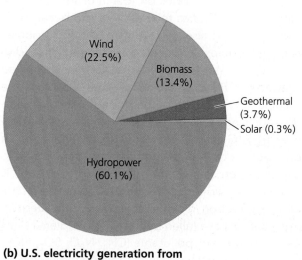

(b) U.S. electricity generation from renewable sources

FIGURE 16.2 ▲ Only 8% of the energy consumed in the United States each year comes from renewable sources. Of this amount **(a)**, most derives from biomass energy and hydropower. Wind power, geothermal energy, and solar energy together account for just 15% of this amount. Similarly, just 10.2% of electricity generated in the United States **(b)** comes from renewable energy sources, predominantly hydropower. Data are for 2010, from Energy Information Administration, U.S. Department of Energy.

able energy sources have received far less in subsidies and tax breaks from governments than have conventional sources. In the United States over the past three decades, renewable energy sources have been granted just one-sixth the public funding for research and development that nuclear energy and fossil fuels have received (**FIGURE 16.3**). Research and development of renewable sources have gone underfunded because fossil fuels continue to be available at inexpensive prices, even though these low prices are enabled in part by government policy that responds to lobbying from fossil fuel interests.

Public policy steps can accelerate our societal transition to renewable energy. Germany's feed-in tariff policy provides

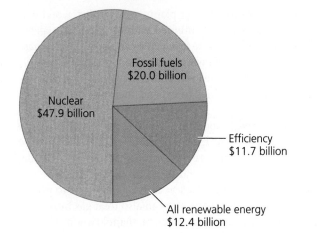

FIGURE 16.3 ▲ Most U.S. research and development funding for energy has gone toward nuclear power and fossil fuels. Between 1974 and 2005, only 13% went toward all renewable energy sources combined. Data from International Energy Agency.

a prime example of an economic policy tool (pp. 106–107) that can hasten the spread of renewable energy by creating financial incentives for businesses and individuals. Renewable energy efforts also received support when many national governments responded to the global financial downturn of 2008–2009 by enacting stimulus packages and boosting spending on green energy programs to help create jobs. As more governments, utilities, corporations, and consumers turn to renewable energy, prices of renewables should continue to fall, further hastening their adoption.

BIOENERGY

Bioenergy—also known as **biomass energy**—is energy obtained from biomass. **Biomass** (p. 69) consists of organic material derived from living or recently living organisms, and it contains chemical energy that originated with sunlight and photosynthesis. We harness bioenergy from many types of plant matter, including wood from trees, charcoal from wood charred in the absence of oxygen, and matter from agricultural crops, as well as from combustible animal waste products such as cattle manure.

The great attraction of bioenergy is that—in principle—it is renewable and releases no net carbon dioxide into the atmosphere. Although burning biomass emits plenty of carbon dioxide, this is balanced by the fact that photosynthesis had pulled this amount of carbon dioxide from the atmosphere to create the biomass just years, months, weeks, or days before. Therefore, in theory, when we replace fossil fuels with bioenergy, we reduce net carbon flux to the atmosphere, helping to alleviate global climate change (Chapter 14). However, in practice it is not so simple, and judging the sustainability of any given bioenergy strategy requires careful consideration of the type of biomass source we are using and the way we gain energy from it.

Bioenergy comes from diverse sources

To a poor farmer in Africa, bioenergy entails cutting wood from trees or collecting livestock manure and burning it to heat and cook for her family. To an industrialized farmer in Iowa, bioenergy means shipping his grain to a high-tech refinery that converts it to liquid fuel to run automobiles. The diversity of sources and approaches involved in bioenergy (**TABLE 16.1**) gives us many ways to address our energy challenges.

Over 1 billion people use wood from trees as their principal energy source. In developing nations, especially in rural areas, families gather fuelwood to burn in their homes for heating, cooking, and lighting (**FIGURE 16.4**). Although fossil fuels are replacing traditional energy sources as developing nations industrialize, fuelwood, charcoal, and manure still account for 35% of energy use in these nations, and up to 90% in the poorest nations.

Fuelwood and other traditional biomass sources constitute nearly 80% of all renewable energy used worldwide. However, biomass is renewable only if it is not overharvested. Harvesting fuelwood at unsustainably rapid rates will lead to deforestation, soil erosion, and desertification (pp. 188, 140), which can damage landscapes, diminish biodiversity, and impoverish human societies.

While much of the world still relies on fuelwood, charcoal, and manure, new bioenergy approaches are being developed using a variety of materials to provide innovative types of energy (see Table 16.1). Some of these materials are

TABLE 16.1 Major Bioenergy Sources
Direct combustion for heating
▶ Wood cut from trees (fuelwood)
▶ Charcoal
▶ Manure from farm animals
Biofuels for powering vehicles
▶ Corn grown for ethanol
▶ Bagasse (sugarcane residue) grown for ethanol
▶ Soybeans, rapeseed, and other crops grown for biodiesel
▶ Used cooking oil for biodiesel
▶ Plant matter treated with enzymes to produce cellulosic ethanol
▶ Algae grown for biofuels
Biopower for generating electricity
▶ Crop residues (such as cornstalks) burned at power plants
▶ Forestry residues (such as wood waste from logging) burned at power plants
▶ Processing wastes (such as waste from sawmills, pulp mills, and paper mills) burned at power plants
▶ "Landfill gas" burned at power plants
▶ Livestock waste from feedlots for gas from anaerobic digesters
▶ Organic components of municipal solid waste from landfills

FIGURE 16.4 ▲ Well over a billion people in developing countries rely on fuelwood for heating and cooking. Wood cut from trees remains the major source of biomass energy used in the world today. In theory, biomass is renewable, but in practice it may not be if forests are overharvested.

burned in power plants to produce **biopower**, generating heat or electricity. Other sources can be converted into **biofuels**, liquid fuels used primarily to power automobiles.

Biopower generates electricity from biomass

We harness biopower by combusting biomass to generate electricity. This can be done using a variety of sources and techniques.

Waste products   The waste products of various industries and processes may be used for biopower. These include woody debris from logging, liquid waste from pulp mills, organic waste from landfills or feedlots, and residue from crops (such as cornstalks and corn husks).

Bioenergy crops   We are beginning to grow certain types of plants as crops to generate biopower. These include fast-growing grasses such as bamboo, fescue, and switchgrass, as well as trees such as specially bred willows and poplars (**FIGURE 16.5**). Many of these plants are also being grown to produce liquid biofuels.

Combustion strategies   At small scales, farmers, ranchers, or villages can operate modular biopower systems that use livestock manure to generate electricity. Small household biodigesters provide portable and decentralized energy production for remote rural areas.

At large scales, power plants built to combust biomass operate like those fired by fossil fuels (see Figure 15.4, p. 330); combustion heats water, creating steam to turn turbines and generators, thereby generating electricity. Much of the biopower produced so far comes from power plants that use cogeneration (p. 344) to generate both electricity and heating. These plants are often located where they can take advantage of forestry waste.

In some coal-fired power plants, wood chips, wood pellets, or other biomass is combined with coal in a specialized boiler in a process called *co-firing*. We can substitute biomass for up to 15% of the coal with only minor equipment modification and no appreciable loss of efficiency. Co-firing is a relatively easy way for utilities to expand their use of renewable energy.

We also harness biopower through *gasification*, in which biomass is vaporized at high temperatures in the absence of oxygen, creating a mixture of hydrogen, carbon monoxide, carbon dioxide, methane, and other gases. This mixture can generate electricity when used to turn a gas turbine to propel a generator in a power plant. We can also treat gas from gasification in various ways to produce methanol (wood alcohol), synthesize a type of diesel fuel, or isolate hydrogen for use in hydrogen fuel cells (p. 375). An alternative method of heating biomass in the absence of oxygen results in *pyrolysis*, which produces a mix of solids, gases, and liquids. This includes a liquid fuel called pyrolysis oil, which can be burned to generate electricity.

Benefits and drawbacks   By enhancing energy efficiency and recycling waste products, biopower helps move our utilities and industries in a sustainable direction. Biopower also helps mitigate climate change by reducing carbon dioxide emissions, and capturing landfill gas reduces emissions of methane, a potent greenhouse gas. When biomass replaces coal in co-firing and direct combustion, biopower reduces emissions of sulfur dioxide because plant matter, unlike coal, contains no appreciable sulfur.

A disadvantage of biopower is that when we burn crops or plant matter for power, we deprive the soil of nutrients it would have gained from the plant matter's decomposition. We essentially draw fertility from the soil and never return it, so that the soil becomes progressively depleted.

Biofuels can power automobiles

Liquid fuels from biomass sources are powering millions of vehicles on today's roads. The two primary biofuels developed so far are ethanol (for gasoline engines) and biodiesel (for diesel engines).

FIGURE 16.5 ▼ Switchgrass, a fast-growing plant native to the North American prairies, provides fuel for biopower now and is being studied as a crop to provide cellulosic ethanol (p. 361).

Ethanol Ethanol is the alcohol in beer, wine, and liquor. It is produced as a biofuel by fermenting biomass, generally from carbohydrate-rich crops, in a process similar to brewing beer. In fermentation, carbohydrates are converted to sugars and then to ethanol. Spurred by the 1990 Clean Air Act amendments and generous government subsidies, ethanol is widely added to gasoline in the United States to reduce automotive emissions. In 2010 in the United States, over 49 billion L (13 billion gal) of ethanol were produced, mostly from corn (**FIGURE 16.6**). This amount is growing rapidly, and nearly 200 U.S. ethanol production facilities are now operating.

(a) Corn grown for ethanol

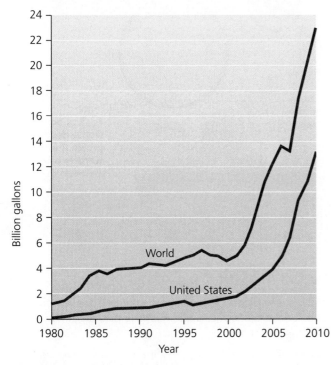

(b) Ethanol production, 1980–2010

FIGURE 16.6 ▲ About 30% of the U.S. corn crop **(a)** is used to produce ethanol, a biofuel that is widely added to gasoline in the United States. Brazil produces most of the rest of the world's ethanol, from bagasse (sugarcane residue). Ethanol production **(b)** has grown rapidly in the last several years. Data from Renewable Fuels Association.

More growth is assured, because the Energy Independence and Security Act passed by Congress in 2007 mandates production and use of 136 billion L (36 billion gal) per year of ethanol by 2022.

Any vehicle with a gasoline engine runs well on gasoline blended with up to 10% ethanol, but automakers are also producing *flexible-fuel vehicles* that run on E-85, a mix of 85% ethanol and 15% gasoline. Over 9 million such cars are on U.S. roads today. In Brazil, half of all new cars are flexible-fuel vehicles, and ethanol from crushed sugarcane residue (called *bagasse*) accounts for 40% of all fuel that Brazil's drivers use.

The enthusiasm for corn-based ethanol shown by U.S policymakers is not widely shared by environmental scientists. Growing corn to produce ethanol exerts considerable impacts on ecosystems, including pesticide use, fertilizer use, fresh water depletion, and other consequences of monocultural industrial agriculture (pp. 136–137). Corn ethanol crops take up precious land (see **ENVISIONIT**, p. 360). If we were to try to produce all the automotive fuel now used in the United States with ethanol from corn, the nation would need to expand its already immense corn acreage by more than 60%, with no loss of productivity and without producing any corn for food. Even at our current level of production, ethanol already competes with food production and drives up food prices.

FAQ

Q: If we substitute ethanol for gasoline, won't that solve most of our problems with oil dependency?

A: In the United States, government subsidies for corn-based ethanol have been politically popular, and many people believe that the more ethanol we produce and substitute for gasoline, the better off we'll be. Increasing the proportion of ethanol in gasoline does indeed help to conserve oil and reduce reliance on foreign imports. However, obtaining the amount of corn ethanol needed to replace gasoline entirely would require that impractically large amounts of land be converted to corn production. Moreover, so much corn would likely be diverted from food to fuel that food prices would rise sharply. This is why researchers are studying other plants as more-efficient sources of ethanol, and trying to develop ways of producing cellulosic ethanol from crop and forestry wastes.

Growing corn for ethanol also requires substantial inputs of fossil fuel energy (for running farm equipment, making petroleum-based pesticides and fertilizers, transporting corn to processing plants, and heating water in refineries to distill ethanol). In fact, corn ethanol yields only a modest amount of energy relative to the energy that needs to be input. The EROI (*energy returned on investment*) ratio (p. 328) for corn-based ethanol is variable, but recent estimates place it around 1.5:1.

In the quest for carbon-neutral alternatives to gasoline, what biofuel could be better, many people have thought, than ethanol made from America's #1 crop, corn?

Ethanol refinery

But growing corn for ethanol requires nearly as much energy as it produces ... and it demands that we convert immense areas of land to industrial farming.

And when fuel crops compete with food crops, that drives food prices up.

Area of corn grown in the U.S. today

Area of corn that would need to be grown if ethanol were to replace all gasoline in U.S.

So, scientists are racing to find more efficient and sustainable biofuels.

STOP!
NOT GASOLINE!
E85 for use in Flexible
Fuel Vehicles (FFVs) only.
Please consult your vehicle

E 85
85% Ethanol

E 85
85% Ethanol

Ethanol at the pump

You Can Make a Difference

➤ Ask your legislators to support university research on non-food-crop biofuels like switchgrass, algae, and cellulosic ethanol.

➤ Urge policymakers to create financial incentives for sustainable biofuels.

➤ Reduce your gasoline consumption by driving less and driving a more fuel-efficient vehicle. You'll save money, too!

This means that to gain 1.5 units of energy from ethanol, we need to expend 1 unit of energy.

Biodiesel Drivers of diesel-fueled vehicles can use **biodiesel**, a fuel produced from vegetable oil, used cooking grease, or animal fat. The oil or fat is mixed with small amounts of ethanol or methanol in the presence of a chemical catalyst. In Europe, where most biodiesel is used, rapeseed oil is the oil of choice, whereas U.S. biodiesel producers generally use soybean oil. Vehicles with diesel engines can run on 100% biodiesel, or biodiesel can be mixed with conventional petrodiesel; a 20% biodiesel mix (called B20) is common.

Biodiesel cuts down on emissions compared with petrodiesel. Its fuel economy is almost as good, it costs just slightly more, and it is nontoxic and biodegradable. Increasing numbers of people are fueling their cars with biodiesel from waste oils (**FIGURE 16.7**). Some buses and recycling trucks now run on biodiesel, and many state and federal fleets use biodiesel blends.

Using waste oil as a biofuel is sustainable, but most biodiesel today, like most ethanol, comes from crops grown specifically for the purpose—and these crops have environmental impacts. Growing soybeans in Brazil (p. 189) or oil palms in Southeast Asia (p. 191) hastens the loss of tropical rainforest. Growing soybeans in the United States and rapeseed in Europe takes up large areas of land as well.

Novel biofuels Because the major crops grown for biodiesel and for ethanol exert heavy impacts on the land, farmers and agricultural scientists are experimenting with a variety of other crops, from wheat, sorghum, cassava, and sugar beets to less-known plants such as hemp, jatropha, and the grass miscanthus. One promising next-generation biofuel crop is algae. Several species of these photosynthetic microorganisms produce large quantities of lipids that can be converted to biodiesel. Alternatively, carbohydrates in algae can be fermented to create ethanol. In fact, a variety of fuels, including jet fuel, can be produced from algae. Algae grow much faster than terrestrial crops, can be harvested every few days, and produce much more oil than other biofuel crops.

Because relying on any monocultural crop for energy may not be a sustainable strategy, researchers are refining techniques to produce **cellulosic ethanol** by using enzymes to produce ethanol from cellulose, which gives structure to all plant material. This would be a substantial advance because ethanol as currently made from corn or sugarcane uses starch, which is valuable to us as food. Cellulose, in contrast, is of no food value to people yet is abundant in all plants. If we can produce cellulosic ethanol in commercially feasible ways, then ethanol could be made from low-value crop waste (such as cornstalks and corn husks), rather than from high-value crops.

Is bioenergy carbon-neutral?

In principle, energy from biomass is carbon-neutral, releasing no net carbon into the atmosphere. This is because burning biomass releases carbon dioxide that plants had recently pulled from the atmosphere during photosynthesis. However, burning biomass for energy is not carbon-neutral if forests are destroyed in order to plant bioenergy crops. Forests sequester more carbon (in vegetation and in soil) than do croplands, so cutting forests to plant crops will increase carbon flux to the atmosphere. Bioenergy also fails to be carbon-neutral if we need to use fossil-fuel energy to produce the biomass (for instance, by driving tractors, using fertilizers, and applying pesticides to grow biofuel crops).

International climate change policy so far has failed to encourage sustainable bioenergy approaches. The Kyoto Protocol (p. 319) required nations to submit data on emissions both from energy use and from land use change (such as deforestation), but only the emissions from energy use were "counted" toward judging nations' performance under the treaty. Negotiators trying to design a follow-up treaty to Kyoto have been trying to address this (p. 319). In the meantime, researchers are busy trying to develop means of using bioenergy that are truly renewable and carbon-neutral. With continued research and careful decision-making, our many bioenergy options may provide promising avenues for sustainable replacement of fossil fuels.

FIGURE 16.7 ▼ At Loyola University Chicago, students and staff produce biodiesel from waste vegetable oil from the dining halls and use it to fuel this biodiesel van. A grant from the U.S. Environmental Protection Agency (EPA) funds them to transport this mini-biodiesel reactor to local high schools to teach students about alternative fuels.

WEIGHING THE ISSUES

Biofuels Do you think producing and using ethanol from corn is a good idea? Do the benefits outweigh the drawbacks? Should we invest billions of dollars into developing next-generation biofuels such as algae and cellulosic ethanol? Can you suggest ways of using biofuels that would minimize environmental impacts?

HYDROELECTRIC POWER

Next to biomass, we draw more renewable energy from the motion of water than from any other resource. In **hydroelectric power**, or **hydropower**, we use the kinetic energy of moving water to turn turbines and generate electricity.

(a) Ice Harbor Dam, Snake River, Washington

(b) Turbine generator inside McNary Dam, Columbia River

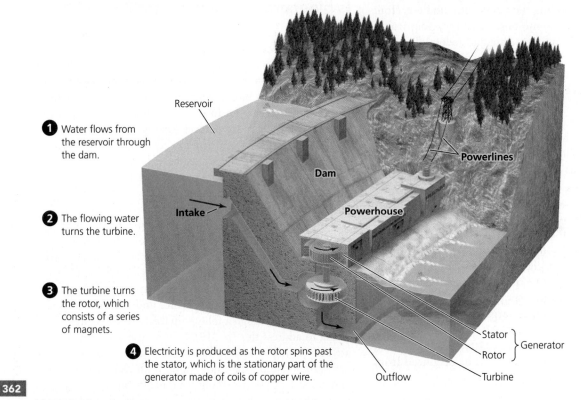

1 Water flows from the reservoir through the dam.

2 The flowing water turns the turbine.

3 The turbine turns the rotor, which consists of a series of magnets.

4 Electricity is produced as the rotor spins past the stator, which is the stationary part of the generator made of coils of copper wire.

Reservoir

Dam

Intake

Powerhouse

Powerlines

Stator } Generator
Rotor
Turbine

Outflow

(c) Hydroelectric power

FIGURE 16.8 ▲ Large dams, such as the Ice Harbor Dam on the Snake River in Washington **(a)**, generate substantial amounts of hydroelectric power. Inside these dams, flowing water is used to turn turbines **(b)** and generate electricity. Water is funneled from the reservoir through a portion of the dam **(c)** to rotate turbines, which turn rotors containing magnets. The spinning rotors generate electricity as their magnets pass coils of copper wire. Electrical current is transmitted away through power lines, and the river's water flows out through the base of the dam.

Modern hydropower uses three approaches

Most of our hydroelectric power today comes from impounding water in reservoirs behind concrete dams that block the flow of river water, and then letting that water pass through the dam. Because immense amounts of water are stored behind dams, this is called the **storage** technique. As reservoir water passes through a dam, it turns the blades of turbines, which cause a generator to generate electricity (**FIGURE 16.8**). Electricity generated in the powerhouse of a dam is transmitted to the electric grid by transmission lines, while the water flows into the riverbed below the dam and continues downriver. By storing water in reservoirs, dam operators can ensure a steady and predictable supply of electricity, even during periods of naturally low river flow.

An alternative approach is the **run-of-river** technique, which generates electricity without greatly disrupting the flow of river water. Several methods can be used; one is to divert a portion of a river's flow through a pipe or channel, passing it through a powerhouse and returning it to the river. Run-of-river systems are useful in areas remote from electrical grids and in regions without the economic resources to build and maintain large dams. This approach cannot guarantee reliable water flow in all seasons, but it minimizes many of the impacts of the storage technique.

To better control the timing of flow, pumped-storage hydropower can be used. In the **pumped-storage** approach, water is pumped from a lower reservoir to a higher reservoir during times when demand for power is weak and prices are low; when demand is strong and prices are high, water is allowed to flow downhill through a turbine, generating electricity. Although energy must be input to pump the water, pumped storage can be profitable, and it also can help even out power supply when paired with intermittent sources such as solar and wind power.

Hydropower is clean and renewable, but also has impacts

Hydropower has two clear advantages over fossil-fuel-generated electricity. First, it is renewable; as long as precipitation falls from the sky and fills rivers and reservoirs, we can use water to turn turbines. Second, no carbon compounds are burned in the production of hydropower, so no carbon dioxide or other pollutants are emitted into the atmosphere. Of course, fossil fuels *are* used in constructing and maintaining dams—and recent evidence indicates that large reservoirs release the greenhouse gas methane as a result of anaerobic decay in deep water. But overall, hydropower accounts for only a small fraction of the greenhouse gas emissions typical of fossil fuel combustion.

In addition, hydropower is efficient. It is thought to have an EROI ratio of 10:1 or more—at least as high as any other modern-day energy source.

Although it is renewable, efficient, and produces little air pollution, hydropower does exert negative impacts. Damming rivers (pp. 261–263) destroys habitat for wildlife as riparian areas above dam sites are submerged and those below dam sites often are starved of water. Because water discharge is regulated to optimize electricity generation, the natural flooding cycles of rivers are disrupted. Suppressing flooding prevents river floodplains from receiving fresh, nutrient-laden sediments. Instead, sediments become trapped behind dams, where they begin filling the reservoir. Dams also cause thermal pollution (p. 269) by changing water temperatures, and this, along with habitat alteration, has diminished or eliminated many native fish populations in dammed waterways. In addition, dams generally block the passage of fish and other aquatic creatures, fragmenting the river and reducing biodiversity in each stretch. These ecological impacts generally translate into negative social and economic impacts on local communities.

Hydroelectric power is widely used, but may not expand much more

Hydropower accounts for 16% of the world's electricity production (see Figure 16.1b). For nations with large amounts of river water and the economic resources to build dams, hydroelectric power has been a keystone of their development and

TABLE 16.2 Top Producers of Hydropower

Nation	Hydropower produced (terawatt-hours)	Percentage of electricity generation from hydropower
China	585	16.9
Canada	383	58.7
Brazil	370	79.8
United States	282	6.5
Russia	167	16.0
Norway	141	98.5
India	114	13.8
Venezuela	87	72.8
Japan	83	7.7
Sweden	69	46.1
Rest of world	1,007	13.6

Data is for 2008, from the International Energy Agency.

wealth. Canada, Brazil, Norway, Austria, Switzerland, Venezuela, and other nations today obtain large amounts of their energy from hydropower (**TABLE 16.2**).

Today the world is witnessing some gargantuan hydroelectric projects. China's recently completed Three Gorges Dam (pp. 261, 263) is the world's largest. However, hydropower is not likely to expand much more. One reason is that most of the world's large rivers are already dammed. Another reason is that people have grown more aware of the ecological impacts of dams, and in some regions residents are resisting dam construction. In the United States, 98% of rivers appropriate for dam construction already are dammed, many of the remaining 2% are protected under the Wild and Scenic Rivers Act, and some people now want to dismantle certain dams and restore river habitats (p. 261). The International Energy Agency forecasts that hydropower's share of electricity generation will decline between now and 2030, whereas the share of other renewable energy sources will triple, from 2% to 6%.

SOLAR ENERGY

The sun releases astounding amounts of energy by converting hydrogen to helium through nuclear fusion. The tiny proportion of this energy that reaches Earth is enough to drive most of the processes in the biosphere, helping to make life possible on our planet. Each day in total, Earth receives enough **solar energy**, or energy from the sun, to power human consumption for a quarter of a century. On average, each square meter of Earth's surface receives about 1 kilowatt of solar energy—17 times the energy of a lightbulb. As a result, a typical home has enough roof area to meet all its power needs with rooftop panels that harness solar energy. However, we are still in the process of developing solar technologies and learning the most effective and cost-efficient ways to put the sun's energy to use.

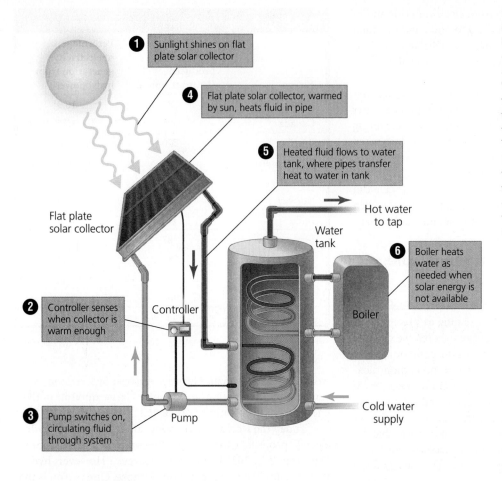

① Sunlight shines on flat plate solar collector

④ Flat plate solar collector, warmed by sun, heats fluid in pipe

⑤ Heated fluid flows to water tank, where pipes transfer heat to water in tank

Flat plate solar collector

Hot water to tap

Water tank

⑥ Boiler heats water as needed when solar energy is not available

② Controller senses when collector is warm enough

Controller

Boiler

③ Pump switches on, circulating fluid through system

Pump

Cold water supply

FIGURE 16.9 ◄ Solar systems for heating water vary in their designs, but typically **①** sunlight is gathered on a flat-plate solar collector, until a controller **②** switches on a pump **③** to circulate fluid through pipes to the collector. The sunlit collector heats the fluid **④**, which flows through pipes **⑤** to a water tank. The hot fluid in the pipes transfers heat to the water in the tank, and this heated water is available for the taps of the home or business. Generally, an external boiler **⑥** kicks in to heat water when solar energy is not available.

We can collect solar energy using passive or active methods

The simplest way to harness solar energy is through **passive solar** energy collection. In this approach, buildings are designed and building materials are chosen to maximize absorption of sunlight in winter and to keep the interior cool in the heat of summer. One such technique involves installing low, south-facing windows to maximize the capture of sunlight in winter. Overhangs shade these windows in summer, when the sun is high in the sky and when cooling, not heating, is desired. Passive solar techniques also may use construction materials that absorb heat, store it, and release it later. Such *thermal mass* (of straw, brick, concrete, or other materials) often makes up floors, roofs, and walls, or can be used in portable blocks. Planting vegetation around a building to buffer the structure from temperature swings is another passive solar approach.

In contrast, **active solar** energy collection makes use of devices to focus, move, or store solar energy. We can use various active solar technologies to heat water and air in our homes and businesses. One common method involves installing *flat-plate solar collectors* on rooftops. These panels generally consist of dark-colored, heat-absorbing metal plates mounted in flat glass-covered boxes. Water, air, or antifreeze runs through tubes that pass through the collectors, transferring heat to the building or its water tank (**FIGURE 16.9**). Heated water can be stored for later use and passed through pipes designed to release the heat into the building.

Over 1.5 million U.S. homes and businesses heat water with solar collectors, although most of this is water for swimming pools. Active solar heating is used more widely in China and also in Europe, where Germans motivated by feed-in tariffs installed 200,000 new systems in 2008 alone.

Concentrating solar rays magnifies energy

We can magnify the intensity of solar energy by gathering sunlight from a wide area and focusing it on a single point. This is the principle behind *solar cookers*, simple portable ovens that use reflectors to focus sunlight onto food and cook it. Such cookers are proving useful in the developing world.

At much larger scales, utilities are using this principle to generate electricity. **Concentrated solar power (CSP)** is being harnessed by several methods in the California desert and elsewhere. In one approach, numerous mirrors concentrate sunlight onto a receiver atop a tall "power tower" (**FIGURE 16.10**). From this central receiver, heat is transported by air or fluids (often molten salts) and piped to a steam-driven generator to create electricity. CSP facilities can harness light from lenses or mirrors spread across large areas of land, and the lenses or mirrors may swivel to track the sun's movement across the sky.

The International Energy Agency estimates that just 260 km² (100 mi²) of Nevada desert could generate enough electricity using CSP to power the entire U.S. economy. Currently, German industrialists and investors are spearheading an effort to create an immense CSP facility in Africa's Sahara Desert. In

FIGURE 16.10 ▲ The Solar Two facility in the southern California desert harnesses concentrated solar power. Hundreds of mirrors reflect sunlight onto a receiver atop a "power tower." The resulting heat is transported through fluid-filled pipes to a steam-driven generator that produces electricity for 10,000 households.

this planned $775 billion project, called Desertec, thousands of mirrors spread across vast areas of desert in Morocco would harness the Sahara's abundant sunlight and transmit electricity to Europe, the Middle East, and North Africa. However, many people are increasingly anxious about the environmen-

tal impacts that such large-scale developments may pose (see **THE SCIENCE BEHIND THE STORY**, pp. 366–367).

Photovoltaic cells generate electricity directly

The most direct way to produce electricity from sunlight involves photovoltaic (PV) systems. **Photovoltaic (PV) cells** convert sunlight to electrical energy when light reaches the PV cell and strikes one of a pair of plates made primarily of silicon, a semiconductor that conducts electricity. The light causes one plate to release electrons, which are attracted by electrostatic forces to the opposing plate. Connecting the two plates with wires enables the electrons to flow back to the original plate, creating an electrical current (direct current, DC), which can be converted into alternating current (AC) and used for residential and commercial electrical power (**FIGURE 16.11**). Small PV cells may already power your watch or your calculator. Atop the roofs of buildings, PV cells are arranged in modules, comprising panels, which can be gathered together in arrays. Arrays of PV panels can be seen on the roofs of the German houses in the photo that opens this chapter (p. 353).

Researchers are experimenting with variations on PV technology, and manufacturers today are already developing

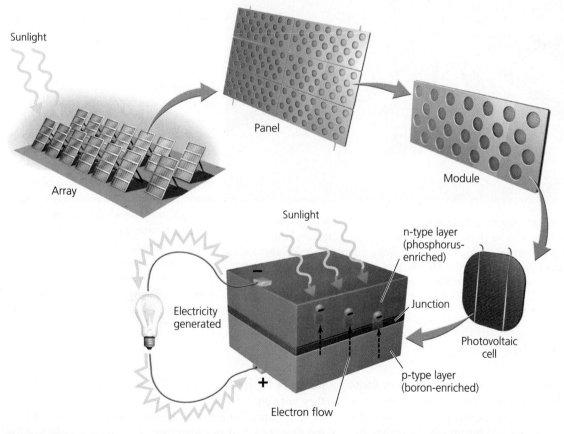

FIGURE 16.11 ▲ A photovoltaic (PV) cell converts sunlight to electrical energy. When sunlight hits the silicon layers of the cell, electrons are knocked loose from some of the silicon atoms and tend to move from the boron-enriched "p-type" layer toward the phosphorus-enriched "n-type" layer. Connecting the two layers with wiring remedies this imbalance as electrical current flows from the n-type layer back to the p-type layer. This direct current (DC) is converted to alternating current (AC) to produce usable electricity. PV cells are grouped in modules, which comprise panels, which can be erected in arrays.

Local residents demonstrate for and against the proposed Cape Wind farm.

THE SCIENCE BEHIND THE STORY

What Are the Impacts of Solar and Wind Development?

Renewable energy sources alleviate many of the negative environmental and social impacts of fossil fuel combustion, and they may one day sustainably fulfill our energy needs. However, this does not mean that renewable energy is a panacea free of costs. As our society decides how to pursue energy sources such as solar power and wind power, we will need to consider their impacts as well as their benefits.

This has become clear in recent years as the Cape Wind project in Massachusetts and a number of large solar projects in California have brought a host of issues that some proponents of clean energy had not considered. Scientific study of these impacts is just getting underway and will be important as energy development proceeds.

Several dozen solar power installations are currently under review for the Mojave Desert and other arid regions of California, and these would, if constructed, cover many thousands of acres of land (**see photo**). Desert environments are particularly sensitive, so researchers say we should expect substantial impacts. Besides altering the pristine appearance of an undeveloped landscape, arrays of thousands of mirrors or panels affect communities of plants and animals by casting shade and altering microclimate. Altered conditions tend to hurt native desert-adapted species while helping invasive weeds. At existing solar facilities, the sites are graded (damaging fragile soils) and sprayed with herbicide. Human presence increases as workers maintain the facilities. Solar power plants also require water for cooling and cleaning, and water is scarce in the arid regions hosting most of these facilities. All these impacts will have consequences for plants, animals, and ecosystems.

Large-scale projects need government approval and are subject to the environmental impact statement process (p. 100). As a result, teams of researchers study the conditions at each site to determine what impacts energy development may have. If impacts are judged to be severe enough, then government agencies can insist that plans be amended. For instance, the California Energy Commission asked for limits on the proposed Calico Solar

This solar plant in Kramer Junction, California, is one of a complex of nine that spread across more than 650 ha (1,600 acres) of the Mojave Desert, providing power for over 230,000 homes. Large-scale solar power farms require vast areas of land and exert substantial environmental impacts, yet researchers estimate that this land use is not appreciably greater than that demanded by fossil fuels. Using coal for energy takes up at least as much land, once one includes the strip mining needed to obtain the coal.

thin-film solar cells, photovoltaic materials compressed into ultra-thin sheets. Although less efficient at converting sunlight to electricity, they are cheaper to produce. Thin-film technologies can be incorporated into roofing shingles and potentially many other types of surfaces, even highways! For these reasons, many people view thin-film solar technologies as a promising direction for the future.

Photovoltaic cells of all types can be connected to batteries that store the accumulated charge until needed. Or, producers of PV electricity can sell power to their local utility if they are connected to the regional electric grid. In parts of 46 U.S. states, homeowners can sell power to their utility in the process called net metering, in which the value of the power the consumer provides is subtracted from the consumer's monthly utility bill. Feed-in tariff systems like Germany's go a step further by paying producers more than the market price of the power, offering producers the hope of turning a profit.

Project in southern California in 2010, after biologists concluded that the project would damage habitat of the desert tortoise and bighorn sheep. The company agreed to reduce the size of its footprint by nearly half, reducing estimated impacts to wildlife by 80%, and the Commission approved the project.

In central California, a solar project underwent 18 months of environmental analysis and was approved only after the Solargen company agreed to purchase and set aside 23,000 acres of preserved land as "mitigation" for the 3,200 acres it was developing. A third California solar project, the Topaz Solar Farm, was scaled back in size after researchers found that scaling back was needed to protect farmland; minimize aesthetic impacts; and lessen disturbance to tule elk, kit foxes, pronghorn antelope, burrowing owls, and seasonal freshwater pools.

On the U.S. Great Plains, researchers from the National Renewable Energy Laboratory are currently studying impacts of solar installations on prairie ecosystems, by comparing a developed site and an undeveloped control site.

Given the impacts of large-scale solar facilities, researchers have determined that installing photovoltaic panels on rooftops of buildings is a low-impact alternative. Simply adding PV panels or roofing tiles to a rooftop has no effect on the landscape. One study, led by five Dutch, German, and American researchers, compared impacts of various ground-based and rooftop PV systems in Germany and in Arizona. The researchers assessed impacts over the systems' entire life cycles (from production to installation through operation). They

found that besides avoiding land use impacts, the rooftop systems also emitted significantly fewer greenhouse gases.

A different study in 2008 measured the amount of energy required by PV cells throughout their life cycles and found that replacing fossil fuel energy with PV solar power would prevent 89–98% of greenhouse gas emissions.

The overall messages from studies so far are that (1) solar power, even with its impacts, is still cleaner and more sustainable than fossil fuel power; and (2) we can minimize the impacts of solar power by using rooftop panels and developing better technologies.

Similar messages are emerging from the scientific study of wind power. One major concern is that birds and bats are killed when they fly into the spinning blades of turbines. At California's Altamont Pass wind farm, turbines killed dozens of golden eagles and other raptors in the 1990s. Studies since then at other sites suggest that bird deaths may be a less severe problem than was initially feared, but uncertainty remains.

For instance, one European study indicated that migrating seabirds fly past offshore turbines without problem, but other data show that resident seabird densities have declined near turbines. On land, the wind industry estimates that about two birds are killed per 1-megawatt-turbine per year. This is far fewer than the hundreds of millions of birds being killed each year by television, radio, and cell phone towers; pesticides; automobiles; glass windows; and domestic cats (**see graph**). If you own a cat and let it outside, you may be killing more wildlife than are most wind farms.

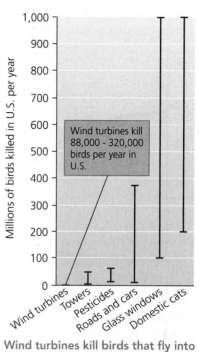

Wind turbines kill birds that fly into them. Yet far more birds are killed by other human causes. Shown are ranges of recent estimates of yearly bird mortality in the United States from several main causes. Habitat alteration is responsible for still more than any of the causes shown. Data from American Bird Conservancy.

At this point, bat mortality appears to be a more severe problem at wind turbines. One key for protecting bats and birds may be selecting sites that are not on migratory flyways or in the midst of prime habitat for species that are likely to fly into the blades. Further research on these questions is urgently needed.

Continued studies on the impacts of wind and solar development should help us find ways to harness renewable energy and attain a sustainable energy future while minimizing the environmental and social impacts of this development.

Solar energy is growing fast

Active solar technology dates from the 18th century, but it was pushed to the sidelines as fossil fuels came to dominate our energy economy. Largely because of a lack of investment, solar energy contributes just 0.15%—15 parts in 10,000—of the U.S. energy supply, and just 0.03% of U.S. electricity generation. Even in Germany, which gets more of its energy from solar than any other nation, its percentage

is just 2%. However, solar energy use has grown by over 30% annually worldwide in the past four decades, a growth rate second only to that of wind power. Solar energy is proving especially attractive in developing countries, many of which are rich in sun but poor in power infrastructure, and where hundreds of millions of people still live without electricity.

PV technology is the fastest-growing power generation technology today, having recently doubled every two years

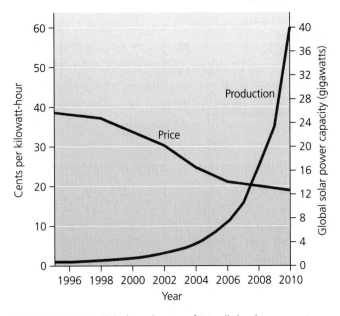

FIGURE 16.12 ▲ Global production of PV cells has been growing exponentially, and minimum prices have fallen rapidly. Data from REN21, 2011. *Renewables 2011: Global status report*. Paris: REN21 Secretariat; European Photovoltaic Industry Association; and U.S. Department of Energy.

(**FIGURE 16.12**). China leads the world in yearly production of PV cells, followed by Germany and Japan. Germany leads the world in installation of PV technology, and German rooftops host over half of all PV cells in the world. Germany's investment began in 1998 when Hermann Scheer spearheaded a "100,000 Rooftops" program to install PV panels atop 100,000 German roofs. The popular program easily surpassed this goal. The United States ranks fifth in production of PV cells. Recent federal tax credits and state-level initiatives may help the United States recover the leadership it lost to other nations in this technology, but China is moving faster and may soon dominate the market.

As production of PV cells increases, prices are falling (see Figure 16.12). At the same time, efficiencies are increasing, making each unit more powerful. Throughout the world, use of solar technology should continue to increase as prices fall, technologies improve, and governments enact economic incentives to spur investment.

Solar energy offers many benefits

The fact that the sun will continue burning for another 4–5 billion years makes it inexhaustible as an energy source for human civilization. Moreover, the amount of solar energy reaching Earth should be enough to power our civilization once we develop technology adequate to harness it. These advantages of solar energy are clear, but the technologies themselves also provide benefits. PV cells and other solar technologies use no fuel, are quiet and safe, contain no moving parts, require little maintenance, and do not require a turbine or generator to create electricity. An average unit can produce energy for 20–30 years.

Solar systems also allow for local, decentralized control over power. Homes, businesses, and isolated communities can use solar power to produce electricity without being near a power plant or connected to a grid. This is especially helpful in developing nations. In contrast, in developed nations, most PV systems are connected to the regional electric grid, and homeowners can sell excess solar energy to their local utility thanks to feed-in tariffs or net metering.

The development and deployment of solar systems is producing many new green-collar jobs. Currently, among major energy sources, PV technology employs the most people per unit energy output.

Finally, a major advantage of solar energy over fossil fuels is that it does not pollute the air with greenhouse gas emissions and other air pollutants. The manufacture of photovoltaic cells *does* currently require fossil fuel use, but once up and running, a PV system produces no emissions.

Location, timing, and cost can be drawbacks

Solar energy currently has three major disadvantages. One is that not all regions are sunny (**FIGURE 16.13**). People in cities such as Seattle might find it difficult to harness enough sunlight most of the year to rely on solar energy. Another is that solar energy is an intermittent resource. Daily or seasonal variation in sunlight can limit stand-alone solar systems if storage capacity in batteries or fuel cells is not adequate or if backup power is not available from a municipal electric grid. Pumped-storage hydropower can sometimes help compensate for periods of low solar power production.

The primary disadvantage of current solar technology is the up-front cost of the equipment. Because of the investment

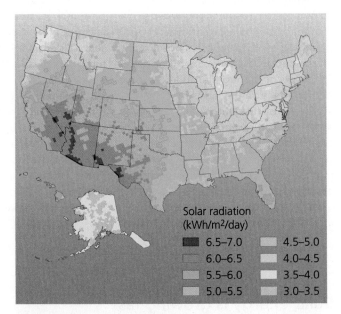

FIGURE 16.13 ▲ Some locations receive more sunlight than others, so harnessing solar energy is more profitable in some areas than in others. In the United States, many areas of Alaska and the Pacific Northwest receive only 3–4 kilowatt-hours per square meter per day, whereas most areas of the Southwest receive 6–7 kilowatt-hours per square meter per day. Data from National Renewable Energy Laboratory, U.S. Department of Energy.

Solar radiation (kWh/m²/day)
- 6.5–7.0
- 6.0–6.5
- 5.5–6.0
- 5.0–5.5
- 4.5–5.0
- 4.0–4.5
- 3.5–4.0
- 3.0–3.5

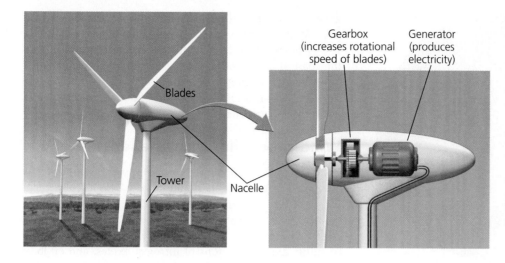

FIGURE 16.14 ◄ A wind turbine converts wind's energy of motion into electrical energy. Wind causes the blades of a wind turbine to spin, turning a shaft that extends into the nacelle that is perched atop the tower. Inside the nacelle, a gearbox converts the rotational speed of the blades, which can be up to 20 revolutions per minute (rpm) or more, into much higher rotational speeds (over 1,500 rpm). These high speeds provide adequate motion for a generator inside the nacelle to produce electricity.

Gearbox (increases rotational speed of blades)

Generator (produces electricity)

Blades

Tower

Nacelle

cost, solar power remains the most expensive way to produce electricity. However, declines in price and improvements in efficiency of solar technologies so far are encouraging, even in the absence of significant funding from government and industry. At their advent in the 1950s, solar technologies had efficiencies of around 6% while costing $600 per watt. Today, PV cells are showing up to 20% efficiency commercially and 40% efficiency in lab research, suggesting that future solar cells could be more efficient than any energy technologies we have today. Solar systems are becoming less expensive and now can sometimes pay for themselves in 10–20 years or less. After that time, they provide energy virtually for free as long as the equipment lasts.

WIND ENERGY

Wind energy—energy derived from the movement of air—is really an indirect form of solar energy because it is the sun's differential heating of air masses on Earth that causes wind to blow. We can harness power from wind by using **wind turbines**, mechanical assemblies that convert wind's kinetic energy (p. 29), or energy of motion, into electrical energy.

Wind turbines convert kinetic energy to electrical energy

Wind blowing into a turbine turns the blades of the rotor, which rotate machinery inside a compartment called a *nacelle*, which sits atop a tall tower (**FIGURE 16.14**). Inside the nacelle are a gearbox and a generator, as well as equipment to monitor and control the turbine's activity. Today's towers average 80 m (260 ft) in height, and the largest are taller than a football field is long. Higher is generally better, to minimize turbulence (and potential damage) while maximizing wind speed. Most rotors consist of three blades and measure 80 m (260 ft) across. Engineers design turbines to yaw, or rotate back and forth in response to changes in wind direction, ensuring that the motor faces into the wind at all times. They also design them to begin turning at specified wind speeds to harness wind energy as efficiently as possible. Turbines are often erected in groups called *wind farms*. The world's largest wind farms contain hundreds of turbines spread across the landscape.

Wind power is growing fast

Like solar energy, wind provides just a small proportion of the world's power needs, but wind power is growing fast—doubling every three years (**FIGURE 16.15**). Five nations account for three-quarters of the world's wind power output (**FIGURE 16.16**), but dozens of nations now produce wind power. Germany had long produced the most, but the United States overtook it in 2008, and China surpassed the United States two years later. Texas accounts for the most wind power generated of all U.S. states. Denmark leads the world in obtaining the greatest percentage of its energy from wind power; in this small European nation, wind farms supply one-fifth of electricity needs.

Experts agree that wind power's rapid growth will continue, because only a small portion of this resource is cur-

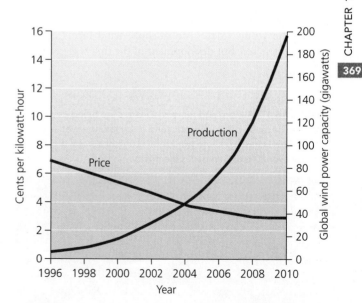

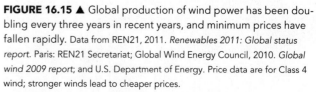

FIGURE 16.15 ▲ Global production of wind power has been doubling every three years in recent years, and minimum prices have fallen rapidly. Data from REN21, 2011. *Renewables 2011: Global status report.* Paris: REN21 Secretariat; Global Wind Energy Council, 2010. *Global wind 2009 report*; and U.S. Department of Energy. Price data are for Class 4 wind; stronger winds lead to cheaper prices.

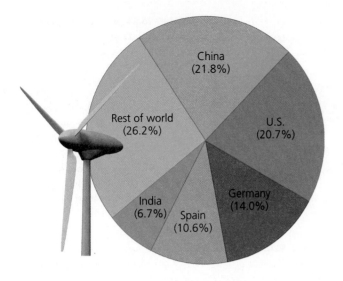

FIGURE 16.16 ▲ Most of the world's fast-growing wind-power generating capacity is concentrated in a handful of countries, led by China, the United States, and Germany. Data from Global Wind Energy Council, 2011. *Global wind 2010 report.* GWEC, Brussels, Belgium.

rently being tapped and because wind power at favorable locations already generates electricity nearly as cheaply as do fossil fuels. A 2008 report by a consortium of government, industry, and environmental experts outlined how the United States could meet fully one-fifth of its electrical demands with wind power by 2030.

Offshore sites hold promise

Wind speeds on average are roughly 20% greater over water than over land. There is also less air turbulence over water. For these reasons, offshore wind turbines are becoming popular (**FIGURE 16.17**). Costs to erect and maintain turbines in water are higher, but the stronger, less turbulent winds make offshore wind potentially more profitable.

In the United States, no offshore wind farms have yet been constructed, but development of the first was given U.S.

FIGURE 16.17 ▼ More and more wind farms are being developed offshore, because offshore winds tend to be stronger yet less turbulent. This Danish offshore wind farm is one of several that help provide over 20% of Denmark's electricity.

government approval in 2010 after nine years of debate. The Cape Wind offshore wind farm, if constructed, will feature 130 turbines rising from Nantucket Sound 8 km (5 mi) off the coast of Cape Cod in Massachusetts. In announcing the government's approval, U.S. Interior Secretary Ken Salazar predicted that it would be "the first of many projects up and down the Atlantic coast."

Wind power has many benefits

Like solar power, wind power produces no emissions once the equipment is manufactured and installed. As a replacement for fossil fuel combustion in the average U.S. power plant, running a 1-megawatt wind turbine for 1 year prevents the release of more than 1,500 tons of carbon dioxide, 6.5 tons of sulfur dioxide, 3.2 tons of nitrogen oxides, and 60 lb of mercury, according to the U.S. Environmental Protection Agency. The amount of carbon pollution that all U.S. wind turbines together prevent from entering the atmosphere is greater than the emissions from 7.5 million cars, or from combusting the cargo of a 600-car freight train of coal each and every day.

Wind power, under optimal conditions, appears considerably more efficient than conventional power sources in its energy returned on investment (EROI; p. 328). One study found that wind turbines produce 23 times more energy than they consume. For nuclear energy, the ratio was 16:1; for coal it was 11:1; and for natural gas it was 5:1.

Wind turbine technology can be used on many scales, from a single tower for local use to farms of hundreds that supply large regions. Small-scale turbine development can help make local areas more self-sufficient, just as solar energy can. Another benefit of wind power is that farmers and ranchers can lease their land for wind development. A single large turbine can bring in $2,000 to $4,500 in annual royalties while occupying just a quarter-acre of land. Because each turbine takes up only a small area, most of the land can still be used for farming or ranching. Royalties from the wind power company provide the farmer or rancher revenue while also increasing property tax income for their rural community.

Lastly, wind power creates job opportunities (**FIGURE 16.18**). The American Wind Energy Association estimates that over 85,000 Americans are now employed in the wind industry. More than 100 colleges and universities offer programs and degrees that train people in the skills needed for jobs in wind power and other renewable energy fields.

Wind power has some downsides

Wind is an intermittent resource; we have no control over when wind will occur. This is a major problem, but is lessened if wind is one of several sources contributing to a utility's power generation. Pumped-storage hydropower can sometimes help compensate during windless times. Moreover, batteries or hydrogen fuel can store energy generated by wind and release it later when needed.

Just as wind varies from time to time, it varies from place to place; some areas are windier than others. Resource planners and wind power companies study wind patterns closely before planning a wind farm. Meteorological research has given us data with which to judge prime areas for locating

FIGURE 16.18 ▲ As the wind power industry expands, it is becoming a major source of new green-collar jobs.

wind farms. A map of average wind speeds across the United States (**FIGURE 16.19A**) reveals that mountainous regions are best, along with areas of the Great Plains. Based on such information, the wind power industry has located much of its generating capacity in states with high wind speeds (**FIGURE 16.19B**) and is seeking to expand in the Great Plains and mountain states.

Good wind resources, however, are not always near population centers that need the energy. Most of North America's people live near the coasts, far from the Great Plains and mountain regions that have the best wind resources. Thus, transmission networks would need to be greatly expanded to get wind power to where people live.

When wind farms *are* proposed near population centers, local residents often oppose them. Turbines are generally located in exposed, conspicuous sites, and some people object to wind farms for aesthetic reasons, feeling that the structures clutter the landscape. Wind turbines also pose a hazard to birds and bats, which are killed when they fly into the rotating blades.

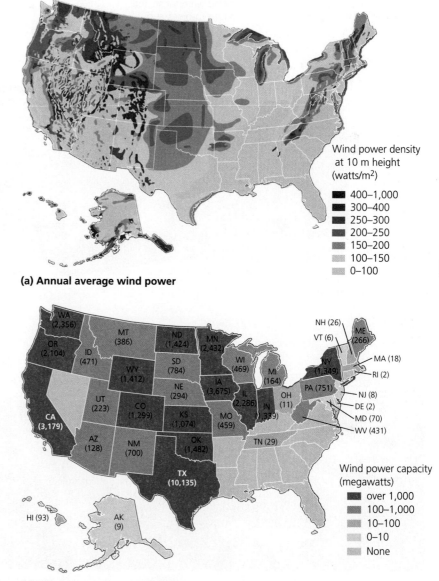

FIGURE 16.19 ▼ Meteorologists have measured wind speed to calculate the potential generating capacity from wind in different areas. The map in **(a)** shows average wind power across the United States, in watts per square meter at a height of 10 m (33 ft) above ground. Such maps are used to help guide placement of wind farms. The development of U.S. wind power so far is summarized in **(b)**, which shows the megawatts of generating capacity developed in each state through April 2011. *Sources:* (a) Elliott, D.L., et al. 1987. *Wind energy resource atlas of the United States.* Golden, CO: Solar Energy Research Institute; (b) American Wind Energy Association.

(a) Annual average wind power

Wind power density at 10 m height (watts/m²)

- 400–1,000
- 300–400
- 250–300
- 200–250
- 150–200
- 100–150
- 0–100

(b) Wind generating capacity, 2011

Wind power capacity (megawatts)

- over 1,000
- 100–1,000
- 10–100
- 0–10
- None

GEOTHERMAL ENERGY

Geothermal energy is thermal energy that arises from beneath Earth's surface. The radioactive decay of elements (p. 27) amid high pressures deep in the interior of our planet generates heat that rises to the surface through magma (molten rock, p. 228) and through cracks and fissures. Where this energy heats groundwater, natural spurts of heated water and steam rise up from below and may erupt through the surface as terrestrial geysers or submarine hydrothermal vents (p. 260). Geothermal energy manifests itself at the surface in these ways only in certain regions.

We harness geothermal energy for heating and electricity

Geothermal energy can be harnessed directly from geysers at the surface, but most often wells must be drilled down hundreds or thousands of meters toward heated groundwater. Hot groundwater can be used directly for heating buildings and for driving industrial processes. The nation of Iceland heats nearly 90% of its homes through direct heating with piped hot water. Such direct use of naturally heated water is efficient and inexpensive, but it is feasible only where geothermal energy sources are readily available. Iceland has a wealth of geothermal resources because it is located along the spreading boundary of two tectonic plates (p. 228–229).

Geothermal power plants harness the energy of naturally heated underground water and steam to generate electricity (**FIGURE 16.20**). Generally, a power plant will bring water at temperatures of 150–370 °C (300–700 °F) or more to the surface and convert it to steam by lowering the pressure in specialized compartments. The steam is then used to turn turbines to generate electricity. The world's largest geothermal power plants, The Geysers in northern California, provide enough electricity to supply 750,000 homes. The United States is the world leader in the use of

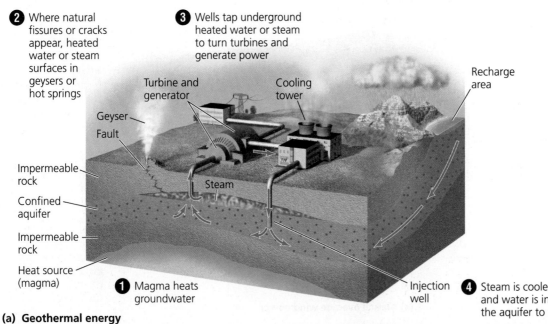

❷ Where natural fissures or cracks appear, heated water or steam surfaces in geysers or hot springs

❸ Wells tap underground heated water or steam to turn turbines and generate power

Turbine and generator

Cooling tower

Recharge area

Geyser

Fault

Impermeable rock

Confined aquifer

Impermeable rock

Heat source (magma)

Steam

❶ Magma heats groundwater

Injection well

❹ Steam is cooled, condensed, and water is injected back into the aquifer to maintain pressure

(a) Geothermal energy

372

(b) Nesjavellir geothermal power station, Iceland

FIGURE 16.20 ◀ With geothermal energy **(a)**, magma heats groundwater deep in the Earth ❶, some of which is let off naturally through surface vents such as geysers ❷. Geothermal facilities tap into heated water below ground and channel steam through turbines in buildings to generate electricity ❸. After being used, the steam is often condensed, and the water is pumped back into the aquifer to maintain pressure ❹. At the Nesjavellir geothermal power station in Iceland **(b)**, steam is piped from wells to a condenser at the plant, where cold water pumped from lakeshore wells is heated. The heated water is sent through an insulated pipeline to the capital city, where residents use it for washing and space heating.

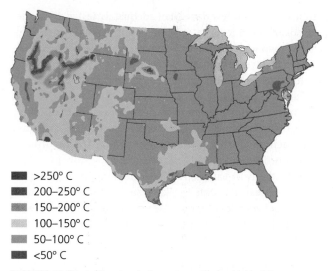

■ >250° C
■ 200–250° C
■ 150–200° C
■ 100–150° C
■ 50–100° C
■ <50° C

FIGURE 16.21 ▲ Geothermal resources in the United States are greatest in the western states. This map shows water temperatures 3 km (1.9 mi) below ground. Although deep subterranean temperatures are greatest in the West, ground-source heat pumps can be used anywhere in the country. Data from Idaho National Laboratory.

geothermal power, but only some U.S. regions have geothermal resources near the surface that can be readily used (**FIGURE 16.21**).

Heat pumps make use of temperature differences above and below ground

Although heated groundwater is available only in certain areas, we can take advantage of the temperature differences that naturally exist between the soil and the air just about anywhere. Soil varies in temperature from season to season

less than air does, and **ground-source heat pumps (GSHPs)** make use of this fact. These pumps provide heating in the winter by transferring heat from the ground into buildings, and they provide cooling in the summer by transferring heat from buildings into the ground. This heat transfer is accomplished with a network of underground plastic pipes that circulate water and antifreeze (**FIGURE 16.22**). Because heat is simply moved from place to place rather than being produced using outside energy inputs, heat pumps can be highly energy-efficient.

More than 600,000 GSHPs are already used to heat U.S. homes. Compared to conventional electric heating and cooling systems, GSHPs heat spaces 50–70% more efficiently, cool them 20–40% more efficiently, can reduce electricity use by 25%–60%, and can reduce emissions by up to 70%.

Geothermal power has benefits and limitations

All forms of geothermal energy greatly reduce emissions relative to fossil fuel combustion. Although geothermally heated water can release dissolved gases, including carbon dioxide, methane, ammonia, and hydrogen sulfide, these are generally in small quantities, and facilities using the latest filtering technologies produce even fewer emissions.

Geothermal energy is renewable in that using it does not affect the amount of thermal energy produced underground. However, not every power plant we build to capture this energy will be able to operate indefinitely. If a geothermal plant uses heated water more quickly than groundwater is recharged, the plant will eventually run out of water. This was occurring at The Geysers in California, so in response, operators began

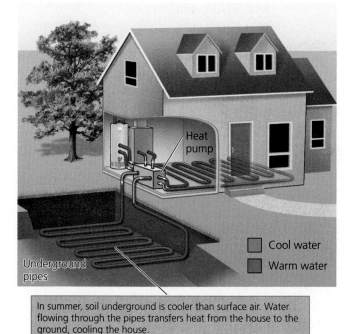

Heat pump

Underground pipes

In summer, soil underground is cooler than surface air. Water flowing through the pipes transfers heat from the house to the ground, cooling the house.

■ Cool water
■ Warm water

Heat pump may warm or cool air in ducts, water in tank, or radiant heating/cooling system under floor.

In winter, soil underground is warmer than surface air. Water flowing through the pipes transfers heat from the ground to the house, warming the house.

FIGURE 16.22 ▲ Ground-source heat pumps provide an efficient way to heat and cool air and water in one's home. A network of plastic pipes filled with water and antifreeze extend underground from the house. Soil is warmer than the air in the winter (**left**), and cooler than the air in the summer (**right**), so by running fluid between the house and the ground, these systems help adjust temperatures inside.

injecting municipal wastewater into the ground to replenish the supply. Moreover, patterns of geothermal activity in Earth's crust shift naturally over time, so an area that produces hot groundwater now may not always do so. In addition, the water of many hot springs is laced with salts and minerals that corrode equipment and pollute the air. These factors may shorten the lifetime of plants, increase maintenance costs, and add to pollution.

The greatest limitation of geothermal electric power is that it is restricted to regions where we can tap the energy from naturally heated groundwater. Iceland, northern California, and Yellowstone National Park are rich in naturally heated groundwater, but most areas of the world are not. Engineers are presently trying to overcome this limitation by developing *enhanced geothermal systems (EGS)*, in which we drill deeply into dry rock, fracture the rock, and pump in cold water. The water becomes heated deep underground and is then drawn up through an outlet well and used to generate power. In theory we could use EGS widely in many locations. For instance, Germany has very little heated groundwater, but feed-in tariffs have enabled an EGS facility to operate profitably here. However, EGS also appears to trigger minor earthquakes. Unless we can develop ways to use EGS safely without causing earthquakes, our use of geothermal power will remain more localized than solar, wind, biomass, or hydropower.

OCEAN ENERGY SOURCES

The oceans are home to several underexploited energy sources stemming from continuous natural processes. Of the four approaches being developed, three involve motion and one involves temperature.

We can harness energy from tides, waves, and currents

Just as dams on rivers use flowing fresh water to generate hydroelectric power, we can use kinetic energy from the natural motion of ocean water to generate electrical power. The rise and fall of ocean tides (p. 257) twice each day moves large amounts of water past any given point on the world's coastlines. Differences in height between low and high tides are especially great in long, narrow bays such as Alaska's Cook Inlet or the Bay of Fundy between New Brunswick and Nova Scotia. Such locations are best for harnessing **tidal energy**, which is done by erecting dams across the outlets of tidal basins. As tidal currents pass through the dam, water turns turbines to generate electricity.

The world's largest tidal generating station, the La Rance facility in France, has operated for over 45 years. Smaller facilities operate in China, Russia, and Canada. San Francisco is seeking to build a tidal energy station under the Golden Gate Bridge, and New York City is considering establishing one in the East River. Tidal stations release few or no pollutant emissions, but they can affect the ecology of estuaries and tidal basins.

People are also working to harness the motion of ocean waves and convert their mechanical energy into electricity. Many designs for machinery to harness **wave energy** have been invented, but few have been adequately tested. Some designs for offshore facilities involve floating devices that move up and down with the waves. Some designs for onshore facilities funnel waves from large areas into narrow channels and elevated reservoirs, from which water then flows out, generating electricity as hydroelectric dams do. Other coastal designs use rising and falling waves to push air into and out of chambers, turning turbines (**FIGURE 16.23**). No commercial wave energy facilities are operating yet, but demonstration projects exist in Europe, Japan, and Oregon.

A third way to harness marine kinetic energy is to use the motion of ocean currents (p. 255), such as the Gulf Stream. Devices that look like underwater wind turbines have been erected in European waters to test this idea.

The ocean stores thermal energy

Each day the tropical oceans absorb an amount of solar radiation equivalent to the heat content of 250 billion barrels of oil—enough to provide 20,000 times the electricity used daily in the United States. The ocean's sun-warmed surface is warmer than its deep water, and **ocean thermal energy conversion (OTEC)** approaches are based on this temperature gradient.

In one approach, warm surface water is piped into a facility to evaporate chemicals, such as ammonia, that boil at low temperatures. These evaporated gases spin turbines to generate electricity. Cold water piped in from ocean depths then condenses the gases so they can be reused. In another approach, warm surface water is evaporated in a vacuum, and its steam turns turbines and then is condensed by cold water. Because ocean water loses salts as it evaporates, the water can be recovered, condensed, and sold as desalinized fresh water

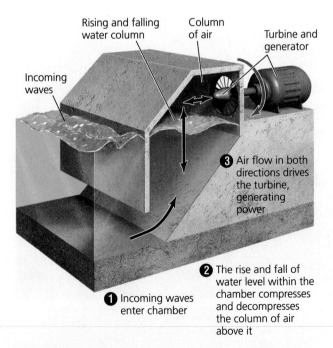

FIGURE 16.23 ▲ Coastal facilities can make use of energy from the motion of ocean waves. In one design, as waves are let into and out of a tightly sealed chamber ❶, the air inside is alternately compressed and decompressed ❷, creating air flow that rotates turbines ❸ to generate electricity.

for drinking or agriculture. Research on OTEC systems has been conducted in Hawaii and elsewhere, but costs remain high, and so far no facility operates commercially.

WEIGHING THE ISSUES

Your Nation's Energy? You are the president of a nation the size of Germany, and your nation's congress is calling on you to propose a national energy policy. Your country is located along a tropical coastline. Your geologists do not yet know whether there are fossil fuel deposits or geothermal resources under your land, but your country gets a lot of sunshine and a fair amount of wind, and broad, shallow shelf regions line its coasts. Your nation's population is moderately wealthy but is growing fast, and importing fossil fuels from other nations is becoming expensive.

What approaches would you propose in your energy policy? Name some specific steps you would urge your congress to fund. Are there trade relationships you would seek to establish with other countries? What questions would you fund your nation's scientists to research?

HYDROGEN

Each of the renewable energy sources we have discussed can be used to generate electricity more cleanly than can fossil fuels. However, electricity cannot be stored easily in large quantities for use when and where it is needed. This is why most vehicles rely on gasoline from oil for their power. The development of fuel cells and of fuel consisting of hydrogen—the universe's simplest and most abundant element—shows promise as a way to store considerable quantities of energy conveniently, cleanly, and efficiently. Like electricity and like batteries, hydrogen is an energy carrier, not a primary energy source. It carries energy that can be converted for use at later times and in different places.

Some yearn for a "hydrogen economy"

Some energy experts envision that hydrogen fuel, together with electricity, could serve as the basis for a clean, safe, and efficient energy system. In such a system, electricity generated from intermittent renewable sources, such as wind and solar energy, could be used to produce hydrogen. Fuel cells (**FIGURE 16.24**) would then use hydrogen to produce electrical energy as needed to power vehicles, computers, cell phones, home heating, and countless other applications.

Basing an energy system on hydrogen could alleviate dependence on foreign fuels and help fight climate change. For these reasons, governments are funding research into hydrogen and fuel-cell technology, and automobile companies have developed vehicles that run on hydrogen. Today, Germany is one of several nations with hydrogen-fueled city buses (**FIGURE 16.25**).

Hydrogen fuel may be produced from water or from other matter

Hydrogen gas (H_2) does not tend to exist freely on Earth. Instead, hydrogen atoms bind to other molecules, becoming incorporated in everything from water to organic compounds. To obtain hydrogen gas for fuel, we must force these substances to release their hydrogen atoms, and this requires an input

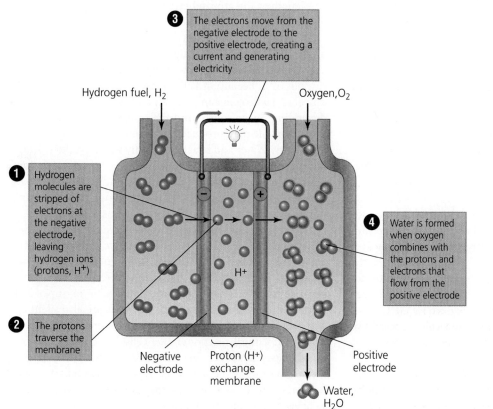

FIGURE 16.24 ◀ Hydrogen fuel drives electricity generation in a fuel cell, creating water as a waste product. First, atoms of hydrogen are split **1** into protons and electrons. The protons, or hydrogen ions **2**, pass through a proton exchange membrane. The electrons, meanwhile, move from a negative electrode to a positive one via an external circuit **3**, creating a current and generating electricity. The protons and electrons then combine with oxygen **4** to form water molecules.

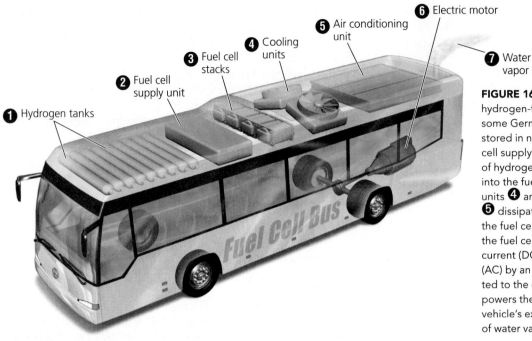

❻ Electric motor

❺ Air conditioning unit

❹ Cooling units

❸ Fuel cell stacks

❷ Fuel cell supply unit

❶ Hydrogen tanks

❼ Water vapor exhaust

FIGURE 16.25 ◄ In one type of hydrogen-fueled bus operating in some German cities, hydrogen is stored in nine fuel tanks ❶. The fuel cell supply unit ❷ controls the flow of hydrogen, air, and cooling water into the fuel cell stacks ❸. Cooling units ❹ and the air conditioning unit ❺ dissipate waste heat produced by the fuel cells. Electricity generated by the fuel cells is changed from direct current (DC) to alternating current (AC) by an inverter, and it is transmitted to the electric motor ❻, which powers the operation of the bus. The vehicle's exhaust ❼ consists simply of water vapor.

of energy. Scientists are studying several potential ways of producing hydrogen. In **electrolysis**, electricity is input to split hydrogen atoms from the oxygen atoms of water molecules:

$$2H_2O \longrightarrow 2H_2 + O_2$$

Electrolysis produces pure hydrogen, and it does so without emitting the carbon- or nitrogen-based pollutants of fossil fuel combustion. However, whether this strategy for producing hydrogen will cause pollution depends on the source of the electricity used for the electrolysis. If coal is burned to generate the electricity, then the process will not reduce emissions compared with reliance on fossil fuels. The "cleanliness" of a future hydrogen economy, therefore, depends largely on the source of electricity used in electrolysis.

The environmental impact of hydrogen production will also depend on the source material for the hydrogen. Besides water, hydrogen can be obtained from biomass and from fossil fuels. Obtaining hydrogen from these sources generally requires less energy input but results in emissions of carbon-based pollutants. For instance, extracting hydrogen from the methane (CH_4) in natural gas entails producing one molecule of the greenhouse gas carbon dioxide for every four molecules of hydrogen gas:

$$CH_4 + 2H_2O \longrightarrow 4H_2 + CO_2$$

Thus, whether a hydrogen-based energy system is cleaner than a fossil fuel system depends on how the hydrogen is extracted.

Once isolated, hydrogen gas can be used as a fuel to produce electricity within fuel cells. The chemical reaction involved in a fuel cell is simply the reverse of that shown for electrolysis; an oxygen molecule and two hydrogen molecules each split so that their atoms can bind and form two water molecules:

$$2H_2 + O_2 \longrightarrow 2H_2O$$

Figure 16.24 shows how this occurs within one common type of fuel cell.

Hydrogen and fuel cells have costs and benefits

One major drawback of hydrogen at this point is a lack of infrastructure. To convert a nation like Germany or the United States to hydrogen would require massive and costly development of facilities to produce, store, transport, and provide the fuel.

Another concern is that some research suggests that leakage of hydrogen from its production, transport, and use could potentially deplete stratospheric ozone (pp. 279, 289–291), and lengthen the atmospheric lifetime of the greenhouse gas methane. Research into these questions is ongoing, because scientists do not want society to switch from fossil fuels to hydrogen without first knowing the risks.

Hydrogen's benefits include the fact that we will never run out of it, because it is the most abundant element in the universe. Hydrogen can be clean and nontoxic to use, and—depending on its source and the source of electricity for its extraction—it may produce few greenhouse gases and other pollutants. Water and heat are the only waste products from a hydrogen fuel cell, along with negligible traces of other compounds. In terms of safety for transport and storage, hydrogen can catch fire and explode, but if kept under pressure, it may not be much more dangerous than gasoline in tanks.

Hydrogen fuel cells are energy-efficient. Depending on the type of fuel cell, 35–70% of the energy released in the reaction can be used. If the system is designed to capture heat as well as electricity, then the energy efficiency of fuel cells can rise to 90%. These rates are comparable or superior to most nonrenewable alternatives. Fuel cells are also silent and nonpolluting. Unlike batteries (which also produce electricity through chemical reactions), fuel cells will generate electricity whenever hydrogen fuel is supplied, without ever needing recharging. For all these reasons, hydrogen fuel cells may soon be used to power cars, much as they are already powering buses operating on the streets of some German cities.

➤ CONCLUSION

Rising concern over air pollution, global climate change, health impacts, and security risks resulting from our reliance on fossil fuels—as well as anxiety over dwindling supplies of oil and natural gas—have convinced many people that we need to shift to renewable energy sources that pollute far less and that will not run out. Bioenergy sources include traditional fuelwood, as well as newer biofuels and various means of generating biopower. These sources can be carbon-neutral but are not all strictly renewable. Hydropower is clean and renewable, but it is nearing its maximal extent of use and can involve substantial ecological impacts. Renewable sources with promise for sustaining our civilization far into the future without greatly degrading our environment include solar energy, wind energy, geothermal energy, and ocean energy sources. By using electricity from renewable sources to produce hydrogen fuel, we may be able to use fuel cells to produce electricity when and where it is needed, helping to create a nonpolluting and renewable transportation sector.

TESTING YOUR COMPREHENSION

1. About how much of our energy now comes from renewable sources? What is the most prevalent form of renewable energy we use? What form of renewable energy is most used to generate electricity?

2. What factors and concerns are causing renewable energy sectors to expand? Which two renewable sources are experiencing the most rapid growth?

3. List five sources of bioenergy. What is the world's most-used source of bioenergy? How does bioenergy use differ between developed and developing nations?

4. Contrast two major approaches to generating hydroelectric power. List one benefit and one negative impact of hydropower.

5. Contrast passive and active solar heating, and give an example of each. Now explain how photovoltaic (PV) cells function and are used.

6. Describe how modern wind turbines generate electricity. What factors affect where we place wind turbines?

7. Define *geothermal energy*, and explain three main ways in which it is obtained and used. Describe one sense in which it is renewable, and one sense in which it is not renewable.

8. List and describe four approaches for obtaining energy from ocean water.

9. For each major type of renewable energy (bioenergy, hydropower, solar, wind, geothermal, and ocean), briefly describe at least one advantage and one disadvantage of its use, relative to use of fossil fuels.

10. How is hydrogen fuel produced? Is this a clean process? What factors determine the amount of pollutants hydrogen production will emit?

SEEKING SOLUTIONS

1. Explain how Germany accelerated its development of renewable energy by establishing a system of feed-in tariffs. Do you think the United States should adopt a similar system? Why or why not?

2. For each source of renewable energy discussed in this chapter, what factors are standing in the way of an expedient transition from fossil fuel use? What could be done in each case to ease a shift to these renewable sources?

3. Do you think we can develop renewable energy resources to replace fossil fuels without great social, economic, and environmental disruption? What steps would we need to take? Will market forces alone suffice to bring about this transition, or will we also need government? Do you think such a shift will be good for our economy? Why or why not?

4. **THINK IT THROUGH** You are an investor looking to invest in alternative energy. You are considering buying stock in companies that (1) build corn ethanol refineries, (2) are developing algae farms for biofuels, (3) construct turbines for hydroelectric dams, (4) produce PV solar panels, (5) install wind turbines, and (6) plan to build a wave energy facility. For each of these companies, what questions would you research before deciding how to invest your money? How do you expect you might apportion your investments, and why?

5. **THINK IT THROUGH** You are the CEO of a company that develops wind farms. Your staff is presenting you with three options, listed below, for sites for your next development. Describe at least one likely advantage and at least one likely disadvantage you would expect to encounter with each option. What further information would you like to know before deciding which to pursue?

   ▶ Option A: A remote rural site in North Dakota
   ▶ Option B: A ridge-top site among the suburbs of Philadelphia
   ▶ Option C: An offshore site off the Florida coast

CALCULATING ECOLOGICAL FOOTPRINTS

Assume that average per capita residential consumption of electricity is 12 kilowatt-hours per day, that photovoltaic cells have an electrical output of 15% incident solar radiation, and that PV panels cost $1,000 per square meter. Now refer to Figure 16.13 on p. 368, and estimate the area and cost of the PV panels needed to provide all of the residential electricity used by each group in the table.

	Area of photovoltaic cells	Cost of photovoltaic cells
You		
A resident of Arizona		
A resident of Alaska		
Total for all U.S. residents		

1. What additional information would you need to increase the accuracy of your estimates for the areas in the table above?

2. Considering the distribution of solar radiation in the United States, where do you think it will be most feasible to greatly increase the percentage of electricity generated from photovoltaic solar cells?

3. The purchase price of a photovoltaic system is considerable. What other costs and benefits should you consider, in addition to the purchase price, when contemplating "going solar"?

Mastering ENVIRONMENTAL SCIENCE™

Go to **www.masteringenvironmentalscience.com** for homework assignments, practice quizzes, Pearson eText, and more.

17 Managing Our Waste

Upon completing this chapter, you will be able to:

➤ Summarize and compare the types of waste we generate

➤ List the major approaches to managing waste

➤ Delineate the scale of the waste dilemma

➤ Describe conventional waste disposal methods: landfills and incineration

➤ Evaluate approaches for reducing waste: source reduction, reuse, composting, and recycling

➤ Discuss industrial solid waste management and principles of industrial ecology

➤ Assess issues in managing hazardous waste

THEN: Fresh Kills Landfill in operation

NOW: Fresh Kills Landfill site today

Fresh Kills Landfill,
Staten Island, New York

Transforming New York's Fresh Kills Landfill

"An extraterrestrial observer might conclude that conversion of raw materials to wastes is the real purpose of human economic activity."
—GARY GARDNER AND PAYAL SAMPAT, WORLDWATCH INSTITUTE

"Recycling is one of the best environmental success stories of the late 20th century."
—U.S. ENVIRONMENTAL PROTECTION AGENCY

The closure of a landfill is not the kind of event that normally draws politicians and the press, but the Fresh Kills Landfill was no ordinary dump. Said to be the largest human-made structure on Earth, Fresh Kills was the primary repository of New York City's garbage for half a century. On March 22, 2001, New York City Mayor Rudolph Giuliani and New York Governor George Pataki were on hand to celebrate as a barge arrived on the shore of Staten Island and dumped the final load of trash at Fresh Kills.

The landfill's closure was a blessing for Staten Island's 450,000 residents, who had long viewed Fresh Kills as a foul-smelling eyesore, health threat, and civic blemish. The 890-ha (2,200-acre) landfill featured six gigantic mounds of trash and soil. The highest, at 69 m (225 ft), was higher than the nearby Statue of Liberty.

New York City had grandiose plans for the site. It planned to transform the old landfill into a world-class public park—a verdant landscape of ball fields, playgrounds, jogging trails, rolling hills, and wetlands teeming with wildlife. Nearly three times bigger than Manhattan's Central Park, the site hosted the region's largest remaining complex of salt marshes and freshwater creeks, while the mounds offered panoramic views of the Manhattan skyline. The city sponsored an international competition to select a landscape architecture firm to design plans for the new park.

Meanwhile, with its only landfill closed, New York City began exporting its waste—and found itself paying contractors exorbitant prices to haul its garbage away one truckload at a time. In the years following the Fresh Kills closure, trucks full of trash rumbled through

neighborhood streets, carrying 12,000 tons of waste each day bound for 26 different landfills and incinerators in New York, New Jersey, Virginia, Pennsylvania, and Ohio. The city sanitation department's expenses doubled, and budget woes caused the city to scale back its recycling program. Some New Yorkers suggested reopening Fresh Kills.

The landfill *was* reopened, but not for a reason anyone could have foreseen. After the September 11, 2001, terrorist attacks, the 1.8 million tons of rubble from the collapsed World Trade Center towers, including unrecoverable human remains, was taken by barge to Fresh Kills. A monument will be erected as part of the new park.

Today, park development is forging ahead. Roads, ball fields, sculptures, and in-line skating rinks are being designed. Wetlands are being restored. People will be able to bicycle on trails alongside the region's largest estuary and reach stunning vistas atop the hills. Recreation areas named Owl Hollow Fields and Schmul Park should open in 2011, and work has begun on a parcel overlooking an adjacent wildlife refuge. The full conversion of Fresh Kills Landfill into Fresh Kills

Park—one of the largest public works projects in the world—will take 30 years. But in the end this longtime symbol of waste will be transformed into a world-class center for recreation and urban ecological restoration.

Meanwhile, the city continues to truck its trash to out-of-state landfills. It has also built a transfer station at Fresh Kills to compact the waste and ship it outward by barge and railroad, at less expense. There is no true "away" for the things we do not reuse or recycle. For New Yorkers—and for the rest of us—the waste we discard needs to go somewhere. ■

APPROACHES TO WASTE MANAGEMENT

As the world's population rises, and as we produce and consume more material goods, we generate more waste. **Waste** refers to any unwanted material or substance that results from a human activity or process.

For management purposes, we divide waste into several main categories. **Municipal solid waste** is nonliquid waste that comes from homes, institutions, and small businesses. **Industrial solid waste** includes waste from production of consumer goods, mining, agriculture, and petroleum extraction and refining. **Hazardous waste** refers to solid or liquid waste that is toxic, chemically reactive, flammable, or corrosive. Another type of waste is wastewater (pp. 270, 272–273), water we use in our households, businesses, industries, or public facilities and drain or flush down our pipes, as well as the polluted runoff from streets and storm drains.

We have several aims in managing waste

Waste can degrade water quality, soil quality, air quality, and human health. Waste is also unpleasant aesthetically. Moreover, waste is a measure of inefficiency, so reducing waste can save money and resources. For all these reasons, waste management has become a vital pursuit.

There are three main components of **waste management**:

1. Minimizing the amount of waste we generate
2. Recovering discarded materials and finding ways to recycle them
3. Disposing of waste safely and effectively

Minimizing waste at its source—called **source reduction**—is the preferred approach. There are several ways to reduce the amount of waste that enters the **waste stream**, the flow of waste as it moves from its sources toward disposal destinations (**FIGURE 17.1**). In this chapter we first examine how the three major approaches are used to manage municipal solid waste, and then we address industrial solid waste and hazardous waste.

MUNICIPAL SOLID WASTE

Municipal solid waste is waste we generate in homes, public facilities, and small businesses. It is what we commonly refer to as "trash" or "garbage."

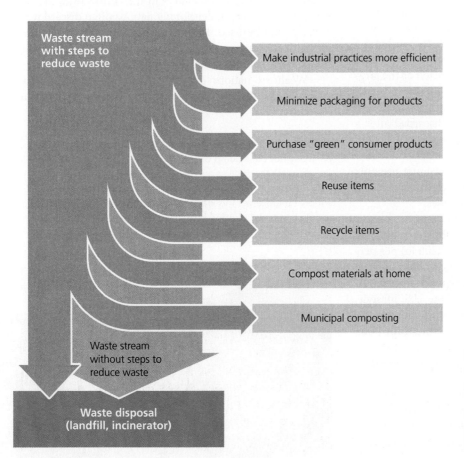

FIGURE 17.1 ◀ The most effective way to manage waste is to minimize the amount of material that enters the waste stream. To do this, manufacturers can increase efficiency, and consumers can buy "green" products that have minimal packaging or are produced in ways that minimize waste.

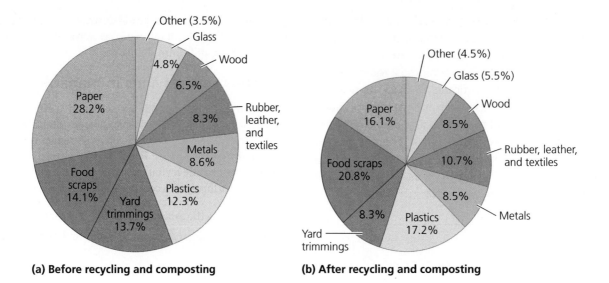

(a) Before recycling and composting

Other (3.5%)
Glass 4.8%
Wood 6.5%
Rubber, leather, and textiles 8.3%
Metals 8.6%
Plastics 12.3%
Yard trimmings 13.7%
Food scraps 14.1%
Paper 28.2%

(b) After recycling and composting

Other (4.5%)
Glass (5.5%)
Wood 8.5%
Rubber, leather, and textiles 10.7%
Metals 8.5%
Plastics 17.2%
Yard trimmings 8.3%
Food scraps 20.8%
Paper 16.1%

FIGURE 17.2 ▲ Paper products comprise the largest component of the municipal solid waste stream in the United States by weight **(a)**, followed by food scraps, yard trimmings, and plastics. After recycling and composting removes many items **(b)**, the waste stream becomes one-third smaller (as shown by the smaller size of the pie chart). Food scraps are now the largest contributor, followed by plastics, because so much paper is recycled and yard waste is composted. Data from U.S. Environmental Protection Agency, 2010. *Municipal solid waste in the United States: 2009 facts and figures.* EPA, Washington, D.C.

Waste generation is rising

In the United States since 1960, waste generation (before recovery) has increased by 2.8 times, and per-person waste generation has risen by 62%. Paper, food scraps, yard trimmings, and plastics are the principal components of municipal solid waste in the United States, together accounting for 68% of what enters the waste stream (**FIGURE 17.2A**). Paper is recycled at a high rate and yard trimmings are composted at a high rate, so as a result, after recycling and composting reduce the waste stream, food scraps and plastics are left as the largest components of U.S. municipal solid waste (**FIGURE 17.2B**).

Most municipal solid waste comes from packaging and nondurable goods (products meant to be discarded after a short period of use). In addition, consumers throw away old durable goods and outdated equipment as they purchase new products. Plastics, which came into wide consumer use only after 1970, have accounted for the greatest relative increase in the waste stream during the last several decades.

As we acquire more goods, we generate more waste. In 2009, U.S. citizens produced 243 million tons of municipal solid waste (before recovery), over 700 kg (1,500 lb) per person. The average American generates 2.0 kg (4.3 lb) of trash per day—considerably more than people of most other developed nations. The relative wastefulness of the U.S. lifestyle, with its excess packaging and reliance on nondurable goods, has caused critics to label the United States "the throwaway society." U.S. waste generation decreased slightly between 2007 and 2009, but this reflected reduced consumption during economic recession.

In developing nations, people consume less and generate considerably less waste. However, consumption is intensifying in developing nations as they become more affluent, and these nations are creating more and more waste. The increase

in waste reflects rising material standards of living, but it also results from an increase in packaging, manufacturing of nondurable goods, and production of inexpensive, poor-quality goods that wear out quickly. As a result, trash is piling up and littering the landscapes of countries from Mexico to Kenya to Indonesia. Like U.S. consumers in the "throwaway society," wealthy consumers in developing nations often discard items that can still be used. In fact, at many dumps and landfills in the developing world, poor people support themselves by selling items they scavenge (**FIGURE 17.3**).

Wealthier nations can afford to invest more in waste collection and disposal, so they are often better able to manage

FIGURE 17.3 ▼ Tens of thousands of people used to scavenge each day from this dump outside Manila in the Philippines, finding items for themselves and selling material to junk dealers for 100–200 pesos (U.S. $2–$4) per day. That so many people could support themselves this way testifies to the immense amount of usable material discarded by wealthier people. The dump was closed in 2000 after an avalanche of trash killed hundreds of people.

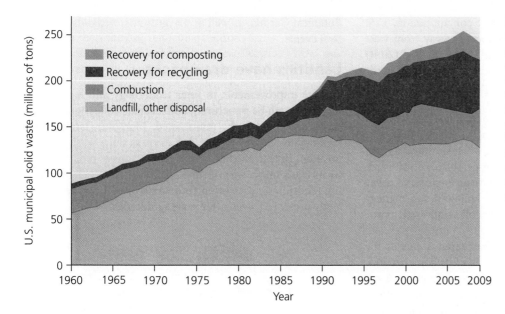

FIGURE 17.4 ◀ Since the 1980s, recycling and composting have grown in the United States, allowing a smaller proportion of waste to go to landfills. As of 2009, 54% of U.S. municipal solid waste was going to landfills and 12% to incinerators, whereas 34% was being recovered for composting and recycling.

Data from U.S. Environmental Protection Agency, 2010. *Municipal solid waste in the United States: 2009 facts and figures.* EPA, Washington, D.C.

their waste and minimize impacts on human health and the environment. Moreover, in many industrialized nations, recovery (recycling and composting) is taking care of an increasingly larger portion of waste (**FIGURE 17.4**). We will examine reduction, reuse, recycling, and composting shortly, but let's first assess how we dispose of waste.

Sanitary landfills are regulated with health and environmental guidelines

In modern **sanitary landfills**, waste is buried in the ground or piled up in large mounds engineered to prevent waste from contaminating the environment and threatening public

health (**FIGURE 17.5**). Most municipal landfills in the United States are regulated locally or by the states, but they must meet national standards set by the U.S. Environmental Protection Agency (EPA) under the federal *Resource Conservation and Recovery Act (RCRA)*, enacted in 1976 and amended in 1984.

In a sanitary landfill, waste is partially decomposed by bacteria and compresses under its own weight to take up less space. Soil is layered along with the waste to speed decomposition, reduce odor, and lessen infestation by pests. Some infiltration of rainwater into the landfill is good, because it encourages biodegradation by aerobic and anaerobic bacteria—yet too much is not good, because contaminants can escape if water carries them out.

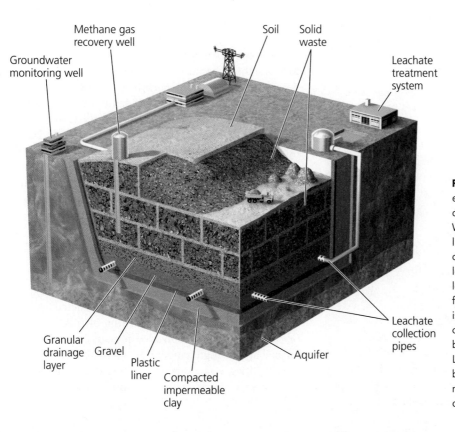

FIGURE 17.5 ◀ Sanitary landfills are engineered to prevent waste from contaminating soil and groundwater. Waste is laid in a large depression lined with plastic and impervious clay designed to prevent liquids from leaching out. Pipes of a leachate collection system draw out these liquids from the bottom of the landfill. Waste is layered along with soil until the depression is filled, and it continues to be built up until the landfill is capped. Landfill gas produced by anaerobic bacteria may be recovered, and waste managers monitor groundwater for contamination.

To protect against environmental contamination, U.S. regulations require that landfills be located away from wetlands and earthquake-prone faults and be at least 6 m (20 ft) above the water table. The bottoms and sides of sanitary landfills must be lined with heavy-duty plastic and 60–120 cm (2–4 ft) of impermeable clay to help prevent contaminants from seeping into aquifers. Sanitary landfills also have systems of pipes, collection ponds, and treatment facilities to collect and treat **leachate**, liquid that results when substances from the trash dissolve in water as rainwater percolates downward. Once a landfill is closed, it is capped with an engineered cover consisting of layers of plastic, gravel, and soil, and managers are required to maintain leachate collection systems for 30 years.

The Fresh Kills Landfill was considered a model for advanced landfill technology at the time of its construction, but it was built before most of the EPA guidelines. As a result, it caused some environmental contamination. However, engineers have retrofitted the landfill with clay liners and a sophisticated leachate collection system. Three of the six mounds have been capped with a "final cover," and the remaining mounds will soon be capped. Because these safeguards need to be maintained for 30 years after closure, designs for a public park at Fresh Kills have had to work around them.

In 1988 the United States had nearly 8,000 landfills, but today it has fewer than 2,000. Waste managers have consolidated the waste stream into fewer landfills of larger size. In many cities, landfills that were closed are now being converted into public parks or other uses (**FIGURE 17.6**). The Fresh Kills endeavor will be the world's largest landfill conversion project, but such efforts date back at least to 1938, when an ash landfill at Flushing Meadows, in Queens, was redeveloped for the 1939 World's Fair. The site subsequently hosted the United Nations and the 1964–1965 World's Fair. Designated a park in 1967, today the site hosts Shea Stadium, the Queens Museum of Art, the New York Hall of Science, and the Queens Botanical Garden, as well as playgrounds, wetlands, and festival events.

Landfills have drawbacks

Despite improvements in liner technology and landfill siting, liners can be punctured, and leachate collection systems eventually cease to be maintained. Moreover, landfills are kept dry to reduce leachate, but dryness slows waste decomposition. In fact, the low-oxygen conditions of most landfills turns trash into a sort of time capsule. Researchers examining landfill contents often find some of their contents perfectly preserved, even after years or decades.

FAQ

Q: How much does garbage decompose in a landfill?

A: You might assume that a banana peel you throw in the trash will soon decay away to nothing in a landfill. However, it might just survive longer than you do! This is because surprisingly little decomposition occurs in landfills. Researcher William Rathje, nicknamed "the Indiana Jones of Solid Waste," made a career out of burrowing into landfills and examining their contents to learn about what we consume and what we throw away. His research teams would routinely come across whole hot dogs, intact pastries that were decades old, and grass clippings that were still green. Newspapers 40 years old are often still legible, and researchers have used them to date layers of trash.

A second challenge with landfills is finding suitable areas to locate them, because most communities do not want them nearby. This *not-in-my-backyard (NIMBY)* reaction is one reason why New York City decided to export its waste—and why residents of states receiving that waste are increasingly protesting. As a result of the NIMBY syndrome, landfills are rarely sited in neighborhoods that are home to wealthy and educated people with the political clout to keep them out. Instead, they are disproportionately sited in poor and minority communities, as environmental justice advocates (pp. 14–15) have frequently pointed out.

Incinerating trash reduces pressure on landfills

Just as sanitary landfills are an improvement over open dumping, incineration in specially constructed facilities is an improvement over open-air burning of trash. **Incineration**, or combustion, is a controlled process in which garbage is burned at very high temperatures (**FIGURE 17.7**). At incineration facilities, waste is generally sorted and metals removed. Metal-free waste is chopped into small pieces and then is

FIGURE 17.6 ▼ Old landfills, once properly capped, can serve other purposes. A number of them, such as Cesar Chavez Park in Berkeley, California, shown here, have been developed into areas for recreation.

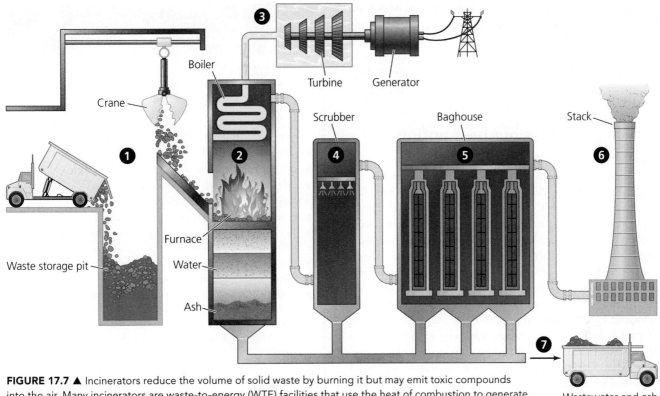

FIGURE 17.7 ▲ Incinerators reduce the volume of solid waste by burning it but may emit toxic compounds into the air. Many incinerators are waste-to-energy (WTE) facilities that use the heat of combustion to generate electricity. In a WTE facility, solid waste ❶ is burned at extremely high temperatures ❷, heating water, which turns to steam. The steam turns a turbine ❸, which powers a generator to create electricity. In an incinerator outfitted with pollution-control technology, toxic gases produced by combustion are mitigated chemically by a scrubber ❹, and airborne particulate matter is filtered physically in a baghouse ❺ before air is emitted from the stack ❻. Ash remaining from the combustion process is disposed of ❼ in a landfill.

Labels in figure: Boiler, Crane, Turbine, Generator, Scrubber, Baghouse, Stack, Furnace, Water, Ash, Waste storage pit, Wastewater and ash for treatment or disposal in landfill

burned in a furnace. Incinerating waste reduces its weight by up to 75% and its volume by up to 90%.

The ash remaining after trash is incinerated contains toxic components and therefore must be disposed of in hazardous waste landfills (p. 395). Moreover, when trash is burned, hazardous chemicals—including dioxins, heavy metals, and PCBs (Chapter 10)—can be created and released into the atmosphere. Such emissions caused a backlash against incineration from citizens concerned about health hazards.

Most developed nations now regulate incinerator emissions, some have banned incineration outright, and engineers have developed technologies to mitigate emissions. Scrubbers (see Figure 13.10, p. 286) chemically treat the gases produced in combustion to remove hazardous components and neutralize acidic gases, such as sulfur dioxide and hydrochloric acid, turning them into water and salt. Scrubbers generally do this either by spraying liquids formulated to neutralize the gases or by passing the gases through dry lime. Particulate matter, called *fly ash*, often contains some of the worst dioxin and heavy metal pollutants in incinerator emissions. To physically remove these tiny particles, facilities may use a system of huge filters known as a *baghouse*. In addition, burning garbage at especially high temperatures can destroy certain pollutants, such as PCBs. Even all these measures, however, do not fully eliminate toxic emissions.

WEIGHING THE ISSUES

Environmental Justice? Do you know where your trash goes? Where is your landfill or incinerator located? Are the people who live closest to the facility wealthy, poor, or middle class? What race or ethnicity are they? Do you know whether the people of this neighborhood protested against the introduction of the landfill or incinerator?

We can gain energy from trash

Incineration reduces the volume of waste, but it often serves to generate electricity as well. Most incinerators now are **waste-to-energy (WTE)** facilities that use the heat produced by waste combustion to boil water, creating steam that drives electricity generation or that fuels heating systems. When burned, waste generates approximately 35% of the energy generated by burning coal.

Combustion in WTE plants is not the only way to gain energy from waste. Deep inside landfills, bacteria decompose waste in an oxygen-deficient environment. This anaerobic decomposition produces **landfill gas**, a mix of gases consisting of roughly half methane (pp. 28, 329). Landfill gas can be collected, processed, and used in the same way as natural gas (pp. 329–330). Today hundreds of landfills are collecting landfill gas and selling it for energy. At Fresh Kills, collection wells pull landfill gas upward through a network of pipes by vacuum pressure. Landfill gas collected from Fresh Kills is sold by the city for $11 million per year and provides energy for 22,000 Staten Island homes.

Reducing waste is our best option

Reducing the amount of material entering the waste stream avoids costs of disposal and recycling, helps conserve resources, minimizes pollution, and can save consumers and businesses money. Recall that diminishing waste generation in this way is known as *source reduction*. One means of source reduction is to lessen the materials used to package goods. Packaging serves worthwhile purposes—preserving freshness, preventing breakage, protecting against tampering, and providing information—but much packaging is extraneous. Consumers can give manufacturers incentive to reduce packaging by choosing minimally packaged goods, buying unwrapped fruit and vegetables, and buying food in bulk. Manufacturers can switch to packaging that is more recyclable. Manufacturers can also reduce the size or weight of goods and materials, as they already have with many items, such as aluminum cans, plastic soft drink bottles, and personal computers. Finally, consumer choice can motivate manufacturers to create goods that last longer.

Some governments have recently taken aim at a major source of waste and litter—plastic grocery bags. These lightweight polyethylene bags can persist for centuries in the environment, choking and entangling wildlife and littering the landscape—yet Americans discard 100 billion of them each year. A number of major cities and over 20 nations have now enacted bans or limits on their use. In 2007, San Francisco became the first U.S. city to ban nonbiodegradable plastic bags. The city's action saves 14,000 bags every day. Financial incentives are also effective. When Ireland began taxing these bags, their use dropped 90%. The Ikea company began charging for them and saw similar drops in usage. Increasing numbers of stores now give discounts if you bring your own reusable bags.

Reuse is a main strategy to reduce waste

To reduce waste, you can save items to use again or substitute disposable goods with durable ones. Habits as simple as bringing your own coffee cup to coffee shops or bringing sturdy reusable cloth bags to the grocery store can, over time, have substantial impact. You can also donate unwanted items and shop for used items yourself at yard sales and resale centers. Over 6,000 reuse centers exist in the United States, including stores run by organizations that resell donated items, such as Goodwill Industries and the Salvation Army. **TABLE 17.1** presents a sampling of actions we all can take to reduce the waste we generate.

Composting recovers organic waste

Composting is the conversion of organic waste into mulch or humus (p. 138) through natural decomposition. People place food and yard waste in compost piles, underground pits, or

TABLE 17.1 Some Everyday Things You Can Do to Reduce and Reuse

- ► Donate used items to charity.
- ► Reuse boxes, paper, plastic wrap, plastic containers, aluminum foil, bags, wrapping paper, fabric, packing material, etc.
- ► Rent or borrow items instead of buying them, when possible . . . and lend your items to friends.
- ► Buy groceries in bulk.
- ► Decline bags at stores when you don't need them.
- ► Bring reusable cloth bags shopping.
- ► Make double-sided photocopies.
- ► Bring your own coffee cup to coffee shops.
- ► Pay a bit extra for durable, long-lasting, reusable goods rather than disposable ones.
- ► Buy rechargeable batteries.
- ► Select goods with less packaging.
- ► Compost kitchen and yard wastes in a compost bin or worm bin (often available from your community or waste hauler).
- ► Buy clothing and other items at resale stores and garage sales.
- ► Use cloth napkins and rags rather than paper napkins and towels.
- ► Write to companies to tell them what you think about their packaging and products.

specially constructed containers. As wastes are added, heat from microbial action builds in the interior, and decomposition proceeds. Banana peels, coffee grounds, grass clippings, autumn leaves, and other organic items can be converted into rich, high-quality compost through the actions of earthworms, bacteria, soil mites, sow bugs, and other detritivores and decomposers (p. 68). The compost is then used to enrich soil. Home composting is a prime example of how we can live more sustainably by mimicking natural cycles and incorporating them into our daily lives.

Municipal composting programs—3,000 across the United States at last count—divert yard debris from the waste stream to central composting facilities, where it decomposes into mulch that community residents can use for gardens and landscaping. Nearly half of U.S. states now ban yard waste from the municipal waste stream, helping to accelerate the drive toward composting. Composting reduces landfill waste, enhances soil biodiversity, helps soil to resist erosion, makes for healthier plants and more pleasing gardens, and reduces the need for chemical fertilizers.

Recycling consists of three steps

Recycling, too, offers many benefits. Recycling involves collecting used items and breaking them down so that their materials can be reprocessed to manufacture new items. The recycling loop includes three basic steps (**FIGURE 17.8**). The first step is collecting and processing used goods and materials.

Communities may designate locations where residents can drop off recyclables or receive money for them. Many of

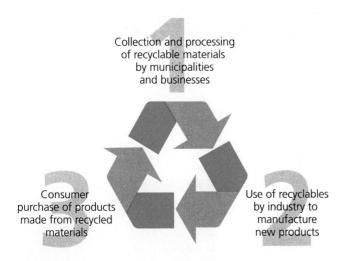

FIGURE 17.8 ▲ The familiar recycling symbol consists of three arrows to represent the three components of a sustainable recycling strategy: collection and processing of recyclable materials, use of the materials in making new products, and consumer purchase of these products.

these have now been replaced by the more convenient option of curbside recycling, in which trucks pick up recyclable items in front of homes, usually in conjunction with municipal trash collection.

Items collected are taken to **materials recovery facilities (MRFs)**, where workers and machines sort items using automated processes including magnetic pulleys, optical sensors, water currents, and air classifiers that separate items by weight and size. The facilities clean the materials, shred them, and prepare them for reprocessing.

Once readied, these materials are used in manufacturing new goods. Newspapers and many other paper products use recycled paper, many glass and metal containers are now made from recycled materials, and some plastic containers are of recycled origin. Some benches, bridges, and walkways in city parks are now made from recycled plastics, and glass is sometimes mixed with asphalt (creating "glassphalt") for paving roads and paths.

If the recycling loop is to function, consumers and businesses must complete the third step in the cycle by purchasing ecolabeled products (p. 109) made from recycled materials. Buying recycled goods provides economic incentive for industries to recycle materials and for new recycling facilities to open or existing ones to expand.

Recycling has grown rapidly

Today 9,000 curbside recycling programs serve nearly half of all Americans. These programs, and the 500 MRFs operating today, have sprung up only in the last 25 years. Recycling in the United States has risen from 6.4% of the waste stream in 1960 to 25.2% in 2009 (and 33.8% if you include composting), according to EPA data (**FIGURE 17.9**).

Recycling rates vary greatly from one product or material type to another—from 7% for plastics to 25% for glass to 62% for paper to 96% for auto batteries. Recycling rates among U.S. states also vary greatly, from 2% to 43%.

Recycling's growth has been propelled in part by economic forces as established businesses see opportunities to save money and as entrepreneurs see opportunities to start new businesses. It has also been driven by the desire of municipalities to reduce waste and by the satisfaction people take in recycling. These latter two forces have driven recycling's rise even when it has not been financially profitable. In fact, many of the increasingly popular municipal recycling programs are run at an economic loss. The expense required to collect, sort, and process recycled goods is often more than recyclables are worth in the market. Furthermore, the more people recycle, the more glass, paper, and plastic is available to manufacturers for purchase, driving down prices. And transporting items to recycling facilities can sometimes involve surprisingly long distances (see **THE SCIENCE BEHIND THE STORY**, pp. 388–389).

Recycling advocates, however, point out that market prices do not take into account external costs (p. 92)—in particular, the environmental and health impacts of *not* recycling. For instance, it has been estimated that globally, recycling saves enough energy to power 6 million households per year. Each year in the United States, recycling and composting together prevent greenhouse gas emissions equal to that of 33 million cars. And recycling aluminum cans saves 95% of the energy required to make the same amount of aluminum from mined virgin bauxite, its source material.

As more manufacturers use recycled products and as more technologies and methods are developed to use recycled materials in new ways, markets should continue to expand and new business opportunities should arise. We are just beginning to shift from an economy that moves linearly

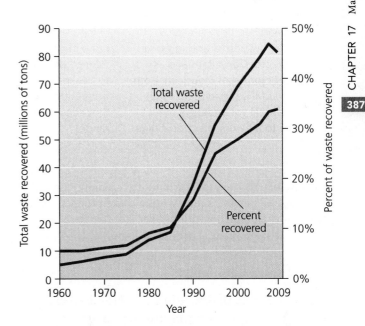

FIGURE 17.9 ▲ Recovery has risen sharply in the United States over the past 50 years. Today over 80 million tons of material is recovered (61 million tons through recycling and 21 million tons through municipal composting), comprising one-third of the waste stream. Data from U.S. Environmental Protection Agency, 2010. *Municipal solid waste in the United States: 2009 facts and figures.* EPA, Washington, D.C.

THE SCIENCE BEHIND THE STORY

Tracking Trash

Where will it go?

Where does your trash go once you throw it away? Where does your recycling go? How far might it travel, and how much energy does it take to get rid of it?

With the help of the latest tracking technology, we can find out. Researchers from the SENSEable City Lab at the Massachusetts Institute of Technology (MIT) are affixing tiny sensors to everyday items in our trash and monitoring them to reveal their hidden travels. By documenting what actually happens to trash and to recyclables, they hope to help make the trash removal process more effective and to encourage better recycling.

The Trash Track project was launched in 2009 in New York City and in Seattle, and is now expanding to other cities. It was inspired by PlaNYC (pp. 411–412), which aims to raise New York City's recycling rate from 30% to 100% by 2030. Early results were unveiled at public exhibitions put on in New York and Seattle in late 2009.

Here's how trash-tracking works: Project director Carlo Ratti and associate director Assaf Biderman, both architects at MIT, organize research teams and local volunteers in the target city to affix tiny electronic tags

(**first figure**) to hundreds of items being thrown away. As each item makes its final journey through the waste stream, its tag calculates its location every few hours and relays the information via the cell phone network to a central server at the MIT lab. A computer plots the movements atop satellite maps, helping the researchers to visualize and interpret the migration of trash.

Each trash tag calculates its position by measuring the signal strength from nearby cell phone towers and comparing this to a map of tower locations. Second-generation tags the project is now using are more accurate, as they combine global positioning system (GPS) technology with better-quality cell network triangulation. To extend battery life to two months or more, the tags are programmed to "sleep" when they are motionless and to "wake up" and report their location frequently when they are moving (when they sense new cell towers coming into range).

Tags used in New York City and Seattle used the cell phone network to calculate their location. Tags now being used in other cities use GPS technology as well.

As an example, a plastic container of liquid soap was tagged on September 5, 2009, and placed in the trash at 457 Madison Avenue in Manhattan (**second figure**). Mapping reveals that the truck that picked it up looped through the city's streets a few times on its route, crossed the Hudson River via the Lincoln Tunnel, and headed to Rutherford, New Jersey. Here it turned south and continued, and was in transit along the Bellevue Turnpike in Kearny, New Jersey, three days later when the tag's battery gave out.

from raw materials to products to waste, to a more sustainable economy that moves circularly using waste products as raw materials for new manufacturing. The steps we have taken in recycling so far are central to this transition.

WEIGHING THE ISSUES

Costs of Recycling and Not Recycling

Should recycling programs be subsidized by governments even if they are run at an economic loss? What types of external costs—costs not reflected in market prices—do you think would be involved in not recycling, say, aluminum cans? Do you feel these costs justify sponsoring recycling programs even when they are not financially self-supporting? Why or why not?

We can recycle material from landfills

With improved technology for sorting rubbish and recyclables, businesses and entrepreneurs are weighing the economic benefits and costs of rummaging through landfills to salvage materials of value that can be recycled. Metals like steel, aluminum, and copper are abundant enough in some landfills to make such salvage operations profitable when market prices for the metals are high enough. For instance, Americans throw out so many aluminum cans that at 2010 prices for aluminum, the nation buries $1.8 billion of this metal in landfills each year. If we could retrieve all the aluminum from U.S. landfills, it would exceed the amount the world produces from a year's worth of mining ore. Besides metals, landfills also offer organic waste that can be mined and sold as premium compost. And old landfill waste can be incinerated

As of 2011, the project had posted mapped data of all its Seattle items on-line, and was preparing to do the same for its New York items.

The results from Seattle reveal some expected patterns but also some odd surprises. Of the 760 trash items tagged, about 200 ended up at the city's Allied Waste Recycling Center and Transfer Station after brief jour-neys. Smaller numbers were trans-ported to other city or regional landfills and recycling centers. But some items followed surprisingly circuitous routes, being transferred from one waste center to another, or back and forth between cities. Some ended up in seemingly random places, perhaps having fallen off a garbage truck along a roadside.

A few items made very long jour-neys. Two printer cartridges were driven down Interstate 5 to California's border with Mexico, perhaps to be disassem-bled at a *maquiladora* (pp. 89, 106) along the border. Two cell phones were transported halfway across the country to Dallas (one flown directly and one making apparent stops at three other cities). A compact fluorescent bulb went to St. Louis after traveling to Portland, Oregon, and back to Seattle. Chicago received a coffee cup that appar-ently made its way east on the nation's interstates. Shipments of batteries were flown 1,500 miles to Minneapolis, 2,500 miles to Pittsburgh, and 2,600 miles to Atlanta.

The longest-traveling piece of trash was a cell phone that was trans-ported over 3,000 miles to the other

corner of the United States, ending up near Ocala, Florida.

Why did these trash items migrate so far? How much gasoline was used to transport them? When we move items such great distances to dispose of them or recycle their parts, does that reflect efficiency from an economy of

scale—or does it indicate an excessive waste of resources? Do we need more recycling and disassembly facilities nearer to each major city? Researchers and waste managers hope to use data from the Trash Track project to address such questions and improve the way we handle waste.

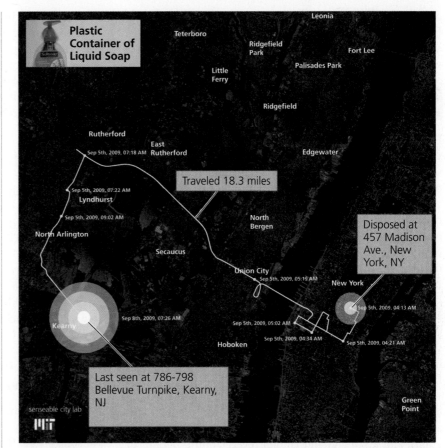

A plastic container of liquid soap put out with the trash on Madison Avenue in New York City looped through midtown Manhattan, crossed the Hudson River, and was last detected traveling down the Bellevue Turnpike in New Jersey.

in newer, cleaner-burning WTE facilities to produce energy. Some companies are even looking into gaining carbon cred-its (p. 322) by harvesting methane leaking from huge open dumps in developing nations in Asia and Africa.

Such approaches are being tried in places from New York to Israel to Sweden to Singapore, and can be profitable when market prices are high enough. The costs of mining landfills and meeting regulatory requirements while commodity pric-es change unpredictably have meant that investing in landfill mining has been risky so far. This could change in the future, though, if prices rise and technologies improve.

Financial incentives help address waste

Waste managers offer consumers economic incentives to reduce the waste stream. In "pay-as-you-throw" garbage collection programs, municipalities charge residents for home trash pickup according to the amount of trash they put out. The less waste the household generates, the less the resident has to pay. Over 7,000 such programs operate in the United States, serving more than one of every four communities.

Bottle bills represent another approach that hinges on financial incentive. In the 11 U.S. states and 17 nations that have these laws, consumers pay a deposit on bottles or cans upon purchase—generally 5 cents per container—and then receive a refund when they return them to stores after use. Where these laws have been enacted, they have proved effec-tive and popular. U.S. states with bottle bills report that their beverage container litter has decreased by 69–84%, their total litter has decreased by 30–64%, and their per capita container recycling rates have risen 2.6-fold. Beverage container recycling rates for states with bottle bills are 2.5 times higher than for states without them. As of 2011, five of the bottle bill states were seeking to expand their programs, and seven other states were considering establishing programs.

One Canadian city showcases reduction and recycling

Edmonton, Alberta, has created one of the world's most advanced waste management programs. Just 40% of the city's waste stream goes to its sanitary landfill, whereas 15% is recycled and an impressive 45% is composted.

When Edmonton's residents put out their trash, city trucks take it to their new co-composting plant—at the size of eight football fields, the largest in North America. The bulk of the waste is mixed with dried sewage sludge for one to two days in immense rotating drums. This mixture travels on a conveyor to a screen that removes nonbiodegradable items. It is aerated for several weeks in the largest stainless steel building in North America (**FIGURE 17.10**). The mix is then passed through a finer screen and is left outside for four to six months. The resulting compost—80,000 tons annually—is made available to area farmers and residents. The facility even filters the air it emits, to eliminate odors.

Edmonton's program includes a state-of-the-art MRF that handles 30,000–40,000 tons of waste annually, a leachate treatment plant, a research center, public education programs, and a wetland and landfill revegetation program. In addition, 100 pipes collect enough landfill gas to power 4,600 homes, bringing thousands of dollars to the city and helping power the new waste management center. Five area businesses reprocess the city's recycled items. Newsprint and magazines are turned into new newsprint and cellulose insulation, and cardboard and paper are converted into building paper and shingles. Household metal is made into rebar and blades for tractors and graders, and recycled glass is used for reflective paint and signs.

390 **FIGURE 17.10** ▼ Edmonton, Alberta, boasts one of North America's most successful waste management programs. Inside the aeration building, which is the size of 14 professional hockey rinks, mixtures of solid waste and sewage sludge are exposed to oxygen and composted for 14–21 days.

Edmonton has just built a new integrated processing and transfer facility to handle both compostable and recyclable waste from homes and businesses, and the city will soon complete a biofuels facility to create ethanol (pp. 359–360) from waste that cannot be recycled or composted. With these new facilities, Edmonton hopes to achieve 90% recovery by 2013.

INDUSTRIAL SOLID WASTE

In the United States, **industrial solid waste** is defined as solid waste that is considered neither municipal solid waste nor hazardous waste under the Resource Conservation and Recovery Act. This includes waste from factories, mining activities, agriculture, petroleum extraction, and more. Each year, U.S. industry generates about 7.6 billion tons of waste, according to the EPA, about 97% of which is wastewater. Thus, very roughly, 230 million tons of solid waste is generated by 60,000 facilities each year—an amount about equal to that of municipal solid waste.

Regulation and economics influence industrial waste generation

Most methods and strategies of waste disposal, reduction, and recycling by industry are similar to those for municipal solid waste. Businesses that dispose of their own waste on site must design and manage their landfills in ways that meet state, local, or tribal guidelines. Other businesses pay to have their waste disposed of at municipal disposal sites. Whereas the federal government regulates municipal solid waste, state or local governments regulate industrial solid waste (with federal guidance). Regulation varies greatly from place to place, but in most cases, state and local regulation of industrial solid waste is less strict than federal regulation of municipal solid waste. In many areas, industries are not required to have permits, install landfill liners or leachate collection systems, or monitor groundwater for contamination.

The amount of waste generated by a manufacturing process is a good measure of its efficiency; the less waste produced per unit or volume of product, the more efficient that process is, from a physical standpoint. However, physical efficiency is not always reflected in economic efficiency. Often it is cheaper for industry to manufacture products or perform services quickly but messily. That is, it can be cheaper to generate waste than to avoid generating waste. In such cases, economic efficiency is maximized, but physical efficiency is not. Because our market system rewards only economic efficiency, all too often industry lacks financial incentive to achieve physical efficiency. The frequent mismatch between these two types of efficiency is a major reason why the output of industrial waste is so great.

Rising costs of waste disposal, however, enhance the financial incentive to decrease waste. Once either government or the market makes the physically efficient use of raw materials also economically efficient, businesses gain financial incentives to reduce their waste.

Industrial ecology seeks to make industry more sustainable

To reduce waste, growing numbers of industries today are experimenting with industrial ecology. A holistic approach that integrates principles from engineering, chemistry, ecology, and economics, **industrial ecology** seeks to redesign industrial systems to reduce resource inputs and to maximize both physical and economic efficiency. Industrial ecologists would reshape industry so that nearly everything produced in a manufacturing process is used, either within that process or in a different one. The larger idea behind industrial ecology is that industrial systems should function more like ecological systems, in which organisms use almost everything that is produced. This principle brings industry closer to the ideal of ecological economists, in which economies function in a circular fashion rather than a linear one (p. 93).

Industrial ecologists pursue their goals in several main ways:

▶ They examine the entire life cycle of a given product—from its origins in raw materials, through its manufacturing, to its use, and finally its disposal—and look for ways to make the process more efficient. This strategy is called **life-cycle analysis**.

▶ They try to identify how waste products from one manufacturing process might be used as raw materials for another one. For instance, used plastic beverage containers

are not refilled because of the potential for contamination, but they can be shredded and reprocessed to make other plastic items, such as benches, tables, and decks.

▶ They examine industrial processes with an eye toward eliminating environmentally harmful products and materials.

▶ They study the flow of materials through industrial systems to look for ways to create products that are more durable, recyclable, or reusable.

Attentive businesses are taking advantage of the insights of industrial ecology to save money while reducing waste. The Swiss Zero Emissions Research and Initiatives (ZERI) Foundation sponsors dozens of innovative projects worldwide that attempt to create goods and services without generating waste. One example involves breweries, currently being pursued in Canada, Sweden, Japan, and Namibia (**FIGURE 17.11**).

Few businesses have taken industrial ecology to heart as much as the carpet tile company Interface, which founder Ray Anderson set on the road to sustainability over a decade ago. Interface asks customers to return used tiles for recycling and for reuse as backing for new carpet. It modified its tile design and its production methods to reduce waste. It adapted its boilers to use landfill gas for its energy needs. Through such steps, Anderson's company cut its waste generation by 80%, its fossil fuel use by 45%, and its water use by 70%—all while saving $30 million per year, holding prices steady for its customers, and raising profits by 49%.

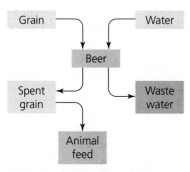

(a) Traditional brewery process

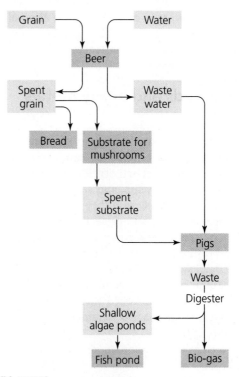

(b) ZERI brewery process

FIGURE 17.11 ▲ Traditional breweries **(a)** produce only beer while generating much waste, some of which goes toward animal feed. ZERI-sponsored breweries **(b)** use their waste grain to make bread and to farm mushrooms. Waste from the mushroom farming, along with brewery wastewater, goes to feed pigs. The pigs' waste is digested in containers that capture natural gas and collect nutrients used to nourish algae for growing fish in fish farms. The brewer derives income from bread, mushrooms, pigs, gas, and fish, as well as beer.

HAZARDOUS WASTE

Hazardous wastes are diverse in their chemical composition and may be liquid, solid, or gaseous. By EPA definition, **hazardous waste** is waste that is either:

► *Ignitable:* Likely to catch fire (for example, natural gas or alcohol)

► *Corrosive:* Apt to corrode metals in storage tanks or equipment

► *Reactive:* Chemically unstable and readily able to react with other compounds, often explosively or by producing noxious fumes

► *Toxic:* Harmful to human health when inhaled, ingested, or touched

Hazardous wastes are diverse

Industry, mining, households, small businesses, agriculture, utilities, and building demolition all create hazardous waste. Industry produces the largest amounts of hazardous waste, but in most developed nations industrial waste generation and disposal is highly regulated. This regulation has limited the amount of hazardous waste entering the environment from industrial activities. As a result, households now are the largest source of unregulated hazardous waste.

Household hazardous waste includes a wide range of items, such as paints, batteries, oils, solvents, cleaning agents, lubricants, and pesticides. U.S. citizens generate 1.6 million tons of household hazardous waste annually, and the average home contains close to 45 kg (100 lb) of it in sheds, basements, closets, and garages.

Many hazardous substances become less hazardous over time as they degrade, but some show especially persistent effects. Radioactive substances are an example, and the disposal of radioactive waste poses a serious dilemma (pp. 349–350). Other types of persistent hazardous substances include organic compounds and heavy metals.

Organic compounds and heavy metals can be hazardous

In our day-to-day lives, we rely on the capacity of synthetic organic compounds and petroleum-derived compounds to resist bacterial, fungal, and insect activity. Plastic containers, rubber tires, pesticides, solvents, and wood preservatives are useful to us precisely because they resist decomposition. We use these substances to protect our buildings from decay, kill pests that attack crops, and keep stored goods intact. However, these compounds' capacity to resist decay is a double-edged sword, for it also makes them persistent pollutants. Many synthetic organic compounds are toxic because they are readily absorbed through the skin and can act as mutagens, carcinogens, teratogens, and endocrine disruptors (p. 212).

Heavy metals such as lead, chromium, mercury, arsenic, cadmium, tin, and copper are used widely in industry for wiring, electronics, metal plating, metal fabrication, pigments, and dyes. Heavy metals enter the environment when paints, electronic devices, batteries, and other materials are disposed of improperly. Lead from fishing weights and hunting ammunition continues to accumulate in rivers, lakes, and forests. In older homes, lead from pipes contaminates drinking water, and lead paint remains a problem, especially for infants. Heavy metals that are fat soluble and break down slowly are prone to bioaccumulate and biomagnify (pp. 216–217).

"E-waste" is growing

Today's proliferation of computers, printers, cell phones, handheld devices, TVs, DVD players, fax machines, MP3 players, and other electronic technology has created a substantial new source of waste (see **ENVISIONIT**, p. 393). These products have short lifetimes before people judge them obsolete, and most are discarded after only a few years. The amount of this **electronic waste**—often called **e-waste**—is growing rapidly, and now comprises 2% of the U.S. solid waste stream. Over 3 billion electronic devices have been sold in the United States since 1980. Of these, half have been disposed of, while about 40% are still being used (or reused) and 10% are in storage. American households discard close to 400 million electronic devices per year—two-thirds of them still in working order (**FIGURE 17.12A**).

Of the electronic items we discard, roughly four of five go to conventional sanitary landfills and incinerators. However, most electronic products contain heavy metals and toxic flame retardants. Recent research suggests that e-waste should instead be treated as hazardous waste, so the EPA and a number of states are now taking steps to do so.

More and more e-waste today is being recycled. Devices are taken apart, and parts and materials are refurbished and reused in new products. As an example of such recycling, the stylish gold, silver, and bronze medals awarded to athletes in the 2010 Winter Olympic Games in Vancouver (**FIGURE 17.12B**) were made from metals recovered from e-waste! According to EPA estimates, however, Americans still recycle

FIGURE 17.12 ▼ Discarded electronic items **(a)** can leach heavy metals and should be considered hazardous waste, researchers say. However, we can recycle this waste and mine it for precious metals. The gold, silver, and bronze medals awarded to athletes at the 2010 Winter Olympic Games in Vancouver **(b)** were manufactured in part from precious metals recycled from discarded e-waste.

(a) Discarded computer monitors   **(b) Vancouver Olympic medals**

Every five minutes, Americans throw away the number of cell phones shown on this page.

The 426,000 cell phones entering the U.S. waste stream daily can leach toxic heavy metals into the environment ...

... or they can be recycled for reuse and for the recovery of valuable metals.

You Can Make a Difference

➤ Recycle your old phone with an approved e-waste recycling service.

➤ Donate your phone to a person or a charity that can reuse it.

➤ Think twice before buying yet another new electronic gadget that you don't really need.

only one-fifth of their e-waste, and so many more items are manufactured each year that the amounts produced and discarded are growing faster than the amount recycled.

Besides keeping toxic substances out of our environment, recycling e-waste is beneficial because a number of trace metals used in electronics are globally rare, so they can be lucrative to recover. A typical cell phone contains close to a dollar's worth of precious metals. Every bit of metal we can recycle from a manufactured item is a bit of metal we don't need to mine from the ground. Thus, "mining" e-waste for precious metals helps reduce environmental impacts from mining. By one estimate, 1 ton of computer scrap contains more gold than 16 tons of mined ore from a gold mine.

There are serious concerns about the health risks that e-waste recycling may pose to workers doing the disassembly. Wealthy nations ship much of their e-waste to developing countries, where low-income workers disassemble the devices and handle toxic materials with minimal safety regulations. These environmental justice concerns need to be resolved if electronics recycling is to be conducted safely and responsibly.

In many North American cities, used electronics are collected by businesses, nonprofit organizations, or municipal services, and are processed for reuse or recycling. So next time you upgrade to a new computer, TV, cell phone, or handheld device, find out what opportunities exist in your area to recycle your old ones.

Several steps precede the disposal of hazardous waste

Many communities designate sites or special collection days to gather household hazardous waste, or designate facilities for the exchange and reuse of substances (**FIGURE 17.13**).

FIGURE 17.13 ▼ Many communities designate collection sites or collection days for household hazardous waste. Here, workers handle waste from an Earth Day collection event near Los Angeles.

394

FIGURE 17.14 ▲ Unscrupulous individuals or businesses sometimes dump hazardous waste illegally to avoid disposal costs.

Once consolidated, the waste is transported for treatment and disposal.

Under the Resource Conservation and Recovery Act, the EPA sets standards by which states manage hazardous waste. RCRA also requires large generators of hazardous waste to obtain permits. Finally, RCRA mandates that hazardous materials be tracked "from cradle to grave." As hazardous waste is generated, transported, and disposed of, the producer, carrier, and disposal facility must each report to the EPA the type and amount of material generated; its location, origin, and destination; and the way it is handled.

Because current U.S. law makes disposing of hazardous waste quite costly, irresponsible companies sometimes illegally dump waste, creating health risks for residents and financial headaches for local governments forced to deal with the mess (**FIGURE 17.14**). Companies from industrialized nations also sometimes dump hazardous waste illegally in developing nations—a major environmental justice issue (p. 15). This occurs despite the Basel Convention, an international treaty to prevent this practice.

High costs of disposal, however, have also encouraged conscientious businesses to invest in reducing their hazardous waste. Many biologically hazardous materials can be broken down by incineration at high temperatures in cement kilns. Others can be treated by exposure to bacteria that break down harmful components and synthesize them into new compounds. Additionally, various plants have been bred or engineered to take up specific contaminants from soil and then break down organic contaminants into safer compounds or concentrate heavy metals in their tissues. The plants are eventually harvested and disposed of.

We use three disposal methods for hazardous waste

We have developed three primary means of hazardous waste disposal: landfills, surface impoundments, and injection wells. These do nothing to lessen the hazards of the substances, but they help keep the waste isolated from people, wildlife, and ecosystems. Design and construction standards for landfills that receive hazardous waste are stricter than those for ordinary sanitary landfills. Hazardous waste landfills must have several impervious liners and leachate removal systems and must be located far from aquifers. Dumping of hazardous waste in ordinary landfills has long been a problem. In New York City, Fresh Kills largely managed to keep hazardous waste out, but most of the city's older landfills were declared to be hazardous sites because of past toxic waste dumping.

Liquid hazardous waste, or waste in dissolved form, may be stored in **surface impoundments**, shallow depressions lined with plastic and an impervious material, such as clay. The liquid or slurry is placed in the impoundment and allowed to evaporate, leaving a residue of solid hazardous waste on the bottom. This process is repeated and eventually the dry residue is removed and transported elsewhere for permanent disposal. Impoundments are not ideal. The underlying layer can crack and leak waste. Some material may evaporate or blow into surrounding areas. Rainstorms may cause waste to overflow and contaminate nearby areas. For these reasons, surface impoundments are used only for temporary storage.

The third method is intended for long-term disposal. In **deep-well injection**, a well is drilled deep beneath the water table into porous rock, and wastes are injected into it (**FIGURE 17.15**). The waste is meant to remain deep underground, isolated from groundwater and human contact. However, wells can corrode and can leak wastes into soil, contaminating aquifers. Roughly 34 billion L (9 billion gal) of hazardous waste are placed in U.S. injection wells each year.

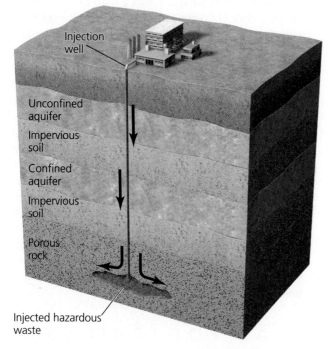

Injection well

Unconfined aquifer

Impervious soil

Confined aquifer

Impervious soil

Porous rock

Injected hazardous waste

FIGURE 17.15 ▲ Liquid hazardous waste may be pumped deep underground by deep-well injection. The well must be drilled below any aquifers, into porous rock separated by impervious clay. The technique is expensive, and waste may leak from the well shaft into groundwater.

Contaminated sites are being cleaned up, slowly

Many thousands of former military and industrial sites remain contaminated with hazardous waste in the United States and virtually every other nation on Earth. For most nations, dealing with these messes is simply too difficult, time-consuming, and expensive. In 1980, however, the U.S. Congress passed the Comprehensive Environmental Response Compensation and Liability Act (CERCLA). This law established a federal program to clean up U.S. sites polluted with hazardous waste. The EPA administers this cleanup program, called the **Superfund**. Under EPA auspices, experts identify polluted sites, take action to protect groundwater, and clean up the pollution. Later laws also charged the EPA with cleaning up **brownfields**, lands whose reuse or development are complicated by the presence of hazardous materials.

Two well-publicized events spurred creation of the Superfund legislation. In *Love Canal*, a residential neighborhood in Niagara Falls, New York, families were evacuated in 1978–1980 after toxic chemicals buried by a company and the city in past decades rose to the surface, contaminating homes and an elementary school. In Missouri, the entire town of *Times Beach* was evacuated and its buildings demolished after being contaminated in the 1970s by dioxin (p. 213) from waste oil sprayed on its roads.

Once a Superfund site is identified, EPA scientists evaluate how close the site is to human habitation, whether wastes are currently confined or likely to spread, and whether the site threatens drinking water supplies. Sites that appear harmful are placed on the National Priorities List, ranked according to the level of risk to human health that they pose. Cleanup proceeds as funds are available. Throughout the process, the EPA is required to hold public hearings to inform area residents of its findings and to receive feedback.

The objective of CERCLA was to charge the polluting parties for cleanup of their sites, according to the *polluter-pays principle* (p. 107). For many sites, however, the responsible parties cannot be found or held liable, and in such cases—roughly 30% so far—cleanups have been covered by taxpayers' funds and from a trust fund established by a federal tax on industries producing petroleum and chemical raw materials. However, Congress let the tax expire and the trust fund went bankrupt in 2004, so taxpayers are now shouldering the entire burden of the program. As funding dwindles and the remaining cleanup jobs become more expensive, fewer cleanups are being completed.

As of 2011, 1,288 Superfund sites remained on the National Priorities List, and only 349 had been cleaned up or otherwise deleted from the list. The average cleanup has cost over $25 million and has taken 15 years. Many sites are contaminated with chemicals we have no effective way to deal with. In such cases, cleanups aim simply to isolate waste from human contact, either by building trenches and clay or concrete barriers around a site or by excavating contaminated material and shipping it to a hazardous waste disposal facility. For all these reasons, the current emphasis in the United States and elsewhere is on preventing hazardous waste contamination in the first place.

➤ CONCLUSION

We have made great strides in addressing our waste problems. Modern methods of waste management are far safer for people and gentler on the environment than past practices of open dumping and open burning. Recycling and composting efforts are advancing steadily, and Americans now divert one-third of all solid waste away from disposal. The continuing growth of recycling, driven by market forces, public policy, and consumer behavior, shows potential to further alleviate our waste problems.

Despite these advances, our prodigious consumption habits have created more waste than ever before. Our waste management efforts are marked by a number of difficult challenges, including the cleanup of Superfund sites, safe disposal of hazardous and radioactive waste, and frequent local opposition to disposal sites. These dilemmas make clear that the best solution is to reduce our generation of waste. Finding ways to reduce, reuse, and efficiently recycle the materials and goods that we use stands as a key challenge for our new century.

TESTING YOUR COMPREHENSION

1. Describe five major methods of managing waste. Why do we practice waste management?

2. Why have some people labeled the United States "the throwaway society"? How much solid waste do Americans generate, and how does this amount compare to that of people from other countries?

3. Name several guidelines by which sanitary landfills are regulated. Describe three problems with landfills.

4. Describe the process of incineration or combustion. What happens to the resulting ash? What is one drawback of incineration?

5. What is composting, and how does it help reduce input to the waste stream?

6. What are the three elements of a sustainable process of recycling?

7. In your own words, describe the goals of industrial ecology.

8. What four criteria are used to define hazardous waste? Why are heavy metals and synthetic organic compounds particularly hazardous?

9. What are the largest sources of hazardous waste? Describe three ways to dispose of hazardous waste.

10. What is the Superfund program? How does it work?

SEEKING SOLUTIONS

1. How much waste do you generate? Look into your waste bin at the end of the day, and categorize and measure the waste there. List all other waste you may have generated in other places throughout the day. How much of this waste could you have avoided generating? How much could have been reused or recycled?

2. Some people have criticized current waste management practices as merely moving waste from one medium to another. How might this criticism apply to the methods now in practice? What are some potential solutions?

3. Of the various waste management approaches covered in this chapter, which ones are your community or campus pursuing, and which are they not pursuing? Would you suggest that your community or campus start pursuing any new approaches? If so, which ones, and why?

4. **THINK IT THROUGH** You are the CEO of a major corporation that produces containers for soft drinks and a wide variety of other consumer products. Your company's

shareholders are asking that you improve the company's image—while not cutting into profits—by taking steps to reduce waste. What steps would you consider taking?

5. **THINK IT THROUGH** You are the president of your college or university. Your trustees want you to engage with local businesses and industries in ways that benefit both the school and the community. Your faculty and students want you to make the school a leader in waste reduction and industrial ecology. Consider the industries and businesses in your community and the ways they interact with facilities on your campus. Bearing in mind the principles of industrial ecology, can you think of any novel ways that your school and local businesses might mutually benefit from one another's services, products, or waste materials? Are there waste products from one business, industry, or campus facility that another might put to good use? Can you design an eco-industrial park that might work on your campus? What steps would you propose to take as president?

CALCULATING ECOLOGICAL FOOTPRINTS

The 17th biennial "State of Garbage in America" survey documents the ability of U.S. residents to generate prodigious amounts of municipal solid waste (MSW). According to the survey, on a per capita basis, Missouri residents generate the least MSW (4.49 lb/day), and Hawaii residents

generate the most (15.84 lb/day). The average for the entire country is 7.01 lb MSW per person per day. Calculate the amount of MSW generated in 1 day and in 1 year by each of the groups indicated, at each of the rates shown in the accompanying table.

Groups generating municipal solid waste	Amount of municipal solid waste generated					
	U.S. average (7.01 lb/day)		Missouri (4.49 lb/day)		Hawaii (15.84 lb/day)	
	Day	Year	Day	Year	Day	Year
You	7.01	2,559				
Your class						
Your state						
United States						
World						

Data from van Haaren, R., et al., 2010. The state of garbage in America. BioCycle 51(10): 16–23.

1. Suppose your town of 50,000 people has just approved construction of a landfill nearby. Estimates are that it will accommodate 1 million tons of MSW. Assuming the landfill is serving only your town, and that your town's residents generate waste at the U.S. average rate, for how many years will it accept waste before filling up? How much longer would a landfill of the same capacity serve a town of the same size in Missouri?

2. One study has estimated that the average world citizen generates 1.47 pounds of trash per day. How many times more does the average U.S. citizen generate?

3. The same study showed that the average resident of a low-income nation generates 1.17 pounds of waste per day and the average resident of a high-income nation generates 2.64 pounds per day. Why do you think U.S. residents generate so much more MSW than people in other "high-income" countries, when standards of living in those countries are comparable?

Mastering ENVIRONMENTAL SCIENCE™

Go to **www.masteringenvironmentalscience.com** for homework assignments, practice quizzes, Pearson eText, and more.

18 The Urban Environment: Creating Sustainable Cities

Upon completing this chapter, you will be able to:

➤ Describe the scale of urbanization

➤ Assess urban and suburban sprawl

➤ Outline city and regional planning and land use strategies

➤ Evaluate mass transit, urban parks, smart growth, and new urbanism

➤ Analyze environmental impacts and advantages of urban centers

➤ Assess urban ecology, green building efforts, and the pursuit of sustainable cities

Pedestrians, mass transit, and greenery in downtown Portland, Oregon

Managing Growth in Portland, Oregon

"Sagebrush subdivisions, coastal condomania, and the ravenous rampage of suburbia in the Willamette Valley all threaten to mock Oregon's status as the environmental model for the nation."
—Oregon Governor Tom McCall, 1973

"We have planning boards. We have zoning regulations. We have urban growth boundaries and 'smart growth' and sprawl conferences. And we still have sprawl."
—Environmental Scientist Donella Meadows

With the fighting words above, Oregon governor Tom McCall challenged his state's legislature in 1973 to take action against runaway sprawling development, which many Oregon residents feared would ruin the communities and landscapes they loved. McCall was echoing the growing concerns of state residents that farms, forests, and open space were being gobbled up and paved over.

Foreseeing a future of subdivisions, strip malls, and traffic jams engulfing the pastoral Willamette Valley, Oregon acted. The state legislature passed Senate Bill 100, a sweeping land use law that would become the focus of acclaim, criticism, and careful study for years afterward by other states and communities trying to manage their own urban and suburban growth.

Oregon's law required every city and county to draw up a comprehensive land use plan in line with statewide guidelines that had gained popular support from the state's electorate. As part of each land use plan, each metropolitan area had to establish an **urban growth boundary (UGB)**, a line on a map intended to separate areas desired to be urban from areas desired to remain rural. Development for housing, commerce, and industry would be encouraged within these urban growth boundaries but restricted beyond them. The intent was to revitalize city centers, prevent suburban sprawl, and protect farmland, forests, and open landscapes around the edges of urbanized areas.

Residents of the area around Portland, the state's largest city, established a new regional planning entity to apportion land in their region. The Metropolitan Service District, or Metro, represents three counties and 25 municipalities. Metro adopted the Portland-area urban growth boundary in 1979 and has tried to focus growth in existing urban centers and to build communities where people can walk or take mass transit between home, work, and shopping. These policies have largely worked as intended. Portland's downtown and older neighborhoods have thrived, regional urban centers are becoming denser and more community oriented, mass transit has expanded, and development has been limited on land beyond the UGB. Portland began attracting international attention for its "livability."

To many Portlanders today, the UGB remains the key to maintaining quality of life in city and countryside alike. In the view of its critics, however, the "Great Wall of Portland" is an elitist and intrusive government regulatory tool. In 2004, Oregon voters approved a ballot measure that threatened to eviscerate the land use rules that most citizens had backed for three decades. Ballot Measure 37 required the state to compensate certain landowners if government regulation had decreased the value of their land. For example, regulations prevent landowners outside UGBs from subdividing their lots and selling them for housing development. Under Measure 37, the state had to pay these landowners to make up for theoretically lost income, or else allow them to ignore the regulations. Because state and local governments did not have enough money to pay such claims, the measure was on track to gut Oregon's zoning, planning, and land use rules.

Landowners filed over 7,500 claims for payments or waivers affecting 296,000 ha (730,000 acres). Although the measure had been promoted to voters as a way to protect the rights of small family landowners, most claims were filed by large developers. Neighbors suddenly found themselves confronting the prospect of massive housing subdivisions, gravel mines, strip malls, or industrial facilities being developed next to their homes—and many who had voted for Measure 37 now had misgivings.

The state legislature, under pressure from opponents and supporters alike, settled on a compromise: to introduce a new ballot measure. Oregon's voters passed Ballot Measure 49 in 2007. It protects the rights of small landowners to gain income from their property by developing small numbers of homes, while restricting large-scale development and development in sensitive natural areas.

In 2010, Metro finalized a historic agreement with representatives and citizens of its region's three counties to determine where urban growth will and will not be allowed over the next 50 years. Metro and the counties apportioned over 121,000 ha (300,000 acres) of undeveloped land into "urban reserves" open for development and "rural reserves" where farmland and forests would be preserved. Boundaries were precisely mapped, to give clarity and direction for landowners and governments alike for half a century.

People are confronting similar issues in communities throughout North America, and debates and negotiations like those in Oregon will determine how our cities and landscapes will change in the future. ∎

OUR URBANIZING WORLD

We live at a turning point in human history. Beginning in 2009, for the first time ever, more people were living in urban areas (cities and suburbs) than in rural areas. As we undergo this historic shift from the countryside into towns and cities—a process called **urbanization**—two pursuits become ever more important. One is to make our urban areas more livable by meeting residents' needs for a safe, clean, healthy urban environment. The other is to make our urban systems sustainable by creating cities that can prosper in the long term while minimizing our ecological footprint and working with natural systems (rather than against them).

Industrialization has driven urbanization

Since 1950, the world's urban population has grown by 4.7 times, whereas the rural population has not quite doubled. Urban populations are growing for two reasons: (1) The human population overall is growing (Chapter 6), and (2) more people are moving from farms to cities than are moving from cities to farms. Industrialization (p. 3) has reduced the need for farm labor while enhancing commerce and jobs in cities. Urbanization, in turn, has bred technological advances that boosted production efficiencies and spurred further industrialization.

By 2050, the United Nations projects that the urban population will nearly double, whereas the rural population will decline by 16%. Trends differ between developed and developing nations, however (**FIGURE 18.1**). In developed nations,

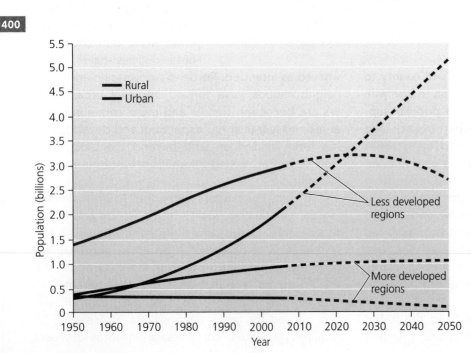

FIGURE 18.1 ◄ In developing nations today, urban populations are growing quickly, whereas rural populations are expected to begin declining. Developed nations are already largely urbanized, so here urban populations are growing more slowly, whereas rural populations are falling. Solid lines in the graph indicate past data, and dashed lines indicate projections of future trends. Data from United Nations Population Division (UNPD), 2010. *World urbanization prospects: The 2009 revision.* New York: UNPD. Reprinted with permission

(a) St. Louis, Missouri

(b) Fort Worth, Texas

FIGURE 18.2 ▲ St. Louis **(a)** is situated on the Mississippi River near its confluence with the Missouri River—and this strategic location for trade drove the city's growth in the 19th and early 20th centuries. Fort Worth, Texas **(b)** grew in the late 20th century as a result of the interstate highway system and a major international airport.

urbanization has slowed because three of every four people already live in cities, towns, and **suburbs**, the smaller communities that ring cities. Back in 1850, the U.S. Census Bureau classified only 15% of U.S. citizens as urban dwellers. That percentage now stands at 80%. Most U.S. urban dwellers reside in suburbs; fully half the U.S. population today is suburban.

In contrast, today's developing nations, where most people still reside on farms, are urbanizing rapidly. In China, India, Pakistan, Nigeria, and most other nations, rural people are streaming to cities in search of jobs and urban lifestyles, or to escape ecological degradation in the countryside. As a result, demographers estimate that urban areas of developing nations will absorb nearly all of the world's population growth from now on.

Environmental factors influence the location of urban areas

Real estate agents use the saying, "Location, location, location," to stress how a home's whereabouts determines its value. Location is vital for urban centers as well. Environmental variables such as climate, topography, and the configuration of waterways influence whether a city will succeed. Think of any major city, and chances are it's situated along a river, seacoast, railroad, or highway—some corridor for trade that has driven economic growth (**FIGURE 18.2**).

Well-located cities often serve as linchpins in trading networks, funneling in resources from agricultural regions, processing them and manufacturing products, and shipping those products to other markets. Portland is situated where the Willamette River joins the Columbia River, and just upriver from where the Columbia flows into the Pacific Ocean. The city grew as it received, processed, and shipped overseas the produce from farms of the river valleys, and as it imported products from other North American ports and from Asia.

Today, powerful technologies and cheap transportation enabled by fossil fuels have allowed cities to thrive even in resource-poor regions. The Dallas–Fort Worth area prospers

from—and relies on—oil-fueled transportation by interstate highways and a major airport. Southwestern cities such as Los Angeles, Las Vegas, and Phoenix flourish in desert regions by appropriating water from distant sources. Whether such cities can sustain themselves as oil and water become increasingly scarce in the future is an important question.

In recent years, many cities in the southern and western United States have grown as people (particularly retirees) have moved south and west in search of warmer weather or more space. Between 1990 and 2009, the population of the Dallas–Fort Worth and Houston metropolitan areas each grew by over 50%; that of the Atlanta area grew by 85%; that of the Phoenix region grew by 94%; and that of the Las Vegas metropolitan area grew by a whopping 122%.

People have moved to suburbs

American cities expanded quickly in the 19th and early 20th centuries, as a result of immigration from abroad and increased trade as the nation spread westward. The bustling economic activity of downtown districts held people in cities despite growing crowding, poverty, and crime. However, by the mid-20th century, many affluent city dwellers were choosing to move outward to cleaner, less crowded, and more parklike suburban communities. These people were pursuing more space, cheaper real estate, less crime, and better schools for their children.

As affluent people moved outward into the expanding suburbs, jobs followed. This hastened the economic decline of downtown districts, and American cities stagnated. Chicago's population declined to 80% of its peak because so many residents moved to its suburbs. Philadelphia's population fell to 76% of its peak, Washington, D.C.'s to 71%, and Detroit's to just 55%.

Portland followed this trajectory, but it also illustrates how some cities have bounced back. After growing rapidly for decades, Portland's population growth stalled in the 1950s to 1970s as crowding and deteriorating economic conditions drove city dwellers to the suburbs. However, subsequent policies to revitalize the city center helped restart Portland's growth (**FIGURE 18.3**).

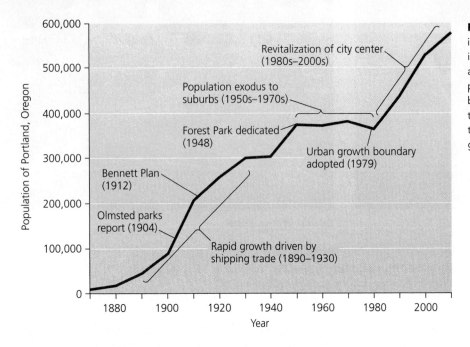

FIGURE 18.3 ◀ Once Portland established its position as a strategically located port, international shipping trade provided jobs and boosted its economy, and the city's population grew rapidly. City residents began leaving for the suburbs in the 1950s to 1970s, but policies designed to enhance the city center then revitalized the city's growth.

The exodus to the suburbs in 20th-century America was enabled by the rise of the automobile, an expanding road network, and inexpensive and abundant oil. Millions of people could now commute by car to downtown workplaces from new homes in suburban "bedroom communities." Automobiles, highway networks, and cheap oil also made it easier for businesses to import and export resources, goods, and waste. The U.S. government's development of the interstate highway system was pivotal in promoting these trends.

In most ways, suburbs have delivered the qualities people sought in them. The wide spacing of homes, with each one on its own plot of land, gives families room and privacy. However, by allotting more space to each person, suburban growth has spread human impact across the landscape. Natural areas have disappeared as housing developments are constructed. Our extensive road networks ease travel, but suburbanites find themselves needing to climb into a car to get anywhere. People commute longer distances to work and spend more time stuck in traffic. The expanding rings of suburbs surrounding cities have grown larger than the cities themselves, and towns are merging into one another. These aspects of suburban growth inspired a new term: *sprawl*.

SPRAWL

The term *sprawl* has become laden with meanings and suggests different things to different people, but we can begin our discussion by giving **sprawl** a simple, nonjudgmental definition: the spread of low-density urban or suburban development outward from an urban center.

Urban areas spread outward

The spatial growth of urban and suburban areas is clear from maps and satellite images of rapidly spreading cities such as

Las Vegas (**FIGURE 18.4**). Another example is Chicago, whose metropolitan area now spreads over a region 40 times the size of the city. All in all, houses and roads supplant over 2,700 ha (6,700 acres) of U.S. land every day.

Several development approaches can lead to sprawl (**FIGURE 18.5**). These approaches allot each person more space than in cities. For example, the average resident of Chicago's suburbs takes up 11 times more space than a resident of the city. As a result, the outward spatial growth of suburbs across the landscape generally outpaces growth in numbers of people. In fact, many researchers define *sprawl* as the physical spread of development at a rate that exceeds the rate of population growth. For instance, the population of Phoenix grew 12 times larger between 1950 and 2000, yet its land area grew 27 times larger. Between 1950 and 1990, the population of 58 major U.S. metropolitan areas rose by 80%, but the land area they covered rose by 305%. Even in 11 metro areas where population declined between 1970 and 1990 (for instance, Rust Belt cities such as Detroit, Cleveland, and Pittsburgh), the amount of land covered increased.

Sprawl has several causes

There are two main components of sprawl. One is human population growth—there are simply more people alive each year. The other is per capita land consumption—each person is taking up more land than in the past, because most people desire space and privacy and dislike congestion. Better highways, inexpensive gasoline, and telecommunications technologies and the Internet have all fostered movement away from city centers by giving workers greater flexibility to live where they desire, and by freeing businesses from dependence on the centralized infrastructure a major city provides.

Economists, politicians, and city boosters have encouraged the unbridled spatial expansion of cities and suburbs. The

(a) Las Vegas, Nevada, 1984

(b) Las Vegas, Nevada, 2009

FIGURE 18.4 ▲ Satellite images show the type of rapid urban and suburban expansion commonly dubbed *sprawl*. Las Vegas, Nevada, is one of the fastest-growing cities in North America. Between 1984 **(a)** and 2009 **(b)**, its population and its developed area each tripled.

conventional assumption has been that growth is always good and that attracting business, industry, and residents will enhance a community's economic well-being, political power, and cultural influence. Today this assumption is being challenged, as growing numbers of people feel the negative effects of sprawl on their lifestyles.

What is wrong with sprawl?

To some people, the word *sprawl* evokes strip malls, traffic jams, homogenous commercial development, and tracts of cookie-cutter houses encroaching on farmland, ranchland, and forests. For other people, sprawl is simply the collective result of choices made by millions of well-meaning individuals trying to make a better life for themselves and their families. What does scientific research tell us about the actual impacts of sprawl?

Transportation Most studies show that sprawl constrains transportation options, essentially forcing people to drive cars. Across the United States, during the 1980s and 1990s the average length of work trips rose by 36%, and total vehicle miles driven rose three times faster than population growth. Besides encouraging traffic congestion, an automobile-oriented culture increases dependence on petroleum, with its economic and environmental consequences (pp. 336–343).

Pollution By promoting automobile use, sprawl increases pollution. Carbon dioxide emissions from vehicles contribute to global climate change (Chapter 14) while nitrogen- and sulfur-containing air pollutants contribute to urban

smog and acid deposition (pp. 287–294). Motor oil and road salt from roads and parking lots run off readily and pollute waterways.

Health Aside from the health impacts of pollution, some research suggests that sprawl promotes physical inactivity and obesity because driving cars largely takes the place of walking during daily errands. A 2003 study found that people from the most-sprawling U.S. counties show higher blood pressure and weigh 2.7 kg (6 lb) more for their height than people from the least-sprawling U.S. counties.

Land use The spread of low-density development means that more land is developed while less is left as forests, fields, farmland, or ranchland. Natural lands and agricultural lands provide vital resources, recreation, aesthetic beauty, wildlife habitat, air and water purification, and other ecosystem services (pp. 2, 36, 90, 94–95).

Economics Sprawl drains tax dollars from communities and funnels money into infrastructure for new development on the fringes of those communities. Funds that could be spent maintaining downtown centers are instead spent on extending the road system, water and sewer system, electricity grid, telephone lines, police and fire service, schools, and libraries. Advocates for sprawling development argue that taxes on new development eventually pay back the investment in infrastructure, but studies have found that in most cases taxpayers continue to subsidize new development unless municipalities pass on infrastructure costs to developers.

(a) Uncentered commercial strip development

(b) Low-density single-use development

(c) Scattered, or leapfrog, development

(d) Sparse street network

FIGURE 18.5 ▲ Several standard approaches to development can result in sprawl. In uncentered commercial strip development **(a)**, businesses are arrayed in a long strip along a roadway, and no attempt is made to create a centralized community with easy access for consumers. In low-density, single-use residential development **(b)**, homes are located on large lots in residential tracts far away from commercial amenities. In scattered or leapfrog development **(c)**, developments are created at great distances from a city center and are not integrated. In developments with a sparse street network **(d)**, roads are far enough apart that moderate-sized areas go undeveloped, but not far enough apart for these areas to function as natural areas or sites for recreation. All these development approaches necessitate frequent automobile use.

THE ISSUES

Sprawl Near You Is there sprawl in the area where you live? Does it bother you, or not? Has development in your area had any of the impacts described above? Do you think your city or town should use its resources to encourage outward growth? Why or why not?

CREATING LIVABLE CITIES

To respond to the challenges that sprawl presents, architects, planners, developers, and policymakers today are trying to restore the vitality of city centers and to plan and manage how urbanizing areas develop. They aim to make cities safer, cleaner, healthier, and more pleasant for city residents.

City and regional planning aim to create livable urban areas

How can we design cities so as to maximize their efficiency, functionality, and beauty? These are the questions central to **city planning** (also known as **urban planning**). City planners advise policymakers on development options, transportation needs, public parks, and other matters.

Washington, D.C., is the earliest example of city planning in the United States. President George Washington hired French architect Pierre Charles L'Enfant in 1791 to design a capital city for the new nation on undeveloped land along the Potomac River. L'Enfant laid out a Baroque-style plan of diagonal avenues cutting across a grid of streets, with plenty of space for majestic public monuments (**FIGURE 18.6**). A century later, a new generation of planners imposed a height restriction on new buildings, which kept the monuments from

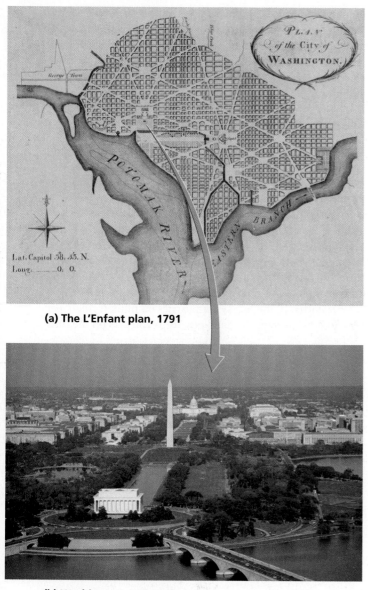

(a) The L'Enfant plan, 1791

(b) Washington, D.C., today

FIGURE 18.6 ▲ Pierre Charles L'Enfant's 1791 plan for Washington, D.C., **(a)** laid out a series of splendid diagonal avenues cutting across gridded streets, allowing plenty of space for the magnificent public monuments **(b)** that grace the city today.

being crowded and dwarfed by modern skyscrapers, thereby preserving the spacious, stately feel of the city.

City planning in North America came into its own at the turn of the 20th century, as urban leaders sought to beautify and impose order on fast-growing, unruly cities. In Portland in 1912, planner Edward Bennett's *Greater Portland Plan* proposed to rebuild the harbor; dredge the river channel; construct new docks, bridges, tunnels, and a waterfront railroad; superimpose wide radial boulevards on the old city street grid; establish civic centers downtown; and greatly expand the number of parks. As the century progressed and other planning efforts were conducted, some ideas, such as establishing a downtown public square, came to fruition.

Your Urban Area Think of your favorite parts of the city you know best. What aspects do you like about them? What do you dislike about some of your least favorite parts of the city? What could this city do to improve the quality of life for its inhabitants?

In today's world of sprawling metropolitan areas, **regional planning** has become at least as important as city planning. Regional planners deal with the same issues as city planners, but they work on broader geographic scales and coordinate their work with multiple municipal governments. In some places, regional planning has been institutionalized in formal government bodies; the Portland area's Metro is the epitome of such a regional planning entity. When Metro and its region's three counties in 2010 announced their collaborative plan apportioning undeveloped land into "urban reserves" and "rural reserves," it marked a historic achievement in regional planning. The agreement designates 11,000 ha (28,000 acres) for urban use and 110,000 ha (272,000 acres) for rural use. It means that homeowners, farmers, developers, and governments can feel informed and secure knowing what kinds of land uses lie in store on and near their land over the next half-century.

Zoning is a key tool for planning

One tool that planners use is **zoning**, the practice of classifying areas for different types of development and land use. For instance, to preserve the cleanliness and tranquility of residential neighborhoods, industrial facilities may be kept out of districts zoned for residential use. By specifying zones for different types of development, planners can guide what gets built where. Zoning also gives home buyers and business owners security, because they know in advance what types of development can and cannot be located nearby.

Zoning involves government restriction on the use of private land and represents a top-down constraint on personal property rights. Yet most people feel that government has a proper role in setting limitations on property rights for the good of the community. Oregon voters sided with private property rights when in 2004 they passed Ballot Measure 37, which shackled government's ability to enforce zoning regulations with landowners who bought their land before the regulations were enacted. However, many Oregonians soon began witnessing new development they did not condone, and in 2007 they passed Ballot Measure 49 to restore public oversight over development. For the most part, people have supported zoning over the years because the common good it produces for communities is widely felt to outweigh the restrictions on private use.

Urban growth boundaries are now widely used

Planners intended Oregon's urban growth boundaries to limit sprawl by containing growth within existing urbanized areas. The UGBs aimed to revitalize downtowns; protect working farms, orchards, ranches, and forests; and ensure urban dwellers access to open space near cities. Since Oregon began its experiment, a number of other states, regions, and cities have adopted UGBs—from Boulder, Colorado, to Lancaster, Pennsylvania, to many California communities.

UGBs save taxpayers money by reducing the amounts that municipalities need to pay for infrastructure. However, UGBs also tend to increase housing prices within their boundaries. In the Portland area, housing has become less affordable, but in most other ways its UGB is working as intended. It has preserved farms and forests outside the UGB while increasing the density of new housing inside the UGB as homes are built on smaller lots and as multistory apartments fulfill a vision of "building up, not out." Downtown employment has grown tremendously as businesses and residents invest anew in the central city.

However, urbanized area still expanded by 101 km² (39 mi²) in the decade after Portland's UGB was established, because 146,000 people were added to the population. This fact suggests that relentless population growth may thwart even the best anti-sprawl efforts. Indeed, Metro has enlarged the Portland-area UGB three dozen times and plans to expand it more in the future.

"Smart growth" and "new urbanism" aim to counter sprawl

As more people feel negative effects of sprawl on their everyday lives, efforts to manage growth with UGBs and other land use policies are springing up throughout North America. Proponents of **smart growth** seek to rejuvenate the older existing communities that so often are drained and impoverished by sprawl (**TABLE 18.1**). Smart growth means "building up, not out"—focusing development and economic investment in existing urban centers and favoring multistory shop-houses and high-rises.

A related approach among architects, planners, and developers is **new urbanism**, which seeks to design walkable neighborhoods with homes, businesses, schools, and other amenities all close together for convenience. The aim is to create functional neighborhoods in which families can meet most

of their needs close to home without using a car. Green spaces, trees, a mix of architectural styles, and creative street layouts add to the visual interest of new urbanist developments. These neighborhoods are often connected to public transit systems, enabling people to travel most places they need to go by train and foot.

Transit options help cities

Traffic jams on roadways cause air pollution, stress, and countless hours of lost personal time. They also cost the U.S. economy an estimated $74 billion each year in fuel and lost productivity. So, a key ingredient in any city planner's recipe is to give citizens alternative transportation options such as public buses, trains, subways, and *light rail* (smaller rail systems powered by electricity). As long as an urban center is large enough to support the infrastructure necessary, both train and bus systems are cheaper, more energy-efficient, and cleaner than roadways choked with cars (**FIGURE 18.7**).

In Portland, the bus system carries 66 million riders per year, and the average Portland bus keeps 250 cars off the road

TABLE 18.1 Ten Principles of "Smart Growth"
▶ Mix land uses
▶ Take advantage of compact building design
▶ Create a range of housing opportunities and choices
▶ Create walkable neighborhoods
▶ Foster distinctive, attractive communities with a strong sense of place
▶ Preserve open space, farmland, natural beauty, and critical environmental areas
▶ Strengthen existing communities and direct development toward them
▶ Provide a variety of transportation choices
▶ Make development decisions predictable, fair, and cost-effective
▶ Encourage community and stakeholder collaboration in development decisions
Source: U.S. Environmental Protection Agency.

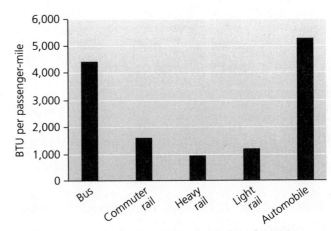

(a) Energy consumption for different modes of transit

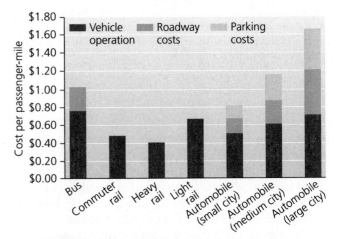

(b) Operating costs for different modes of transit

FIGURE 18.7 ▲ Rail transit consumes far less energy per passenger mile **(a)** than bus or automobile transit. Rail transit involves fewer costs per passenger mile **(b)** than bus or automobile transit. Data from Litman, T., 2005. *Rail transit in America: A comprehensive evaluation of benefits.* Victoria, BC: Victoria Transport Policy Institute.

FIGURE 18.8 ▲ The "bullet trains" of the high-speed rail systems of Japan and Europe can travel at 150–220 mph. China, too, is investing in high-speed trains, such as this one speeding through the city of Qingdao. Plans to establish such lines in the United States are just getting underway.

each day. Portland also boasts downtown streetcars and one of the nation's leading light rail systems. America's most-used train systems are the extensive heavy rail systems in large cities, including New York City's subways, Washington, D.C.'s Metro, the T in Boston, and the San Francisco Bay area's BART. Each of these carries more than one-fourth of its city's daily commuters.

In general, however, the United States lags behind most nations in mass transit. Many countries, rich and poor alike, have extensive bus systems that ferry citizens within and between towns and cities cheaply and effectively. Curitiba, Brazil, a metropolis of 2.5 million people, has an outstanding bus system that is used each day by three-quarters of the population. Japan, China, and many European nations have developed modern high-speed "bullet" trains (**FIGURE 18.8**), whereas the United States has long starved its only national passenger rail network, Amtrak, of funding.

The United States chose to invest instead in road networks for cars and trucks largely because population density was low and gasoline was cheap. As energy costs and population rise, however, mass transit becomes increasingly appealing, and citizens begin to clamor for train and bus systems in their communities. The United States may even see its first bullet train before long; the 2009 stimulus bill passed by Congress set aside $8 billion to develop high-speed rail, and the Obama administration identified 10 potential corridors for such trains.

To make urban transportation more efficient, policymakers can take a variety of other actions. They can raise fuel taxes, tax inefficient modes of transport, reward carpoolers with carpool lanes, encourage bicycle use and bus ridership, and charge trucks for road damage.

Urban residents need parklands

City dwellers often desire some escape from the noise, commotion, and stress of urban life. Natural lands, public parks, and open space provide greenery, scenic beauty, freedom of movement, and places for recreation. These lands also keep ecological processes functioning by helping

to regulate climate, purify air and water, and provide wildlife habitat.

America's city parks began to arise in the late 19th century, as politicians and citizens yearning to make their crowded and dirty cities more livable established public spaces using aesthetic ideals borrowed from European parks and gardens. The lawns, shaded groves, curved pathways, and pastoral vistas of many American city parks originated with these European ideals, as interpreted by America's leading landscape architect, Frederick Law Olmsted. Olmsted designed New York's Central Park (**FIGURE 18.9**) and many other urban park systems.

Portland's quest for parks began in 1900, when city leaders created a parks commission and hired Olmsted's son, John Olmsted, to design a park system. His 1904 plan recommended acquiring land to ring the city generously with parks, but no action was taken. A full 44 years later, citizens pressured city leaders to create Forest Park along a large forested ridge on the northwest side of the city. At 11 km (7 mi) long, it is today the largest city park in the United States.

Large city parks are vital to a healthy urban environment, but even small spaces can make a big difference. Playgrounds provide places where children can be active outdoors and interact with their peers. Community gardens allow people to grow vegetables and flowers in a neighborhood setting. "Greenways" along rivers, streams, or canals can provide walking trails, protect water quality, boost property values, and serve as corridors for the movement of wildlife.

FIGURE 18.9 ▼ Central Park in Manhattan was one of the first American city parks to be developed, and it remains one of the largest and finest.

Green buildings bring benefits

Although we need parklands, we spend most of our lives indoors, so our health is affected by the buildings in which we live and work. Moreover, buildings consume 40% of the energy and 70% of the electricity we use, contributing to fossil fuel consumption and the greenhouse gas emissions that drive climate change. As a result, today there is a thriving movement to design **green buildings**, structures that use technologies and approaches to minimize the ecological footprint (p. 4) of their construction and operation. Green buildings are built from sustainable materials, limit their use of energy and water, minimize health impacts on their occupants, control their pollution, and recycle their waste (**FIGURE 18.10**). The U.S. Green Building Council promotes these efforts by running the **Leadership in Energy and Environmental Design (LEED)**

certification program. Buildings (new buildings or renovation projects) apply for certification, and depending on their performance, may be granted silver, gold, or platinum status.

Green building techniques add expense to construction, but the added cost is generally less than 3% for a LEED-silver building and 10% for a LEED-platinum building. LEED certification is booming throughout North America. One example in Portland is the Rosa Parks Elementary School. Completed in 2007 and certified LEED-gold, the school was built with locally sourced and nontoxic materials, diverted nearly all its construction waste from the landfill, and uses 24% less energy and water than comparable buildings. Schoolchildren learn about renewable energy by watching a display of the electricity produced by their school's solar photovoltaic (PV) system. Green buildings are becoming especially popular at colleges and universities (pp. 414–415) across North America.

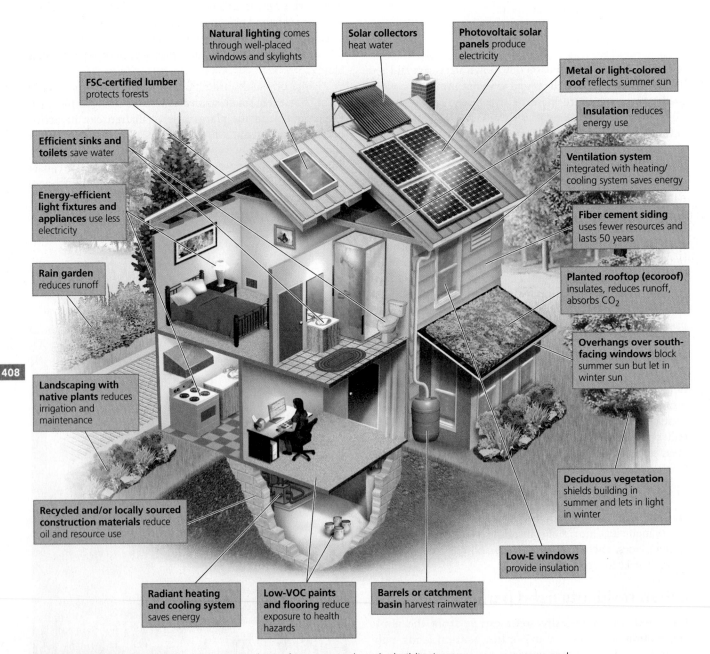

FIGURE 18.10 ▲ Green buildings incorporate design features to reduce the building's energy use, water use, and ecological footprint. Made from sustainable materials, they are built to be healthy for their occupants, to limit pollution, and to recycle waste.

URBAN SUSTAINABILITY

Most of our efforts to make cities more livable also help to make them more sustainable. A sustainable city is one that can function and prosper over the long term, providing generations of residents a good quality of life far into the future.

Urban centers have a mix of environmental effects

Making a city sustainable entails minimizing its impacts on the natural systems and resources that nourish it. You might guess that urban living has more environmental impact than rural living. However, the picture is not so simple. Urban centers exert a complex mix of positive and negative environmental impacts.

Resource use and efficiency Cities and towns are sinks (p. 37) for resources, having to import from source areas beyond their borders most of what they need to feed, clothe, and house their inhabitants and power their commerce. Indeed, for their day-to-day survival, people in cities such as New York, Boston, San Francisco, and Los Angeles depend on water pumped in from faraway watersheds. Thus, urban areas rely on large expanses of land elsewhere for resources and ecosystem services, and they also burn fossil fuels to import resources and goods.

However, imagine that the world's 3.6 billion urban residents were instead spread evenly across the landscape. What would the transportation requirements be, then, to move resources and goods around to all those people? A world without cities would likely require *more* transportation to provide people the same level of access to resources and goods. Moreover, once resources arrive in a city, it maximizes efficiency in distributing goods and services. The density of cities facilitates the provision of electricity, medical care, education, water and sewer systems, waste disposal, and public transportation. Thus, although a city has a large ecological footprint (p. 4) in total, it may have a moderate or small footprint in per capita terms.

Land preservation Because people pack densely together in cities, more land outside cities is left undeveloped. Indeed, this is the idea behind urban growth boundaries. If, instead, all 7 billion of us were evenly spread across the planet's land area, no large blocks of land would be left uninhabited, and we would have far less room for agriculture, wilderness, biodiversity, or privacy. The fact that half the human population is concentrated in discrete locations helps allow room for natural ecosystems to continue functioning and provide the ecosystem services on which all of us, urban and rural, depend.

Pollution Just as cities import resources, they export wastes, either passively through pollution or actively through trade. In so doing, urban centers transfer the costs of their activities to other regions—and mask the costs from their own residents. Citizens of Indianapolis, Columbus, or Buffalo may not recognize that pollution from nearby coal-fired power plants worsens acid rain hundreds of miles to the east. Citizens of New York City may not realize how much garbage their city produces if it is shipped elsewhere for disposal.

However, not all waste and pollution leaves the city. Urban residents are exposed to toxic industrial compounds, photochemical and industrial smog, fossil fuel emissions, noise pollution, and light pollution. City residents even suffer thermal pollution, in the form of the **urban heat island effect** (**FIGURE 18.11**). Pollution and the health risks it poses are not evenly shared among residents. As environmental justice advocates point out (pp. 14–15), those who receive the brunt of pollution are often those who are too poor to live in cleaner areas.

Innovation Cities promote a flourishing cultural life and, by mixing diverse people and influences, spark innovation and creativity. The urban environment can promote education and scientific research, and cities have long been viewed as engines of technological and artistic inventiveness. This inventiveness can lead to solutions to societal problems, including ways to reduce environmental impacts.

95°F(35°C)

Heat emanates from urban areas

88°F(31°C)

88°F(31°C)

Farmland Suburbs City center Suburbs Forest

Vegetation keeps forests, farms, and parkland relatively cool

Pavement and other surfaces in cities absorb sunlight and re-radiate heat at night

Cars, buildings, industry, and people radiate heat in urban areas

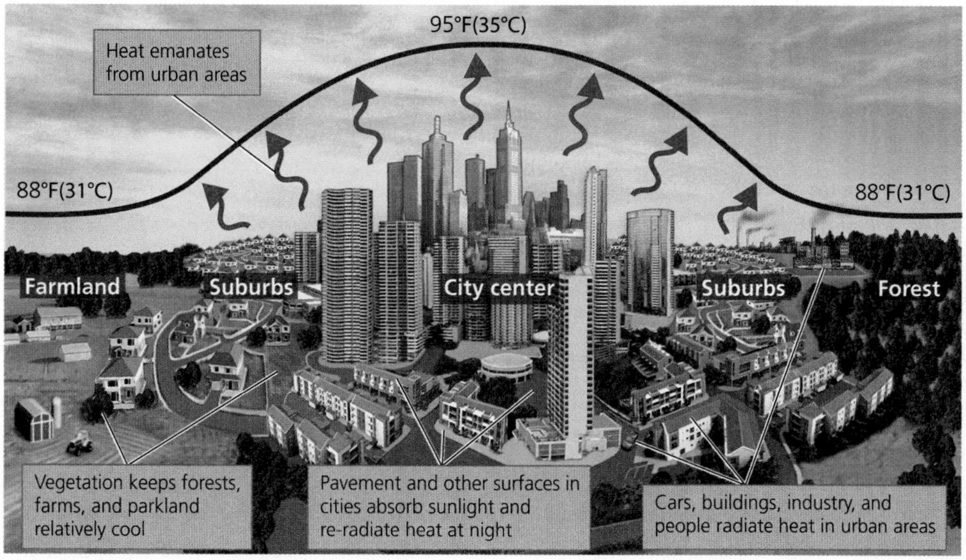

FIGURE 18.11 ◀ Cities are often several degrees warmer than surrounding areas, creating the *urban heat island* effect. People, buildings, vehicles, and factories generate heat, and buildings and dark paved surfaces absorb daytime heat and then release it slowly at night, warming the air and disrupting convective circulation patterns that would otherwise cool the city.

An urban ecologist samples water under an overpass in Baltimore.

Baltimore and Phoenix Showcase Urban Ecology

Researchers in urban ecology examine how ecosystems function in cities and suburbs, how natural systems respond to urbanization, and how people interact with the urban environment. Today, Baltimore and Phoenix are centers for urban ecology.

These two cities are very different: Baltimore is an Atlantic port city on Chesapeake Bay with a long history, whereas Phoenix is a young and fast-growing southwestern metropolis sprawling across the desert. Each was picked by the U.S. National Science Foundation to serve as a research site in its prestigious Long Term Ecological Research (LTER) program, which funds multi-decade ecological research. Since 1997, hundreds of researchers have studied Baltimore and Phoenix explicitly as ecosystems, examining nutrient cycling, biodiversity, air and water quality, how people react to environmental health threats, and more.

Research teams in both cities are combining old maps, aerial photos, and new remote sensing satellite data to reconstruct the history of landscape change. In Phoenix, one team showed how urban development spread across the desert in a "wave of advance," affecting soils, vegetation, and microclimate as it went. In Baltimore, mapping efforts showed that development fragmented the area's forest into smaller patches over the past 100 years, even while the overall amount of forest remained the same.

The study regions designated for each city encompass both heavily urbanized central city areas and rural and natural areas on the urban fringe. To detect the impacts of urbanization, many research projects compare conditions in these two types of areas.

Baltimore scientists can see ecological effects of urbanization just by comparing the urban lower end of their site's watershed with its less-developed upper end. In the lower end, pavement, rooftops, and compacted soil prevent rain from infiltrating the soil, so water runs off quickly into streams. The rapid flow cuts streambeds deeply into the earth, while leaving the surrounding soil drier. As a result, wetland-adapted trees and shrubs are vanishing, replaced by dry-adapted upland trees and shrubs.

The fast flow of water also worsens pollution. In natural areas, streams and wetlands filter pollution by breaking down nitrogen compounds. But in urban areas, where wetlands dry up and runoff from pavement creates flash floods, streams lose their filtering ability. In Baltimore, the resulting pollution ends up in Chesapeake Bay, which suffers eutrophication and a large hypoxic dead zone (pp. 21–22). Baltimore scientists studying nutrient cycling (p. 36–41) found that urban and suburban watersheds have far more nitrate pollution than natural forests (see first figure).

Baltimore research has also revealed that applying salt to icy roads in winter has major environmental impacts. The salt makes its way into streams, which become up to 100 times saltier, even in summer when road salt is not applied. Such high salinity kills organisms

FAQ

Q: Aren't cities bad for the environment?

A: Stand in the middle of a big city and look around. You see concrete, cars, and pollution. Environmentally bad, right? Not necessarily. Cities have a mix of consequences, but the widespread impression that urban living is less environmentally friendly than rural living is largely a misconception. Consider that in a city you can walk to the grocery store instead of driving. You can take the bus or the train. Police, fire, and medical services are close at hand. Water and electricity are easily supplied to your entire neighborhood, and waste is easily collected. In contrast, if you live in the country, resources must be used to transport all these services long distances across the landscape, or you have to burn gasoline and time traveling to reach them. By clustering people together, cities allow us to distribute resources efficiently, while also preserving natural lands outside the city. In many ways, each person can receive more, with less environmental impact, in a city than in the country.

Urban ecology helps cities toward sustainability

Cities that import all their resources and export all their wastes have a linear, one-way metabolism. Linear models of production and consumption tend to destabilize environmental systems. Researchers in **urban ecology** urge us to

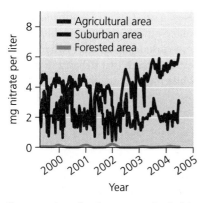

Streams in suburban areas in Baltimore had nitrate levels much higher than nearby forested areas, but lower than nearby agricultural areas, where fertilizers are used liberally. Adapted from Groffman, P.M., et al., 2004. Nitrogen fluxes and retention in urban watershed ecosystems. *Ecosystems* 7: 393-403; and Baltimore Ecosystem Study, www.beslter.org/frame4-page_3f_05.html.

(see second figure), degrades habitat and water quality, and impairs streams' ability to remove nitrate.

To study contamination of groundwater and drinking water, researchers are using isotopes (pp. 26–27) in a forensics approach to trace where salts in the most polluted streams are coming from. Baltimore is now improving water quality substantially by spending $900 million upgrading its sewer system.

Urbanization also affects species and ecological communities. Cities and suburbs facilitate the spread of non-native species, because people introduce exotic ornamental plants and because

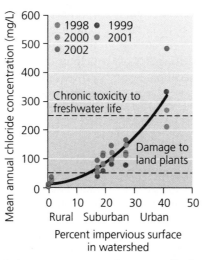

Salt concentrations (from runoff of road salt) in Baltimore-area streams were high enough to damage plants in the suburbs and to kill aquatic animals in urban areas. Adapted from Kaushal, S.S., et al., 2005. Increased salinization of fresh water in the northeastern United States. *Proc. Natl. Acad. Sci. USA* 102: 13517–13520.

impacts on the soil, climate, and landscape favor weedy generalist species over more specialized native ones.

Compared with natural landscapes, cities offer steady and reliable food resources—think of people's bird feeders, or food scraps from dumpsters. Growing seasons are extended and seasonal variation is buffered in cities, as well. The urban heat island effect (p. 409) raises nighttime temperatures and makes temperatures more similar year-round. Buildings and ornamental vegetation shelter animals from extreme conditions,

and irrigation in yards and gardens provides water. In a desert city like Phoenix, watering boosts primary productivity and lowers daytime temperatures. Together, all these changes lead to higher population densities of animals but lower species diversity as generalists thrive and displace specialists.

Urban ecologists in Phoenix and Baltimore are studying social and demographic aspects of the urban environment as well. Some studies measure how natural amenities affect property values. For example, one study found that proximity to a park increases a home's property values—unless crime is pervasive. If the robbery rate surpasses 6.5 times the national average (as it does in Baltimore) then proximity to a park begins to depress property values.

Other studies focus on environmental justice concerns (pp. 14–15). These have repeatedly found that sources of industrial pollution tend to be located in neighborhoods that are less affluent and that are home to people of racial and ethnic minorities. Phoenix researchers mapped patterns of air pollution and toxic chemical releases and found that minorities and the poor are exposed to a greater share of these hazards. As a result, they suffer from higher rates of childhood asthma.

Whether addressing the people, natural communities, or changing ecosystems of the urban environment, studies on urban ecology like those in Phoenix and Baltimore will be vitally informative in our ever more urban world.

develop circular systems akin to those found in nature, which recycle materials and use renewable sources of energy. These researchers hold that the fundamentals of ecosystem ecology and systems science (Chapter 3) apply to the urban environment. Major urban ecology projects are ongoing in Baltimore and Phoenix, where scientists are studying these cities explicitly as ecological systems (see **THE SCIENCE BEHIND THE STORY**, above). Urban sustainability advocates suggest that cities follow an ecosystem-centered model by striving to:

▶ Maximize efficient use of resources.

▶ Recycle as much as possible (pp. 386–390).

▶ Develop environmentally friendly technologies.

▶ Account fully for external costs (p. 92).

▶ Offer tax incentives to encourage sustainable practices.

▶ Use locally produced resources.

▶ Use organic waste and wastewater to fertilize soil.

▶ Encourage urban agriculture.

More and more cities are adopting these strategies. Urban agriculture is thriving in many places, from Portland to Cuba to Japan. Municipal recycling programs continue to grow. Cities like Curitiba, Brazil, provide mass transit, environmental education, job training for the poor, and free health care.

In 2007, New York City unveiled an ambitious plan that Mayor Michael Bloomberg hoped would make it "the first environmentally sustainable 21st-century city." PlaNYC is a 127-item program to reduce greenhouse gas emissions, improve mass transit, plant trees, clean up polluted land and rivers, and enhance access to parkland—to make New York City a better place to live as it accommodates 1 million more people by 2030. As of 2011, accomplishments included completing 100 energy efficiency projects in city buildings, planting 430,000 trees, opening or renovating 180 school playgrounds, acquiring 29,000 acres to protect the upstate water supply, installing 200 miles of bike lanes and 5,000 bike racks, retrofitting the Staten Island Ferry to reduce pollution, converting 30% of the taxi fleet to hybrid vehicles, and reducing greenhouse gas emissions by 13%.

Success stories from New York City to Curitiba to Portland suggest that we can make cities more sustainable. Continuing experimentation in urban areas everywhere will help us determine how best to ensure that urbanization improves our quality of life while maintaining the quality of our environment.

➤ CONCLUSION

As half the human population has shifted from rural to urban lifestyles, the nature of our environmental impact has changed. As urban and suburban dwellers, our impacts are less direct but more far-reaching. Making urban and suburban areas more sustainable is vital for our future. Fortunately, we are developing sustainable solutions while making urban areas better places to live. Planning and zoning entail long-term vision, so they can be powerful forces for sustaining urban communities. Smart growth and new urbanism cut down on energy consumption, helping us address the looming challenges of peak oil (pp. 333–335) and climate change (Chapter 14). Mass transit systems reduce fuel use and carbon emissions. Parks promote health and offer ecosystem services. And green buildings bring a diversity of environmental and health benefits. Indeed, because urban centers affect the environment in some positive ways and have the potential for efficient resource use, they are a key element in achieving progress toward global sustainability.

TESTING YOUR COMPREHENSION

1. What factors lie behind the shift of population from rural areas to urban areas? What types of cities and countries are experiencing the fastest urban growth today, and why?

2. Why have so many city dwellers in the United States, Canada, and other nations moved into suburbs?

3. Give two definitions of *sprawl*. Describe five negative impacts that have been suggested to result from sprawl.

4. What are city planning and regional planning? Contrast planning with zoning. Give examples of some of the suggestions typically made by early planners such as Edward Bennett in Portland.

5. How are some people trying to prevent or slow sprawl? Describe some key elements of "smart growth." What effects, positive and negative, do urban growth boundaries tend to have?

6. Describe several apparent benefits of rail transit systems. What is a potential drawback?

7. How are city parks thought to make urban areas more livable? Give three examples of types of parks or public spaces.

8. What is a green building? Describe several features a LEED-certified building may have.

9. Describe the connection between urban ecology and sustainable cities. List three actions a city can take to enhance its sustainability.

10. Name two positive effects of urban centers on the natural environment.

SEEKING SOLUTIONS

1. Evaluate the causes of the spread of suburbs and assess the environmental, social, and economic impacts of sprawl. Overall, do you think the spread of urban and suburban development that is commonly labeled *sprawl* is predominantly a good thing or a bad thing? Do you think it is inevitable? Give reasons for your answers.

2. Would you personally want to live in a neighborhood developed in the new-urbanist style? Would you like to live in a region with an urban growth boundary? Why or why not?

3. Consider the variety of approaches used to construct or renovate a building using LEED-certified green building techniques. Are there any LEED-certified buildings on your campus? If so, how do they differ from conventional buildings? Think about a building on your campus that you think is unhealthy or environmentally wasteful. Name three to five specific ways in which green building techniques might be used to improve this building.

4. **THINK IT THROUGH** You are the president of your college or university, and students are clamoring for you to help make your campus into the world's first fully sustainable campus. Considering what you have learned about enhancing livability and sustainability in cities, what lessons might you try to apply to your college or university? You are scheduled to give a speech to the campus community about your plans and will need to name five specific actions you plan to take to pursue a sustainable campus. What will they be, and what will you say about each choice to describe its importance?

5. **THINK IT THROUGH** After you earn your college degree, you are offered three equally desirable jobs, in three very different locations. If you take the first, you will live in the midst of a vibrant, densely populated city. If you

accept the second, you will live in a suburb where you'll have more space but where sprawl may soon surround you. If you select the third, you will live in a peaceful rural area with plenty of space but few cultural amenities.

You are a person who aims to live in the most ecologically sustainable way you can. Where would you choose to live? Why? What considerations will you factor into your decision?

CALCULATING ECOLOGICAL FOOTPRINTS

One way to reduce your ecological footprint is to consider transportation alternatives. Each gallon of gasoline is converted during combustion to approximately 20 lb of carbon dioxide (CO_2), which is released into the atmosphere. The table lists typical amounts of CO_2 released per person per mile, using various forms of transportation, assuming typical fuel efficiencies.

For an average North American who travels 12,000 miles per year, calculate and record in the table the CO_2 emitted yearly for each transportation option, and the reduction in CO_2 emission that one could achieve by relying solely on each option.

	CO_2 per person per mile	CO_2 per person per year	CO_2 emission reduction	Your estimated mileage per year	Your CO_2 emissions per year
Automobile (driver only)	0.825 lb	9,900 lb	0		
Automobile (2 persons)	0.413 lb				
Automobile (4 persons)	0.206 lb				
Vanpool (8 persons)	0.103 lb				
Bus	0.261 lb				
Walking	0.082 lb				
Bicycle	0.049 lb				
				Total = 12,000	

1. Which transportation option provides the most miles traveled per unit of carbon dioxide emitted?

2. In the last two columns, estimate what proportion of the 12,000 annual miles you think that you personally travel by each method, and then calculate the CO_2 emissions that you are responsible for generating over the course

of a year. Which transportation option accounts for the most emissions for you?

3. Examining this data, how could you reduce your CO_2 emissions? How many pounds of emissions do you think you could realistically eliminate over the course of the next year by making changes in your transportation decisions?

Go to **www.masteringenvironmentalscience.com** for homework assignments, practice quizzes, Pearson eText, and more.

EPILOGUE
Sustainable Solutions

The notion of sustainability has run throughout this book. With one issue after another, we have seen how society has depleted or degraded resources, endangering their availability for future generations. Yet in case after case, we have seen that dedicated people today are helping to devise creative solutions to our dilemmas. Humanity's challenge is to continue developing solutions that enhance our quality of life while protecting and restoring the natural environment that supports us.

Sustaining our society in a healthy and functional condition requires sustaining our natural environment in a healthy and functional condition. Our civilization wholly depends on the contributions of biodiversity (pp. 172–174) and ecosystem goods and services (pp. 2, 36, 90, 94–95). As more and more people come to appreciate Earth's limited capacity to accommodate our rising population and consumption, we are finding ways to adjust our behaviors, institutions, and technologies in order to sustain our civilization and the natural environment on which it depends.

We can develop sustainably

As people have learned how our quality of life depends on environmental quality, they have also recognized that environ-

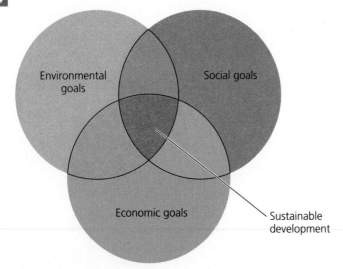

FIGURE E.1 ▲ Modern conceptions of sustainable development hold that sustainability occurs where three sets of goals overlap: social, economic, and environmental goals.

mental degradation often affects society's poorer people the most. As a result, advocates for environmental protection, economic development, and social justice are cooperating in promoting sustainable development (p. 17). Today we recognize that sustainability does not mean simply protecting the environment against the ravages of human development. It means finding ways to promote social justice, economic well-being, and environmental quality at the same time (**FIGURE E.1**).

Students are leading sustainability efforts on campus

In today's quest for sustainable solutions, students at colleges and universities are playing a crucial role. They are creating models for the wider world by leading sustainability initiatives on their campuses (see **ENVISIONIT**, p. 415).

Students are advancing **campus sustainability** efforts by running recycling programs, promoting "green" transportation options on campus, planting trees and restoring native plants, growing organic gardens, fostering sustainable dining halls, and pushing for curricular changes. They are working with faculty and administrators to improve energy efficiency and water conservation in campus buildings and to ensure that new buildings meet certification guidelines for sustainable construction. To help address global climate change, students are urging their institutions to reduce greenhouse gas emissions and to use and invest in renewable energy. In response, nearly 700 university presidents have signed on to the American College and University Presidents' Climate Commitment, a public pledge to inventory emissions, set target dates and milestones for becoming carbon-neutral, and take immediate steps to lower emissions, while also integrating sustainability into the curriculum.

These diverse efforts by students, faculty, staff, and administrators are beginning to reduce the ecological footprints of college and university campuses. In the process, students prepare themselves for sustainability efforts in the wider world, while serving as positive models for their peers on campus.

Environmental protection can enhance economic opportunity

For students and nonstudents alike, economic concerns play a major role in our day-to-day decisions. For too long, we

Installing solar panels at Arizona State

Students everywhere are helping to make their campuses more sustainable.

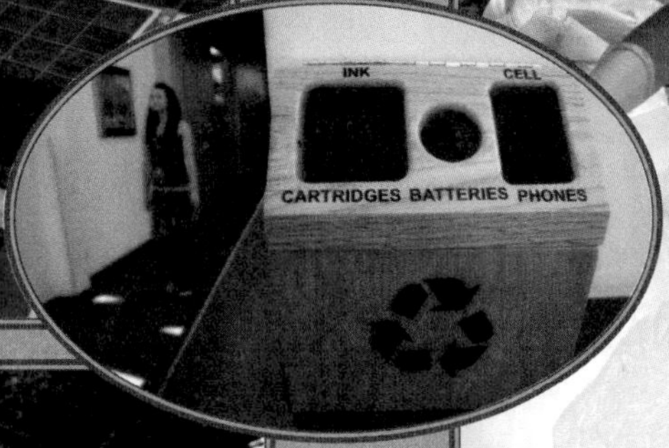

INK CELL
CARTRIDGES BATTERIES PHONES

Recycling at Davidson

Bike-sharing at U. of Rhode Island

URIde
CELS Sustainability Initiative
401-874-4947

Every item you recycle, each light you turn off, and each trip made by bike instead of car makes a difference — and serves as a positive example for your peers.

E-waste drive at UT-Austin

18 million American college students are a powerful force for change!

Biodiesel at U. of Virginia

You Can Make a Difference

➤ Join the student environmental club on your campus . . . or start one!

➤ Register to vote, and help with voter registration drives on campus.

➤ Talk to your institution's president about signing the American College and University Presidents' Climate Commitment.

415

FIGURE E.2 ▲ As we progress toward a sustainable economy, new job opportunities open up. Green-collar jobs, such as employment as a wind-power technician, are already increasing in the United States and other nations.

have labored under the misconception that economic well-being and environmental protection are in conflict. In reality, our well-being depends on a healthy environment, and protecting environmental quality can improve our economic bottom line.

For individuals, businesses, and institutions, reducing resource consumption and waste often saves money. For society, promoting environmental quality can enhance economic opportunity by generating new types of employment. As we transition to a more sustainable economy, some resource-extraction industries will decline, but a variety of recycling-oriented and high-technology industries will spring up to take their place. As we reduce our dependence on fossil fuels, green-collar jobs (p. 355) and novel investment opportunities are blossoming in renewable energy sectors such as wind power, solar power, and fuel-cell technology (**FIGURE E.2**).

Moreover, people desire to live in areas with clean air, clean water, intact forests, public parks, and open space. Environmental protection enhances a region's attractiveness, drawing residents and increasing property values and the tax revenues that fund social services. As a result, regions that safeguard their environments tend to retain and increase their wealth and quality of life. In all these ways, environmental protection enhances economic opportunity. Indeed, a recent U.S. government review concluded that the economic benefits of environmental regulations greatly exceed their economic costs. Both the U.S. economy and the global economy have expanded rapidly in the past 50 years, the very period during which environmental protection measures have proliferated.

We can improve our lives while consuming less

Environmental protection can enhance our economic well-being, but well-being is not the same as growth (pp. 92–94). Economic growth is driven largely by consumption, the purchase of material goods and services (and thus the use of resources involved in their manufacture) by consumers (**FIGURE E.3**).

Our tendency to believe that more, bigger, and faster are always better is reinforced by advertisers seeking to sell us more goods more quickly. The United States, with less than 5% of the world's population, consumes 30% of the world's energy resources and 40% of total global resources. U.S. homes are larger than ever, gas-guzzling vehicles remain popular, and many Americans have more material belongings than they know what to do with.

Our lavishly consumptive lifestyles are a brand-new phenomenon on Earth. We are enjoying the greatest material prosperity in human history—yet if we do not find ways to make our wealth sustainable, the party may not last much longer. Many of Earth's natural resources are limited and nonrenewable, so if we do not shift to sustainable practices of resource use, consumption will decline for rich and poor alike as resources dwindle.

Fortunately, material consumption alone does not reflect a person's happiness or quality of life. We can enhance our quality of life while reducing our consumption—squeezing more from less—in at least three ways. One way is to improve the technology of materials and the efficiency of manufacturing processes, so that industry produces more goods using fewer natural resources. Another way is to develop manufacturing systems that are circular and recycle, in which the waste from one process becomes raw material for input into others (p. 391). A third way is to adjust our attitudes and lifestyles to reduce consumption and slow down the frantic pace of our busy lives.

Population growth will eventually cease

Just as continued growth in consumption is not sustainable, neither is growth in the human population. We have seen (pp. 58–59) that populations may grow exponentially for a time but that they eventually encounter limiting factors and level off or decline. We have used technology to increase Earth's carrying capacity for our species, but sooner or later, our population will stop growing. The question is how: through war, plagues, and famine, or through voluntary means? Thanks to urbani-

FIGURE E.3 ▼ Americans consume more than the people of any other nation. Unless we find ways to increase the sustainability of our manufacturing processes, our rising rate of consumption cannot be sustained in the long run.

zation, wealth, education, and the empowerment of women, the demographic transition (pp. 122–123) is already far along in many developed nations. If today's developing nations also pass through the demographic transition, then humanity may be able to rein in its population growth while creating a more prosperous and equitable society.

Technology can help us toward sustainability

Technologies—developed with the agricultural revolution, the industrial revolution, and advances in medicine and health—have facilitated our population increase. Most technologies have magnified our impacts on Earth's environmental systems. However, new sustainable technologies can also give us ways to reduce our impact. Catalytic converters on cars have reduced emissions, as have scrubbers on industrial smokestacks. Recycling technology and advances in wastewater treatment are reducing our waste output. Solar, wind, and geothermal energy technologies are producing cleaner renewable energy. Technological advances such as these help explain why people of the United States and western Europe today enjoy cleaner environments—although they consume far more—than people of eastern Europe or rapidly industrializing nations such as China.

We can follow several strategies for sustainability

Truly lasting win-win solutions for humanity and for Earth's environmental systems are numerous, and we have seen specific examples throughout this book. Let's now summarize 10 broad strategies or approaches that can spawn sustainable solutions (**TABLE E.1**).

We have already touched on four of these approaches: redefining our priorities on economic growth and quality of life, reducing unnecessary consumption, limiting population growth, and encouraging sustainable technologies. Other economic strategies are to implement green taxes (p. 107), phase out harmful subsidies (p. 108), and incorporate external costs (p. 92) into the market prices of goods and services.

As industries today seek to develop green technologies and sustainable practices, they have an excellent model: nature

itself. Environmental systems tend to operate in cycles featuring feedback loops and the circular flow of materials (Chapter 2). Some forward-thinking industrialists are making their manufacturing processes sustainable by transforming linear pathways into circular ones, in which waste is recycled and reused (p. 391).

Encouraging local self-sufficiency is an important element of building sustainable societies. When people are tied closely to the region where they live, they tend to value the region and seek to sustain its environment and human community. Moreover, relying on locally made products cuts down on fossil fuel use from long-distance transport. This argument is frequently made regarding the cultivation and distribution of food, in promoting local organic or sustainable agriculture (pp. 155, 158).

At the same time, globalization allows people of the world's diverse cultures to communicate and learn about one another, making us more likely to respect and celebrate, rather than fear, differences among cultures. In addition, a globalized free-market system holds a great asset for sustainability: Consumers can exercise influence through what they choose to buy. When goods produced sustainably are ecolabeled (p. 109), we all can "vote with our wallets" by purchasing these products. Consumer choice already drives sales of everything from recycled paper to organic produce to "dolphin-safe" tuna.

Many sustainable solutions require policymakers to usher them through, and policymakers respond to whoever exerts influence. Corporations and interest groups employ lobbyists to sway politicians all the time. Citizens in a democratic republic have power as well, if they choose to exercise it. You can exercise your power at the ballot box, by attending public hearings, by donating money or time to advocacy groups that promote positions you favor, and by writing letters and making phone calls to office-holders. The environmental and consumer-protection laws we all benefit from today came about because ordinary citizens pressured their government representatives to act. As we enjoy today's cleaner air, cleaner water, and greater prosperity, we owe a debt to the people who fought hard for the legislation that enabled these advances in the 1960s and 1970s. In turn, we owe it to future generations to engage ourselves now so that they have a better world in which to live.

However we pursue the solutions we seek, truly sustainable solutions require that we think in the long term (**FIGURE E.4**). Policymakers in democracies often act for short-term good because they compete to produce quick results that will help them be reelected. Yet many environmental dilemmas are cumulative, worsen gradually, and can be resolved only over long periods. Often the costs for addressing environmental problems are short term, whereas the benefits are long term. In such a situation, citizen pressure on policymakers is especially vital.

Finally, we each can magnify our influence by helping to educate others and by serving as role models through our actions. The discipline of environmental science provides information we all can use to make wise decisions about environmental issues and about our own lives. By promoting scientific research and by educating the public about environmental science, we can all assist in the pursuit of sustainable solutions.

TABLE E.1 Some Major Approaches to Sustainability

▶ Rethink our assumptions on economic growth
▶ Consume less while maintaining quality of life
▶ Limit population growth
▶ Encourage "green" technologies
▶ Mimic natural systems by promoting closed-loop industrial processes
▶ Enhance local self-sufficiency, yet embrace some aspects of globalization
▶ Vote with our wallets
▶ Be politically active
▶ Think in the long term
▶ Promote research and education

FIGURE E.4 ▲ Sustainable solutions require thinking in the long term. These Sri Lankan children are planting tree seedlings on deforested and eroded hillsides around their village. In doing so, they are investing in their own future.

Time is precious

We can bring sustainable solutions within reach, but time is getting short, and many human impacts continue to intensify, including deforestation, overfishing, wetland loss, resource extraction, biodiversity loss, and climate change. Even if we can visualize sustainable solutions to our many challenges, how can we find the time to implement them before we do irreparable damage to our environment and our future?

In 1961, U.S. President John F. Kennedy announced that within a decade the United States would be "landing a man on the moon and returning him safely to the Earth." It was a bold and astonishing statement; the technology to achieve this almost unimaginable feat did not yet exist. Yet just eight years later, astronauts walked on the moon. America accomplished this historic milestone by harnessing public enthusiasm for a goal and by supporting scientists and engineers so they could develop methods and technologies to meet the goal.

Today humanity faces a challenge more important than any previous one—achieving sustainability. Attaining sustainability is a larger and more complex process than traveling to the moon. However, it is one to which every person on Earth can contribute; in which government, industry, and citizens can all cooperate; and toward which all nations can work together. If America was able to reach the moon in a mere eight years, then certainly humanity can begin down the road to sustainability with comparable speed. We have the ingenuity;

we now need to rally public resolve and engage our governments, institutions, and entrepreneurs in the race.

Fortunately, in our global society today we have thousands of scientists who study Earth's processes and resources. Thanks to their efforts, we are amassing a detailed knowledge and an ever-developing understanding of our dynamic planet, what it offers us, and what impacts it can bear. This science, this study of Earth and of ourselves, offers us hope for our future.

Earth is an island

We began this book with the vision of Earth as an island, and indeed it is (**FIGURE E.5**). Islands can be paradise, as Easter Island (pp. 6–7) likely was when the Polynesians first reached it. Yet when Europeans arrived there, they witnessed the aftermath of a civilization that had depleted its resources, degraded its environment, and collapsed. People of the once-mighty culture had cut trees unsustainably, kicking the base out from beneath their prosperous civilization.

As Easter Island's trees disappeared, some individuals must have spoken out for conservation and for finding ways to live sustainably amid dwindling resources. Others likely ignored those calls and went on extracting more than the land could bear, assuming that somehow things would turn out all right. Indeed, whoever cut down the last tree from atop the most remote mountaintop could have looked out across the island and seen that it was the last tree. And yet that person cut it down.

It would be tragic folly to let such a fate occur to our planet as a whole. By recognizing this, by shifting our individual behavior and our cultural institutions in ways that encourage sustainable practices, and by employing science to help us achieve these ends, we may yet be able to live happily and sustainably on our wondrous island, Earth.

FIGURE E.5 ▼ This photo of Earth, taken by astronauts orbiting the moon, shows our planet as it truly is—an island in space. Everything we know, need, love, and value comes from and resides on this small sphere—so we had best treat it well!

resenting data in ways that help make trends and patterns visually apparent is a vital part of the scientific endeavor. For scientists, businesspeople, policymakers, and others, the primary tool for expressing patterns in data is the graph. Thus, the ability to interpret graphs is a skill that you will want to cultivate. This appendix guides you in how to read graphs, introduces a few vital conceptual points, and surveys the most common types of graphs, giving rationales for their use.

Navigating a Graph

A graph is a diagram that shows relationships among *variables*, which are factors that can change in value. The most common types of graphs relate values of a *dependent variable* to those of an *independent variable*. As explained in Chapter 1 (p. 9), a dependent variable is so named because its values "depend on" the values of an independent variable. In other words, as the values of an independent variable change, the values of the dependent variable change in response. In a manipulative experiment (pp. 9–10), changes that a researcher specifies in the value of the independent variable *cause* changes in the value of the dependent variable. In observational studies, there may be no causal relationship, and scientists may plot a correlation (p. 10). In a positive correlation, values of one variable go up or down along with values of another. In a negative correlation, values of one variable go up when values of the other go down, or else go down when the others go up. Whether we are graphing a correlation or a causal relationship, the values of the independent variable are known or specified by the researcher, and the values of the dependent variable are unknown until the research has taken place. The latter are what we are interested in observing or measuring.

By convention, independent variables are generally represented on the horizontal axis, or *x axis*, of a graph, while dependent variables are represented on the vertical axis, or *y axis*. Numerical values of variables generally become larger as one proceeds rightward on the *x* axis or upward on the *y* axis. Note that the tick marks along the axes must be uniformly spaced so that when the data are plotted, the graph gives an accurate visual representation of the scale of quantitative change in the data.

In many cases, independent variables are not numbers, but categories. For example, in a graph that presents population sizes of several nations, the nations would comprise a categorical independent variable, whereas population size would be a numerical dependent variable.

As a simple example, **FIGURE A.1** shows data from a classic early lab experiment that measured population growth among yeast cells. The *x* axis shows values of the independent variable, which in this case was time, expressed in units of hours. The researcher was interested in how many yeast cells would propagate over time, so the dependent variable, presented on the *y* axis, is the number of yeast cells present. For each hour at which data were measured during the experiment, a data point on the graph is plotted to show the number of yeast cells present. In this particular graph, a line (red curve) was then drawn through the actual data points (orange dots), showing how closely the empirical data matched the logistic growth curve (p. 58), a theoretical phenomenon of importance in ecology.

Now that you're familiar with the basic building blocks of a graph, let's survey the most common types of graphs you'll see, and examine a few vital concepts in graphing.

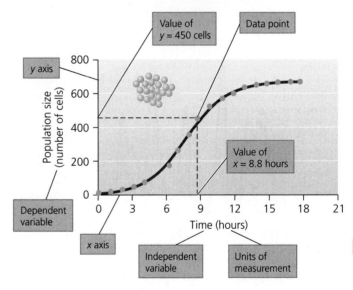

FIGURE A.1 ▲ Logistic population growth, demonstrated by the growth of yeast cells over time in a classic lab experiment. (Figure 3.12a, p. 59)

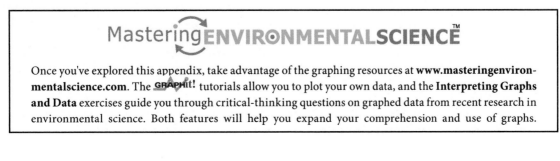

Once you've explored this appendix, take advantage of the graphing resources at **www.masteringenviron-mentalscience.com**. The GRAPHit! tutorials allow you to plot your own data, and the **Interpreting Graphs and Data** exercises guide you through critical-thinking questions on graphed data from recent research in environmental science. Both features will help you expand your comprehension and use of graphs.

GRAPH TYPE: Line Graph

A line graph is used when a data set involves a sequence of some kind, such as a series of values that occur one by one and change through time or across distance. In a line graph, a line runs from one data point to the next. Line graphs are most appropriate when the *y* axis expresses a continuous numerical variable and the *x* axis expresses either continuous numerical data or discrete sequential categories (such as years). **FIGURE A.2** shows values for the size of the ozone hole over Antarctica in recent years. Note how the data show that the size increases until 1987, when the Montreal Protocol (pp. 290–291) came into force, and then begins to stabilize afterwards.

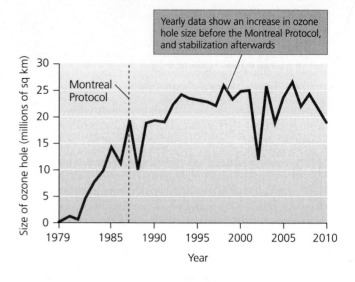

FIGURE A.2 ▲ Size of the Antarctic ozone hole before and after a treaty intended to address it. (Figure 13.17, p. 291)

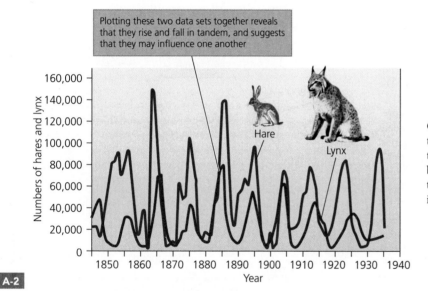

FIGURE A.3 ▲ Fluctuations in recorded numbers of hare and lynx. (Figure 4.4, p. 67)

One useful technique is to plot two or more data sets together on the same graph. This allows us to compare trends in the data sets to see whether and how they may be related. In **FIGURE A.3**, recorded numbers of a predator species rise and fall immediately following those of its prey, suggesting a possible connection.

KEY CONCEPT: Projections

Besides showing observed data, we can use graphs to show data that are predicted for the future. Such *projections* of data are based on models, simulations, or extrapolations from past data, but they are only as good as the information that goes into them—and future trends may not hold if conditions change in unforeseen ways. Thus, in this textbook, projected future data on a line graph are shown with dashed lines, as in **FIGURE A.4**, to indicate that they are less certain than data that have already been observed. Be careful when interpreting graphs in the popular media, however; often newspapers, magazines, and ads will show projected future data in the same way as known past data!

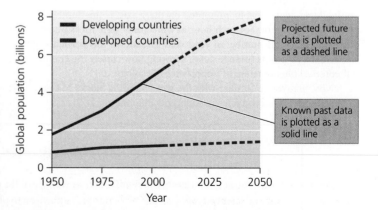

FIGURE A.4 ▲ Past and projected population growth for developing and developed countries. (Figure 6.17, p. 126)

GRAPH TYPE: Bar Chart

A bar chart is most often used when one variable is a category and the other is a number. In such a chart, the height (or length) of each bar represents the numerical value of a given category. Higher or longer bars mean larger values. In **FIGURE A.5**, the bar for the category "Automobile" is higher than that for "Light rail," indicating that automobiles use more energy per passenger-mile (the numerical variable on the *y* axis) than light rail systems do.

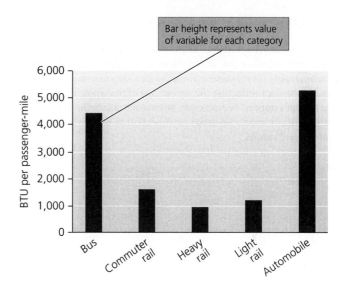

Bar height represents value of variable for each category

FIGURE A.5 ▲ Energy consumption for different modes of transit. (Figure 18.7a, p. 406)

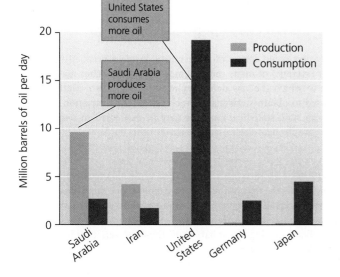

United States consumes more oil

Saudi Arabia produces more oil

FIGURE A.6 ▲ Oil production and consumption by selected nations. (Figure 15.13, p. 342)

It is often instructive to graph two or more data sets together to reveal patterns and relationships. A bar chart such as **FIGURE A.6** lets us compare two data sets (oil production and oil consumption) both within and among nations. A graph that does double duty in this way allows for higher-level analysis (in this case, suggesting which nations depend on others for petroleum imports). Most bar charts in this book illustrate multiple types of information at once in this manner.

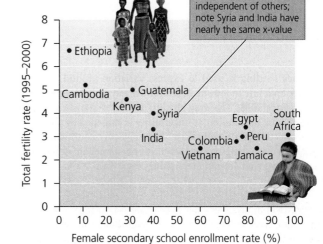

Each data point is independent of others; note Syria and India have nearly the same x-value

GRAPH TYPE: Scatter Plot

A scatter plot is often used when data are not sequential, and when a given *x*-axis value could have multiple *y*-axis values. A scatter plot allows us to visualize a broad positive or negative correlation between variables. **FIGURE A.7** shows a negative correlation (that is, one value goes up while the other goes down): Nations with higher rates of school enrollment for girls tend to have lower fertility rates. Jamaica has high enrollment and low fertility, whereas Ethiopia has low enrollment and high fertility.

FIGURE A.7 ▲ Fertility rate and female education. (Figure 6.13, p. 124)

GRAPH TYPE: Pie Chart

A pie chart is used when we wish to compare the numerical proportions of some whole that are taken up by each of several categories. Each category is represented visually like a slice from a pie, with the size of the slice reflecting the percentage of the whole that is taken up by that category. For example, **FIGURE A.8** shows the percentages of genetically modified crops worldwide that are soybeans, corn, cotton, and canola.

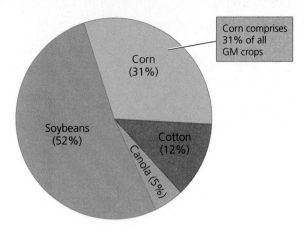

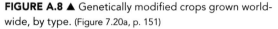

FIGURE A.8 ▲ Genetically modified crops grown worldwide, by type. (Figure 7.20a, p. 151)

KEY CONCEPT: Statistical Uncertainty

Most data sets involve some degree of uncertainty. When a graphed value represents the *mean* (average) of many measurements, the researcher may want to show the degree to which the raw data vary around this mean. Mathematical techniques are used to obtain statistically precise degrees of variation around a mean. Results from such statistical analyses may be expressed in a number of ways, and the two graphs here show methods used in this book.

In a bar chart (**FIGURE A.9**), thin black lines called *error bars* may be shown extending above and below the tops of the bars, or simply above them. Longer error bars indicate more uncertainty or variation, whereas short error bars mean we can place high confidence in the mean value. In this example of woody debris remaining after salvage logging, error bars show the most variation in measurements at "Burned & logged" sites.

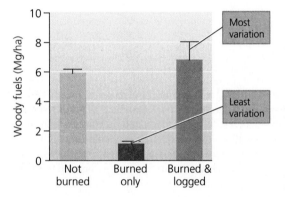

FIGURE A.9 ▲ Fine-scale woody debris left after treatments in salvage logging study. (Figure 9.SBS, p. 199)

Sometimes shading is used to express variation around a mean. The black data line in **FIGURE A.10** shows mean global temperature readings since 1850. The gray shading indicates statistical variation. Note how the amount of uncertainty is exceeded by the sheer scale of the temperature increase. This gives us confidence that globally warming temperatures are a real phenomenon, despite the uncertainty we find around mean values each year.

The statistical analysis of data is critically important in science. In this book we provide a broad and streamlined introduction to many topics, so we often omit error bars from our graphs and details of statistical significance from our discussions. Bear in mind that this is for clarity of presentation only; the research we discuss analyzes its data in far more depth than any textbook can cover.

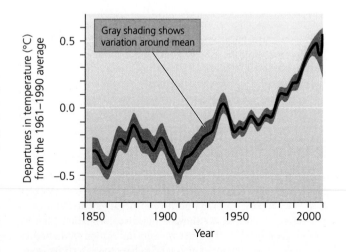

FIGURE A.10 ▲ Change in global temperature, measured since 1850. (Figure 14.10a, p. 308)

APPENDIX B
Metric System

Measurement	Unit and Abbreviation	Metric Equivalent	Metric to English Conversion Factor	English to Metric Conversion Factor
Length	1 kilometer (km)	$= 1{,}000\ (10^3)$ meters	1 km = 0.62 mile	1 mile = 1.61 km
	1 meter (m)	$= 100\ (10^2)$ centimeters	1 m = 1.09 yards	1 yard = 0.914 m
		= 1,000 millimeters	1 m = 3.28 feet	1 foot = 0.305 m
			1 m = 39.37 inches	
	1 centimeter (cm)	$= 0.01\ (10^{-2})$ meter	1 cm = 0.394 inch	1 foot = 30.5 cm
				1 inch = 2.54 cm
	1 millimeter (mm)	$= 0.001\ (10^{-3})$ meter	1 mm = 0.039 inch	
Area	1 square meter (m²)	= 10,000 square centimeters	1 m² = 1.1960 square yards	1 square yard = 0.8361 m²
			1 m² = 10.764 square feet	1 square foot = 0.0929 m²
	1 square centimeter (cm²)	= 100 square millimeters	1 cm² = 0.155 square inch	1 square inch = 6.4516 cm²
Mass	1 metric ton (t)	= 1,000 kilograms	1 t = 1.103 ton	1 ton = 0.907 t
	1 kilogram (kg)	= 1,000 grams	1 kg = 2.205 pounds	1 pound = 0.4536 kg
	1 gram (g)	= 1,000 milligrams	1 g = 0.0353 ounce	1 ounce = 28.35 g
	1 milligram (mg)	= 0.001 gram		
Volume (solids)	1 cubic meter (m³)	= 1,000,000 cubic centimeters	1 m³ = 1.3080 cubic yards	1 cubic yard = 0.7646 m³
			1 m³ = 35.315 cubic feet	1 cubic foot = 0.0283 m³
	1 cubic centimeter (cm³ or cc)	= 0.000001 cubic meter	1 cm³ = 0.0610 cubic inch	1 cubic inch = 16.387 cm³
		= 1 milliliter		
	1 cubic millimeter (mm³)	= 0.000000001 cubic meter		
Volume (liquids and gases)	1 kiloliter (kl or kL)	= 1,000 liters	1 kL = 264.17 gallons	1 gallon = 3.785 L
	1 liter (l or L)	= 1,000 milliliters	1 L = 0.264 gallons	1 quart = 0.946 L
			1 L = 1.057 quarts	
	1 milliliter (ml or mL)	= 0.001 liter	1 ml = 0.034 fluid ounce	1 quart = 946 ml
		= 1 cubic centimeter	1 ml = approximately $\frac{1}{4}$ teaspoon	1 pint = 473 ml
				1 fluid ounce = 29.57 ml
				1 teaspoon = approx. 5 ml
Time	1 millisecond (ms)	= 0.001 second		
Temperature	Degrees Celsius (°C)		$°C = \frac{5}{9}(°F - 32)$	$°F = \frac{9}{5}°C + 32$
Energy and Power	1 kilowatt-hour	= 34,113 BTUs = 860,421 calories		
	1 watt	= 3.413 BTU/hr		
		= 14.34 calorie/min		
	1 calorie	= the amount of heat necessary to raise the temperature of 1 gram (1 cm³) of water 1 degree Celsius		
	1 horsepower	$= 7.457 \times 10^2$ watts		
	1 joule	$= 9.481 \times 10^{-4})$ BTU		
		= 0.239 cal		
		$= 2.778 \times 10^{-7}$ kilowatt-hour		
Pressure	1 pound per square inch (psi)	= 6894.757 pascal (Pa)		
		= 0.068045961 atmosphere (atm)		
		= 51.71493 millimeters of mercury (mm hg = Torr)		
		= 68.94757 millibars (mbar)		
		= 6.894757 kilopascal (kPa)		
	1 atmosphere (atm)	= 101.325 kilopascal (kPa)		

Periodic Table of the Elements

Representative (main group) elements

IA	IIA	IIIB	IVB	VB	VIB	VIIB	VIIIB	VIIIB	VIIIB	IB	IIB	IIIA	IVA	VA	VIA	VIIA	VIIIA
1 **H** 1.0079 Hydrogen																	2 **He** 4.003 Helium
3 **Li** 6.941 Lithium	4 **Be** 9.012 Beryllium											5 **B** 10.811 Boron	6 **C** 12.011 Carbon	7 **N** 14.007 Nitrogen	8 **O** 15.999 Oxygen	9 **F** 18.998 Fluorine	10 **Ne** 20.180 Neon
11 **Na** 22.990 Sodium	12 **Mg** 24.305 Magnesium											13 **Al** 26.982 Aluminum	14 **Si** 28.086 Silicon	15 **P** 30.974 Phosphorus	16 **S** 32.066 Sulfur	17 **Cl** 35.453 Chlorine	18 **Ar** 39.948 Argon
19 **K** 39.098 Potassium	20 **Ca** 40.078 Calcium	21 **Sc** 44.956 Scandium	22 **Ti** 47.88 Titanium	23 **V** 50.942 Vanadium	24 **Cr** 51.996 Chromium	25 **Mn** 54.938 Manganese	26 **Fe** 55.845 Iron	27 **Co** 58.933 Cobalt	28 **Ni** 58.69 Nickel	29 **Cu** 63.546 Copper	30 **Zn** 65.39 Zinc	31 **Ga** 69.723 Gallium	32 **Ge** 72.61 Germanium	33 **As** 74.922 Arsenic	34 **Se** 78.96 Selenium	35 **Br** 79.904 Bromine	36 **Kr** 83.8 Krypton
37 **Rb** 85.468 Rubidium	38 **Sr** 87.62 Strontium	39 **Y** 88.906 Yttrium	40 **Zr** 91.224 Zirconium	41 **Nb** 92.906 Niobium	42 **Mo** 95.94 Molybdenum	43 **Tc** 98 Technetium	44 **Ru** 101.07 Ruthenium	45 **Rh** 102.906 Rhodium	46 **Pd** 106.42 Palladium	47 **Ag** 107.868 Silver	48 **Cd** 112.411 Cadmium	49 **In** 114.82 Indium	50 **Sn** 118.71 Tin	51 **Sb** 121.76 Antimony	52 **Te** 127.60 Tellurium	53 **I** 126.905 Iodine	54 **Xe** 131.29 Xenon
55 **Cs** 132.905 Cesium	56 **Ba** 137.327 Barium	57 **La** 138.906 Lanthanum	72 **Hf** 178.49 Hafnium	73 **Ta** 180.948 Tantalum	74 **W** 183.84 Tungsten	75 **Re** 186.207 Rhenium	76 **Os** 190.23 Osmium	77 **Ir** 192.22 Iridium	78 **Pt** 195.08 Platinum	79 **Au** 196.967 Gold	80 **Hg** 200.59 Mercury	81 **Tl** 204.383 Thallium	82 **Pb** 207.2 Lead	83 **Bi** 208.980 Bismuth	84 **Po** 209 Polonium	85 **At** 210 Astatine	86 **Rn** 222 Radon
87 **Fr** 223 Francium	88 **Ra** 226.025 Radium	89 **Ac** 227.028 Actinium	104 **Rf** 261 Rutherfordium	105 **Db** 262 Dubnium	106 **Sg** 263 Seaborgium	107 **Bh** 262 Bohrium	108 **Hs** 265 Hassium	109 **Mt** 266 Meitnerium	110 **Ds** 269 Darmstadtium	111 **Rg** 272 Roentgenium	112 **Cn** 277 Copernicium		114		116		

Transition metals — groups IIIB–IIB. *Rare earth elements* — Lanthanides and Actinides.

Lanthanides

58 **Ce** 140.115 Cerium	59 **Pr** 140.908 Praseodymium	60 **Nd** 144.24 Neodymium	61 **Pm** 145 Promethium	62 **Sm** 150.36 Samarium	63 **Eu** 151.964 Europium	64 **Gd** 157.25 Gadolinium	65 **Tb** 158.925 Terbium	66 **Dy** 162.5 Dysprosium	67 **Ho** 164.93 Holmium	68 **Er** 167.26 Erbium	69 **Tm** 168.934 Thulium	70 **Yb** 173.04 Ytterbium	71 **Lu** 174.967 Lutetium

Actinides

90 **Th** 232.038 Thorium	91 **Pa** 231.036 Protactinium	92 **U** 238.029 Uranium	93 **Np** 237.048 Neptunium	94 **Pu** 244 Plutonium	95 **Am** 243 Americium	96 **Cm** 247 Curium	97 **Bk** 247 Berkelium	98 **Cf** 251 Californium	99 **Es** 252 Einsteinium	100 **Fm** 257 Fermium	101 **Md** 258 Mendelevium	102 **No** 259 Nobelium	103 **Lr** 262 Lawrencium

The periodic table arranges elements by atomic number and atomic weight into horizontal rows called periods and vertical columns called groups.

Elements of each group in Class A have similar chemical and physical properties. This reflects the fact that members of a particular group have the same number of valence shell electrons, which is indicated by the group's number. For example, group IA elements have one valence shell electron, group IIA elements have two, and group VA elements have five. In contrast, as you progress across a period from left to right, properties of the elements change, varying from the very metallic properties of groups IA and IIA to the nonmetallic properties of group VIIA to the inert elements (noble gases) in group VIIIA. This reflects changes in the number of valence shell electrons.

Class B elements, or transition elements, are metals, and generally have one or two valence shell electrons. In these elements, some electrons occupy more distant electron shells before the deeper shells are filled.

In this periodic table, elements with symbols printed in black exist as solids under standard conditions (25 °C and 1 atmosphere of pressure), whereas elements in red exist as gases, and those in dark blue as liquids. Elements with symbols in green do not exist in nature and must be created by some type of nuclear reaction.

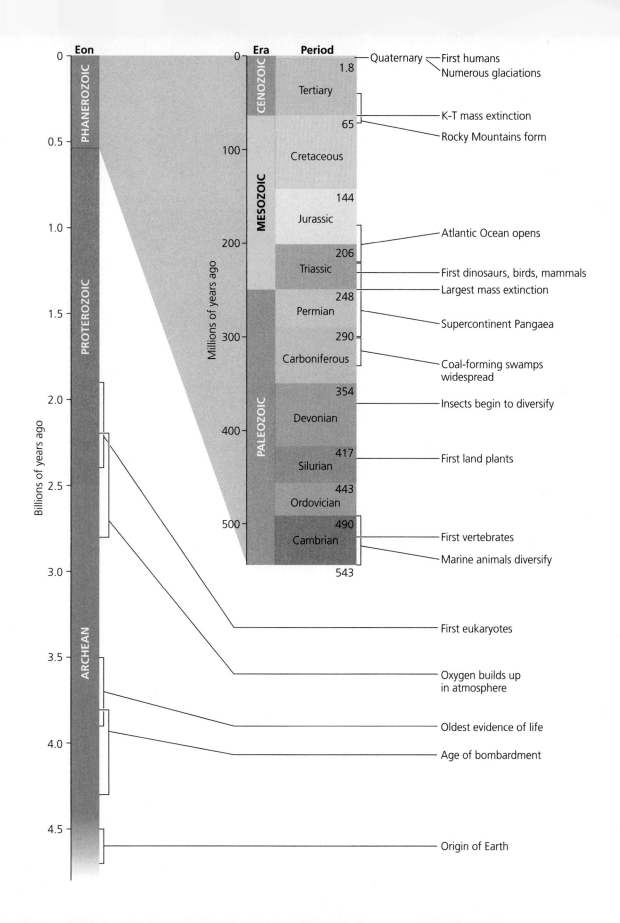

APPENDIX D Geologic Time Scale

D-1

GLOSSARY

abiotic   Nonliving. Compare *biotic*.

acid deposition   The settling of acidic or acid-forming pollutants from the *atmosphere* onto Earth's surface. This can take place by precipitation, fog, gases, or the deposition of dry particles. Compare *acid rain; acid precipitation*.

acid drainage   A process in which sulfide minerals in newly exposed rock surfaces react with *oxygen* and rainwater to produce sulfuric acid, which causes chemical *runoff* as it *leaches* metals from the rocks. Although acid drainage is a natural phenomenon, mining can greatly accelerate its rate by exposing many new rock surfaces at once.

acidic   The property of a *solution* in which the concentration of *hydrogen* (H⁺) *ions* is greater than the concentration of hydroxide (OH⁻) ions. Compare *basic*.

acid rain   *Acid deposition* that takes place through rain. Compare *acid precipitation*.

activation   A rare process in which enzymes that detoxify harmful substances within organisms convert nontoxic substances into toxic compounds.

active solar energy collection   An approach in which technological devices are used to focus, move, or store solar energy. Compare *passive solar energy collection*.

acute exposure   Exposure to a *toxicant* occurring in high amounts for short periods of time. Compare *chronic exposure*.

adaptation   (1) See *adaptive trait*. (2) A response to *global climate change*, consisting of an attempt to minimize its impacts on us. The goal is to adapt to change. Compare *mitigation*.

adaptive management   The systematic testing of different management approaches to improve methods over time.

adaptive trait   A trait that confers greater likelihood that an individual will reproduce.

aerosols   Very fine liquid droplets or solid particles aloft in the atmosphere.

affluenza   Term coined by social critics to describe the failure of material goods to bring happiness to people who have the financial means to afford them.

age distribution   The relative numbers of organisms of each age within a *population*. Age distributions can have a strong effect on rates of population growth or decline and are often expressed as a ratio of age classes, consisting of organisms (1) not yet mature enough to reproduce, (2) capable of reproduction, and (3) beyond their reproductive years.

age structure   See *age distribution*.

age structure diagram (population pyramid)   A diagram *demographers* use to show the age structure of a population. The width of each horizontal bar represents the relative number of individuals in each age class.

agricultural revolution   The shift around 10,000 years ago from a hunter-gatherer lifestyle to an agricultural way of life in which people began to grow their own crops and raise domestic animals. Compare *industrial revolution*.

agriculture   The practice of cultivating *soil*, producing crops, and raising livestock for human use and consumption.

A horizon   A layer of *soil* found in a typical *soil profile*. It forms the top layer or lies below the *O horizon* (if one exists). It consists of mostly inorganic mineral components such as *weathered* substrate, with some organic matter and *humus* from above mixed in. The A horizon is often referred to as *topsoil*. Compare *B horizon; C horizon; E horizon; R horizon*.

air pollutants   Gases and particulate material added to the atmosphere that can affect *climate* or harm people or other organisms.

air pollution   The act of polluting the air, or the condition of being polluted by *air pollutants*.

airshed   The geographic area that produces air pollutants likely to end up in a waterway.

albedo   The capacity of a surface to reflect light. Higher albedo values refer to greater reflectivity.

allergen   A *toxicant* that overactivates the immune system, causing an immune response when one is not necessary.

allopatric speciation   Species formation due to the physical separation of populations over some geographic distance. Compare *sympatric speciation*.

alloy   A mixture of a metal with another metal or with a nonmetallic substance.

alpine tundra   *Tundra* that occurs at the tops of mountains.

ambient air pollution   See *outdoor air pollution*.

amino acids   Organic molecules that join in long chains to form *proteins*.

anaerobic   Occurring in an *environment* that has little or no *oxygen*. The conversion of organic matter to *fossil fuels (crude oil, coal, natural gas)* at the bottom of a deep lake, swamp, or shallow sea is an example of anaerobic decomposition. The opposite of *aerobic*.

anthropocentrism   A human-centered view of our relationship with the *environment*.

anthropogenic   Caused by humans.

application   An applied use of science, such as a new technology, policy decision, or resource management strategy.

aquaculture   The raising of aquatic organisms for food in controlled *environments*.

aquifer   An underground water reservoir.

artificial selection   *Natural selection* conducted under human direction. Examples include the *selective breeding* of crop plants, pets, and livestock.

asbestos   Any of several types of *mineral* that form long, thin microscopic fibers—a structure that allows asbestos to insulate buildings for heat, muffle sound, and resist fire. When inhaled and lodged in lung tissue, asbestos scars the tissue and may eventually lead to lung cancer or *asbestosis*.

asbestosis   A disorder resulting from lung tissue scarred by acid following prolonged inhalation of *asbestos*.

asthenosphere   A layer of the upper *mantle*, just below the *lithosphere*, consisting of especially soft rock.

atmosphere   The thin layer of gases surrounding planet Earth. Compare *biosphere; hydrosphere; lithosphere*.

atmospheric deposition   The wet or dry deposition on land of a wide variety of pollutants, including mercury, nitrates, organochlorines, and others. *Acid deposition* is one type of atmospheric deposition.

atom   The smallest component of an *element* that maintains the chemical properties of that element.

atomic number   The number of *protons* in a given *atom*.

autotroph (producer)   An organism that uses energy from sunlight to produce its own food by *photosynthesis*. Includes green plants, algae, and cyanobacteria.

Bacillus thuringiensis (Bt)   A naturally occurring *soil* bacterium that produces a protein that kills many pests, including caterpillars and the larvae of some flies and beetles.

background rate of extinction   The average rate of *extinction* that occurred before the appearance of humans. For example, the *fossil record* indicates that for both birds and mammals, one *species* in the world typically became extinct every 500 to 1,000 years. Compare *mass extinction event*.

bagasse   Crushed sugarcane residue, whose sugars are used in Brazil to make *ethanol* that helps powers millions of vehicles.

baghouse   A system of large filters that physically removes *particulate matter* from *incinerator emissions*.

basic   The property of a *solution* in which the concentration of hydroxide (OH⁻) *ions* is greater than the concentration of *hydrogen* (H⁺) ions. Compare *acidic*.

bedrock   The continuous mass of solid rock that makes up Earth's *crust*.

benthic   Of, relating to, or living on the bottom of a water body. Compare *pelagic*.

benthic zone   The bottom layer of a water body. Compare *littoral zone; limnetic zone*.

B horizon   The layer of *soil* that lies below the *E horizon* and above the *C horizon*. *Minerals* that leach out of the E horizon are carried down into the B horizon (or subsoil) and accumulate there. Sometimes called the "zone of accumulation" or "zone of deposition." Compare *A horizon; O horizon; R horizon*.

bioaccumulation   The buildup of *toxicants* in the tissues of an animal.

biocentrism   A philosophy that ascribes relative values to actions, entities, or properties on the basis of their effects on living things or on the *biotic* realm in general.

biochemical blocker A *toxicant* that interrupts vital chemical processes in organisms, causing injury or death. Examples include cyanide (which interrupts chemical pathways in mitochondria) and the herbicide atrazine (which blocks biochemical pathways in photosynthesis).

biodiesel Diesel fuel produced by mixing vegetable oil, used cooking grease, or animal fat with small amounts of *ethanol* or methanol (wood alcohol) in the presence of a chemical catalyst.

biodiversity (biological diversity) The variety of life across all levels of biological organization, including the diversity of *species*, their *genes*, their *populations*, and their *communities*.

biodiversity hotspot An area that supports an especially great diversity of *species*, particularly species that are *endemic* to the area.

bioenergy (biomass energy) *Energy* harnessed from plant and animal matter, including wood from trees, charcoal from burned wood, and combustible animal waste products, such as cattle manure.

biofuel Fuel produced from *biomass energy* sources and used primarily to power automobiles.

biogenic Type of *natural gas* created at shallow depths by the anaerobic decomposition of organic matter by bacteria. Consists of nearly pure *methane*. Compare *thermogenic*.

biogeochemical cycle See *nutrient cycle*.

biological control (biocontrol) The attempt to battle pests and weeds with organisms that prey on or parasitize them, rather than by using *pesticides*.

biological diversity See *biodiversity*.

biological hazard Human health hazards that result from ecological interactions among organisms. These include *parasitism* by viruses, bacteria, or other *pathogens*. Compare *infectious disease; chemical hazard; cultural hazard; physical hazard*.

biomagnification The magnification of the concentration of *toxicants* in an organism caused by its consumption of other organisms in which toxicants have *bioaccumulated*.

biomass (1) In ecology, organic material that makes up living organisms; the collective mass of living matter in a given place and time. (2) In energy, organic material derived from living or recently living organisms, containing chemical energy that originated with *photosynthesis*.

biomass energy See *bioenergy*.

biome A major regional complex of similar plant *communities*; a large *ecological* unit defined by its dominant plant type and vegetation structure.

biophilia An instinctive love for nature; an emotional bond people feel with other living things.

biopower The burning of *biomass energy* sources to generate electricity.

biosphere The sum total of all the planet's living organisms and the *abiotic* portions of the *environment* with which they interact.

biosphere reserve A tract of land with exceptional *biodiversity* that couples preservation with *sustainable development* to benefit local people. Designated by UNESCO (the *United Nations* Educational, Scientific, and Cultural Organization) following application by local stakeholders.

biotechnology The material application of biological *science* to create products derived from organisms. The creation of *transgenic* organisms is one type of biotechnology.

biotic Living. Compare *abiotic*.

biotic potential An organism's capacity to produce offspring.

birth control The effort to control the number of children one bears, particularly by reducing the frequency of pregnancy. Compare *contraception*.

bisphenol A (BPA) A substance widely used in plastics and to line food and drink cans, which has raised health concerns because it is an estrogen mimic.

bitumen A thick and heavy form of *petroleum* rich in *carbon* and poor in *hydrogen*.

bog A type of *wetland* in which a pond is thoroughly covered with a thick floating mat of vegetation. Compare *freshwater marsh; swamp*.

boreal forest A *biome* of northern coniferous forest that stretches in a broad band across much of Canada, Alaska, Russia, and Scandinavia. Also known as *taiga*, boreal forest consists of a limited number of *species* of evergreen trees, such as black spruce, that dominate large regions of forests interspersed with occasional bogs and lakes.

Borlaug, Norman (1914–2009) American agricultural scientist who introduced specially bred crops to developing nations in the 20th century, helping to spur the *Green Revolution*.

bottle bill A law that allows consumers to return bottles and cans to stores after their use and receive a monetary refund. Bottle bills have proven highly successful in reducing litter and boosting recycling rates.

bottom-trawling Fishing practice that involves dragging weighted nets across the seafloor to catch *benthic* organisms. Trawling crushes many organisms in its path and leaves long swaths of damaged sea bottom.

breakdown product A *compound* that results from the degradation of a toxicant.

brownfield An area of land whose redevelopment or reuse is complicated by the presence or potential presence of hazardous material.

bycatch That portion of a commercial fishing catch consisting of animals caught unintentionally. Bycatch kills many thousands of fish, sharks, marine mammals, and birds each year.

Calvin cycle In *photosynthesis*, a series of chemical reactions in which *carbon atoms* from *carbon dioxide* are linked together to manufacture sugars.

campus sustainability A term encompassing a wide variety of efforts by students, faculty, staff, and administrators of colleges and universities to make campus operations more sustainable. Includes efforts toward energy efficiency, water efficiency, emission reductions, transportation improvements, sustainable dining, landscaping improvements, renewable energy, curricular changes, and more.

cap-and-trade A *permit trading* system in which government determines an acceptable level of *pollution* and then issues polluting parties permits to pollute. A polluting party receives credit for amounts it does not emit and can then sell this credit to other parties. A type of *emissions trading system*.

capitalist market economy An *economy* in which buyers and sellers interact to determine which *goods* and *services* to produce, how much of them to produce, and how to distribute them. Compare *centrally planned economy*.

captive breeding The practice of capturing members of threatened and endangered *species* so that their young can be bred and raised in controlled *environments* and subsequently reintroduced into the wild.

carbohydrate An *organic compound* consisting of *atoms* of *carbon*, *hydrogen*, and *oxygen*.

carbon The chemical *element* with six protons and six neutrons. A key element in *organic compounds*.

carbon capture Technologies or approaches that remove *carbon dioxide* from power plant or other emissions, in an effort to mitigate *global climate change*.

carbon cycle A major *nutrient cycle* consisting of the routes that *carbon atoms* take through the nested networks of environmental *systems*.

carbon dioxide (CO_2) A colorless gas used by plants for *photosynthesis*, given off by *respiration*, and released by burning *fossil fuels*. A primary *greenhouse gas* whose buildup contributes to *global climate change*.

carbon footprint The cumulative amount of carbon, or *carbon dioxide*, that a person or institution emits, and is indirectly responsible for emitting, into the *atmosphere*, contributing to *global climate change*. Compare *ecological footprint*.

carbon monoxide (CO) A colorless, odorless gas produced primarily by the incomplete combustion of fuel. An EPA *criteria pollutant*.

carbon-neutrality The state in which an individual, business, or institution emits no net carbon to the atmosphere. This may be achieved by reducing carbon emissions and/or employing *carbon offsets* to offset emissions.

carbon offset A voluntary payment to another entity intended to enable that entity to reduce the *greenhouse gas* emissions that one is unable or unwilling to reduce oneself. The payment thus offsets one's own emissions.

carbon sequestration Technologies or approaches to sequester, or store, *carbon dioxide* from industrial emissions (e.g., underground under pressure in locations where it will not seep out) in an effort to mitigate *global climate change*. We are still a long way from developing adequate technology and secure storage space to accomplish this.

carbon storage See *carbon sequestration*.

carbon tax A fee charged to entities that pollute by emitting *carbon dioxide*. A carbon tax gives polluters a financial incentive to reduce pollution, and is thus foreseen as a way to address *global climate change*.

carcinogen A chemical or type of radiation that causes cancer.

carrying capacity The maximum *population size* that a given *environment* can sustain.

categorical imperative An *ethical standard* described by Immanuel Kant, which roughly approximates Christianity's "golden rule": to treat others as you would prefer to be treated yourself.

cell The most basic organizational unit of organisms.

cellular respiration The process by which a *cell* uses the chemical reactivity of *oxygen* to split glucose into its constituent parts, water and *carbon dioxide*, and thereby release chemical energy that can be used to form chemical bonds or to perform other tasks within the cell. Compare *photosynthesis*.

cellulosic ethanol *Ethanol* produced by treating the cellulose in plant tissues with enzymes. Techniques for producing cellulosic ethanol are under development because of the desire to make ethanol from low-value crop waste (residues such as corn stalks and husks), rather than from the sugars of high-value crops.

centrally planned economy An *economy* in which a nation's government determines how to allocate resources in a top-down manner. Also called a "state socialist economy." Compare *capitalist market economy.*

chaparral A *biome* consisting mostly of densely thicketed evergreen shrubs occurring in limited small patches. Its "Mediterranean" *climate* of mild, wet winters and warm, dry summers is induced by oceanic influences. In addition to ringing the Mediterranean Sea, chaparral occurs along the coasts of California, Chile, and southern Australia.

chemical energy *Potential energy* held in the bonds between atoms.

chemical formula A shorthand way to indicate the type and number of atoms in a molecule using numbers and chemical symbols.

chemical hazard Chemicals that pose human health hazards. These include *toxins* produced naturally, as well as many of the disinfectants, *pesticides*, and other synthetic chemicals that our society produces. Compare *biological hazard; cultural hazard; physical hazard.*

Chernobyl Site of a nuclear power plant in Ukraine (then part of the Soviet Union), where in 1986 an explosion caused the most severe *nuclear reactor* accident the world has yet seen. As with *Three Mile Island*, the term is often used to denote the accident itself.

chlorofluorocarbon (CFC) One of a group of human-made *organic compounds* derived from simple *hydrocarbons*, such as ethane and methane, in which *hydrogen atoms* are replaced by chlorine, bromine, or fluorine. CFCs deplete the protective *ozone layer* in the *stratosphere.*

chlorophyll The light-absorbing pigment that enables *photosynthesis* and makes plants green.

chloroplast A cell organelle containing *chlorophyll* in which *photosynthesis* occurs.

C horizon The layer of *soil* that lies below the *B horizon* and above the *R horizon*. It contains rock particles that are larger and less *weathered* than the layers above. It consists of *parent material* that has been altered only slightly or not at all by the process of *soil* formation. Compare *A horizon; E horizon; O horizon.*

chronic exposure Exposure for long periods of time to a *toxicant* occurring in low amounts. Compare *acute exposure.*

circum-Pacific belt A 40,000-km (25,000-mi) arc of *subduction* zones and *fault* systems that encircles much of the Pacific Ocean basin. Popularly called the "ring of fire," 90% of *earthquakes* and over half the world's *volcanoes* occur here.

city planning The professional pursuit that attempts to design cities in such a way as to maximize their efficiency, functionality, and beauty.

classical economics Founded by *Adam Smith*, the study of the behavior of buyers and sellers in a free-market *economy*. Holds that individuals acting in their own self-interest may benefit society, provided that their behavior is constrained by the rule of law and by private property rights and operates within competitive markets. See also *neoclassical economics.*

Clean Air Act U.S. federal *legislation* to control *air pollution* that funds research into pollution control, sets standards for air quality, imposes limits on emissions from new stationary and mobile sources, enables citizens to sue parties violating the standards, and introduces an *emissions trading* program for *sulfur dioxide*. First enacted in 1963 and amended multiple times, particularly in 1970 and 1990.

clean coal technologies A wide array of techniques, equipment, and approaches that seek to remove chemical contaminants (such as sulfur) during the process of generating electricity from coal.

clear-cutting The harvesting of timber by cutting all the trees in an area, leaving only stumps. Although it is the most cost-efficient method, clear-cutting is also the most damaging to the *environment.*

climate The pattern of atmospheric conditions found across large geographic regions over long periods of time. Compare *weather.*

climate diagram (climatograph) A visual representation of a region's average monthly temperature and *precipitation.*

climate model A computer program that combines what is known about weather patterns, atmospheric circulation, atmosphere–ocean interactions, and feedback mechanisms to simulate *climate* processes.

climax community In the traditional view of ecological *succession*, a *community* that remains in place with little modification until disturbance restarts the successional process. Today, ecologists recognize that community change is more variable and less predictable than originally thought, and that assemblages of species may instead form complex mosaics in space and time.

cloud forests Moist *forests*, generally at high elevations in the tropics and subtropics, that derive much of their moisture from low-moving clouds.

clumped distribution Distribution pattern in which organisms arrange themselves in patches, generally according to the availability of the resources they need.

coal A *fossil fuel* composed of organic matter that was compressed under very high pressure to form a dense, solid *carbon* structure.

coalbed methane *Methane* that emanates from *coal* seams, which commonly leaks to the *atmosphere* during coal mining. To avoid this waste and reduce methane emissions, engineers are trying to capture more of this gas for energy.

coevolution Process by which two or more species evolve in response to one another. Parasites and hosts may coevolve, as may flowers and their pollinators.

co-firing A process in which *biomass* is combined with *coal* in coal-fired power plants. Can be a relatively easy and inexpensive way for *fossil-fuel*-based utilities to expand their use of *renewable energy.*

cogeneration A practice in which excess heat produced in electricity generation is captured and used to heat nearby workplaces and homes and to produce other kinds of power.

colony collapse disorder An undiagnosed cause of mass die-offs of honeybees (*Apis mellifera*) in recent years.

command-and-control An approach to protecting the *environment* in which government sets strict legal limits and threatens punishment for violations of those limits.

communicable disease See *infectious disease.*

community An assemblage of *populations* of organisms that live in the same place at the same time.

community-based conservation The practice of engaging local people to protect land and wildlife in their own region.

community ecology The study of the interactions among *species*, from one-to-one interactions to complex interrelationships involving entire *communities.*

community-supported agriculture A system in which consumers pay farmers in advance for a share of their yield, usually in the form of weekly deliveries of produce.

competition A relationship in which multiple organisms seek the same limited resource.

compost A mixture produced when decomposers break down organic matter, including food and crop waste, in a controlled environment.

composting The conversion of organic *waste* into mulch or *humus* by encouraging, in a controlled manner, the natural biological processes of decomposition.

compound A *molecule* whose *atoms* are composed of two or more *elements.*

concentrated animal feeding operation See *feedlot.*

concentrated solar power (CSP) An array of technologies by which energy from the sun is harnessed from a large area and focused onto a small area in order to generate electricity.

concession The right to extract a resource, granted by a government to a corporation. Sometimes conservation organizations purchase concessions in order to preserve habitat.

conservation biology A scientific discipline devoted to understanding the factors, forces, and processes that influence the loss, protection, and restoration of *biological diversity* within and among *ecosystems.*

conservation ethic An *ethic* holding that humans should put *natural resources* to use but also have a responsibility to manage them wisely. Compare *preservation ethic.*

conservation geneticist A scientist who studies genetic attributes of organisms, generally to infer the status of their *populations* in order to help conserve them.

Conservation Reserve Program U.S. policy in farm bills since 1985 that pays farmers to stop cultivating highly erodible cropland and instead place it in conservation reserves planted with grasses and trees.

conservation tillage *Agriculture* that limits the amount of tilling (plowing, disking, harrowing, or chiseling) of *soil*. Compare *no-till.*

consumer See *heterotroph.*

consumptive use Use of *fresh water* whereby water is removed from a particular *aquifer* or surface water body and is not returned to it. *Irrigation* for *agriculture* is an example of consumptive use. Compare *nonconsumptive use.*

continental collision The meeting of two tectonic plates of continental *lithosphere* at a *convergent plate boundary*, wherein the continental *crust* on both sides resists *subduction* and instead crushes together, bending, buckling, and deforming layers of rock and forcing portions of the buckled crust upward, often creating mountain ranges.

contour farming The practice of plowing furrows sideways across a hillside, perpendicular to its slope, to help prevent the formation of rills and gullies. The technique is so named because the furrows follow the natural contours of the land.

contraception The deliberate attempt to prevent pregnancy despite sexual intercourse. Compare *birth control*.

control The portion of an *experiment* in which a *variable* has been left unmanipulated. Serves as a point of comparison with the *treatment*.

control rods Rods made of a metallic alloy that absorbs *neutrons*, which are placed in a *nuclear reactor* among the water-bathed *fuel rods* of uranium. Engineers move these control rods into and out of the water to maintain the *fission* reaction at the desired rate.

controlled experiment An *experiment* in which the effects of all *variables* are held constant, except the one whose effect is being tested by comparison of *treatment* and *control* conditions.

convective circulation A circular *current* (of air, water, magma, etc.) driven by temperature differences. In the atmosphere, warm air rises into regions of lower *atmospheric pressure*, where it expands and cools and then descends and becomes denser, replacing warm air that is rising. The air picks up heat and moisture near ground level and prepares to rise again.

conventional law International law that arises from *conventions*, or treaties, that nations agree to enter into. Compare *customary law*.

Convention on Biological Diversity A 1993 international treaty that aims to conserve *biodiversity*, use biodiversity in a *sustainable* manner, and ensure the fair distribution of biodiversity's benefits.

Convention on International Trade in Endangered Species of Wild Fauna and Flora (CITES) A 1973 international treaty that protects endangered *species* by banning the international transport of their body parts.

convergent evolution Evolutionary change in populations that occurs when natural selection causes distantly related species to converge in appearance, generally as a result of similar selective pressures in similar environments.

convergent plate boundary Area where tectonic plates converge or come together. Can result in *subduction* or *continental collision*. Compare *divergent plate boundary* and *transform plate boundary*.

coral Tiny marine animals that build *coral reefs*. Corals attach to rock or existing reef and capture passing food with stinging tentacles. They also derive nourishment from photosynthetic symbiotic algae known as *zooxanthellae*.

coral reef A mass of calcium carbonate composed of the skeletons of tiny colonial marine organisms called *corals*.

core The innermost part of the Earth, made up mostly of iron, that lies beneath the *crust* and *mantle*.

Coriolis effect The apparent deflection of north-south air *currents* to a partly east-west direction, caused by the faster spin of regions near the equator than of regions near the poles as a result of Earth's rotation.

corporate average fuel efficiency (CAFE) standards Miles-per-gallon fuel efficiency standards set by the U.S. Congress for auto manufacturers to meet, by a sales-weighted average of all models of the manufacturer's fleet.

correlation A relationship among *variables*.

corridor A passageway of protected land established to allow animals to travel between islands of protected *habitat*.

corrosive Able to corrode metals. One criterion for defining *hazardous waste*.

cost-benefit analysis A method commonly used by *neoclassical economists*, in which estimated costs for a proposed action are totaled and then compared to the sum of benefits estimated to result from the action.

covalent bond A chemical bond formed by *atoms* "sharing" electrons.

criteria pollutants Six *air pollutants*—*carbon monoxide, sulfur dioxide, nitrogen dioxide, tropospheric ozone, particulate matter*, and *lead*—for which the *Environmental Protection Agency* has established maximum allowable concentrations in ambient outdoor air because of the threats they pose to human health.

cropland Land that humans use to raise plants for food and fiber.

crop rotation The practice of alternating the kind of crop grown in a particular field from one season or year to the next.

crude oil (petroleum) A *fossil fuel* produced by the conversion of *organic compounds* by heat and pressure. Crude oil is a mixture of hundreds of different types of *hydrocarbon* molecules characterized by *carbon* chains of different length.

crust The lightweight outer layer of the Earth, consisting of rock that floats atop the malleable *mantle*, which in turn surrounds a mostly iron core.

cultural hazard Human health hazards that result from the place we live, our socioeconomic status, our occupation, or our behavioral choices. These include choosing to smoke cigarettes, or living or working with people who do. Compare *biological hazard; chemical hazard; physical hazard*.

current The flow of a liquid or gas in a certain direction.

customary law International law that arises from long-standing practices, or customs, held in common by most *cultures*. Compare *conventional law*.

cyclone A cyclonic storm that forms over the ocean but can do damage upon its arrival on land.

dam Any obstruction placed in a river or stream to block the flow of water so that water can be stored in a reservoir. Dams are built to prevent floods, provide drinking water, facilitate *irrigation*, and generate electricity.

Darwin, Charles (1809–1882) English naturalist who proposed the concept of *natural selection* as a mechanism for *evolution* and as a way to explain the great variety of living things. See also *Wallace, Alfred Russel*.

data Information, generally quantitative information.

deciduous Term describing trees that lose their leaves each fall and remain dormant during winter, when hard freezes would endanger leaves.

decomposer An organism, such as a fungus or bacterium, that breaks down leaf litter and other nonliving matter into simple constituents that can be taken up and used by plants. Compare *detritivore*.

Deepwater Horizon The British Petroleum offshore drilling platform that sank in 2010, creating the largest oil spill in U.S. history.

deep-well injection A *hazardous waste* disposal method in which a well is drilled deep beneath an area's *water table* into porous rock below an impervious *soil* layer. Wastes are then injected into the well, so that they will be absorbed into the porous rock and remain deep underground, isolated from *groundwater* and human contact. Compare *surface impoundment*.

deforestation The clearing and loss of *forests*.

demand The amount of a product people will buy at a given price if free to do so.

demographer A scientist who studies human populations.

demographic transition A theoretical *model* of economic and cultural change that explains the declining death rates and birth rates that occurred in Western nations as they became industrialized. The model holds that industrialization caused these rates to fall naturally by decreasing mortality and by lessening the need for large families. Parents would thereafter choose to invest in quality of life rather than quantity of children.

demography A *social science* that applies the principles of *population ecology* to the study of statistical change in human *populations*.

denitrifying bacteria Bacteria that convert the nitrates in *soil* or water to gaseous *nitrogen* and release it back into the *atmosphere*.

density-dependent factor A *limiting factor* whose effects on a *population* increase or decrease depending on the *population density*. Compare *density-independent factor*.

density-independent factor A *limiting factor* whose effects on a *population* are constant regardless of *population density*. Compare *density-dependent factor*.

deoxyribonucleic acid See *DNA*.

dependent variable The *variable* that is affected by manipulation of the *independent variable*.

deposition The arrival of eroded *soil* at a new location. Compare *erosion*.

desalination (desalinization) The removal of salt from seawater.

desert The driest *biome* on Earth, with annual *precipitation* of less than 25 cm. Because deserts have relatively little vegetation to insulate them from temperature extremes, sunlight readily heats them in the daytime, but daytime heat is quickly lost at night, so temperatures vary widely from day to night and in different seasons.

desertification A form of land degradation in which more than 10% of productivity is lost due to *erosion, soil* compaction, forest removal, *overgrazing*, drought, *salinization*, climate change, water depletion, or other factors. Severe desertification can result in the expansion of desert areas or creation of new ones. Compare *soil degradation*.

detritivore An organism, such as a millipede or soil insect, that scavenges the waste products or dead bodies of other community members. Compare *decomposer*.

dike A long raised mound of earth erected along a river bank to protect against floods by holding rising water in the main channel.

directional drilling Modern methods of drilling underground for *oil* and *natural gas* in which the drill is bent or curved as it descends, allowing many areas to be reached from a single drill pad and thus reducing the environmental impact of drilling on the surface.

distillation The removal of substances dissolved in water by evaporating the water and condensing its purified vapor.

divergent plate boundary Area where tectonic plates push apart from one another as *magma* rises upward to the surface, creating new *lithosphere* as it cools and spreads. A prime example is the Mid-Atlantic Ridge. Compare *convergent plate boundary* and *transform plate boundary*.

DNA (deoxyribonucleic acid) A double-stranded *nucleic acid* composed of four nucleotides, each of which contains a sugar (deoxyribose), a phosphate group, and a nitrogenous base. DNA carries the hereditary information for living organisms and is responsible for passing traits from parents to offspring. Compare *RNA*.

dose The amount of *toxicant* a test animal receives in a dose-response test. Compare *response*.

dose–response analysis A set of experiments that measure the *response* of test animals to different *doses* of a *toxicant*. The response is generally quantified by measuring the proportion of animals exhibiting negative effects.

dose–response curve A curve that plots the *response* of test animals to different *doses* of a *toxicant*, as a result of *dose-response analysis*.

downwelling In the ocean, the flow of warm surface water toward the ocean floor. Downwelling occurs where surface *currents* converge. Compare *upwelling*.

driftnet Fishing net that spans large expanses of water, arrayed strategically to drift with currents so as to capture passing fish, and held vertical by floats at the top and weights at the bottom. Driftnetting captures substantial *bycatch* of dolphins, seals, sea turtles, and non-target fish.

Dust Bowl An area that loses huge amounts of *topsoil* to wind *erosion* as a result of drought and/or human impact; first used to name the region in the North American Great Plains severely affected by drought and topsoil loss in the 1930s. The term is now also used to describe that historical event and others like it.

earthquake A release of energy occurring as Earth relieves accumulated pressure between masses of lithosphere, and which results in shaking at the surface.

ecocentrism A philosophy that considers actions in terms of their damage or benefit to the integrity of whole ecological *systems*, including both *biotic* and *abiotic* elements.

ecofeminism A philosophy holding that the patriarchal (male-dominated) structure of society is a root cause of both social and environmental problems. Ecofeminists hold that a *worldview* traditionally associated with women, which interprets the world in terms of interrelationships and cooperation, is more in tune with nature than a worldview traditionally associated with men, which interprets the world in terms of hierarchies and competition.

ecolabeling The practice of designating on a product's label how the product was grown, harvested, or manufactured, so that consumers buying it are aware of the processes involved and can differentiate between brands that use processes believed to be *environmentally* beneficial (or less harmful than others) and those that do not.

ecological economics A developing school of *economics* that applies the principles of *ecology* and *systems* thinking to the description and analysis of *economies*. Compare *environmental economics; neoclassical economics*.

ecological footprint The cumulative amount of land and water required to provide the raw materials a person or *population* consumes and to dispose of or *recycle* the *waste* that is produced.

ecological modeling The practice of constructing and testing *models* that aim to explain and predict how ecological systems function.

ecological restoration Efforts to reverse the effects of human disruption of ecological systems and to restore *communities* to their condition before the disruption. The practice that applies principles of *restoration ecology*.

ecology The *science* that deals with the distribution and abundance of organisms, the interactions among them, and the interactions between organisms and their *abiotic environments*.

economically recoverable Extractable such that income from a resource's sale exceeds the costs of extracting it. Applied to *fossil fuel* deposits.

economics The study of how we decide to use scarce resources to satisfy the demand for *goods* and *services*.

economy A social *system* that converts resources into *goods* and *services*.

ecosystem All organisms and nonliving entities that occur and interact in a particular area at the same time.

ecosystem-based management The attempt to manage the harvesting of resources in ways that minimize impact on the *ecosystems* and ecological processes that provide the resources.

ecosystem diversity The number and variety of ecosystems in a particular area. One way to express *biodiversity*. Related concepts consider the geographic arrangement of *habitats, communities*, or *ecosystems* at the landscape level, including the sizes, shapes, and interconnectedness of patches of these entities.

ecosystem service An essential service an *ecosystem* provides that supports life and makes *economic* activity possible. For example, ecosystems naturally purify air and water, cycle *nutrients*, provide for plants to be *pollinated* by animals, and serve as receptacles and *recycling* systems for the *waste* generated by our economic activity.

ecotone A transitional zone where *ecosystems* meet.

ecotourism Visitation of natural areas for tourism and recreation. Most often involves tourism by affluent people, which may generate *economic* benefits for less-affluent communities near natural areas and thus provide economic incentives for conservation of natural areas.

ED$_{50}$ (effective dose–50%) The amount of a *toxicant* it takes to affect 50% of a *population* of test animals. Compare *threshold dose; LD$_{50}$*.

edge effect An impact on organisms, populations, or communities that results because conditions along the edge of a habitat fragment differ from conditions in the interior.

effluent Water that flows out of a facility such as a wastewater treatment plant or power plant.

E horizon The layer of *soil* that lies below the *A horizon* and above the *B horizon*. The letter "E" stands for "eluviation," meaning "loss," and the E horizon is characterized by the loss of certain *minerals* through *leaching*. It is sometimes called the "zone of leaching." Compare *C horizon; O horizon; R horizon*.

electricity A secondary form of energy that can be transferred over long distances and applied for a variety of uses.

electrolysis A process in which electrical current is passed through a *compound* to release *ions*. Electrolysis offers one way to produce *hydrogen* for use as fuel: Electrical current is passed through water, splitting the water *molecules* into hydrogen and *oxygen atoms*.

electron A negatively charged particle that surrounds the nucleus of an *atom*.

electronic waste (e-waste) Discarded electronic products such as computers, monitors, printers, DVD players, cell phones, and other devices. *Heavy metals* in these products mean that this waste may be judged hazardous.

element A fundamental type of matter; a chemical substance with a given set of properties, which cannot be broken down into substances with other properties. Chemists currently recognize 92 elements that occur in nature, as well as more than 20 others that have been artificially created.

El Niño Strong warming of the eastern Pacific Ocean that occurs every 2 to 7 years, depressing local fish and bird *populations* by altering the marine *food web* in the area, and having widespread climatic consequences. Compare *La Niña*. See also *El Niño–Southern Oscillation*.

El Niño–Southern Oscillation (ENSO) A systematic shift in atmospheric pressure, sea surface temperature, and ocean circulation in the tropical Pacific Ocean. ENSO cycles give rise to *El Niño* and *La Niña* conditions.

emigration The departure of individuals from a *population*.

emissions trading system A *permit trading* system in which a government issues marketable emissions permits to businesses, industries, or utilities to emit pollutants. Under a *cap-and-trade* system, the government determines an acceptable level of *pollution* and then issues permits to pollute. A polluting party receives credit for amounts it does not emit and can then sell this credit to other parties. Compare *cap-and-trade*.

Endangered Species Act (ESA) The primary *legislation*, enacted in 1973, for protecting *biodiversity* in the United States. It forbids the government and private citizens from taking actions (such as developing land) that would destroy endangered *species* or their *habitats*, and it prohibits trade in products made from endangered species.

endemic Native or restricted to a particular geographic region. An endemic species occurs in one region and nowhere else on Earth.

endocrine disruptor A *toxicant* that interferes with the *endocrine (hormone) system*.

endocrine system The body's *hormone* system.

energy An intangible phenomenon that can change the position, physical composition, or temperature of matter.

energy conservation The practice of reducing *energy* use as a way of extending the lifetime of our *fossil fuel* supplies, of being less wasteful, and of reducing our impact on the *environment*. Conservation can result from behavioral decisions or from technologies that demonstrate *energy efficiency*.

energy efficiency The ability to obtain a given result or amount of output while using less energy input. Technologies permitting greater energy efficiency are one main route to *energy conservation*.

energy returned on investment (EROI) The ratio determined by dividing the quantity of *energy* returned from a process by the quantity of energy invested in the process. Higher EROI ratios mean that more energy is produced from each unit of energy invested. Compare *net energy*.

enhanced geothermal systems A new approach whereby engineers drill deeply into rock, fracture it, pump in water, and then pump it out once it is heated below ground. This approach could potentially enable us to obtain *geothermal energy* in any location.

entropy The degree of disorder in a substance, *system*, or process. See *second law of thermodynamics*.

environment The sum total of our surroundings, including all of the living things and nonliving things with which we interact.

environmental economics A developing school of *economics* that modifies the principles of *neoclassical economics* to address environmental challenges. An environmental economist believes that we can attain *sustainability* within our current economic *systems*. Whereas ecological economists call for revolution, environmental economists call for reform. Compare *ecological economics; neoclassical economics*.

environmental ethics The application of *ethical standards* to environmental questions.

environmental health Environmental factors that influence human health and quality of life and the health of *ecological* systems essential to environmental quality and long-term human well-being.

environmental impact statement (EIS) A report of results from detailed studies that assess the potential effects on the *environment* that would likely result from development projects or other actions undertaken by the government.

environmentalism A social movement dedicated to protecting the natural world.

environmental justice A movement based on a moral sense of fairness and equality that seeks to expand society's domain of ethical concern from men to women, from humans to nonhumans, from rich to poor, and from majority races and ethnic groups to minority ones.

environmental policy *Public policy* that pertains to human interactions with the *environment*. It generally aims to regulate resource use or reduce *pollution* to promote human welfare and/or protect natural systems.

Environmental Protection Agency (EPA) A U.S. administrative agency charged with monitoring environmental quality, conducting and evaluating research, setting standards, enforcing those standards, assisting the states in meeting standards and goals, and educating the public.

environmental resistance The collective force of limiting factors, which together stabilize a population size at its carrying capacity.

environmental science The study of how the natural world functions and how humans and the *environment* interact.

environmental studies An academic *environmental science* program that heavily incor-porates the social sciences as well as the natural sciences.

environmental toxicology The study of *toxicants* that come from or are discharged into the *environment*, including the study of health effects on humans, other animals, and *ecosystems*.

enzyme A chemical that catalyzes a chemical reaction.

epidemiological study A study that involves large-scale comparisons among groups of people, usually contrasting a group known to have been exposed to some *toxicant* and a group that has not.

EROI See *energy returned on investment*.

erosion The removal of material from one place and its transport to another by the action of wind or water.

estuary An area where a river flows into the ocean, mixing *fresh water* with salt water.

ethanol The alcohol in beer, wine, and liquor, produced as a *biofuel* by fermenting biomass, generally from *carbohydrate*-rich crops such as corn.

ethical standard A criterion that helps differentiate right from wrong.

ethics The study of good and bad, right and wrong. The term can also refer to a person's or group's set of moral principles or values.

European Union (EU) Political and economic organization formed after World War II to promote Europe's economic and social progress. As of 2011, the EU consisted of 27 member nations.

eutrophic Term describing a water body that has high-nutrient and low-oxygen conditions. Compare *oligotrophic*.

eutrophication The process of *nutrient* enrichment, increased production of organic matter, and subsequent *ecosystem* degradation in a water body.

evaporation The conversion of a substance from a liquid to a gaseous form.

even-aged Condition of timber plantations—generally *monocultures* of a single *species*—in which all trees are of the same age. Most *ecologists* view plantations of even-aged stands more as crop *agriculture* than as ecologically functional *forests*. Compare *uneven-aged*.

evenness See *relative abundance*.

evolution Genetically based change in the appearance, functioning, and/or behavior of organisms across generations, often by the process of *natural selection*.

evolutionary arms race A duel of escalating adaptations between species. Like rival nations racing to stay ahead of one another in military technology, *host* and *parasite* may repeatedly evolve new responses to the other's latest advance.

e-waste See *electronic waste*.

exotic Non-native to an area (as, an exotic organism).

experiment An activity designed to test the validity of a *hypothesis* by manipulating *variables*. See *manipulative experiment* and *natural experiment*.

exploitative interaction A species interaction in which one participant benefits while another is harmed; that is, one species exploits the other. Such interactions include *predation*, *parasitism*, and *herbivory*.

exploratory drilling Drilling that takes place after a *fossil fuel* deposit has been identified, in order to gauge how much of the fuel exists and whether extraction will prove worthwhile. Involves drilling small holes that descend to great depths.

exponential growth The increase of a *population* (or of anything) by a fixed percentage each year.

external cost A negative *externality*; a cost borne by someone not involved in an economic transaction. Examples include harm to citizens from *water pollution* or *air pollution* discharged by nearby factories.

extinction The disappearance of an entire *species* from the face of the Earth. Compare *extirpation*.

extirpation The disappearance of a particular *population* from a given area, but not the entire *species* globally. Compare *extinction*.

extrusive Term for igneous rock formed when magma is ejected from a *volcano* and cools quickly (e.g., basalt).

factory farm See *feedlot*.

family planning The effort to plan the number and spacing of one's children, so as to offer children and parents the best quality of life possible.

farmers' markets Markets at which local farmers and food producers sell fresh locally grown items.

fault A fracture in Earth's crust, at which earthquakes may occur.

fee-and-dividend A program of *carbon taxes* in which proceeds from the taxes are paid to consumers as a tax refund or "dividend." This way, if polluters pass their costs along to consumers, consumers will not lose money.

feedback loop A circular process in which a *system*'s output serves as input to that same system. See *negative feedback loop; positive feedback loop*.

feed-in tariff A system whereby utilities are mandated to buy power from anyone who can generate power from renewable energy sources and feed it into the electric grid. When prices are set at a premium, this can act as a powerful mechanism to promote the spread of renewable energy technologies. Compare *net metering*.

feedlot A huge barn or outdoor pen designed to deliver *energy*-rich food to animals living at extremely high densities. Also called a *factory farm* or concentrated animal feeding operation (CAFO).

Ferrel cell One of a pair of cells of *convective circulation* between 30° and 60° north and south latitude that influence global *climate* patterns. Compare *Hadley cell; polar cell*.

fertilizer A substance that promotes plant growth by supplying essential *nutrients* such as *nitrogen* or *phosphorus*.

first law of thermodynamics Physical law stating that *energy* can change from one form to another but cannot be created or lost. The total energy in the universe remains constant and is said to be conserved.

flagship species A *species* that has wide appeal with the public and that can be used to promote conservation efforts that also benefit other less charismatic species.

flat-plate solar collectors See *solar panels*.

flexible fuel vehicle A vehicle that runs on fuel that is a mixture of *ethanol* and gasoline, such as E-85, a mix of 85% ethanol and 15% gasoline.

flooding The spillage of water over a river's banks due to heavy rain or snowmelt.

floodplain The region of land over which a river has historically wandered and periodically floods.

flux The movement of nutrients among *pools* or *reservoirs* in a *nutrient cycle*.

food chain A linear series of feeding relationships. As organisms feed on one another, energy is transferred from lower to higher *trophic levels*. Compare *food web*.

food security An adequate, reliable, and available food supply to all people at all times.

food web A visual representation of feeding interactions within an *ecological community* that shows an array of relationships between organisms at different *trophic levels*. Compare *food chain*.

forensic science The scientific analysis of evidence to make an identification or answer a question, most often relating to a crime or accident. *Conservation biologists* are now employing forensic science to protect *species* from illegal harvesting.

forensics See *forensic science*.

forest Any ecosystem characterized by a high density of trees.

forester Professionals who manage *forests* through the practice of *forestry*.

forestry The professional management of *forests*.

forest type Categories of *forest* defined by the predominant tree *species*.

fossil The remains, impression, or trace of an animal or plant of past geological ages that has been preserved in rock or *sediments*.

fossil fuel A *nonrenewable natural resource*, such as *crude oil*, *natural gas*, or *coal*, produced by the decomposition and compression of organic matter from ancient life.

fossil record The cumulative body of *fossils* worldwide, which paleontologists study to infer the history of past life on Earth.

Fourth Assessment Report A 2007 report from the *Intergovernmental Panel on Climate Change (IPCC)* that summarizes thousands of scientific studies, documenting observed trends in surface temperature, precipitation patterns, snow and ice cover, sea levels, storm intensity, and other factors. It also predicts future changes in these phenomena under a range of emission scenarios; addresses impacts of *climate change* on wildlife, ecosystems, and human societies; and discusses strategies we might pursue in response. The Fourth Assessment represents the consensus of scientific climate research from around the world.

free rider A party that fails to invest in controlling *pollution* or carrying out other environmentally responsible activities and instead relies on the efforts of other parties to do so. For example, a factory that fails to control its emissions gets a "free ride" on the efforts of other factories that do make the sacrifices necessary to reduce emissions.

fresh water Water that is relatively pure, holding very few dissolved salts.

freshwater marsh A type of *wetland* in which shallow water allows plants such as cattails to grow above the water surface. Compare *swamp; bog*.

fuel rods Rods of uranium that supply the fuel for nuclear *fission*, and are kept bathed in a *moderator* in a *nuclear reactor*.

fungicide A type of chemical *pesticide* that kills fungi.

gasification A process in which *biomass* is vaporized at extremely high temperatures in the absence of *oxygen*, creating a gaseous mixture including *hydrogen*, *carbon monoxide*, and *methane*, in order to produce *biopower* or *biofuels*. In coal gasification, *coal* is converted to a syngas by reacting it with oxygen and steam at a high temperature.

gene A stretch of *DNA* that represents a unit of hereditary information.

genera Plural of *genus*.

General Land Ordinances of 1785 and 1787 Laws that gave the U.S. government the right to manage Western lands and created a grid system for surveying them and readying them for private ownership.

General Mining Act of 1872 U.S. law that legalized and promoted mining by private individuals on public lands for just $5 per acre, subject to local customs, with no government oversight.

generalist A *species* that can survive in a wide array of *habitats* or use a wide array of resources. Compare *specialist*.

genetically modified (GM) organism An organism that has been *genetically engineered* using a technique called *recombinant DNA* technology.

genetic diversity A measurement of the differences in *DNA* composition among individuals within a given *species*.

genetic engineering Any process scientists use to manipulate an organism's genetic material in the lab by adding, deleting, or changing segments of its *DNA*.

genus A taxonomic level in the Linnaean classification system that is above *species* and below family. A genus is made up of one or more closely related species.

geographic information system (GIS) Computer software that overlays multiple types of data (for instance, on geology, hydrology, vegetation, animal species, and human development) onto a common set of geographic coordinates. The idea is to create a complete picture of a landscape and to analyze how elements of the datasets are arrayed spatially and how they may be correlated. A common tool of geographers, landscape ecologists, resource managers, and conservation biologists.

geologic hazards Natural hazards to human life and property that result from geologic processes. Examples include *earthquakes*, *volcanoes*, and *mass wasting* events.

geology The scientific study of Earth's physical features, processes, and history.

geothermal energy Renewable *energy* that is generated deep within Earth. The radioactive decay of elements amid the extremely high pressures and temperatures at depth generate heat that rises to the surface in magma and through fissures and cracks. Where this energy heats *groundwater*, heated water and steam may erupt from below.

geyser A natural spurt of heated groundwater and steam pressurized and sent up from below ground that erupts through the surface.

glaciation The extension of ice sheets from the polar regions far into Earth's temperate zones during cold periods of Earth's history.

global climate change Any change in aspects of Earth's *climate*, such as temperature, *precipitation*, and storm intensity. Generally refers today to the current warming trend in global

temperatures and associated climatic changes. Compare *global warming*.

global warming An increase in Earth's average surface temperature. The term is most frequently used in reference to the pronounced warming trend of recent years and decades. Global warming is one aspect of *global climate change*, and in turn drives other components of climate change.

global warming potential A quantity that specifies the ability of one *molecule* of a given *greenhouse gas* to contribute to atmospheric warming, relative to *carbon dioxide*.

good A material commodity manufactured for and bought by individuals and businesses.

green building (1) A structure that minimizes the *ecological footprint* of its construction and operation by using sustainable materials, using minimal energy and water, reducing health impacts, limiting pollution, and recycling waste. (2) The pursuit of constructing or renovating such buildings.

green-collar job A job in an industry dedicated to new sustainable technologies or practices. Examples include jobs in the design, installation, and management of renewable energy facilities.

greenhouse effect The warming of Earth's surface and *atmosphere* (especially the *troposphere*) caused by the *energy* emitted by *greenhouse gases*.

greenhouse gas A gas that absorbs infrared radiation released by Earth's surface and then warms the surface and *troposphere* by emitting *energy*, thus giving rise to the *greenhouse effect*. Greenhouse gases include *carbon dioxide* (CO_2), water vapor, ozone (O_3), nitrous oxide (N_2O), halocarbon gases, and *methane* (CH_4).

green manure Organic *fertilizer* comprised of freshly dead plant material.

Green Revolution An intensification of the industrialization of *agriculture* in the developing world in the latter half of the 20th century that has dramatically increased crop yields produced per unit area of farmland. Practices include devoting large areas to *monocultures* of crops specially bred for high yields and rapid growth; heavy use of *fertilizers, pesticides,* and *irrigation* water; and sowing and harvesting on the same piece of land more than once per year or per season.

green tax A levy on *environmentally* harmful activities and products aimed at providing a market-based incentive to correct for *market failure*. Compare *subsidy*.

greenwashing A public-relations effort, generally by a corporation or business, to mislead customers or the public into thinking it is acting more sustainably than it actually is.

gross primary production The *energy* that results when *autotrophs* convert solar energy (sunlight) to energy of chemical bonds in sugars through *photosynthesis*. Autotrophs use a portion of this production to power their own metabolism, which entails oxidizing *organic compounds* by *cellular respiration*. Compare *net primary production*.

ground-source heat pump A pump that harnesses *geothermal energy* from near-surface sources of earth and water, and that can help heat and cool buildings. Operates on the principle that temperatures below ground are more stable than temperatures above ground.

groundwater Water held in aquifers underground. Compare *surface water*.

growth rate The net change in a *population's* size, per 1,000 individuals. Calculated by adding the crude birth rate to the *immigration* rate and then subtracting the crude death rate and the *emigration* rate, each expressed as the number per 1,000 individuals per year.

habitat The specific *environment* in which an organism lives, including both *biotic* and *abiotic factors*.

habitat conservation plan A cooperative agreement that allows landowners to harm threatened or endangered species in some ways if they voluntarily improve habitat for the species in others.

habitat fragmentation The process by which an expanse of natural *habitat* becomes broken up into discontinuous fragments, often as a result of farming, logging, road-building, and other types of human development and land use.

habitat selection The process by which organisms select *habitats* from among the range of options they encounter.

habitat use The process by which organisms use *habitats* from among the range of options they encounter.

Hadley cell One of a pair of cells of *convective circulation* between the equator and 30° north and south latitude that influence global *climate* patterns. Compare *Ferrel cell; polar cell*.

half-life The amount of time it takes for one-half the atoms of a *radioisotope* to emit radiation and decay. Different radioisotopes have different half-lives, ranging from fractions of a second to billions of years.

harmful algal bloom A *population* explosion of toxic algae caused by excessive *nutrient* concentrations.

hazardous waste *Waste* that is toxic, chemically reactive, flammable, or corrosive. Compare *industrial solid waste; municipal solid waste*.

herbicide A type of chemical *pesticide* that kills plants.

herbivory The consumption of plants by animals.

heterotroph (consumer) An organism that consumes other organisms. Includes most animals, as well as fungi and microbes that decompose organic matter.

horizon A distinct layer of *soil*. See *A horizon; B horizon; C horizon; E horizon; O horizon; R horizon*.

hormone A chemical messenger that travels though the bloodstream to stimulate growth, development, and sexual maturity; and regulate brain function, appetite, sexual drive, and many other aspects of physiology and behavior.

host The organism in a parasitic relationship that suffers harm while providing the *parasite* nourishment or some other benefit.

Hubbert's peak The peak in production of *crude oil* in the United States, which occurred in 1970 just as Shell Oil geologist M. King Hubbert had predicted in 1956.

humus A dark, spongy, crumbly mass of material made up of complex organic compounds, resulting from the partial decomposition of organic matter.

hurricane A cyclonic storm that forms over the ocean but can do damage upon its arrival on land. A type of *cyclone* or *typhoon* that usually forms over the Atlantic Ocean.

hydrocarbon An *organic compound* consisting solely of *hydrogen* and *carbon atoms*.

hydroelectric power (hydropower) The generation of electricity using the *kinetic energy* of moving water.

hydrogen The chemical *element* with one proton. The most abundant element in the universe. Also a possible fuel for our future economy.

hydrologic cycle The flow of water—in liquid, gaseous, and solid forms—through our *biotic* and *abiotic environment*.

hydropower See *hydroelectric power*.

hydrosphere All water—salt or fresh, liquid, ice, or vapor—in surface bodies, underground, and in the *atmosphere*. Compare *biosphere; lithosphere*.

hypothesis A statement that attempts to explain a phenomenon or answer a *scientific* question. Compare *theory*.

hypoxia The condition of extremely low dissolved *oxygen* concentrations in a body of water.

igneous rock One of the three main categories of rock. Formed from cooling *magma*. Granite and basalt are examples of igneous rock. Compare *metamorphic rock; sedimentary rock*.

ignitable Easily able to catch fire. One criterion for defining *hazardous waste*.

immigration The arrival of individuals from outside a *population*.

inbreeding depression A state that occurs in a *population* when genetically similar parents mate and produce weak or defective offspring as a result.

incineration A controlled process of burning solid waste for disposal in which mixed garbage is combusted at very high temperatures. Compare *sanitary landfill*.

independent variable The *variable* that the scientist manipulates in a *manipulative experiment*.

indoor air pollution *Air pollution* that occurs indoors.

industrial agriculture A form of *agriculture* that uses large-scale mechanization and *fossil fuel* combustion, enabling farmers to replace horses and oxen with faster and more powerful means of cultivating, harvesting, transporting, and processing crops. Other aspects include *irrigation* and the use of *inorganic fertilizers*. Use of chemical *herbicides* and *pesticides* reduces *competition* from weeds and *herbivory* by insects. Compare *traditional agriculture*.

industrial ecology A holistic approach to industry that integrates principles from engineering, chemistry, *ecology*, *economics*, and other disciplines and seeks to redesign industrial *systems* in order to reduce resource inputs and minimize inefficiency.

industrial fixation *Nitrogen fixation* performed by people to produce *fertilizers* and industrial chemicals.

industrial revolution The shift in the mid-1700s from rural life, animal-powered agriculture, and manufacturing by craftsmen to an urban society powered by *fossil fuels* such as *coal* and *crude oil*. Compare *agricultural revolution*.

industrial smog Gray-air *smog* caused by the incomplete combustion of *coal* or oil when burned. Compare *photochemical smog*.

industrial solid waste Nonliquid *waste* that is not especially hazardous and that comes from production of consumer goods, mining, *petroleum* extraction and *refining*, and *agriculture*.

Compare *hazardous waste; municipal solid waste*.

industrial stage The third stage of the *demographic transition* model, characterized by falling birth rates that close the gap with falling death rates and reduce the rate of *population* growth. Compare *pre-industrial stage; post-industrial stage; transitional stage*.

infectious disease A disease in which a pathogen attacks a host.

infiltration The flow of *surface water* downward through *rock* and *soil* to recharge *groundwater reservoirs*.

inorganic compound A *compound* that does not contain *carbon atoms* joined to other carbon atoms by *covalent bonds*. Compare *organic compound*.

inorganic fertilizer A *fertilizer* that consists of mined or synthetically manufactured mineral supplements.

insecticide A type of chemical *pesticide* that kills insects.

in-situ recovery See *solution mining*.

integrated pest management (IPM) The use of multiple techniques in combination to achieve long-term suppression of pests, including *biocontrol*, use of *pesticides*, close monitoring of *populations*, *habitat* alteration, *crop rotation*, *transgenic* crops, alternative tillage methods, and mechanical pest removal.

intercropping Planting different types of crops in alternating bands or other spatially mixed arrangements.

interdisciplinary field A field that borrows techniques from several more traditional fields of study and brings together research results from these fields into a broad synthesis.

Intergovernmental Panel on Climate Change (IPCC) An international panel of *climate* scientists and government officials established in 1988 by the *United Nations Environment Programme* and the World Meteorological Organization. The IPCC's mission is to assess and synthesize scientific research on *global climate change* and to offer guidance to the world's policymakers. The IPCC's 2007 *Fourth Assessment Report* summarizes current and projected future global trends in climate, and represents the consensus of climate scientists around the world.

interspecific competition *Competition* that takes place among members of two or more different *species*. Compare *intraspecific competition*.

intertidal Of, relating to, or living along shorelines between the highest reach of the highest *tide* and the lowest reach of the lowest tide.

intraspecific competition *Competition* that takes place among members of the same *species*. Compare *interspecific competition*.

intrusive Term for igneous rock formed when magma cools slowly while it is well below Earth's surface (e.g., granite).

invasive species A *species* that spreads widely and rapidly becomes dominant in a *community*, interfering with the community's normal functioning.

inversion layer In a *temperature inversion*, the band of air in which temperature rises with altitude (instead of falling with altitude, as temperature does normally).

ion An electrically charged *atom* or combination of atoms.

ionic bond A chemical bond formed by the attraction between oppositely-charged *ions*.

IPAT model A formula that represents how humans' total impact (I) on the *environment* results from the interaction among three factors: *population* (P), affluence (A), and technology (T).

irrigation The artificial provision of water to support *agriculture*.

isotope One of several forms of an *element* having differing numbers of *neutrons* in the nucleus of its *atoms*. Chemically, isotopes of an element behave almost identically, but they have different physical properties because they differ in mass.

kelp Large brown algae or seaweed that can form underwater "forests," providing habitat for marine organisms.

keystone species A *species* that has an especially far-reaching effect on a *community*.

kinetic energy *Energy* of motion. Compare *potential energy*.

K–selected Term denoting a *species* with low *biotic potential* whose members produce a small number of offspring and take a long time to gestate and raise each of their young, but invest heavily in promoting the survival and growth of these few offspring. *Populations* of K–selected species are generally regulated by *density-dependent factors*. Compare *r–selected*.

kwashiorkor A form of *malnutrition* that results from a high-*starch* diet with inadequate *protein* or *amino acids*. In children, causes bloating of the abdomen, deterioration and discoloration of hair, mental disability, immune suppression, developmental delays, anemia, and reduced growth.

Kyoto Protocol An international agreement drafted in 1997 that calls for reducing, by 2012, emissions of six *greenhouse gases* to levels lower than their levels in 1990. Although the United States refused to ratify the protocol, it came into force in 2005 once Russia ratified it, the 127th nation to do so.

lahar A large flow of destabilized mud racing downslope.

landfill gas A mix of gases that consists of roughly half *methane* produced by anaerobic decomposition deep inside *landfills*.

landscape ecology The study of how landscape structure affects the abundance, distribution, and interaction of organisms. This approach to the study of organisms and their *environments* at the landscape scale focuses on broad geographical areas that include multiple *ecosystems*.

landslide The collapse and downhill flow of large amounts of rock or soil. A severe and sudden form of *mass wasting*.

land trust A private organization, generally local or regional, that preserves lands valued by its members. In most cases, land trusts purchase land outright with the aim of preserving it in its natural condition.

La Niña Strong cooling of surface water in the equatorial Pacific Ocean that occurs every 2 to 7 years and has widespread climatic consequences. Compare *El Niño*. See also *El Niño–Southern Oscillation*.

latitudinal gradient The tendency in many groups of organisms for *species richness* to increase as one approaches the equator.

lava *Magma* that is released from the *lithosphere* and flows or spatters across Earth's surface.

law of conservation of matter Physical law stating that *matter* may be transformed from one type of substance into others, but that it cannot be created or destroyed.

LD$_{50}$ (lethal dose–50%) The amount of a *toxicant* it takes to kill 50% of a *population* of test animals. Compare *ED$_{50}$; threshold dose*.

leachate Liquids that seep through liners of a *sanitary landfill* and leach into the *soil* underneath.

leaching The process by which solid materials such as *minerals* are dissolved in a liquid (usually water) and transported to another location.

lead A heavy metal that may be ingested through water or paint, or that may enter the *atmosphere* as a particulate pollutant through combustion of leaded gasoline or other processes. Atmospheric lead deposited on land and water can enter the *food chain*, accumulate within body tissues, and cause *lead poisoning* in animals and people. An EPA *criteria pollutant*.

Leadership in Energy and Environmental Design (LEED) The leading set of standards for sustainable building. Compare *green building*.

lead poisoning Poisoning by ingestion or inhalation of the heavy metal *lead*, causing an array of maladies including damage to the brain, liver, kidney, and stomach; learning problems and behavioral abnormalities; anemia; hearing loss; and even death. Lead poisoning can result from drinking water that passes through old lead pipes or ingesting dust or chips of old lead-based paint.

legislation Statutory law.

Leopold, Aldo (1887–1949) American scientist, scholar, philosopher, and author. His book *The Land Ethic* argued that humans should view themselves and the land itself as members of the same *community* and that humans are obligated to treat the land *ethically*.

levee See *dike*.

life-cycle analysis In *industrial ecology*, the examination of the entire life cycle of a given product—from its origins in raw materials, through its manufacturing, to its use, and finally its disposal—in an attempt to identify ways to make the process more *ecologically* efficient.

life expectancy The average number of years that individuals in particular age groups are likely to continue to live.

lifestyle hazard See *cultural hazard*.

light rail Public transit systems consisting of rail cars powered by electricity.

light reactions In *photosynthesis*, a series of chemical reactions in which water molecules are split and react to form *hydrogen ions* (H$^+$), molecular *oxygen* (O$_2$), and small, high-energy molecules used to fuel the *Calvin cycle*.

limiting factor A physical, chemical, or biological characteristic of the *environment* that restrains *population* growth.

limnetic zone In a water body, the layer of open water through which sunlight penetrates. Compare *littoral zone; benthic zone*.

liquefied natural gas (LNG) *Natural gas* that has been converted to a liquid at low temperatures and that can be shipped long distances in refrigerated tankers.

lithification The formation of rock through the processes of compaction, binding, and crystallization.

lithosphere The outer layer of Earth, consisting of *crust* and uppermost *mantle*, and located just above the *asthenosphere*. More generally, the solid part of the Earth, including the rocks, *sediment*, and *soil* at the surface and extending down many miles underground. Compare *atmosphere; biosphere; hydrosphere*.

littoral See *intertidal*.

littoral zone The region ringing the edge of a water body. Compare *benthic zone; limnetic zone*.

Living Planet Index A metric that summarizes trends in the *populations* of over 2,500 *species* that are well enough monitored to provide reliable data. Developed by scientists at the World Wildlife Fund and the *United Nations Environment Programme* to give an overall idea of how natural populations are faring. Between 1970 and 2007, this index fell by roughly 30%.

lobbying The expenditure of time or money in an attempt to influence an elected official.

logistic growth curve A plot that shows how the initial *exponential growth* of a *population* is slowed and finally brought to a standstill by *limiting factors*.

longline fishing Fishing practice that involves setting out extremely long lines with up to several thousand baited hooks spaced along their lengths. Kills turtles, sharks, and an estimated 300,000 seabirds each year in *bycatch*.

Love Canal A residential neighborhood in Niagara Falls, New York, from which families were evacuated after buried toxic chemicals rose to the surface, contaminating homes and an elementary school. The well-publicized event helped spark passage of the *Superfund* legislation. Compare *Times Beach*.

low-input agriculture *Agriculture* that uses smaller amounts of *pesticides, fertilizers*, growth hormones, water, and *fossil fuel* energy than are used in *industrial agriculture*. Compare *sustainable agriculture; organic agriculture*.

macromolecule A very large molecule, such as a *protein, nucleic acid, carbohydrate*, or lipid.

magma Molten, liquid rock.

malnutrition The condition of lacking *nutrients* the body needs, including a complete complement of vitamins and minerals.

Malthus, Thomas (1766–1834) British economist who maintained that rising human *population* would eventually deplete the available food supply until starvation, war, or disease arose and reduced the population.

mangrove A tree with roots that curve upward to obtain *oxygen*, which is lacking in the mud in which they grow, and that serve as stilts to support the tree in changing water levels. Mangrove forests grow on the coastlines of the tropics and subtropics.

manipulative experiment An *experiment* in which the researcher actively chooses and manipulates the *independent variable*. Compare *natural experiment*.

mantle The malleable layer of rock that lies beneath Earth's *crust* and surrounds a mostly iron *core*.

maquiladora A U.S.-owned factory on the Mexican side of the U.S.-Mexico border.

marasmus A form of *malnutrition* that results from *protein* deficiency together with a lack of calories, causing wasting or shriveling among millions of children in the developing world.

marine protected area (MPA) An area of the ocean set aside to protect marine life from fishing pressures. An MPA may be protected from some human activities but be open to others. Compare *marine reserve*.

marine reserve An area of the ocean designated as a "no-fishing" zone, allowing no extractive activities. Compare *marine protected area*.

market failure The failure of markets to take into account the *environment*'s positive effects on *economies* (for example, *ecosystem services*) or to reflect the negative effects of economic activity on the environment and thereby on people (*external costs*).

mass extinction event The extinction of a large proportion of the world's *species* in a very short time period due to some extreme and rapid change or catastrophic event. Earth has seen five mass extinction events in the past half-billion years.

mass number The combined number of *protons* and *neutrons* in an *atom*.

mass wasting The downslope movement of soil and rock due to gravity. Compare *landslide*.

materials recovery facility (MRF) A *recycling* facility where items are sorted, cleaned, shredded, and prepared for reprocessing into new items.

matter All material in the universe that has mass and occupies space. See *law of conservation of matter*.

maximum sustainable yield The maximal harvest of a particular *renewable natural resource* that can be accomplished while still keeping the resource available for the future.

meltdown The accidental melting of the uranium fuel rods inside the core of a *nuclear reactor*, causing the release of radiation.

metal A type of chemical element, or a mass of such an element, that typically is lustrous, opaque, malleable, and can conduct heat and electricity.

metamorphic rock One of the three main categories of rock. Formed by great heat and/or pressure that reshapes crystals within the rock and changes its appearance and physical properties. Common metamorphic rocks include marble and slate. Compare *igneous rock*; *sedimentary rock*.

methane (CH_4) A colorless gas produced primarily by *anaerobic* decomposition. The major constituent of *natural gas*, and a *greenhouse gas* that is molecule-for-molecule more potent than *carbon dioxide*.

methane hydrate An ice-like solid consisting of molecules of *methane* embedded in a crystal lattice of water molecules. Methane hydrates are being investigated as a potential new source of *energy* from *fossil fuels*.

Milankovitch cycle One of three types of variations in Earth's rotation and orbit around the sun that result in slight changes in the relative amount of solar radiation reaching Earth's surface at different latitudes. As the cycles proceed, they change the way solar radiation is distributed over Earth's surface and contribute to changes in *atmospheric* heating and circulation that have triggered the ice ages and other *climate* changes.

Millennium Development Goals A program of targets for *sustainable development* set by the international community through the *United Nations* at the turn of this century.

Millennium Ecosystem Assessment The most comprehensive scientific assessment of the present condition of the world's ecological systems and their ability to continue supporting our civilization. Prepared by over 2,000 of the world's leading environmental scientists from nearly 100 nations, and completed in 2005.

mineral A naturally occurring solid *element* or *inorganic compound* with a crystal structure, a specific chemical composition, and distinct physical properties. Compare *ore* and *rock*.

minimum viable population size The *population* size for a given population of a *species* that should assure its continued existence indefinitely into the future. Calculated and used by *conservation geneticists*, population biologists, and wildlife managers.

mining (1) In the broad sense, the extraction of any resource that is nonrenewable on the timescale of our society (such as fossil fuels or groundwater). (2) In relation to mineral resources, the systematic removal of rock, soil, or other material for the purpose of extracting minerals of economic interest.

mitigation A response to *global climate change*, consisting of an attempt to reduce *greenhouse gas* emissions so as to lessen the severity of climate change. Compare *adaptation*.

mixed economy An economy that combines elements of a *capitalist market economy* and a *centrally planned economy*.

model A simplified representation of a complex natural process, designed by scientists to help understand how the process occurs and to make predictions.

moderator Within a *nuclear reactor*, a substance, most often water or graphite, that slows the *neutrons* bombarding uranium so that *fission* can begin.

molecule A combination of two or more *atoms*.

monoculture The uniform planting of a single crop over a large area. Characterizes *industrial agriculture*. Compare *polyculture*.

Montreal Protocol International treaty ratified in 1987 in which 180 signatory nations agreed to restrict production of *chlorofluorocarbons (CFCs)* in order to forestall stratospheric ozone depletion.

mortality Rate of death within a *population*.

mosaic In *landscape ecology*, a spatial configuration of *patches* arrayed across a landscape.

mountaintop removal mining A large-scale form of *coal* mining in which entire mountaintops are leveled. Supplies much of U.S. coal, but exerts severe environmental impacts on surrounding ecosystems and human residents.

Muir, John (1838–1914) Scottish immigrant to the United States who eventually settled in California and made the Yosemite Valley his wilderness home. Today, he is most strongly associated with the *preservation ethic*. He argued that nature deserved protection for its own inherent values (an *ecocentrist* argument) but also claimed that nature played a large role in human happiness and fulfillment (an *anthropocentrist* argument).

municipal solid waste Nonliquid *waste* that is not especially hazardous and that comes from homes, institutions, and small businesses. Compare *hazardous waste*; *industrial solid waste*.

mutagen A *toxicant* that causes *mutations* in the *DNA* of organisms.

mutation An accidental change in *DNA* that may range in magnitude from the deletion, substitution, or addition of a single nucleotide to a change affecting entire sets of chromosomes. Mutations provide the raw material for evolutionary change.

mutualism A relationship in which all participating organisms benefit from their interaction. Compare *parasitism*.

nacelle Compartment in a *wind turbine* containing machinery for generating power.

NAFTA See *North American Free Trade Agreement*.

natality Rate of birth within a *population*.

National Environmental Policy Act (NEPA) A 1970 U.S. law that created the Council on Environmental Quality and required that an *environmental impact statement* be prepared for any major federal action.

national forest Public lands consisting of 191 million acres (more than 8% of the nation's land area) in many tracts spread across all but a few states.

National Forest Management Act *Legislation* passed by the U.S. Congress in 1976, mandating that each *national forest* draw up plans for renewable resource management, based on the concepts of *multiple use* and *sustainable development* and subject to broad public participation.

national park A scenic area set aside for recreation and enjoyment by the public. The national park system today numbers 392 sites totaling 84 million acres and includes national historic sites, national recreation areas, national wild and scenic rivers, and other types of areas.

national wildlife refuge An area set aside to serve as a haven for wildlife and also sometimes to encourage hunting, fishing, wildlife observation, photography, environmental education, and other public uses.

natural experiment An *experiment* in which the researcher cannot directly manipulate the *variables* and therefore must observe nature, comparing conditions in which variables differ, and interpret the results. Compare *manipulative experiment*.

natural gas A *fossil fuel* composed primarily of methane (CH_4), produced as a by-product when bacteria decompose organic material under *anaerobic* conditions.

natural rate of population change The rate of change in a *population*'s size resulting from birth and death rates alone, excluding migration.

natural resource Any of the various substances and *energy* sources we need in order to survive.

Natural Resources Conservation Service U.S. agency that promotes soil conservation, as well as water quality protection and pollution control. Prior to 1994, known as the Soil Conservation Service.

natural science An academic discipline that studies the natural world. Compare *social science*.

natural selection The process by which traits that enhance survival and reproduction are passed on more frequently to future generations of organisms than those that do not, thus altering the genetic makeup of populations through time. Natural selection acts on genetic variation and is a primary driver of *evolution*.

negative feedback loop A *feedback loop* in which output of one type acts as input that moves the *system* in the opposite direction. The input and output essentially neutralize each other's effects, stabilizing the system. Compare *positive feedback loop*.

neoclassical economics A *theory* of economics that explains market prices in terms of consumer preferences for units of particular commodities. Buyers desire the lowest possible price, whereas sellers desire the highest possible price. This conflict between buyers and sellers results in a compromise price being reached and the "right" quantity of commodities being bought and sold. Compare *ecological economics; environmental economics*.

net energy The quantitative difference between *energy* returned from a process and energy invested in the process. Positive net energy values mean that a process produces more energy than is invested. Compare *energy returned on investment*.

net metering Process by which owners of houses with *photovoltaic* systems or *wind turbines* can sell their excess *solar energy* or *wind energy* to their local power utility. Whereas *feed-in tariffs* award producers with prices above market rates, net metering offers market-rate prices.

net primary production The *energy* or biomass that remains in an ecosystem after *autotrophs* have metabolized enough for their own maintenance through *cellular respiration*. Net primary production is the energy or biomass available for consumption by *heterotrophs*. Compare *gross primary production*.

net primary productivity The rate at which *net primary production* is produced. See *productivity; gross primary production; net primary production*.

neurotoxin A *toxicant* that assaults the nervous system. Neurotoxins include heavy metals, *pesticides*, and some chemical weapons developed for use in war.

neutral The property of a *solution* in which the concentration of hydroxide *ions* (OH⁻) is equal to the concentration of *hydrogen ions* (H⁺).

neutron An electrically neutral (uncharged) particle in the nucleus of an *atom*.

new urbanism An approach among architects, planners, and developers that seeks to design neighborhoods in which homes, businesses, schools, and other amenities are within walking distance of one another. Proponents of new urbanism aim to counter *sprawl* by creating functional neighborhoods in which families can meet most needs close to home without use of a car.

niche The functional role of a *species* in a *community*.

NIMBY See *not-in-my-backyard*.

nitrification The conversion by bacteria of ammonium ions (NH₄⁺) first into nitrite ions (NO₂⁻) and then into nitrate ions (NO₃⁻).

nitrogen The chemical *element* with seven *protons* and seven *neutrons*. The most abundant element in the *atmosphere*, a key element in *macromolecules*, and a crucial plant *nutrient*.

nitrogen cycle A major *nutrient cycle* consisting of the routes that *nitrogen atoms* take through the nested networks of environmental *systems*.

nitrogen dioxide (NO₂) A foul-smelling reddish brown gas that contributes to *smog* and *acid deposition*. It results when atmospheric *nitrogen* and *oxygen* react at the high temperatures created by combustion engines. An EPA *criteria pollutant*.

nitrogen fixation The process by which inert *nitrogen* gas combines with *hydrogen* to form ammonium ions (NH₄⁺), which are chemically

and biologically active and can be taken up by plants.

nitrogen-fixing Term describing bacteria that live in a *mutualistic* relationship with many types of plants and provide *nutrients* to the plants by converting *nitrogen* to a usable form.

nitrogen oxide (NOₓ) One of a family of chemical compounds containing a nitrogen atom bonded to one or more oxygen atoms. *Nitrogen dioxide* is one example.

nonconsumptive use Use of *fresh water* whereby the water from a particular *aquifer* or surface water body either is not removed or is removed only temporarily and then returned. The use of water to generate electricity in hydroelectric *dams* is an example. Compare *consumptive use*.

nongovernmental organization (NGO) An organization unaffiliated with any government or corporation that exists to promote an issue or agenda. Many operate internationally, and some exert influence over environmental policy.

nonmarket value A value that is not usually included in the price of a *good* or *service*.

non-point source A diffuse source of *pollutants*, often consisting of many small sources. Compare *point source*.

nonrenewable natural resource A *natural resource* that is in limited supply and is formed much more slowly than we use it. Compare *renewable natural resource*.

North American Free Trade Agreement (NAFTA) A 1994 treaty among Canada, Mexico, and the United States to promote commerce and cross-border trade. Among many other provisions, NAFTA allows industries and corporations to weaken environmental protection laws if they view them as barriers to trade.

North Atlantic Deep Water (NADW) The deep portion of the *thermohaline circulation* in the northern Atlantic Ocean.

no-till *Agriculture* that does not involve tilling (plowing, disking, harrowing, or chiseling) the *soil*. The most extreme form of *conservation tillage*.

not-in-my-backyard (NIMBY) Syndrome in which people do not want something (e.g., a polluting facility) near where they live, even if they may want or need the thing to exist somewhere.

nuclear energy The *energy* that holds together *protons* and *neutrons* within the nucleus of an *atom*. Several processes, each of which involves transforming *isotopes* of one *element* into isotopes of other elements, can convert nuclear energy into thermal energy, which is then used to generate electricity. See also *nuclear fission; nuclear reactor*.

nuclear fission The conversion of the *energy* within an *atom's* nucleus to usable thermal energy by splitting apart atomic nuclei. Compare *nuclear fusion*.

nucleic acid A *macromolecule* that directs the production of *proteins*. Includes *DNA* and *RNA*.

nutrient An *element* or *compound* that organisms consume and require for survival.

nutrient cycle The comprehensive set of cyclical pathways by which a given *nutrient* moves through the *environment*.

ocean acidification The process by which today's oceans are becoming more *acidic* (attaining lower *pH*), as a result of increased car-

bon dioxide concentrations in the atmosphere. Ocean acidification occurs as ocean water absorbs CO₂ from the air and forms carbonic acid. Ocean acidification threatens to dissolve *coral reefs*, which would have strong impacts on marine biodiversity.

ocean thermal energy conversion (OTEC) A potential *energy* source that involves harnessing the solar radiation absorbed by tropical oceans in the tropics. See *closed cycle; open cycle*.

O horizon The top layer of *soil* in some *soil profiles*, made up of organic matter, such as decomposing branches, leaves, crop residue, and animal waste. Compare *A horizon; B horizon; C horizon; E horizon; R horizon*.

oil See *petroleum*.

oil sands (tar sands) Deposits that can be mined from the ground, consisting of moist sand and clay containing 1–20% *bitumen*, that some envision as a replacement for *crude oil* as this resource is depleted. Oil sands represent crude oil deposits that have been degraded and chemically altered by water *erosion* and bacterial decomposition.

oil shale *Sedimentary rock* filled with *kerogen* that can be processed to produce liquid *petroleum*. Oil shale is formed by the same processes that form *crude oil*, but occurs when kerogen was not buried deeply enough or subjected to enough heat and pressure to form oil.

oligotrophic Term describing a water body that has low-nutrient and high-oxygen conditions. Compare *eutrophic*.

open-pit mining A *mining* technique that involves digging a gigantic hole and removing the desired *ore*, along with waste rock that surrounds the ore.

ore A *mineral* or grouping of minerals from which we extract *metals*.

organic agriculture *Agriculture* that uses no synthetic *fertilizers* or *pesticides* but instead relies on biological approaches such as *composting* and *biocontrol*. Compare *low-input agriculture; sustainable agriculture*.

organic compound A *compound* made up of *carbon atoms* (and, generally, *hydrogen* atoms) joined by *covalent bonds* and sometimes including other *elements*, such as *nitrogen, oxygen*, sulfur, or *phosphorus*. The unusual ability of carbon to build elaborate *molecules* has resulted in millions of different organic compounds showing various degrees of complexity. Compare *inorganic compound*.

organic fertilizer A *fertilizer* made up of natural materials (largely the remains or *wastes* of organisms), including animal manure, crop residues, fresh vegetation, and *compost*. Compare *inorganic fertilizer*.

Organization of Petroleum Exporting Countries (OPEC) Cartel of predominantly Arab nations that in 1973 embargoed oil shipments to the United States and other nations supporting Israel, setting off the nation's first oil shortage.

outdoor air pollution *Air pollution* that occurs outdoors.

overgrazing The consumption by too many animals of plant cover, impeding plant regrowth and the replacement of *biomass*. Overgrazing can exacerbate damage to *soils*, natural *communities*, and the land's productivity for further grazing.

overnutrition A condition of excessive food intake in which people receive more than their daily caloric needs.

overshoot The amount by which humanity has surpassed Earth's long-term carrying capacity for our species.

oxygen The chemical *element* with eight *protons* and eight *neutrons*. A key element in the *atmosphere* that is produced by *photosynthesis*.

ozone A *molecule* consisting of three atoms of *oxygen*. Absorbs ultraviolet radiation in the *stratosphere*. Compare *ozone layer; tropospheric ozone*.

ozone-depleting substances Airborne chemicals, such as *halocarbons*, that destroy *ozone* molecules and thin the *ozone layer* in the *stratosphere*.

ozone hole Term popularly used to describe the thinning of the stratospheric *ozone layer* that occurs over Antarctica each year, as a result of *ozone-depleting substances*.

ozone layer A portion of the *stratosphere*, roughly 17–30 km (10–19 mi) above sea level, that contains most of the *ozone* in the *atmosphere*.

paper park A term referring to parks that are protected "on paper" but not in reality.

paradigm A dominant philosophical and theoretical framework within a scientific discipline.

parasite The organism in a parasitic relationship that extracts nourishment or some other benefit from the *host*.

parasitism A relationship in which one organism, the *parasite*, depends on another, the *host*, for nourishment or some other benefit while simultaneously doing the host harm. Compare *mutualism*.

parasitoid An insect that parasitizes other insects, generally causing eventual death of the *host*. Compare *parasite*.

parent material The base geological material in a particular location.

particulate matter Solid or liquid particles small enough to be suspended in the *atmosphere* and able to damage respiratory tissues when inhaled. Includes *primary pollutants* such as dust and soot as well as *secondary pollutants* such as sulfates and nitrates. An EPA *criteria pollutant*.

passive solar energy collection An approach in which buildings are designed and building materials are chosen to maximize their direct absorption of sunlight in winter, even as they keep the interior cool in the summer. Compare *active solar energy collection*.

patch In *landscape ecology*, spatial areas within a landscape. Depending on a researcher's perspective, patches may consist of habitat for a particular organism, or communities, or ecosystems. An array of patches forms a *mosaic*.

pathogen A microbe that causes disease.

peak oil Term used to describe the point of maximum production of petroleum in the world (or for a given nation), after which oil production declines. This is also expected to be roughly the midway point of extraction of the world's oil supplies. The term is generally used in contexts suggesting that our society will face tremendous challenges once the peak is past. Compare *Hubbert's peak*.

peer review The process by which a manuscript submitted for publication in an academic journal is examined by other specialists in the field, who provide comments and criticism (generally anonymously), and judge whether the work merits publication in the journal.

pelagic Of, relating to, or living between the surface and floor of the ocean. Compare *benthic*.

periodic table of the elements (see Appendix C) Standard table in chemistry that summarizes information on the *elements*.

permafrost In *tundra*, underground soil that remains more or less permanently frozen.

permit trading The practice of buying and selling government-issued marketable emissions permits to conduct environmentally harmful activities. Under a *cap-and-trade* system, the government determines an acceptable level of *pollution* and then issues permits to pollute. A company receives credit for amounts it does not emit and can then sell this credit to other companies. Compare *emissions trading system*.

peroxyacyl nitrate A chemical created by the reaction of NO_2 with hydrocarbons that can induce further reactions that damage living tissues in animals and plants.

pest A pejorative term for any organism that damages crops that are valuable to us. The term is subjective and defined by our own economic interest, and is not biologically meaningful. Compare *weed*.

pesticide An artificial chemical used to kill insects (*insecticide*), plants (*herbicide*), or fungi (*fungicide*).

pesticide drift Airborne transport of *pesticides*.

petroleum See *crude oil*.

pH A measure of the concentration of *hydrogen ions* in a *solution*. The pH scale ranges from 0 to 14: A solution with a pH of 7 is neutral; solutions with a pH below 7 are *acidic*, and those with a pH higher than 7 are *basic*. Because the pH scale is logarithmic, each step on the scale represents a tenfold difference in hydrogen ion concentration.

phase shift A fundamental shift in the overall character of an ecological community, generally occurring after some extreme disturbance, and after which the community may not return to its original state. Also known as a *regime shift*.

phosphorus The chemical *element* with 15 *protons* and 15 *neutrons*. An abundant element in the *lithosphere*, a key element in *macromolecules*, and a crucial plant *nutrient*.

phosphorus cycle A major *nutrient cycle* consisting of the routes that *phosphorus atoms* take through the nested networks of environmental *systems*.

photic zone In the ocean or a freshwater body, the well-lit top layer of water where *photosynthesis* occurs.

photobiological Term describing a process that produces fuel by utilizing light and living organisms.

photochemical smog Brown-air *smog* caused by light-driven reactions of *primary pollutants* with normal atmospheric *compounds* that produce a mix of over 100 different chemicals, ground-level ozone often being the most abundant among them. Compare *industrial smog*.

photosynthesis The process by which *autotrophs* produce their own food. Sunlight powers a series of chemical reactions that convert *carbon dioxide* and water into sugar (glucose), thus transforming low-quality *energy* from the sun into high-quality energy the organism can use. Compare *cellular respiration*.

photovoltaic (PV) cell A device designed to collect sunlight and directly convert it to electrical energy. When light strikes one of a pair of metal plates within the cell, it causes the release of electrons, which are attracted by electrostatic forces to the opposing plate. The flow of electrons from one plate to the other creates an electrical current.

phthalates Hormone-disrupting chemicals used to soften plastics and enhance fragrances in children's toys, cosmetics, perfumes, and other consumer products.

phylogenetic tree A treelike diagram that represents the history of divergence of *species* or other taxonomic groups of organisms.

physical hazard Physical processes that occur naturally in our environment and pose human health hazards. These include discrete events such as *earthquakes*, *volcanic eruptions*, fires, floods, blizzards, *landslides*, hurricanes, and droughts, as well as ongoing natural phenomena such as ultraviolet radiation from sunlight. Compare *biological hazard; chemical hazard; cultural hazard*.

phytoplankton Microscopic photosynthetic algae, protists, and cyanobacteria that drift near the surface of water bodies and generally form the first *trophic level* in an aquatic *food chain*. Compare *zooplankton*.

Pinchot, Gifford (1865–1946) The first professionally trained American *forester*, Pinchot helped establish the U.S. Forest Service. Today, he is the person most closely associated with the *conservation ethic*.

pioneer species A *species* that arrives earliest, beginning the ecological process of *succession* in a terrestrial or aquatic *community*.

placer mining A *mining* technique that involves sifting through material in modern or ancient riverbed deposits, generally using running water to separate lightweight mud and gravel from heavier minerals of value.

plastics Synthetic (human-made) *polymers* used in numerous manufactured products.

plate tectonics The process by which Earth's surface is shaped by the extremely slow movement of tectonic plates, or sections of *crust*. Earth's surface includes about 15 major tectonic plates. Their interaction gives rise to processes that build mountains, cause *earthquakes*, and otherwise influence the landscape.

point source A specific spot—such as a factory's smokestacks—where large quantities of *pollutants* are discharged. Compare *non-point source*.

polar cell One of a pair of cells of *convective circulation* between the poles and 60° north and south latitude that influence global *climate* patterns. Compare *Ferrel cell; Hadley cell*.

polar stratospheric clouds High-altitude icy clouds containing condensed nitric acid that enhance the destruction of stratospheric ozone in the spring.

polar vortex Circular wind currents that trap air over Antarctica during winter, worsening ozone depletion over the continent.

policy A rule or guideline that directs individual, organizational, or societal behavior.

pollination An interaction in which one organism (for example, bees) transfers pollen (male sex cells) from one flower to the ova (female cells) of another, fertilizing the female flower and leading to production of fruit and seeds.

polluter-pays principle Principle specifying that the party responsible for producing *pollution* should pay the costs of cleaning up the pollution or mitigating its impacts.

pollution Any matter or *energy* released into the *environment* that causes undesirable impacts on the health and well-being of humans or other organisms. Pollution can be physical, chemical, or biological, and can affect water, air, or soil.

polybrominated diphenyl ethers (PBDEs) Synthetic compounds that provide fire-retardant properties and are used in a diverse array of consumer products, including computers, televisions, plastics, and furniture. Released during production, disposal, and use of products, these chemicals persist and accumulate in living tissue, and appear to be *endocrine disruptors*.

polyculture The planting of multiple crops in a mixed arrangement or in close proximity. An example is some traditional Native American farming that mixed maize, beans, squash, and peppers. Compare *monoculture*.

polymer A chemical *compound* or mixture of compounds consisting of long chains of repeated *molecules*. Some polymers play key roles in the building blocks of life.

pool A location in which nutrients in a *biogeochemical cycle* remain for a period of time before moving to another pool. Can be living or nonliving entities. Synonymous with *reservoir*. Compare *flux; residence time*.

population A group of organisms of the same *species* that live in the same area. Species are often composed of multiple populations.

population density The number of individuals within a *population* per unit area. Compare *population size*.

population dispersion See *population distribution*.

population distribution The spatial arrangement of organisms within a particular area.

population ecology The study of the quantitative dynamics of population change and the factors that affect the distribution and abundance of members of a population.

population pyramid See *age structure diagram*.

population size The number of individual organisms present at a given time.

positive feedback loop A *feedback loop* in which output of one type acts as input that moves the *system* in the same direction. The input and output drive the system further toward one extreme or another. Compare *negative feedback loop*.

post-industrial stage The fourth and final stage of the *demographic transition* model, in which both birth and death rates have fallen to a low level and remain stable there, and *populations* may even decline slightly. Compare *industrial stage; pre-industrial stage; transition stage*.

potential energy *Energy* of position. Compare *kinetic energy*.

precautionary principle The idea that one should not undertake a new action until the ramifications of that action are well understood.

precipitation Water that condenses out of the *atmosphere* and falls to Earth in droplets or crystals.

predation The process in which one *species* (the *predator*) hunts, tracks, captures, and ultimately kills its *prey*.

predator An organism that hunts, captures, kills, and consumes individuals of another species, the *prey*.

prediction A specific statement, generally arising from a *hypothesis*, that can be tested directly and unequivocally.

pre-industrial stage The first stage of the *demographic transition* model, characterized by conditions that defined most of human history. In pre-industrial societies, both death rates and birth rates are high. Compare *industrial stage; post-industrial stage; transitional stage*.

prescribed burning The practice of burning areas of *forest* or grassland under carefully controlled conditions to improve the health of *ecosystems*, return them to a more natural state, and help prevent uncontrolled catastrophic fires.

preservation ethic An ethic holding that we should protect the natural *environment* in a pristine, unaltered state. Compare *conservation ethic*.

prey An organism that is killed and consumed by a *predator*.

primary consumer An organism that consumes *producers* and feeds at the second *trophic level*.

primary extraction The initial drilling and pumping of the most easily accessible *crude oil*. Compare *secondary extraction*.

primary forest *Forest* uncut by people. Compare *second-growth*.

primary pollutant A hazardous substance, such as soot or *carbon monoxide*, that is emitted into the *troposphere* in a form that is directly harmful. Compare *secondary pollutant*.

primary production The conversion of solar energy to the energy of chemical bonds in sugars during *photosynthesis*, performed by *autotrophs*.

primary succession A stereotypical series of changes as an *ecological community* develops over time, beginning with a lifeless substrate. In terrestrial *systems*, primary succession begins when a bare expanse of rock, *sand*, or *sediment* becomes newly exposed to the atmosphere and *pioneer species* arrive. Compare *secondary succession*.

primary treatment A stage of *wastewater* treatment in which contaminants are physically removed. Wastewater flows into tanks in which sewage solids, grit, and particulate matter settle to the bottom. Greases and oils float to the surface and can be skimmed off. Compare *secondary treatment*.

probability A quantitative description of the likelihood of a certain outcome.

producer See *autotroph*.

productivity The rate at which plants convert solar *energy* (sunlight) to *biomass*. Ecosystems whose plants convert solar energy to biomass rapidly are said to have high productivity. See *net primary productivity; gross primary production; net primary production*.

protein A *macromolecule* made up of long chains of amino acids.

proton A positively charged particle in the nucleus of an *atom*.

proven recoverable reserve The amount of a given *fossil fuel* in a deposit that is technologically and economically feasible to remove under current conditions.

proxy indicator A type of indirect evidence that serves as a proxy, or substitute, for direct

measurement, and that sheds light on conditions of the past. For example, pollen from *sediment* cores and air bubbles from ice cores provide data on the past *climate*.

public policy *Policy* that is made by governments, including those at the local, state, federal, and international levels; it consists of *legislation, regulations*, orders, incentives, and practices intended to advance societal welfare. See also *environmental policy*.

public-private partnership A combined effort of government and a for-profit entity, generally intended to use the efficiency of the private sector to help achieve a public policy goal.

pumped storage An approach to even power supply and demand whereby water is pumped uphill when electricity demand is low and allowed to flow downhill through a turbine when electricity demand is high.

pyroclastic flow A fast-moving cloud of toxic gas, ash, and rock fragments that races down the slopes of a *volcano*.

pyrolysis The chemical breakdown of organic matter (such as *biomass, oil shale*, and so on) by heating in the absence of *oxygen*, which often produces materials that can be more easily converted to usable energy. A number of variations on this process exist.

quarry A pit used to extract mineral resources such as clay, gravel, sand, and stone such as limestone, granite, marble, and slate.

radiative forcing The amount of change in *energy* that a given factor (such as *aerosols, albedo*, or *greenhouse gases*) exerts over Earth's energy balance. Positive radiative forcing warms the surface, whereas negative radiative forcing cools it.

radioactive The quality by which some *isotopes* "decay," changing their chemical identity as they shed atomic particles and emit high-energy radiation.

radioisotope A radioactive *isotope* that emits subatomic particles and high-*energy* radiation as it "decays" into progressively lighter isotopes until becoming a *stable isotope*.

radon A highly *toxic*, radioactive, colorless gas that seeps up from the ground in areas with certain types of bedrock and can build up inside basements and homes with poor air circulation.

random distribution Distribution pattern in which individuals are located haphazardly in space in no particular pattern (often when needed resources are spread throughout an area and other organisms do not strongly influence where individuals settle).

rangeland Land used for grazing livestock.

REACH Program of the European Union that shifts the burden of proof for testing chemical safety from national governments to industry, and requires that chemical substances produced or imported in amounts of over 1 metric ton per year be registered with a new European Chemicals Agency. REACH, which stands for Registration, Evaluation, Authorisation and Restriction of Chemicals, went into effect in 2007.

reactive Chemically unstable and readily able to react with other compounds, often explosively or by producing noxious fumes. One criterion for defining *hazardous waste*.

reclamation The act of restoring a *mining* site to an approximation of its pre-mining condition. To reclaim a site, companies are required to remove buildings and other structures used for mining, replace overburden, fill in shafts, and replant the area with vegetation.

recombinant DNA *DNA* that has been patched together from the DNA of multiple organisms in an attempt to produce desirable traits (such as rapid growth, disease and pest resistance, or higher nutritional content) in organisms lacking those traits.

recovery Waste management strategy composed of *recycling* and *composting*.

recycling The collection of materials that can be broken down and reprocessed to manufacture new items.

Red List An updated list of *species* facing unusually high risks of *extinction*. The list is maintained by the World Conservation Union.

red tide A *harmful algal bloom* consisting of algae that produce reddish pigments that discolor surface waters.

Reducing Emissions from Deforestation and Forest Degradation (REDD) A program being developed to mitigate *climate change*, whereby wealthy industrialized nations would pay poorer developing nations to conserve *forest* while the wealthy nations would receive credits for *carbon offsets*.

regime shift See *phase shift*.

regional planning *City planning* done on broader geographic scales, generally involving multiple municipal governments.

regulation A specific rule issued by an administrative agency, based on the more broadly written statutory law passed by Congress and enacted by the president.

relative abundance The extent to which numbers of individuals of different species are equal or skewed. One way to express species diversity. See *evenness*; compare *species richness*.

relativist An ethicist who maintains that *ethics* do and should vary with social context. Compare *universalist*.

renewable natural resource A *natural resource* that is virtually unlimited or that is replenished by the *environment* over relatively short periods of hours to weeks to years. Compare *nonrenewable natural resource*.

replacement fertility The *total fertility rate (TFR)* that maintains a stable *population* size.

reproductive window The period of a woman's life, beginning with sexual maturity and ending with menopause, during which she may become pregnant.

reserves-to-production ratio (R/P ratio) The total remaining reserves of a *fossil fuel* divided by the annual rate of production (extraction and processing).

reservoir (1) An artificial water body behind a dam that stores water for human use. (2) See *pool*.

residence time In a biogeochemical cycle, the amount of time a nutrient remains in a given *pool* or *reservoir* before moving to another. Compare *flux*.

resilience The ability of an ecological *community* to change in response to disturbance but later return to its original state. Compare *resistance*.

resistance The ability of an ecological *community* to remain stable in the presence of a disturbance. Compare *resilience*.

Resource Conservation and Recovery Act (RCRA) Congressional *legislation* (enacted

in 1976 and amended in 1984) that specifies, among other things, how to manage *sanitary landfills* to protect against environmental contamination.

resource management Strategic decision making about who should extract resources and in what ways, so that resources are used wisely and not wasted.

resource partitioning The process by which *species* adapt to *competition* by evolving to use slightly different resources, or to use their shared resources in different ways, thus minimizing interference with one another.

response The type or magnitude of negative effects an animal exhibits in response to a *dose* of *toxicant* in a *dose-response analysis*. Compare *dose*.

restoration ecology The study of the historical conditions of *ecological communities* as they existed before humans altered them. Principles of restoration ecology are applied in the practice of *ecological restoration*.

reverse osmosis A process that removes dissolved substances by forcing a liquid through a membrane at high pressure. Water molecules pass through pores in the membrane because of their relatively small size, whereas salts and other larger molecules cannot.

R horizon The bottommost layer of *soil* in a typical *soil profile*. Also called *bedrock*. Compare *A horizon; B horizon; C horizon; E horizon; O horizon*.

ribonucleic acid See *RNA*.

riparian Relating to a river or the area along a river.

risk The mathematical probability that some harmful outcome (for instance, injury, death, *environmental* damage, or *economic* loss) will result from a given action, event, or substance.

risk assessment The quantitative measurement of *risk*, together with the comparison of risks involved in different activities or substances.

risk management The process of considering information from scientific *risk assessment* in light of economic, social, and political needs and values, in order to make decisions and design strategies to minimize *risk*.

RNA (ribonucleic acid) A usually single-stranded *nucleic acid* composed of four nucleotides, each of which contains a sugar (ribose), a phosphate group, and a nitrogenous base. RNA carries the hereditary information for living organisms and is responsible for passing traits from parents to offspring. Compare *DNA*.

roadless rule A 2001 Clinton Administration executive order that put 31% of national forest land off-limits to road construction or maintenance.

rock A solid aggregation of *minerals*.

rock cycle The very slow process in which *rocks* and the *minerals* that make them up are heated, melted, cooled, broken, and reassembled, forming *igneous*, *sedimentary*, and *metamorphic* rocks.

rotation time The number of years that pass between the time a *forest* stand is cut for timber and the next time it is cut.

r–selected Term denoting a *species* with high *biotic potential* whose members produce a large number of offspring in a relatively short time but do not care for their young after birth. *Populations* of r–selected species are generally regulated by *density-independent factors*. Compare *K–selected*.

runoff The water from *precipitation* that flows into streams, rivers, lakes, and ponds, and (in many cases) eventually to the ocean.

run-of-river Any of several methods used to generate *hydroelectric power* without greatly disrupting the flow of river water. Run-of-river approaches eliminate much of the *environmental* impact of large *dams*. Compare *storage*.

safe harbor agreement A cooperative agreement that allows landowners to harm threatened or endangered species in some ways if they voluntarily improve habitat for the species in others.

salinization The buildup of salts in surface *soil* layers.

salt marsh Flat land that is intermittently flooded by the ocean where the *tide* reaches inland. Salt marshes occur along temperate coastlines and are thickly vegetated with grasses, rushes, shrubs, and other herbaceous plants.

salvage logging The removal of dead trees following a natural disturbance. Although it may be economically beneficial, salvage logging can be ecologically destructive, because the dead trees provide food and shelter for a variety of insects and wildlife and because removing timber from recently burned land can cause severe *erosion* and damage to *soil*.

sanitary landfill A site at which solid waste is buried in the ground or piled up in large mounds for disposal, designed to prevent the waste from contaminating the *environment*. Compare *incineration*.

savanna A *biome* characterized by grassland interspersed with clusters of acacias and other trees. Found across parts of Africa, South America, Australia, India, and other dry tropical regions.

science A systematic process for learning about the world and testing our understanding of it.

scientific method A formalized method for testing ideas with observations that involves several assumptions and a more or less consistent series of interrelated steps.

scrubber Technology to chemically treat gases produced in combustion to remove hazardous components and neutralize acidic gases, such as *sulfur dioxide* and hydrochloric acid, turning them into water and salt, in order to reduce smokestack emissions.

secondary consumer An organism that consumes *primary consumers* and feeds at the third *trophic level*.

secondary extraction The extraction of *crude oil* remaining after *primary extraction* by using solvents or by flushing underground rocks with water or steam. Compare *primary extraction*.

secondary forest *Forest* that has grown back after *primary forest* has been cut. Consists of *second-growth* trees.

secondary pollutant A hazardous substance produced through the reaction of substances added to the *atmosphere* with chemicals normally found in the atmosphere. Compare *primary pollutant*.

secondary succession A stereotypical series of changes as an *ecological community* develops over time, beginning when some event disrupts or dramatically alters an existing community. Compare *primary succession*.

secondary treatment A stage of *wastewater* treatment in which biological means are used to remove contaminants remaining after pri-

mary treatment. Wastewater is stirred up in the presence of bacteria, which degrade organic pollutants in the water. The wastewater then passes to another settling tank, where remaining solids drift to the bottom. Compare *primary treatment.*

second-growth Term describing trees that have sprouted and grown to partial maturity after virgin timber has been cut.

second law of thermodynamics Physical law stating that the nature of *energy* tends to change from a more-ordered state to a less-ordered state; that is, *entropy* increases.

sediment The eroded remains of rocks.

sedimentary rock One of the three main categories of rock. Formed when dissolved *minerals* seep through *sediment* layers and act as a kind of glue, crystallizing and binding sediment particles together. Sandstone and shale are examples of sedimentary rock. Compare *igneous rock; metamorphic rock.*

seed bank A storehouse for samples of the world's crop diversity.

selective breeding See *artificial selection.*

septic system A *wastewater* disposal method, common in rural areas, consisting of an underground tank and series of drainpipes. Wastewater runs from the house to the tank, where solids precipitate out. The water proceeds downhill to a drain field of perforated pipes laid horizontally in gravel-filled trenches, where microbes decompose the remaining waste.

service Work done for others as a form of business.

sex ratio The proportion of males to females in a *population.*

shelterbelt A row of trees or other tall perennial plants that are planted along the edges of farm fields to break the wind and thereby minimize wind *erosion.*

sick-building syndrome A *building-related illness* produced by indoor *pollution* in which the specific cause is not identifiable.

sink In a *nutrient cycle*, a *pool* that accepts more *nutrients* than it releases.

SLOSS (Single Large or Several Small) dilemma The debate over whether it is better to make reserves large in size and few in number or many in number but small in size.

smart growth A *city planning* concept in which a community's growth is managed in ways that limit *sprawl* and maintain or improve residents' quality of life. It involves guiding the rate, placement, and style of development such that it serves the *environment*, the *economy*, and the community.

smelting A process in which ore is heated beyond its melting point and combined with other metals or chemicals, in order to form metal with desired characteristics. Steel is created by smelting iron ore with carbon, for example.

Smith, Adam (1723–1790) Scottish philosopher known today as the father of *classical economics.* He believed that when people are free to pursue their own economic self-interest in a competitive marketplace, the marketplace will behave as if guided by "an invisible hand" that ensures that their actions will benefit society as a whole.

smog Term popularly used to describe unhealthy mixtures of air *pollutants* that often form over urban areas. See *industrial smog; photochemical smog.*

snag A dead or dying tree that is still standing. Snags are valuable for wildlife.

social science An academic discipline that studies human interactions and institutions. Compare *natural science.*

soil A complex plant-supporting *system* consisting of disintegrated rock, organic matter, air, water, *nutrients*, and microorganisms.

soil degradation A deterioration of soil quality and decline in soil productivity, resulting primarily from forest removal, cropland agriculture, and *overgrazing* of livestock. Compare *desertification.*

soil profile The cross-section of a *soil* as a whole, from the surface to the *bedrock.*

solar cooker A simple portable oven that uses reflectors to focus sunlight onto food and cook it.

solar energy Energy from the sun. It is perpetually renewable and may be harnessed in multiple ways.

solution A homogenous mixture of substances in which elements, molecules, or compounds come together without chemically bonding. Most often a liquid, but sometimes a gas or solid.

solution mining A *mining* technique in which a narrow borehole is drilled deep into the ground to reach a *mineral* deposit, and water, acid, or another liquid is injected down the borehole to leach the resource from the surrounding rock and dissolve it in the liquid. The resulting solution is then sucked out, and the desired resource is isolated.

source In a *nutrient cycle*, a *pool* that releases more *nutrients* than it accepts.

source reduction The reduction of the amount of material that enters the *waste stream* to avoid the costs of disposal and *recycling*, help conserve resources, minimize *pollution*, and save consumers and businesses money.

specialist A *species* that can survive only in a narrow range of *habitats* that contain very specific resources. Compare *generalist.*

speciation The process by which new *species* are generated.

species A *population* or group of populations of a particular type of organism whose members share certain characteristics and can breed freely with one another and produce fertile offspring. Different biologists may have different approaches to diagnosing species boundaries.

species diversity The number and variety of *species* in the world or in a particular region.

species richness The number of species in a particular region. One way to express species diversity. Compare *evenness; relative abundance.*

sprawl The unrestrained spread of urban or *suburban* development outward from a city center and across the landscape. Sometimes specified as growth in which the area of development outpaces *population* growth.

stable isotope An *isotope* that is not *radioactive.*

steady-state economy An *economy* that does not grow or shrink but remains stable.

Stockholm Convention on Persistent Organic Pollutants A 2004 international *treaty* that aims to end the use of 12 persistent organic *pollutants* nicknamed the "dirty dozen."

storage Technique used to generate *hydroelectric power*, in which large amounts of water are impounded in a reservoir behind a concrete

dam and then passed through the dam to turn *turbines* that generate electricity. Compare *run-of-river.*

storm surge A temporary and localized rise in sea level brought on by the high tides and winds associated with storms.

stratosphere The layer of the *atmosphere* above the *troposphere* and below the *mesosphere*; it extends from 11 km (7 mi) to 50 km (31 mi) above sea level.

strip mining The use of heavy machinery to remove huge amounts of earth to expose *coal* or *minerals*, which are mined out directly. Compare *subsurface mining.*

subduction The *plate tectonic* process by which denser *crust* slides beneath lighter crust at a *convergent plate boundary.* Often results in *volcanism.*

subsidy A government incentive (a giveaway of cash or publicly owned resources, or a *tax break*) intended to encourage a particular industry or activity. Compare *green tax.*

subsistence economy A survival *economy*, one in which people meet most or all of their daily needs directly from nature and do not purchase or trade for most of life's necessities.

subspecies *Populations* of a *species* that occur in different geographic areas and vary from one another in some characteristics. Subspecies are formed by the same processes that drive *speciation*, but result when divergence does not proceed far enough to create separate species.

subsurface mining Method of mining underground *coal* deposits, in which shafts are dug deeply into the ground and networks of tunnels are dug or blasted out to follow coal seams. Compare *strip mining.*

suburb A smaller community that rings a city.

succession A stereotypical series of changes in the composition and structure of an *ecological community* through time. See *primary succession; secondary succession.*

sulfur dioxide (SO₂) A colorless gas that may result from volcanic eruptions or from the combustion of *coal.* In the *atmosphere*, it may react to form sulfur trioxide and sulfuric acid, which may return to Earth in *acid deposition.* An EPA *criteria pollutant.*

Superfund A program administered by the *Environmental Protection Agency* in which experts identify sites polluted with hazardous chemicals, protect *groundwater* near these sites, and clean up the *pollution.* Established by the 1980 Comprehensive Environmental Response Compensation and Liability Act (CERCLA).

supply The amount of a product offered for sale at a given price.

surface impoundment (1) A disposal method for *hazardous waste* or mining waste in which waste in liquid or slurry form is placed into a shallow depression lined with impervious material, such as clay. (2) The site of such disposal. Compare *deep-well injection.*

Surface Mining Control and Reclamation Act A 1977 U.S. law that mandates efforts to reclaim (restore) *mining* sites after use, requiring companies to post bonds to cover *reclamation* costs before mining can be approved.

surface water Water located atop Earth's surface. Compare *groundwater.*

sustainability A guiding principle of *environmental science* that requires us to live in such a way as to maintain Earth's systems and its *natural resources* for the foreseeable future.

sustainable agriculture *Agriculture* that does not deplete *soils* faster than they form, or reduce the clean water and genetic diversity essential to long-term crop and livestock production.

sustainable development Development that satisfies our current needs without compromising the future availability of *natural resources* or our future quality of life.

sustainable forest certification A form of *ecolabeling* that identifies timber products that have been produced using *sustainable* methods. The Forest Stewardship Council and several other organizations issue such certification.

swamp A type of *wetland* consisting of shallow water rich with vegetation, occurring in a forested area. Compare *bog; freshwater marsh*.

swidden The traditional form of *agriculture* in tropical forested areas, in which the farmer cultivates a plot for one year to a few years and then moves on to clear another plot, leaving the first to grow back to *forest*. When the forest is burned, this may be called "slash-and-burn" agriculture.

symbiosis A *parasitic* or *mutualistic* relationship between different *species* of organisms that live in close physical proximity.

synergistic effect An interactive effect (as of *toxicants*) that is more than or different from the simple sum of their constituent effects.

system A network of relationships among a group of parts, elements, or components that interact with and influence one another through the exchange of *energy*, matter, and/or information.

taiga See *boreal forest*.

tailings Portions of *ore* left over after *metals* have been extracted in *mining*.

tar sands See *oil sands*.

tax break A common form of *subsidy* in which government reduces or eliminates taxes required of a business or an individual, for the purpose of promoting industries or activities deemed desirable.

taxonomist A scientist who classifies species, using an organism's physical appearance and/or genetic makeup, and who groups species by their similarity into a hierarchy of categories meant to reflect evolutionary relationships.

technically recoverable Extractable using current technology. Applied to *fossil fuel* deposits.

temperate deciduous forest A *biome* consisting of midlatitude *forests* characterized by broad-leafed trees that lose their leaves each fall and remain dormant during winter. These forests occur in areas where *precipitation* is spread relatively evenly throughout the year: much of Europe, eastern China, and eastern North America.

temperate grassland A *biome* whose vegetation is dominated by grasses and features more extreme temperature differences between winter and summer and less *precipitation* than *temperate deciduous forests*.

temperate rainforest A *biome* consisting of tall coniferous trees, cooler and less *species*-rich than *tropical rainforest* and milder and wetter than *temperate deciduous forest*.

temperature (thermal) inversion A departure from the normal temperature distribution in the *atmosphere*, in which a pocket of relatively cold air occurs near the ground, with warmer air above it. The cold air, denser than the air above it, traps *pollutants* near the ground and causes a buildup of *smog*.

teratogen A *toxicant* that causes harm to the unborn, resulting in birth defects.

terracing The cutting of level platforms, sometimes with raised edges, into steep hillsides to contain water from *irrigation* and *precipitation*. Terracing transforms slopes into series of steps like a staircase, enabling farmers to cultivate hilly land while minimizing their loss of *soil* to water *erosion*.

tertiary consumer An organism that consumes *secondary consumers* and feeds at the fourth *trophic level*.

theory A widely accepted, well-tested explanation of one or more cause-and-effect relationships that has been extensively validated by a great amount of research. Compare *hypothesis*.

thermal inversion See *temperature inversion*.

thermal mass Construction materials that absorb heat, store it, and release it later, for use in *passive solar energy* approaches.

thermogenic Type of *natural gas* created by compression and heat deep underground. Contains *methane* and small amounts of other *hydrocarbon* gases. Compare *biogenic*.

thermohaline circulation A worldwide system of ocean currents in which warmer, fresher water moves along the surface and colder, saltier water (which is denser) moves deep beneath the surface.

thin-film solar cells Photovoltaic cells made of material compressed into ultra-thin sheets. Can be incorporated into roofing shingles and is widely viewed as a promising energy source for the future.

Three Mile Island Nuclear power plant in Pennsylvania that in 1979 experienced a partial *meltdown*. The term is often used to denote the accident itself, the most serious *nuclear reactor* malfunction that the United States has thus far experienced.

threshold dose The amount of a *toxicant* at which it begins to affect a *population* of test animals. Compare ED_{50}; LD_{50}.

tidal creek A channel in a *salt marsh* through which the *tide* flows in and out.

tidal energy *Energy* harnessed by erecting a *dam* across the outlet of a tidal basin. Water flowing with the incoming or outgoing *tide* through sluices in the dam turns turbines to generate *electricity*.

tide The periodic rise and fall of the ocean's height at a given location, caused by the gravitational pull of the moon and sun.

Times Beach A town in Missouri whose residents were evacuated and whose buildings were demolished after being contaminated by *dioxin* from waste oil sprayed on its roads. The well-publicized event helped spark passage of the *Superfund* legislation. Compare *Love Canal*.

topsoil That portion of the *soil* that is most nutritive for plants and is thus of the most direct importance to *ecosystems* and to *agriculture*. Also known as the *A horizon*.

tornado A type of cyclonic storm in which funnel clouds pick up soil and objects and can do great damage to structures.

total fertility rate (TFR) The average number of children born per female member of a *population* during her lifetime.

toxic Poisonous; able to harm health of people or other organisms when a substance is inhaled, ingested, or touched. One criterion for defining *hazardous waste*.

toxic air pollutant *Air pollutant* that is known to cause cancer, reproductive defects, or neurological, developmental, immune system, or respiratory problems in humans, and/or to cause substantial *ecological* harm by affecting the health of nonhuman animals and plants. The *Clean Air Act of 1990* identifies 188 toxic air pollutants, ranging from the heavy metal mercury to *volatile organic compounds* such as benzene and methylene chloride.

toxicant A substance that acts as a poison to humans or wildlife.

toxicity The degree of harm a chemical substance can inflict.

toxicology The scientific field that examines the effects of poisonous chemicals and other agents on humans and other organisms.

Toxic Substances Control Act A 1976 U.S. law that directs the *Environmental Protection Agency* to monitor thousands of industrial chemicals and gives the EPA authority to regulate and ban substances found to pose excessive risk.

toxin A *toxic* chemical stored or manufactured in the tissues of living organisms. For example, a chemical that plants use to ward off *herbivores* or that insects use to deter *predators*.

traditional agriculture Biologically powered *agriculture*, in which human and animal muscle power, along with hand tools and simple machines, perform the work of cultivating, harvesting, storing, and distributing crops. Compare *industrial agriculture*.

transboundary park A reserve of protected land that overlaps national borders.

transform plate boundary Area where two tectonic plates meet and slip and grind alongside one another, creating *earthquakes*. For example, the Pacific Plate and the North American Plate rub against each other along California's San Andreas Fault. Compare *convergent plate boundary* and *divergent plate boundary*.

transgene A *gene* that has been extracted from the *DNA* of one organism and transferred into the DNA of an organism of another *species*.

transgenic Term describing an organism that contains *DNA* from another *species*.

transitional stage The second stage of the *demographic transition* model, which occurs during the transition from the *pre-industrial stage* to the *industrial stage*. It is characterized by declining death rates but continued high birth rates. See also *post-industrial stage*. Compare *industrial stage; post-industrial stage; pre-industrial stage*.

transmissible disease See *infectious disease*.

transpiration The release of water vapor by plants through their leaves.

trawling Fishing method that entails dragging immense cone-shaped nets through the water, with weights at the bottom and floats at the top to keep the nets open. Compare *bottom-trawling*.

treatment The portion of an *experiment* in which a *variable* has been manipulated in order to test its effect. Compare *control*.

tributary A smaller river that flows into a larger one.

triple bottom line An approach to sustainability that attempts to meet environmental, economic, and social goals simultaneously.

trophic cascade A series of changes in the *population* sizes of organisms at different *trophic levels* in a *food chain*, occurring when *predators* at high trophic levels indirectly promote populations of organisms at low trophic levels by keeping species at intermediate trophic levels in check. Trophic cascades may become apparent when a top predator is eliminated from a system.

trophic level Rank in the feeding hierarchy of a *food chain*. Organisms at higher trophic levels consume those at lower trophic levels.

tropical dry forest A *biome* that consists of deciduous trees and occurs at tropical and subtropical latitudes where wet and dry seasons each span about half the year. Widespread in India, Africa, South America, and northern Australia.

tropical rainforest A *biome* characterized by year-round rain and uniformly warm temperatures. Found in Central America, South America, Southeast Asia, west Africa, and other tropical regions. Tropical rainforests have dark, damp interiors; lush vegetation; and highly diverse *biotic communities*.

tropopause The boundary between the *troposphere* and the *stratosphere*. Acts like a cap, limiting mixing between these atmospheric layers.

troposphere The bottommost layer of the *atmosphere*; it extends to 11 km (7 mi) above sea level. Compare *stratosphere*.

tropospheric ozone *Ozone* that occurs in the *troposphere*, where it is a *secondary pollutant* created by the interaction of sunlight, heat, *nitrogen oxides*, and volatile *carbon*-containing chemicals. A major component of *smog*, it can injure living tissues and cause respiratory problems. An EPA *criteria pollutant*.

tsunami An immense swell, or wave, of ocean water triggered by an *earthquake*, *volcano*, or *landslide*, that can travel long distances across oceans and inundate coasts.

tundra A *biome* that is nearly as dry as *desert* but is located at very high latitudes along the northern edges of Russia, Canada, and Scandinavia. Extremely cold winters with little daylight and moderately cool summers with lengthy days characterize this landscape of lichens and low, scrubby vegetation.

typhoon See *cyclone*.

umbrella species A *species* for which meeting its *habitat* needs automatically helps meet those of many other species. Umbrella species generally are species that require large areas of habitat.

undernutrition A condition of insufficient *nutrition* in which people receive less than 90% of their daily caloric needs.

uneven-aged Term describing stands of trees in timber plantations that are of different ages. Uneven-aged stands more closely approximate a natural *forest* than do *even-aged* stands.

uniform distribution Distribution pattern in which individuals are evenly spaced (as when individuals hold territories or otherwise compete for space).

United Nations (U.N.) Organization founded in 1945 to promote international peace and to cooperate in solving international economic, social, cultural, and humanitarian problems.

United Nations Environment Programme (UNEP) Agency within the United Nations that deals with *environmental policy*. Created in 1972.

U.N. Framework Convention on Climate Change (FCCC) International agreement to reduce *greenhouse gas* emissions to 1990 levels by the year 2000, signed by nations represented at the 1992 Earth Summit convened in Rio de Janeiro by the *United Nations*. The FCCC called for a voluntary, nation-by-nation approach. Its imminent failure sparked introduction of the *Kyoto Protocol*.

universalist An *ethicist* who maintains that there exist objective notions of right and wrong that hold across cultures and situations. Compare *relativist*.

upwelling In the ocean, the flow of cold, deep water toward the surface. Upwelling occurs in areas where surface *currents* diverge. Compare *downwelling*.

urban ecology A scientific field that views cities explicitly as *ecosystems*. Researchers in this field seek to apply the fundamentals of *ecosystem ecology* and *systems* science to urban areas.

urban growth boundary (UGB) In *city planning*, a geographic boundary intended to separate areas desired to be urban from areas desired to remain rural. Development for housing, commerce, and industry are encouraged within urban growth boundaries, but beyond them such development is severely restricted.

urban heat island effect The phenomenon whereby a city becomes warmer than outlying areas because of the concentration of heat-generating buildings, vehicles, and people, and because buildings and dark paved surfaces absorb heat.

urbanization The shift from rural to city and *suburban* living.

urban planning See *city planning*.

utility An *ethical standard*, elaborated by British philosophers Jeremy Bentham and John Stuart Mill, holding that something is right when it produces the greatest practical benefits for the most people.

variable In an *experiment*, a condition that can change. See *dependent variable* and *independent variable*.

vector An organism that transfers a *pathogen* to its *host*. An example is a mosquito that transfers the malaria pathogen to humans.

vernal pool A seasonal wetland that forms in spring from rain and snowmelt, then dries up later in the year.

volatile organic compound (VOC) One of a large group of potentially harmful organic chemicals used in industrial processes.

volcano A site where molten rock, hot gas, or ash erupt through Earth's surface, often creating a mountain over time as cooled *lava* accumulates.

Wallace, Alfred Russel (1823–1913) English naturalist who proposed, independently of *Charles Darwin*, the concept of *natural selection* as a mechanism for *evolution* and as a way to explain the great variety of living things.

waste Any unwanted product that results from a human activity or process.

waste management Strategic decision making to minimize the amount of *waste* generated and to dispose of waste safely and effectively.

waste stream The flow of *waste* as it moves from its sources toward disposal destinations.

waste-to-energy (WTE) facility An incinerator that uses heat from its furnace to boil water to create steam that drives electricity generation or that fuels heating systems.

wastewater Any water that is used in households, businesses, industries, or public facilities and is drained or flushed down pipes, as well as the polluted *runoff* from streets and storm drains.

wastewater treatment plant A centralized facility that treats *wastewater* from municipal and/or industrial sources, then releases the treated *effluent* into the environment.

waterlogging The saturation of *soil* by water, in which the *water table* is raised to the point that water bathes plant roots. Waterlogging deprives roots of access to gases, essentially suffocating them and eventually damaging or killing the plants.

water pollution The act of polluting water, or the condition of being polluted by water pollutants.

watershed The entire area of land from which water drains into a given river.

water table The upper limit of *groundwater* held in an *aquifer*.

wave energy *Energy* harnessed from the motion of wind-driven waves at the ocean's surface. Many designs for machinery to harness wave energy have been invented, but few have been adequately tested.

weather The local physical properties of the *troposphere*, such as temperature, pressure, humidity, cloudiness, and wind, over relatively short time periods. Compare *climate*.

weathering The physical, chemical, and biological processes that break down rocks and *minerals*, turning large particles into smaller particles.

weed A pejorative term for any plant that competes with our crops. The term is subjective and defined by our own economic interest, and is not biologically meaningful. Compare *pest*.

wetland A system in which the soil is saturated with water and which generally features shallow standing water with ample vegetation. These biologically productive systems include *freshwater marshes*, *swamps*, *bogs*, and seasonal wetlands such as *vernal pools*.

wilderness area Federal land that is designated off-limits to development of any kind but is open to public recreation, such as hiking, nature study, and other activities that have minimal impact on the land.

wildland-urban interface A region where urban or suburban development meets forested or undeveloped lands.

windbreak See *shelterbelt*.

wind energy *Energy* from the motion of wind. In this source of renewable energy, the passage of wind through *wind turbines* is used to generate *electricity*.

wind farm A development involving a group of *wind turbines*.

wind turbine A mechanical assembly that converts the wind's *kinetic energy*, or energy of motion, into electrical energy.

World Bank Institution founded in 1944 that serves as one of the globe's largest sources of funding for *economic* development, including such major projects as *dams*, *irrigation* infrastructure, and other undertakings.

world heritage site A location internationally designated by the *United Nations* for its cultural or natural value. There are over 900 such sites worldwide.

World Trade Organization (WTO) Organization based in Geneva, Switzerland, that represents multinational corporations and promotes free trade by reducing obstacles to international commerce and enforcing fairness among nations in trading practices.

worldview A way of looking at the world that reflects a person's (or a group's) beliefs about the meaning, purpose, operation, and essence of the world.

xeriscaping Landscaping using plants adapted to arid conditions.

zoning The practice of classifying areas for different types of development and land use.

zooplankton Tiny aquatic animals that feed on *phytoplankton* and generally comprise the second *trophic level* in an aquatic *food chain*. Compare *phytoplankton*.

zooxanthellae Symbiotic algae that inhabit the bodies of *corals* and produce food through *photosynthesis*.

PHOTO CREDITS

Contents—1: NASA/Johnson Space Center; **2**: Edwin Remsberg/Alamy; **3**: Michael & Patricia Fogden/Corbis; **4**: Poelzer Wolfgang; **5**: Newscom; **6**: Redlink/Corbis; **7**: Curt Maas/AGE Fotostock, insets: USDA/Natural Resources Conservation Service; **8**: Charles McClean/Alamy; **9**: Newpage Corporation; **10**: Olinchuk/Shutterstock; **11**: Eyedea/ZUMA Press/Gamma; **12**: William Lang; **13**: Morteza Nikoubazi/Reuters Limited, inset: Patrick Baz/AFP/Getty Images; **14**: wennoddpix/Newscom; **15**: Newscom; **16**: Daniel Schoenen/easyfotostock/AGE Fotostock; **17**: left: Ray Pfortner/Photolibrary, right: Irvin Silverstain/Landov Media; **18**:David L. Moore – Oregon/Alamy.

Chapter 1—Opening Photo: NASA/Johnson Space Center; **1.2a**: Charles O'Rear/Corbis; **1.6b**: Jay Withgott; **1.6a**: Reuters NewMedia/Corbis; **1.8**: Brian J. Skerry/National Geographic Stock; **1.11**: Library of Congress; **1.12**: Corbis; **1.13**: Corbis; **1.14**: Corbis; **1.15**: Mario Tama/Getty Images; **1.18**: AP Images; **1.SBS**, top inset: Andrzej Gibasiewicz/Shutterstock; **1.SBS**: vasen/Shutterstock.

Chapter 2—Opening Photo: Edwin Remsberg/Alamy; **2.11a**: Mcpics/Dreamstime; **2.11b**: Jupiterimages/Thinkstock; **2.SBS**, top inset: Photo by Russell P. Burke, Courtesy of David M. Schulte; **2.SBS.2**: Photo by Andy Reid, US Army Corps. of Engineers.

Chapter 3—Opening Photo: Michael & Patricia Fogden/Corbis; **3.1a**: Nicholas Athanas/Tropical Birding; **3.1b**: Oliver Gerhard/Alamy; **3.1c**: Michael & Patricia Fogden/Corbis; **3.1d**: Michael Fogden/Photolibrary; **3.2b1**: Philippe Clement/Nature Picture Library; **3.2b2**: A Jagel/AGE Fotostock; **3.6**: Colin Keates/Dorling Kindersley Media Library; **3.8a**: G. I. Bernard/Photo Researchers; **3.8b**: Wisconsin Historical Society; **3.9a**: Kennan Ward/Corbis; **3.9b**: Art Wolfe/The Image Bank/Getty Images; **3.9c**: Photodisc/Getty Images; **3.10a**: Mandar Khadilkar/Alamy; **3.13**: Matthias Clamer/Stone/Getty Images; **3.SBS**, top inset: Photographer: Italo G. Tapia. Courtesy of J. Alan Pounds.

Chapter 4—Opening Photo: Poelzer Wolfgang; **4.1**: Peter Yates/Photo Researchers; **4.3a**: Michael & Patricia Fogden/Corbis; **4.3b**: James L. Amos/National Geographic Stock; **4.3c**: Andrew Darrington/Alamy Images; **4.5a**: Peter Johnson/Corbis; **4.5b**: Luca Bertolli/Shutterstock; **4.5c**: Photoshot/Alamy; **4.6**: Michael & Patricia Fogden/Corbis; **4.12**: Asgeir Helgestad/Nature Picture Library; **4.14**: Newscom; **4.17a**: JayKay57/iStockphoto; **4.18a**: Philip Gould/Corbis; **4.19a**: Charles Mauzy/Corbis; **4.20a**: David Samuel Robbins/Corbis; **4.21a**: O. Alamany & E. Vicens/Corbis; **4.22a**: Wolfgang Kaehler/Corbis; **4.23a**: Getty Images/Digital Vision; **4.24a**: Radius Image/Alamy; **4.26a**: Gary Nafis; **4.SBS**, top inset: Photo by Heather Malcom. Courtesy David Strayer.

Chapter 5—Opening Photo: Newscom; **5.1a**: Jamie Kum/Ocean Imaging; **5.4**: Annie Griffiths Belt/Corbis; **5.6a**: thinair28/iStockphoto; **5.6b**: Belizar/Dreamstime; **5.6c**: BigWest1/iStockphoto; **5.6d**: Corbis; **5.6e**: Frans Lanting/Corbis; **5.6f**: Image Source/Corbis; **5.6g**: Pgiam/iStockphoto; **5.8**: Walmart; **5.10a**: Bettmann/Corbis; **5.10b**: Museum of History & Industry/Corbis; **5.10c**: University of Washington Libraries; **5.11**: Erich Hartmann/Magnum Photos; **5.12**: Bettmann/Corbis; **5.14**: 350.org; **5.19**: Kristin Piljay; **5.SBS**, top inset: Alan Decker/The New York Times/Redux Pictures; **5.SBS**: AP Images.

Chapter 6—Opening Photo: Redlink/Corbis; **6.1**: EPA/Corbis; **6.ENV**: all images: Skyscan Photolibrary/Alamy; **6.10c**: SETBOUN/Corbis; **6.10d**: AFP/Getty Images; **6.18a**: Peter Menzel/menzelphoto.com; **6.18b**: Peter Ginter; **6.SBS**, top inset: Karen Tweedy; **6.SBS.2**: Mark Edwards/Still Pictures/Peter Arnold/Photolibrary.

Chapter 7—Opening Photo: Curt Maas/AGE Fotostock, insets: USDA/Natural Resources Conservation Service; **7.3**: Peter Trunley/Corbis; **7.4**: Art Rickerby/Time Life Pictures/Getty Images; **7.7a**, inset: Fletcher & Baylis/Photo Researchers; **7.7a**: Ron Giling/Peter Arnold/Photolibrary; **7.7b**, inset: Biological Resources Division, U.S. Geological Survey; **7.7b**: Philip Gould/Corbis; **7.8**: Lynn Betts, USDA/Natural Resources Conservation Service; **7.9a**: Finney County Historical Society; **7.10a**: Scott Sinklier/AGE Fotostock; **7.10b**: Kevin Horan/Stone/Getty Images; **7.10c**: Keren Su/Stone/Getty Images; **7.10d**: Ron Giling/Lineair/Peter Arnold/Photolibrary; **7.10e**: Yann Arthus-Bertrand/Corbis; **7.10f**: U.S. Department of Agriculture; **7.13a**: Photodisc/Getty Images; **7.13b**: Carol Cohen/Corbis; **7.16a, b**: Department of Natural Resources, Queensland, Australia; **7.17a**: Photodisc/Getty Images; **7.17b**: Bob Rowan, Progressive Image/Corbis; **7.24a**: IndexStock/SuperStock; **7.24b**: Joseph Sohm/Visions of America/AGE Fotostock; **7.SBS**, top inset, **7.SBS.2a,b**: FIBL Switzerland; **7.ENV**: top: Lyza Gardner; middle left: Miichael Siluk/The Image Works; bottom left: David R. Frazier Photolibrary/Alamy; bottom right: Alamy.

Chapter 8—Opening Photo: Charles McClean/Alamy; **8.9a**: Photobank/Shutterstock; **8.11**: Konrad Wothe/Photolibrary; **8.13**: Jan Martin Will/Shutterstock; **8.17**: Ant/NHPA/Photoshot; **8.18**: Jeremy Holden; **8.ENV**: top left: Phillippa Psaila/Photo Researchers; top right: Frans Lanting/National Geographic Stock; middle left: Martin Harvey/Alamy; middle right: Hong Kong's Agricultural Fisheries and Conservation Department/AP Images; bottom left: Jordi Bas Casas/NHPA/Photoshot; bottom right: Marvin Dembinsky Photo Associates/Alamy; **8.19a**: Drnickburton/Dreamstime; **8.20**: Corbis News; **8.21**: Alamy; **8.23**: Gilbert Kent/AP Images; **8.SBS**, top inset: Scott Baker; **8.SBS.1**: Neil Beer/Corbis.

Chapter 9—Opening Photo: Newpage Corporation; **9.2a**: John Shaw/NHPA/Photoshot; **9.2b**: Larry Richardson/Danita Delimont/Alamy; **9.2c**: Clint Farlinger/Alamy; **9.2d**: Tetra Images/Alamy; **9.ENV**: top left and right: NASA Earth Observing System; middle left: Michael Nichols/National Geographic Stock; middle right: Rodrigo Baleia/Liaison/Getty Images; bottom: Ron Giling/APIS/Peter Arnold/Photolibrary; **9.6a**: Frans Lanting/www.lanting.com; **9.10**: Rob Badger/Getty Images; **9.12**: Weyerhaeuser Company; **9.14**: Karl Mondon/MCT/Landov Media; **9.15a**: Tracy Ferrero/Alamy; **9.15b**, inset: Dion Manastryrski/British Columbia Province of B.C.-Ministry of Citizens' Services; **9.16**: David Daut, Fountains Forestry; **9.17**: SashaBuzko/iStockphoto; **9.18b**: Erika Vohman, www.mayanutinstitute.org; **9.19a**: Gary Braasch/Corbis; **9.19c**: James Zipp/Photo Researchers; **9.20a**: Courtesy of R.O. Bierregaard Jr.; **9.SBS**, top inset: Ross Hamilton/The Oregonian.

Chapter 10—Opening Photo: Olinchuk/Shutterstock; **10.1**: Gary Porter/Milwaukee Journal Sentinel/MCT/Newsom; **10.2a**: Ingram Publishing/Alamy; **10.2b**: Spencer Grant/PhotoEdit; **10.2c**: Knorre/Dreamstime; **10.2d**: thinkstock/iStockphoto; **10.5**: Bettmann/Corbis; **10.9a**: Howard K. Suzuki; **10.9b**: Peg Skorpinski Photography; **10.SBS**, top inset: Patricia Hunt; **10.SBS.1a, b**: Reproduced by permission from PA Hunt, KE Koehler, M Susiarjo, CA Hodges, A Ilagan, RC Voight, S Thomas, BF Thomas and TJ Hassold. Bisphenol A exposure causes meiotic aneuploidy in the female mouse. Current Biology 13, 546–553. Copyright by Elsevier Science Ltd.

Chapter 11—Opening Photo: Eyedea/ZUMA Press/Gamma; **11.41**: Julien Grondin/Shutterstock; **11.42**: Joy M. Prescott/Shutterstock; **11.6**: Julie Jacobson/AP Images; **11.7b**: Weatherstock/Peter Arnold/Photolibrary; **11.8**: zumalive/Newscom; **11.9**: ainichi Newspaper/Aflo/Newscom; **11.12a**: DEA/R Appiani/AGE Fotostock; **11.12b**: Jim Coon Studio; **11.12c**: Renn Sminkey, Creative Digital Visions, Pearson Science; **11.15**: Michael Williamson/Washington Post Writers Group/Getty Images; **11.16**: David R. Frazier/The Image Works; **11.17**: Eyedea/ZUMA Press/Gamma; **11.18**: George Steinmetz/Corbis; **11.19**: Greenshoots Communications/Alamy; **11.21**: Monty Rakusen/Cultura Creative/Alamy; **11.SBS**, top inset: Rob Crandall/Stock Connection Blue/Alamy.

Chapter 12—Opening Photo: William Lang; **12.1.c**: NASA Earth Observing System; **12.6**: N.D. Smith; **12.8**: Ed Reschke/Peter Arnold/Photolibrary; **12.13**: Genevieve Vallee/Alamy; **12.14**: Kevin M. Kerfoot/Shutterstock; **12.15**: Mark A. Johnson/Corbis; **12.17**: Ralph A. Clevenger/Corbis; **12.18a**: Jeff Hunter/Image Bank/Getty Images; **12.18b**: Stephen Frink/Corbis; **12.19**: Laguna Design/SPL/Photo Researchers; **12.21**: Margaret Croft/The News-Star/AP Images; **12.23a**: AP Images; **12.23b**: Ian Berry/Magnum Photos; **12.24c**: Gilles Saussier; **12.24a, b**: NASA Earth Observing System; **12.27a**: Sean Gardner/MCT/Newscom; **12.SBS**, top inset: Chad Hunt/Corbis; **12.ENV** – top: Michael Pitts/Nature Picture Library; middle left: Matt Cramer/Algalita Marine Research Foundation; middle right: Chris Jordan/Algalita Marine Research Foundation; bottom: Bill Macdonald/Algalita Marine Research Foundation.

Chapter 13—Opening Photo: Morteza Nikoubazi/Reuters Limited, inset: Patrick Baz/AFP/Getty Images; **13.5a**: NASA GSFC Visualization Analysis Laboratory; **13.5b**: Eric Nguyen/Corbis; **13.6a**: David Ponton/Design Pics/Alamy; **13.6b**: Biological Resources Division, U.S. Geological Survey; **13.6c**: NASA Earth Observing System; **13.11**: Jennifer Brown/Corbis; **13.12b**: Pittsburgh Post-Gazette; **13.13b**: Allen Russell/Photolibrary; **13.16**: NASA; **13.20**: Ton Koene/Picture Contact BV/Alamy; **13.SBS**, top inset: AP Images.

Chapter 14—Opening Photo: wennoddpix/Newscom; **14.6a**: Ted Spiegel/Corbis; **14.6b**: CSIRO/Photo Researchers; **14.13a, b**: T. J. Hileman -1938, Carl Key -1981, Dan Fagre/USGS -1998, Blase Reardon/USGS - 25/Courtesy of Glacier National Park Archives/AP Images; **14.15a**: Ron Lacey/iStockphoto; **14.15b**: Thomas Kitchin, Victoria Hurst/Alamy; **14.16a**: Jan Martin Will/Shutterstock; **14.16b**: Brian and Cherry Alexander/Photo Researchers; **14.16c**: Ashley Cooper/Woodfall Wild Images/Photoshot; **14.20**: Bob Strong/Reuters Limited; **14.21**: Imaginechina/AP Images; **14.22a**: Dustin Snipes/AP Images; **14.22b**: Dane Andrew/ZUMA/Corbis; **14.SBS**, top inset: USDA/NRCS/Natural Resources Conservation Service; **14.SBS.1**: USDA/NRCS/Natural Resources Conservation Service.

Chapter 15—Opening Photo: Newscom; **15.1b**: Zhang Jun/Xinhua/Photoshot/Newscom; **15.6**: Bob Fleumer/Corbis; **15.7a**: Corbis; **15.10**: Larry Macdougal/First Light/Alamy; **15.21b**: Caroline Penn/Corbis; **15.ENV**: top left: Charlie Varley/Newscom; top right: Eric Gay/AP Images; middle left: Julie Dermansky/Corbis; middle right: Sean Gardner/Reuters Limited; bottom: Dave Martin/AP Images; **15.21a**: AP Images; **15.SBS**, top inset: Stan Honda/AFP/Getty Images/Newscom.

Chapter 16—Opening Photo: Daniel Schoenen/easyfotostock/AGE Fotostock; **16.4**: Chris Steele-Perkins/Magnum Photos; **16.5**: Wolfgang Hoffmann/USDA/ARS/Agricultural Research Service; **16.6a**: Curt Maas/AGE Fotostock; **16.ENV**: top left: iStockphoto; top right: Jeffrey Zalesny/Fotolia; bottom left: Rick Wilking/Corbis; bottom right (2): Curt Maas/AGE Fotostock; **16.7**: Loyola University Chicago; **16.8a**: Earl Roberge/Photo Researchers; **16.8b**: Michael Melford/The Image Bank/Getty Images; **16.10**: George Steinmetz/Corbis; **16.17**: Jorgen Schytte/Still Pictures/Peter Arnold/Photolibrary; **16.18**: Greg Smith/Corbis; **16.20b**: Simon Fraser/Photo Researchers; **16.SBS**, top inset: Julia Cumes/AP Images; **16.SBS.1**: Sandia National Laboratory.

Chapter 17—Opening Photos: left: Ray Pfortner/Photolibrary, right: Irvin Silverstain/Landov Media; **17.ENV**: Chris Jordan Photographic Arts; **17.3**: Nigel Dickinson; **17.6**: Lawrence Migdale/Pix; **17.10**: City of Edmonton; **17.12a**: Walter Bieri, Keystone/AP Images; **17.12b**: International Olympic Committee; **17.13**: Joe Sohm/Alamy; **17.14**: Robert Brook/Photo Researchers; **17.SBS**, top inset, **17.SBS.1**, **17SBS.2**: Senseable City Laboratory.

Chapter 18—Opening Photo: David L. Moore - Oregon/Alamy; **18.2a**: Chad Palmer/Shutterstock; **18.2b**: JupiterImages/Thinkstock/Alamy; **18.4a, b**: NASA Earth Observing System; **18.5a**: Lester Lefkowitz/Corbis; **18.5b**: Bob Krist/Corbis; **18.5c**: David R. Frazier Photolibrary; **18.5d**: Aldo Torelli/Getty Images; **18.6a**: Library of Congress; **18.6b**: Stock Connection Blue/Alamy; **18.8**: Wu Hong, U.S. Environmental Protection Agency Headquarters; **18.9**: Peter Scholey/Robert Harding World Imagery/Getty Images; **18.SBS**, top inset: Steward Pickett/Cary Institute of Ecosystems Studies.

EPILOGUE—EP.ENV: top left: Rick D'Elia/Corbis; top middle: Jenna Close/P2 Photography; top right: Davidson College; bottom left: Jason O. Watson Photography; middle left: Victoria Arocho/AP Images; bottom right: Bob Daemmrich/Corbis; **EP.2**: Tim Pannell/Corbis; **EP.3**: Costco Wholesale; **EP.4**: Mark Edwards/Peter Arnold/Photolibrary; **EP.5**: NASA/Langley Research Center.

INDEX